What is Materials Science and Engineering?

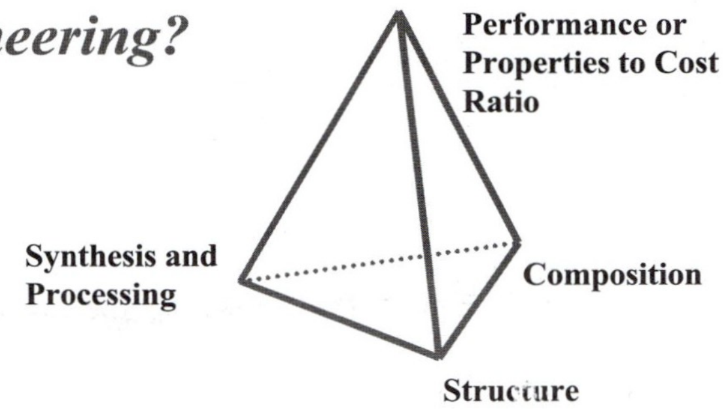

Macro-Scale Structure
Engine Block
≅ upto 1 meter

Performance Criteria
- **Power generated**
- **Efficiency**
- **Durability**
- **Cost**

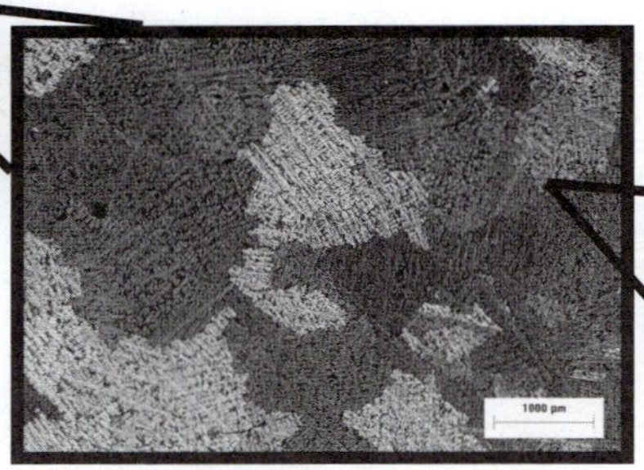

Microstructure
- Grains
≅ 1 – 10 millimeters

Properties affected
- High cycle fatigue
- Ductility

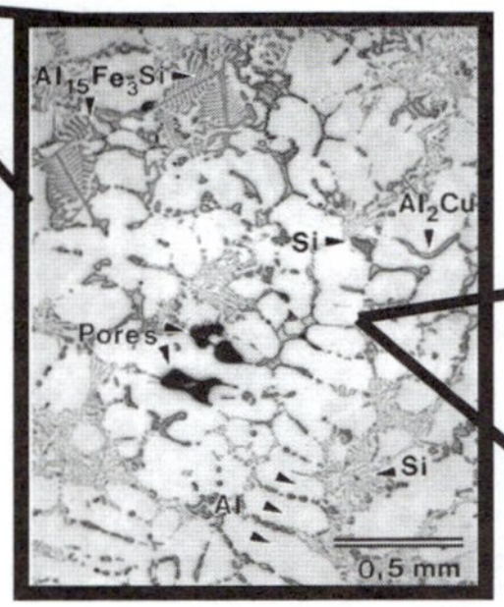

Microstructure
- Dendrites & Phases
≅ 50 – 500 micrometers

Properties affected
- Yield strength
- Ultimate tensile strength
- High cycle fatigue
- Low cycle fatigue
- Thermal Growth
- Ductility

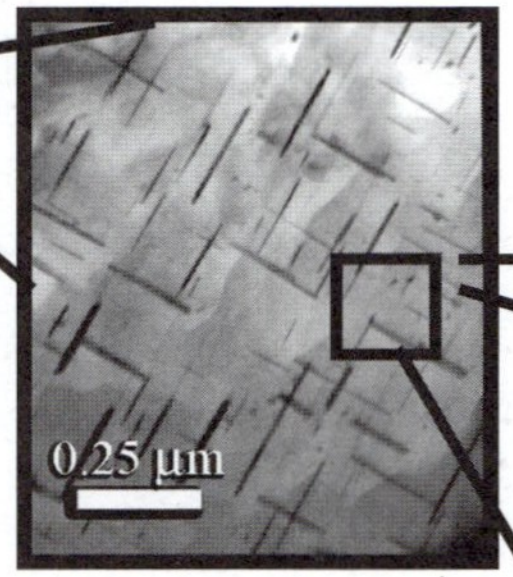

Nano-structure
- Precipitates
≅ 3-100 nanometers

Properties affected
- Yield strength
- Ultimate tensile strength
- Low cycle fatigue
- Ductility

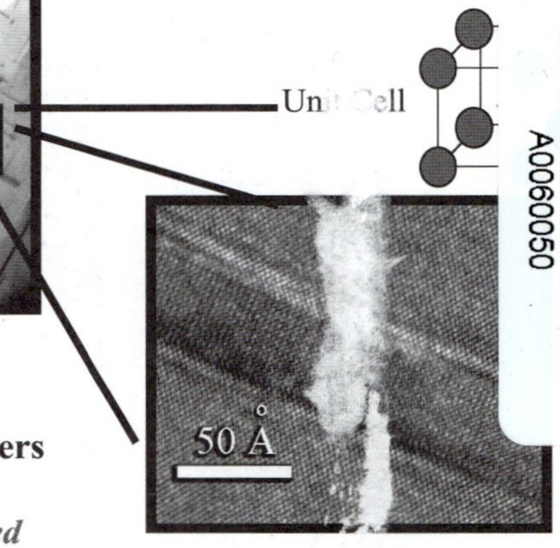

Atomic-scale structur
≅ 1-100 Angstroms

Properties affected
- Young's modulus
- Thermal Growth

A real-world example of important microstructural features at different length-scales, resulting from the sophisticated synthesis and processing used, and the properties they influence. The atomic, nano, micro, and macro-scale structures of cast aluminum alloys (for engine blocks) in relation to the properties affected and performance are shown. The materials science and engineeri (MSE) tetrahedron that represents this approach is shown in the upper right corner.

(Illustrations Courtesy of John Allison and William Donlon, Ford Motor Company)

Essentials of Materials Science and Engineering

Second Edition, SI

Donald R. Askeland
University of Missouri—Rolla, Emeritus

Pradeep P. Fulay
University of Pittsburgh

SI Edition prepared by

D. K. Bhattacharya
Solid State Physics Laboratories
New Delhi

CENGAGE Learning™

Australia • Brazil • Japan • Korea • Mexico • Singapore • Spain • United Kingdom • United States

CENGAGE Learning™

Essentials of Materials Science and Engineering, Second Edition, SI
Donald R. Askeland and Pradeep P. Fulay
SI Edition prepared by D. K. Bhattacharya

Director, Global Engineering Program: Chris Carson
Senior Developmental Editor: Hilda Gowans
Development Editor: Swati Meherishi
Marketing Specialist: Lauren Betsos
Content Project Manager: Abigail Greshik
Production Service: Pre-Press PMG, RPK Editorial Services, Inc.
Copy Editor: Pat Daly
Proofreader: Martha McMaster
Indexer: Shelly Gerger-Knechtl
Compositor: Pre-Press PMG
Senior Art Director: Michelle Kunkler
Internal Design: RPK Editorial Services
Cover Design: Andrew Adams
Cover Image: © Lorelyn Medina/iStockphoto; © Ivan Dinev/iStockphoto; © Jankaliciak/Dreamstime
Senior First Print Buyer: Doug Wilke

Library of Congress Control Number: 2008942016
ISBN-13: 978-0-495-43850-2
ISBN-10: 0-495-43850-2

Cengage Learning
200 First Stamford Place, Suite 400
Stamford, CT 06902
USA

Cengage Learning is a leading provider of customized learning solutions with office locations around the globe, including Singapore, the United Kingdom, Australia, Mexico, Brazil, and Japan. Locate your local office at: **international.cengage.com/region**

Cengage Learning products are represented in Canada by Nelson Education Ltd.

For your course and learning solutions, visit **www.cengage.com/engineering**

Purchase any of our products at your local college store or at our preferred online store **www.ichapters.com**

Printed in Canada
1 2 3 4 5 6 7 8 12 11 10 09

Contents

Preface

This book, *Essentials of Materials Science and Engineering Second Edition*, is a direct result of the success of the *Fifth Edition* of *The Science and Engineering of Materials*, published in 2006. We received positive feedback on both the contents and the integrated approach we used to develop materials science and engineering foundations by presenting the student with real-world applications and problems.

This positive feedback gave us the inspiration to develop *Essentials of Materials Science and Engineering*. The main objective of this book is to provide a *concise* overview of the principles of materials science and engineering for undergraduate students in varying engineering and science disciplines. This *Essentials* text contains the same integrated approach as the *Fifth Edition*, using real-world applications to present and then solve fundamental material science and engineering problems.

The contents of the *Essentials of Materials Science and Engineering* book have been carefully selected such that the reader can develop key ideas that are essential to a solid understanding of materials science and engineering. This book also contains several new examples of modern applications of advanced materials such as those used in information technology, energy technology, nanotechnology microelectromechanical systems (MEMS), and biomedical technology.

The concise approach used in this book will allow instructors to complete an introductory materials science and engineering course in one semester.

We feel that while reading and using this book, students will find materials science and engineering very interesting, and they will clearly see the relevance of what they are learning. We have presented many examples of modern applications of materials science and engineering that impact students' lives. Our feeling is that if students recognize that many of today's technological marvels depend on the availability of engineering materials they will be more motivated and remain interested in learning about how to apply the essentials of materials science and engineering.

Audience and Prerequisites

This book has been developed to cater to the needs of students from different engineering disciplines and backgrounds other than materials science and engineering (e.g., mechanical, industrial, manufacturing, chemical, civil, biomedical, and electrical engineering). At the same time, a conscious effort has been made so that the contents are very well suited for undergraduates majoring in materials science and engineering and closely related disciplines (e.g., metallurgy, ceramics, polymers, and engineering physics). In this sense, from a technical and educational perspective, the book has not been "watered down" in any way. The subjects presented in this text are a careful selection of topics based on our analysis of the needs and feedback from reviewers. Many of the topics related to electronic, magnetic, thermal, and optical properties have not been included in this book to keep the page length down. For instructors and students who wish to develop these omitted concepts, we suggest using the *Fifth Edition* of *The Science and Engineering of Materials.*

This text is intended for engineering students who have completed courses in general physics, chemistry, physics, and calculus. Completion of a general introduction to Engineering or Engineering Technology will be helpful, but not necessary. The text does not presume that the students have had any engineering courses related to statics, dynamics, or mechanics of materials.

Features

We have many unique features to this book.

Have You Ever Wondered? Questions Each chapter opens with a section entitled "*Have You Ever Wondered*?" These questions are designed to arouse the reader's interest, put things in perspective, and form the framework for what the reader will learn in that chapter.

Examples Many real-world Examples have been integrated to accompany the chapter discussions. These Examples specifically cover design considerations, such as operating temperature, presence of corrosive material, economic considerations, recyclability, and environmental constraints. The examples also apply to theoretical material and numeric calculations to further reinforce the presentation.

Glossary All of the Glossary terms that appear in the chapter are set in boldface type the first time they appear within the text. This provides an easy reference to the definitions provided in the end of each chapter Glossary.

Answers to Selected Problems The answers to the selected problems are provided at the end of the text to help the student work through the end-of-chapter problems.

Appendices and Endpapers Appendix A provides a listing of selected physical properties of metals and Appendix B presents the atomic and ionic radii of selected elements. The Endpapers include SI Conversion tables and Selected Physical Properties of elements.

Strategies for Teaching from the Book

Most of the material presented here can be covered in a typical one-semester course. By selecting the appropriate topics, however, the instructor can emphasize the desired materials (i.e., metals, alloys, ceramics, polymers, composites, etc.), provide an overview of materials, concentrate on behavior, or focus on physical properties. In addition, the text provides the student with a useful reference for subsequent courses in manufacturing, design, and materials selection. For students specializing in materials science and engineering, or closely related disciplines, sections related to synthesis and processing could be discussed in greater detail.

Supplements

Supplements for the instructor include:

- The Instructor's Solutions Manual that provides complete, worked-out solutions to selected text problems and additional text items.
- Power Point slides of all figures from the textbook available from the book website at http://academic.cengage.com/engineering.

Acknowledgments

It takes a team of many people and a lot of hard work to create a quality textbook. We are indebted to all of the people who provided the assistance, encouragement, and constructive criticism leading to the preparation of this book.

First, we wish to acknowledge the many instructors who have provided helpful feedback of both *The Science and Engineering of Materials* and *Essentials of Materials Science and Engineering*.

C. Maurice Balik, North Carolina State University

the late Deepak Bhat, University of Arkansass, Fayetteville

Brian Cousins, University of Tasmania

Raymond Cutler, Ceramatec Inc.

Arthur F. Diaz, San Jose State University

Phil Guichelaar, Western Michigan University

Richard S. Harmer, University of Dayton

Prashant N. Kumta, Carnegie Mellon University

Rafael Manory, Royal Melbourne Institute of Technology

Sharon Nightingale, University of Wollongong, Australia

Christopher K. Ober, Cornell University

David Poirier, University of Arizona

Ramurthy Prabhakaran, Old Dominion University

Lew Rabenberg, The Unviersity of Texas at Austin

Wayne Reitz, North Dakota State University

John Schlup, Kansas State University

Robert L. Snyder, Georgia Institute of Technology

J. Rasty, Texas Tech University

Lisa Friis, University of Kansas

Blair London, California Polytechnic State University, San Luis Obispo

Yu-Lin Shen, University of New Mexico

Stephen W. Stafford, University of Texas at El Paso

Rodney Trice, Purdue University

David S. Wilkinson, McMaster University

Indranath Dutta, Naval Postgraduate School

Richard B. Griffin, Texas A&M University

F. Scott Miller, Missouri University of Science and Technology

Amod A. Ogale, Clemson University

Martin Pugh, Concordia University

Thanks most certainly to everyone at Cengage Learning for their encouragement, knowledge, and patience in seeing this text to fruition.

We wish to thank three people, in particular, for their diligent efforts: Many thanks to Chris Carson, our publisher, who set the tone for excellence and who provided the vision, expertise, and leadership to create such a quality product; to Hilda Gowans, our developmental editor and to Rose Kernan, our production editor, who worked long hours to improve our prose and produce this quality text from the first pages of manuscript to the final, bound product.

Pradeep Fulay would like specifically to thank his wife, Dr. Jyotsna Fulay and children, Aarohee and Suyash, for their patience, understanding, and encouragement. Pradeep Fulay would also like to thank his parents Prabhakar and Pratibha Fulay for their support and encouragement. Thanks are also due to Professor S.H. Risbud, University of California–Davis, for his advice and encouragement and to all of our colleagues who provided many useful illustrations.

Donald R. Askeland
University of Missouri–Rolla, Emeritus

Pradeep P. Fulay
University of Pittsburgh

About the Authors

Donald R. Askeland is a Distinguished Teaching Professor Emeritus of Metallurgical Engineering at the University of Missouri–Rolla. He received his degrees from the Thayer School of Engineering at Dartmouth College and the University of Michigan prior to joining the faculty at the University of Missouri–Rolla in 1970. Dr. Askeland taught a number of courses in materials and manufacturing engineering to students in a variety of engineering and science curricula. He received a number of awards for excellence in teaching and advising at UMR. He served as a Key Professor for the Foundry Educational Foundation and received several awards for his service to that organization. His teaching and research were directed primarily to metals casting and joining, in particular lost foam casting, and resulted in over 50 publications and a number of awards for service and best papers from the American Foundry Society.

Dr. Pradeep Fulay has been a Professor of Materials Science and engineering in the Department of Mechanical Engineering and Materials Science for almost 19 years. Currently, Dr. Fulay serves as the Program Director (PD) for the Electronic, Photonic Devices Technology Program (EPDT) at the National Science Foundation (NSF). He joined the University of Pittsburgh in 1989, was promoted to Associate Professor in 1994, and then to full professor in 1999. Dr. Fulay received a Ph.D. in Materials Science and Engineering from the University of Arizona (1989) and a B. Tech (1983) and M. Tech (1984) in Metallurgical Engineering from the Indian Institute of Technology Bombay (Mumbai) India.

He has authored close to 60 publications and has two U.S. patents issued. He has received the Alcoa Foundation and Ford Foundation research awards.

He has been an outstanding teacher and educator and was listed on the Faculty Honor Roll at the University of Pittsburgh (2001) for outstanding services and assistance. From 1992–1999, he was the William Kepler Whiteford Faculty Fellow at the University of Pittsburgh. From August to December 2002, Dr. Fulay was a visiting scientist at the Ford Scientific Research Laboratory in Dearborn, MI.

Dr. Fulay's primary research areas are chemical synthesis and processing of ceramics, electronic ceramics and magnetic materials, development of smart materials and systems. He was the President of Ceramic Educational Council (2003–2004) and a Member of the Program Committee for the Electronics Division of the American ceramic society since 1996.

He has also served as an Associate Editor for the *Journal of the American Ceramic Society* (1994–2000). He has been the lead organizer for symposia on ceramics for sol-gel processing, wireless communications, and smart structures and sensors. In 2002, Dr. Fulay was elected as a Fellow of the American Ceramic Society. Dr. Fulay's research has been supported by National Science Foundation (NSF) and other organizations.

1

Introduction to Materials Science and Engineering

Have You Ever Wondered?

- *Why do jewellers add copper to gold?*
- *How sheet steel can be processed to produce a high-strength, lightweight, energy absorbing, malleable material used in the manufacture of car chassis?*
- *Can we make flexible and lightweight electronic circuits using plastics?*
- *What is a "smart material?"*
- *What is a superconductor?*

In this chapter, we will introduce you to the field of materials science and engineering (MSE) using different real-world examples. We will then provide an introduction to the classification of materials. Materials science underlies most technological advances. Understanding the basics of materials and their applications will not only make you a better engineer, but will help you during the design process. In order to be a good designer, you must learn what materials will be appropriate to use in different applications. The most important aspect of materials is that they are *enabling*; materials make things happen. For example, in the history of civilization, materials such as stone, iron, and bronze played a key role in mankind's development. In today's fast-paced world, the discovery of silicon single crystals and an understanding of their properties have enabled the information age.

In this chapter and throughout the book, we will provide compelling examples of real-world applications of engineered materials. The diversity of applications and the unique uses of materials illustrate why an engineer needs to thoroughly understand and know how to apply the principles of materials science and engineering. In each chapter, we begin with a section entitled *Have You Ever Wondered?* These questions are designed to pique your curiosity, put things in perspective, and form a framework for what you will learn in that chapter.

1-1 What is Materials Science and Engineering?

Materials science and engineering (MSE) is an interdisciplinary field concerned with inventing new materials and improving previously known materials by developing a deeper understanding of the microstructure-composition-synthesis-processing relationships. The term **composition** means the chemical make-up of a material. The term **structure** means a description of the arrangement of atoms, as seen at different levels of detail. Materials scientists and engineers not only deal with the development of materials, but also with the **synthesis** and **processing** of materials and manufacturing processes related to the production of components. The term "synthesis" refers to how materials are made from naturally occurring or man-made chemicals. The term "processing" means how materials are shaped into useful components. One of the most important functions of materials scientists and engineers is to establish the relationships between the properties of a material and its performance. In **materials science**, the emphasis is on the underlying relationships between the synthesis and processing, structure, and properties of materials. In **materials engineering**, the focus is on how to translate or transform materials into a useful device or structure.

One of the most fascinating aspects of materials science involves the investigation into the structure of a material. The structure of materials has a profound influence on many properties of materials, even if the overall composition does not change! For example, if you take a pure copper wire and bend it repeatedly, the wire not only becomes harder but also becomes increasingly brittle! Eventually, the pure copper wire becomes so hard and brittle that it will break rather easily. The electrical resistivity of wire will also increase as we bend it repeatedly. In this simple example, note that we did not change the material's composition (i.e., its chemical make up). The changes in the material's properties are often due to a change in its internal structure. If you examine the wire after bending using an optical microscope, it will look the same as before (other than the bends, of course). However, its structure has been changed at a very small or microscopic scale. The structure at this microscopic scale is known as **microstructure**. If we can understand what has changed at a micrometer level, we can begin to discover ways to control the material's properties.

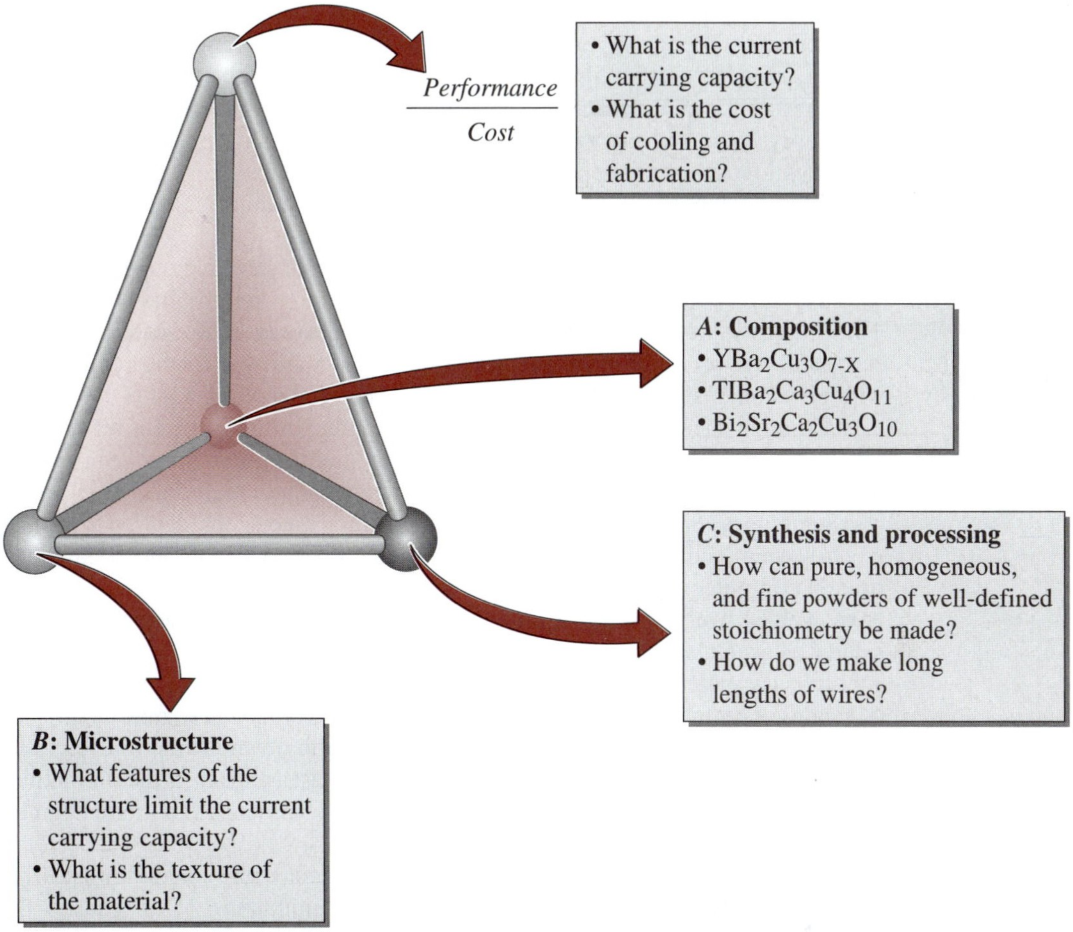

Figure 1-1 Application of the tetrahedron of materials science and engineering to ceramic superconductors. Note that the microstructure-synthesis and processing-composition are all interconnected and affect the performance-to-cost ratio.

Let's put the materials science and engineering tetrahedron in perspective by examining a sample product–ceramic superconductors invented in 1986 (Figure 1-1). You may be aware that ceramic materials usually do not conduct electricity. Scientists found, serendipitously, that certain ceramic compounds based on yttrium barium copper oxides (known as *YBCO*) can actually carry electrical current without any resistance under certain conditions. Based on what was known then about metallic superconductors and the electrical properties of ceramics, superconducting behavior in ceramics was not considered as a strong possibility. Thus, the first step in this case was the *discovery* of superconducting behavior in ceramic materials. These materials were discovered through some experimental research. A limitation of these materials is that they can superconduct only at low temperatures (<150 K).

The next step was to determine how to make these materials better. By "better" we mean: How can we retain superconducting behavior in these materials at higher temperatures, or how can we transport a large amount of current over a long distance? This involves materials processing and careful structure-property studies. Materials scientists wanted to know how the composition and microstructure affect the superconducting behavior. They also want to know if there are other compounds that exhibited superconductivity. Through experimentation, the scientists developed controlled *synthesis* of ultrafine powders or thin films that are used to create useful devices.

An example of approaching this from a *materials engineering* perspective will be to find a way to make long wires for power transmission. In applications, we ultimately want to know if we can make reliable and reproducible long lengths of superconducting wires that are superior to the current copper and aluminum wires. Can we produce such wires in a cost-effective way?

The next challenge was to make long lengths of ceramic superconductor wires. Ceramic superconductors are brittle, so making long lengths of wires was difficult. Thus, *materials processing* techniques had to be developed to create these wires. One successful way of creating these superconducting wires was to fill hollow silver tubes with powders of superconductor ceramic and then draw wires.

Although the discovery of ceramic superconductors did cause a lot of excitement, the path toward translating that discovery into useful products has been met by many challenges related to the synthesis and processing of these materials.

Sometimes, discoveries of new materials, phenomena, or devices are heralded as *revolutionary*. Today, as we look back, the 1948 discovery of the silicon-based transistor used in computer chips is considered revolutionary. On the other hand, materials that have evolved over a period of time can be just as important. These materials are considered as *evolutionary*. Many alloys based on iron, copper, and the like are examples of evolutionary materials. Of course, it is important to recognize that what are considered as evolutionary materials now, did create revolutionary advances many years back. It is not uncommon for materials or phenomena to be discovered first and then for many years to go by before commercial products or processes appear in the marketplace. The transition from the development of novel materials or processes to useful commercial or industrial applications can be slow and difficult.

Let's examine another example using the materials science and engineering tetrahedron. Let's look at "sheet steels" used in the manufacture of car chassis. Steels, as you may know, have been used in manufacturing for more than a hundred years. Earlier steels probably existed in a crude form during the Iron Age, thousands of years ago. In the manufacture of automobile chassis, a material is needed that possesses extremely high strength but is easily formed into aerodynamic contours. Another consideration is fuel-efficiency, so the sheet steel must also be thin and lightweight. The sheet steels should also be able to absorb significant amounts of energy in the event of a crash, thereby increasing vehicle safety. These are somewhat contradictory requirements.

Thus, in this case, materials scientists are concerned with the sheet steel's

- composition;
- strength;
- density;
- energy absorption properties; and
- ductility (formability).

Materials scientists would examine steel at a microscopic level to determine if its properties can be altered to meet all of these requirements. They also would have to process this material into a car chassis in a cost-effective way. Will the shaping process itself affect the mechanical properties of the steel? What kind of coatings can be developed to make the steel corrosion-resistant? We also need to know if these steels could be welded easily. From this discussion, you can see that many issues need to be considered during the design and materials selection for any product.

1-2 Classification of Materials

There are different ways of classifying materials. One way is to describe five groups (Table 1-1):

1. **metals** and **alloys**;
2. ceramics, **glasses**, and **glass-ceramics**;
3. polymers (plastics);
4. semiconductors; and
5. composite materials.

Materials in each of these groups possess different structures and properties. The differences in strength, which are compared in Figure 1-2, illustrate the wide range of properties engineers can select from. Since metallic materials are extensively used for

TABLE 1-1 ■ ***Representative examples, applications, and properties for each category of materials***

	Examples of Applications	Properties
Metals and Alloys		
Copper	Electrical conductor wire	High electrical conductivity, good formability
Gray cast iron	Automobile engine blocks	Castable, machinable, vibration-damping
Alloy steels	Wrenches, automobile chassis	Significantly strengthened by heat treatment
Ceramics and Glasses		
SiO_2-Na_2O-CaO	Window glass or soda-lime glass	Optically transparent, thermally insulating
Al_2O_3, MgO, SiO_2	Refractories (i.e., heat-resistant lining of furnaces) for containing molten metal	Thermally insulating, withstand high temperatures, relatively inert to molten metal
Barium titanate	Capacitors for microelectronics	High ability to store charge
Silica	Optical fibers for information technology	Refractive index, low optical losses
Polymers		
Polyethylene	Food packaging	Easily formed into thin, flexible, airtight film
Epoxy	Encapsulation of integrated circuits	Electrically insulating and moisture-resistant
Phenolics	Adhesives for joining plies in plywood	Strong, moisture resistant
Semiconductors		
Silicon (Si)	Transistors and integrated circuits	Unique electrical behavior
GaAs	Optoelectronic systems	Converts electrical signals to light, lasers, laser diodes, etc.
Composites		
Graphite-epoxy	Aircraft components	High strength-to-weight ratio
Tungsten carbide-cobalt (WC-Co)	Carbide cutting tools for machining	High hardness, yet good shock resistance
Titanium-clad steel	Reactor vessels	Low cost and high strength of steel, with the corrosion resistance of titanium

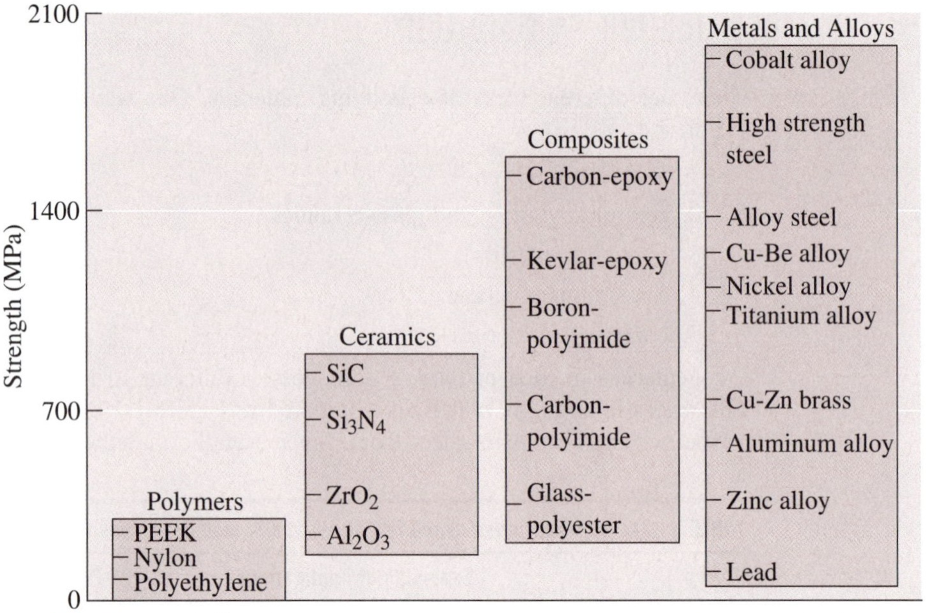

Figure 1-2 Representative strengths of various categories of materials. The strength of ceramics is under a compressive stress.

load-bearing applications, their mechanical properties are of great practical interest. We briefly introduce these here. The term "stress" refers to load or force per unit area. "Strain" refers to elongation or change in dimension divided by original dimension. Application of "stress" causes "strain." If the strain goes away after the load or applied stress is removed, the strain is said to be "elastic." If the strain remains after the stress is removed, the strain is said to be "plastic." When the deformation is elastic, stress and strain are linearly related, the slope of the stress-strain diagram is known as the elastic or Young's modulus. A level of stress needed to initiate plastic deformation is known as "yield strength." The maximum percent deformation we can get is a measure of the ductility of a metallic material. These concepts are discussed further in Chapter 6.

Metals and Alloys These include steels, aluminum, magnesium, zinc, cast iron, titanium, copper, and nickel. In general, metals have good electrical and thermal conductivity. Metals and alloys have relatively high strength, high stiffness, ductility or formability, and shock resistance. They are particularly useful for structural or load-bearing applications. Although pure metals are occasionally used, combinations of metals called alloys provide improvement in a particular desirable property or permit better combinations of properties. The cross section of a jet engine shown in Figure 1-3 illustrates the use of metallic materials for a number of critical applications.

Ceramics Ceramics can be defined as inorganic crystalline materials. Ceramics are probably the most "natural" materials. Beach sand and rocks are examples of naturally occurring ceramics. Advanced ceramics are materials made by refining naturally occurring ceramics and other special processes. Advanced ceramics are used in substrates that house computer chips, sensors and actuators, capacitors, spark plugs, inductors, and electrical insulation. Some ceramics are used as thermal-barrier coatings to protect metallic substrates in turbine engines. Ceramics are also used in such consumer products as paints, plastics, tires, and for industrial applications such as the tiles for the space

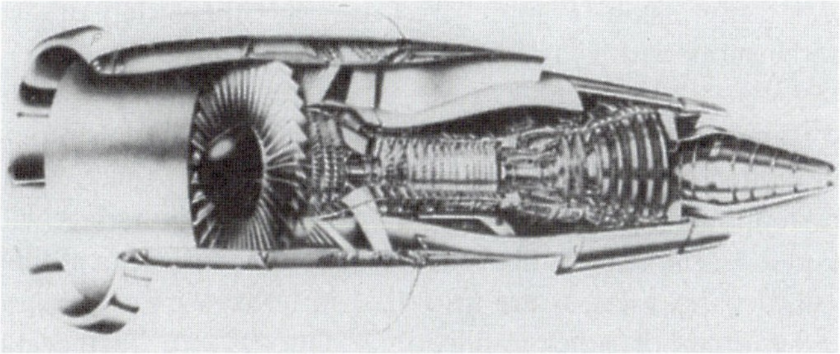

Figure 1-3 A section through a jet engine. The forward compression section operates at low to medium temperatures, and titanium parts are often used. The rear combustion section operates at high temperatures and nickel-based superalloys are required. The outside shell experiences low temperatures, and aluminum and composites are satisfactory. (*Courtesy of GE Aircraft Engines.*)

shuttle, a catalyst support, and oxygen sensors used in cars. Traditional ceramics are used to make bricks, tableware, sanitaryware, refractories (heat-resistant material), and abrasives. In general, due to the presence of porosity (small holes), ceramics tend to be brittle. Ceramics must also be heated to very high temperatures before they can melt. Ceramics are strong and hard, but also very brittle. We normally prepare fine powders of ceramics and convert these into different shapes. New processing techniques make ceramics sufficiently resistant to fracture that they can be used in load-bearing applications, such as impellers in turbine engines (Figure 1-4). Ceramics have exceptional

Figure 1-4 A variety of complex ceramic components, including impellers and blades, which allow turbine engines to operate more efficiently at higher temperatures. (*Courtesy of Certech, Inc.*)

strength under compression (Figure 1-2). Can you believe that the weight of an entire fire truck can be supported using four ceramic coffee cups?

Glasses and Glass-Ceramics Glass is an amorphous material, often, but not always, derived from molten silica. The term "amorphous" refers to materials that do not have a regular, periodic arrangement of atoms. Amorphous materials will be discussed in detail in Chapter 3. The fiber optics industry is founded on optical fibers made by using high-purity silica glass. Glasses are also used in houses, cars, computer and television screens, and hundreds of other applications. Glasses can be thermally treated (tempered) to make them stronger. Forming glasses and nucleating (creating) small crystals within them by a special thermal process creates materials that are known as glass-ceramics. Zerodur™ is an example of a glass-ceramic material that is used to make the mirror substrates for large telescopes (e.g., the Chandra and Hubble telescopes). Glasses and glass-ceramics are usually processed by melting and casting.

Polymers Polymers are typically organic materials produced using a process known as **polymerization**. Polymeric materials include rubber (elastomers) and many types of adhesives. Many polymers have very good electrical resistivity. They can also provide good thermal insulation. Although they have lower strength, polymers have a very good **strength-to-weight ratio**. They are typically not suitable for use at high temperatures. Many polymers have very good resistance to corrosive chemicals. Polymers have thousands of applications ranging from bulletproof vests, compact disks (CDs), ropes, and liquid crystal displays (LCDs) to clothes and coffee cups. **Thermoplastic** polymers, in which the long molecular chains are not rigidly connected, have good ductility and formability; **thermosetting** polymers are stronger but more brittle because the molecular chains are tightly linked (Figure 1-5). Polymers are used in many applications, including electronic devices. Thermoplastics are made by shaping their molten form. Thermosets are typically cast into molds. The term **plastics** is used to describe polymeric materials containing additives that enhance their properties.

Semiconductors Silicon, germanium, and gallium arsenide-based semiconductors are part of a broader class of materials known as electronic materials. The electrical conductivity of semiconducting materials is between that of ceramic insulators and metallic conductors. **Semiconductors** have enabled the information age. In semiconductors, the

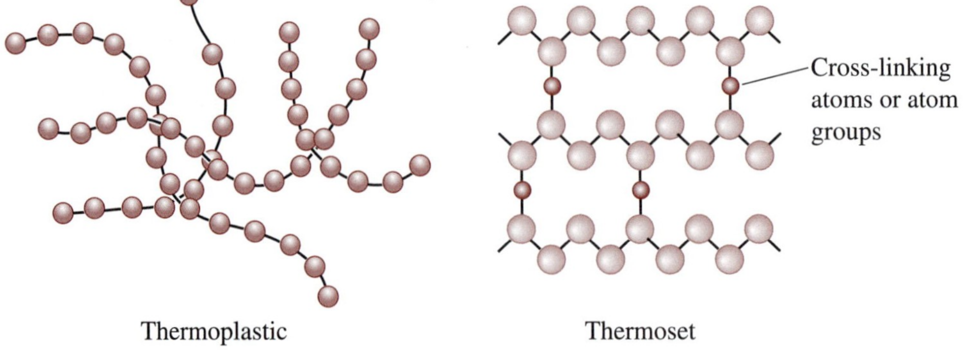

Figure 1-5 Polymerization occurs when small molecules, represented by the circles, combine to produce larger molecules, or polymers. The polymer molecules can have a structure that consists of many chains that are entangled but not connected (thermoplastics) or can form three-dimensional networks in which chains are cross-linked (thermosets).

level of conductivity is controlled to enable their use in electronic devices such as transistors, diodes, etc., that are used to build integrated circuits. In many applications, we need large single crystals of semiconductors. These are grown from molten materials. Often, thin films of semiconducting materials are also made using specialized processes.

Composite Materials The main idea in developing **composites** is to blend the properties of different materials. The composites are formed from two or more materials, producing properties not found in any single material. Concrete, plywood, and fiberglass are examples of composite materials. Fiberglass is made by dispersing glass fibers in a polymer matrix. The glass fibers make the polymer matrix stiffer, without significantly increasing its density. With composites we can produce lightweight, strong, ductile, high temperature-resistant materials or we can produce hard, yet shock-resistant, cutting tools that would otherwise shatter. Advanced aircraft and aerospace vehicles rely heavily on composites such as carbon-fiber-reinforced polymers. Sports equipment such as bicycles, golf clubs, tennis rackets, and the like also make use of different kinds of composite materials that are light and stiff.

1-3 Functional Classification of Materials

We can classify materials based on whether the most important function they perform is mechanical (structural), biological, electrical, magnetic, or optical. This classification of materials is shown in Figure 1-6. Some examples of each category are shown. These categories can be broken down further into subcategories.

Aerospace Light materials such as wood and an aluminum alloy (that accidentally strengthened the alloy used for making the engine even more by picking up copper from the mold used for casting) were used in the Wright brothers' historic flight. Aluminum alloys, plastics, silica for space shuttle tiles, carbon-carbon composites, and many other materials belong to this category.

Biomedical Our bones and teeth are made, in part, from a naturally formed ceramic known as hydroxyapatite. A number of artificial organs, bone replacement parts, cardiovascular stents, orthodontic braces, and other components are made using different plastics, titanium alloys, and nonmagnetic stainless steels. Ultrasonic imaging systems make use of ceramics known as *PZT* (lead zirconium titanate). Magnets used for magnetic resonance imaging make use of metallic niobium tin-based superconductors.

Electronic Materials As mentioned before, semiconductors, such as those made from silicon, are used to make integrated circuits for computer chips. Barium titanate ($BaTiO_3$), tantalum oxide (Ta_2O_5), and many other dielectric materials are used to make ceramic capacitors and other devices. Superconductors are used in making powerful magnets. Copper, aluminum, and other metals are used as conductors in power transmission and in microelectronics.

Energy Technology and Environmental Technology The nuclear industry uses materials such as uranium dioxide and plutonium as fuel. Numerous other materials, such as glasses and stainless steels, are used in handling nuclear materials and managing radioactive waste. New technologies related to batteries and fuel cells make use of many ceramic materials such as zirconia (ZrO_2) and polymers. The battery technology has

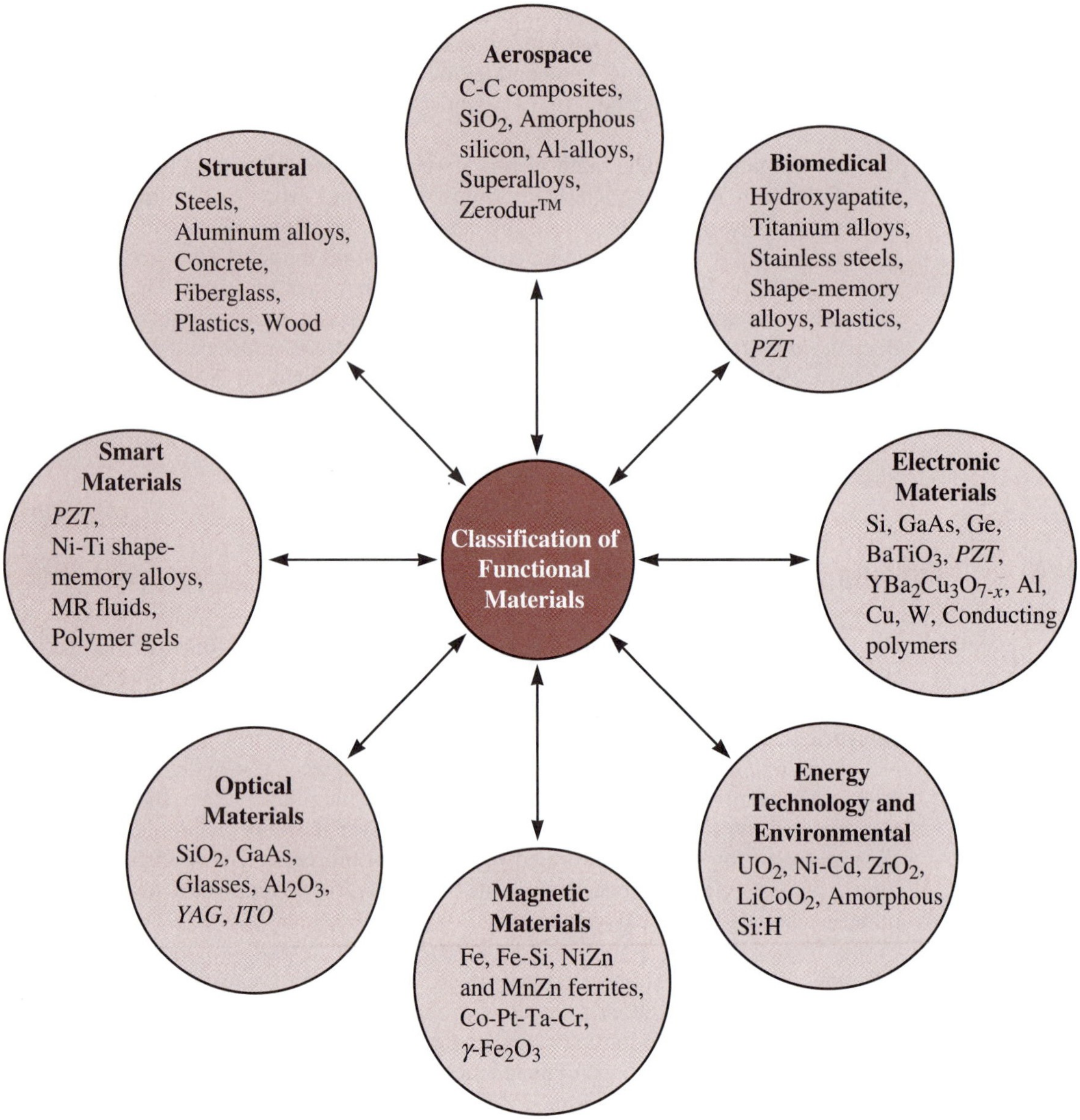

Figure 1-6 Functional classification of materials. Notice that metals, plastics, and ceramics occur in different categories. A limited number of examples in each category are provided.

gained significant importance owing to the need for many electronic devices that require longer lasting and portable power. Fuel cells are also being used in some cars. The oil and petroleum industry widely uses zeolites, alumina, and other materials as catalyst substrates. They use Pt, Pt/Rh and many other metals as catalysts. Many membrane technologies for purification of liquids and gases make use of ceramics and plastics. Solar power is generated using materials such as crystalline Si and amorphous silicon (a:Si:H).

Magnetic Materials Computer hard disks and audio and video cassettes make use of many ceramic, metallic, and polymeric materials. For example, particles of a special form of iron oxide, known as gamma iron oxide (γ-Fe_2O_3) are deposited on a polymer substrate to make audio cassettes. High-purity iron particles are used for making videotapes. Computer hard disks are made using alloys based on cobalt-platinum-tantalum-chromium (Co-Pt-Ta-Cr) alloys. Many magnetic ferrites are used to make inductors and components for wireless communications. Steels based on iron and silicon are used to make transformer cores.

Photonic or Optical Materials Silica is used widely for making optical fibers. Almost ten million kilometers of optical fiber have been installed around the world. Optical materials are used for making semiconductor detectors and lasers used in fiber optic communications systems and other applications. Similarly, alumina (Al_2O_3) and yttrium aluminum garnets (*YAG*) are used for making lasers. Amorphous silicon is used to make solar cells and photovoltaic modules. Polymers are used to make liquid crystal displays (*LCDs*).

Smart Materials A **smart material** can sense and respond to an external stimulus such as a change in temperature, the application of a stress, or a change in humidity or chemical environment. Usually a smart-material-based system consists of sensors and actuators that read changes and initiate an action. An example of a passively smart material is lead zirconium titanate (*PZT*) and shape-memory alloys. When properly processed, *PZT* can be subjected to a stress and a voltage is generated. This effect is used to make such devices as spark generators for gas grills and sensors that can detect underwater objects such as fish and submarines. Other examples of smart materials include magnetorheological or MR fluids. These are magnetic paints that respond to magnetic fields and are being used in suspension systems of automobiles. Other examples of smart materials and systems are photochromic glasses and automatic dimming mirrors based on electrochromic materials.

Structural Materials These materials are designed for carrying some type of stress. Steels, concrete, and composites are used to make buildings and bridges. Steels, glasses, plastics, and composites are also used widely to make automotives. Often in these applications, combinations of strength, stiffness, and toughness are needed under different conditions of temperature and loading.

1-4 Classification of Materials Based on Structure

As mentioned before, the term "structure" means the arrangement of a material's atoms; the structure at a microscopic scale is known as "microstructure." We can view these arrangements at different scales, ranging from a few angstrom units to a millimeter. We will learn in Chapter 3 that some materials may be **crystalline** (where the material's atoms are arranged in a periodic fashion) or they may be amorphous (where the material's atoms do not have a long-range order). Some crystalline materials may be in the form of one crystal and are known as **single crystals**. Others consist of many crystals or **grains** and are known as **polycrystalline**. The characteristics of crystals or grains (size, shape, etc.) and that of the regions between them, known as the **grain boundaries**, also affect the properties of materials. We will further discuss these concepts in later chapters. A micrograph of a stainless steel sample (showing grains and grain boundaries) is shown in Figure 1-7. For this sample, each grain reflects the light differently and this produces a contrast between the grains.

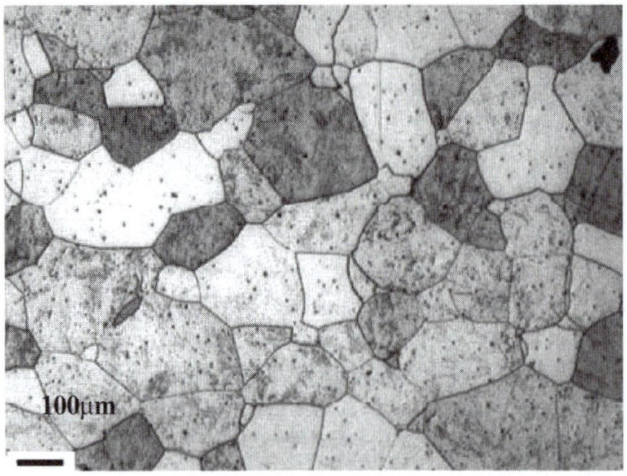

Figure 1-7
Micrograph of stainless steel showing grains and grain boundaries. (*Courtesy Dr. Hua and Dr. DeArdo—University of Pittsburgh.*)

1-5 Environmental and Other Effects

The structure-property relationships in materials fabricated into components are often influenced by the surroundings to which the material is subjected during use. This can include exposure to high or low temperatures, cyclical stresses, sudden impact, corrosion or oxidation. These effects must be accounted for in design to ensure that components do not fail unexpectedly.

Temperature Changes in temperature dramatically alter the properties of materials (Figure 1-8). Metals and alloys that have been strengthened by certain heat treatments or forming techniques will lose their strength when heated. A tragic reminder of this is the collapse of the steel beams used in the World Trade Center towers on September 11, 2001.

High temperatures change the structure of ceramics and cause polymers to melt or char. Very low temperatures, at the other extreme, may cause a metal or polymer to fail in a brittle manner, even though the applied loads are low. This low temperature em-

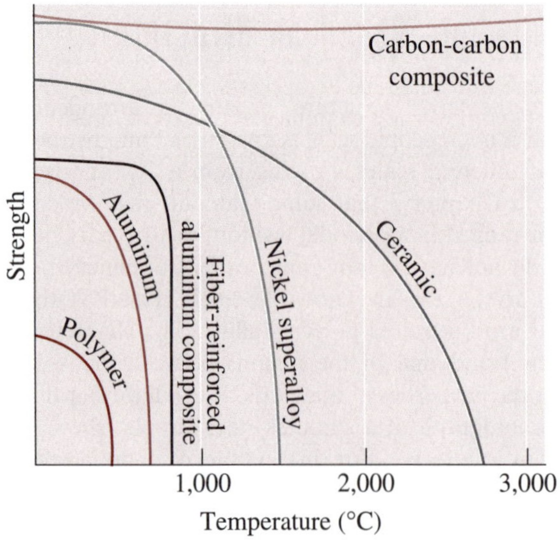

Figure 1-8
Increasing temperature normally reduces the strength of a material. Polymers are suitable only at low temperatures. Some composites, such as carbon-carbon composites, special alloys, and ceramics, have excellent properties at high temperatures.

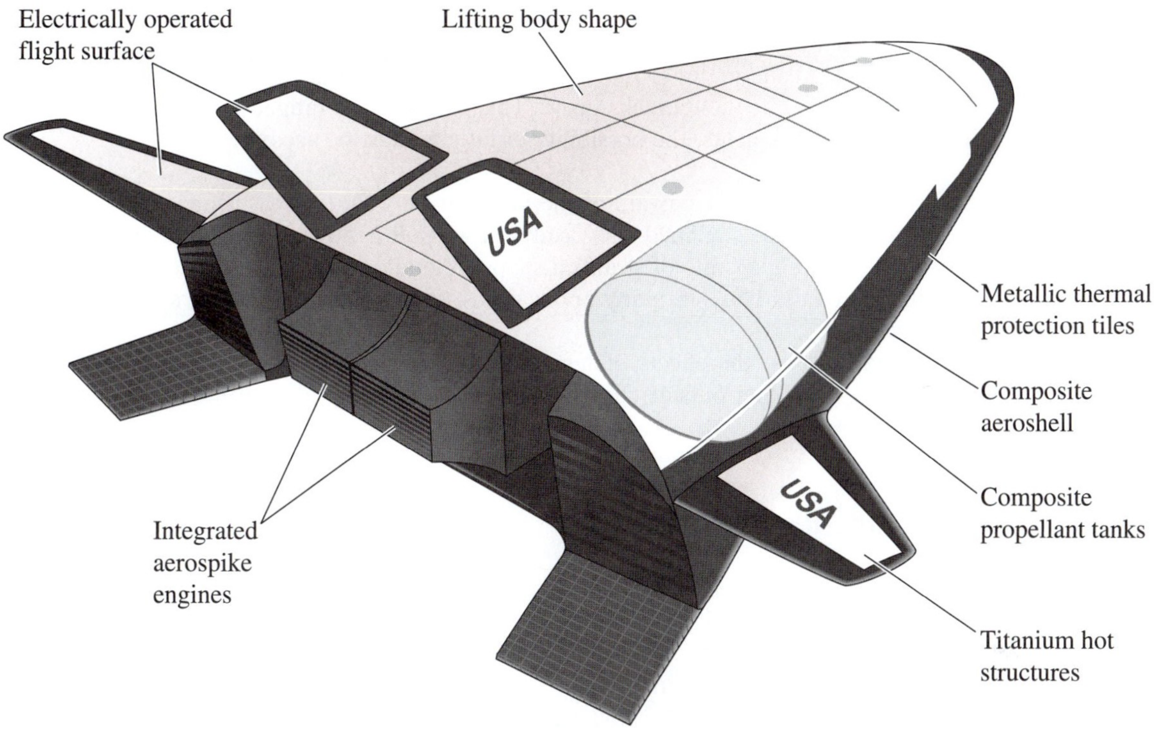

Figure 1-9 Schematic of a *X-33* plane prototype. Notice the use of different materials for different parts. This type of vehicle will test several components for the *Venturestar* (*From "A Simpler Ride into Space," by T.K. Mattingly, October, 1997,* Scientific American, *p. 125. Copyright © 1997 Slim Films.*)

brittlement was a factor that caused the *Titanic* to fracture and sink. Similarly, the 1986 *Challenger* accident, in part, was due to embrittlement of rubber O-rings. The reasons why some polymers and metallic materials become brittle are different. We will discuss these concepts in later chapters.

The design of materials with improved resistance to temperature extremes is essential in many technologies related to aerospace. As faster speeds are attained, more heating of the vehicle skin occurs because of friction with the air. At the same time, engines operate more efficiently at higher temperatures. So, in order to achieve higher speed and better fuel economy, new materials have gradually increased allowable skin and engine temperatures. But materials engineers are continually faced with new challenges. The *X-33* and *Venturestar* are examples of advanced reusable vehicles intended to carry passengers into space using a single stage of rocket engines. Figure 1-9 shows a schematic of the *X-33* prototype. The development of even more exotic materials and processing techniques is necessary in order to tolerate the high temperatures that will be encountered.

Corrosion Most of the time, failure of materials occurs as a result of corrosion and some form of tensile overload. Most metals and polymers react with oxygen or other gases, particularly at elevated temperatures. Metals and ceramics may disintegrate and polymers and nonoxide ceramics may oxidize. Materials are also attacked by corrosive liquids, leading to premature failure. The engineer faces the challenge of selecting materials or coatings that prevent these reactions and permit operation in extreme environments. In space applications, we may have to consider the effects of the presence of radiation, the presence of atomic oxygen, and the impact from debris.

Fatigue In many applications, components must be designed such that the load on the material may not be enough to cause permanent deformation. However, when we do load and unload the material thousands of times, small cracks may begin to develop and materials fail as these cracks grow. This is known as **fatigue failure**. In designing load-bearing components, the possibility of fatigue must be accounted for.

Strain Rate You may be aware of the fact that Silly Putty®, a silicone- (not silicon-) based plastic, can be stretched significantly if we pull it slowly (small rate of strain). If you pull it fast (higher rate of strain) it snaps. A similar behavior can occur with many metallic materials. Thus, in many applications, the level and rate of strain have to be considered.

In many cases, the effects of temperature, fatigue, stress, and corrosion may be interrelated, and other outside effects could affect the material's performance.

1-6 Materials Design and Selection

When a material is designed for a given application, a number of factors must be considered. The material must possess the desired physical and mechanical properties. It must be capable of being processed or manufactured into the desired shape, and must provide an economical solution to the design problem. Satisfying these requirements in a manner that protects the environment—perhaps by encouraging recycling of the materials—is also essential. In meeting these design requirements, the engineer may have to make a number of tradeoffs in order to produce a serviceable, yet marketable, product.

As an example, material cost is normally calculated on a cost-per-kg basis. We must consider the **density** of the material, or its weight-per-unit volume, in our design and selection (Table 1-2). Aluminum may cost more per kg than steel, but it is only one-third the weight of steel. Although parts made from aluminum may have to be thicker, the aluminum part may be less expensive than the one made from steel because of the weight difference.

TABLE 1-2 ■ ***Strength-to-weight ratios of various materials***

Material	Strength (kg/m^2)	Density (gm/cm^3)	Strength-to-weight ratio (cm)
Polyethylene	70×10^4	0.83	8.43×10^4
Pure aluminum	455×10^4	2.71	16.79×10^4
Al_2O_3	21×10^6	3.16	0.66×10^6
Epoxy	105×10^5	1.38	7.61×10^5
Heat-treated alloy steel	17×10^7	7.75	0.22×10^7
Heat-treated aluminum alloy	60×10^6	2.71	2.21×10^6
Carbon-carbon composite	42×10^6	1.80	2.33×10^6
Heat-treated titanium alloy	12×10^7	4.43	0.27×10^7
Kevlar-epoxy composite	46×10^6	1.47	3.13×10^6
Carbon-epoxy composite	56×10^6	1.38	4.06×10^6

In some instances, particularly in aerospace applications, the weight issue is critical, since additional vehicle weight increases fuel consumption and reduces range. By using materials that are lightweight but very strong, aerospace or automobile vehicles can be designed to improve fuel efficiency. Many advanced aerospace vehicles use composite materials instead of aluminum alloys. These composites, such as carbon-epoxy, are more expensive than the traditional aluminum alloys; however, the fuel savings yielded by the higher strength-to-weight ratio of the composite (Table 1-2) may offset the higher initial cost of the aircraft. The body of one of the latest Boeing aircrafts known as the Dreamliner is made almost entirely from carbon-carbon composite materials. There are literally thousands of applications in which similar considerations apply. Usually the selection of materials involves trade-offs between many properties.

By this point of our discussion, we hope that you can appreciate that the properties of materials depend not only on composition, but also on how the materials are made (synthesis and processing) and, most importantly, their internal structure. This is why it is not a good idea for an engineer to simply refer to a handbook and select a material for a given application. The handbooks may be a good starting point. A good engineer will consider: the effects of how the material is made, what exactly is the composition of the candidate material for the application being considered, any processing that may have to be done for shaping the material or fabricating a component, the structure of the material after processing into a component or device, the environment in which the material will be used, and the cost-to-performance ratio. The knowledge of principles of materials science and engineering will empower you with the fundamental concepts. These will allow you to make technically sound decisions in designing with engineered materials.

EXAMPLE 1-1 *Materials for a Bicycle Frame*

Bicycle frames are made using steel, aluminum alloys, titanium alloys containing aluminum and vanadium, and carbon-fiber composites (Figure 1-10). (a) If a steel-frame bicycle weighs 14 kg, what will be the weight of the frame assuming we use aluminum, titanium, and a carbon-fiber composite to make the frame in such a way that the volume of frame (the diameter of the tubes) is constant? (b) What other considerations can come into play in designing bicycle frames?

Figure 1-10
Bicycle frames need to be lightweight, stiff, and corrosion resistant (for Example 1-1). (*Courtesy of Chris harve/StockXpert.*)

Note: The densities of steel, aluminum alloy, titanium alloy, and carbon-fiber composite can be assumed to be 7.8, 2.7, 4.5, and 1.85 g/cm^3.

SOLUTION

(a) The weight of the bicycle frame made from steel is stated to be 14 kg. The volume of this frame will be

$$V_{\text{frame}} = 14000 \text{ g}/(7.8) = 1795 \text{ cm}^3$$

For aluminum frame the weight will be

$$W_{\text{al}} = (1795 \text{ cm}^3) \times (2.7 \text{ g/cm}^3) = 4846.5 \text{ grams}$$

Another and simpler way to arrive at this answer is to take the ratio of densities, since the volume is assumed constant.

The weight of the aluminum alloy frame

$$W_{\text{alloy}} = (\text{density of aluminum alloy/density of steel})$$
$$\times (\text{wt. of the steel frame})$$
$$= (2.7/7.8) \times 14 \text{ kg} = 4.846 \text{ kg}$$

Thus, the aluminum frame weighs roughly one-third of the steel frame.

Similarly, the weight of titanium frame will be

$$W_{\text{Ti}} = (\text{density of titanium alloy/density of steel})$$
$$\times (\text{wt. of the steel frame})$$
$$= (4.5/7.8) \times 14 \text{ kg} = 8.08 \text{ kg}$$

Finally, the weight of the frame made using carbon-fiber composite will be

$$W_{\text{cf}} = (\text{density of carbon fiber composite/density of steel})$$
$$\times (\text{wt. of the steel frame})$$
$$= (1.85/7.8) \times 14 \text{ kg} = 3.32 \text{ kg}$$

As can be seen, substantial reduction in weight is possible using materials other than steel.

(b) One of the other factors that comes into play is the stiffness of the structure. This is related to the elastic modulus of the material (Chapter 2). For example, for the same tube dimensions, an aluminum tube will be not as stiff as steel. This will make the aluminum frame bicycle ride "soft." This effect can be compensated for by making the aluminum tubes larger in diameter and the walls of the tubes thicker. Some other factors to consider are the toughness of each of the materials. For example, even though a carbon-fiber frame is very light, it is relatively brittle. Additional considerations would be the ability to weld or join the frame to other parts of the bicycle, corrosion resistance, and of course, cost.

EXAMPLE 1-2 Ceramic-Carbon-Fiber Brakes for Cars

Car breaks are typically made using cast iron and weigh about 9 kg. What other materials can be used to make brakes that would last long and weigh less?

SOLUTION

The brakes could be made using other lower density materials, such as aluminum or titanium. Cost and wear resistance are clearly important. Titanium alloys will be very expensive, and both titanium and aluminum will wear out more easily.

We could make the brakes out of ceramics, such as alumina (Al_2O_3) or silicon carbide (SiC), since both have densities lower than cast iron. However, ceramics are too brittle, and even though they have very good resistance, they will fracture easily.

We can use a material that is a composite of carbon fibers and ceramics, such as SiC. This composite material will provide the lightweight and wear-resistance necessary, so that the brakes do not have to be replaced often. Some companies are already producing such ceramic-carbon-fiber brakes.

SUMMARY

- The properties of engineered materials depend upon their composition, structure, synthesis, and processing. An important performance index for materials or devices is their cost-to-performance ratio.
- The structure at a microscopic level is known as the microstructure (length scale 10 nm to 1000 nm).
- Many properties of materials depend strongly on the structure, even if the composition of the material remains the same. This is why the structure-property or microstructure-property relationships in materials are extremely important.
- Materials are often classified as metals and alloys, ceramics, glasses, and glass ceramics, composites, polymers, and semiconductors.
- Metals and alloys have good strength, good ductility, and good formability. Pure metals have good electrical and thermal conductivity. Metals and alloys play an indispensable role in many applications such as automotives, buildings, bridges, aerospace, and the like.
- Ceramics are inorganic, crystalline materials. They are strong, serve as good electrical and thermal insulators, are often resistant to damage by high temperatures and corrosive environments, but are mechanically brittle. Modern ceramics form the underpinnings of many of the microelectronic and photonic technologies.
- Polymers have relatively low strength; however, the strength-to-weight ratio is very favorable. Polymers are not suitable for use at high temperatures. They have very good corrosion resistance, and—like ceramics—provide good electrical and thermal insulation. Polymers may be either ductile or brittle, depending on structure, temperature, and the strain rate.
- Materials can also be classified as crystalline or amorphous. Crystalline materials may be single crystal or polycrystalline.

- Selection of a material having the needed properties and the potential to be manufactured economically and safely into a useful product is a complicated process requiring the knowledge of the structure-property-processing-composition relationships.

GLOSSARY

Alloy A metallic material that is obtained by chemical combinations of different elements (e.g., steel is made from iron and carbon). Typically, alloys have better mechanical properties than pure metals.

Ceramics Crystalline inorganic materials characterized by good strength in compression, and high melting temperatures. Many ceramics are very good electrical insulators and have good thermal insulation behavior.

Composition The chemical make-up of a material.

Composites A group of materials formed from metals, ceramics, or polymers in such a manner that unusual combinations of properties are obtained (e.g., fiberglass).

Crystal structure The arrangement of the atoms in a crystalline material.

Crystalline material A material comprised of one or many crystals. In each crystal atoms or ions show a long-range periodic arrangement.

Density Mass per unit volume of a material, usually expressed in units of g/cm^3.

Fatigue failure Failure of a material due to repeated loading and unloading.

Glass An amorphous material derived from the molten state, typically, but not always, based on silica.

Glass-ceramics A special class of crystalline materials obtained by forming a glass and then heat treating it to form small crystals.

Grains Crystals in a polycrystalline material.

Grain boundaries Regions between grains of a polycrystalline material.

Materials engineering An engineering oriented field that focuses on how to translate or transform materials into a useful device or structure.

Materials science and engineering (MSE) An interdisciplinary field concerned with inventing new materials and improving previously known materials by developing a deeper understanding of the microstructure-composition-synthesis-processing relationships between different materials.

Materials science A field of science that emphasizes studies of relationships between the internal or microstructure, synthesis and processing and the properties of materials.

Materials science and engineering tetrahedron A tetrahedron diagram showing how the performance-to-cost ratio of materials depends upon the composition, microstructure, synthesis, and processing.

Mechanical properties Properties of a material, such as strength, that describe how well a material withstands applied forces, including tensile or compressive forces, impact forces, cyclical or fatigue forces, or forces at high temperatures.

Metal An element that has metallic bonding and generally good ductility, strength, and electrical conductivity.

Microstructure The structure of a material at a length scale of 10 nm to 1000 nm (1 μm).

Physical properties Describe characteristics such as color, elasticity, electrical or thermal conductivity, magnetism, and optical behavior that generally are not significantly influenced by forces acting on a material.

Polycrystalline material A material comprised of many crystals (as opposed to a single-crystal material that has only one crystal). The crystals are also known as grains.

Polymerization The process by which organic molecules are joined into giant molecules, or polymers.

Polymers A group of materials normally obtained by joining organic molecules into giant molecular chains or networks. Polymers are characterized by low strengths, low melting temperatures, and poor electrical conductivity.

Plastics These are polymeric materials consisting of other additives that enhance their properties.

Processing Different ways for shaping materials into useful components or changing their properties.

Semiconductors A group of materials having electrical conductivity between metals and typical ceramics (e.g., Si, GaAs).

Single crystal A crystalline material that is made of only one crystal (there are no grain boundaries).

Smart material A material that can sense and respond to an external stimulus such as change in temperature, application of a stress, or change in humidity or chemical environment.

Strength-to-weight ratio The strength of a material divided by its density; materials with a high strength-to-weight ratio are strong but lightweight.

Structure Description of the arrangements of atoms or ions in a material. The structure of materials has a profound influence on many properties of materials, even if the overall composition does not change!

Synthesis The process by which materials are made from naturally occurring or other chemicals.

Thermoplastics A special group of polymers in which molecular chains are entangled but not interconnected. They can be easily melted and formed into useful shapes. Normally, these polymers have a chainlike structure (e.g., polyethylene).

Thermosets A special group of polymers that decompose rather than melt upon heating. They are normally quite brittle due to a relatively rigid, three-dimensional network structure comprising chains that are bonded to one another (e.g., polyurethane).

PROBLEMS

Section 1-1 What is Materials Science and Engineering?

1-1 Define Material Science and Engineering (MSE).

1-2 Define the following terms: **(a)** composition, **(b)** structure, **(c)** synthesis, **(d)** processing, and **(e)** microstructure.

1-3 Explain the difference between the terms materials science and materials engineering.

1-4 Name one revolutionary discovery of a material. Name one evolutionary discovery of a material.

Section 1-2 Classification of Materials

Section 1-3 Functional Classification of Materials

Section 1-4 Classification of Materials Based on Structure

Section 1-5 Environmental and Other Effects

1-5 Steel is often coated with a thin layer of zinc if it is to be used outside. What characteristics do you think the zinc provides to this coated, or galvanized, steel? What precautions should be considered in producing this product? How will the recyclability of the steel be affected as a result of the galvanization?

1-6 We would like to produce a transparent canopy for an aircraft. If we were to use a ceramic (that is, traditional window glass) canopy, rocks or birds might cause it to shatter. Design a material that would minimize damage or at least keep the canopy from breaking into pieces.

1-7 Coiled springs ought to be very strong and stiff. Silicon nitride (Si_3N_4) is a strong, stiff material. Would you select this material for a spring? Explain.

1-8 Temperature indicators are sometimes produced from a coiled metal strip that uncoils a specific amount when the temperature increases. How does this work; from what kind of material would the indicator be made; and what are the important properties that the material in the indicator must possess?

Section 1-6 Materials Design and Selection

1-9 You would like to design an aircraft that can be flown by human power nonstop for a distance of 30 km. What types of material properties would you recommend? What materials might be appropriate?

1-10 You would like to place a 90-cm diameter microsatellite into orbit. The satellite will contain delicate electronic equipment that will send and receive radio signals from earth. Design the outer shell within which the electronic equipment is contained. What properties will be required, and what kind of materials might be considered?

1-11 What properties should the head of a carpenter's hammer possess? How would you manufacture a hammer head?

1-12 The hull of the space shuttle consists of ceramic tiles bonded to an aluminum skin. Discuss the design requirements of the shuttle hull that led to the use of this combination of materials. What problems in producing the hull might the designers and manufacturers have faced?

1-13 You would like to select a material for the electrical contacts in an electrical switching device which opens and closes frequently and forcefully. What properties should the contact material possess? What type of material might you recommend? Would Al_2O_3 be a good choice? Explain.

1-14 Aluminum has a density of 2.7 g/cm^3. Suppose you would like to produce a composite material based on aluminum having a density of 1.5 g/cm^3. Design a material that would have this density. Would introducing beads of polyethylene, with a density of 0.95 g/cm^3, into the aluminum be a likely possibility? Explain.

1-15 You would like to be able to identify different materials without resorting to chemical analysis or lengthy testing procedures. Describe some possible testing and sorting techniques you might be able to use based on the physical properties of materials.

1-16 You would like to be able to physically separate different materials in a scrap recycling plant. Describe some possible methods that might be used to separate materials such as polymers, aluminum alloys, and steels from one another.

1-17 Some pistons for automobile engines might be produced from a composite material containing small, hard silicon carbide particles in an aluminum alloy matrix. Explain what benefits each material in the composite may provide to the overall part. What problems might the different properties of the two materials cause in producing the part?

1-18 Look up information on materials known as Geofoam. How are these materials used to reinforce ground that may be otherwise unstable?

1-19 An airplane made using primarily aluminum alloys weighs 2300 kg. What will be the weight of this airplane if it is made using primarily carbon-fiber composites?

1-20 Ladders can be made using aluminum alloy, fiber-glass, and wood. What will be the pros and cons of using each of these materials? One thing to keep in mind is that aluminum alloys are good conductors of electricity.

1-21 Replacing about half of the steel-based materials in a car can reduce the weight of the car by almost 60%. This can lead to nearly a 30% increase in fuel efficiency. What kinds of materials could replace steel in cars? What would be the advantages and disadvantages in using these materials?

2

Atomic Structure

Have You Ever Wondered?

- *What is nanotechnology?*
- *Why carbon, in the form of diamond, is one of the hardest materials known, but as graphite is very soft and can be used as a solid lubricant?*
- *How silica, which forms the main chemical in beach sand, is used in an ultrapure form to make optical fibers?*

The goal of this chapter is to describe the underlying physical concepts related to the structure of matter. You will learn that the structure of atoms affects the types of bonds that exist in different types of materials. These different types of bonds directly affect suitability of materials for real-world engineering applications.

Both the **composition** and the **structure** of a material have a profound influence on its properties and behavior. Engineers and scientists who study and develop materials must understand their atomic structure. The properties of materials are controllable and can be tailored to the needs of a given application by controlling their structure and composition.

We can examine and describe the structure of materials at five different levels:

1. macrostructure;
2. microstructure;
3. nanostructure;
4. short- and long-range atomic arrangements; and
5. atomic structure.

Engineers and scientists concerned with development and practical applications of advanced materials need to understand the **microstructure** and **macrostructure** of various materials and the ways of controlling them. Microstructure is the structure of material at a **length-scale** of $\sim$10 to 1000 nm. Length-scale is a characteristic length or range of dimensions over which we are describing the properties of a material or the phenomena occurring in materials. Microstructure typically includes such features as average grain size, grain size distribution, grain shape, grain orientation, and other features related to defects in materials. A grain is a small crystal of the material within which the arrangement of atoms and repeats in a particular fashion in all three dimensions. Macrostructure is the structure of a material at a macroscopic level where the length-scale is $\sim>$100,000 nm. Features that constitute macrostructure include porosity, surface coatings, and such features as internal or external micro-cracks.

It is also important to understand **atomic structure** and how the atomic bonds lead to different atomic or ionic arrangements in materials. The atomic structure includes all atoms and their arrangements, which constitute the building blocks of matter. It is from these building blocks that all the nano, micro, and macrolevels of structures emerge. The insights gained by understanding atomic structure and bonding configurations of atoms and molecules are essential for the proper selection of engineering materials, as well as for developing new, advanced materials.

A close examination of atomic arrangement allows us to distinguish between materials that are **amorphous** or **crystalline** (those that exhibit periodic arrangements of atoms or ions). Amorphous materials have only **short-range atomic arrangements** while crystalline materials have short- and **long-range arrangements**. In short-range atomic arrangements, the atoms or ions show a particular order only over relatively short distances. For crystalline materials, the long-range atomic order is in the form of atoms or ions arranged in a three-dimensional pattern that repeats over much larger distances (from $\sim>$100 nm to up to few cm).

2-1 The Structure of Materials: Technological Relevance

In today's world, information technology (IT), biotechnology, energy technology, environmental technology, and many other areas require smaller, lighter, faster, portable, more efficient, reliable, durable, and inexpensive devices. We want batteries that are smaller, lighter, and longer lasting. We need cars that are affordable, lightweight, safe, highly fuel efficient, and "loaded" with many advanced features, ranging from global positioning systems (GPS) to sophisticated sensors for airbag deployment.

Some of these needs have generated considerable interest in **nanotechnology** and **micro-electro-mechanical systems** (MEMS). A real-world example of the MEMS technology, Figure 2-1 shows a small accelerometer sensor obtained by the micromachining of silicon (Si). This sensor is used to measure acceleration in automobiles. The information

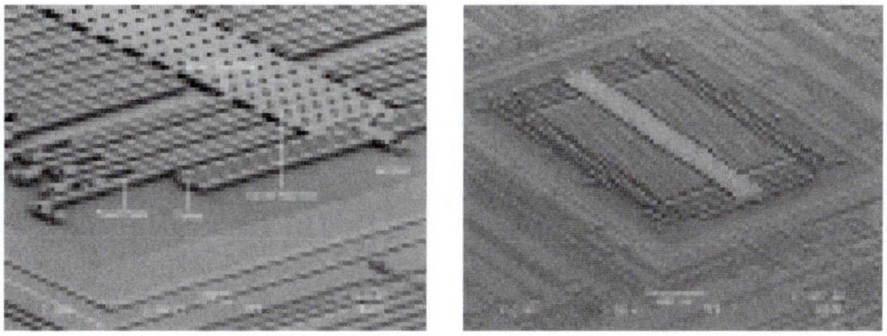

(a) (b)

Figure 2-1 Micro-machined silicon sensors used in automotives to control airbag deployment. (*Courtesy of Dr. John Yasaitis, Analog Devices, Inc.*)

is processed by a central computer and then used for controlling airbag deployment. Properties and behavior of materials at these "micro" levels can vary greatly when compared to those in their "macro" or bulk state. As a result, understanding the structure at nano-scale or **nanostructure** (i.e., the structure and properties of materials at a **nano-scale** or ~length-scale 1–100 nm) and microstructure are areas that have received considerable attention. The term "nanotechnology" is used to describe a set of technologies that are based on physical, chemical, and biological phenomena occuring at a nano-scale.

The applications shown in Table 2-1 and accompanying figures (Figures 2-2 through 2-7) illustrate how important the different levels of structure are to the material behavior. The applications illustrated are broken out by their levels of structure and their length-scales (the approximate characteristic length that is important for a given application). Examples of how such an application would be used within industry, as well as an illustration, are also provided.

We now turn our attention to the details concerning the structure of atoms, the bonding between atoms, and how these form a foundation for the properties of materials. Atomic structure influences how atoms are bonded together. An understanding of this helps categorize materials as metals, semiconductors, ceramics, or polymers. It also permits us to draw some preliminary conclusions concerning the general mechanical properties and physical behaviors of these four classes of materials.

2-2 The Structure of the Atom

An atom is composed of a nucleus surrounded by electrons. The nucleus contains neutrons and positively charged protons and carries a net positive charge. The negatively charged electrons are held around the nucleus by an electrostatic attraction. The electrical charge q carried by each electron and proton is 1.60×10^{-19} coulomb (C). Because the numbers of electrons and protons in the atom are equal, the atom as a whole is electrically neutral.

The **atomic number** of an element is equal to the number of electrons or protons in each atom. Thus, an iron atom, which contains 26 electrons and 26 protons, has an atomic number of 26.

Most of the mass of the atom is contained within its nucleus. The mass of each proton and neutron is 1.67×10^{-24} g, but the mass of each electron is only 9.11×10^{-28} g. The **atomic mass** M, which is equal to the average number of protons and

TABLE 2-1 ■ *Levels of structures*

Level of Structure	Example of Technologies
Atomic Structure	*Diamond*: Diamond is based on carbon-carbon (C-C) covalent bonds. Materials with this type of bonding are expected to be relatively hard. Thin films of diamonds are used for providing a wear-resistant edge in cutting tools.
Atomic Arrangements: Long-Range Order (LRO)	*Lead-zirconium-titanate [$Pb(Zr_x Ti_{1-x})O_3$] or PZT:* When ions in this material are arranged such that they exhibit tetragonal and/or rhombohedral crystal structures, the material is piezoelectric (i.e., it develops a voltage when subjected to pressure or stress). PZT ceramics are used widely for many applications including gas igniters, ultrasound generation, and vibration control.
Atomic Arrangements: Short-Range Order (SRO)	*Ions in silica-based (SiO_2) glasses* exhibit only a short-range order in which Si^{+4} and O^{-2} ions are arranged in a particular way (each Si^{+4} is bonded with 4 O^{-2} ions in a tetrahedral coordination). This order, however, is not maintained over long distances, thus making silica-based glasses amorphous. Amorphous glasses based on silica and certain other oxides form the basis for the entire fiber optical communications industry.

	Approximate Length-Scale
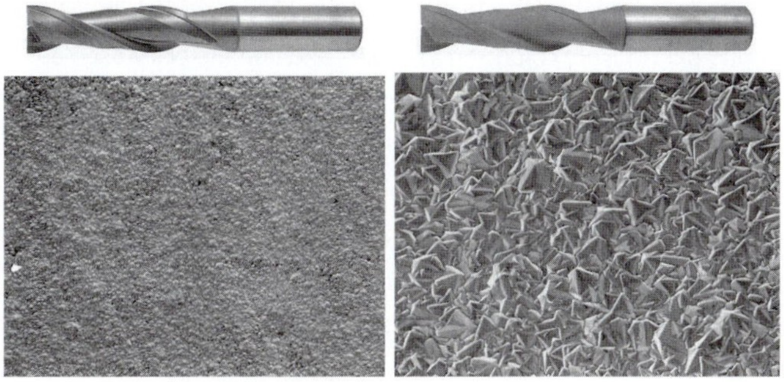 **Figure 2-2** Diamond-coated cutting tools. (*Courtesy of OSG Tap & Die, Inc.*)	~Up to 10^{-10} m (1 A)
 **Figure 2-3** Piezoelectric PZT-based gas igniters. When the piezoelectric material is stressed (by applying a pressure) a voltage develops and a spark is created between the electrodes. (*Courtesy of Morgan Electro Ceramics, Inc.*)	~10^{-10} to 10^{-9} m (1 to 10 A), ordering can exist up to a few cm in larger crystals
 **Figure 2-4** Optical fibers based on a form of silica that is amorphous. (*Courtesy of Corning Incorporated.*)	~10^{-10} to 10^{-9} m (1 to 10 A)

TABLE 2-1 (continued)

Level of Structure	Example of Technologies
Nanostructure	Nano-sized particles (~5–10 nm) of iron oxide are used in ferrofluids or liquid magnets. These nano-sized iron oxide particles are dispersed in liquids and commercially used as ferrofluids. An application of these liquid magnets is as a cooling (heat transfer) medium for loudspeakers.
Microstructure	The mechanical strength of many metals and alloys depends very strongly on the grain size. The grains and grain boundaries in this accompanying micrograph of steel are part of the microstructural features of this crystalline material. In general, at room temperature a finer grain size leads to higher strength. Many important properties of materials are sensitive to the microstructure.
Macrostructure	Relatively thick coatings, such as paints on automobiles and other applications, are used not only for aesthetics, but also to provide corrosion resistance.

	Approximate Length-Scale
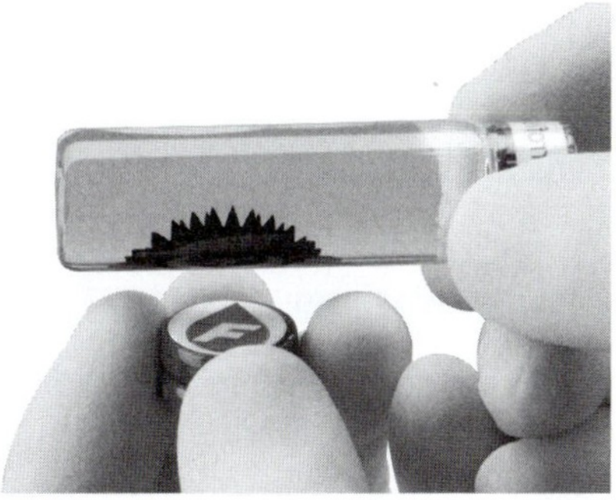 **Figure 2-5** Ferrofluid nanoparticles responding to a magnet. (*Courtesy of Ferro Tec, Inc.*)	~10^{-9} to 10^{-7} m (1 to 100 nm)
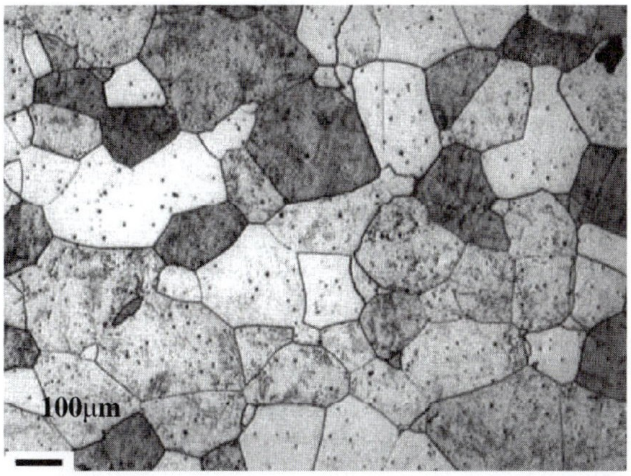 **Figure 2-6** Micrograph of stainless steel showing grains and grain boundaries. (*Courtesy of Dr. Hua and Dr. DeArdo—University of Pittsburgh.*)	~>10^{-8} to 10^{-6} m (10 nm to 1000 nm)
 **Figure 2-7** A number of organic and inorganic coatings protect the steel in the car from corrosion and provide a pleasing appearance. (*Courtesy of Ford Motor Company.*)	~>10^{-4} m (100,000 nm)

neutrons in the atom, is also the mass in grams of the **Avogadro number** N_A of atoms. The quantity $N_A = 6.02 \times 10^{23}$ atoms/mol is the number of atoms or molecules in a mole. Therefore, the atomic mass has units of g/mol. An alternate unit for atomic mass is the **atomic mass unit**, or amu, which is 1/12 the mass of carbon 12 (i.e., carbon atom with 12 protons). As an example, one mole of iron contains 6.02×10^{23} atoms and has a mass of 55.847 g, or 55.847 amu. Calculations including a material's atomic mass and Avogadro's number are helpful to understanding more about the structure of a material. Example 2-1 illustrates applications to magnetic materials.

EXAMPLE 2-1 *Nano-Sized Iron-Platinum Particles For Information Storage*

Scientists are considering using nano-particles of such magnetic materials as iron-platinum (Fe-Pt) as a medium for ultrahigh density data storage. Arrays of such particles potentially can lead to storage of trillions of bits of data per square inch—a capacity that will be 10 to 100 times higher than any other devices such as computer hard disks. Consider iron (Fe) particles that are 3 nm in diameter, what will be the number of atoms in one such particle?

SOLUTION

Let us assume the magnetic particles are spherical in shape.

The radius of a particle is 1.5 nm.

$$\text{Volume of each iron magnetic nano-particle} = (4/3)\pi(1.5 \times 10^{-7}\ \text{cm})^3$$
$$= 1.4137 \times 10^{-20}\ \text{cm}^3$$

Density of iron = 7.8 g/cm^3. Atomic mass of iron is 56 g/mol.

$$\text{Mass of each iron nano-particle} = 7.8\ \text{g/cm}^3 \times 1.4137 \times 10^{-20}\ \text{cm}^3$$
$$= 1.102 \times 10^{-19}\ \text{g.}$$

One mole or 56 g of Fe contains 6.023×10^{23} atoms, therefore, the number of atoms in one Fe nano-particle will be 1186. This is a very small number of atoms. Compare this with the number of atoms in an iron particle that is 10 micrometers in diameter. Such larger iron particles often are used in breakfast cereals, vitamin tablets, and other applications.

2-3 The Electronic Structure of the Atom

For some students, reviewing some of these concepts from introductory Chemistry courses will be useful. Electrons occupy discrete energy levels within the atom. Each electron possesses a particular energy, with no more than two electrons in each atom having the same energy. This also implies that there is a discrete energy difference between any two energy levels.

Quantum Numbers The energy level to which each electron belongs is determined by four **quantum numbers**. Quantum numbers are the numbers in an atom that assign electrons to discrete energy levels. The four quantum numbers are the principal **quantum**

number n, the **Azimuthal quantum number** l, the **magnetic quantum number** m_l, and the **spin quantum number** m_s. Azimuthal quantum numbers describe the energy levels in each quantum shell. The spin quantum number (m_s) is assigned values $+1/2$ and $-1/2$ and reflects the different electron spins. The number of possible energy levels is determined by the first three quantum numbers.

The shorthand notation frequently used to denote the electronic structure of an atom combines the numerical value of the principal quantum number, the lowercase letter notation for the azimuthal quantum number, and a superscript showing the number of electrons in each orbital. The shorthand notation for germanium (Ge), which has an atomic number of 32, is:

$$1s^2 2s^2 2p^6 3s^2 3p^6 3d^{10} 4s^2 4p^2$$

Deviations from Expected Electronic Structures The orderly building up of the electronic structure is not always followed, particularly when the atomic number is large and the d and f levels begin to fill. For example, we would expect the electronic structure of iron atom, atomic number 26, to be:

$$1s^2 2s^2 2p^6 3s^2 3p^6 \ \boxed{3d^8}$$

The actual electronic structure, however, is:

$$1s^2 2s^2 2p^6 3s^2 3p^6 \ \boxed{3d^6 4s^2}$$

The unfilled $3d$ level causes the magnetic behavior of iron.

Valence The **valence** of an atom is the number of electrons in an atom that participate in bonding or chemical reactions. Usually, the valence is the number of electrons in the outer s and p energy levels. The valence of an atom is related to the ability of the atom to enter into chemical combination with other elements. Examples of the valence are:

$$\text{Mg: } 1s^2 2s^2 2p^6 \ \boxed{3s^2} \qquad \text{valence} = 2$$

$$\text{Al: } 1s^2 2s^2 2p^6 \ \boxed{3s^2 3p^1} \qquad \text{valence} = 3$$

$$\text{Ge: } 1s^2 2s^2 2p^6 3s^2 3p^6 3d^{10} \ \boxed{4s^2 4p^2} \qquad \text{valence} = 4$$

Valence also depends on the immediate environment surrounding the atom or the neighboring atoms available for bonding. Phosphorus has a valence of five when it combines with oxygen. But the valence of phosphorus is only three—the electrons in the $3p$ level—when it reacts with hydrogen. Manganese may have a valence of 2, 3, 4, 6, or 7!

Atomic Stability and Electronegativity If an atom has a valence of zero, the element is inert (non-reactive). An example is argon (Ar), which has the electronic structure:

$$1s^2 2s^2 2p^6 \ \boxed{3s^2 3p^6}$$

Other atoms prefer to behave as if their outer s and p levels are either completely full, with eight electrons, or completely empty. Aluminum has three electrons in its outer s and p levels. An aluminum atom readily gives up its outer three electrons to empty the $3s$ and $3p$ levels. The atomic bonding and the chemical behavior of aluminum are determined by the mechanism through which these three electrons interact with surrounding atoms.

On the other hand, chlorine contains seven electrons in the outer $3s$ and $3p$ levels. The reactivity of chlorine is caused by its desire to fill its outer energy level by accepting an electron.

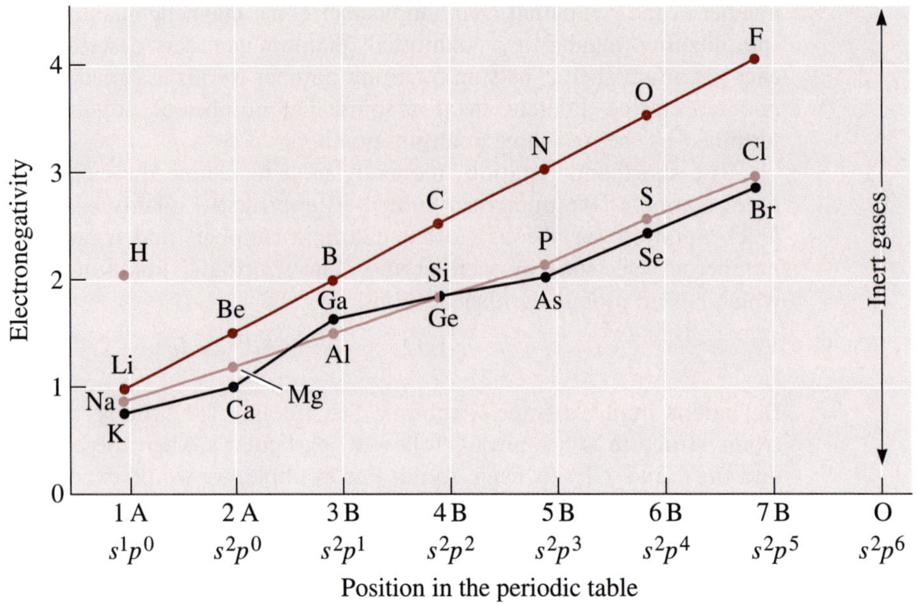

Figure 2-8 The electronegativities of selected elements relative to the position of the elements in the periodic table.

Electronegativity describes the tendency of an atom to gain an electron. Atoms with almost completely filled outer energy levels—such as chlorine—are strongly electronegative and readily accept electrons. However, atoms with nearly empty outer levels—such as sodium—readily give up electrons and have low electronegativity. High atomic number elements also have low electronegativity because the outer electrons are at a greater distance from the positive nucleus, so that they are not as strongly attracted to the atom. Electronegativities for some elements are shown in Figure 2-8. *Note*: The symbol O on the *x*-axis is group zero and not for oxygen.

2-4 The Periodic Table

The periodic table contains valuable information about specific elements, and can also help identify trends in atomic size, melting point, chemical reactivity, and other properties. The familiar periodic table (Figure 2-9) is constructed in accordance with the electronic structure of the elements. Not all elements in the periodic table are naturally occurring. Rows in the periodic table correspond to quantum shells, or principal quantum numbers. Columns typically refer to the number of electrons in the outermost *s* and *p* energy levels and correspond to the most common valence. In engineering, we are mostly concerned with:

(a) polymers (plastics) (primarily based on carbon, which appears in group 4B);

(b) ceramics (typically based on combinations of many elements appearing in Groups 1 through 5B, and such elements as oxygen, carbon, and nitrogen); and

(c) metallic materials (typically based on elements in Groups 1, 2 and transition metal elements).

Many technologically important semiconductors appear in group 4B (e.g., silicon (Si), diamond (C), germanium (Ge)). Semiconductors also can be combinations of

Key: Atomic number — Electron configuration — Symbol — Melting point [°C] — Boiling point [°C] — Atomic mass (mean relative) — Oxidation states — Name — Density [g/cm3], for gases [g/l] (0° C, 1013mbar) — Electronegativity — * Radioactive

Example: 43 (98.91) [Kr] $4d^5 5s^2$ 7 Tc Technetium 2140 11.5 5030 * 1.9

Period	Group	Atomic number	Symbol	Name	Atomic mass	Electron configuration	Oxidation states	Melting point [°C]	Boiling point [°C]	Density	Electronegativity
1	1 1A	1	H	Hydrogen	1.01	$1s$	1, −1	−259	−253	0.09	2.1
1	18 0	2	He	Helium	4.00	$1s^2$		—	−269	0.18	—
2	1 1A	3	Li	Lithium	6.94	[He] $2s$	1	181	1330	0.53	1.0
2	2 2A	4	Be	Beryllium	9.01	[He] $2s^2$	2	1277	2970	1.85	1.5
2	13 3B	5	B	Boron	10.81	[He] $2s^2 2p$	3	(2030)	2550	2.35	2.0
2	14 4B	6	C	Carbon	12.01	[He] $2s^2 2p^2$	4, 2, −4	(3550)	4830	2.2	2.5
2	15 5B	7	N	Nitrogen	14.01	[He] $2s^2 2p^3$	5, 4, 3, 2, −3	−210	−196	1.25	3.0
2	16 6B	8	O	Oxygen	16.00	[He] $2s^2 2p^4$	−2, −1	−219	−183	1.43	3.5
2	17 7B	9	F	Fluorine	19.00	[He] $2s^2 2p^5$	−1	−220	−188	1.7	4.0
2	18 0	10	Ne	Neon	20.18	[He] $2s^2 3p^5$		−249	−246	0.9	—
3	1 1A	11	Na	Sodium	22.99	[Ne] $3s$	1	98	892	0.97	0.9
3	2 2A	12	Mg	Magnesium	24.31	[Ne] $3s^2$	2	650	1107	1.74	1.2
3	13 3B	13	Al	Aluminum	26.98	[Ne] $3s^2 3p$	3	660	2450	2.70	1.5
3	14 4B	14	Si	Silicon	28.09	[Ne] $3s^2 3p^2$	4, −4	1410	2680	2.33	1.8
3	15 5B	15	P	Phosphorus	30.97	[Ne] $3s^2 3p^3$	5, 3, −3	44	280	1.82	2.1
3	16 6B	16	S	Sulfur	32.06	[Ne] $3s^2 3p^4$	6, 4, 2, −2	119	445	2.07	2.5
3	17 7B	17	Cl	Chlorine	35.45	[Ne] $3s^2 3p^5$	7, 5, 3, 1, −1	−101	−35	3.2	3.0
3	18 0	18	Ar	Argon	39.95	[Ne] $3s^2 3p^6$		−189	−183	1.78	—
4	1 1A	19	K	Potassium	39.10	[Ar] $4s$	1	64	760	0.86	0.8
4	2 2A	20	Ca	Calcium	40.08	[Ar] $4s^2$	2	838	1440	1.55	1.0
4	3 3A	21	Sc	Scandium	44.96	[Ar] $3d\,4s^2$	3	1539	2730	3.0	1.3
4	4 4A	22	Ti	Titanium	47.88	[Ar] $3d^2 4s^2$	4, 3	1668	3260	4.54	1.5
4	5 5A	23	V	Vanadium	50.94	[Ar] $3d^3 4s^2$	5, 4, 3, 2, 0	1900	3450	6.1	1.5
4	6 6A	24	Cr	Chromium	52.00	[Ar] $3d^5 4s$	6, 3, 2, 0	1875	2200	7.19	1.6
4	7 7A	25	Mn	Manganese	54.94	[Ar] $3d^5 4s^2$	7, 6, 4, 3, 2, 0, −1	1245	2097	7.43	1.5
4	8 8	26	Fe	Iron	55.85	[Ar] $3d^6 4s^2$	6, 3, 2, 0, −2	1536	3000	7.86	1.8
4	9 8	27	Co	Cobalt	58.93	[Ar] $3d^7 4s^2$	3, 2, 0, −1	1495	2900	8.9	1.9
4	10 8	28	Ni	Nickel	58.70	[Ar] $3d^8 4s^2$	3, 2, 0	1453	2730	8.9	1.9
4	11 1B	29	Cu	Copper	63.55	[Ar] $3d^{10} 4s$	2, 1	1080	2562	8.9	1.9
4	12 2B	30	Zn	Zinc	65.38	[Ar] $3d^{10} 4s^2$	2	420	906	7.13	1.6
4	13 3B	31	Ga	Gallium	69.72	[Ar] $3d^{10} 4s^2 4p$	3	30	2237	5.91	1.6
4	14 4B	32	Ge	Germanium	72.59	[Ar] $3d^{10} 4s^2 4p^2$	4	937	2830	5.32	1.8
4	15 5B	33	As	Arsenic	74.91	[Ar] $3d^{10} 4s^2 4p^3$	5, 3, −3	Subl.	—	5.72	2.0
4	16 6B	34	Se	Selenium	74.91	[Ar] $3d^{10} 4s^2 4p^4$	6, 4, 2	217	685	4.79	2.4
4	17 7B	35	Br	Bromine	79.90	[Ar] $3d^{10} 4s^2 4p^5$	7, 5, 3, 1, −1	−7	58	3.12	2.8
4	18 0	36	Kr	Krypton	83.80	[Ar] $3d^{10} 4s^2 4p^6$		−157	−152	3.7	—
5	1 1A	37	Rb	Rubidium	85.47	[Kr] $5s$	1	39	688	1.53	0.8
5	2 2A	38	Sr	Strontium	87.62	[Kr] $5s^2$	2	768	1380	2.6	1.0
5	3 3A	39	Y	Yttrium	88.91	[Kr] $4d\,5s^2$	3	1509	2927	4.47	1.2
5	4 4A	40	Zr	Zirconium	91.22	[Kr] $4d^2 5s^2$	4	1852	3580	6.49	1.4
5	5 5A	41	Nb	Niobium	91.22	[Kr] $4d^4 5s$	5, 3	2468	4927	8.57	1.6
5	6 6A	42	Mo	Molybdenum	95.94	[Kr] $4d^5 5s$	6, 5, 4, 3, 2, 0	2610	5560	10.2	1.8
5	7 7A	43	Tc	Technetium	(98.91)	[Kr] $4d^5 5s^2$	7	2140	5030 *	11.5	1.9
5	8 8	44	Ru	Ruthenium	101.07	[Kr] $4d^7 5s$	8, 6, 4, 3, 2, 0, −2	2500	3900	12.4	2.2
5	9 8	45	Rh	Rhodium	102.91	[Kr] $4d^8 5s$	5, 4, 3, 1, 2, 0	1966	3730	12.4	2.2
5	10 8	46	Pd	Palladium	106.42	[Kr] $4d^{10}$	4, 2, 0	1552	3140	12.0	2.2
5	11 1B	47	Ag	Silver	107.87	[Kr] $4d^{10} 5s$	2, 1	961	2210	10.5	1.9
5	12 2B	48	Cd	Cadmium	112.41	[Kr] $4d^{10} 5s^2$	2	321	765	8.65	1.7
5	13 3B	49	In	Indium	114.82	[Kr] $4d^{10} 5s^2 5p$	3	156	2080	7.31	1.7
5	14 4B	50	Sn	Tin	118.69	[Kr] $4d^{10} 5s^2 5p^2$	4, 2	232	2270	7.30	1.8
5	15 5B	51	Sb	Antimony	121.75	[Kr] $4d^{10} 5s^2 5p^3$	5, 3, −3	631	1380	6.69	1.9
5	16 6B	52	Te	Tellurium	127.60	[Kr] $3d^{10} 5s^2 5p^4$	6, 4, −2	450	990	6.24	2.1
5	17 7B	53	I	Iodine	126.90	[Kr] $4d^{10} 5s^2 5p^5$	7, 5, 1, −1	114	183	4.94	2.5
5	18 0	54	Xe	Xenon	131.29	[Kr] $4d^{10} 5s^2 5p^6$		−112	−108	5.89	—
6	1 1A	55	Cs	Cesium	132.91	[Xe] $6s$	1	29	690	1.90	0.7
6	2 2A	56	Ba	Barium	137.33	[Xe] $6s^2$	2	714	1640	3.76	0.9
6	3 3A	71	Lu	Lutetium	174.97	[Xe] $4f^{14} 5d\,6s^2$	3	1652	3327	9.84	1.2
6	4 4A	72	Hf	Hafnium	178.49	[Xe] $4f^{14} 5d^2 6s^2$	4	2222	5400	13.31	1.3
6	5 5A	73	Ta	Tantalum	180.95	[Xe] $4f^{14} 5d^3 6s^2$	5	2996	5425	16.5	1.5
6	6 6A	74	W	Tungsten	183.85	[Xe] $4f^{14} 5d^4 6s^2$	6, 5, 4, 3, 2, 0	3410	5930	19.3	1.7
6	7 7A	75	Re	Rhenium	186.21	[Xe] $4f^{14} 5d^5 6s^2$	7, 6, 4, 2, −1	3180	5900	21.0	1.9
6	8 8	76	Os	Osmium	190.20	[Xe] $4f^{14} 5d^6 6s^2$	8, 6, 4, 3, 2, 0, −2	3050	5500	22.6	2.2
6	9 8	77	Ir	Iridium	192.22	[Xe] $4f^{14} 5d^7 6s^2$	6, 4, 3, 2, 1, 0, −1	2454	4500	22.7	2.2
6	10 8	78	Pt	Platinum	195.08	[Xe] $4f^{14} 5d^9 6s^2$	4, 2, 0	1769	3830	21.4	2.2
6	11 1B	79	Au	Gold	196.97	[Xe] $4f^{14} 5d^{10} 6s$	3, 1	1063	2970	19.3	2.4
6	12 2B	80	Hg	Mercury	200.59	[Xe] $4f^{14} 5d^{10} 6s^2$	2, 1	−38	357	13.6	1.9
6	13 3B	81	Tl	Thallium	204.38	[Xe] $4f^{14} 5d^{10} 6s^2 6p$	3, 1	303	1457	11.85	1.8
6	14 4B	82	Pb	Lead	207.20	[Xe] $4f^{14} 5d^{10} 6s^2 6p^2$	4, 2	327	1725	11.4	1.9
6	15 5B	83	Bi	Bismuth	208.98	[Xe] $4f^{14} 5d^{10} 6s^2 6p^3$	5, 3	271	1560	9.8	1.9
6	16 6B	84	Po	Polonium	(209)	[Xe] $4f^{14} 5d^{10} 6s^2 6p^4$	6, 4, 2	254	962	9.3	2.0
6	17 7B	85	At	Astatine	(210)	[Xe] $4f^{14} 5d^{10} 6s^2 6p^5$	7, 5, 3, 1, −1	(302)	337	—	2.2
6	18 0	86	Rn	Radon	(222)	[Xe] $4f^{14} 5d^{10} 6s^2 6p^6$		(−71)	−62	9.73	—
7	1 1A	87	Fr	Francium	(223.0)	[Rn] $7s$	1	(27)	677 *	—	0.7
7	2 2A	88	Ra	Radium	226.03	[Rn] $7s^2$	2	700	1140 *	5.0	0.9
7	3 3A	103	Lr	Lawrencium	(262.1)	[Rn] $5f^{14} 6d\,7s^2$	3	—	— *	—	—
7	4 4A	104	Rf	Rutherfordium	(261.1)	[Rn] $5f^{14} 6d^2 7s^2$		—	— *	—	—
7	5 5A	105	Db	Dubnium	(262.1)	[Rn] $5f^{14} 6d^3 7s^2$		—	— *	—	—
7	6 6A	106	Sg	Seaborgium	(263.1)	[Rn] $5f^{14} 6d^4 7s^2$		—	— *	—	—
7	7 7A	107	Bh	Bohrium	(264.1)	[Rn] $5f^{14} 6d^5 7s^2$		—	— *	—	—
7	8 8	108	Hs	Hassium	(265.1)	[Rn] $5f^{14} 6d^6 7s^2$		—	— *	—	—
7	9 8	109	Mt	Meitnerium	(268)	[Rn] $5f^{14} 6d^7 7s^2$		—	— *	—	—
7	10 8	110	Uun		(269)	[Rn] $5f^{14} 6d^8 7s^2$		—	— *	—	—
7	11 1B	111	Uuu		(272)	[Rn] $5f^{14} 6d^{10} 7s^2$		—	— *	—	—
7	12 2B	112	Uub		(277)			—	— *	—	—
7	14 4B	114	Uuq		(289)			—	— *	—	—
7	16 6B	116	Uuh		(289)			—	— *	—	—
7	18 0	118	Uuo		(293)			—	— *	—	—
6	Lanthanoids	57	La	Lanthanum	138.91	[Xe] $5d\,6s^2$	3	920	3470	6.17	1.1
6	Lanthanoids	58	Ce	Cerium	140.12	[Xe] $4f\,5d\,6s^2$	4, 3	795	3468	6.67	1.1
6	Lanthanoids	59	Pr	Praseodymium	140.91	[Xe] $4f^3 6s^2$	4, 3	935	3127	6.77	1.1
6	Lanthanoids	60	Nd	Neodymium	144.24	[Xe] $4f^4 6s^2$	3	1024	3027	7.00	1.2
6	Lanthanoids	61	Pm	Promethium	(145)	[Xe] $4f^5 6s^2$	3	(1027)	2460 *	7.22	—
6	Lanthanoids	62	Sm	Samarium	150.36	[Xe] $4f^6 6s^2$	3, 2	1072	1790	7.54	1.2
6	Lanthanoids	63	Eu	Europium	151.96	[Xe] $4f^7 6s^2$	3, 2	828	1439	5.26	—
6	Lanthanoids	64	Gd	Gadolinium	157.25	[Xe] $4f^7 5d\,6s^2$	3	1312	3000	7.89	1.1
6	Lanthanoids	65	Tb	Terbium	158.93	[Xe] $4f^9 6s^2$	4, 3	1356	2800	8.27	1.2
6	Lanthanoids	66	Dy	Dysprosium	162.50	[Xe] $4f^{10} 6s^2$	3	1407	2600	8.54	—
6	Lanthanoids	67	Ho	Holmium	164.93	[Xe] $4f^{11} 6s^2$	3	1461	2600	8.80	1.2
6	Lanthanoids	68	Er	Erbium	167.26	[Xe] $4f^{12} 6s^2$	3	1497	2900	9.05	1.2
6	Lanthanoids	69	Tm	Thulium	168.93	[Xe] $4f^{13} 6s^2$	3, 2	1545	1727	9.33	1.2
6	Lanthanoids	70	Yb	Ytterbium	173.04	[Xe] $4f^{14} 6s^2$	3, 2	824	1196	6.98	1.1
7	Actinoids	89	Ac	Actinium	(227)	[Rn] $6d\,7s^2$	3	1050	— *	10.1	1.1
7	Actinoids	90	Th	Thorium	232.04	[Rn] $6d^2 7s^2$	4	1750	3850 *	11.7	1.3
7	Actinoids	91	Pa	Protactinium	231.04	[Rn] $5f^2 6d\,7s^2$	5, 4	(1230)	— *	15.4	1.4
7	Actinoids	92	U	Uranium	238.03	[Rn] $5f^3 6d\,7s^2$	6, 5, 4, 3	1132	3818 *	19.07	1.4
7	Actinoids	93	Np	Neptunium	(237.05)	[Rn] $5f^4 6d\,7s^2$	6, 5, 4, 3	637	3900 *	19.5	1.3
7	Actinoids	94	Pu	Plutonium	(244)	[Rn] $5f^6 7s^2$	6, 5, 4, 3	640	3235 *	19.81	1.3
7	Actinoids	95	Am	Americium	(243)	[Rn] $5f^7 7s^2$	6, 5, 4, 3	994	— *	13.7	1.3
7	Actinoids	96	Cm	Curium	(247)	[Rn] $5f^7 6d\,7s^2$	4, 3	(1340)	3100 *	13.51	—
7	Actinoids	97	Bk	Berkelium	(247)	[Rn] $5f^9 7s^2$	4, 3	—	— *	—	—
7	Actinoids	98	Cf	Californium	(251)	[Rn] $5f^{10} 7s^2$	4, 3	—	— *	—	—
7	Actinoids	99	Es	Einsteinium	(254)	[Rn] $5f^{11} 7s^2$	3	—	— *	—	—
7	Actinoids	100	Fm	Fermium	(257)	[Rn] $5f^{12} 7s^2$	3	—	— *	—	—
7	Actinoids	101	Md	Mendelevium	(258)	[Rn] $5f^{13} 7s^2$	3	—	— *	—	—
7	Actinoids	102	No	Nobelium	(259)	[Rn] $5f^{14} 7s^2$	3, 2	—	— *	—	—

Figure 2-9 Periodic table of elements.

elements from groups 2B and 6B (e.g., cadmium selenide (CdSe), based on cadmium (Cd) from group 2 and seleneium (Se) based on Group 6). These are known as **II–VI** (two-six) **semiconductors**. Similarly, gallium arsenide (GaAs) is a **III–V** (three-five) **semiconductor** based on gallium (Ga) from group 3B and arsenic (As) from group 5B. Many **transition elements** (e.g., titanium (Ti), vanadium (V), iron (Fe), nickel (Ni), cobalt (Co), etc.) are particularly useful for magnetic and optical materials due to their electronic configuration that allows multiple valencies.

Trends in Properties The periodic table contains a wealth of useful information (e.g., atomic mass, atomic number of different elements, etc.). It also points to trends in atomic size, melting points, and chemical reactivity. For example, carbon (in its diamond form) has the highest melting point (3550°C). Melting points of the elements below carbon decrease (i.e., silicon (Si) (1410°C), germanium (Ge) (937°C), tin (Sn) (232°C), and lead (Pb) (327°C). Note that the melting temperature of Pb is higher than that of Sn. What we can conclude is that the trends are not exact variations in properties.

We also can discern trends in other properties from the periodic table. Diamond (carbon), a group 4B element, is a material with a very large bandgap (i.e., it is not a very effective conductor of electricity). This is consistent with the fact that it has the highest melting point among group 4 elements, which suggests the interatomic forces are strong (see Section 2-6). As we move down the column, the bandgap decreases (the bandgaps of semiconductors Si and Ge are 1.11 and 0.67 eV, respectively). Moving further down column 4, one form of tin is a semiconductor. Another form of tin is metallic. If we look at group 1A, we see that lithium is highly **electropositive** (i.e., an element whose atoms want to participate in chemical interactions by donating electrons and are therefore highly reactive). Likewise, if we move down column 1A, we can see that the chemical reactivity of elements decreases.

2-5 Atomic Bonding

There are four important mechanisms by which atoms are bonded in engineered materials. These are:

1. **metallic bond**;
2. **covalent bond**;
3. **ionic bond**; and
4. **van der Waals bond**.

In the first three of these mechanisms, bonding is achieved when the atoms fill their outer *s* and *p* levels. These bonds are relatively strong and are known as **primary bonds** (relatively strong bonds between adjacent atoms resulting from the transfer or sharing of outer orbital electrons). The van der Waals bonds are secondary bonds and originate from a different mechanism and are relatively weaker. Let's look at each of these types of bonds.

The Metallic Bond The metallic elements have more electropositive atoms that donate their valence electrons to form a "sea" of electrons surrounding the atoms (Figure 2-10). Aluminum, for example, gives up its three valence electrons, leaving behind a core consisting of the nucleus and inner electrons. Since three negatively charged electrons are missing from this core, it has a positive charge of three. The valence electrons move

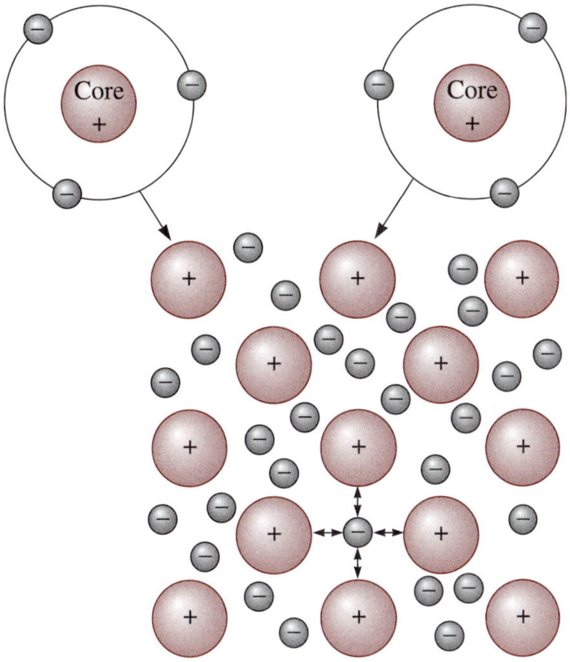

Figure 2-10
The metallic bond forms when atoms give up their valence electrons, which then form an electron sea. The positively charged atom cores are bonded by mutual attraction to the negatively charged electrons.

freely within the electron sea and become associated with several atom cores. The positively charged ion cores are held together by mutual attraction to the electron, thus producing a strong metallic bond.

Because their valence electrons are not fixed in any one position, most pure metals are good electrical conductors of electricity at relatively low temperatures ($\sim T <$ 300 K). Under the influence of an applied voltage, the valence electrons move, causing a current to flow if the circuit is complete.

Materials with metallic bonding exhibit relatively high Young's modulus since the bonds are strong. Metals also show good ductility since the metallic bonds are non-directional. There are other important reasons related to microstructure that can explain why metals actually exhibit *lower strengths* and *higher ductility* than what we may anticipate from their bonding. **Ductility** refers to the ability of materials to be stretched or bent without breaking. We will discuss these concepts in greater detail in Chapter 6. In general, the melting points of metals are relatively high. From an optical properties viewpoint, metals make good reflectors of visible radiation. Owing to their electropositive character, many metals such as iron tend to undergo corrosion or oxidation. Many pure metals are good conductors of heat and are effectively used in many heat transfer applications. We emphasize that metallic bonding is *one of the factors* in our efforts to rationalize the trends in observed properties of metallic materials. As we will see in some of the following chapters, there are other factors related to microstructure that also play a crucial role in determining the properties of metallic materials.

The Covalent Bond Materials with **covalent bonding** are characterized by bonds that are formed by sharing of valence electrons among two or more atoms. For example, a silicon atom, which has a valence of four, obtains eight electrons in its outer energy shell by sharing its electrons with four surrounding silicon atoms (Figure 2-11). Each instance of sharing represents one covalent bond; thus, each silicon atom is bonded to four neighboring atoms by four covalent bonds. In order for the covalent bonds to be

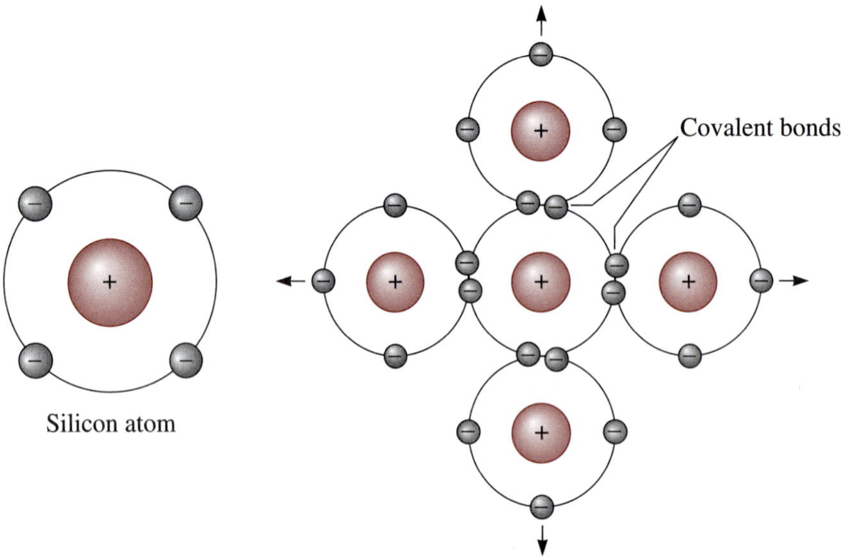

Figure 2-11 Covalent bonding requires that electrons be shared between atoms in such a way that each atom has its outer *sp* orbital filled. In silicon, with a valence of four, four covalent bonds must be formed for every atom.

formed, the silicon atoms must be arranged so the bonds have a fixed **directional relationship** with one another. A directional relationship is formed when the bonds between atoms in a covalently bonded material form specific angles, depending on the material. In the case of silicon, this arrangement produces a tetrahedron, with angles of 109.5° between the covalent bonds (Figure 2-12).

Covalent bonds are very strong. As a result, covalently bonded materials are very strong and hard. For example, diamond (C), silicon carbide (SiC), silicon nitride (Si_3N_4), and boron nitride (BN) all exhibit covalency. These materials also exhibit very high melting points, which means they could be useful for high-temperature applications. On the other hand, the temperature resistance of these materials presents challenges in their processing. The materials bonded in this manner typically have limited ductility because the bonds tend to be directional. The electrical conductivity of many covalently bonded materials (i.e., silicon, diamond, and many ceramics) is not high since the valence electrons are locked in bonds between atoms and are not readily available for conduction. With some of these materials, such as Si, we can get useful and controlled levels of electrical conductivity by deliberately introducing small levels of other elements known as dopants. Conductive polymers are also a good example of covalently bonded materials that can be turned into semiconducting materials. The development of conducting polymers that are lightweight has captured the attention of many scientists and engineers for developing flexible electronic components.

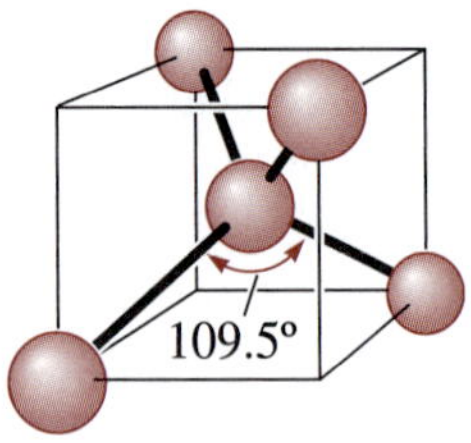

Figure 2-12
Covalent bonds are directional. In silicon, a tetrahedral structure is formed, with angles of 109.5° required between each covalent bond.

We cannot simply predict whether or not a material will be high or low strength, ductile or brittle, simply based on the nature of bonding! We need additional information on the atomic, microstructure, and macrostructure of the material. However, the nature of bonding does point to a trend for materials with certain types of bonding and chemical compositions. Example 2-2 explores how one such bond of oxygen and silicon join to form silica.

EXAMPLE 2-2 *How Do Oxygen and Silicon Atoms Join to Form Silica?*

Assuming that silica (SiO_2) has 100% covalent bonding, describe how oxygen and silicon atoms in silica (SiO_2) are joined.

SOLUTION

Silicon has a valence of four and shares electrons with four oxygen atoms, thus giving a total of eight electrons for each silicon atom. However, oxygen has a valence of six and shares electrons with two silicon atoms, giving oxygen a total of eight electrons. Figure 2-13 illustrates one of the possible structures. The arrows indicate to what other atom is a particular electron bonded with. Similar to silicon (Si), a tetrahedral structure is produced. We will discuss later in this chapter how to account for the ionic and covalent nature of bonding in silica.

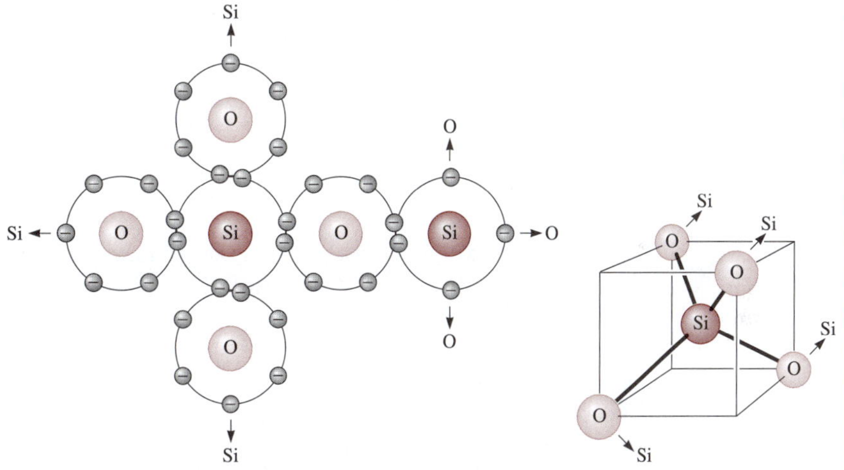

Figure 2-13 The tetrahedral structure of silica (SiO_2), which contains covalent bonds between silicon and oxygen atoms (for Example 2-2).

The Ionic Bond When more than one type of atoms are present in a material, one atom may donate its valence electrons to a different atom, filling the outer energy shell of the second atom. Both atoms now have filled (or emptied) outer energy levels, but both have acquired an electrical charge and behave as ions. The atom that contributes the electrons is left with a net positive charge and is called a **cation**, while the atom that accepts the electrons acquires a net negative charge and is called an **anion**. The oppositely charged ions are then attracted to one another and produce the **ionic bond**. For example, the attraction between sodium and chloride ions (Figure 2-14) produces sodium chloride (NaCl), or table salt.

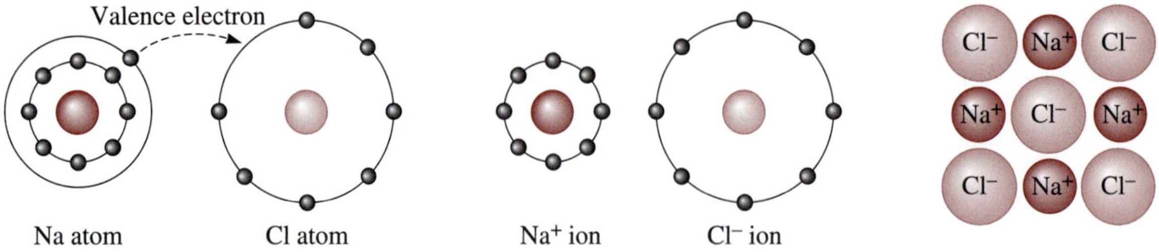

Figure 2-14 An ionic bond is created between two unlike atoms with different electronegativities. When sodium donates its valence electron to chlorine, each atom becomes an ion, and the ionic bond is formed.

EXAMPLE 2-3 *Describing the Ionic Bond Between Magnesium and Chlorine*

Describe the ionic bonding between magnesium and chlorine.

SOLUTION

The electronic structures and valences are

$$\text{Mg: } 1s^2 2s^2 2p^6 \; \boxed{3s^2} \qquad \text{valence electrons} = 2$$

$$\text{Cl: } 1s^2 2s^2 2p^6 \; \boxed{3s^2 3p^5} \qquad \text{valence electrons} = 7$$

Each magnesium atom gives up its two valence electrons, becoming a Mg^{2+} ion. Each chlorine atom accepts one electron, becoming a Cl^- ion. To satisfy the ionic bonding, there must be twice as many chloride ions as magnesium ions present, and a compound, $MgCl_2$, is formed.

Solids that exhibit considerable ionic bonding are also often mechanically strong because of the strength of the bonds. Electrical conductivity of ionically bonded solids is very limited. A large fraction of the electrical current is transferred via the movement of ions (Figure 2-15). Owing to their size, ions typically do not move as easily as electrons. However, in many technological applications we make use of the electrical conduction that can occur via movement of ions as a result of increased temperature, chemical potential gradient, or an electrochemical driving force. Examples of these include lithium ion batteries that make use of lithium cobalt oxide, conductive indium tin oxide (ITO) coatings on glass for touch sensitive displays, and solid oxide fuel cells (SOFC) based on compositions based on zirconia (ZrO_2).

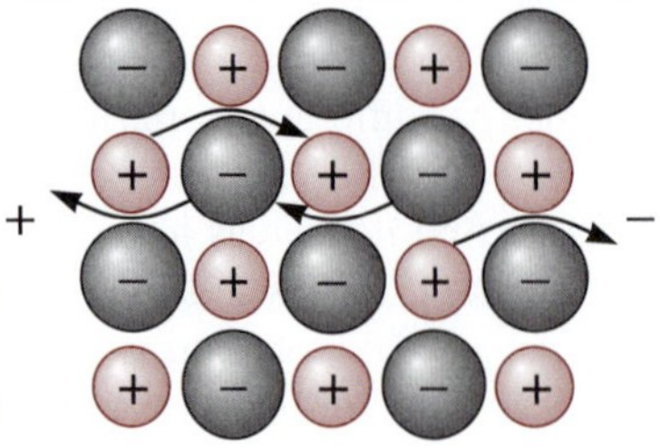

Figure 2-15 When voltage is applied to an ionic material, entire ions must move to cause a current to flow. Ion movement is slow and the electrical conductivity is poor at low temperatures (for Example 2-3).

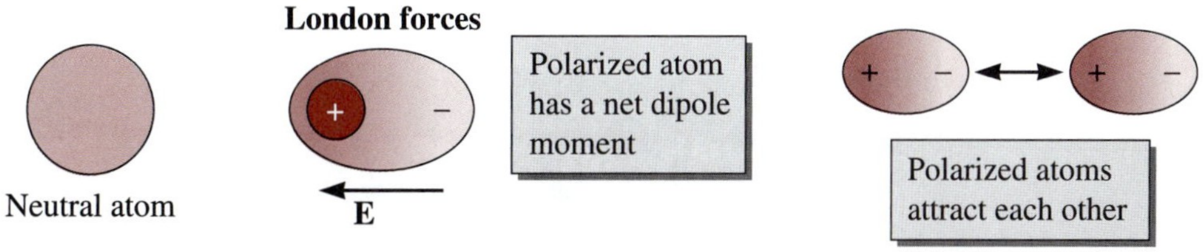

Figure 2-16 Illustration of London forces, a type of a van der Waals force, between atoms.

Van der Waals Bonding The origin of van der Waals forces between atoms and molecules is quantum mechanical in nature and a detailed discussion is beyond the scope of this book. We present here a simplified picture. If two electrical charges $+q$ and $-q$ are separated by a distance d, the arrangement is called a dipole and the dipole moment is defined as $q \times d$. Atoms are electrically neutral. Also, the centers of the positive charge (nucleus) and negative charge (electron cloud) coincide. Therefore, a neutral atom has no dipole moment. When a neutral atom is exposed to an internal or external electric field the atom gets polarized (i.e., the centers of positive and negative charges separate). This creates or induces a dipole moment (Figure 2-16). In some molecules, the dipole moment does not have to be induced—it exists by virtue of the direction of bonds and the nature of atoms. These molecules are known as **polar molecules**. An example of such a molecule that has a permanently built-in dipole moment is water (Figure 2-17).

There are three types of **van der Waals** interactions, namely London forces, Keesom forces, and Debye forces. If the interactions are between two dipoles that are induced in atoms or molecules, we refer to them as **London forces** (e.g., carbon tetrachloride) (Figure 2-16). When an induced dipole (that is, a dipole that is induced in what is otherwise a non-polar atom or molecule) interacts with a molecule that has a permanent dipole moment, we refer to this interaction as a **Debye interaction**. An example of Debye interaction would be forces between water molecules and those of carbon tetrachloride.

If the interactions are between molecules that are permanently polarized (e.g., water molecules attracting other water molecules or other polar molecules), we refer to these as **Keesom interactions**. The attraction between the positively charged regions of one water molecule and the negatively charged regions of a second water molecule provides an attractive bond between the two water molecules (Figure 2-17).

The bonding between molecules that have a permanent dipole moment, known as the Keesom force, is often referred to as the **hydrogen bond**, where hydrogen atoms represent one of the polarized regions. Thus, hydrogen bonding is essentially a Keesom force and is a type of van der Waals force.

Note that van der Waals bonds are **secondary bonds**, which means bond energies are smaller. However, the atoms within the molecule or group of atoms are joined by strong covalent or ionic bonds. Thus, heating water to the boiling point breaks the van

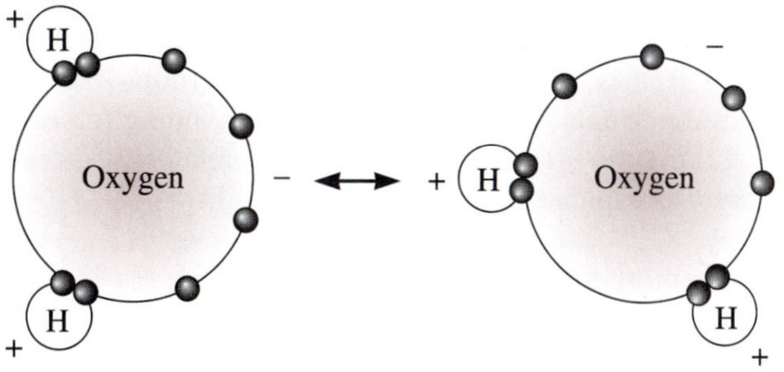

Figure 2-17
The Keesom interactions are formed as a result of polarization of molecules or groups of atoms. In water, electrons in the oxygen tend to concentrate away from the hydrogen. The resulting charge difference permits the molecule to be weakly bonded to other water molecules.

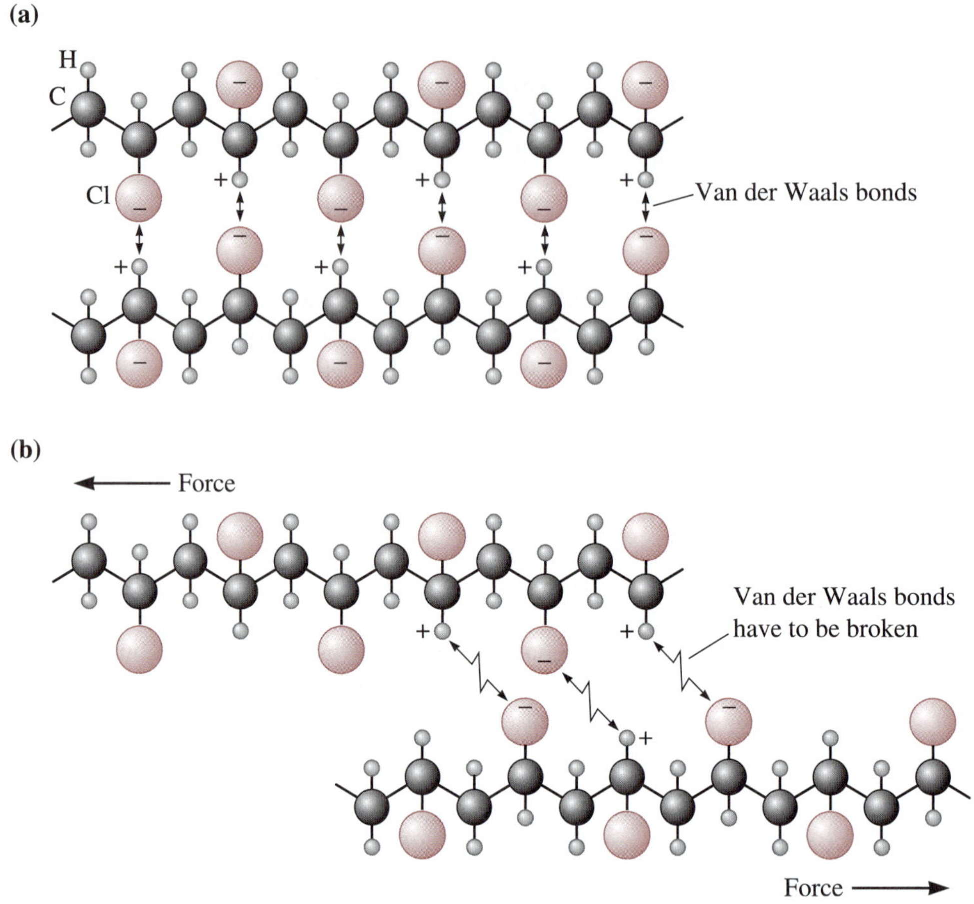

Figure 2-18 (a) In polyvinyl chloride (PVC), the chlorine atoms attached to the polymer chain have a negative charge and the hydrogen atoms are positively charged. The chains are weakly bonded by van der Waals bonds. This additional bonding makes PVC stiffer. (b) When a force is applied to the polymer, the van der Waals bonds are broken and the chains slide past one another.

der Waals bonds and changes water to steam, but much higher temperatures are required to break the covalent bonds joining oxygen and hydrogen atoms.

Although termed "secondary," based on the bond energies, van der Waals forces play a very important role in many areas of engineering. Van der Waals forces between atoms and molecules play a vital role in determining the surface tension and boiling points of liquids.

Van der Waals bonds can change dramatically the properties of certain materials. For example, graphite and diamond have very different mechanical properties. In many plastic materials, molecules contain polar parts or side groups (e.g., cotton or cellulose, PVC, Teflon). Van der Waals forces provide an extra binding force between the chains of these polymers (Figure 2-18). This makes PVC relatively more brittle; materials known as plasticizers are added to enhance PVC ductility.

Mixed Bonding In most materials, bonding between atoms is a mixture of two or more types. Iron, for example, is bonded by a combination of metallic and covalent bonding that prevents atoms from packing as efficiently as we might expect.

Compounds formed from two or more metals (**intermetallic compounds**) may be bonded by a mixture of metallic and ionic bonds, particularly when there is a large difference in electronegativity between the elements. Because lithium has an electronegativity of 1.0 and aluminum has an electronegativity of 1.5, we would expect AlLi to have a combination of metallic and ionic bonding. On the other hand, because both aluminum and vanadium have electronegativities of 1.5, we would expect Al_3V to be bonded primarily by metallic bonds.

Many ceramic and semiconducting compounds, which are combinations of metallic and nonmetallic elements, have a mixture of covalent and ionic bonding. As the electronegativity difference between the atoms increases, the bonding becomes more ionic. The fraction of bonding that is covalent can be estimated from the following equation:

$$\text{Fraction covalent} = \exp(-0.25\Delta E^2) \tag{2-1}$$

where ΔE is the difference in electronegativities.

Example 2-4 explores the nature of the bonds in silica.

EXAMPLE 2-4 *Determine if Silica is Ionically or Covalently Bonded*

In a previous example, we used silica (SiO_2) as an example of a covalently bonded material. In reality, silica exhibits ionic and covalent bonding. What fraction of the bonding is covalent? Give examples of applications in which silica is used.

SOLUTION

From Figure 2-8, we estimate the electronegativity of silicon to be 1.8 and that of oxygen to be 3.5. The fraction of the bonding that is covalent is:

$$\text{Fraction covalent} = \exp[-0.25(3.5 - 1.8)^2] = \exp(-0.72) = 0.486$$

Although the covalent bonding represents only about half of the bonding, the directional nature of these bonds still plays an important role in the structure of SiO_2.

Silica has many applications. Silica is used for making glasses and optical fibers. We add nano-sized particles of silica to tires to enhance the stiffness of the rubber. High-purity silicon (Si) crystals for computer chips are made by reducing silica to silicon.

EXAMPLE 2-5 *Thermal Expansion of Silicon for Computer Chips*

Silicon crystals cut into thin wafers are widely used to make computer chips. The coefficient of expansion of a single crystal of silicon is $\alpha = 2.5 \times 10^{-6}$ K^{-1}. (a) On this silicon wafer, a thin layer of silica (SiO_2) is grown by heating the silicon wafer to high temperatures (e.g., 900°C). (See Figure 2-19 on the next page.) The thermal expansion coefficient of silica is 0.5×10^{-6} K^{-1}. Will the silica layer experience a compressive or tensile stress when it cools down to room temperature? (b) If an aluminum film ($\alpha = 24 \times 10^{-6}$ K^{-1}) is grown on silicon, what type of stress (compressive or tensile) will be expected in the aluminum film? Assume that there are no chemical reactions occurring between the film and the substrate.

SOLUTION

(a) The silica layer forms on silicon at high temperatures. When the silicon wafer is at a high temperature, it has expanded compared to its original dimensions. The silica layer is formed at this high temperature and has the same width as that of silicon.

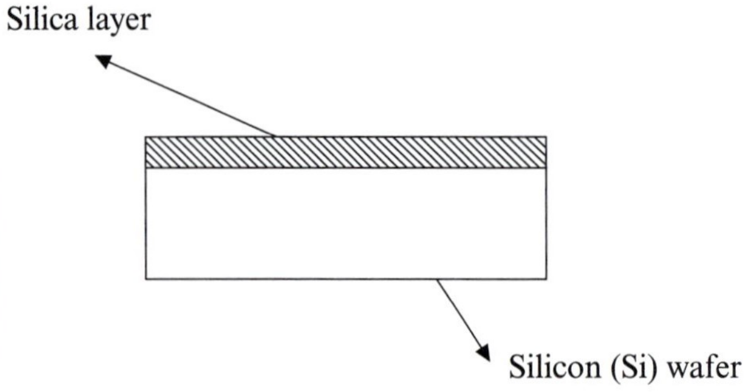

Figure 2-19 Silica layer on silicon (for Example 2-5).

During cooling, the silicon wafer likes to shrink more compared to silica. This is because silicon has a higher thermal expansion coefficient. Silica, on the other hand, does not shrink as much ($\alpha_{SiO2} < \alpha_{Si}$). However, since it is in contact with silicon, it is forced to shrink with silicon (i.e., the silica layer is subjected to a compressive stress). Since the *difference* in thermal expansion coefficients between the two materials is not very high, the silica layer remains adherent. Since the stress in silica film is compressive, it also will be able to withstand any mechanical shock or load more easily. This is one of the major reasons why silica films can be used as insulators and in other applications for processing silicon chips.

(b) When aluminum is deposited onto silicon, during cooling the aluminum would like to shrink its dimensions more than the silicon. This is because of $\alpha_{Al} \gg \alpha_{Si}$. However, the aluminum film can not shrink its dimensions as much because it is attached to the silicon wafer. Thus, when processing is finished, aluminum film will be subjected to a residual tensile stress (i.e., it would have liked to shrink more but is constrained by the underlying silicon wafer). In the case of aluminum, it is able to tolerate this tensile stress. Other materials (such as if we were to attempt depositing a ceramic material with coefficient of thermal expansion higher than that of the substrate) will just fracture during cooling because of the thermal stress generated by the mismatch in thermal expansion values.

2-6 Binding Energy and Interatomic Spacing

Interatomic Spacing The equilibrium distance between atoms results from a balance between repulsive and attractive forces. In the metallic bond, for example, the attraction between the electrons and the ion cores is balanced by the repulsion between ion cores. Equilibrium separation occurs when the total interatomic energy (IAE) of the pair of

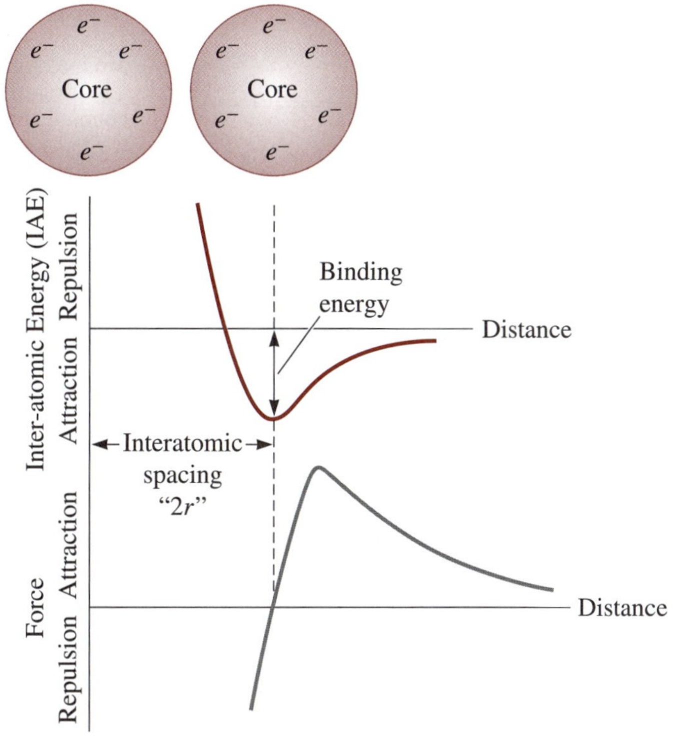

Figure 2-20
Atoms or ions are separated by an equilibrium spacing that corresponds to the minimum interatomic energy for a pair of atoms or ions (or when zero force is acting to repel or attract the atoms or ions).

atoms is at a minimum, or when no net force is acting to either attract or repel the atoms (Figure 2-20).

The **interatomic spacing** in a solid metal is *approximately* equal to the atomic diameter, or twice the atomic radius *r*. We cannot use this approach for ionically bonded materials, however, since the spacing is the sum of the two different ionic radii. Atomic and ionic radii for the elements are listed in Appendix B and will be used in the next chapter.

The minimum energy in Figure 2-20 is the **binding energy**, or the energy required to create or break the bond. Consequently, materials having a high binding energy also have a high strength and a high melting temperature. Ionically bonded materials have a particularly large binding energy (Table 2-2) because of the large difference in electronegativities between the ions. Metals have lower binding energies because the electronegativities of the atoms are similar.

Other properties can be related to the force-distance and energy-distance expressions in Figure 2-21. For example, the **modulus of elasticity** of a material (the slope of

TABLE 2-2 ■ ***Binding energies for the four bonding mechanisms***

Bond	Binding Energy (kcal/mol)
Ionic	150–370
Covalent	125–300
Metallic	25–200
Van der Waals	<10

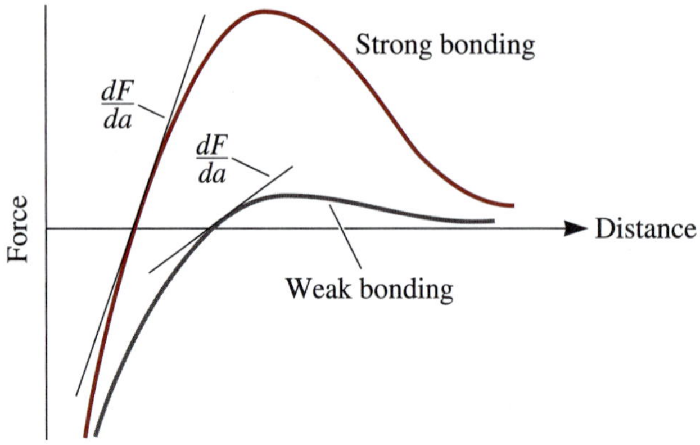

Figure 2-21
The force-distance curve for two materials, showing the relationship between atomic bonding and the modulus of elasticity. A steep dF/da slope gives a high modulus.

the stress-strain curve in the elastic region (E), also known as Young's modulus) is related to the slope of the force-distance curve (Figure 2-21). A steep slope, which correlates with a higher binding energy and a higher melting point, means that a greater force is required to stretch the bond; thus, the material has a high modulus of elasticity.

An interesting point that needs to be made is that not all properties of engineered materials are highly microstructure sensitive. Modulus of elasticity is one such property. If we have two aluminum samples that have essentially the same chemical composition but different grain size, we can expect that the modulus of elasticity of these samples will be about the same. However, the **yield strength**, a level of stress at which the material begins to deform easily with increasing stress, of these samples will be quite different. The yield strength, therefore, is a microstructure sensitive property. We will learn in subsequent chapters that, compared to other mechanical properties such as yield strength and tensile strength, the modulus of elasticity does not depend strongly on the microstructure. The modulus of elasticity can be linked directly to the strength of bonds between atoms. Thus, the modulus of elasticity depends primarily on the atoms that make up the material.

Another property that can be linked to the binding energy or interatomic force-distance curves is the **coefficient of thermal expansion** (CTE) defined as $\alpha = (1/L)(dL/dT)$, where the overall dimensions of the material in a given direction, L, will increase with increasing temperature, T. The CTE describes how much a material expands or contracts when its temperature is changed. It is also related to the strength of the atomic bonds. In order for the atoms to move from their equilibrium separation, energy must be supplied to the material. If a very deep interatomic energy (IAE) trough caused by strong atomic bonding is characteristic of the material (Figure 2-22), the atoms separate to a lesser degree resulting in a low, linear coefficient of thermal expansion. Materials with a low coefficient of thermal expansion maintain their dimensions more accurately when the temperature changes. Note that there are microstructural features (e.g., anistropy, or varying properties, in thermal expansion along with different crystallographic directions and other effects) that also have a significant effect on the overall thermal expansion coefficient of an engineered material. Indeed, materials scientists and engineers have developed materials that show very low, zero or negative coefficients of thermal expansion.

Materials that have very low expansion are useful in many applications where the components are expected to repeatedly undergo relatively rapid heating and cooling. For example, cordierite ceramics (used as catalyst support in catalytic converters in cars), ultra-low expansion (ULE) glasses, Visionware^{TM}, and other glass-ceramics developed by Corning, have very low thermal expansion coefficients.

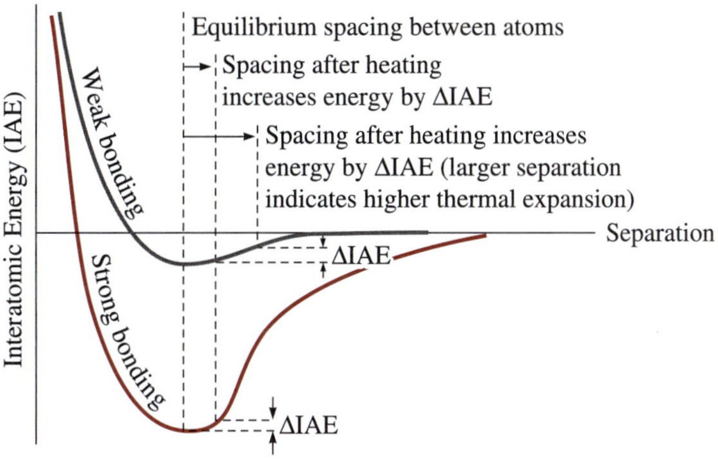

Figure 2-22 The interatomic energy (IAE)–separation curve for two atoms. Materials that display a steep curve with a deep trough have low linear coefficients of thermal expansion.

EXAMPLE 2-6 *Steel Beams and Expansion Joints for Construction*

A steel beam used in a bridge is 10 meters long at 30°C. (a) On a winter day, the temperature changes to −20°C. What will be the change in the length of the beam in millimeters? (b) If the elastic modulus of the steel used is 200×10^9 N/m^2, what will be the level of stress generated in the steel beam if it is not allowed to contract? (c) Is this stress compressive or tensile? Assume that the thermal coefficient of expansion for steel used in the beam is 13×10^{-6}/C.

SOLUTION

(a) The coefficient of thermal expansion is defined as

$$\alpha = \left(\frac{1}{L}\right)\left(\frac{dL}{dT}\right)$$

In this case, the original length of the beam is 10 meters. The temperature change will be 50°C. Therefore,

$$13 \times 10^{-6} = \left(\frac{1}{10\text{ m}}\right)\left(\frac{dL}{30-(-20)}\right)$$

$$(13 \times 10^{-6}) \times (10\text{ m}) \times (50) = dL$$

Thus, the change in length is 65×10^{-4} m or 6.5 mm. Thus, the steel beam would contract slightly more than one-quarter inch.

(b) The elastic modulus of steel is 200×10^9 N/m^2. This is the ratio of stress to strain.

$$200 \times 10^9\text{ N/m}^2 = \frac{\text{stress}}{\text{strain}}$$

Strain is defined as change in length divided by the original length. Thus,

$$\text{Strain} = \frac{dL}{L} = \frac{65 \times 10^{-4}\ \text{m}}{10\ \text{m}} = 65 \times 10^{-5}$$

You can also show from the definitions of strain and α that strain is simply $\alpha \times (dT)$.

Thus, the stress generated in the steel beam if it is not allowed to contract will be given by

$$\text{Stress} = (200 \times 10^9\ \text{N/m}^2) \times (65 \times 10^{-5}) = 130 \times 10^6\ \text{N/m}^2$$

The stress generated in the steel beam will be $130 \times 10^6\ \text{N/m}^2$ or 130 mega-Pascals (MPa). As will be seen in Chapter 6, this level of stress is typically not enough to cause a permanent deformation in the steel beam.

(c) The steel beam wants to shrink, however, if it is anchored between rigid supports, it will not be allowed to do so. Thus, the stress generated in the beam is actually tensile (i.e., the beam acts as if it has been pulled to a longer length).

To accommodate the effect of such strains generated by thermal expansion and contraction of steels, civil engineers use expansion joints on bridges (Figure 2-23).

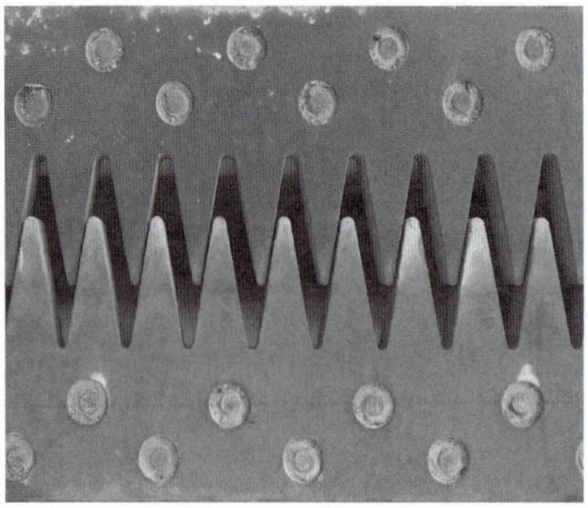

Figure 2-23
Expansion joint on a bridge.
(*Courtesy of CheriJon/Stockphoto.*)

SUMMARY

- Similar to composition, structure of a material has a profound influence on the properties of a material.
- Structure of materials can be understood at various levels: atomic structure, long- and short-range atomic arrangements, nanostructure, microstructure, and macrostructure. Engineers concerned with practical applications need to understand the structure at both the micro and macro levels. Given that atoms and atomic arrangements constitute the building blocks of advanced materials, we need to understand the structure at an atomic level. There are many emerging novel devices centered on micro-electro-mechanical systems (MEMS) and nanotechnology. As a result, understanding the structure of materials at nano-scale is also very important for some applications.

- Atomic bonding is determined partly by how the valence electrons associated with each atom interact. Types of bonds include metallic, covalent, ionic, and van der Waals. Most engineered materials exhibit mixed bonding.
- A metallic bond is formed as a result of atoms of low electronegativity elements donating their valence electrons and leading to the formation of a "sea" of electrons. Metallic bonds are non-directional and relatively strong. Metals are good conductors of heat and electricity and reflect visible light.
- A covalent bond is formed between two atoms when each atom donates an electron that is needed in the bond formation. Covalent bonds are found in many polymeric and ceramic materials. These bonds are strong, most inorganic materials with covalent bonds exhibit high levels of strength, hardness, and limited ductility. Most covalently bonded materials tend to be relatively good electrical insulators. Some materials such as Si and Ge behave as semiconductors.
- The ionic bonding found in many ceramics is produced when an electron is "donated" from one electropositive atom to an electronegative atom, creating positively charged cations and negatively charged anions. As in covalently bonded materials, these materials tend to be mechanically strong and hard, but brittle. Melting points of ionically bonded materials are relatively high. These materials are typically electrical insulators. In some cases, though, the microstructure of these materials can be tailored so that significant ionic conductivity is obtained.
- The van der Waals bonds are formed when atoms or groups of atoms have a nonsymmetrical electrical charge. The asymmetry in the charge is a result of dipoles that are induced or dipoles that are permanent.
- The binding energy is related to the strength of the bonds and is particularly high in ionically and covalently bonded materials. Materials with a high binding energy often have a high-melting temperature, a high modulus of elasticity, and a low coefficient of thermal expansion.

GLOSSARY

Amorphous material A material that does not have a long-range order of atoms (e.g., silica glass).

Anion A negatively charged ion produced when an atom, usually of a nonmetal, accepts one or more electrons.

Atomic mass The mass of the Avogadro number of atoms, g/mol. Normally, this is the average number of protons and neutrons in the atom. Also called the atomic weight.

Atomic mass unit The mass of an atom expressed as 1/12 of the mass of a carbon atom.

Atomic number The number of protons or electrons in an atom.

Atomic structure All atoms and their arrangements that constitute the building blocks of matter.

Avogadro number The number of atoms or molecules in a mole. The Avogadro number is 6.02×10^{23} per mole.

Azimuthal quantum number (l) A quantum number that designates different energy levels in principal shells.

Binding energy The energy required to separate two atoms from their equilibrium spacing to an infinite distance apart. Alternately, the binding energy is the strength of the bond between two atoms.

Cation A positively charged ion produced when an atom, usually of a metal, gives up its valence electrons.

Coefficient of thermal expansion (CTE) The amount by which a material changes its dimensions when the temperature changes. A material with a low coefficient of thermal expansion tends to retain its dimensions when the temperature changes.

Composition The chemical make-up of a material.

Covalent bond The bond formed between two atoms when the atoms share their valence electrons.

Crystalline materials Materials in which atoms are arranged in a periodic fashion exhibiting a long-range order.

Debye interactions Van der Waals forces that occur between two molecules, with only one of them with a permanent dipole moment.

Directional relationship The bonds between atoms in covalently bonded materials form specific angles, depending on the material.

Ductility The ability of materials to be stretched or bent without breaking.

Electronegativity The relative tendency of an atom to accept an electron and become an anion. Strongly electronegative atoms readily accept electrons.

Hydrogen bond A Keesom interaction (a type of van der Waals bond) between molecules in which a hydrogen atom is involved (e.g., bonds *between* water molecules).

Interatomic spacing The equilibrium spacing between the centers of two atoms. In solid elements, the interatomic spacing equals the apparent diameter of the atom.

Intermetallic compound A compound such as Al_3V formed by two or more metallic atoms; bonding is typically a combination of metallic and ionic bonds.

Ionic bond The bond formed between two different atom species when one atom (the cation) donates its valence electrons to the second atom (the anion). An electrostatic attraction binds the ions together.

Keesom interactions Van der Waals forces that occur between molecules that have a permanent dipole moment.

Length-scale A relative distance or range of distances used to describe materials-related structure, properties or phenomena.

London forces Van der Waals forces that occur between molecules that do not have a permanent dipole moment.

Long-range atomic arrangements Repetitive three-dimensional patterns with which atoms or ions are arranged in crystalline materials.

Magnetic quantum number (M_l) A quantum number that describes energy levels for each azimuthal quantum number.

Macrostructure Structure of a material at a macroscopic level. The length-scale is ~>100,000 nm. Typical features include porosity, surface coatings, and internal or external micro-cracks.

Metallic bond The electrostatic attraction between the valence electrons and the positively charged ion cores.

Micro-electro-mechanical systems (MEMS) These consist of miniaturized devices typically prepared by micromachining.

Microstructure Structure of a material at a length-scale of ~10 to 1000 nm. This typically includes such features as average grain size, grain size distribution, grain orientation and those related to defects in materials.

Modulus of elasticity The slope of the stress-strain curve in the elastic region (E). Also known as Young's modulus.

Nano-scale A length scale of 1–100 nm.

Nanostructure Structure of a material at a nano-scale (~length-scale 1–100 nm).

Nanotechnology An emerging set of technologies based on nano-scale devices, phenomena, and materials.

Polarized molecules Molecules that have developed a dipole moment by virtue of an internal or external electric field.

Primary bonds Strong bonds between adjacent atoms resulting from the transfer or sharing of outer orbital electrons.

Quantum numbers The numbers that assign electrons in an atom to discrete energy levels. The four quantum numbers are the principal quantum number n, the azimuthal quantum number l, the magnetic quantum number m_l, and the spin quantum number m_s.

Quantum shell A set of fixed energy levels to which electrons belong. Each electron in the shell is designated by four quantum numbers.

Secondary bond Weak bonds, such as van der Waals bonds, that typically join molecules to one another.

Short-range atomic arrangements Atomic arrangements up to a distance of a few nm.

Spin quantum number (m_s) A quantum number that indicates spin of an electron.

Structure Description of spatial arrangements of atoms or ions in a material.

Transition elements A set of elements whose electronic configurations are such that their inner d and f levels begin to fill up. These elements usually exhibit multiple valence and are useful for electronic, magnetic and optical applications.

Three-five (III-V) semiconductor A semiconductor that is based on group 3A and 5B elements (e.g., GaAs).

Two-six (II-VI) semiconductor A semiconductor that is based on group 2B and 6B elements (e.g., CdSe).

Valence The number of electrons in an atom that participate in bonding or chemical reactions. Usually, the valence is the number of electrons in the outer s and p energy levels.

Van der Waals bond A secondary bond developed between atoms and molecules as a result of interactions between dipoles that are induced or permanent.

Yield strength The level of stress above which a material begins to show permanent deformation.

PROBLEMS

Section 2-1 The Structure of Materials—An Introduction

2-1 What is meant by the term *composition* of a material?

2-2 What is meant by the term *structure* of a material?

2-3 What are the different levels of structure of a material?

2-4 Why is it important to consider the structure of a material while designing and fabricating engineering components?

2-5 What is the difference between the microstructure and the macrostructure of a material?

Section 2-2 The Structure of the Atom

2-6 **(a)** Aluminum foil used for storing food weighs about 0.05 g/cm^2. How many atoms of aluminum are contained in one square inch of the foil?

(b) Using the densities and atomic weights given in Appendix A, calculate and compare the number of atoms per cubic centimeter in **(i)** lead and **(ii)** lithium.

2-7 **(a)** Using data in Appendix A, calculate the number of iron atoms in 1000 kg of iron.

(b) Using data in Appendix A, calculate the volume in cubic centimeters occupied by one mole of boron.

2-8 In order to plate a steel part having a surface area of 1250 cm^2 with a 0.005 cm thick layer of nickel: **(a)** How many atoms of nickel are required? **(b)** How many moles of nickel are required?

Section 2-3 The Electronic Structure of the Atom

2-9 Suppose an element has a valence of 2 and an atomic number of 27. Based only on the quantum numbers, how many electrons must be present in the 3*d* energy level?

2-10 Indium, which has an atomic number of 49, contains no electrons in its 4*f* energy levels. Based only on this information, what must be the valence of indium?

Section 2-4 The Periodic Table

2-11 The periodic table of elements can help us better rationalize trends in properties of elements and compounds based on elements from different groups. Search the literature and obtain the coefficients of thermal expansions of elements from group 4B. Establish a trend and see if it correlates with the melting temperatures and other properties (e.g., bandgap) of these elements.

2-12 Bonding in the intermetallic compound Ni_3Al is predominantly metallic. Explain why there will be little, if any, ionic bonding component. The electronegativity of nickel is about 1.9.

2-13 Plot the melting temperatures of elements in the 4A to 8–10 columns of the periodic table versus atomic number (i.e., plot melting temperatures of Ti through Ni, Zr through Pd, and Hf through Pt). Discuss these relationships, based on atomic bonding and binding energies: **(a)** as the atomic number increases in each row of the periodic table and **(b)** as the atomic number increases in each column of the periodic table.

2-14 Plot the melting temperature of the elements in the 1A column of the periodic table versus atomic number (i.e., plot melting temperatures of Li through Cs). Discuss this relationship, based on atomic bonding and binding energy.

Section 2-5 Atomic Bonding

2-15 Methane (CH_4) has a tetrahedral structure similar to that of SiO_2, with a carbon atom of radius 0.77×10^{-8} cm at the center and hydrogen atoms of radius 0.46×10^{-8} cm at four of the eight corners. Calculate the size of the tetrahedral cube for methane.

2-16 The compound aluminum phosphide (AlP) is a compound semiconductor having mixed ionic and covalent bonding. Calculate the fraction of the bonding that is ionic.

2-17 Calculate the fraction of bonding of MgO that is ionic.

2-18 What is the type of bonding in diamond? Are the properties of diamond commensurate with the nature of bonding?

2-19 Such materials as silicon carbide (SiC) and silicon nitride (Si_3N_4) are used for grinding and polishing applications. Rationalize the choice of these materials for this application.

2-20 Explain the role of van der Waals forces in determining the properties of PVC plastic.

2-21 Calculate the fractions of ionic bonds in silicon carbide (SiC) and in silicon nitride (Si_3N_4).

2-22 One particular form of boron nitride (BN) known as cubic born nitride (CBN) is a very hard material and is used in grinding applications. Calculate the fraction of covalent bond character in this material.

2-23 Another form of boron nitride (BN) known as hexagonal boron nitride (HBN) is used as a solid lubricant. Explain how this may be possible by comparing this situation with that encountered in two forms of carbon, namely diamond and graphite.

Section 2-6 Binding Energy and Interatomic Spacing

2-24 Beryllium and magnesium, both in the 2A column of the periodic table, are lightweight metals. Which would you expect to have the higher modulus of elasticity? Explain, considering binding energy and atomic radii and using appropriate sketches of force versus interatomic spacing.

2-25 Boron has a much lower coefficient of thermal expansion than aluminum, even though both are in the 3B column of the periodic table. Explain, based on binding energy, atomic size, and the energy well, why this difference is expected.

2-26 Would you expect MgO or magnesium (Mg) to have the higher modulus of elasticity? Explain.

2-27 Would you expect Al_2O_3 or aluminum (Al) to have the higher coefficient of thermal expansion? Explain.

2-28 Aluminum and silicon are side-by-side in the periodic table. Which would you expect to have the higher modulus of elasticity (E)? Explain.

2-29 Explain why the modulus of elasticity of simple thermoplastic polymers, such as polyethylene and polystyrene, is expected to be very low compared with that of metals and ceramics.

2-30 Steel is coated with a thin layer of ceramic to help protect against corrosion. What do you expect to happen to the coating when the temperature of the steel is increased significantly? Explain.

2-31 Why is the modulus of elasticity considered a relatively structure insensitive property?

2-32 An aluminum-alloy bar of length 2 meters at room temperature (300 K) is exposed to a temperature of 100°C ($\alpha = 23 \times 10^{-6}$ K^{-1}). What will be the length of this bar at 100°C?

2-33 If the elastic modulus of the aluminum alloy in the previous example is 70×10^9 N/m^2 (or Pa), what will be stress generated in the aluminum-alloy bar heated to 100°C if the bar was constrained between rigid supports and thus not allowed to expand? Will this stress be compressive or tensile in nature?

Design Problems

2-34 You wish to introduce ceramic fibers into a metal matrix to produce a composite material, which is subjected to high forces and large temperature changes. What design parameters might you consider to ensure that the fibers will remain intact and provide strength to the matrix? What problems might occur?

2-35 Turbine blades used in jet engines can be made from such materials as nickel-based superalloys. We can, in principle, even use ceramic materials such as zirconia or other alloys based on steels. In some cases, the blades also may have to be coated with a thermal barrier coating (TBC) to minimize exposure of the blade material to high temperatures. What design parameters would you consider in selecting a material for the turbine blade and for the coating that would work successfully in a turbine engine. Note that different parts of the engine are exposed to different temperatures, and not all blades are exposed to relatively high operating temperatures. What problems might occur? Consider the factors such as temperature and humidity in the environment that the turbine blades must function.

2-36 You want to design a material for making a mirror for a telescope that will be launched in space. Given that the temperatures in space can change considerably, what material will you consider using? Remember that this material should not expand or contract at all, if possible. It also should be as strong and as low a density as possible, and one should be able to coat it so that it can serve as a mirror.

2-37 You want to use a material that can be used for making a catalytic converter substrate. The job of this material is to be a carrier for the nanoparticles of metals (such as platinum and palladium), which are the actual catalysts. The main considerations are that this catalyst-support material must be able to withstand the constant, cyclic heating and cooling that it will be exposed to. (*Note*: The gases from automobile exhaust reach temperatures up to 500°C, and the material will get heated up to high temperatures and then cool down when the car is not being used.) What kinds of materials can be used for this application?

2-38 *Solid-Oxide Fuel-Cell Materials.* A solid-oxide fuel cell is made using a thin film of yttria stabilized zirconia (ZrO_2) (known as YSZ). The film is deposited onto a ceramic tube of a material called

3

Atomic and Ionic Arrangements

Have You Ever Wondered?

- *What is amorphous silicon and how is it different from the silicon used to make computer chips?*
- *What are liquid crystals?*
- *If you were to pack a cubical box with uniform-sized spheres, what is the maximum packing possible?*
- *How can we calculate the density of different materials?*

Arrangements of atoms and ions play an important role in determining the microstructure and properties of a material. The main objectives of this chapter are to:

(a) learn classification of materials based on atomic/ionic arrangements; and

(b) describe the arrangements in crystalline solids based on **lattice**, **basis**, and **crystal structure**.

For crystalline solids, we will illustrate the concepts of Bravais lattices, unit cells, crystallographic directions, and planes by examining the arrangements of atoms or ions in many technologically important materials. These include metals (e.g., Cu, Al, Fe, W, Mg, etc.), semiconductors (e.g., Si, Ge, GaAs, etc.), advanced ceramics (e.g., ZrO_2, Al_2O_3, $BaTiO_3$, diamond, etc.), and other materials. We will develop the necessary nomenclature used to characterize atomic or ionic arrangements in crystalline materials. We will present an overview of different types of amorphous materials such as amorphous silicon, metallic glasses, polymers, and inorganic glasses.

3-1 Short-Range Order versus Long-Range Order

In different states of matter, we can define four types of atomic or ionic arrangements (Figure 3-1).

No Order In monoatomic gases, such as argon (Ar) or plasma created in a fluorescent tubelight, atoms or ions have no orderly arrangement. These materials randomly fill up whatever space is available to them.

Short-Range Order (SRO) A material displays **short-range order (SRO)** if the special arrangement of the atoms extends only to the atom's nearest neighbors. Each water molecule in steam has a short-range order due to the covalent bonds between the

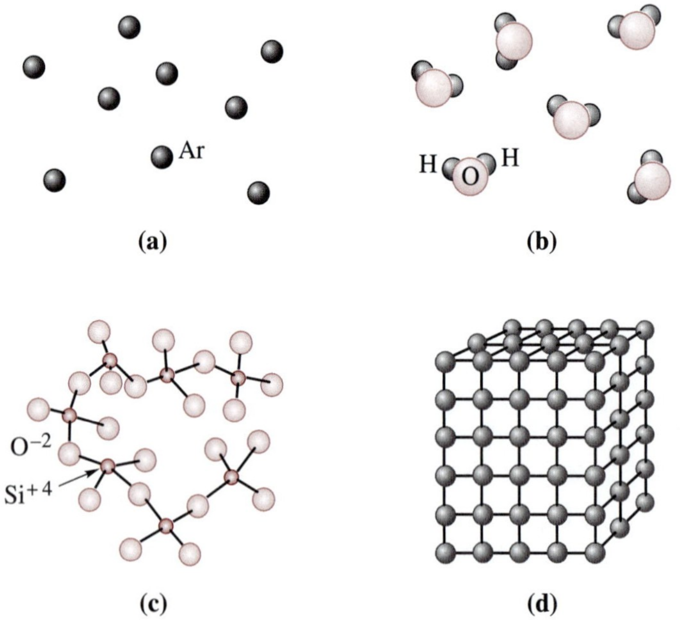

Figure 3-1 Levels of atomic arrangements in materials: (a) Inert monoatomic gases have no regular ordering of atoms. (b,c) Some materials, including water vapor, nitrogen gas, amorphous silicon and silicate glass have short-range order. (d) Metals, alloys, many ceramics and some polymers have regular ordering of atoms/ions that extends through the material.

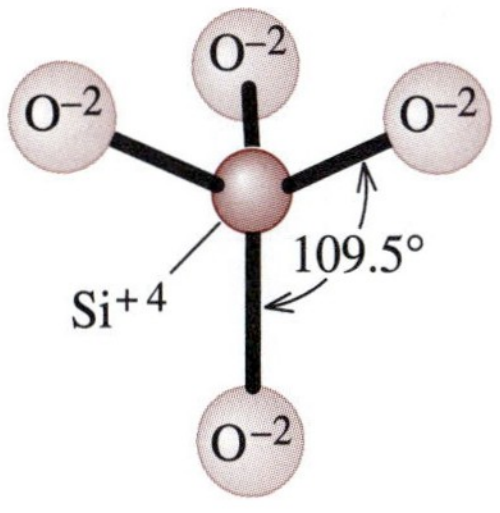

Figure 3-2
Basic Si-O tetrahedron in silicate glass.

hydrogen and oxygen atoms; that is, each oxygen atom is joined to two hydrogen atoms, forming an angle of 104.5° between the bonds. However, the water molecules in steam have no special arrangement with respect to each other's position.

A similar situation exists in materials known as inorganic glasses. In Chapter 2, we described the **tetrahedral structure** in silica that satisfies the requirement that four oxygen ions be bonded to each silicon ion. However, beyond the basic unit of a $(SiO_4)^{4-}$ tetrahedron (Figure 3-2), there is no periodicity in the way these tetrahedra are connected. In contrast, in quartz or other forms of crystalline silica, the silicate $(SiO_4)^{4-}$ tetrahedra are indeed connected in different periodic arrangements.

Many polymers also display short-range atomic arrangements that closely resemble the silicate glass structure.

Long-Range Order (LRO) Most metals and alloys, semiconductors, ceramics, and some polymers have a crystalline structure in which the atoms or ions display **long-range order (LRO)**; the special atomic arrangement extends over much larger length scales $\sim>100$ nm. The atoms or ions in these materials form a regular repetitive, grid-like pattern, in three dimensions. We refer to these materials as **crystalline materials**. If a crystalline material consists of only one crystal, we refer to it as a **single crystal material**. Single crystal materials are useful in many electronic and optical applications. For example, computer chips are made from silicon in the form of large (up to 12-inch diameter) single crystals [Figure 3-3(a)]. A **polycrystalline material** is comprised of many crystals with varying orientations in space. These crystals in a polycrystalline material are known as **grains**. A polycrystalline material is similar to a collage of several tiny single crystals. The borders between tiny crystals, where the crystals are in misalignment and are known as **grain boundaries**. Figure 3-3(b) shows the microstructure of a polycrystalline stainless steel material. Many crystalline materials we deal with in engineering applications are polycrystalline (e.g., steels used in construction, aluminum alloys for aircrafts, etc.). We will learn in later chapters that many properties of polycrystalline materials depend upon the physical and chemical characteristics of both grains and grain boundaries. The properties of single crystal materials depend upon the chemical composition and specific directions within the crystal (known as the crystallographic directions). Long-range order in crystalline materials can be detected and measured using techniques such as **x-ray diffraction** or **electron diffraction** (Section 3-9).

Liquid crystals (LCs) are polymeric materials that have a special type of order. Liquid crystal polymers behave as amorphous materials (liquid-like) in one state. However, when an external stimulus (such as an electric field or a temperature change) is provided, some polymer molecules undergo alignment and form small regions that are crystalline, hence the name "liquid crystals."

The Nobel Prize in Physics for 2001 went to Eric A. Cornell, Wolfgang Ketterle, and Carl E. Wieman. These scientists have verified a new state of matter known as the Bose-Einstein condensate (BEC).

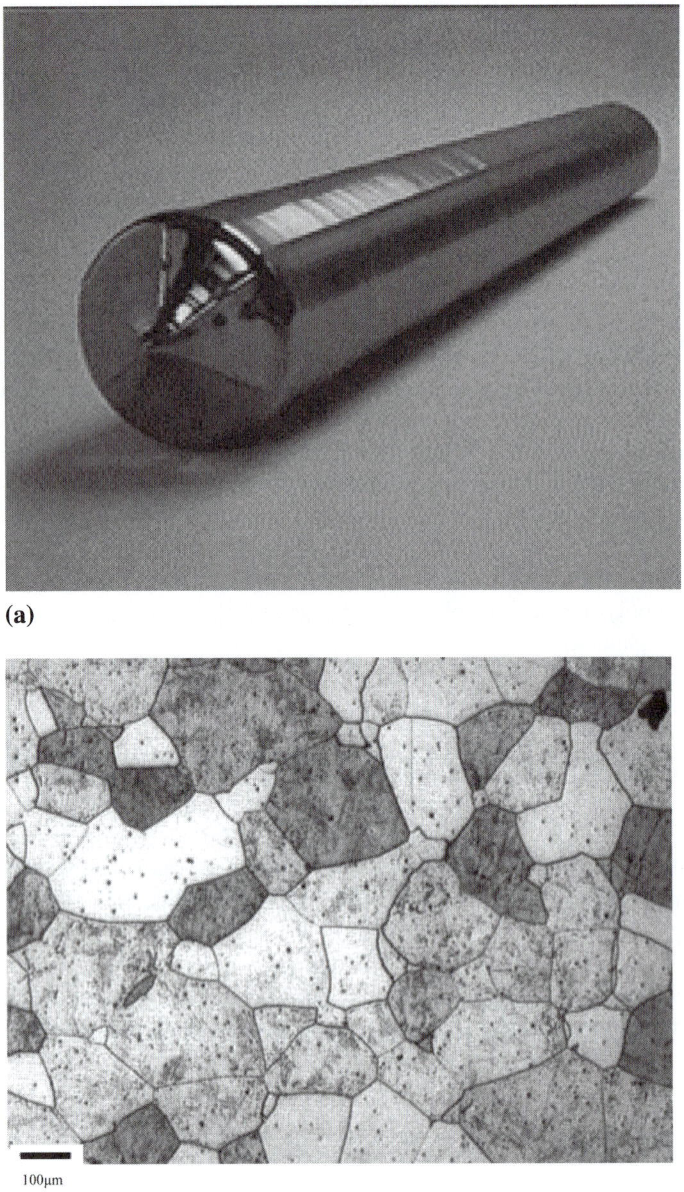

Figure 3-3
(a) Photograph of a silicon single crystal. (b) Micrograph of a polycrystalline stainless steel showing grains and grain boundaries (*Courtesy of Dr. M. Hua, Dr. I. Garcia, and Dr. A.J. DeArdo.*)

3-2 Amorphous Materials: Principles and Technological Applications

Any material that exhibits only a short-range order of atoms or ions is an **amorphous material**; that is, a noncrystalline one. In general, most materials want to form periodic arrangements since this configuration maximizes the thermodynamic stability of the material. Amorphous materials tend to form when, for one reason or other, the kinetics of the process by which the material was made did not allow for the formation of periodic arrangements. **Glasses**, which typically form in ceramic and polymer systems, are good examples of amorphous materials. Similarly, certain types of polymeric or colloidal gels, or gel-like materials, are also considered amorphous. Amorphous materials

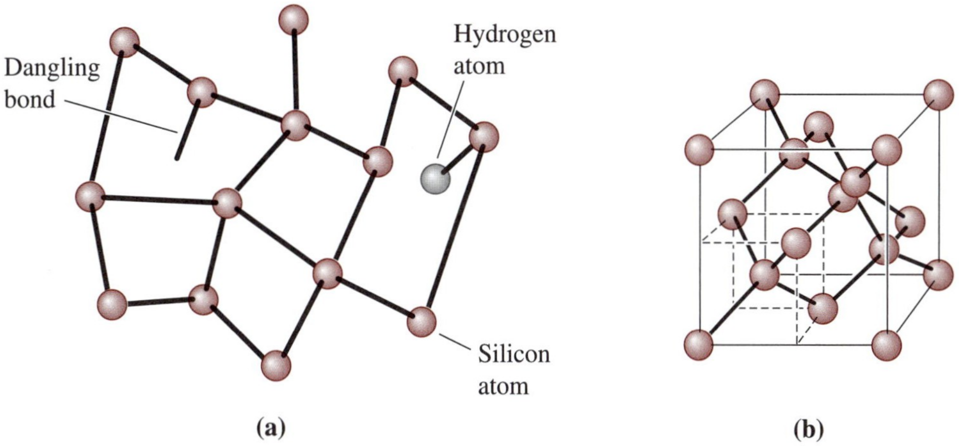

Figure 3-4 Atomic arrangements in amorphous silicon and crystalline silicon. (a) Amorphous silicon. (b) Crystalline silicon. Note the variation in the interatomic distance for amorphous silicon.

often offer a unique and unusual blend of properties since the atoms or ions are not assembled into their "regular" and periodic arrangements.

Similar to inorganic glasses, many plastics are also amorphous. They do contain small portions of material that are crystalline. During processing, relatively large chains of polymer molecules get entangled with each other, like spaghetti. Entangled polymer molecules do not organize themselves into crystalline materials.

Compared to plastics and inorganic glasses, metals and alloys tend to form crystalline materials rather easily. As a result, special efforts must be made to quench the metals and alloys quickly; a cooling rate of $>10^6$C/s is required to form **metallic glasses**. This process of cooling materials at a high rate is called **rapid solidification**.

Amorphous silicon, denoted a:Si-H, is another important example of a material that has the basic short-range order of crystalline silicon (Figure 3-4). The H in the symbol tells us that this material also contains some hydrogen. In amorphous silicon, the silicon tetrahedra are not connected to each other in the periodic arrangement seen in crystalline silicon. Also, some bonds are incomplete or "dangling." Thin films of amorphous silicon are used to make transistors for active matrix displays in computers. Amorphous silicon and polycrystalline silicon are both widely used for such applications as solar cells and solar panels.

3-3 Lattice, Unit Cells, Basis, and Crystal Structures

A **lattice** is a collection of points called **lattice points** that are arranged in a periodic pattern so that the surroundings of each point in the lattice are identical. A lattice may be one, two, or three dimensional. In materials science and engineering, we use the concept of "lattice" to describe arrangements of atoms or ions. A group of one or more atoms, located in a particular way with respect to each other and associated with each lattice point, is known as the **motif** or **basis**. We obtain a **crystal structure** by adding the lattice and basis (i.e., crystal structure = lattice + basis).

The **unit cell** is the subdivision of a lattice that still retains the overall characteristics of the entire lattice. Unit cells are shown in Figure 3-5. By stacking identical unit cells, the entire lattice can be constructed. There are seven unique arrangements, known as

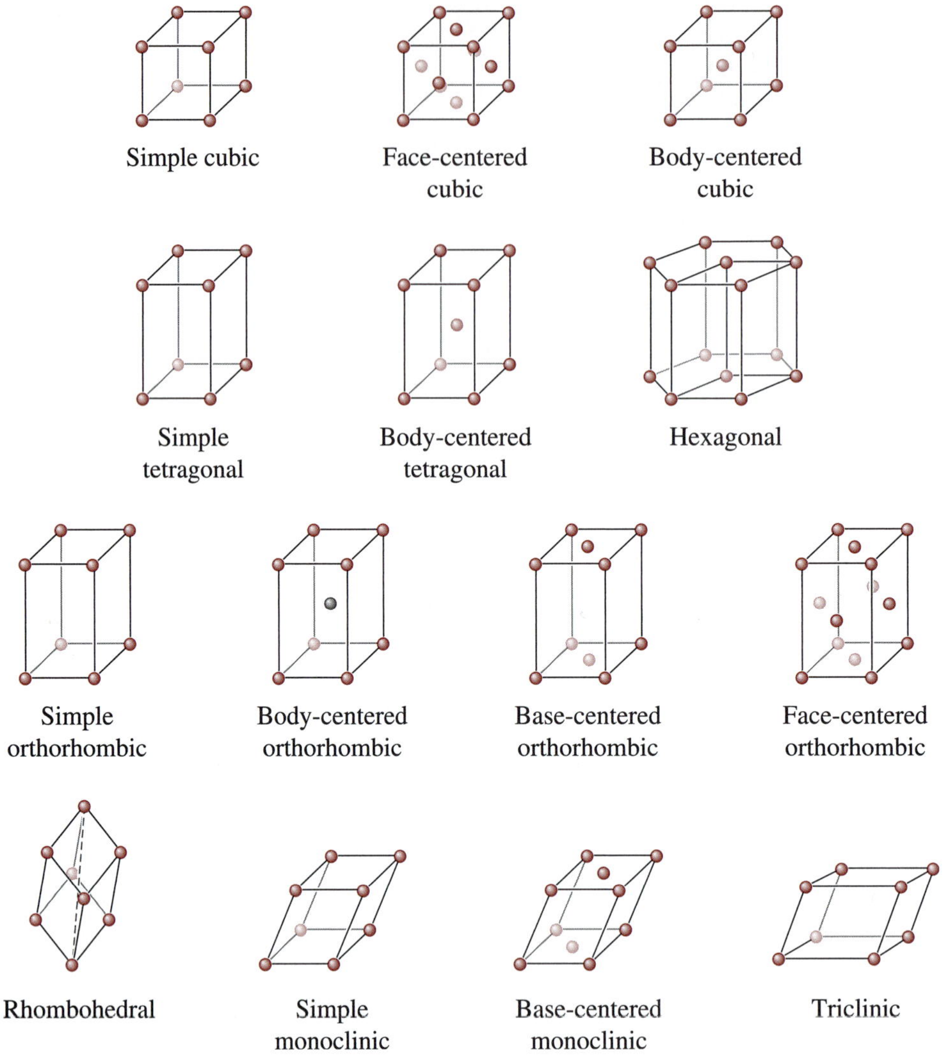

Figure 3-5 The fourteen types of Bravais lattices grouped in seven crystal systems. The actual unit cell for a hexagonal system is shown in Figures 3-6 and 3-10.

crystal systems, which can be used to fill up a three-dimensional space. These are cubic, tetragonal, orthorhombic, rhombohedral (also known as trigonal), hexagonal, monoclinic, and triclinic. Although there are seven crystal systems, we have a total of 14 distinct arrangements of lattice points. These unique arrangements of lattice points are known as the **Bravais lattices** (Figure 3-5 and Table 3-1). Lattice points are located at the corners of the unit cells and, in some cases, at either faces or the center of the unit cell. Note that for the cubic crystal system we have simple cubic (SC), face-centered cubic (FCC), and body-centered cubic (BCC) Bravais lattices. Similarly, for the tetragonal crystal system, we have simple tetragonal and body centered tetragonal Bravais lattices. Any other arrangement of atoms can be expressed using these 14 Bravais lattices. Note that the concept of a lattice is mathematical and does not mention atoms, ions or molecules. It is only when we take a Bravais lattice and begin to define the basis (i.e., one or more atoms associated with each lattice point) that we can describe a crystal structure. For example, if we take the face-centered cubic lattice and assume that at each lattice point we have one atom, then we get a face-centered cubic crystal structure.

TABLE 3-1 ■ *Characteristics of the seven crystal systems*

Structure	Axes	Angles between Axes	Volume of the Unit Cell
Cubic	$a = b = c$	All angles equal 90°	a^3
Tetragonal	$a = b \neq c$	All angles equal 90°	a^2c
Orthorhombic	$a \neq b \neq c$	All angles equal 90°	abc
Hexagonal	$a = b \neq c$	Two angles equal 90°. One angle equals 120°.	$0.866a^2c$
Rhombohedral or trigonal	$a = b = c$	All angles are equal and none equals 90°	$a^3\sqrt{1 - 3\cos^2\alpha + 2\cos^3\alpha}$
Monoclinic	$a \neq b \neq c$	Two angles equal 90°. One angle (β) is not equal to 90°	$abc \sin\beta$
Triclinic	$a \neq b \neq c$	All angles are different and none equals 90°	$abc\sqrt{1 - \cos^2\alpha - \cos^2\beta - \cos^2\gamma + 2\cos\alpha\cos\beta\cos\gamma}$

Note that although we have only 14 Bravais lattices, we can have many more bases. Since crystal structure is derived by adding lattice and basis, we have hundreds of different crystal structures. Many different materials can have the same crystal structure. For example, copper and nickel have the face-centered cubic crystal structure. In this book, for the sake of simplicity, we will assume that each lattice point has only one atom (i.e., the basis is one), unless otherwise stated. This assumption allows us to refer to the terms lattice and the crystal structure interchangeably. Let's look at some of the characteristics of a lattice or unit cell.

Lattice Parameter The **lattice parameters**, which describe the size and shape of the unit cell, include the dimensions of the sides of the unit cell and the angles between the sides (Figure 3-6). In a cubic crystal system, only the length of one of the sides of the cube

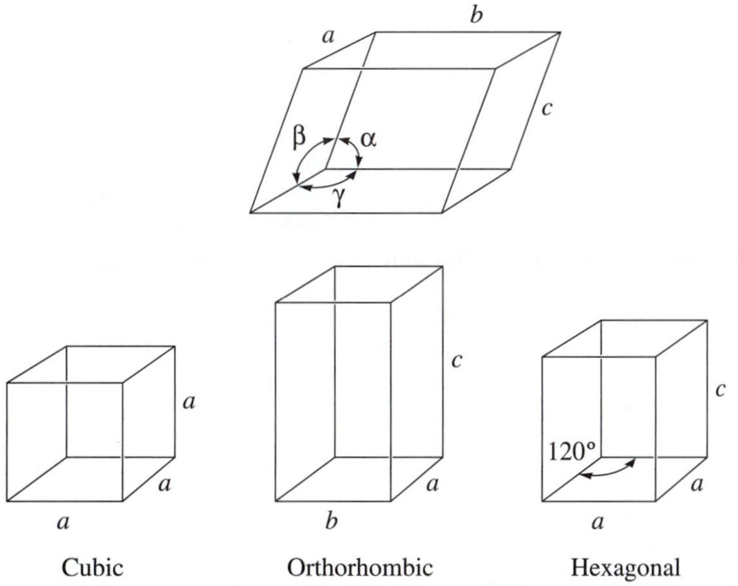

Figure 3-6 Definition of the lattice parameters and their use in cubic, orthorhombic, and hexagonal crystal systems.

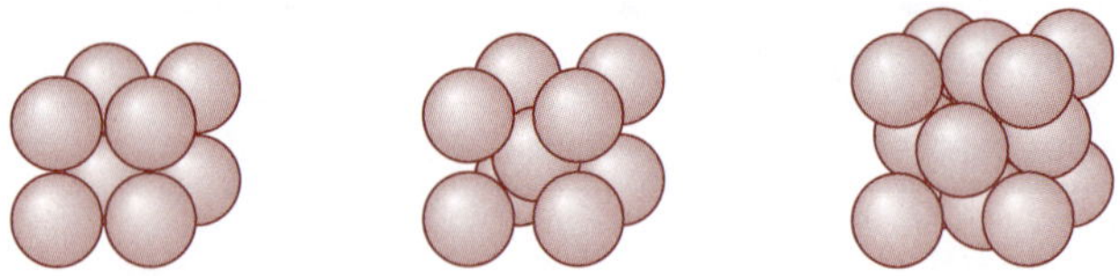

Figure 3-7 The models for simple cubic (SC), body-centered cubic (BCC), and face-centered cubic (FCC) unit cells, assuming only one atom per lattice point.

is necessary to completely describe the cell (angles of 90° are assumed unless otherwise specified). This length is the lattice parameter a (some times designated as a_0). The length is often given in nanometers (nm) or Angstrom (Å) units, where:

$$1 \text{ nanometer (nm)} = 10^{-9} \text{ m} = 10^{-7} \text{ cm} = 10 \text{ Å}$$

$$1 \text{ angstrom (Å)} = 0.1 \text{ nm} = 10^{-10} \text{ m} = 10^{-8} \text{ cm}$$

Several lattice parameters are required to define the size and shape of complex unit cells. For an orthorhombic unit cell, we must specify the dimensions of all three sides of the cell: a_0, b_0, and c_0. Hexagonal unit cells require two dimensions, a_0 and c_0, and the angle of 120° between the a_0 axes. The most complicated cell, the triclinic cell, is described by three lengths and three angles.

Number of Atoms per Unit Cell A specific number of lattice points defines each of the unit cells. For example, the corners of the cells are easily identified, as are the body-centered (center of the cell) and face-centered (centers of the six sides of the cell) positions (Figure 3-5). When counting the number of lattice points belonging to each unit cell, we must recognize that lattice points may be shared by more than one unit cell. A lattice point at a corner of one unit cell is shared by seven adjacent unit cells (thus a total of eight cells); only one-eighth of each corner lattice point belongs to one particular cell. Thus, the number of lattice points from all of the corner positions in one unit cell is:

$$\left(\frac{1}{8}\frac{\text{lattice point}}{\text{corner}}\right)\left(8\frac{\text{corners}}{\text{cell}}\right) = 1\frac{\text{lattice point}}{\text{unit cell}}$$

The number of atoms per unit cell is the product of the number of atoms per lattice point and the number of lattice points per unit cell. In most metals, one atom is located at each lattice point. The structures of simple cubic (SC), body-centered cubic (BCC), and face-centered cubic (FCC) unit cells, with one atom located at each lattice point, are shown in Figure 3-7. Example 3-1 illustrates how to determine the number of lattice points in cubic crystal systems.

EXAMPLE 3-1 ***Determining the Number of Lattice Points in Cubic Crystal Systems***

Determine the number of lattice points per cell in the cubic crystal systems. If there is only one atom located at each lattice point, calculate the number of atoms per unit cell.

SOLUTION

In the SC unit cell, lattice points are located only at the corners of the cube:

$$\frac{\text{lattice point}}{\text{unit cell}} = (8 \text{ corners})\left(\frac{1}{8}\right) = 1$$

In BCC unit cells, lattice points are located at each corners and with one at the center of the cube:

$$\frac{\text{lattice point}}{\text{unit cell}} = (8 \text{ corners})\left(\frac{1}{8}\right) + (1 \text{ center})(1) = 2$$

In FCC unit cells, lattice points are located at all corners and all faces of the cube:

$$\frac{\text{lattice point}}{\text{unit cell}} = (8 \text{ corners})\left(\frac{1}{8}\right) + (6 \text{ faces})\left(\frac{1}{2}\right) = 4$$

Since we are assuming there is only one atom located at each lattice point, the number of *atoms* per unit cell would be 1, 2, and 4, for the simple cubic, body-centered cubic, and face-centered cubic, unit cells, respectively.

Atomic Radius versus Lattice Parameter Directions in the unit cell along which atoms are in continuous contact are **close-packed directions**. In simple structures, particularly those with only one atom per lattice point, we use these directions to calculate the relationship between the apparent size of the atom and the size of the unit cell. By geometrically determining the length of the direction relative to the lattice parameters, and then adding the number of **atomic radii** along this direction, we can determine the desired relationship. Example 3-2 illustrates how the relationships between lattice parameters and atomic radius are determined.

EXAMPLE 3-2 ***Determining the Relationship between Atomic Radius and Lattice Parameters***

Determine the relationship between the atomic radius (r) and the lattice parameter (a_0) in SC, BCC, and FCC structures when one atom is located at each lattice point.

SOLUTION

If we refer to Figure 3-8, we find that atoms touch along the edge of the cube in an SC structure. The corner atoms are centered on the corners of the cube, so:

$$a_0 = 2r \tag{3-1}$$

In a BCC structure, atoms touch along the body diagonal, which is $\sqrt{3}a_0$ in length. There are two atomic radii from the center atom and one atomic radius from each of the corner atoms on the body diagonal, so

$$a_0 = \frac{4r}{\sqrt{3}} \tag{3-2}$$

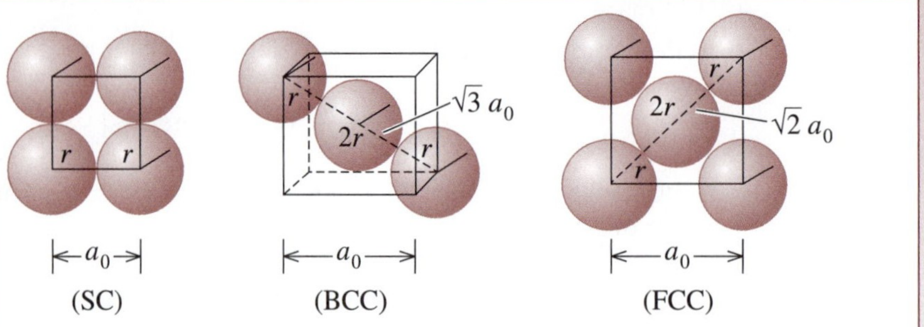

Figure 3-8 The relationships between the atomic radius and the lattice parameter in cubic systems (for Example 3-2).

In an FCC structure, atoms touch along the face diagonal of the cube, which is $\sqrt{2}a_0$ in length. There are four atomic radii along this length—two radii from the face-centered atom and one radius from each corner, so:

$$a_0 = \frac{4r}{\sqrt{2}} \tag{3-3}$$

Coordination Number The **coordination number** is the number of atoms touching a particular atom, or the number of nearest neighbors for that particular atom. This is one indication of how tightly and efficiently atoms are packed together. For ionic solids, the coordination number of cations is defined as the number of nearest anions. The coordination number of anions is the number of nearest cations. We will discuss the crystal structures of different ionic solids and other materials in Section 3-7.

In cubic structures containing only one atom per lattice point, atoms have a coordination number related to the lattice structure. By inspecting the unit cells in Figure 3-9, we see that each atom in the SC structure has a coordination number of six, while each atom in the BCC structure has eight nearest neighbors. In Section 3-5, we will show that each atom in the FCC structure has a coordination number of 12, which is the maximum.

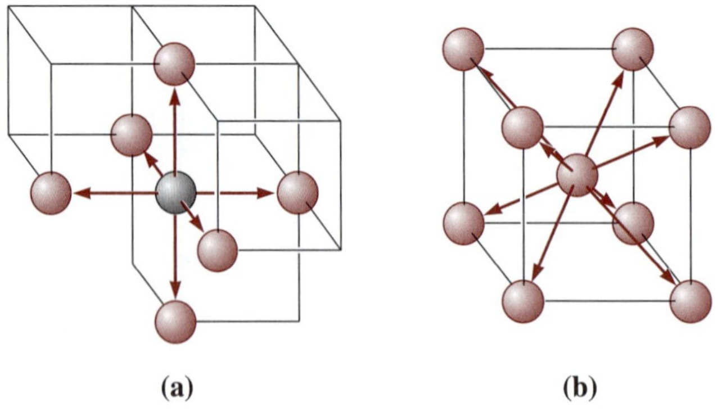

Figure 3-9 Illustration of coordinations in (a) SC and (b) BCC unit cells. Six atoms touch each atom in SC, while eight atoms touch each atom in the BCC unit cell.

Packing Factor The **packing factor** is the fraction of space occupied by atoms, assuming that atoms are hard spheres sized so that they touch their closest neighbor. The general expression for the packing factor is:

$$\text{Packing factor} = \frac{(\text{number of atoms/cell})(\text{volume of each atom})}{\text{volume of unit cell}} \quad (3\text{-}4)$$

Example 3-3 illustrates how to calculate the packing factor for a FCC cell.

EXAMPLE 3-3 *Calculating the Packing Factor*

Calculate the packing factor for the FCC cell.

SOLUTION

In a FCC cell, there are four lattice points per cell; if there is one atom per lattice point, there are also four atoms per cell. The volume of one atom is $4\pi r^3/3$ and the volume of the unit cell is a_0^3.

$$\text{Packing factor} = \frac{(4\text{ atoms/cell})\left(\frac{4}{3}\pi r^3\right)}{a_0^3}$$

Since, for FCC unit cells, $a_0 = 4r/\sqrt{2}$:

$$\text{Packing factor} = \frac{(4)\left(\frac{4}{3}\pi r^3\right)}{(4r/\sqrt{2})^3} = \frac{\pi}{\sqrt{18}} \cong 0.74$$

The packing factor of $\pi/\sqrt{18} \cong 0.74$ in the FCC unit cell is the most efficient packing possible. BCC cells have a packing factor of 0.68 and SC cells have a packing factor of 0.52. Notice that the packing factor is independent of the radius of atoms, as long as we assume that all atoms have a fixed radius.

The FCC arrangement represents a **close-packed structure** (CP) (i.e., the packing fraction is the highest possible with atoms of one size). The SC and BCC structures are relatively open. We will see in the next section that it is possible to have a hexagonal structure that has the same packing efficiency as the FCC structure. This structure is known as the **hexagonal close-packed structure** (HCP). Metals with only metallic bonding are packed as efficiently as possible. Metals with mixed bonding, such as iron, may have unit cells with less than the maximum packing factor. No commonly encountered engineering metals or alloys have the SC structure, although this structure is found in ceramic materials.

Density The theoretical **density** of a material can be calculated using the properties of the crystal structure. The general formula is:

$$\text{Density } \rho = \frac{(\text{number of atoms/cell})(\text{atomic mass})}{(\text{volume of unit cell})(\text{Avogadro's number})} \quad (3\text{-}5)$$

If a material is ionic and consists of different types of atoms or ions, this formula will have to be modified to reflect these differences. Example 3-4 illustrates how to determine the density of BCC iron.

EXAMPLE 3-4 *Determining the Density of BCC Iron*

Determine the density of BCC iron, which has a lattice parameter of 0.2866 nm.

SOLUTION

For a BCC cell,

$$\text{Atoms/cell} = 2$$

$$a_0 = 0.2866 \text{ nm} = 2.866 \times 10^{-8} \text{ cm}$$

$$\text{Atomic mass of iron} = 55.847 \text{ g/mol}$$

$$\text{Volume of unit cell} = a_0^3 = (2.866 \times 10^{-8} \text{ cm})^3 = 23.54 \times 10^{-24} \text{ cm}^3\text{/cell}$$

$$\text{Avogadro's number } N_A = 6.02 \times 10^{23} \text{ atoms/mol}$$

$$\text{Density } \rho = \frac{(\text{number of atoms/cell})(\text{atomic mass of iron})}{(\text{volume of unit cell})(\text{Avogadro's number})}$$

$$\rho = \frac{(2)(55.847)}{(23.54 \times 10^{-24})(6.02 \times 10^{23})} = 7.882 \text{ g/cm}^3$$

The measured density is 7.870 g/cm^3. The slight discrepancy between the theoretical and measured densities is a consequence of defects in the material. As mentioned before, the term "defect" in this context means imperfections with regard to the atomic arrangement.

The Hexagonal Close-Packed Structure A special form of the hexagonal structure, the hexagonal close-packed structure (HCP), is shown in Figure 3-10. The unit cell is the skewed prism, shown separately. The HCP structure has one lattice point per cell—one from each of the eight corners of the prism—but two atoms are associated with each lattice point. One atom is located at a corner, while the second is located within the unit cell. Thus, the basis is 2.

In metals with an ideal HCP structure, the a_0 and c_0 axes are related by the ratio $c_0/a_0 = 1.633$. Most HCP metals, however, have c_0/a_0 ratios that differ slightly from the ideal value because of mixed bonding. Because the HCP structure, like the FCC structure, has the most efficient packing factor of 0.74 and a coordination number of 12, a number of metals possess this structure. Table 3-2 summarizes the characteristics of crystal structures of some metals.

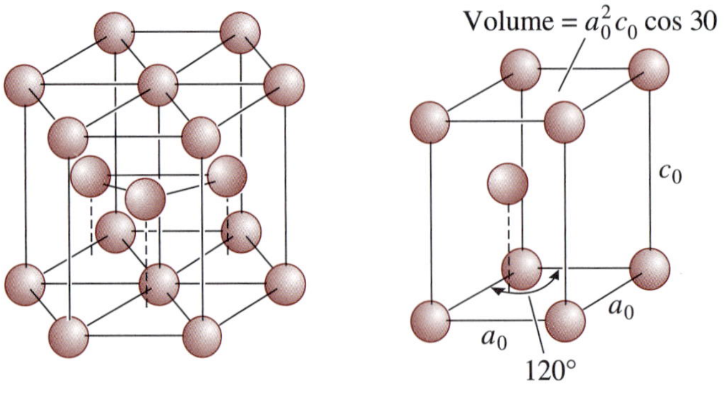

Figure 3-10 The hexagonal close-packed (HCP) structure (left) and its unit cell.

TABLE 3-2 ■ *Crystal structure characteristics of some metals*

Structure	a_0 versus r	Atoms per Cell	Coordination Number	Packing Factor	Examples
Simple cubic (SC)	$a_0 = 2r$	1	6	0.52	Polonium (Po), α-Mn
Body-centered cubic	$a_0 = 4r/\sqrt{3}$	2	8	0.68	Fe, Ti, W, Mo, Nb, Ta, K, Na, V, Zr, Cr
Face-centered cubic	$a_0 = 4r/\sqrt{2}$	4	12	0.74	Fe, Cu, Au, Pt, Ag, Pb, Ni
Hexagonal close-packed	$a_0 = 2r$ $c_0 \approx 1.633a_0$	2	12	0.74	Ti, Mg, Zn, Be, Co, Zr, Cd

Structures of ionically bonded materials can be viewed as formed by the packing (cubic or hexagonal) of anions. Cations enter into the interstitial sites or holes that remain after the packing of anions. Section 3-7 discusses this in greater detail.

3-4 Allotropic or Polymorphic Transformations

Materials with more than one crystal structure are called allotropic or polymorphic. The term **allotropy** is normally reserved for this behavior in pure elements, while the term **polymorphism** is used for compounds. You may have noticed in Table 3-2 that some metals, such as iron and titanium, have more than one crystal structure. At low temperatures, iron has the BCC structure, but at higher temperatures, iron transforms to an FCC structure. These transformations result in changes in properties of materials and form the basis for the heat treatment of steels and many other alloys.

Many ceramic materials, such as silica (SiO_2) and zirconia (ZrO_2), also are polymorphic. A volume change may accompany the transformation during heating or cooling; if not properly controlled, this volume change causes the ceramic material to crack and fail.

Polymorphism is also of central importance to several other applications. The properties of some materials can depend quite strongly on the type of polymorph. For example, the dielectric properties of such materials as *PZT* (Chapter 2) and $BaTiO_3$ depend upon the particular polymorphic form. Example 3-5 illustrates how to calculate volume changes in polymorphs of zirconia.

EXAMPLE 3-5 *Calculating Volume Changes in Polymorphs of Zirconia (ZrO_2)*

Calculate the percent volume change as zirconia (ZrO_2) transforms from a tetragonal to a monoclinic structure. The lattice constants for the monoclinic unit cells are: $a = 5.156$, $b = 5.191$, and $c = 5.304$ Å, respectively. The angle β for the monoclinic unit cell is 98.9°. The lattice constants for the tetragonal unit cell are $a = 5.094$ and $c = 5.304$ Å, respectively. Does the zirconia expand or contract during this transformation? What is the implication of this transformation on the mechanical properties of zirconia ceramics?

SOLUTION

The volume of a tetragonal unit cell is given by $V = a^2c = (5.094)^2(5.304) = 134.33$ Å^3.

The volume of a monoclinic unit cell is given by $V = abc \sin \beta = (5.156)(5.191)(5.304) \sin(98.9) = 140.25$ Å^3.

Thus, there is an expansion of the unit cell as ZrO_2 transforms from a tetragonal to monoclinic form.

The percent change in volume = (final volume − initial volume)/(initial volume) $*$ 100 = (140.25 − 134.33 Å^3)/140.25 Å^3 $*$ 100 = 4.21%.

Most ceramics are very brittle and cannot withstand more than a 0.1% change in volume. (We will discuss mechanical behavior of materials in Chapters 6 and 7.) The conclusion here is that ZrO_2 ceramics cannot be used in their monoclinic form since, when zirconia does transform to the tetragonal form, it will most likely fracture. Therefore, ZrO_2 is often stabilized in a cubic form using different additives such as CaO, MgO, and Y_2O_3. On the other hand, the expansion associated with the tetragonal to monoclinic form is used to create transformation toughened ceramics. When small crystals of tetragonal zirconia are subjected to a stress they become monoclinic. The expansion creates a compressive stress near a crack and toughens the ceramic material.

3-5 Points, Directions, and Planes in the Unit Cell

Coordinates of Points We can locate certain points, such as atom positions, in the lattice or unit cell by constructing the right-handed coordinate system in Figure 3-11. Distance is measured in terms of the number of lattice parameters we must move in each of the x, y, and z coordinates to get from the origin to the point in question. The coordinates are written as the three distances, with commas separating the numbers.

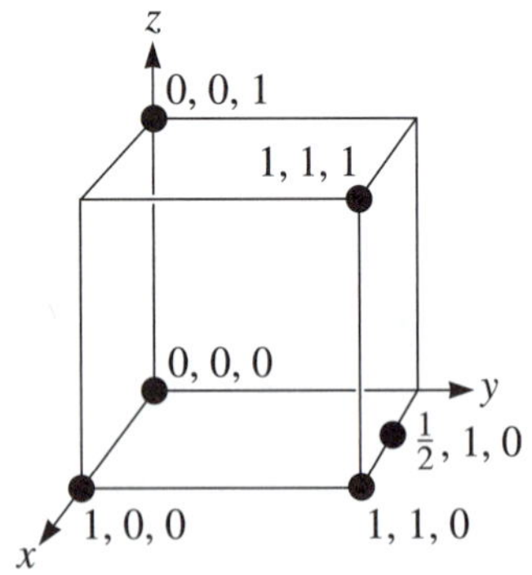

Figure 3-11
Coordinates of selected points in the unit cell. The number refers to the distance from the origin in terms of lattice parameters.

Directions in the Unit Cell Certain directions in the unit cell are of particular importance. **Miller indices** for directions are the shorthand notation used to describe these directions. The procedure for finding the Miller indices for directions is as follows:

1. Using a right-handed coordinate system, determine the coordinates of two points that lie on the direction.
2. Subtract the coordinates of the "tail" point from the coordinates of the "head" point to obtain the number of lattice parameters traveled in the direction of each axis of the coordinate system.
3. Clear fractions and/or reduce the results obtained from the subtraction to lowest integers.
4. Enclose the numbers in square brackets [ ]. If a negative sign is produced, represent the negative sign with a bar over the number.

Example 3-6 illustrates a way of determining the Miller indices of direction.

EXAMPLE 3-6 *Determining Miller Indices of Directions*

Determine the Miller indices of directions A, B, and C in Figure 3-12.

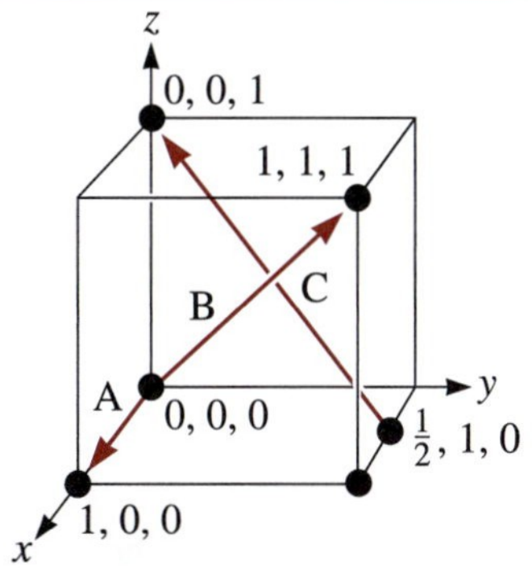

Figure 3-12
Crystallographic directions and coordinates (for Example 3-6).

SOLUTION

Direction *A*

1. Two points are 1, 0, 0, and 0, 0, 0
2. 1, 0, 0 −0, 0, 0 = 1, 0, 0
3. No fractions to clear or integers to reduce
4. [100]

Direction *B*

1. Two points are 1, 1, 1 and 0, 0, 0
2. 1, 1, 1 −0, 0, 0 = 1, 1, 1
3. No fractions to clear or integers to reduce
4. [111]

Direction *C*

1. Two points are 0, 0, 1 and $\frac{1}{2}$, 1, 0
2. 0, 0, 1 $-\frac{1}{2}$, 1, 0 = $-\frac{1}{2}$, −1, 1
3. $2(-\frac{1}{2}, -1, 1) = -1, -2, 2$
4. $[\bar{1}\bar{2}2]$

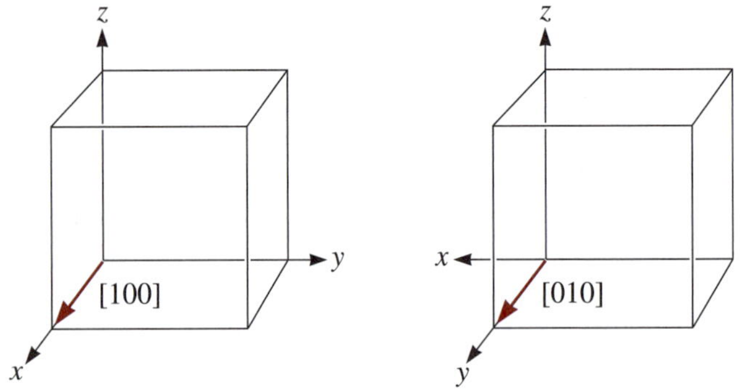

Figure 3-13 Equivalency of crystallographic directions of a form in cubic systems.

TABLE 3-3 ■ *Directions of the form ⟨110⟩ in cubic systems*

⟨110⟩ =	[110]	$[\bar{1}\bar{1}0]$
	[101]	$[\bar{1}0\bar{1}]$
	[011]	$[0\bar{1}\bar{1}]$
	$[1\bar{1}0]$	$[\bar{1}10]$
	$[10\bar{1}]$	$[\bar{1}01]$
	$[01\bar{1}]$	$[0\bar{1}1]$

Several points should be noted about the use of Miller indices for directions:

1. Because directions are vectors, a direction and its negative are not identical; [100] is not equal to $[\bar{1}00]$. They represent the same line, but opposite directions.
2. A direction and its multiple are *identical*; [100] is the same direction as [200]. We just forgot to reduce to lowest integers.
3. Certain groups of directions are *equivalent*; they have their particular indices because of the way we construct the coordinates. For example, in a cubic system, a [100] direction is a [010] direction if we redefine the coordinate system as shown in Figure 3-13. We may refer to groups of equivalent directions as **directions of a form**. The special brackets ⟨ ⟩ are used to indicate this collection of directions. All of the directions of the form ⟨110⟩ are shown in Table 3-3. We would expect a material to have the same properties in each of these 12 directions of the form ⟨110⟩.

Significance of Crystallographic Directions Crystallographic directions are used to indicate a particular orientation of a single crystal or of an oriented polycrystalline material. Knowing how to describe these can be useful in many applications. Metals deform more easily, for example, in directions along which atoms are in closest contact. Another real-world example is the dependence of the magnetic properties of iron and other magnetic materials on the crystallographic directions. It is much easier to magnetize iron in the [100] direction compared to [111] or [110] directions. This is why the grains in Fe-Si steels used in magnetic applications (e.g., transformer cores) are oriented in the [100] or equivalent directions. In the case of magnetic materials used for recording media, we have to make sure the grains are aligned in a particular crystallographic direction such that the stored information is not erased easily. Similarly, crystals used for making turbine blades are aligned along certain directions for better mechanical properties.

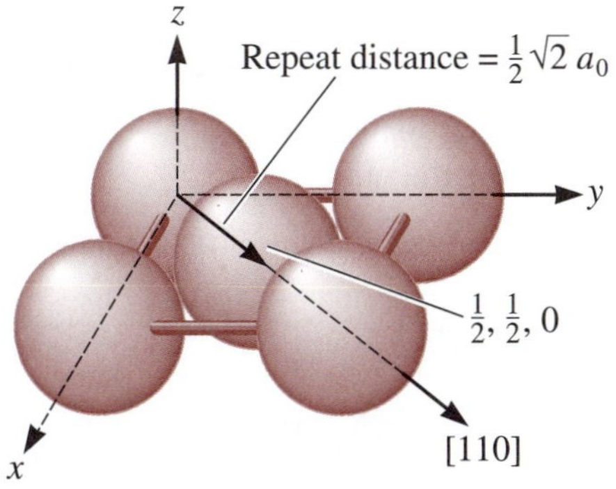

Figure 3-14
Determining the repeat distance, linear density, and packing fraction for a [110] direction in FCC copper.

Repeat Distance, Linear Density, and Packing Fraction Another way of characterizing directions is by the **repeat distance** or the distance between lattice points along the direction. For example, we could examine the [110] direction in an FCC unit cell (Figure 3-14); if we start at the 0, 0, 0 location, the next lattice point is at the center of a face, or a 1/2, 1/2, 0 site. The distance between lattice points is therefore one-half of the face diagonal, or $\frac{1}{2}\sqrt{2}a_0$. In copper, which has a lattice parameter of 0.36151 nm, the repeat distance is 0.2556 nm.

The **linear density** is the number of lattice points per unit length along the direction. In copper, there are two repeat distances along the [110] direction in each unit cell; since this distance is $\sqrt{2}a_0 = 0.51125$ nm, then:

$$\text{Linear density} = \frac{2 \text{ repeat distances}}{0.51125 \text{ nm}} = 3.91 \text{ lattice points/nm}$$

Note that the linear density is also the reciprocal of the repeat distance.

Finally, we could compute the **packing fraction** of a particular direction, or the fraction actually covered by atoms. For copper, in which one atom is located at each lattice point, this fraction is equal to the product of the linear density and twice the atomic radius. For the [110] direction in FCC copper, the atomic radius $r = \sqrt{2}a_0/4 =$ 0.12781 nm. Therefore, the packing fraction is:

$$\begin{aligned}\text{Packing fraction} &= (\text{linear density})(2r)\\ &= (3.91)(2)(0.12781)\\ &= (1.0)\end{aligned}$$

Atoms touch along the [110] direction, since the [110] direction is close-packed in FCC metals.

Planes in the Unit Cell Certain planes of atoms in a crystal also carry particular significance. For example, metals deform easily along planes of atoms that are most tightly packed together. The surface energy of different faces of a crystal depends upon the particular crystallographic planes. This becomes important in crystal growth. In thin film growth of certain electronic materials (e.g., Si or GaAs), we need to be sure the substrate is oriented in such a way that the thin film can grow on a particular crystallographic plane.

Miller indices are used as a shorthand notation to identify these important planes, as described in the following procedure

1. Identify the points at which the plane intercepts the *x*, *y*, and *z* coordinates in terms of the number of lattice parameters. If the plane passes through the origin, the origin of the coordinate system must be moved!

2. Take reciprocals of these intercepts.
3. Clear fractions but do *not* reduce to lowest integers.
4. Enclose the resulting numbers in parentheses ( ). Again, negative numbers should be written with a bar over the number.

The following example shows how Miller indices of planes can be obtained.

EXAMPLE 3-7 *Determining Miller Indices of Planes*

Determine the Miller indices of planes *A*, *B*, and *C* in Figure 3-15.

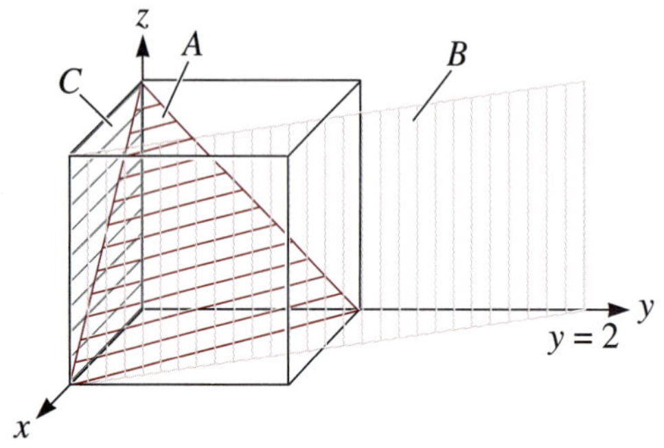

Figure 3-15 Crystallographic planes and intercepts (for Example 3-7).

SOLUTION

Plane *A*

1. $x = 1,\ y = 1,\ z = 1$
2. $\frac{1}{x} = 1,\ \frac{1}{y} = 1,\ \frac{1}{z} = 1$
3. No fractions to clear
4. (111)

Plane *B*

1. The plane never intercepts the z axis, so $z = \infty$, other intercepts are $x = 1$ and $y = 2$
2. $\frac{1}{x} = 1,\ \frac{1}{y} = \frac{1}{2},\ \frac{1}{z} = 0$
3. Clear fractions: $\frac{1}{x} = 2,\ \frac{1}{y} = 1,\ \frac{1}{z} = 0$
4. (210)

Plane *C*

1. We must move the origin, since the plane passes through 0, 0, 0. Let's move the origin one lattice parameter in the y-direction. Then, $x = \infty,\ y = -1$, and $z = \infty$
2. $\frac{1}{x} = 0,\ \frac{1}{y} = -1,\ \frac{1}{z} = 0$
3. No fractions to clear.
4. $(0\bar{1}0)$

Several important aspects of the Miller indices for planes should be noted:

1. Planes and their negatives are identical (this was not the case for directions). Therefore, $(020) = (0\bar{2}0)$.
2. Planes and their multiples are not identical (again, this is the opposite of what we found for directions). We can show this by defining planar densities and planar packing fractions. The **planar density** is the number of atoms per unit area whose centers lie on the plane; the packing fraction is the fraction of the area of that plane actually covered by these atoms. Example 3-8 shows how these can be calculated.
3. In each unit cell, **planes of a form** represent groups of equivalent planes that have their particular indices because of the orientation of the coordinates. We represent these groups of similar planes with the notation { }. The planes of a form {110} in cubic systems are (110), (101), (011), $(1\bar{1}0)$, $(10\bar{1})$, and $(01\bar{1})$.
4. In cubic systems, a direction that has the same indices as a plane is perpendicular to that plane.

EXAMPLE 3-8 *Calculating the Planar Density and Packing Fraction*

Calculate the planar density and planar packing fraction for the (010) and (020) planes in simple cubic polonium, which has a lattice parameter of 0.334 nm.

SOLUTION

The two planes are drawn in Figure 3-16. On the (010) plane, the atoms are centered at each corner of the cube face, with 1/4 of each atom actually in the face of the unit cell. Thus, the total atoms on each face is one. The planar density is:

$$\text{Planar density (010)} = \frac{\text{atom per face}}{\text{area of face}} = \frac{1 \text{ atom per face}}{(0.334)^2}$$

$$= 8.96 \text{ atoms/nm}^2 = 8.96 \times 10^{14} \text{ atoms/cm}^2$$

The planar packing fraction is given by:

$$\text{Packing fraction (010)} = \frac{\text{area of atoms per face}}{\text{area of face}} = \frac{(1 \text{ atom})(\pi r^2)}{(a_0)^2}$$

$$= \frac{\pi r^2}{(2r)^2} = 0.79$$

However, no atoms are centered on the (020) planes. Therefore, the planar density and the planar packing fraction are both zero. The (010) and (020) planes are not equivalent!

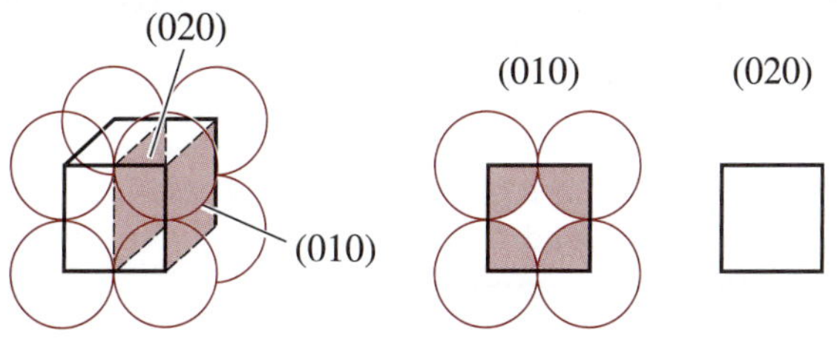

Figure 3-16 The planar densities of the (010) and (020) planes in SC unit cells are not identical (for Example 3-8).

Construction of Directions and Planes To construct a direction or plane in the unit cell, we simply work backwards. Example 3-9 shows how we might do this.

EXAMPLE 3-9 *Drawing Direction and Plane*

Draw (a) the $[1\bar{2}1]$ direction and (b) the $(\bar{2}10)$ plane in a cubic unit cell.

SOLUTION

a. Because we know that we will need to move in the negative y-direction, let's locate the new origin at 0, +1, 0. The "tail" of the direction will be located at this new origin. A second point on the direction can be determined by moving +1 in the x-direction, −2 in the y-direction, and +1 in the z-direction [Figure 3-17(a)].

b. To draw in the $(\bar{2}10)$ plane, first take reciprocals of the indices to obtain the intercepts, that is:

$$x = \frac{1}{-2} = -\frac{1}{2} \quad y = \frac{1}{1} = 1 \quad z = \frac{1}{0} = \infty$$

Since the x-intercept is in a negative direction, and we wish to draw the plane within the unit cell, let's move the new origin +1 in the x-direction to 1, 0, 0. Then we can locate the x-intercept at −1/2 and the y-intercept at +1. The plane will be parallel to the z-axis [Figure 3-17(b)].

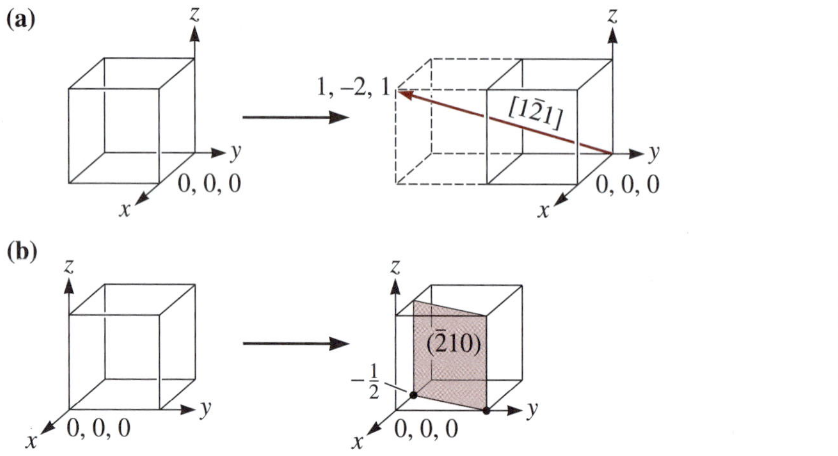

Figure 3-17 Construction of a (a) direction and (b) plane within a unit cell (for Example 3-9).

Miller Indices for Hexagonal Unit Cells A special set of **Miller-Bravais indices** has been devised for hexagonal unit cells because of the unique symmetry of the system (Figure 3-18). The coordinate system uses four axes instead of three, with the a_3 axis being redundant. The procedure for finding the indices of planes is exactly the same as before, but four intercepts are required, giving indices of the form $(hkil)$. Because of the redundancy of the a_3 axis and the special geometry of the system, the first three integers in the designation, corresponding to the a_1, a_2, and a_3 intercepts, are related by $h + k = -i$.

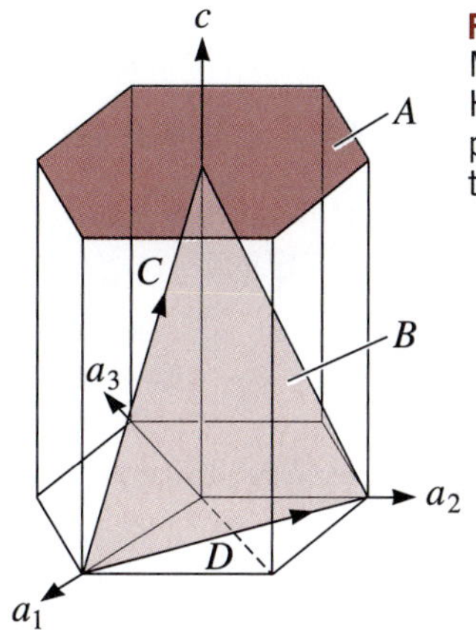

Figure 3-18
Miller-Bravais indices are obtained for crystallographic planes in HCP unit cells by using a four-axis coordinate system. The planes labeled *A* and *B* and the directions labeled *C* and *D* are those discussed in Example 3-10.

Directions in HCP cells are denoted with either the three-axis or four-axis system. With the three-axis system, the procedure is the same as for conventional Miller indices; examples of this procedure are shown in Example 3-10. A more complicated procedure, by which the direction is broken up into four vectors, is needed for the four-axis system. We determine the number of lattice parameters we must move in each direction to get from the "tail" to the "head" of the direction, while for consistency still making sure that $h + k = -i$. This is illustrated in Figure 3-19, showing that the [010] direction is the same as the $[\bar{1}2\bar{1}0]$ direction.

We can also convert the three-axis notation to the four-axis notation for directions by the following relationships, where h', k', and l' are the indices in the three-axis system:

$$\left.\begin{aligned} h &= \frac{1}{3}(2h' - k') \\ k &= \frac{1}{3}(2k' - h') \\ i &= -\frac{1}{3}(h' + k') \\ l &= l' \end{aligned}\right\} \tag{3-6}$$

After conversion, the values of h, k, i, and l may require clearing of fractions or reducing to lowest integers.

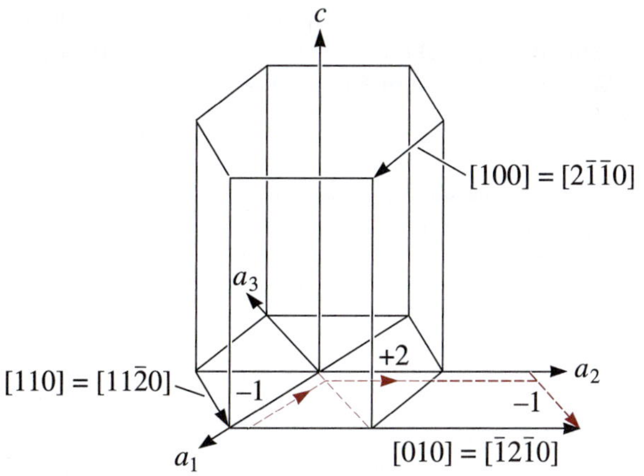

Figure 3-19
Typical directions in the HCP unit cell, using both three- and four-axis systems. The dashed lines show that the $[\bar{1}2\bar{1}0]$ direction is equivalent to a [010] direction.

EXAMPLE 3-10 *Determining the Miller-Bravais Indices for Planes and Directions*

Determine the Miller-Bravais indices for planes *A* and *B* and directions *C* and *D* in Figure 3-18.

SOLUTION

Plane *A*

1. $a_1 = a_2 = a_3 = \infty$, $c = 1$
2. $\frac{1}{a_1} = \frac{1}{a_2} = \frac{1}{a_3} = 0$, $\frac{1}{c} = 1$
3. No fractions to clear
4. (0001)

Plane *B*

1. $a_1 = 1$, $a_2 = 1$, $a_3 = -\frac{1}{2}$, $c = 1$
2. $\frac{1}{a_1} = 1$, $\frac{1}{a_2} = 1$, $\frac{1}{a_3} = -2$, $\frac{1}{c} = 1$
3. No fractions to clear.
4. $(11\bar{2}1)$

Direction *C*

1. Two points are 0, 0, 1 and 1, 0, 0.
2. 0, 0, 1 −1, 0, 0 = −1, 0, 1
3. No fractions to clear or integers to reduce.
4. $[\bar{1}01]$ or $[\bar{2}113]$

Direction *D*

1. Two points are 0, 1, 0 and 1, 0, 0.
2. 0, 1, 0 −1, 0, 0 = −1, 1, 0
3. No fractions to clear or integers to reduce.
4. $[\bar{1}10]$ or $[\bar{1}100]$

Close-Packed Planes and Directions In examining the relationship between atomic radius and lattice parameter, we looked for close-packed directions, where atoms are in continuous contact. We can now assign Miller indices to these close-packed directions, as shown in Table 3-4.

TABLE 3-4 ■ *Close-packed planes and directions*

Structure	Directions	Planes
SC	$\langle 100 \rangle$	None
BCC	$\langle 111 \rangle$	None
FCC	$\langle 110 \rangle$	{111}
HCP	$\langle 100 \rangle$, $\langle 110 \rangle$ or $\langle 11\bar{2}0 \rangle$	(0001), (0002)

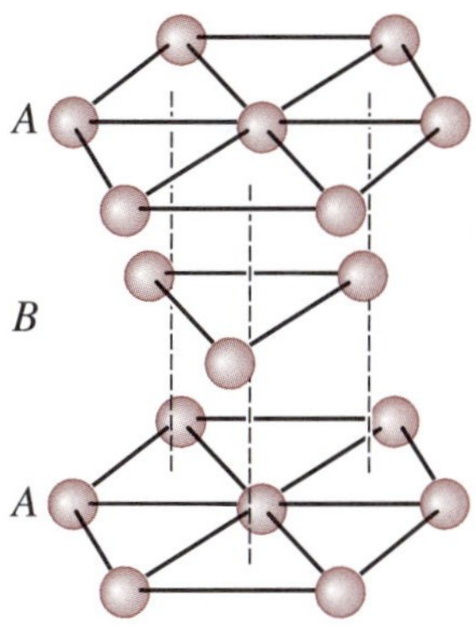

Figure 3-20
The *ABABAB* stacking sequence of close-packed planes produces the HCP structure.

We can also examine FCC and HCP unit cells more closely and discover that there is at least one set of close-packed planes in each. Close-packed planes are shown in Figure 3-20. Notice that a hexagonal arrangement of atoms is produced in two dimensions. The close-packed planes are easy to find in the HCP unit cell; they are the (0001) and (0002) planes of the HCP structure and are given the special name **basal planes**. In fact, we can build up an HCP unit cell by stacking together close-packed planes in an ... *ABABAB* ... **stacking sequence** (Figure 3-20). Atoms on plane *B*, the (0002) plane, fit into the valleys between atoms on plane *A*, the bottom (0001) plane. If another plane identical in orientation to plane *A* is placed in the valleys of plane *B*, the HCP structure is created. Notice that all of the possible close-packed planes are parallel to one another. Only the basal planes—(0001) and (0002)—are close-packed.

From Figure 3-20, we find the coordination number of the atoms in the HCP structure. The center atom in a basal plane is touched by six other atoms in the same plane. Three atoms in a lower plane and three atoms in an upper plane also touch the same atom. The coordination number is 12.

In the FCC structure, close-packed planes are of the form {111} (Figure 3-21). When parallel (111) planes are stacked, atoms in plane *B* fit over valleys in plane *A* and atoms in plane *C* fit over valleys in both planes *A* and *B*. The fourth plane fits directly over atoms in plane *A*. Consequently, a stacking sequence ... *ABCABCABC* ... is produced using the (111) plane. Again, we find that each atom has a coordination number of 12.

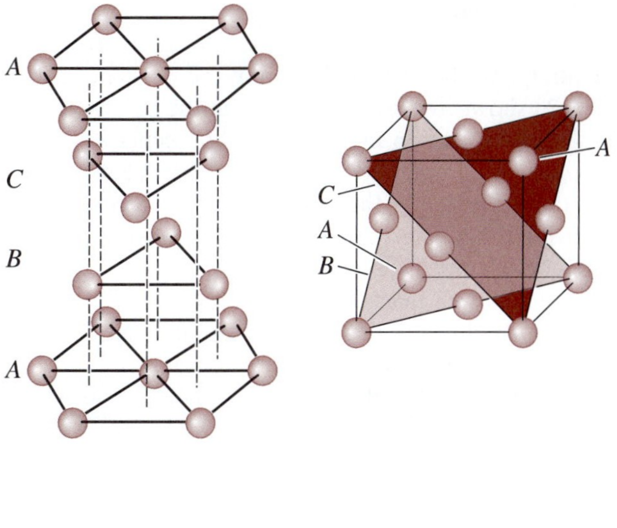

Figure 3-21
The *ABCABCABC* stacking sequence of close-packed planes produces the FCC structure.

Unlike the HCP unit cell, there are four sets of nonparallel close-packed planes—(111), $(11\bar{1})$, $(1\bar{1}1)$, and $(\bar{1}11)$—in the FCC cell. This difference between the FCC and HCP unit cells—the presence or absence of intersecting close-packed planes—affects the behavior of metals with these structures.

Isotropic and Anisotropic Behavior Because of differences in atomic arrangement in the planes and directions within a crystal, some properties also vary with direction. A material is crystallographically **anisotropic** if its properties depend on the crystallographic direction along which the property is measured. For example, the modulus of elasticity of aluminum is 75.9 GPa in $\langle 111 \rangle$ directions, but only 63.4 GPa in $\langle 100 \rangle$ directions. If the properties are identical in all directions, the material is crystallographically **isotropic**. Note that a material such as aluminum, which is crystallographically anisotropic, may behave as an isotropic material if it is in a polycrystalline form. This is because the random orientations of different crystals in a polycrystalline material will mostly cancel out any effect of the anisotropy as a result of crystal structure. In general, most polycrystalline materials will exhibit isotropic properties. Materials that are single crystals or in which many grains are oriented along certain directions (natural or deliberately obtained by processing) will typically have anisotropic mechanical, optical, magnetic, and dielectric properties.

Interplanar Spacing The distance between two adjacent parallel planes of atoms with the same Miller indices is called the **interplanar spacing** (d_{hkl}). The interplanar spacing in *cubic* materials is given by the general equation

$$d_{hkl} = \frac{a_0}{\sqrt{h^2 + k^2 + l^2}} \tag{3-7}$$

where a_0 is the lattice parameter and h, k, and l represent the Miller indices of the adjacent planes being considered. The interplanar spacings for non-cubic materials are given by more complex expressions.

3-6 Interstitial Sites

In any of the crystal structures that have been described, there are small holes between the usual atoms into which smaller atoms may be placed. These holes in the crystal structure are called **interstitial sites**.

An atom, when placed into an interstitial site, touches two or more atoms in the lattice. This interstitial atom has a coordination number equal to the number of atoms it touches. Figure 3-22 shows interstitial locations in the SC, BCC, and FCC structures. The **cubic site**, with a coordination number of eight, occurs in the SC structure. **Octahedral sites** give a coordination number of six (not eight). They are known as octahedral sites because the atoms contacting the interstitial atom form an octahedron with the larger atoms occupying the regular lattice points. **Tetrahedral sites** give a coordination number of four. As an example, the octahedral sites in BCC unit cells are located at faces of the cube; a small atom placed in the octahedral site touches the four atoms at the corners of the face, the atom in the center of the unit cell, plus another atom at the center of the adjacent unit cell, giving a coordination number of six. In FCC unit cells, octahedral sites occur at the center of each edge of the cube, as well as in the center of the unit cell.

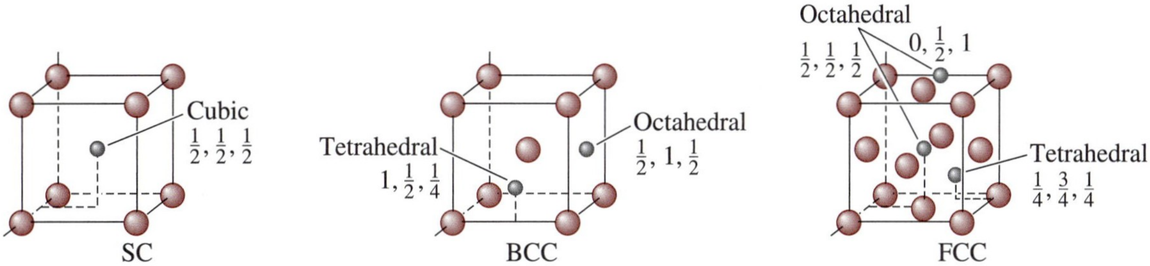

Figure 3-22 The location of the interstitial sites in cubic unit cells. Only representative sites are shown.

EXAMPLE 3-11 *Calculating Octahedral Sites*

Calculate the number of octahedral sites that *uniquely* belong to one FCC unit cell.

SOLUTION

The octahedral sites include the centers of the 12 edges of the unit cell, with the coordinates

$$\tfrac{1}{2},0,0 \quad \tfrac{1}{2},1,0 \quad \tfrac{1}{2},0,1 \quad \tfrac{1}{2},1,1$$
$$0,\tfrac{1}{2},0 \quad 1,\tfrac{1}{2},0 \quad 1,\tfrac{1}{2},1 \quad 0,\tfrac{1}{2},1$$
$$0,0,\tfrac{1}{2} \quad 1,0,\tfrac{1}{2} \quad 1,1,\tfrac{1}{2} \quad 0,1,\tfrac{1}{2}$$

plus the center position, 1/2, 1/2, 1/2. Each of the sites on the edge of the unit cell is shared between four unit cells, so only 1/4 of each site belongs uniquely to each unit cell. Therefore, the number of sites belonging uniquely to each cell is:

$$(12 \text{ edges})\left(\tfrac{1}{4} \text{ per cell}\right) + 1 \text{ center location} = 4 \text{ octahedral sites}$$

Interstitial atoms or ions whose radii are slightly larger than the radius of the interstitial site may enter that site, pushing the surrounding atoms slightly apart. However, atoms whose radii are smaller than the radius of the hole are not allowed to fit into the interstitial site, because the ion would "rattle" around in the site. If the interstitial atom becomes too large, it prefers to enter a site having a larger coordination number (Table 3-5). Therefore, an atom whose radius ratio is between 0.225 and 0.414 enters a tetrahedral site; if its radius ratio is somewhat larger than 0.414, it enters an octahedral site instead. When atoms have the same size, as in pure metals, the radius ratio is one and the coordination number is 12, which is the case for metals with the FCC and HCP structures.

Many ionic crystals (Section 3-7) can be viewed as being generated by close packing of larger anions. Cations are viewed as smaller ions that fit into the interstitial sites of the close packed anions. Thus, the radius ratios described in Table 3-5 also apply to the ratios of radius of the cation to that of the anion. The packing in ionic crystals is not as tight as that in FCC or HCP metals.

TABLE 3-5 ■ *The coordination number and the radius ratio*

Coordination Number	Location of Interstitial	Radius Ratio	Representation
2	Linear	0–0.155	
3	Center of triangle	0.155–0.225	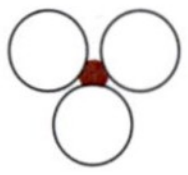
4	Center of tetrahedron	0.225–0.414	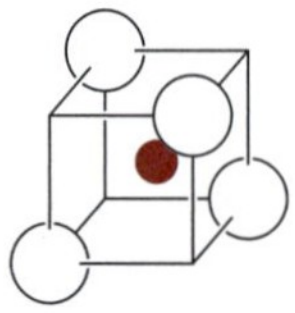
6	Center of octahedron	0.414–0.732	
8	Center of cube	0.732–1.000	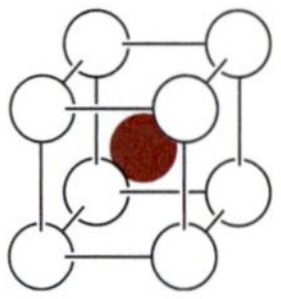

3-7 Crystal Structures of Ionic Materials

Many ceramic materials (Chapter 15) contain considerable fraction of ionic bonds between the anions and cations. These ionic materials must have crystal structures that assure electrical neutrality and stoichiometry, yet permit ions of different sizes to be packed efficiently. As mentioned before, ionic crystal structures can be viewed as close-packed structures of anions. Anions form tetrahedra or octahedra, allowing the cations to fit into their appropriate interstitial sites. In some cases, it may be easier to visualize coordination polyhedra of cations with anions going to the interstial sites. Some typical structures of ionic materials are discussed here.

Visualization of Crystal Structures Using Computers Before we begin to describe different crystal structures, it is important to note that there are many new software programs and tools that have become available recently. These are quite effective in better understanding the crystal structure concepts as well as other concepts discussed in Chapter 4. One example of a useful program is the CaRIne^{TM} software.

Cesium Chloride Structure Cesium chloride (CsCl) is simple cubic, with the "cubic" interstitial site filled by the Cl anion [Figure 3-23(a)]. The radius ratio,

$$\frac{r_{Cs^+}}{r_{Cl^{-1}}} = \frac{0.167 \text{ nm}}{0.181 \text{ nm}} = 0.92$$

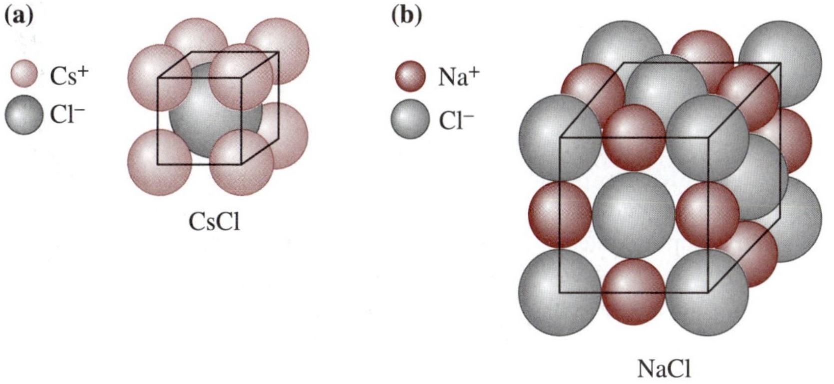

Figure 3-23 (a) The cesium chloride structure, a SC unit cell with two ions (Cs^+ and Cl^-) per lattice point. (b) The sodium chloride structure, a FCC unit cell with two ions (Na^+ and Cl^-) per lattice point. *Note*: Ion sizes not to scale.

dictates that cesium chloride has a coordination number of eight. We can characterize the structure as a simple cubic structure with two ions—one Cs^+ and one Cl^{-1}—associated with each lattice point (or a basis of 2). This structure is possible when the anion and the cation have the same valence.

Sodium Chloride Structure The radius ratio for sodium and chloride ions is $r_{Na^+}/r_{Cl^-} = 0.097 \text{ nm}/0.181 \text{ nm} = 0.536$; the sodium ion has a charge of $+1$; the chloride ion has a charge of -1. Therefore, based on the charge balance and radius ratio, each anion and cation must have a coordination number of six. The FCC structure, with Cl^{-1} ions at FCC positions and Na^+ at the four octahedral sites, satisfies these requirements [Figure 3-23(b)]. We can also consider this structure to be FCC with two ions—one Na^{+1} and one Cl^{-1}—associated with each lattice point. Many ceramics, including magnesium oxide (MgO), calcium oxide (CaO), and iron oxide (FeO) have this structure.

EXAMPLE 3-12 *Illustrating a Crystal Structure and Calculating Density*

Show that MgO has the sodium chloride crystal structure and calculate the density of MgO.

SOLUTION

From Appendix B, $r_{Mg^{+2}} = 0.066$ nm and $r_{O^{-2}} = 0.132$ nm, so:

$$\frac{r_{Mg^{+2}}}{r_{O^{-2}}} = \frac{0.066}{0.132} = 0.50$$

Since $0.414 < 0.50 < 0.732$, the coordination number for each ion is six, and the sodium chloride structure is possible.

The atomic masses are 24.312 and 16 g/mol for magnesium and oxygen, respectively. The ions touch along the edge of the cube, so:

$$a_0 = 2r_{Mg^{+2}} + 2r_{O^{-2}} = 2(0.066) + 2(0.132) = 0.396 \text{ nm} = 3.96 \times 10^{-8} \text{ cm}$$

$$\rho = \frac{(4Mg^{+2})(24.312) + (4O^{-2})(16)}{(3.96 \times 10^{-8} \text{ cm})^3(6.02 \times 10^{23})} = 4.31 \text{ g/cm}^3$$

Zinc Blende Structure Although the Zn ions have a charge of +2 and S ions have a charge of −2, zinc blende (ZnS) cannot have the sodium chloride structure because $\frac{r_{Zn^{+2}}}{r_{S^{-2}}} = 0.074\ \text{nm}/0.184\ \text{nm} = 0.402$. This radius ratio demands a coordination number of four, which in turn means that the sulfide ions enter tetrahedral sites in a unit cell, as indicated by the small "cubelet" in the unit cell (Figure 3-24). The FCC structure, with Zn cations at the normal lattice points and S anions at half of the tetrahedral sites, can accommodate the restrictions of both charge balance and coordination number. A variety of materials, including the semiconductor GaAs and many other III–V semiconductors (Chapter 2) have this structure.

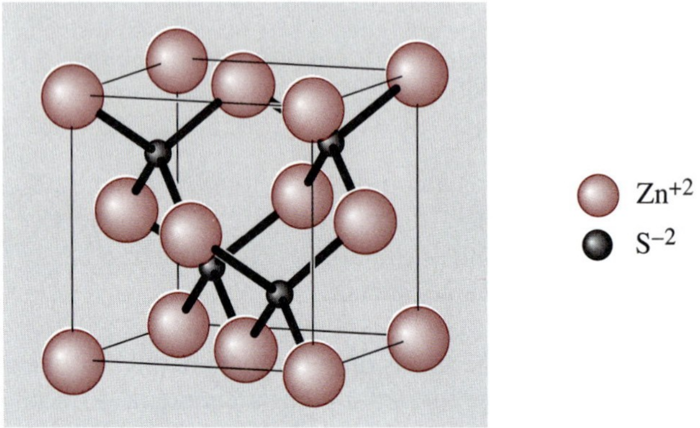

Figure 3-24
The zinc blende unit cell.
Note: Ion sizes not to scale.

Fluorite Structure The fluorite structure is FCC, with anions located at all eight of the tetrahedral positions (Figure 3-25). One of the polymorphs of ZrO_2 known as cubic zirconia exhibits this crystal structure. Other compounds that exhibit this structure include UO_2, ThO_2, and CeO_2.

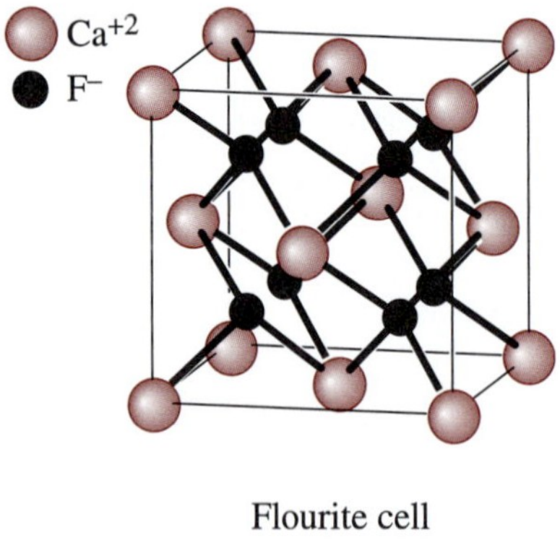

Figure 3-25
Fluorite unit cell. *Note*: Ion sizes not to scale.

Perovskite Structure This is the crystal structure of $CaTiO_3$ and $BaTiO_3$ (Figure 3-26). In $CaTiO_3$, oxygen anions occupy the face centers of the perovskite unit cell, the corners or the A-sites are occupied by the Ca^{+2} ions and the octahedral B-site at the cube center is occupied by the Ti^{+4} ions. Billions of capacitors for electronic applications are made using formulations based on $BaTiO_3$.

Corundum Structure This is one of the crystal structures of alumina known as alpha alumina (α-Al_2O_3). In alumina, the oxygen anions pack in a hexagonal arrangement

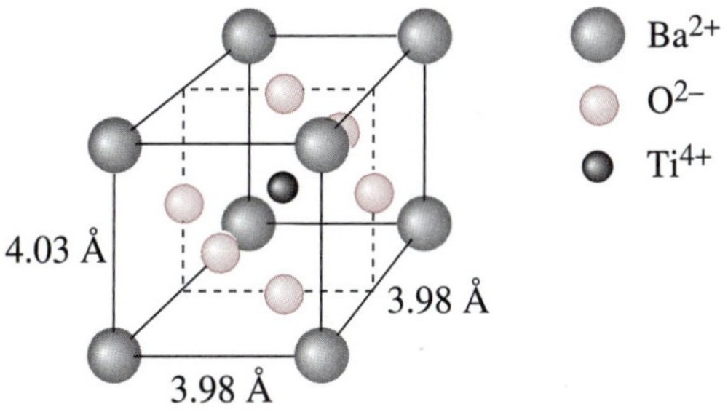

Figure 3-26
The perovskite unit cell showing the arrangement of different ions. *Note*: Ion sizes not to scale.

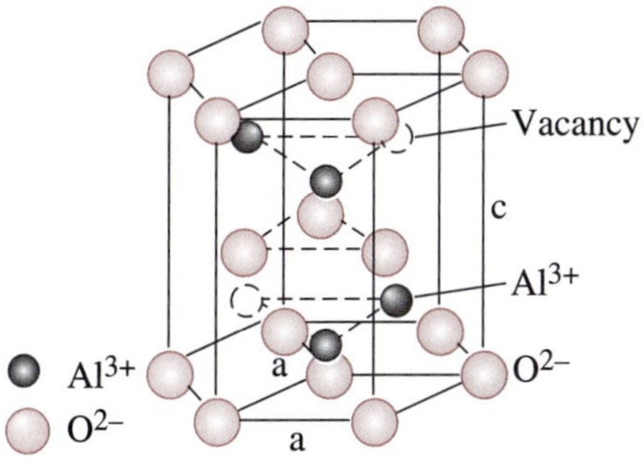

Figure 3-27
Corundum structure of alpha-alumina (α-Al_2O_3). *Note*: Ion sizes not to scale.

and the aluminum cations occupy some of the available octahedral positions (Figure 3-27). Alumina is probably the most widely used ceramic material. Applications include, but are not limited to, spark plugs, refractories, electronic packaging substrates, and abrasives.

3-8 Covalent Structures

Covalently bonded materials frequently have complex structures in order to satisfy the directional restraints imposed by the bonding.

Diamond Cubic Structure Elements such as silicon, germanium (Ge), α-Sn, and carbon (in its diamond form) are bonded by four covalent bonds and produce a tetrahedron. The coordination number for each silicon atom is only four, because of the nature of the covalent bonding. (See Figure 3-28.)

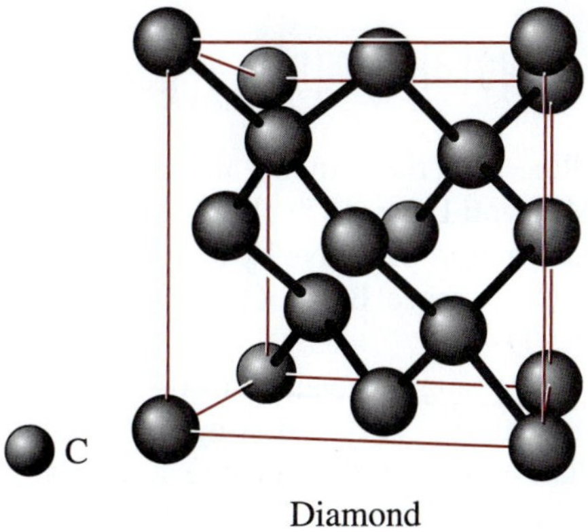

Figure 3-28
The diamond cubic (DC) unit cell. This open structure is produced because of the requirements of covalent bonding.

EXAMPLE 3-13 ***Determining the Packing Factor for Diamond Cubic Silicon***

Determine the packing factor for diamond cubic silicon.

SOLUTION

We find that atoms touch along the body diagonal of the cell (Figure 3-29). Although atoms are not present at all locations along the body diagonal, there are voids that have the same diameter as atoms. Consequently:

$$\sqrt{3}a_0 = 8r$$

$$\text{Packing factor} = \frac{(8\text{ atoms/cell})\left(\frac{4}{3}\pi r^3\right)}{a_0^3}$$

$$= \frac{(8)\left(\frac{4}{3}\pi r^3\right)}{(8r/\sqrt{3})^3}$$

$$= 0.34$$

Compared to close packed structures this is a relatively open structure. In Chapter 5, we will learn that openness of the structure is one of the factors that affects the rate at which different atoms can diffuse inside a given material.

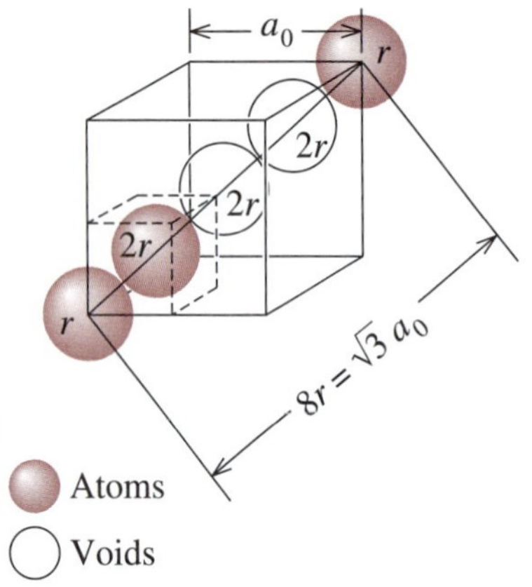

Figure 3-29 Determining the relationship between lattice parameter and atomic radius in a diamond cubic cell (for Example 3-13).

3-9 Diffraction Techniques for Crystal Structure Analysis

A crystal structure of a crystalline material can be analyzed using **x-ray diffraction** (XRD) or electron diffraction. Max von Laue (1879–1960) won the Nobel Prize in 1912 for his discovery related to the diffraction of x-rays by a crystal. William Henry Bragg (1862–1942) and his son William Lawrence Bragg (1890–1971) won the 1915 Nobel Prize for their contributions to XRD.

When a beam of x-rays having a single wavelength (on the same order of magnitude as the atomic spacing in the material) strikes that material, x-rays are scattered in all directions. Most of the radiation scattered from one atom cancels out radiation

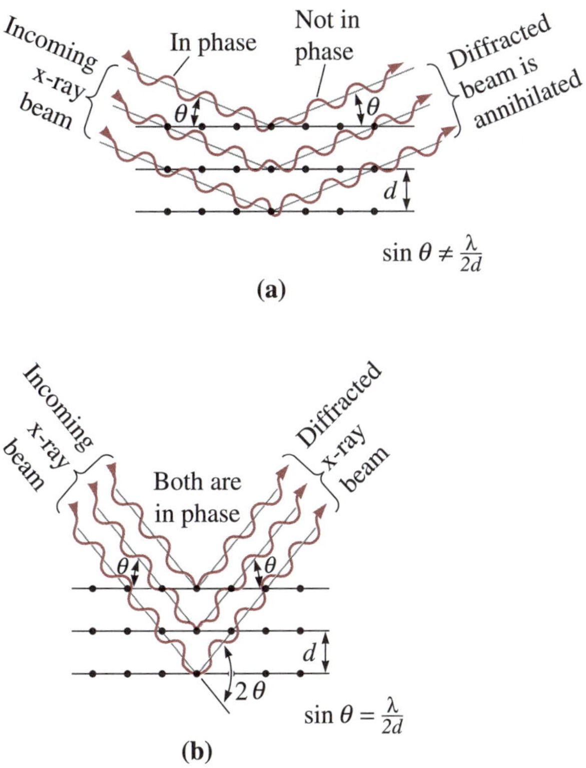

Figure 3-30
(a) Destructive and (b) reinforcing interactions between x-rays and the crystalline material. Reinforcement occurs at angles that satisfy Bragg's law.

scattered from other atoms. However, x-rays that strike certain crystallographic planes at specific angles are reinforced rather than annihilated. This phenomenon is called **diffraction**. The x-rays are diffracted, or the beam is reinforced, when conditions satisfy **Bragg's law**,

$$\sin \theta = \frac{\lambda}{2d_{hkl}} \tag{3-8}$$

where the angle θ is half the angle between the diffracted beam and the original beam direction, λ is the wavelength of the x-rays, and d_{hkl} is the interplanar spacing between the planes that cause constructive reinforcement of the beam (see Figure 3-30).

When the material is prepared in the form of a fine powder, there are always at least some powder particles (tiny crystals or aggregates of tiny crystals) whose planes (hkl) are oriented at the proper θ angle to satisfy Bragg's law. Therefore, a diffracted beam, making an angle of 2θ with the incident beam, is produced. In a **diffractometer**, a moving x-ray detector records the 2θ angles at which the beam is diffracted, giving a characteristic diffraction pattern. If we know the wavelength of the x-rays, we can determine the interplanar spacings and, eventually, the identity of the planes that cause the diffraction.

Electron Diffraction and Microscopy In **electron diffraction**, we make use of high-energy (~100,000 to 400,000 eV) electrons. These electrons are diffracted from electron transparent samples of materials. The electron beam that exits from the sample is also used to form an image of the sample. Both **transmission electron microscopy** (TEM) and electron diffraction are used for imaging microstructural features and determining crystal structures.

EXAMPLE 3-14 Barium Titanate ($BaTiO_3$) Lattice Constant

Barium titanate ($BaTiO_3$) is a ceramic material used to make capacitors that store electrical charge. The lattice constant for the cubic crystal structure is to be determined. This material was analyzed using copper K-α radiation of wavelength 1.54 A. It was seen that the value of 2θ at which the (111) reflection from the diffracted x-rays was at 39°. What is the lattice constant a_0 for the cubic form of $BaTiO_3$?

SOLUTION

We will use Bragg's law:

$$\sin \theta = \lambda / d_{hkl}$$

For the plane (111) in cubic $BaTiO_3$,

$$d_{111} = 1.54 \text{ A}/\sin(19.5) = 1.54/0.3338 = 4.61 \text{ A}$$

Note that we used the value of the angle θ and not that of 2θ. Now,

$$d_{111} = \frac{a_0}{\sqrt{(h^2 + k^2 + l^2)}} = \frac{a_0}{\sqrt{1+1+1}} = \frac{a_0}{\sqrt{3}}$$

Substituting 4.61 A for d_{111}, we get $a_0 = 4.00$ A. This is very close to the experimentally observed value of the lattice constant for cubic form of $BaTiO_3$.

SUMMARY

- Atoms or ions may be arranged in solid materials with either a short-range or long-range order.
- Amorphous materials, such as silicate glasses, metallic glasses, amorphous silicon and many polymers, have only a short-range order. Amorphous materials form whenever the kinetics of a process involved in the fabrication of a material do not allow the atoms or ions to assume the equilibrium positions. These materials often offer very novel and unusual properties.
- Crystalline materials, including metals and many ceramics, have both long- and short-range order. The long-range periodicity in these materials is described by the crystal structure.
- The atomic or ionic arrangements of crystalline materials are described by seven general crystal systems, which include 14 specific Bravais lattices. Examples include simple cubic, body-centered cubic, face-centered cubic, and hexagonal lattices.
- A lattice is a collection of points organized in a unique manner. The basis or motif refers to one or more atoms associated with each lattice point. Crystal structure is derived by adding lattice and basis. Although there are only 14 Bravais lattices there are hundreds of crystal structures.
- A crystal structure is characterized by the lattice parameters of the unit cell, which is the smallest subdivision of the crystal structure that still describes the overall structure of the lattice.

- Allotropic, or polymorphic, materials have more than one possible crystal structure. The properties of materials can depend strongly on the type of particular polymorph or allotrope.
- The atoms of metals having the face-centered cubic and hexagonal close-packed crystal structures are closely packed; atoms are arranged in a manner that occupies the greatest fraction of space. The FCC and HCP structures achieve the closest packing by different stacking sequences of close-packed planes of atoms.
- Points, directions, and planes within the crystal structure can be identified in a formal manner by the assignment of coordinates and Miller indices.
- Interstitial sites, or holes in a crystal structure, can be filled by other atoms or ions. The crystal structure of many ceramic materials can be understood by considering how these sites are occupied. Atoms or ions located in interstitial sites play an important role in strengthening materials, influencing the physical properties of materials, and controlling the processing of materials.
- Crystal structures of many ionic materials form by the packing of anions (e.g., oxygen ions (O^{-2})). Cations fit into coordination polyhedra formed by anions.
- Crystal structures of covalently bonded materials tend to be open. Examples include diamond cubic (e.g., Si, Ge).
- XRD and electron diffraction are used for the determination of the crystal structure of crystalline materials. Transmission electron microscopy can also be used for imaging of microstructural features in materials at smaller length scales.

GLOSSARY

Allotropy The characteristic of an element being able to exist in more than one crystal structure, depending on temperature and pressure.

Amorphous materials Materials, including glasses, that have no long-range order, or crystal structure.

Anisotropic Having different properties in different directions.

Atomic radius The apparent radius of an atom, typically calculated from the dimensions of the unit cell, using close-packed directions (depends upon coordination number).

Basal plane The special name given to the close-packed plane in hexagonal close-packed unit cells.

Basis A group of atoms associated with a lattice point (same as motif).

Bragg's law The relationship describing the angle at which a beam of x-rays of a particular wavelength diffracts from crystallographic planes of a given interplanar spacing.

Bravais lattices The fourteen possible lattices that can be created using lattice points.

Close-packed directions Directions in a crystal along which atoms are in contact.

Close-packed structure Structures showing a packing fraction of 0.74 (FCC and HCP).

Coordination number The number of nearest neighbors to an atom in its atomic arrangement.

Crystal structure The arrangement of the atoms in a material into a regular repeatable lattice.

Crystal systems Cubic, tetragonal, orthorhombic, hexagonal, monoclinic, rhombohedral and triclinic arrangements of points in space that lead to 14 Bravais lattices and hundreds of crystal structures.

Crystalline materials Materials comprised of one or many small crystals or grains.

Crystallization The process responsible for the formation of crystals, typically in an amorphous material.

Cubic site An interstitial position that has a coordination number of eight. An atom or ion in the cubic site touches eight other atoms or ions.

Density Mass per unit volume of a material, usually in units of g/cm^3.

Diamond cubic (DC) A special type of face-centered cubic crystal structure found in carbon, silicon, and other covalently bonded materials.

Diffraction The constructive interference, or reinforcement, of a beam of x-rays or electrons interacting with a material. The diffracted beam provides useful information concerning the structure of the material.

Directions of a form Crystallographic directions that all have the same characteristics, although their "sense" is different. Denoted by $\langle\ \rangle$ brackets.

Electron diffraction A method to determine the level of crystallinity at relatively smaller length scale. Based on the diffraction of electrons typically involving use of a transmission electron microscope.

Glasses Non-crystalline materials (typically derived from the molten state) that have only short-range atomic order.

Grain A crystal in a polycrystalline material.

Grain boundaries Regions between grains of a polycrystalline material.

Interplanar spacing Distance between two adjacent parallel planes with the same Miller indices.

Interstitial sites Locations between the "normal" atoms or ions in a crystal into which another—usually different—atom or ion is placed. Typically, the size of this interstitial location is smaller than the atom or ion that is to be introduced.

Isotropic Having the same properties in all directions.

Lattice A collection of points that divide space into smaller equally sized segments.

Lattice parameters The lengths of the sides of the unit cell and the angles between those sides. The lattice parameters describe the size and shape of the unit cell.

Lattice points Points that make up the lattice. The surroundings of each lattice point are identical anywhere in the material.

Linear density The number of lattice points per unit length along a direction.

Liquid crystals Polymeric materials that are typically amorphous but can become partially crystalline when an external electric field is applied. The effect of the electric field is reversible. Such materials are used in liquid crystal displays.

Long-range order (LRO) A regular repetitive arrangement of atoms in a solid which extends over a very large distance.

Metallic glass Amorphous metals or alloys obtained using rapid solidification.

Miller-Bravais indices A special shorthand notation to describe the crystallographic planes in hexagonal close-packed unit cells.

Miller indices A shorthand notation to describe certain crystallographic directions and planes in a material. Denoted by [ ] brackers. A negative number is represented by a bar over the number.

Motif A group of atoms affiliated with a lattice point (same as basis).

Octahedral site An interstitial position that has a coordination number of six. An atom or ion in the octahedral site touches six other atoms or ions.

Packing factor The fraction of space in a unit cell occupied by atoms.

Packing fraction The fraction of a direction (linear-packing fraction) or a plane (planar-packing factor) that is actually covered by atoms or ions. When one atom is located at each lattice point, the linear packing fraction along a direction is the product of the linear density and twice the atomic radius.

Planar density The number of atoms per unit area whose centers lie on the plane.

Planes of a form Crystallographic planes that all have the same characteristics, although their orientations are different. Denoted by { } braces.

Polycrystalline material A material comprised of many grains.

Polymorphism Compounds exhibiting more than one type of crystal structure.

Rapid solidification A technique used to cool metals and alloys very quickly.

Repeat distance The distance from one lattice point to the adjacent lattice point along a direction.

Short-range order The regular and predictable arrangement of the atoms over a short distance—usually one or two atom spacings.

Stacking sequence The sequence in which close-packed planes are stacked. If the sequence is *ABABAB*, a hexagonal close-packed unit cell is produced; if the sequence is *ABCABCABC*, a face-centered cubic structure is produced.

Transmission electron microscopy (TEM) A technique for imaging and analysis of microstructures using a high-energy electron beam.

Tetrahedral site An interstitial position that has a coordination number of four. An atom or ion in the tetrahedral site touches four other atoms or ions.

Unit cell A subdivision of the lattice that still retains the overall characteristics of the entire lattice.

X-ray diffraction (XRD) A technique for analysis of crystalline materials using a beam of x-rays.

PROBLEMS

Section 3-1 Short-Range Order versus Long-Range Order

Section 3-2 Amorphous Materials: Principles and Technological Applications

Section 3-3 Lattice, Unit Cells, Basis, and Crystal Structures

3-1 Define the terms lattice, unit cell, basis, and crystal structure.

3-2 Explain why there is no face-centered tetragonal Bravais lattice.

3-3 Calculate the atomic radius in cm for the following:

(a) BCC metal with $a_0 = 0.3294$ nm and one atom per lattice point; and
(b) FCC metal with $a_0 = 4.0862$ Å and one atom per lattice point.

3-4 Determine the crystal structure for the following:

(a) a metal with $a_0 = 4.9489$ Å, $r = 1.75$ Å, and one atom per lattice point; and
(b) a metal with $a_0 = 0.42906$ nm, $r = 0.1858$ nm, and one atom per lattice point.

3-5 The density of potassium, which has the BCC structure and one atom per lattice point, is 0.855 g/cm^3. The atomic weight of potassium is 39.09 g/mol. Calculate

(a) the lattice parameter; and
(b) the atomic radius of potassium.

3-6 The density of thorium, which has the FCC structure and one atom per lattice point, is 11.72 g/cm^3. The atomic weight of thorium is 232 g/mol. Calculate

(a) the lattice parameter; and
(b) the atomic radius of thorium.

3-7 A metal has a cubic structure with a density of 2.6 g/cm^3, an atomic weight of 87.62 g/mol, and a lattice parameter of 6.0849 Å. One atom is associated with each lattice point. Determine the crystal structure of the metal.

3-8 A metal has a cubic structure with a density of 1.892 g/cm^3, an atomic weight of 132.91 g/mol, and a lattice parameter of 6.13 Å. One atom is associated with each lattice point. Determine the crystal structure of the metal.

3-9 Indium has a tetragonal structure, with $a_0 = 0.32517$ nm and $c_0 = 0.49459$ nm. The density is 7.286 g/cm^3 and the atomic weight is 114.82 g/mol. Does indium have the simple tetragonal or body-centered tetragonal structure?

3-10 Gallium has an orthorhombic structure, with $a_0 = 0.45258$ nm, $b_0 = 0.45186$ nm, and $c_0 = 0.76570$ nm. The atomic radius is 0.1218 nm. The density is 5.904 g/cm^3 and the atomic weight is 69.72 g/mol. Determine

(a) the number of atoms in each unit cell; and
(b) the packing factor in the unit cell.

3-11 Beryllium has a hexagonal crystal structure, with $a_0 = 0.22858$ nm and $c_0 = 0.35842$ nm. The atomic radius is 0.1143 nm, the density is 1.848 g/cm^3, and the atomic weight is 9.01 g/mol. Determine

(a) the number of atoms in each unit cell; and
(b) the packing factor in the unit cell.

3-12 A typical paper clip weighs 0.59 g. Assume that it is made from BCC iron. Calculate

(a) the number of unit cells; and
(b) the number of iron atoms in the paper clip. (See Appendix A for required data.)

3-13 Aluminum foil used to package food is approximately 0.0025 cm thick. Assume that all of the unit cells of the aluminum are arranged so that a_0 is perpendicular to the foil surface. For a 10 cm × 10 cm square of the foil, determine

(a) the total number of unit cells in the foil; and
(b) the thickness of the foil in number of unit cells. (See Appendix A.)

Section 3-4 Allotropic or Polymorphic Transformations

3-14 What is the difference between an allotrope and a polymorph?

3-15 Above 882°C, titanium has a BCC crystal structure, with $a = 0.332$ nm. Below this temperature, titanium has a HCP structure with $a = 0.2978$ nm and $c = 0.4735$ nm. Determine the percent volume change when BCC titanium transforms to HCP titanium. Is this a contraction or expansion?

3-16 α-Mn has a cubic structure with $a_0 = 0.8931$ nm and a density of 7.47 g/cm^3. β-Mn has a different cubic structure with $a_0 = 0.6326$ nm and a density of 7.26 g/cm^3. The atomic weight of manganese is 54.938 g/mol and the atomic radius is 0.112 nm. Determine the percent volume change that would occur if α-Mn transforms to β-Mn.

3-17 What are the two allotropes of iron?

Section 3-5 Points, Directions, and Planes in the Unit Cell

3-18 Explain the significance of crystallographic directions using an example of an application.

3-19 Determine the Miller indices for the directions in the cubic unit cell shown in Figure 3-31.

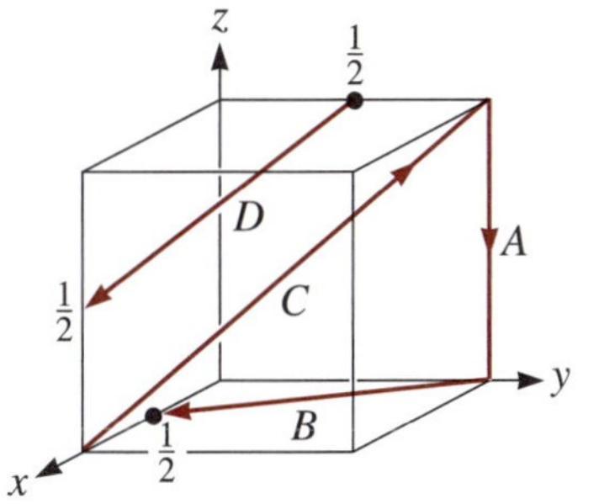

Figure 3-31 Directions in a cubic unit cell for Problem 3-19.

3-20 Determine the indices for the directions in the cubic unit cell shown in Figure 3-32.

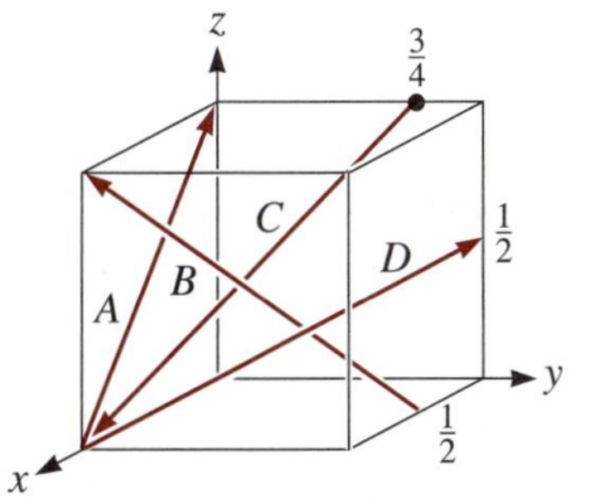

Figure 3-32 Directions in a cubic unit cell for Problem 3-20.

3-21 Determine the indices for the planes in the cubic unit cell shown in Figure 3-33.

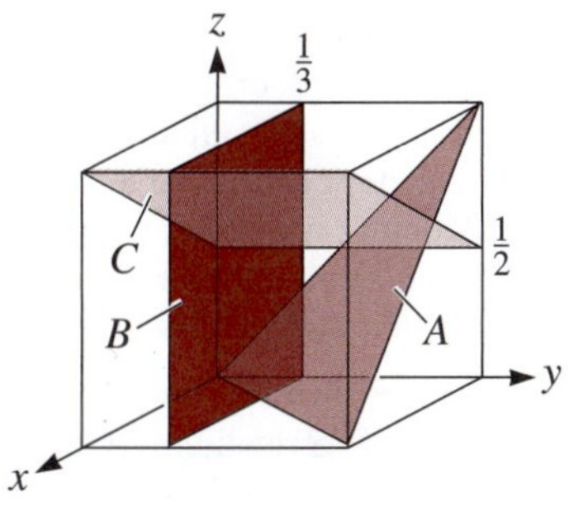

Figure 3-33 Planes in a cubic unit cell for Problem 3-21.

3-22 Determine the indices for the planes in the cubic unit cell shown in Figure 3-34.

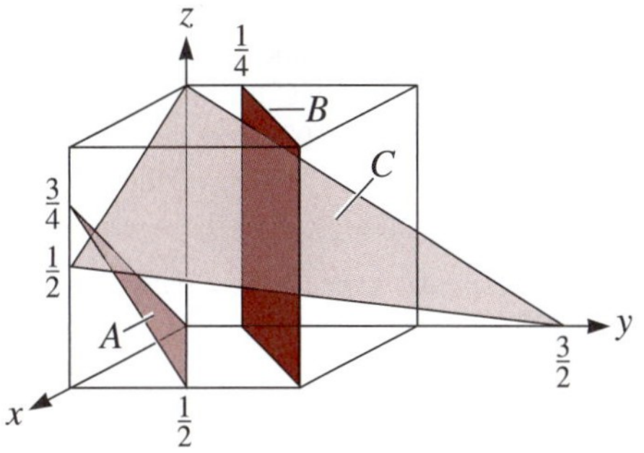

Figure 3-34 Planes in a cubic unit cell for Problem 3-22.

3-23 Sketch the following planes and directions within a cubic unit cell:

(a) $[101]$ **(b)** $[0\bar{1}0]$ **(c)** $[12\bar{2}]$ **(d)** $[301]$
(e) $[\bar{2}01]$ **(f)** $[2\bar{1}3]$ **(g)** $(0\bar{1}\bar{1})$ **(h)** (102)
(i) (002) **(j)** $(1\bar{3}0)$ **(k)** $(\bar{2}12)$ **(l)** $(3\bar{1}\bar{2})$

3-24 Sketch the following planes and directions within a cubic unit cell:

(a) $[1\bar{1}0]$ **(b)** $[\bar{2}\bar{2}1]$ **(c)** $[410]$ **(d)** $[0\bar{1}2]$
(e) $[\bar{3}\bar{2}1]$ **(f)** $[1\bar{1}1]$ **(g)** $(11\bar{1})$ **(h)** $(01\bar{1})$
(i) (030) **(j)** $(\bar{1}21)$ **(k)** $(11\bar{3})$ **(l)** $(0\bar{4}1)$

3-25 What are the indices of the six directions of the form $\langle 110 \rangle$ that lie in the $(11\bar{1})$ plane of a cubic cell?

3-26 What are the indices of the four directions of the form $\langle 111 \rangle$ that lie in the $(\bar{1}01)$ plane of a cubic cell?

3-27 Determine the number of directions of the form $\langle 110 \rangle$ in a tetragonal unit cell and compare to the number of directions of the form $\langle 110 \rangle$ in an orthorhombic unit cell.

3-28 Determine the angle between the $[110]$ direction and the (110) plane in a tetragonal unit cell; then determine the angle between the $[011]$ direction and the (011) plane in a tetragonal cell. The lattice parameters are $a_0 = 4.0$ Å and $c_0 = 5.0$ Å. What is responsible for the difference?

3-29 Determine the Miller indices of the plane that passes through three points having the following coordinates:

(a) 0, 0, 1; 1, 0, 0; and 1/2, 1/2, 0
(b) 1/2, 0, 1; 1/2, 0, 0; and 0, 1, 0
(c) 1, 0, 0; 0, 1, 1/2; and 1, 1/2, 1/4
(d) 1, 0, 0; 0, 0, 1/4; and 1/2, 1, 0

3-30 Determine the repeat distance, linear density, and packing fraction for FCC nickel, which has a lattice parameter of 0.35167 nm, in the [100], [110], and [111] directions. Which of these directions is close packed?

3-31 Determine the repeat distance, linear density, and packing fraction for BCC lithium, which has a lattice parameter of 0.35089 nm, in the [100], [110], and [111] directions. Which of these directions is close packed?

3-32 Determine the planar density and packing fraction for FCC nickel in the (100), (110), and (111) planes. Which, if any, of these planes is close-packed?

3-33 Determine the planar density and packing fraction for BCC lithium in the (100), (110), and (111) planes. Which, if any, of these planes is close packed?

3-34 Suppose that FCC rhodium is produced as a 1-mm thick sheet, with the (111) plane parallel to the surface of the sheet. How many (111) interplanar spacings d_{111} thick is the sheet? See Appendix A for necessary data.

3-35 What are the Miller indices of the plane shown in the Figure 3-35?

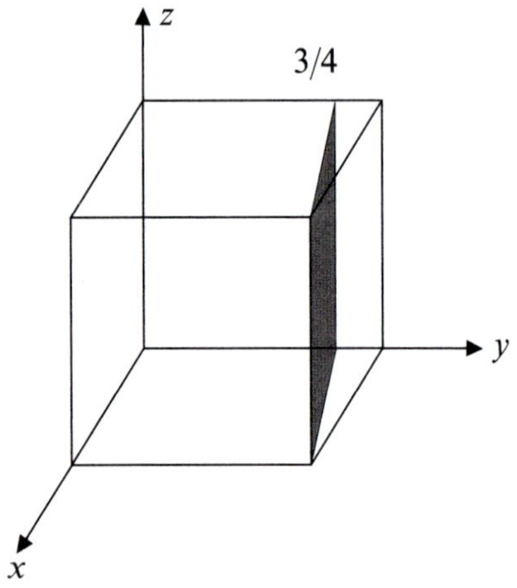

Figure 3-35 Plane in a cubic unit cell for Problem 3-35.

Section 3-6 Interstitial Sites

3-36 Determine the minimum radius of an atom that will just fit into:

(a) the tetrahedral interstitial site in FCC nickel; and

(b) the octahedral interstitial site in BCC lithium.

3-37 What are the coordination numbers for octahedral and tetrahedral sites?

Section 3-7 Crystal Structures of Ionic Materials

3-38 What is meant by coordination polyhedra?

3-39 Is the radius of an atom or ion fixed? Explain.

3-40 Explain why we consider anions to form the close-packed structures and cations to enter the interstitial sites?

3-41 What is the coordination number for the titanium ion in the perovskite crystal structure?

3-42 What is the radius of an atom that will just fit into the octahedral site in FCC copper without disturbing the crystal structure?

3-43 Would you expect NiO to have the cesium chloride, sodium chloride, or zinc blende structure? Based on your answer, determine

(a) the lattice parameter;
(b) the density; and
(c) the packing factor.

3-44 Would you expect UO_2 to have the sodium chloride, zinc blende, or fluorite structure? Based on your answer, determine

(a) the lattice parameter;
(b) the density; and
(c) the packing factor.

3-45 Would you expect BeO to have the sodium chloride, zinc blende, or fluorite structure? Based on your answer, determine

(a) the lattice parameter;
(b) the density; and
(c) the packing factor.

3-46 Would you expect CsBr to have the sodium chloride, zinc blende, fluorite, or cesium chloride structure? Based on your answer, determine

(a) the lattice parameter;
(b) the density; and
(d) the packing factor.

3-47 Recently, gallium nitride (GaN) material has been used to make light-emitting diodes (LEDs) that emit a blue or ultraviolet light. Such LEDs are used in DVD players and other electronic devices. This material has two crystal structures. One form is the zinc-blende crystal structure (lattice constant $a_0 = 0.450$ nm), which has a density of 6.1 g/cm^3 at 300 K. Calculate the number of Ga and N atoms per unit cell of this form of GaN.

3-48 The theoretical density of germanium (Ge) is 5.323 g/cm^3 at 300 K. Germanium has the same crystal structure as diamond. What is the lattice constant of germanium at 300 K?

3-49 The lattice constant of zinc selenide (ZnSe) is 0.567 nm. The crystal structure is that of zinc blende. Show that the theoretical density for ZnSe should be 5.26 g/cm^3.

Section 3-8 Covalent Structures

3-50 Calculate the theoretical density of α-Sn. Assume diamond cubic structure and obtain the radius information from Appendix B.

3-51 What are the different polymorphs of carbon?

Section 3-9 Diffraction Techniques for Crystal Structure Analysis

3-52 Explain the principle of XRD.

3-53 A sample of cubic SiC was analyzed using XRD. It was found that the (111) peak was located at 2θ of 16°. The wavelength (λ) of the x-ray radiation used in this experiment was 0.6975 A. Show that the lattice constant (a_0) of this form of SiC is 4.0867 A.

3-54 For the cubic phase of $BaTiO_3$, a diffraction peak is seen at a value of $2\theta = 45°$. What crystallographic plane does this peak correspond to if the XRD analysis was done using Cu K-α x-rays ($\lambda = 1.54$ A)?

3-55 The lattice constant of $BaTiO_3$, a ceramic material used to make capacitors, for the cubic crystal structure is 4 A. This material is analyzed using copper K-α radiation of wavelength 1.54 A. What will be the value of 2θ at which the (200) reflection from the diffracted x-rays can be expected?

Design Problems

3-56 An oxygen sensor is to be made to measure dissolved oxygen in a large vessel containing molten steel. What kind of material would you choose for this application? Explain.

3-57 You would like to sort iron specimens, some of which are FCC and others BCC. Design an x-ray diffraction method by which this can be accomplished.

4

Imperfections in the Atomic and Ionic Arrangements

Have You Ever Wondered?

- *Why silicon crystals, used in the manufacture of semiconductor chips, contain trace amounts of dopants, such as phosphorous or boron?*
- *What makes steel considerably harder and stronger than pure iron?*
- *Why do we use very high-purity copper as a conductor in electrical applications?*
- *Why FCC metals (such as copper and aluminum) tend to be more ductile than BCC and HCP metals?*

The arrangement of the atoms or ions in engineered materials contains imperfections or defects. These **defects** often have a profound effect on the properties of materials. In this chapter, we introduce the three basic types of imperfections: point defects, line defects (or dislocations), and surface defects. These imperfections only represent defects in or deviations from the perfect or ideal atomic or ionic arrangements expected in a given crystal structure. The material is not considered defective from an application viewpoint. In many applications, the presence of such defects actually is useful. There are a few applications, though, where we will strive to minimize a particular type of defect. For example, defects known as dislocations are useful in increasing the strength of metals and alloys. However, in single crystal silicon, used for manufacturing computer chips, the presence of dislocations is undesirable. Often the "defects" are created intentionally to produce a desired set of electronic, magnetic, optical, and mechanical properties. For example, pure iron is relatively soft, yet, when we add a small amount of carbon, we create defects in the crystalline arrangement of iron and turn it into a plain carbon steel that exhibits considerably higher strength. Similarly, a crystal of pure alumina (Al_2O_3) is transparent and colorless, but, when we add a small amount of chromium (Cr), it creates a special defect, resulting in a beautiful red ruby crystal.

Grain boundaries, regions between different grains of a polycrystalline material, represent one type of defect that can control properties. For example, the new ceramic superconductors, under certain conditions, can conduct electricity without any electrical resistance. Materials scientists and engineers have made long wires or tapes of such materials. They have also discovered that, although the current flows quite well within the grains of a polycrystalline superconductor, there is considerable resistance to the flow of current from one grain onto another—across the grain boundary. On the other hand, the presence of grain boundaries actually helps strengthen metallic materials. In later chapters, we will show how we can control the concentrations of these defects through tailoring of composition or processing techniques. In this chapter, we explore the nature and effects of different types of defects.

4-1 Point Defects

Point defects are localized disruptions in an otherwise perfect atomic or ionic arrangements in a crystal structure. Even though we call them point defects, the disruption affects a region involving several of the surrounding atoms or ions. These imperfections, shown in Figure 4-1, may be introduced by movement of the atoms or ions when they gain energy by heating, during processing of the material or by introduction of other atoms. The distinction between an impurity and a dopant is as follows: Typically, **impurities** are elements or compounds that are present from raw materials or processing. For example, silicon single crystals grown in quartz crucibles contain oxygen as an impurity. **Dopants**, on the other hand, are elements or compounds that are deliberately added, in known concentrations, at specific locations in the microstructure, with an intended beneficial effect on properties or processing. In general, the effect of impurities is deleterious, whereas the effect of dopants on the properties of materials is useful. Phosphorus (P) and boron (B) are examples of dopants that are added to silicon crystals to improve or alter the electrical properties of pure silicon (Si).

A point defect typically involves one atom or ion, or a pair of atoms or ions, and thus is different from **extended defects,** such as dislocations, grain boundaries, etc. An important "point" about point defects is that although the defect occurs at one or two sites, their presence is "felt" over much larger distances in a crystalline material.

Vacancies A **vacancy** is produced when an atom or an ion is missing from its normal site in the crystal structure, as in Figure 4-1(a). When atoms or ions are missing (i.e., when vacancies are present), the overall randomness or entropy of the material increases, which increases the thermodynamic stability of a crystalline material. All

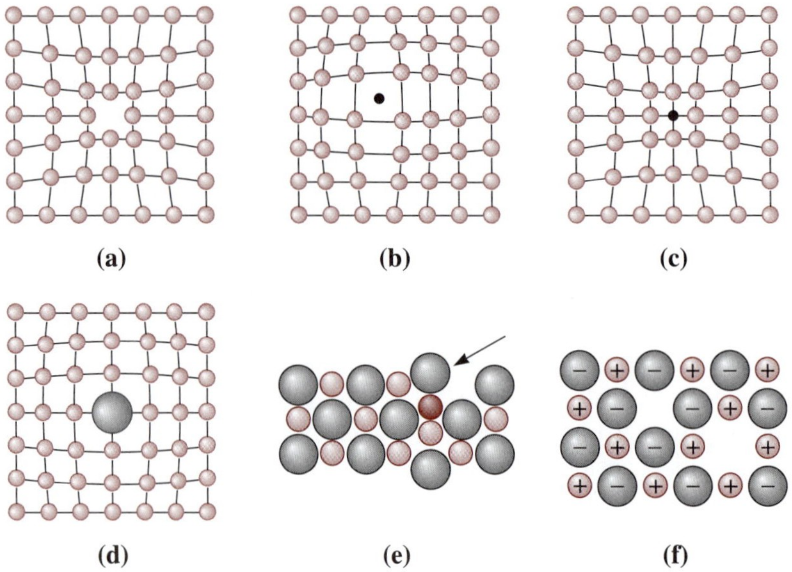

Figure 4-1 Point defects: (a) vacancy, (b) interstitial atom, (c) small substitutional atom, (d) large substitutional atom, (e) Frenkel defect, and (f) Schottky defect. All of these defects disrupt the perfect arrangement of the surrounding atoms and create a strain in the crystal structure.

crystalline materials have vacancy defects. Vacancies are introduced into metals and alloys during solidification, at high temperatures, or as a consequence of radiation damage. Vacancies play an important role in determining the rate at which atoms or ions can move around, or diffuse in a solid material, especially in pure metals. We will see this effect in greater detail in Chapter 5. In some other applications, we make use of the vacancies created in a ceramic material to tune its electrical properties. This includes many ceramics that are used as conductive and transparent oxides such as indium tin oxide (*ITO*) and zirconia oxygen (ZrO_2) sensors.

At room temperature (~300 K), the concentration of vacancies is small, but the concentration of vacancies increases exponentially as we increase the temperature, as shown by the following Arrhenius type behavior:

$$n_v = n \exp\left(\frac{-Q_v}{RT}\right) \tag{4-1}$$

where

n_v is the number of vacancies per cm^3;

n is the number of atoms per cm^3;

Q_v is the energy required to produce one mole of vacancies, in cal/mol or Joules/mol;

R is the gas constant, $1.987 \dfrac{\text{cal}}{\text{mol} - \text{K}}$ or $8.31 \dfrac{\text{Joules}}{\text{mol} - \text{K}}$; and

T is the temperature in degrees Kelvin.

Due to the large thermal energy atoms have near the melting temperature of a material, there may be as many as one vacancy per 1000 atoms. Note that this equation provides for *equilibrium* concentration of vacancies at a given temperature. It is also possible to retain a *non-equilibrium* concentration of vacancies produced at a high temperature by quenching the material rapidly. Thus, in many situations the concentration of vacancies observed at room temperature is *not* the equilibrium concentration predicted by Equation 4-1.

EXAMPLE 4-1 *The Effect of Temperature on Vacancy Concentrations*

Calculate the concentration of vacancies in copper at 25°C. What temperature will be needed to heat treat copper such that the concentration of vacancies produced will be 1000 times more than the equilibrium concentration of vacancies at room temperature? Assume that 20,000 cal are required to produce a mole of vacancies in copper.

SOLUTION

The lattice parameter of FCC copper is 0.36151 nm. The basis is 1, therefore, the number of copper atoms, or lattice points, per cm^3 is:

$$n = \frac{4 \text{ atoms/cell}}{(3.6151 \times 10^{-8} \text{ cm})^3} = 8.47 \times 10^{22} \text{ copper atoms/cm}^3$$

At 25°C, $T = 25 + 273 = 298$ K:

$$n_v = n \exp\left(\frac{-Q_v}{RT}\right)$$

$$= \left(8.47 \times 10^{22} \frac{\text{atoms}}{\text{cm}^3}\right) \cdot \exp\left(\frac{-20{,}000 \frac{\text{cal}}{\text{mol}}}{1.987 \frac{\text{cal}}{\text{mol}-\text{K}} \times 298 \text{ K}}\right)$$

$$= 1.815 \times 10^8 \text{ vacancies/cm}^3$$

We wish to find a heat treatment temperature that will lead to a concentration of vacancies which is 1000 times higher than this number, or $n_v = 1.815 \times 10^{11}$ vacancies/cm^3.

We could do this by heating the copper to a temperature at which this number of vacancies forms:

$$n_v = 1.815 \times 10^{11} = n \exp\left(\frac{-Q_v}{RT}\right)$$

$$= (8.47 \times 10^{22}) \exp(-20{,}000/(1.987 \times T))$$

$$\exp\left(\frac{-20{,}000}{1.987 \times T}\right) = \frac{1.815 \times 10^{11}}{8.47 \times 10^{22}} = 0.214 \times 10^{-11}$$

$$\frac{-20{,}000}{1.987T} = \ln(0.214 \times 10^{-11}) = -26.87$$

$$T = \frac{20{,}000}{(1.987)(26.87)} = 375 \text{ K} = 102°\text{C}$$

By heating the copper slightly above 100°C, until equilibrium is reached, and then rapidly cooling the copper back to room temperature, the number of vacancies trapped in the structure may be one thousand times greater than the equilibrium number of vacancies at room temperature. Thus, vacancy concentrations encountered in relatively pure materials are often dictated by both the thermodynamic and kinetic factors.

EXAMPLE 4-2 *Vacancy Concentrations in Iron*

Determine the number of vacancies needed for a BCC iron crystal to have a density of 7.87 g/cm^3. The lattice parameter of BCC iron is 2.866×10^{-8} cm.

SOLUTION

The expected theoretical density of iron can be calculated from the lattice parameter and the atomic mass. Since the iron is BCC, two iron atoms are present in each unit cell.

$$\rho = \frac{(2 \text{ atoms/cell})(55.847 \text{ g/mol})}{(2.866 \times 10^{-8} \text{ cm})^3(6.02 \times 10^{23} \text{ atoms/mol})} = 7.8814 \text{ g/cm}^3$$

We would like to produce iron with a density of 7.87 g/cm^3. We could do this by intentionally introducing vacancies into the crystal. Let's calculate the number of iron atoms and vacancies that would be present in each unit cell for the required density of 7.87 g/cm^3:

$$\rho = \frac{(X \text{ atoms/cell})(55.847 \text{ g/mol})}{(2.866 \times 10^{-8} \text{ cm})^3(6.02 \times 10^{23} \text{ atoms/mol})} = 7.87 \text{ g/cm}^3$$

$$X \text{ atoms/cell} = \frac{(7.87)(2.866 \times 10^{-8})^3(6.02 \times 10^{23})}{55.847} = 1.9971$$

Or, there should be $2.00 - 1.9971 = 0.0029$ vacancies per unit cell. The number of vacancies per cm^3 is:

$$\text{Vacancies/cm}^3 = \frac{0.0029 \text{ vacancies/cell}}{(2.866 \times 10^{-8} \text{ cm})^3} = 1.23 \times 10^{20}$$

We assume that introduction of vacancies does not change the lattice constant. If additional information, such as the energy required to produce a vacancy in iron, is available, we might be able to design a heat treatment (as we did in Example 4-1) to produce this concentration of vacancies.

Interstitial Defects An **interstitial defect** is formed when an extra atom or ion is inserted into the crystal structure at a normally unoccupied position, as in Figure 4-1(b). The interstitial sites were illustrated in Table 3-5. Interstitial atoms or ions, although much smaller than the atoms or ions located at the lattice points, are still larger than the interstitial sites that they occupy. Consequently, the surrounding crystal region is compressed and distorted. Interstitial atoms such as hydrogen are often present as impurities; whereas carbon atoms are intentionally added to iron to produce steel. For small concentrations, carbon atoms occupy interstitial sites in the iron crystal structure, introducing a stress in the localized region of the crystal in their vicinity. If there are dislocations in the crystals trying to move around these types of defects, they face a resistance to their motion, making it difficult to create permanent deformation in metals and alloys. This is one important way of increasing the strength of metallic materials. Unlike vacancies, once introduced, the number of interstitial atoms or ions in the structure remains nearly constant, even when the temperature is changed.

EXAMPLE 4-3 *Interstitial Sites for Carbon in Iron*

In FCC iron, carbon atoms are located at *octahedral* sites at the center of each edge of the unit cell (1/2, 0, 0) and at the center of the unit cell (1/2, 1/2, 1/2). In BCC iron, carbon atoms enter *tetrahedral* sites, such as 1/4, 1/2, 0. The lattice parameter is 0.3571 nm for FCC iron and 0.2866 nm for BCC iron. Assume that carbon atoms have a radius of 0.071 nm. (1) Would we expect a greater distortion of the crystal by an interstitial carbon atom in FCC or BCC iron? (2) What would be the atomic percentage of carbon in each type of iron if all the interstitial sites were filled?

SOLUTION

1. We could calculate the size of the interstitial site at the 1/4, 1/2, 0 location with the help of Figure 4-2(a). The radius R_{BCC} of the iron atom is:

$$R_{BCC} = \frac{\sqrt{3}a_0}{4} = \frac{(\sqrt{3})(0.2866)}{4} = 0.1241 \text{ nm}$$

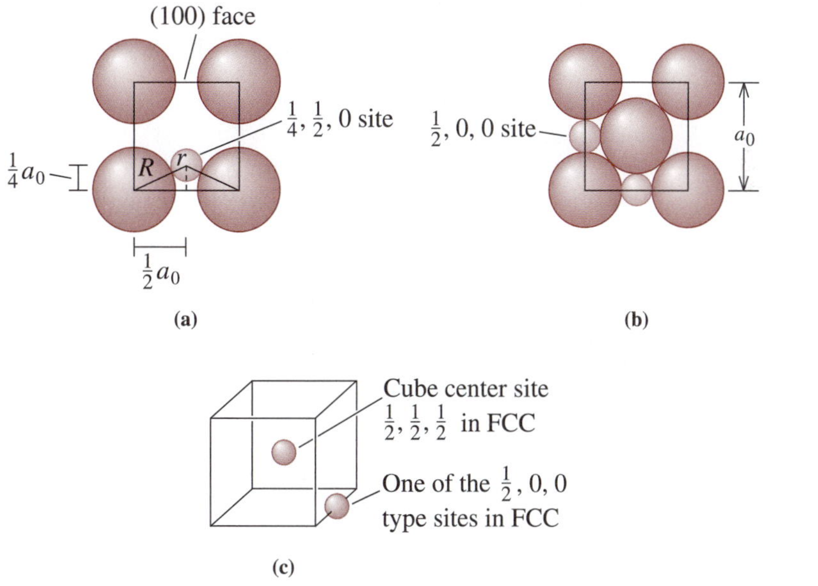

Figure 4-2 (a) The location of the $\frac{1}{4}, \frac{1}{2}$, 0 interstitial site in BCC metals, showing the arrangement of the normal atoms and the interstitial atom (b) $\frac{1}{2}$, 0, 0 site in FCC metals. (c) Edge centers and cube centers are some of the interstitial sites in the FCC structure. (For Example 4-3.)

From Figure 4-2(a), we find that:

$$\left(\frac{1}{2}a_0\right)^2 + \left(\frac{1}{4}a_0\right)^2 = (r_{\text{interstitial}} + R_{\text{BCC}})^2$$

$$(r_{\text{interstitial}} + R_{\text{BCC}})^2 = 0.3125a_0^2 = (0.3125)(0.2866 \text{ nm})^2 = 0.02567$$

$$r_{\text{interstitial}} = \sqrt{0.02567} - 0.1241 = 0.0361 \text{ nm}$$

For FCC iron, the interstitial site such as the 1/2, 0, 0 lies along ⟨100⟩ directions. Thus, the radius of the iron atom and the radius of the interstitial site are [Figure 4-2(b)]:

$$R_{\text{FCC}} = \frac{\sqrt{2}a_0}{4} = \frac{(\sqrt{2})(0.3571)}{4} = 0.1263 \text{ nm}$$

$$2r_{\text{interstitial}} + 2R_{\text{FCC}} = a_0$$

$$r_{\text{interstitial}} = \frac{0.3571 - (2)(0.1263)}{2} = 0.0522 \text{ nm}$$

The interstitial site in the BCC iron is smaller than the interstitial site in the FCC iron. Since both are smaller than the size of the carbon atom, carbon distorts the BCC crystal structure more than the FCC crystal. As a result, fewer carbon atoms are expected to enter interstitial positions in BCC iron than those in FCC iron.

2. In BCC iron, two iron atoms are expected in each unit cell. We can find a total of 24 interstitial sites of the type 1/4, 1/2, 0; however, since each site is located at a face of the unit cell, only half of each site belongs uniquely to a single cell. Thus,

$$(24 \text{ sites})\left(\frac{1}{2}\right) = 12 \text{ interstitial sites per unit cell}$$

If all of the interstitial sites were filled, the atomic percentage of carbon contained in the iron would be

$$\text{at \% C} = \frac{12 \text{ C atoms}}{12 \text{ C atoms} + 2 \text{ Fe atoms}} \times 100 = 86\%$$

In FCC iron, four iron atoms are expected in each unit cell, and the number of octahedral interstitial sites is:

$$(12 \text{ edges})\left(\frac{1}{4}\right) + 1 \text{ center} = 4 \text{ interstitial sites per unit cell [Figure 4-2(c)]}$$

Again, if all the octahedral interstitial sites were filled, the atomic percentage of carbon in the FCC iron would be:

$$\text{at \% C} = \frac{4 \text{ C atoms}}{4 \text{ C atoms} + 4 \text{ Fe atoms}} \times 100 = 50\%$$

As we will see in a later chapter, the maximum atomic percentage of carbon present in the two forms of iron under equilibrium conditions is:

BCC: 1.0% FCC: 8.9%

Because of the strain imposed on the iron crystal structure by the interstitial atoms—particularly in the BCC iron—the fraction of the interstitial sites that can be occupied is quite small.

Substitutional Defects A **substitutional defect** is introduced when one atom or ion is replaced by a different type of atom or ion as in Figure 4-1(c) and (d). The substitutional atoms or ions occupy the normal lattice sites. Substitutional atoms or ions may either be larger than the normal atoms or ions in the crystal structure, in which case the surrounding interatomic spacings are reduced, or smaller causing the surrounding atoms to have larger interatomic spacings. In either case, the substitutional defects alter the interatomic distances in the surrounding crystal. Again, the substitutional defect can be introduced either as an impurity or as a deliberate alloying addition, and, once introduced, the number of defects is relatively independent of temperature.

Examples of substitutional defects include incorporation of dopants such as phosphorus (P) or boron (B) into Si. Similarly, if we added copper to nickel, copper atoms will occupy crystallographic sites where nickel atoms would normally be present. The substitutional atoms will often increase the strength of the metallic material. Substitutional defects also appear in ceramic materials. For example, if we add MgO to NiO, Mg^{+2} ions occupy Ni^{+2} sites and O^{-2} ions from MgO occupy O^{-2} sites of NiO. Whether atoms or ions added go into interstitial or substitutional sites depends upon the size and valence of guest atoms or ions compared to the size and valence of host ions. The size of the available sites also plays a role in this.

4-2 Other Point Defects

An **interstitialcy** is created when an atom identical to those at the normal lattice points is located in an interstitial position. These defects are most likely to be found in crystal structures having a low packing factor.

A **Frenkel defect** is a vacancy-interstitial pair formed when an ion jumps from a normal lattice point to an interstitial site, as in Figure 4-1(e), leaving behind a vacancy. Although this is described for an ionic material, a Frenkel defect can occur in metals and covalently bonded materials. A **Schottky defect**, Figure 4-1(f), is unique to ionic materials and is commonly found in many ceramic materials. In this defect vacancies occur in an ionically bonded material; a stoichiometric number of anions and cations must be missing from the crystal if electrical neutrality is to be preserved in the crystal. For example, one Mg^{+2} and one O^{-2} missing in MgO constitute a Schottky pair. In ZrO_2, for one missing zirconium ion there will be two oxygen ions missing.

An important substitutional point defect occurs when an ion of one charge replaces an ion of a different charge. This might be the case when an ion with a valence of +2 replaces an ion with a valence of +1 (Figure 4-3). In this case, an extra positive charge is introduced into the structure. To maintain a charge balance, a vacancy might be

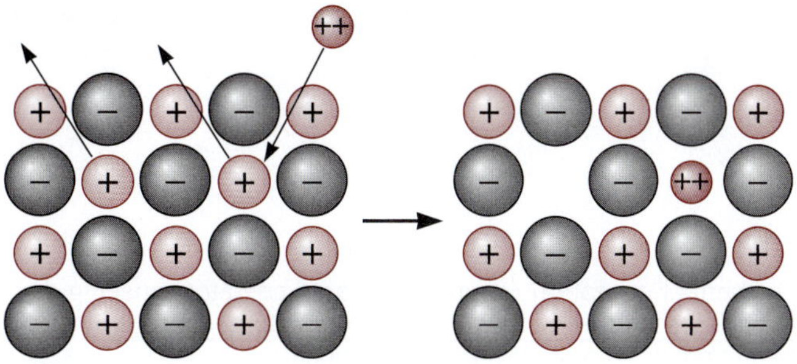

Figure 4-3
When a divalent cation replaces a monovalent cation, a second monovalent cation must also be removed, creating a vacancy.

created where a +1 cation normally would be located. Again, this imperfection is observed in materials that have pronounced ionic bonding.

Thus, in ionic solids, when point defects are introduced the following rules have to be observed:

(a) a charge balance must be maintained so that the crystalline material as a whole is electrically neutral;

(b) a mass balance must be maintained; and

(c) the number of crystallographic sites must be conserved.

For example, in nickel oxide (NiO) if one oxygen ion is missing, it creates an oxygen ion vacancy (designated as $V_{\ddot{O}}$). Each dot (.) on the subscript indicates an *effective* positive charge of one. To maintain stoichiometry, mass balance and charge balance we must also create a vacancy of nickel ion (designated as V''_{Ni}). Each accent (′) in the superscript indicates an *effective* charge of negative 1.

The Kröger-Vink notation is used to write the defect chemistry equations.

4-3 Dislocations

Dislocations are line imperfections in an otherwise perfect crystal. They are introduced typically into the crystal during solidification of the material or when the material is deformed permanently. Although dislocations are present in all materials, including ceramics and polymers, *they are particularly useful in explaining deformation and strengthening in metallic materials.* We can identify three types of dislocations: the screw dislocation, the edge dislocation, and the mixed dislocation.

Screw Dislocations The **screw dislocation** (Figure 4-4) can be illustrated by cutting partway through a perfect crystal, then skewing the crystal one atom spacing. If we follow a crystallographic plane one revolution around the axis on which the crystal was skewed, starting at point *x* and traveling equal atom spacings in each direction, we finish one atom spacing below our starting point (point *y*). The vector required to complete the loop and return us to our starting point is the **Burgers vector b**. If we continued our rotation, we would trace out a spiral path. The axis, or line around which we trace out this path, is the screw dislocation. The Burgers vector is parallel to the screw dislocation.

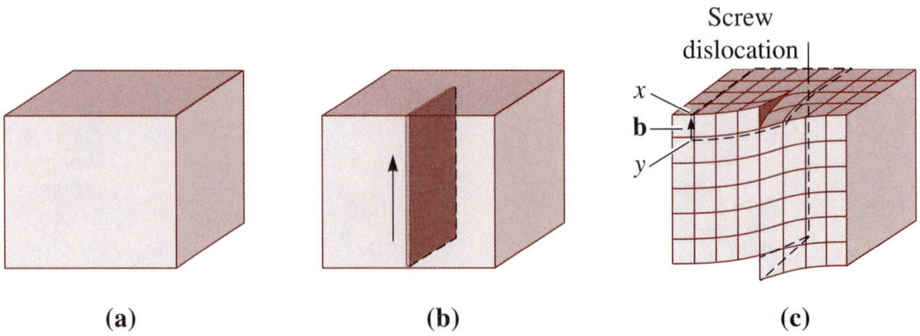

Figure 4-4 The perfect crystal (a) is cut and sheared one atom spacing, (b) and (c). The line along which shearing occurs is a screw dislocation. A Burgers vector **b** is required to close a loop of equal atom spacings around the screw dislocation.

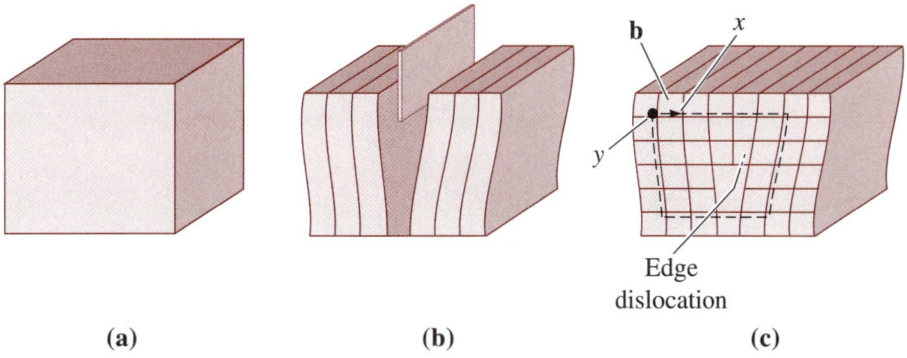

Figure 4-5 The perfect crystal in (a) is cut and an extra plane of atoms is inserted (b). The bottom edge of the extra plane is an edge dislocation (c). A Burgers vector **b** is required to close a loop of equal atom spacings around the edge dislocation. (*Adapted from J.D. Verhoeven*, Fundamentals of Physical Metallurgy, *Wiley, 1975.*)

Edge Dislocations An edge dislocation (Figure 4-5) can be illustrated by slicing partway through a perfect crystal, spreading the crystal apart, and partly filling the cut with an extra plane of atoms. The bottom edge of this inserted plane represents the edge dislocation. If we describe a clockwise loop around the edge dislocation, starting at point x and going an equal number of atoms spacings in each direction, we finish, at point y, one atom spacing from the starting point. The vector required to complete the loop is, again, the Burgers vector. In this case, the Burgers vector is perpendicular to the dislocation. By introducing the dislocation, the atoms above the dislocation line are squeezed too closely together, while the atoms below the dislocation are stretched too far apart. The surrounding region of the crystal has been disturbed by the presence of the dislocation. [This is illustrated later on in Figure 4-8(b).] Unlike an **edge dislocation**, a screw dislocation cannot be visualized as an extra half plane of atoms.

Mixed Dislocations As shown in Figure 4-6, **mixed dislocations** have both edge and screw components, with a transition region between them. The Burgers vector, however, remains the same for all portions of the mixed dislocation.

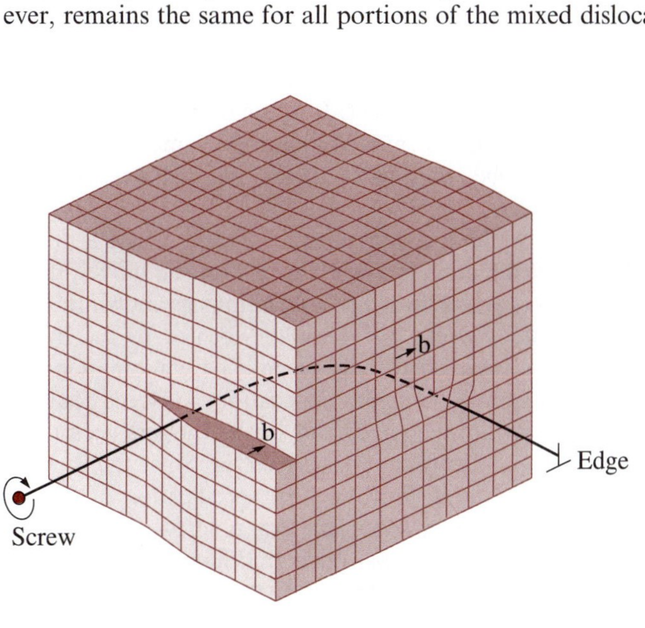

Figure 4-6
A mixed dislocation. The screw dislocation at the front face of the crystal gradually changes to an edge dislocation at the side of the crystal. (*Adapted from W.T. Read*, Dislocations in Crystals. *McGraw-Hill, 1953.*)

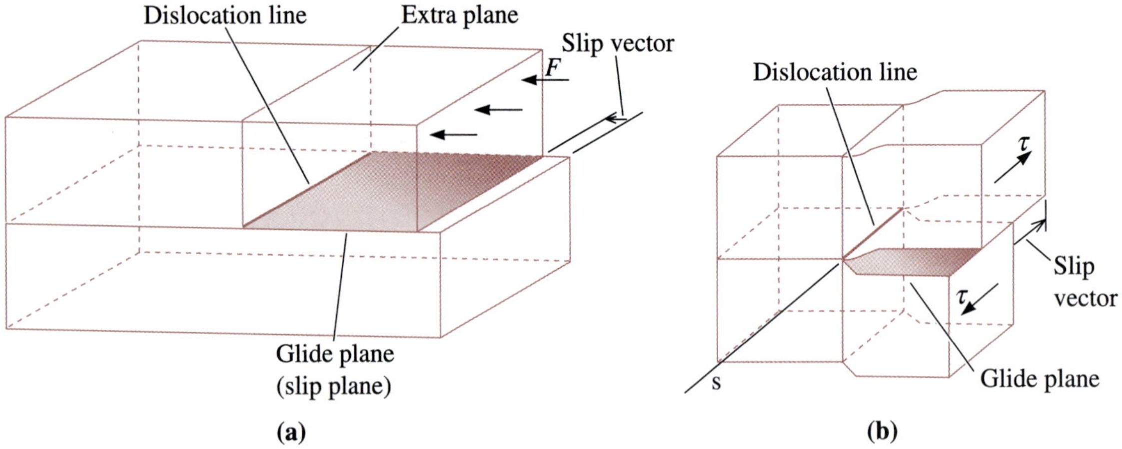

Figure 4-7 Schematic of slip line, slip plane, and slip (Burgers) vector for (a) an edge dislocation and (b) for a screw dislocation. (*Adapted from J.D. Verhoeven,* Fundamentals of Physical Metallurgy, *Wiley, 1975.*)

A schematic for slip line, slip plane, and slip vector (Burgers vector) for an edge and a screw dislocation are shown in Figure 4-7. The Burgers vector and the plane are helpful in explaining how materials deform.

When a shear force acting in the direction of the Burgers vector is applied to a crystal containing a dislocation, the dislocation can move by breaking the bonds between the atoms in one plane. The cut plane is shifted slightly to establish bonds with the original partial plane of atoms. This shift causes the dislocation to move one atom spacing to the side, as shown in Figure 4-8(a). If this process continues, the dislocation moves through the crystal until a step is produced on the exterior of the crystal; the crystal has then been deformed. Another analogy is the motion by which a caterpillar moves [Figure 4-8(d)]. A caterpillar will lift some of its legs at any given time and use that motion to move from one place to another rather than lifting all the legs at one time. The speed with which dislocations move in materials is close to or greater than the speed of sound! Another way to visualize dislocation motion is to think about how a fold or crease in a carpet would move if we were trying to remove it by pushing it across rather than by lifting the carpet. If dislocations could be introduced continually into one side of the crystal and moved along the same path through the crystal, the crystal would eventually be cut in half.

Slip The process by which a dislocation moves and causes a metallic material to deform is called **slip**. The direction in which the dislocation moves, the **slip direction**, is the direction of the Burgers vector for edge dislocations as shown in Figure 4-8(b). During slip, the edge dislocation sweeps out the plane formed by the Burgers vector and the dislocation. This plane is called the **slip plane**. The combination of slip direction and slip plane is the **slip system**. A screw dislocation produces the same result; the dislocation moves in a direction perpendicular to the Burgers vector, although the crystal deforms in a direction parallel to the Burgers vector. Since the Burgers vector of a screw dislocation is parallel to the dislocation line, specification of Burgers vector and dislocation line does not define a slip plane for a screw dislocation. As mentioned in Chapter 3, there are new software packages that have been developed and use of these can be very effective in visualizing some of these concepts.

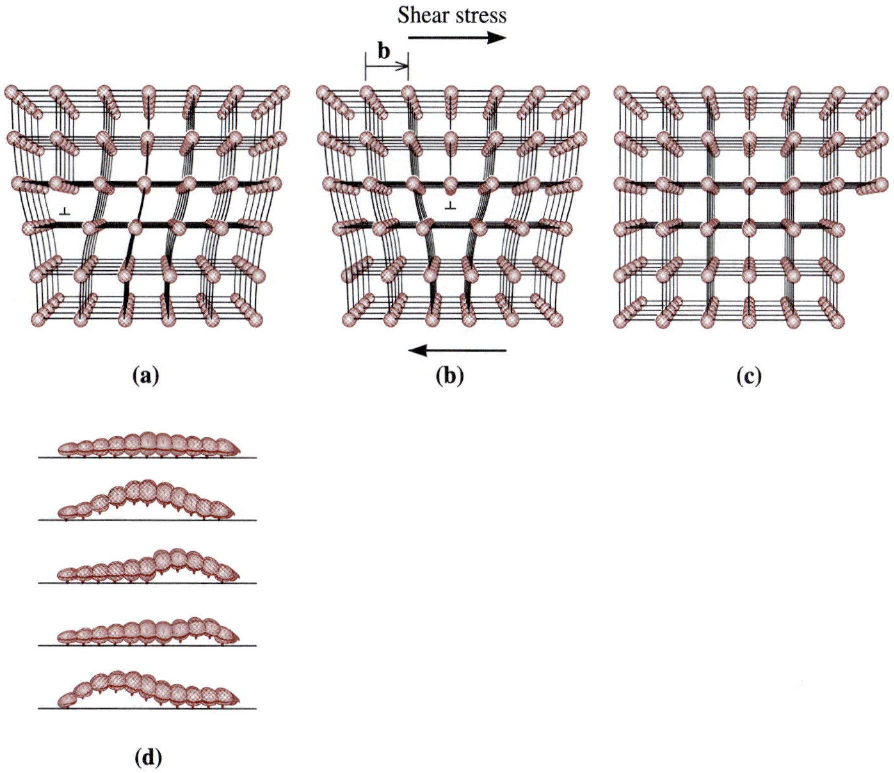

Figure 4-8 (a) When a shear stress is applied to the dislocation in (a), the atoms are displaced, causing the dislocation to move one Burgers vector in the slip direction (b). Continued movement of the dislocation eventually creates a step (c), and the crystal is deformed. (*Adapted from A.G. Guy,* Essentials of Materials Science, *McGraw-Hill, 1976.*) (d) Motion of caterpillar is analogous to the motion of a dislocation.

During slip, a dislocation moves from one set of surroundings to an identical set of surroundings. The **Peierls-Nabarro stress** (Equation 4-2) is required to move the dislocation from one equilibrium location to another,

$$\tau = c \exp(-kd/b) \tag{4-2}$$

where τ is the shear stress required to move the dislocation, d is the interplanar spacing between adjacent slip planes, b is the magnitude of the Burgers vector, and both c and k are constants for the material. The dislocation moves in a slip system that requires the least expenditure of energy. Several important factors determine the most likely slip systems that will be active:

1. The stress required to cause the dislocation to move increases exponentially with the length of the Burgers vector. Thus, the slip direction should have a small repeat distance or high linear density. The close-packed directions in metals and alloys satisfy this criterion and are the usual slip directions.
2. The stress required to cause the dislocation to move decreases exponentially with the interplanar spacing of the slip planes. Slip occurs most easily between planes of atoms that are smooth (so there are smaller "hills and valleys" on the surface) and between planes that are far apart (or have a relatively large interplanar spacing). Planes with a high planar density fulfill this requirement. Therefore,

TABLE 4-1 ■ *Slip planes and directions in metallic structures*

Crystal Structure	Slip Plane	Slip Direction
BCC metals	{110}	⟨111⟩
	{112}	
	{123}	
FCC metals	{111}	⟨110⟩
HCP metals	{0001}	⟨100⟩
	$\{11\bar{2}0\}$ See Note	⟨110⟩
	$\{10\bar{1}0\}$ See Note	or $\langle 11\bar{2}0\rangle$
	$\{10\bar{1}1\}$ See Note	
MgO, NaCl (ionic)	{110}	⟨110⟩
Silicon (covalent)	{111}	⟨110⟩

Note: These planes are active in some metals and alloys or at elevated temperatures.

the slip planes are typically close-packed planes or those as closely packed as possible. Common slip systems in several materials are summarized in Table 4-1.

3. Dislocations do not move easily in materials such as silicon or polymers, which have covalent bonds. Because of the strength and directionality of the bonds, the materials typically fail in a brittle manner before the force becomes high enough to cause appreciable slip. In many engineering polymers dislocations play a relatively minor role in their deformation. The deformation in polymers occurs mainly as polymer chains become untangled and then are stretched.
4. Materials with ionic bonding, including many ceramics such as MgO, also are resistant to slip. Movement of a dislocation disrupts the charge balance around the anions and cations, requiring that bonds between anions and cations be broken. During slip, ions with a like charge must also pass close together, causing repulsion. Finally, the repeat distance along the slip direction, or the Burgers vector, is larger than that in metals and alloys. Again, brittle failure of ceramic materials typically occurs owing to the presence of flaws such as small holes (pores) before the applied level of stress is sufficient to cause dislocations to move. It is possible to obtain some ductility in certain ceramics using special processing techniques.

The following examples illustrate the calculation of the magnitude of the Burgers vector and identification of slip planes.

EXAMPLE 4-4 *Dislocations in Ceramic Materials*

A sketch of a dislocation in magnesium oxide (MgO), which has the sodium chloride crystal structure and a lattice parameter of 0.396 nm, is shown in Figure 4-9. Determine the length of the Burgers vector.

SOLUTION

In Figure 4-9, we begin a clockwise loop around the dislocation at point x, then move equal atom spacings to finish at point y. The vector **b** is the Burgers vector. Because **b** is a [110] direction, it must be perpendicular to {110} planes.

The length of **b** is the distance between two adjacent (110) planes. From Equation 3-7,

$$d_{110} = \frac{a_0}{\sqrt{h^2 + k^2 + l^2}} = \frac{0.396}{\sqrt{1^2 + 1^2 + 0^2}} = 0.280 \text{ nm}$$

The Burgers vector is a $\langle 110 \rangle$ direction that is 0.280 nm in length. Note, however, that two extra half-planes of atoms make up the dislocation-one composed of oxygen ions and one of magnesium ions (Figure 4-9). Note that this formula for calculating the magnitude of the Burgers vector will not work for non-cubic systems. It is better to consider the magnitude of the Burgers vector as equal to the repeat distance in the slip direction.

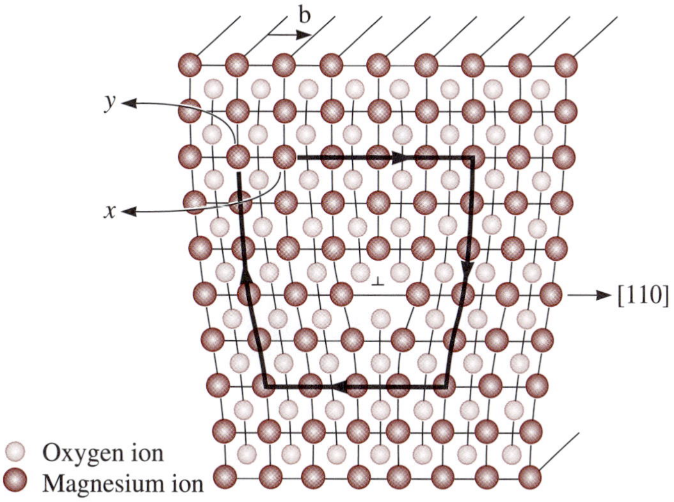

Figure 4-9 An edge dislocation in MgO showing the slip direction and Burgers vector (for Example 4-4). (*Adapted from W.D. Kingery, H.K. Bowen, and D.R. Uhlmann,* Introduction to Ceramics, *John Wiley, 1976.*)

EXAMPLE 4-5 Burgers Vector Calculation

Calculate the length of the Burgers vector in copper.

SOLUTION

Copper has an FCC crystal structure. The lattice parameter of copper (Cu) is 0.36151 nm. The close-packed directions, or the directions of the Burgers vector, are of the form $\langle 110 \rangle$. The repeat distance along the $\langle 110 \rangle$ directions is one-half the face diagonal, since lattice points are located at corners and centers of faces [Figure 4-10(a)].

$$\text{Face diagonal} = \sqrt{2}a_0 = (\sqrt{2})(0.36151) = 0.51125 \text{ nm}$$

The length of the Burgers vector, or the repeat distance, is:

$$b = \tfrac{1}{2}(0.51125 \text{ nm}) = 0.25563 \text{ nm}$$

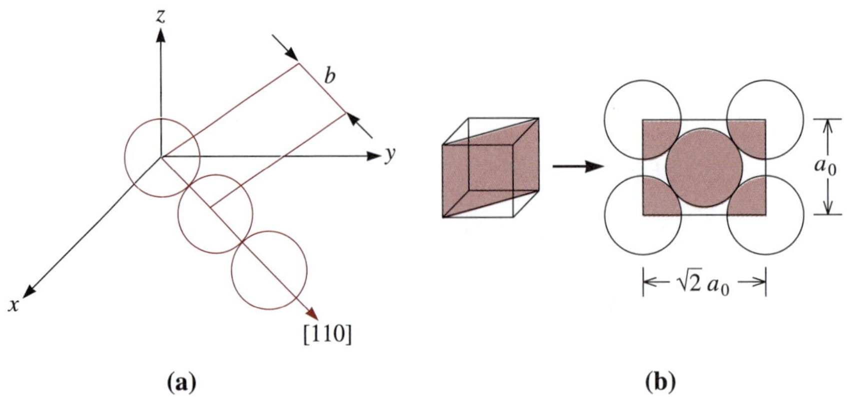

Figure 4-10 (a) Burgers vector for FCC copper. (b) The atom locations on a (110) plane in a BCC unit cell (for Examples 4-5 and 4-6, respectively).

EXAMPLE 4-6 *Identification of Preferred Slip Planes*

The planar density of the (112) plane in BCC iron is 9.94×10^{14} atoms/cm^2. Calculate (1) the planar density of the (110) plane and (2) the interplanar spacings for both the (112) and (110) planes. On which plane would slip normally occur?

SOLUTION

The lattice parameter of BCC iron is 0.2866 nm or 2.866×10^{-8} cm. The (110) plane is shown in Figure 4-10(b), with the portion of the atoms lying within the unit cell being shaded. Note that one-fourth of the four corner atoms plus the center atom lie within an area of a_0 times $\sqrt{2}a_0$.

1. The planar density is:

$$\text{Planar density (110)} = \frac{\text{atoms}}{\text{area}} = \frac{2}{(\sqrt{2})(2.866 \times 10^{-8}\ \text{cm})^2}$$

$$= 1.72 \times 10^{15}\ \text{atoms/cm}^2$$

$$\text{Planar density (112)} = 0.994 \times 10^{15}\ \text{atoms/cm}^2\ \text{(from problem statement)}$$

2. The interplanar spacings are:

$$d_{110} = \frac{2.866 \times 10^{-8}}{\sqrt{1^2 + 1^2 + 0}} = 2.0266 \times 10^{-8}\ \text{cm}$$

$$d_{112} = \frac{2.866 \times 10^{-8}}{\sqrt{1^2 + 1^2 + 2^2}} = 1.17 \times 10^{-8}\ \text{cm}$$

The planar density and interplanar spacing of the (110) plane are larger than those for the (112) plane; therefore, the (110) plane would be the preferred slip plane.

4-4 Significance of Dislocations

First, dislocations are most significant in metals and alloys since they provide a mechanism for plastic deformation, which is the cumulative effect of slip of numerous dislocations. **Plastic deformation** refers to irreversible deformation or change in shape that occurs when the force or stress that caused it is removed. This is because the applied stress causes dislocation motion that in turn causes permanent deformation. When we use the term "plastic deformation" the implication is that it is caused by dislocaiton motion. There are, however, other mechanisms that cause permanent deformation. We will see these in later chapters. The plastic deformation is to be distinguished from **elastic deformation**, which is a temporary change in shape that occurs while a force or stress remains applied to a material. In elastic deformation, the shape change is a result of stretching of interatomic bonds, however, no dislocation motion occurs. Slip can occur in some ceramics and polymers. However, other factors (e.g., porosity in ceramics, disentanglement of chains in polymers, etc.) dominate the near room-temperature mechanical behavior of polymers and ceramics. Amorphous materials such as silicate glasses do not have a periodic arrangement of ions and hence do not contain dislocations. *The slip process, therefore, is particularly important in understanding the mechanical behavior of metals.* First, slip explains why the strength of metals is much *lower* than the value predicted from the metallic bond. If slip occurs, only a tiny fraction of all of the metallic bonds across the interface need to be broken at any one time, and the force required to deform the metal is small. It can be shown that the actual strength of metals is 10^3 to 10^4 times *smaller* than that expected from the strength of metallic bonds.

Second, slip provides ductility in metals. If no dislocations were present, an iron bar would be brittle and the metal could not be shaped by metalworking processes, such as forging, into useful shapes.

Third, we control the mechanical properties of a metal or alloy by interfering with the movement of dislocations. An obstacle introduced into the crystal prevents a dislocation from slipping unless we apply higher forces. Thus, the presence of dislocations helps strengthen metallic materials.

Enormous numbers of dislocations are found in materials. The **dislocation density**, or total length of dislocations per unit volume, is usually used to represent the amount of dislocations present. Dislocation densities of 10^6 cm/cm^3 are typical of the softest metals, while densities up to 10^{12} cm/cm^3 can be achieved by deforming the material.

Dislocations also influence electronic and optical properties of materials. For example, the resistance of pure copper increases with increasing dislocation density. We mentioned previously that the resistivity of pure copper also depends strongly on small levels of impurities.

4-5 Schmid's Law

We can understand the differences in behavior of metals that have different crystal structures by examining the force required to initiate the slip process. Suppose we apply a unidirectional tensile stress to a cylinder of metal that is a single crystal (Figure 4-11). We can orient the slip plane and slip direction to the applied force (F) by defining the angles λ, and ϕ. The angle between the slip direction and the applied force is λ, and ϕ is the angle between the normal to the slip plane and the applied force. Note that the sum of angles ϕ and λ can be but does not have to be 90°.

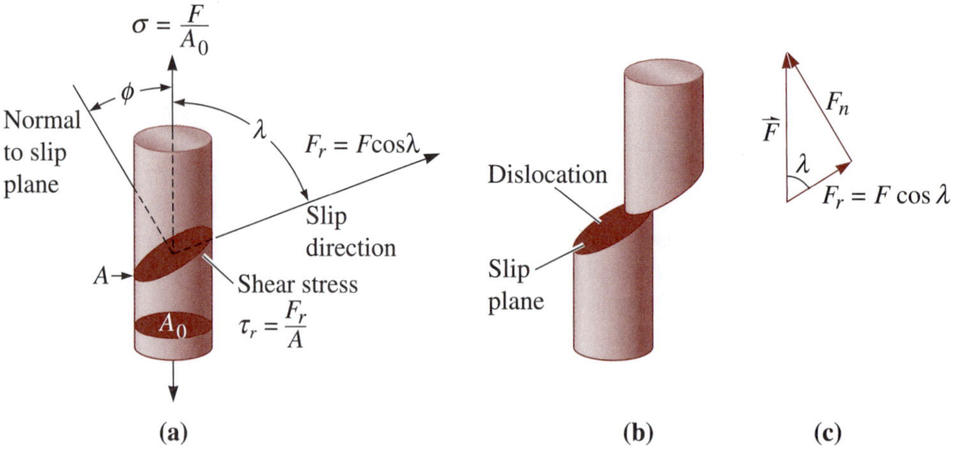

Figure 4-11 (a) A resolved shear stress τ is produced on a slip system. (Note: $(\phi + \lambda)$ does not have to be 90°.) (b) Movement of dislocations on the slip system deforms the material. (c) Resolving the force.

In order for the dislocation to move in its slip system, a shear force acting in the slip direction must be produced by the applied force. This resolved shear force F_r is given by:

$$F_r = F \cos \lambda$$

If we divide the equation by the area of the slip plane, $A = A_0/\cos \phi$, we obtain the following equation known as **Schmid's law**,

$$\tau_r = \sigma \cos \phi \cos \lambda, \tag{4-3}$$

where

$$\tau_r = \frac{F_r}{A} = \text{resolved shear } \textit{stress} \text{ in the slip direction}$$

$$\sigma = \frac{F}{A_0} = \text{unidirectional } \textit{stress} \text{ applied to the cylinder}$$

The example that follows illustrates an application of Schmid's law.

EXAMPLE 4-7 *Calculation of Resolved Shear Stress*

Apply Schmid's law for a situation in which the single crystal is at an orientation so that the slip plane is perpendicular to the applied tensile stress.

SOLUTION

Suppose the slip plane is perpendicular to the applied stress σ, as in Figure 4-12. Then, $\phi = 0°$, $\lambda = 90°$, $\cos \lambda = 0$, and therefore $\tau_r = 0$. As noted before, the angles ϕ and λ can but do not always add up to 90°. Even if the applied stress σ is enormous, no resolved shear stress develops along the slip direction and the dislocation cannot move. (You could perform a simple experiment to demonstrate this with a deck of cards. If you push on the deck at an angle, the cards slide over one another, as in the slip process. If you push perpendicular to the deck, however, the cards do not slide.) Slip cannot occur if the slip system is oriented so that either λ or ϕ is 90°.

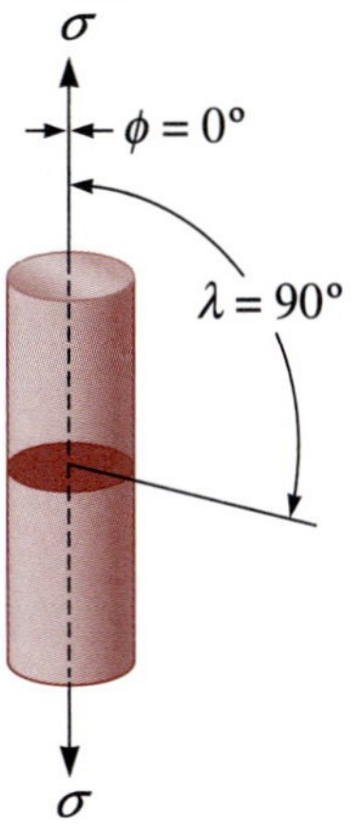

Figure 4-12
When the slip plane is perpendicular to the applied stress σ, the angle λ is 90° and no shear stress is resolved. (For Example 4-7.)

EXAMPLE 4-8 *Schmid's Law Calculation of Resolved Shear Stress*

Consider the (111) slip plane and $[0\bar{1}1]$ slip direction for a single crystal of copper. A tensile stress (σ) of 21 MPa is applied to this crystal along the [001] direction. What is the resolved shear stress (τ_r) along the slip direction?

SOLUTION

We will use Schmid's law:

$$\tau_r = \sigma \cos(\lambda) \cos(\phi)$$

We calculate the angle (λ) between the tensile stress direction [001] and the slip direction $[0\bar{1}1]$ from the dot product as follows:

$$\cos\lambda = \frac{[001]\cdot[0\bar{1}1]}{\|001\| \times \|0\bar{1}1\|} = \frac{[(0\times 0)+(0\times(-1))+(1\times 1)]}{(\sqrt{0^2+0^2+1^2})\times(\sqrt{0^2+(-1)^2+(1)^2})} = \frac{1}{\sqrt{1}\sqrt{2}}$$

The symbol $\|\ \|$ stands for the magnitude of the vector which is given by

$$\sqrt{h^2+k^2+1^2}$$

Since $\cos\lambda = \frac{1}{\sqrt{2}}$, the angle λ will be 45°.

The normal to the slip plane (111) is [111]. Thus, the angle ϕ between the tensile stress direction [001] and normal to the slip plane (i.e., [111] is given as follows).

$$\cos\phi = \frac{[001]\cdot[111]}{\|001\| \times \|111\|} = \frac{[(0\times 1)+(0\times 1)+(1\times 1)]}{(\sqrt{0^2+0^2+1^2})\times(\sqrt{1^2+(1)^2+(1)^2})} = \frac{1}{\sqrt{3}}$$

$\cos\phi = \frac{1}{\sqrt{3}}$, the angle ϕ will be 54.73°.

Thus, the resolved shear stress will be

$$\tau_r = (21 \text{ MPa})\frac{1}{\sqrt{2}}\frac{1}{\sqrt{3}} = 8.6 \text{ MPa}$$

If we had a combination of slip systems (i.e., different slip planes and slip directions), we could calculate the values of critical resolved shear stress for each one. The direction that had the highest resolved shear stress would become active first (i.e., dislocation in that particular system will begin to move first).

The **critical resolved shear stress** τ_{crss} is the shear stress required to break enough metallic bonds in order for slip to occur. Thus slip occurs, causing the metal to plastically deform, when the *applied* stress (σ) produces a *resolved* shear stress (τ_r) that equals the *critical* resolved shear stress.

$$\tau_r = \tau_{crss} \qquad (4\text{-}4)$$

4-6 Influence of Crystal Structure

We can use Schmid's law to compare the properties of metals having BCC, FCC, and HCP crystal structures. Table 4-2 lists three important factors that we can examine. We must be careful to note, however, that this discussion describes the behavior of nearly perfect single crystals. Most engineered materials are seldom single crystals and always contain large numbers of defects. Since different crystals or grains are oriented in different random directions, we can not apply Schmid's law to predict the mechanical behavior of polycrystalline materials.

Critical Resolved Shear Stress If the critical resolved shear stress in a metal is very high, the applied stress σ must also be high in order for τ_r to equal τ_{crss}. A higher τ_{crss} implies a higher stress is necessary to deform a metal, which in turn indicates the metal has a high strength! In FCC metals, which have close-packed $\{111\}$ planes, the critical resolved shear stress is low—about 0.34 to 0.69 MPa in a perfect crystal; FCC metals

TABLE 4-2 ■ *Summary of factors affecting slip in metallic structures*

Factor	FCC	BCC	HCP $\left(\frac{c}{a} > 1.633^*\right)$
Critical resolved shear stress (MPa)	0.34–0.69	34–69	0.34–0.69[a]
Number of slip systems	12	48	3[b]
Cross-slip	Can occur	Can occur	Cannot occur[b]
Summary of properties	Ductile	Strong	Relatively brittle

[a] *For slip on basal planes.*

[b] *By alloying or heating to elevated temperatures, additional slip systems are active in HCP metals, permitting cross-slip to occur and thereby improving ductility.*

* *Note: For most elements $c/a < 1.633$, slip occurs on planes other than (0001) and τ_{crss} is high.*

tend to have low strengths. On the other hand, BCC crystal structures contain no close-packed planes and we must exceed a higher critical resolved shear stress—on the order of 69 MPa in perfect crystals—before slip occurs; therefore, BCC metals tend to have high strengths and lower ductilities.

We would expect the HCP metals, because they contain close-packed basal planes, to have low critical resolved shear stresses. In fact, in HCP metals such as zinc that have a c/a ratio greater than or equal to the theoretical ratio of 1.633, the critical resolved shear stress is less than 0.69 MPa, just as in FCC metals. As noted in Table 4-2, for most HCP elements $c/a < 1.633$. The slip occurs on non-basal planes and τ_{crss} is high. For example, in HCP titanium, the c/a ratio is less than 1.633; the close-packed planes are spaced too closely together. Slip now occurs on planes such as $(10\bar{1}0)$, the "prism" planes or faces of the hexagon, and the critical resolved shear stress is then as great as or greater than in BCC metals.

Number of Slip Systems If at least one slip system is oriented to give the angles λ and ϕ near 45°, then τ_r equals τ_{crss} at low applied stresses. Ideal HCP metals possess only one set of parallel close-packed planes, the (0001) planes, and three close-packed directions, giving three slip systems. Consequently, the probability of the close-packed planes and directions being oriented with λ and ϕ near 45° is very low. The HCP crystal may fail in a brittle manner without a significant amount of slip. However, in HCP metals with a low c/a ratio, or when HCP metals are properly alloyed, or when the temperature is increased, other slip systems become active, making these metals less brittle than expected.

On the other hand, FCC metals contain four nonparallel close-packed planes of the form $\{111\}$ and three close-packed directions of the form $\langle 110 \rangle$ within each plane, giving a total of 12 slip systems. At least one slip system is favorably oriented for slip to occur at low applied stresses, permitting FCC metals to have high ductilities.

Finally, BCC metals have as many as 48 slip systems that are nearly close-packed. Several slip systems are always properly oriented for slip to occur, allowing BCC metals to also have ductility.

Cross-Slip Consider a screw dislocation moving on one slip plane that encounters an obstacle and is blocked from further movement. This dislocation can shift to a second intersecting slip system, also properly oriented, and continue to move. This is called **cross-slip**. In many HCP metals, no cross-slip can occur because the slip planes are parallel (i.e., not intersecting). Therefore, polycrystalline HCP metals tend to be brittle. Fortunately, additional slip systems become active when HCP metals are alloyed or heated, thus improving ductility. Cross-slip is possible in both FCC and BCC metals because a number of intersecting slip systems are present. Consequently, cross-slip helps maintain ductility in these metals.

4-7 Surface Defects

Surface defects are the boundaries, or planes, that separate a material into regions, each region having the same crystal structure but different orientations.

Material Surface The exterior dimensions of the material represent surfaces at which the crystal abruptly ends. Each atom at the surface no longer has the proper coordination number and atomic bonding is disrupted. This is very often an important factor in making silicon based microelectronic devices. The exterior surface may also be very

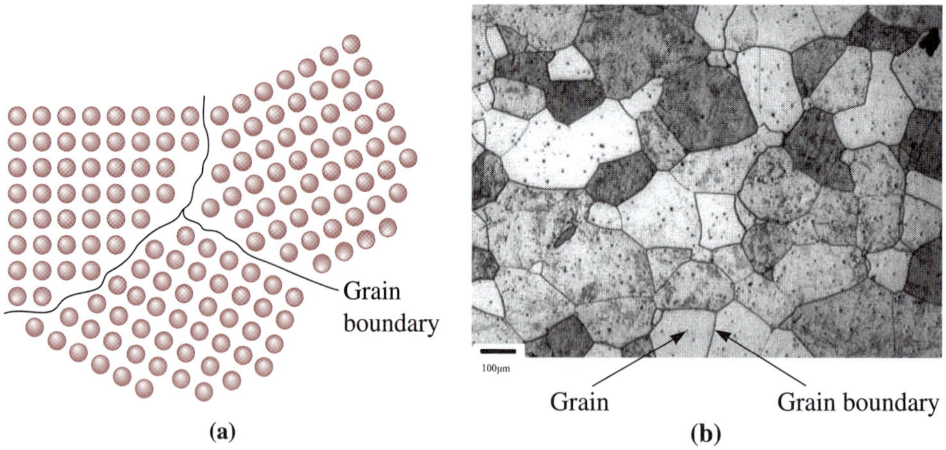

Figure 4-13 (a) The atoms near the boundaries of the three grains do not have an equilibrium spacing or arrangement. (b) Grains and grain boundaries in a stainless steel sample. (*Courtesy of Dr. A. DeArdo.*)

rough, may contain tiny notches, and may be much more reactive than the bulk of the material.

In nano-structured materials, the ratio of the number of atoms or ions at the surface to that in the bulk is very high. As a result, these materials have a large surface area per unit mass. Therefore, surface defects play an important role on their properties.

Grain Boundaries The microstructure of many engineered ceramic and metallic materials consists of many grains. A **grain** is a crystalline portion of the material within which the arrangement of the atoms is nearly identical. However, the orientation of the atom arrangement, or crystal structure, is different for each adjoining grain. Three grains are shown schematically in Figure 4-13(a); the arrangement of atoms in each grain is identical but the grains are oriented differently. A **grain boundary**, the surface that separates the individual grains, is a narrow zone in which the atoms are not properly spaced. That is to say, the atoms are so close together at some locations in the grain boundary that they cause a region of compression, and in other areas they are so far apart that they cause a region of tension. Figure 4-13(b), a micrograph of a stainless steel sample, shows grains and grain boundaries.

One method of controlling the properties of a material is by controlling the grain size. By reducing the grain size, we increase the number of grains and, hence, increase the amount of grain boundary area. Any dislocation moves only a short distance before encountering a grain boundary and being stopped, and the strength of the metallic material is increased. The **Hall-Petch equation** relates the grain size to the **yield strength** (σ_y),

$$\sigma_y = \sigma_0 + \mathrm{K}d^{-1/2} \tag{4-5}$$

where d is the average diameter of the grains, and σ_0 and K are constants for the metal. Recall from Chapter 1 that yield strength (σ_y) of a metallic material is the minimum level of stress that is needed to initiate plastic (permanent) deformation. Figure 4-14 shows this relationship in steel. The Hall-Petch equation is not valid for materials with unusally large or ultrafine grains. In the chapters that follow, we will describe how the grain size of metals and alloys can be controlled through solidification, alloying, and heat treatment. The following example illustrates an application of the Hall-Petch equation.

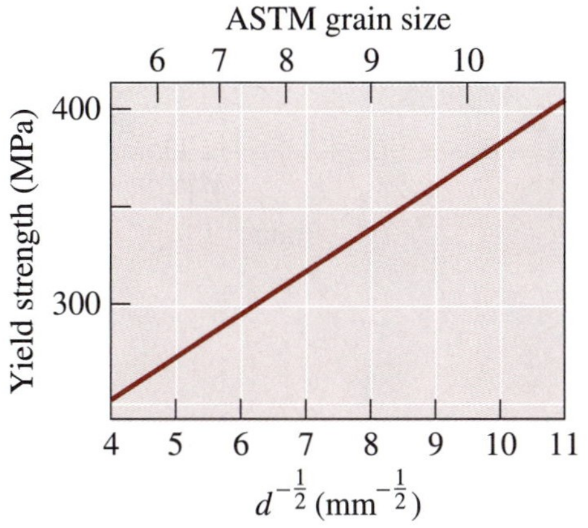

Figure 4-14
The effect of grain size on the yield strength of steel at room temperature.

EXAMPLE 4-9 Design of a Mild Steel

The yield strength of mild steel with an average grain size of 0.05 mm is 138 MPa. The yield stress of the same steel with a grain size of 0.007 mm is 276 MPa. What will be the average grain size of the same steel with a yield stress of 207 MPa? Assume the Hall-Petch equation is valid and that changes in the observed yield stress are due to changes in grain size.

SOLUTION

$$\sigma_y = \sigma_0 + Kd^{-1/2}$$

Thus, for a grain size of 0.05 mm the yield stress is 138 MPa.
Using the Hall-Petch equation

$$138 = \sigma_0 + \frac{K}{\sqrt{0.05}}$$

For the grain size of 0.007 mm, the yield stress is 276 MPa. Therefore, again using the Hall-Petch equation:

$$276 = \sigma_0 + \frac{K}{\sqrt{0.007}}$$

Solving these two equations K = 18.43 MPa-mm$^{1/2}$, and $\sigma_0 = 55.5$ MPa. Now we have the Hall-Petch equation as

$$\sigma_y = 55.5 + 18.43 \times d^{-1/2}$$

If we want a yield stress of 207 MPa, the grain size will be 0.0148 mm or 14.8 μm.

Optical microscopy is one technique that is used to reveal microstructural features such as grain boundaries that require less than about 2000 magnification. The process of preparing a metallic sample and observing or recording its microstructure is called

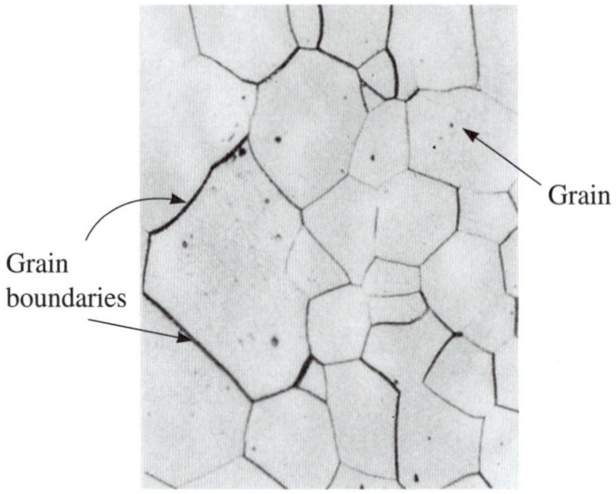

Figure 4-15
Microstructure of palladium (×100). (*From* ASM Handbook, *Vol. 9, Metallography and Microstructure (1985), ASM International, Materials Park, OH 44073.*)

metallography. A sample of the material is sanded and polished to a mirror-like finish. The surface is then exposed to chemical attack, or *etching*, with grain boundaries being attacked more aggressively than the remainder of the grain. Light from an optical microscope is reflected or scattered from the sample surface, depending on how the surface is etched. Since more light is scattered from deeply etched features such as the grain boundaries, these features appear dark (Figure 4-15). In ceramic samples, a technique known as **thermal grooving** is often used to observe grain boundaries. It involves polishing and heating, for a short time, a ceramic sample to temperatures below the sintering temperature.

One way to specify grain size is by using the **ASTM grain size number** (ASTM is the **American Society for Testing and Materials**). The number of grains per square inch (N) is determined from a micrograph of the metal taken at magnification ×100. The number of grains per square inch N is entered into Equation 4-6 and the ASTM grain size number n is calculated:

$$N = 2^{n-1} \tag{4-6}$$

A large ASTM number indicates many grains, or a finer grain size, and correlates with high strengths for metals and alloys.

When describing a microstructure, whenever possible, it is preferable to use a micrometer marker or some other scale on the micrograph, instead of stating the magnification. A number of sophisticated **image analysis** programs are also available. These programs make it easy to determine the average grain size and grain-size distribution. The following example illustrates the calculation of the ASTM grain size number.

EXAMPLE 4-10 *Calculation of ASTM Grain Size Number*

Suppose we count 16 grains per square inch in a photomicrograph taken at magnification ×250. What is the ASTM grain size number?

SOLUTION

If we count 16 grains per square inch at magnification ×250, then at magnification ×100 we must have:

$$N = \left(\frac{250}{100}\right)^2 (16) = 100 \text{ grains/in.}^2 = 2^{n-1}$$

$$\log 100 = (n - 1) \log 2$$

$$2 = (n - 1)(0.301)$$

$$n = 7.64$$

The approach discussed above to compute the ASTM grain size number is based on English units. In countries that use the metric system of units, alternate formulae have been developed to calculate the grain size number:

$$N_m = 8(2^n) \tag{4-7}$$

where N_m is the number of grains per mm^2 at magnification $\times$ 1.

The value of n calculated from Equation 4-7 is slightly greater than that calculated using Equation 4-6, but the difference is negligible. The Swedish, Italian, Russian, French, and ISO standards use Equation 4-7. The German standard is also based on metric units but uses a different formulation:

$$n = 3.7 + 3.33 \log(N_{mG}) \tag{4-8}$$

where N_{mG} is the number of grains per cm^2 at magnification $\times$ 100.

Small Angle Grain Boundaries A **small angle grain boundary** is an array of dislocations that produces a small misorientation between the adjoining crystals. Because the energy of the small-angle grain boundaries is less than that of a regular grain boundary, the small angle grain boundaries are not as effective in resisting slip.

Stacking Faults **Stacking faults**, which occur in FCC metals, represent an error in the stacking sequence of close-packed planes. Normally, a stacking sequence of *ABC ABC ABC* is produced in a perfect FCC crystal. But, suppose the following sequence is produced:

ABC ABAB CABC

—∨—

In the portion of the sequence indicated, a type *A* plane replaces where a type *C* plane would normally be located. This small region, which has a HCP stacking sequence instead of the FCC stacking sequence, represents a stacking fault. Stacking faults interfere with the slip process.

Twin Boundaries A **twin boundary** is a plane across which there is a special mirror image misorientation of the crystal structure (Figure 4-16). Twins can be produced when a shear force, acting along the twin boundary, causes the atoms to shift out of position. Twinning occurs during deformation or heat treatment of certain metals or alloys. The twin boundaries interfere with the slip process and increase the strength of the metal. Movement of twin boundaries can also cause a metal to deform. Figure 4-16 (c) shows that the formation of a twin has changed the shape of the metal. Twinning also occurs in some ceramic materials as well.

The effectiveness of the surface defects in interfering with the slip process can be judged from the surface energies (see Table 4-3 on the next page). The high-energy grain boundaries are much more effective in blocking dislocations than either stacking faults or twin boundaries.

TABLE 4-3 ■ ***Energies of surface imperfections in selected metals***

Surface Imperfection (ergs/cm^2)	Al	Cu	Pt	Fe
Stacking fault	200	75	95	—
Twin boundary	120	45	195	190
Grain boundary	625	645	1000	780

4-8 Importance of Defects

Extended and point defects play a major role in influencing mechanical, electrical, optical and magnetic properties of engineered materials. In this section, we recapitulate the importance of defects on properties of materials.

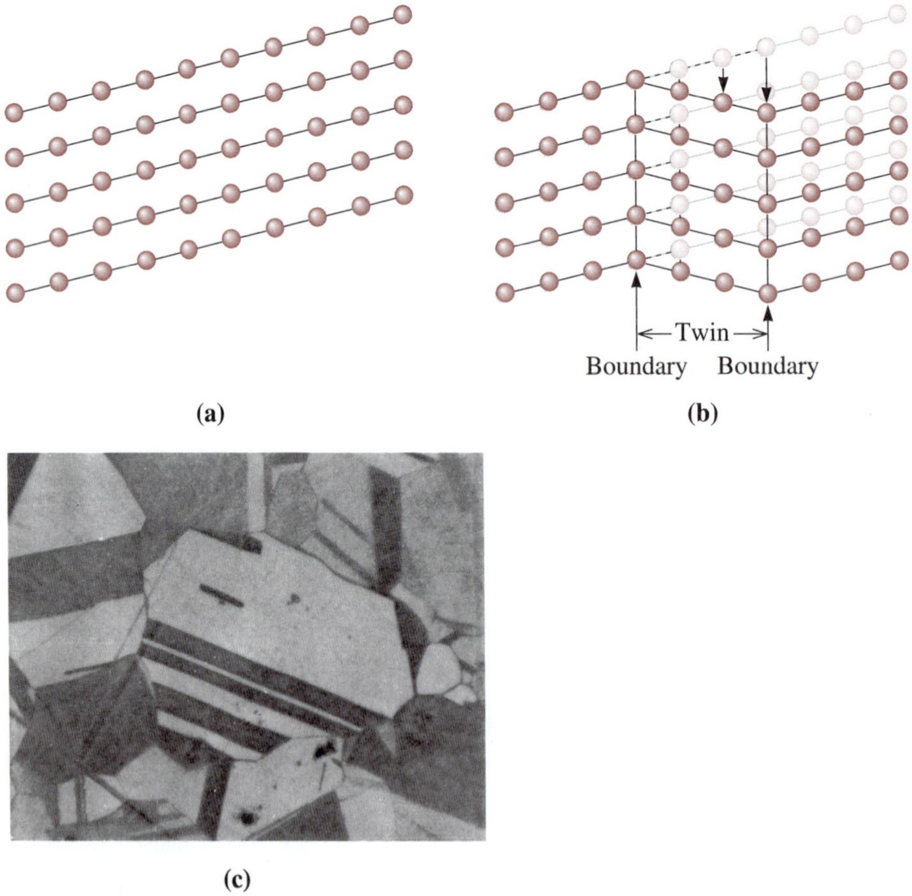

Figure 4-16 Application of a stress to the perfect crystal (a) may cause a displacement of the atoms, (b) causing the formation of a twin. Note that the crystal has deformed as a result of twinning. (c) A micrograph of twins within a grain of brass (×250).

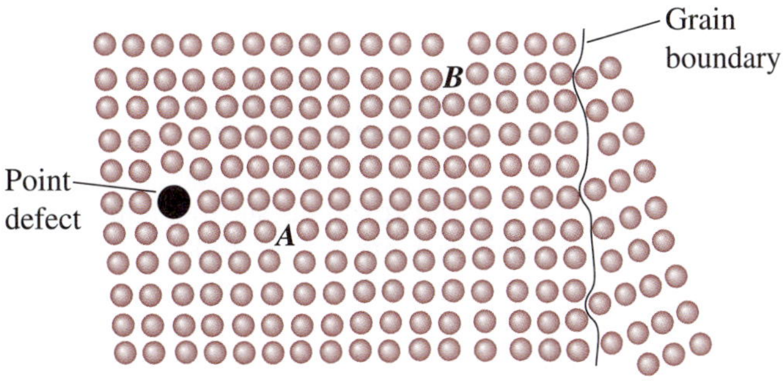

Figure 4-17 If the dislocation at point *A* moves to the left, it is blocked by the point defect. If the dislocation moves to the right, it interacts with the disturbed lattice near the second dislocation at point *B*. If the dislocation moves farther to the right, it is blocked by a grain boundary.

Effect on Mechanical Properties via Control of the Slip Process Any imperfection in the crystal raises the internal energy at the location of the imperfection. The local energy is increased because, near the imperfection, the atoms either are squeezed too closely together (compression) or are forced too far apart (tension).

One dislocation in an otherwise perfect metallic crystal can move easily through the crystal if the resolved shear stress equals the critical resolved shear stress. However, if the dislocation encounters a region where the atoms are displaced from their usual positions, a higher stress is required to force the dislocation past the region of high local energy; thus, the material is stronger. *Defects in materials, such as dislocations, point defects, and grain boundaries, serve as "stop signs" for dislocations.* We can control the strength of a metallic material by controlling the number and type of imperfections. Three common strengthening mechanisms are based on the three categories of defects in crystals. Since dislocation motion is relatively easy in metals and alloys, these mechanisms typically work best for metallic materials.

Strain Hardening Dislocations disrupt the perfection of the crystal structure. In Figure 4-17, the atoms below the dislocation line at point *B* are compressed, while the atoms above dislocation *B* are too far apart. If dislocation *A* moves to the right and passes near dislocation *B*, dislocation *A* encounters a region where the atoms are not properly arranged. Higher stresses are required to keep the second dislocation moving; consequently, the metal must be stronger. Increasing the number of dislocations further increases the strength of the material since increasing the dislocation density causes more stop signs for dislocation motion. The dislocation density can be shown to increase markedly as we strain or deform a material. This mechanism of increasing the strength of a material by deformation is known as **strain hardening**, which is discussed formally in Chapter 7. We can also show that dislocation densities can be reduced substantially by heating a metallic material to a relatively high temperature and holding it there for a long period of time. This heat treatment is known as **annealing** and is used to impart ductility to metallic materials. Thus, controlling the dislocation density is an important way of controlling the strength and ductility of metals and alloys.

Solid-Solution Strengthening Any of the point defects also disrupt the perfection of the crystal structure. A solid solution is formed when atoms or ions of a guest element or compound are assimilated completely into the crystal structure of the host material. This is similar to the way salt or sugar in small concentrations dissolve in water.

If dislocation *A* moves to the left (Figure 4-17), it encounters a disturbed crystal caused by the point defect; higher stresses are needed to continue slip of the dislocation. By intentionally introducing substitutional or interstitial atoms, we cause *solid-solution strengthening*, which is discussed further in Chapter 10.

Grain-Size Strengthening Surface imperfections such as grain boundaries disturb the arrangement of atoms in crystalline materials. If dislocation *B* moves to the right (Figure 4-17), it encounters a grain boundary and is blocked. By increasing the number of grains or reducing the grain size, **grain-size strengthening** is achieved in metallic materials. This is the basis for the Hall-Petch equation (Equation 4-5).

Effects on Electrical, Optical, and Magnetic Properties In previous sections, we stated how profound is the effect of point defects on the electrical properties of semiconductors. The entire microelectronics industry critically depends upon the successful incorporation of substitutional dopants such as P, As, B, and Al in Si and other semiconductors. These dopant atoms allow us to have a significant control of the electrical properties of the semiconductors.

SUMMARY

- Imperfections, or defects, in a crystalline material are of three general types: point defects, line defects or dislocations, and surface defects.
- The number of vacancies depends on the temperature of the material; interstitial atoms (located at interstitial sites between the normal atoms) and substitutional atoms (which replace the host atoms at lattice points) are often deliberately introduced and are typically unaffected by changes in temperature.
- Frenkel and Schottky defects are commonly seen in ionic materials.
- Dislocations are line defects which, when a force is applied to a metallic material, move and cause a metallic material to deform.
- The critical resolved shear stress is the stress required to move a dislocation.
- The dislocation moves in a slip system, composed of a slip plane and a slip direction. The slip direction, or Burgers vector, is typically along a close-packed direction. The slip plane is also normally close packed or nearly close packed.
- In metallic crystals, the number and type of slip directions and slip planes influence the properties of the metal. In FCC metals, the critical resolved shear stress is low and an optimum number of slip planes are available; consequently, FCC metals tend to be ductile. In BCC metals, no close-packed planes are available and the critical resolved shear stress is high; thus, the BCC metals tend to be strong. The number of slip systems in HCP metals is limited, causing these metals to behave in a brittle manner.
- Point defects, which include vacancies, interstitial atoms, and substitutional atoms, introduce compressive or tensile strain fields that disturb the atomic arrangements in the surrounding crystal. As a result, dislocations cannot easily slip in the vicinity of point defects and the strength of the metallic material is increased.
- Surface defects include grain boundaries. Producing a very small grain size increases the amount of grain boundary area; because dislocations cannot easily pass through a grain boundary, the material is strengthened (Hall-Petch equation).

- The number and type of crystal defects control the ease of movement of dislocations and, therefore, directly influence the mechanical properties of the material.
- Strain hardening is obtained by increasing the number of dislocations; solid-solution strengthening involves the introduction of point defects; and grain-size strengthening is obtained by producing a small grain size.
- Annealing is a heat treatment used to undo the effects of strain hardening by reducing the dislocation density. This leads to increased ductility in metallic materials.

GLOSSARY

ASTM American Society for Testing and Materials.

ASTM grain size number (*n*) A measure of the size of the grains in a crystalline material obtained by counting the number of grains per square inch at magnification ×100.

Annealing A heat treatment that typically involves heating a metallic material to a high temperature for an extended period of time, conducted with a view to lower the dislocation density and hence impart ductility.

Burgers vector The direction and distance that a dislocation moves in each step, also known as the slip vector.

Critical resolved shear stress The shear stress required to cause a dislocation to move and cause slip.

Cross-slip A change in the slip system of a dislocation.

Defect A microstructural feature representing a disruption in the perfect periodic arrangement of atoms/ions in a crystalline material. This term is not used to convey the presence of a flaw in the material.

Dislocation A line imperfection in a crystalline material. Movement of dislocations helps explain how metallic materials deform. Interference with the movement of dislocations helps explain how metallic materials are strengthened.

Dislocation density The total length of dislocation line per cubic centimeter in a material.

Domain boundaries Region between domains in a material.

Dopants Elements or compounds typically added, in known concentrations and appearing at specific places within the microstructure, to enhance the properties or processing of a material.

Edge dislocation A dislocation introduced into the crystal by adding an "extra half plane" of atoms.

Elastic deformation Deformation that is fully recovered when the stress causing it is removed.

Extended defects Defects that involve several atoms/ions and thus occur over a finite volume of the crystalline material (e.g., dislocations, stacking faults, etc.).

Frenkel defect A pair of point defects produced when an ion moves to create an interstitial site, leaving behind a vacancy.

Grain One of the crystals present in a polycrystalline material.

Grain boundary A surface defect representing the boundary between two grains. The crystal has a different orientation on either side of the grain boundary.

Hall-Petch equation The relationship between yield strength and grain size in a metallic material—that is, $\sigma_y = \sigma_0 + Kd^{-1/2}$.

Image analysis A technique that is used to analyze images of microstructures to obtain quantitative information on grain size, shape, grain size distribution, etc.

Impurities Elements or compounds that find their way into a material, often originating from processing or raw materials and typically having a deleterious effect on the properties or processing of a material.

Interstitial defect A point defect produced when an atom is placed into the crystal at a site that is normally not a lattice point.

Interstitialcy A point defect caused when a "normal" atom occupies an interstitial site in the crystal.

Line defects Defects such as dislocations in which atoms or ions are missing in a row.

Metallography Preparation of a metallic sample of a material by polishing and etching so that the structure can be examined using a microscope.

Mixed dislocation A dislocation that contains partly edge components and partly screw components.

Peierls-Nabarro stress The shear stress, which depends on the Burgers vector and the interplanar spacing, required to cause a dislocation to move—that is, $\tau = c \exp(-kd/b)$.

Point defects Imperfections, such as vacancies, that are located typically at one (in some cases a few) sites in the crystal.

Schmid's law The relationship between shear stress, the applied stress, and the orientation of the slip system—that is, $\tau_r = \sigma \cos \lambda \cos \phi$.

Schottky defect A point defect in ionically bonded materials. In order to maintain a neutral charge, a stoichiometric number of cation and anion vacancies must form.

Screw dislocation A dislocation produced by skewing a crystal so that one atomic plane produces a spiral ramp about the dislocation.

Slip Deformation of a metallic material by the movement of dislocations through the crystal.

Slip direction The direction in the crystal in which the dislocation moves. The slip direction is the same as the direction of the Burgers vector.

Slip plane The plane swept out by the dislocation line during slip. Normally, the slip plane is a close-packed plane, if one exists in the crystal structure.

Slip system The combination of the slip plane and the slip direction.

Small angle grain boundary An array of dislocations causing a small misorientation of the crystals across the surface of the imperfection.

Stacking fault A surface defect in FCC metals caused by the improper stacking sequence of close-packed planes.

Substitutional defect A point defect produced when an atom is removed from a regular lattice point and replaced with a different atom, usually of a different size.

Surface defects Imperfections, such as grain boundaries, that form a two-dimensional plane within the crystal.

Thermal grooving A technique used for observing microstructures in ceramic materials, involves heating, for a short time, a polished sample to a temperature slightly below the sintering temperature.

Twin boundary A plane across which there is a special misorientation of the crystal structure.

Vacancy An atom or an ion missing from its regular crystallographic site.

PROBLEMS

Section 4-1 Point Defects

4-1 Calculate the number of vacancies per cm^3 expected in copper at 1080°C (just below the melting temperature). The energy for vacancy formation is 20,000 cal/mol.

4-2 The fraction of lattice points occupied by vacancies in solid aluminum at 660°C is 10^{-3}. What is the energy required to create vacancies in aluminum?

4-3 The density of a sample of FCC palladium is 11.98 g/cm^3 and its lattice parameter is 3.8902 Å Calculate

(a) the fraction of the lattice points that contain vacancies; and
(b) the total number of vacancies in a cubic centimeter of Pd.

4-4 The density of a sample of HCP beryllium is 1.844 g/cm^3 and the lattice parameters are $a_0 =$ 0.22858 nm and $c_0 = 0.35842$ nm. Calculate

(a) the fraction of the lattice points that contain vacancies; and
(b) the total number of vacancies in a cubic centimeter.

4-5 BCC lithium has a lattice parameter of 3.5089×10^{-8} cm and contains one vacancy per 200 unit cells. Calculate

(a) the number of vacancies per cubic centimeter; and
(b) the density of Li.

4-6 FCC lead (Pb) has a lattice parameter of 0.4949 nm and contains one vacancy per 500 Pb atoms. Calculate

(a) the density; and
(b) the number of vacancies per gram of Pb.

4-7 A niobium alloy is produced by introducing tungsten substitutional atoms in the BCC structure; eventually an alloy is produced that has a lattice parameter of 0.32554 nm and a density of 11.95 g/cm^3. Calculate the fraction of the atoms in the alloy that are tungsten.

4-8 Tin atoms are introduced into a FCC copper crystal, producing an alloy with a lattice parameter of 3.7589×10^{-8} cm and a density of 8.772 g/cm^3. Calculate the atomic percentage of tin present in the alloy.

4-9 We replace 7.5 atomic percent of the chromium atoms in its BCC crystal with tantalum. X-ray diffraction shows that the lattice parameter is 0.29158 nm. Calculate the density of the alloy.

4-10 Suppose we introduce one carbon atom for every 100 iron atoms in an interstitial position in BCC iron, giving a lattice parameter of 0.2867 nm. For the Fe-C alloy, find the density and the packing factor.

4-11 The density of BCC iron is 7.882 g/cm^3 and the lattice parameter is 0.2866 nm when hydrogen atoms are introduced at interstitial positions. Calculate

(a) the atomic fraction of hydrogen atoms; and
(b) number of unit cells on average that contain hydrogen atoms.

Section 4-2 Other Point Defects

4-12 Suppose one Schottky defect is present in every tenth unit cell of MgO. MgO has the sodium chloride crystal structure and a lattice parameter of 0.396 nm. Calculate

(a) the number of anion vacancies per cm^3; and
(b) the density of the ceramic.

4-13 ZnS has the zinc blende structure. If the density is 3.02 g/cm^3 and the lattice parameter is 0.59583 nm, determine the number of Schottky defects

(a) per unit cell; and
(b) per cubic centimeter.

Section 4-3 Dislocations

4-14 What are the Miller indices of the slip directions:

(a) on the (111) plane in an FCC unit cell?
(b) on the (011) plane in a BCC unit cell?

4-15 What are the Miller indices of the slip planes in FCC unit cells that include the [101] slip direction?

4-16 What are the Miller indices of the {110} slip planes in BCC unit cells that include the [111] slip direction?

4-17 Calculate the length of the Burgers vector in the following materials:

(a) BCC niobium;
(b) FCC silver; and
(c) diamond cubic silicon.

4-18 Determine the interplanar spacing and the length of the Burgers vector for slip on the expected slip systems in FCC aluminum. Repeat, assuming that the slip system is a (110) plane and a $[1\bar{1}1]$ direction. What is the ratio between the shear stresses required for slip for the two systems? Assume that k = 2 in Equation 4-2.

4-19 Determine the interplanar spacing and the length of the Burgers vector for slip on the $(110)/[1\bar{1}1]$ slip system in BCC tantalum. Repeat, assuming that the slip system is a $(111)/[1\bar{1}0]$ system. What is the ratio between the shear stresses required for slip for the two systems? Assume that k = 2 in Equation 4-2.

Section 4-4 Significance of Dislocations

4-20 How many grams of aluminum, with a dislocation density of 10^{10} cm/cm^3, are required to give a total dislocation length that would stretch from New York City to Los Angeles (4860 km)?

4-21 Compare the c/a ratios for the following HCP metals, determine the likely slip processes in each, and estimate the approximate critical resolved shear stress. Explain. (See data in Appendix A.)

(a) zinc **(b)** magnesium **(c)** titanium
(d) zirconium **(e)** rhenium **(f)** beryllium

Section 4-5 Schmid's Law

4-22 A single crystal of an FCC metal is oriented so that the [001] direction is parallel to an applied stress of 35 MPa. Calculate the resolved shear stress acting on the (111) slip plane in the $[\bar{1}10]$, $[0\bar{1}1]$, and $[10\bar{1}]$ slip directions. Which slip system (s) will become active first?

4-23 A single crystal of a BCC metal is oriented so that the [001] direction is parallel to the applied stress. If the critical resolved shear stress required for slip is 83 MPa, calculate the magnitude of the applied stress required to cause slip to begin in the $[1\bar{1}1]$ direction on the (110), (011), and $(10\bar{1})$ slip planes.

Section 4-6 Influence of Crystal Structure

4-24 Why is it that single crystal and polycrystalline copper are both ductile, however, single crystal, but not polycrystalline, zinc can exhibit considerable ductility?

4-25 Why is it that cross slip in BCC and FCC metals is easier than that in HCP metals? How does this influence the ductility of BCC, FCC, and HCP metals?

Section 4-7 Surface Defects

4-26 The strength of titanium is found to be 448 MPa when the grain size is 17×10^{-6} m and 565 MPa when the grain size is 0.8×10^{-6} m. Determine

(a) the constants in the Hall-Petch equation; and
(b) the strength of the titanium when the grain size is reduced to 0.2×10^{-6} m.

4-27 A copper-zinc alloy has the following properties

Grain Diameter (mm)	Strength (MPa)
0.015	170 MPa
0.025	158 MPa
0.035	151 MPa
0.050	145 MPa

Determine

(a) the constants in the Hall-Petch equation; and
(b) the grain size required to obtain a strength of 200 MPa.

4-28 For an ASTM grain size number of 8, calculate the number of grains per square inch

(a) at a magnification of 100 and
(b) with no magnification.

4-29 Determine the ASTM grain size number if 20 grains/square inch are observed at a magnification of 400.

4-30 Determine the ASTM grain size number if 25 grains/square inch are observed at a magnification of 50.

4-31 Determine the ASTM grain size number for the materials in: Figure 4-15 and Figure 4-18.

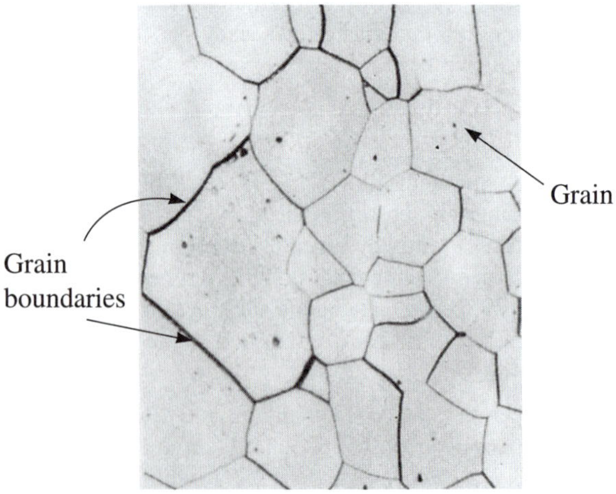

Figure 4-15 (Repeated for Problem 4-31) Microstructure of palladium (×100). (*From* ASM Handbook, *Vol. 9, Metallography and Microstructure (1985), ASM International, Materials Park, OH 44073.)*

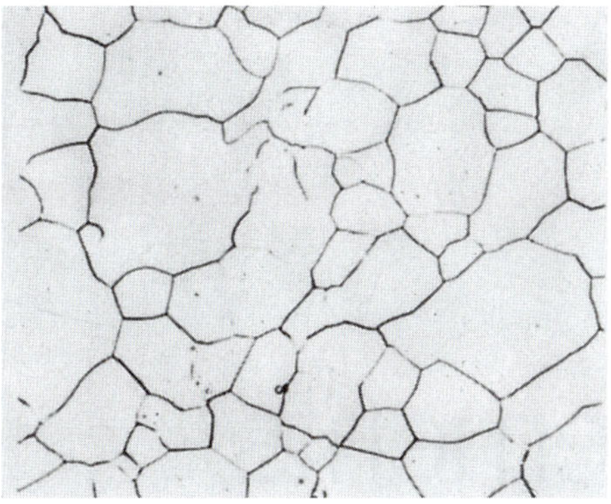

Figure 4-18 Microstructure of iron, for Problem 4-31 (×500). (*From* ASM Handbook, *Vol. 9, Metallography and Microstructure (1985), ASM International, Materials Park, OH 44073.*)

4-32 The yield stress of several samples of a steel containing 0.12% carbon with different grain sizes was measured. The data are shown here.

Sample ID	Grain-Size Inverse Square Root ($d^{-1/2}$)	Yield Stress (MPa)
A	24	500
B	19	420
C	12	320
D	10	250
E	6	190

(a) Calculate the grain size of each steel sample in micrometers.

(b) Which sample has the grain size of 27.7 μm?

(c) Fit these data to a straight line and calculate the constants σ_0 and K for the Hall-Petch equation.

(d) What is the grain size of the sample that has the highest yield strength?

(e) A sample of this steel with 15 μm grain size is produced. What will be the yield stress of this sample?

4-33 A researcher working in the nano-science area develops a sample of 0.12% carbon steel such that the value of $d^{-1/2}$ is 110. What will be the grain size of this steel? Can she use the Hall-Petch relationship developed for this steel in the previous problem to predict the yield stress of this sample?

Section 4-8 Importance of Defects

4-34 What is meant by the term strain hardening?

4-35 Which mechanism of strengthening is the Hall-Petch equation related to?

4-36 Pure copper is strengthened by addition of small concentration of Be. What mechanism of strengthening is this related to?

Design Problems

4-37 The density of pure aluminum calculated from crystallographic data is expected to be 2.69955 g/cm³.

(a) Design an aluminum alloy that has a density of 2.6450 g/cm³.

(b) Design an aluminum alloy that has a density of 2.7450 g/cm³.

4-38 You would like to use a metal plate with good weldability. During the welding process, the metal next to the weld is heated almost to the melting temperature and, depending on the welding parameters, may remain hot for some period of time. Design an alloy that will minimize the loss of strength in this "heat-affected zone" during the welding process.

5

Atom and Ion Movements in Materials

Have You Ever Wondered?

- *Aluminum oxidizes more easily than iron, so why do we say aluminum does not "rust?"*
- *What kind of plastic is used to make carbonated beverage bottles?*
- *How are the surfaces of certain steels hardened?*
- *Why do we encase optical fibers using a polymeric coating?*
- *What is galvanized steel?*
- *How does a tungsten filament in a light bulb fail?*

In Chapter 4, we learned that the atomic and ionic arrangements in materials are never perfect. We also saw that most engineered materials are not pure elements; they are alloys or blends of different elements or compounds. Different types of atoms or ions typically "diffuse", or move within the material, so the differences in their concentration are minimized. Diffusion refers to an observable net flux of atoms or other species. Diffusion depends upon the initial concentration gradient and temperature. Just as water flows from a mountain towards the sea to minimize its gravitational potential energy, atoms and ions have a tendency to move in a predictable fashion so as to eliminate concentration differences and produce homogeneous, uniform compositions that make the material thermodynamically more stable.

In this chapter, we will learn that temperature influences the kinetics of diffusion and that the concentration difference contributes to the overall net flux of diffusing species. The goal of this chapter is to examine the principles and applications of diffusion in materials. We'll illustrate the concept of diffusion through examples of several real-world technologies dependent on the diffusion of atoms, ions, or molecules.

We will present an overview of Fick's laws that describe the diffusion process quantitatively. We will also see how the relative openness of different crystal structures and the size of atoms or ions, temperature, and concentration of diffusing species affect the rate at which diffusion occurs. We will discuss specific examples of how diffusion is used in the synthesis and processing of advanced materials as well as manufacturing of components using advanced materials.

5-1 Applications of Diffusion

Diffusion refers to the net flux of any species, such as ions, atoms, electrons, holes, and molecules. The magnitude of this flux depends upon the initial concentration gradient and temperature. The process of diffusion is central to a large number of today's important technologies. In materials processing technologies, control over the diffusion of atoms, ions, molecules, or other species is key. There are hundreds of applications and technologies that depend on either enhancing or limiting diffusion. The following are just a few examples.

Carburization for Surface Hardening of Steels Let's say we want a surface, such as the teeth of a gear, to be hard. However, we do not want the entire gear to be hard. Carburization processes are used to increase surface hardness. In carburization, a source of carbon, such as a graphite powder or gaseous phase containing carbon, is diffused into steel components such as gears (Figure 5-1). In later chapters, you will learn how increased carbon concentration on the surface of the steel increases the steel's hardness.

Dopant Diffusion for Semiconductor Devices The entire microelectronics industry, as we know it today, would not exist if we did not have a very good understanding of the diffusion of different atoms into silicon or other semiconductors.

Conductive Ceramics Diffusion of ions, electrons, or holes also plays an important role in the electrical conductivity of many conductive ceramics, such as partially or fully stabilized zirconia (ZrO_2) or indium tin oxide (*ITO*). Lithium cobalt oxide ($LiCoO_2$) is an example of an ionically conductive material that is used in lithium ion batteries. These ionically conductive materials are used for such products as oxygen sensors in cars, touch-screen displays, fuel cells, and batteries.

Manufacturing of Plastic Beverage Bottles The occurrence of diffusion may not always be beneficial. In some applications, we may want to limit the occurrence of diffusion for certain species. For example, in the creation of certain plastic bottles, the diffusion of carbon dioxide (CO_2) must be minimized. This is one of the major reasons why

Figure 5-1 Furnace for heat treating steel using the carburization process. (*Courtesy of Cincinnati Steel Treating.*)

we use polyethylene terephthalate (PET) to make bottles which ensure that the carbonated beverages they contain will not loose their fizz for a reasonable period of time!

Oxidation of Aluminum You may have heard or know that aluminum does not "rust." In reality, aluminum oxidizes (rusts) more easily than iron. However, the aluminum oxide (Al_2O_3) forms a very protective but thin coating on the aluminum's surface preventing any further diffusion of oxygen and hindering further oxidation of the underlying aluminum. The oxide coating does not have a color and is thin and, hence, invisible.

Coatings and Thin Films Coatings and thin films are often used in the manufacturing process to limit the diffusion of water vapor, oxygen, or other chemicals.

Optical Fibers and Microelectronic Components Another example of diffusion is that of the coatings around optical fibers. Therefore, optical fibers made from silica (SiO_2) are coated with polymeric materials to prevent diffusion of water molecules. Water reacts with the glass-fiber surface and degrades the ability of optical fibers to carry information.

Drift and Diffusion In some engineering applications, we are concerned with the movement of small particles as a result of forces other than concentration gradient and temperature. We distinguish between the movement of atoms, molecules, ions, electrons, holes, etc. as a result of concentration gradient and temperature (diffusion) or some other **driving force**, such as gradients in density, electric field, or magnetic field gradients. The movement of particles, atoms, ions, electrons, holes, etc., under driving forces other than the concentration gradient is called **drift**. The following example illustrates the principle of diffusion.

EXAMPLE 5-1 *Diffusion of Ar/He and Cu/Ni*

Consider a box containing an impermeable partition that divides the box into equal volumes (Figure 5-2). On one side, we have pure argon (Ar) gas; on the other side, we have pure helium (He) gas. Explain what will happen when the partition is opened? What will happen if we replace the Ar side with a Cu single crystal and the He side with a Ni single crystal?

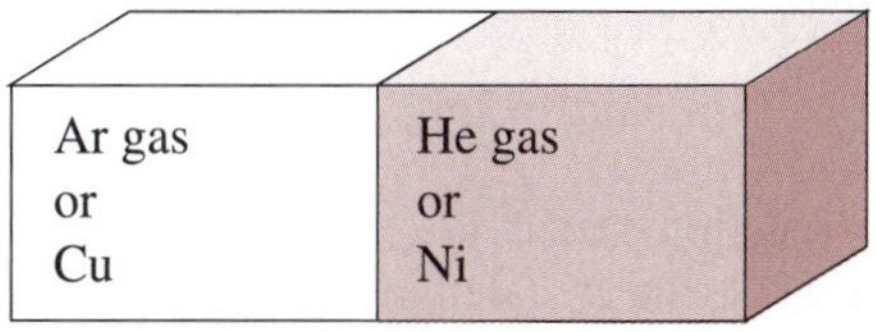

Figure 5-2
Illustration for Diffusion of Ar/He and Cu/Ni (for Example 5-1).

SOLUTION

Before the partition is opened, one compartment has no argon and the other has no helium (i.e., there is a concentration gradient of Ar and He). When the partition is opened, Ar atoms will diffuse toward the He side, and vice versa. This diffusion will continue until the entire box has a uniform concentration of both gases. There may be some density gradient driven convective flows as well. If we took random samples of different regions in this box after a few hours, we would get statistically uniform concentration of Ar and He. Note that Ar and He atoms will continue to move around in the box as a result of thermal; however, there will be no concentration gradients.

If we open the hypothetical partition between the Ni and Cu single crystals at room temperature, we would find that, similar to the Ar/He situation, the concentration gradients exist but the temperature is too low to see any significant diffusion of Cu atoms into Ni single crystal and vice-versa. This is an example of a situation in which there exists a concentration gradient for diffusion. However, because of the lower temperature the kinetics for diffusion are not favorable. Certainly, if we increase the temperature (say to 600°C) and waited for a longer period of time (e.g., ~24 hours), we would see diffusion of copper atoms into Ni single crystal and vice versa. After a very long time (say a few months), the entire solid will have a uniform concentration of Ni and Cu atoms.

5-2 Stability of Atoms and Ions

Atoms or ions in their normal positions in the crystal structures are not stable or at rest. Instead, the atoms or ions possess thermal energy and they will move. For instance, an atom may move from a normal crystal structure location to occupy a nearby vacancy.

An atom may also move from one interstitial site to another. Atoms or ions may jump across a grain boundary, causing the grain boundary to move.

The ability of atoms and ions to diffuse increases as the temperature, or thermal energy, possessed by the atoms and ions increases. The rate of atom or ion movement is related to temperature or thermal energy by the *Arrhenius equation*:

$$\text{Rate} = c_0 \exp\left(\frac{-Q}{RT}\right) \tag{5-1}$$

where c_0 is a constant, R is the gas constant $\left(1.987 \frac{\text{cal}}{\text{mol} \cdot \text{K}}\right)$, T is the absolute temperature (K), and Q is the **activation energy** (cal/mol) required to cause an Avogadro's number of atoms or ions to move. This equation is derived from a statistical analysis of the probability that the atoms will have the extra energy Q needed to cause movement. The rate is related to the number of atoms that move.

We can rewrite the equation by taking natural logarithms of both sides:

$$\ln(\text{rate}) = \ln(c_0) - \frac{Q}{RT} \tag{5-2}$$

If we plot ln(rate) of some reaction versus $1/T$ (Figure 5-3), the slope of the curve will be $-Q/R$ and, consequently, Q can be calculated. The constant c_0 corresponds to the intercept at ln c_0 when $1/T$ is zero.

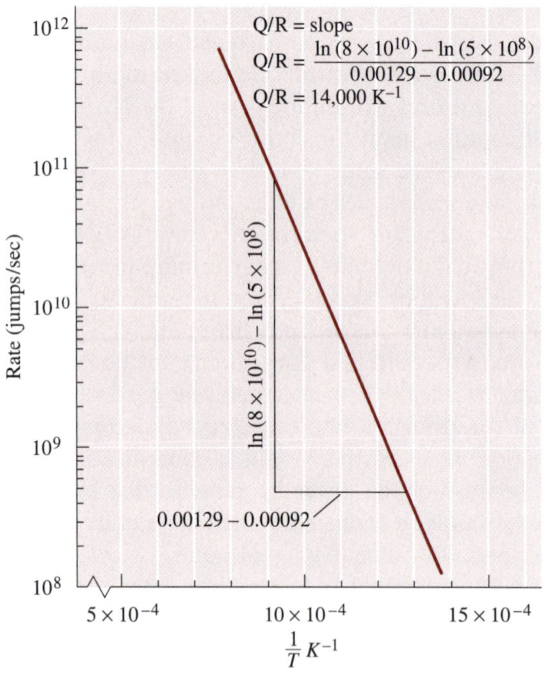

Figure 5-3 The Arrhenius plot of ln (rate) versus $1/T$ can be used to determine the activation energy required for a reaction.

EXAMPLE 5-2 *Activation Energy for Interstitial Atoms*

Suppose that interstitial atoms are found to move from one site to another at the rates of 5×10^8 jumps/s at 500°C and 8×10^{10} jumps/s at 800°C. Calculate the activation energy Q for the process.

SOLUTION

Figure 5-3 represents the data on a ln(rate) versus $1/T$ plot; the slope of this line, as calculated in the figure, gives $Q/R = 14{,}000\ \text{K}^{-1}$, or $Q = 27{,}880$ cal/mol. Alternately, we could write two simultaneous equations:

$$\text{Rate}\left(\frac{\text{jumps}}{\text{s}}\right) = c_0 \exp\left(\frac{-Q}{RT}\right)$$

$$5 \times 10^8\left(\frac{\text{jumps}}{\text{s}}\right) = c_0\left(\frac{\text{jumps}}{\text{s}}\right)\exp\left[\frac{-Q\left(\frac{\text{cal}}{\text{mol}}\right)}{\left[1.987\left(\frac{\text{cal}}{\text{mol}\cdot\text{K}}\right)\right](500+273)\text{K}}\right]$$

$$= c_0 \exp(-0.000651Q)$$

$$8 \times 10^{10}\left(\frac{\text{jumps}}{\text{s}}\right) = c_0\left(\frac{\text{jumps}}{\text{s}}\right)\exp\left[\frac{-Q\left(\frac{\text{cal}}{\text{mol}}\right)}{\left[1.987\left(\frac{\text{cal}}{\text{mol}\cdot\text{K}}\right)\right](800+273)\text{K}}\right]$$

$$= c_0 \exp(-0.000469Q)$$

Note the temperatures were converted into K.

Since

$$c_0 = \frac{5 \times 10^8}{\exp(-0.000651Q)}\left(\frac{\text{jumps}}{\text{s}}\right),$$

then

$$8 \times 10^{10} = \frac{(5 \times 10^8)\exp(-0.000469Q)}{\exp(-0.000651Q)}$$

$$160 = \exp[(0.000651 - 0.000469)Q] = \exp(0.000182Q)$$

$$\ln(160) = 5.075 = 0.000182Q$$

$$Q = \frac{5.075}{0.000182} = 27{,}880\ \text{cal/mol}$$

5-3 Mechanisms for Diffusion

The disorder vacancies create (i.e., increased entropy) helps minimize the free energy and, therefore, enhances the thermodynamic stability of a crystalline material. Crystalline materials also contain other types of defects. In materials containing vacancies, atoms move or "jump" from one lattice position to another. This process, known as **self-diffusion**, can be detected by using radioactive tracers. As an example, suppose we were to introduce a radioactive isotope of gold (Au^{198}) onto the surface of normal gold (Au^{197}). After a period of time, the radioactive atoms would move into the normal gold. Eventually, the radioactive atoms would be uniformly distributed throughout the entire regular gold sample. Although self-diffusion occurs continually in all materials, its effect on the material's behavior is generally not significant.

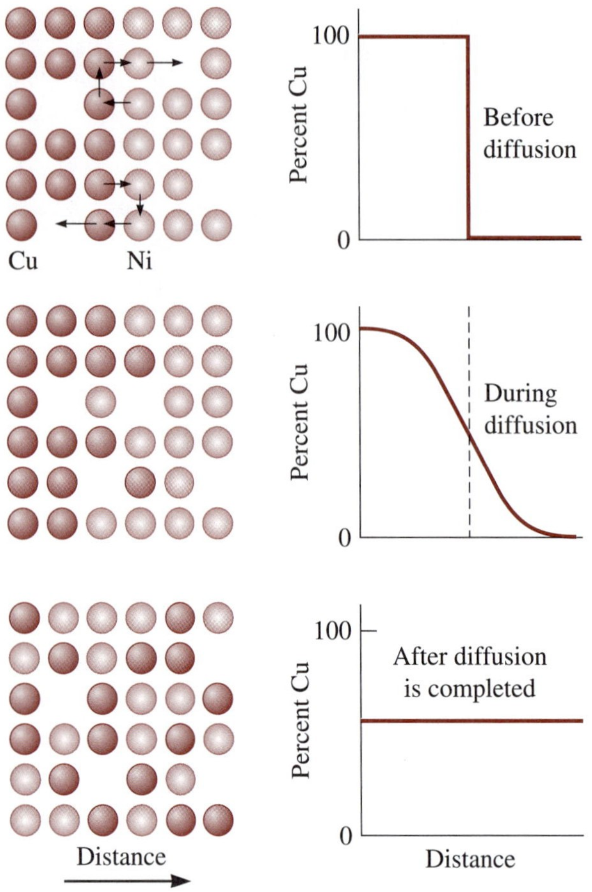

Figure 5-4
Diffusion of copper atoms into nickel. Eventually, the copper atoms are randomly distributed throughout the nickel.

Diffusion of unlike atoms in materials also occurs (Figure 5-4). Consider a nickel sheet bonded to a copper sheet. At high temperatures, nickel atoms gradually diffuse into the copper and copper atoms migrate into the nickel. Again, the nickel and copper atoms eventually are uniformly distributed. There are two important mechanisms by which atoms or ions can diffuse (Figure 5-5).

Vacancy Diffusion In self-diffusion and diffusion involving substitutional atoms, an atom leaves its lattice site to fill a nearby vacancy (thus creating a new vacancy at the original lattice site). As diffusion continues, we have countercurrent flows of atoms and vacancies, called **vacancy diffusion**. The number of vacancies, which increases as the temperature increases, helps determine the extent of both self-diffusion and diffusion of substitutional atoms.

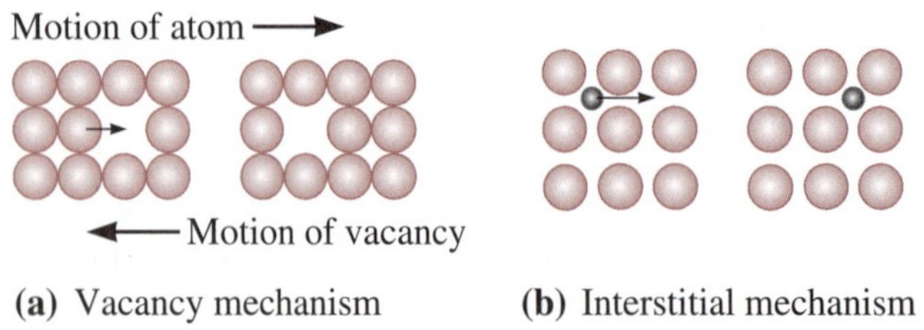

Figure 5-5
Diffusion mechanisms in materials: (a) vacancy or substitutional atom diffusion and (b) interstitial diffusion.

Interstitial Diffusion When a small interstitial atom or ion is present in the crystal structure, the atom or ion moves from one interstitial site to another. No vacancies are required for this mechanism. Partly because there are many more interstitial sites than vacancies, interstitial diffusion occurs more easily than vacancy diffusion. Also, interstitial atoms that are relatively smaller can diffuse faster. In Chapter 3, we have seen that for many ceramics with ionic bonding, the structure can be considered as close packing of anions with cations in the interstitial sites. In these materials, smaller cations often diffuse faster than larger anions.

5-4 Activation Energy for Diffusion

A diffusing atom must squeeze past the surrounding atoms to reach its new site. In order for this to happen, energy must be supplied to force the atom to its new position, as is shown schematically for vacancy and interstitial diffusion in Figure 5-6. The atom is originally in a low-energy, relatively stable location. In order to move to a new location, the atom must overcome an energy barrier. The energy barrier is the activation energy Q. The thermal energy supplies atoms or ions with the energy needed to exceed this barrier. Note that the symbol Q is often used for activation energies for different processes (rate at which atoms jump, a chemical reaction, energy needed to produce vacancies, etc.).

Normally, less energy is required to squeeze an interstitial atom past the surrounding atoms; consequently, activation energies are lower for interstitial diffusion than for vacancy diffusion (Figure 5-6). A good analogy for this is as follows. In basketball, relatively smaller and shorter players can quickly "diffuse by" taller and bigger players and score baskets. Typical values for activation energies for diffusion of different atoms in different host materials are shown in Table 5-1. We use the term **diffusion couple** to indicate a combination of an atom of a given element (e.g., carbon) diffusing in a host material (e.g., BCC Fe). A low-activation energy indicates easy diffusion. In self-diffusion, the activation energy is equal to the energy needed to create a vacancy and to

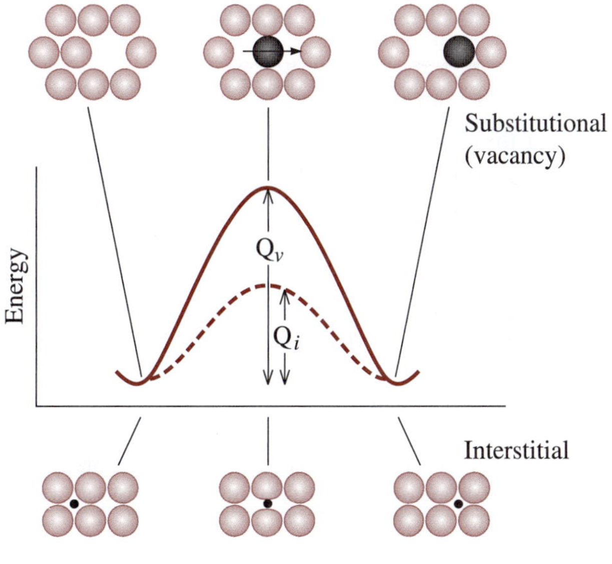

Figure 5-6 A high energy is required to squeeze atoms past one another during diffusion. This energy is the activation energy Q. Generally more energy is required for a substitutional atom than for an interstitial atom.

TABLE 5-1 ■ ***Diffusion data for selected materials***

Diffusion Couple	Q (cal/mol)	D_0 (cm^2/s)
Interstitial diffusion:		
C in FCC iron	32,900	0.23
C in BCC iron	20,900	0.011
N in FCC iron	34,600	0.0034
N in BCC iron	18,300	0.0047
H in FCC iron	10,300	0.0063
H in BCC iron	3,600	0.0012
Self-diffusion (vacancy diffusion):		
Pb in FCC Pb	25,900	1.27
Al in FCC Al	32,200	0.10
Cu in FCC Cu	49,300	0.36
Fe in FCC Fe	66,700	0.65
Zn in HCP Zn	21,800	0.1
Mg in HCP Mg	32,200	1.0
Fe in BCC Fe	58,900	4.1
W in BCC W	143,300	1.88
Si in Si (covalent)	110,000	1800.0
C in C (covalent)	163,000	5.0
Heterogeneous diffusion (vacancy diffusion):		
Ni in Cu	57,900	2.3
Cu in Ni	61,500	0.65
Zn in Cu	43,900	0.78
Ni in FCC iron	64,000	4.1
Au in Ag	45,500	0.26
Ag in Au	40,200	0.072
Al in Cu	39,500	0.045
Al in Al_2O_3	114,000	28.0
O in Al_2O_3	152,000	1900.0
Mg in MgO	79,000	0.249
O in MgO	82,100	0.000043

Data from several sources, including Adda, Y. and Philibert, J., La Diffusion dans les Solides, Vol. 2, 1966.

cause the movement of the atom. Table 5-1 also shows values of D_0, the pre-exponent term, and the constant c_0 from Equation 5-1 where the rate process is diffusion. We will see later that D_0 is the diffusion coefficient when $1/T = 0$.

5-5 Rate of Diffusion (Fick's First Law)

The rate at which atoms, ions, particles or other species diffuse in a material can be measured by the **flux** (J). Here we are mainly concerned with diffusion of ions or atoms. The flux is defined as the number of atoms passing through a plane of unit area per unit time (Figure 5-7). **Fick's first law** explains the net flux of atoms:

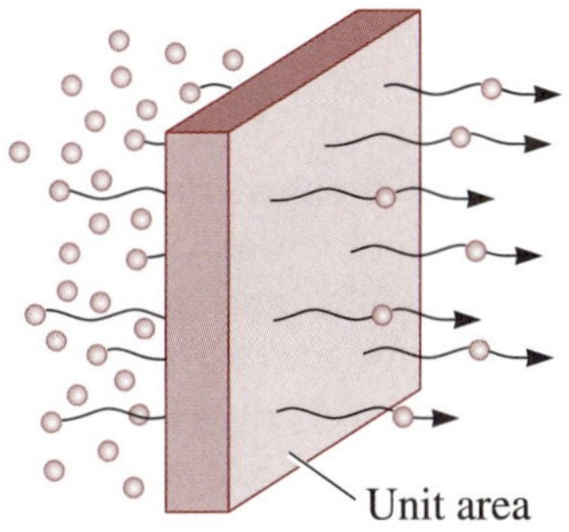

Figure 5-7
The flux during diffusion is defined as the number of atoms passing through a plane of unit area per unit time.

$$J = -D\frac{dc}{dx} \tag{5-3}$$

where J is the flux, D is the **diffusivity** or **diffusion coefficient** $\left(\frac{\text{cm}^2}{\text{s}}\right)$, and dc/dx is the **concentration gradient** $\left(\frac{\text{atoms}}{\text{cm}^3 \cdot \text{cm}}\right)$. Depending upon the situation, concentration may be expressed as atom percent (at%), weight percent (wt%), mole percent (mol%), atom fraction, or mole fraction. The units of concentration gradient and flux will also change respectively.

Several factors affect the flux of atoms during diffusion. If we are dealing with diffusion of ions, electrons, holes, etc., the units of J, D, and $\frac{dc}{dx}$ will reflect the appropriate species that are being considered. The negative sign in Equation 5-3 tells us that the flux of diffusing species is from higher to lower concentrations, making the $\frac{dc}{dx}$ term negative, and hence J will be positive. Thermal energy associated with atoms, ions etc. causes the random movement of atoms. At a microscopic scale the thermodynamic driving force for diffusion is concentration gradient. A net or an observable flux is created depending upon temperature and concentration gradient.

Concentration Gradient The concentration gradient shows how the composition of the material varies with distance: Δc is the difference in concentration over the distance Δx (Figure 5-8). The concentration gradient may be created when two materials of different composition are placed in contact, when a gas or liquid is in contact with a solid material, when nonequilibrium structures are produced in a material due to processing, and from a host of other sources.

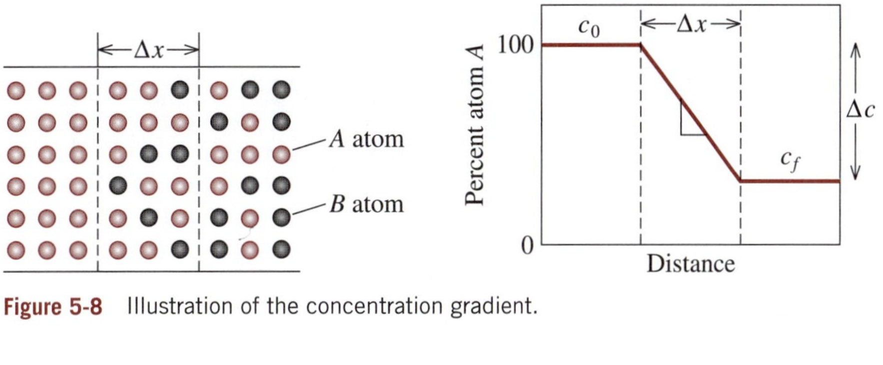

Figure 5-8 Illustration of the concentration gradient.

The flux at a particular temperature is constant only if the concentration gradient is also constant—that is, the compositions on each side of the plane in Figure 5-7 remain unchanged. In many practical situations, however, these compositions vary as atoms are redistributed, and thus the flux also changes. Often, we find that the flux is initially high and then gradually decreases as the concentration gradient is reduced by diffusion. The example that follows illustrates calculations of flux and concentration gradients for diffusion of dopants in semiconductors, but only for the case of constant concentration gradient. Later in this chapter, we will consider non-steady state diffusion with the aid of Fick's second law.

EXAMPLE 5-3 *Semiconductor Doping*

One way to manufacture transistors, which amplify electrical signals, is to diffuse impurity atoms into a semiconductor material such as silicon (Si). Suppose a silicon wafer 0.1 cm thick, which originally contains one phosphorus atom for every 10 million Si atoms, is treated so that there are 400 phosphorous (P) atoms for every 10 million Si atoms at the surface (Figure 5-9). Calculate the concentration gradient (a) in atomic percent/cm and (b) in $\frac{\text{atoms}}{\text{cm}^3 \cdot \text{cm}}$. The lattice parameter of silicon is 5.4307 Å.

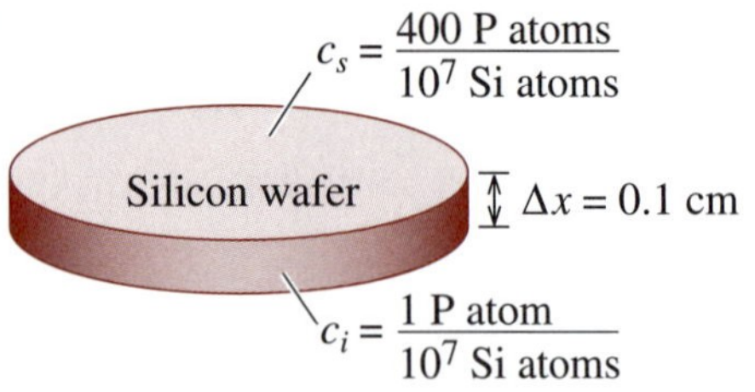

Figure 5-9 Silicon wafer showing variation in concentration of P atoms (for Example 5-3).

SOLUTION

First, let's calculate the initial and surface compositions in atomic percent:

$$c_i = \frac{1 \text{ P atom}}{10^7 \text{ atoms}} \times 100 = 0.00001 \text{ at\% P}$$

$$c_s = \frac{400 \text{ P atoms}}{10^7 \text{ atoms}} \times 100 = 0.004 \text{ at\% P}$$

$$\frac{\Delta c}{\Delta x} = \frac{0.00001 - 0.004 \text{ at\% P}}{0.1 \text{ cm}} = -0.0399 \frac{\text{at\% P}}{\text{cm}}$$

To find the gradient in terms of $\frac{\text{atoms}}{\text{cm}^3 \cdot \text{cm}}$, we must find the volume of the unit cell:

$$V_{\text{cell}} = (5.4307 \times 10^{-8} \text{ cm})^3 = 1.6 \times 10^{-22} \frac{\text{cm}^3}{\text{cell}}$$

The volume occupied by 10^7 Si atoms, which are arranged in a diamond cubic (DC) structure with 8 atoms/cell, is:

$$V = \left[\frac{10^7 \text{ atoms}}{8 \frac{\text{atoms}}{\text{cell}}}\right]\left[1.6 \times 10^{-22}\left(\frac{\text{cm}^3}{\text{cell}}\right)\right] = 2 \times 10^{-16} \text{ cm}^3$$

The compositions in atoms/cm^3 are:

$$c_i = \frac{1 \text{ P atom}}{2 \times 10^{-16} \text{ cm}^3} = 0.005 \times 10^{18} \text{ P}\left(\frac{\text{atoms}}{\text{cm}^3}\right)$$

$$c_s = \frac{400 \text{ P atoms}}{2 \times 10^{-16} \text{ cm}^3} = 2 \times 10^{18} \text{ P}\left(\frac{\text{atoms}}{\text{cm}^3}\right)$$

$$\frac{\Delta c}{\Delta x} = \frac{0.005 \times 10^{18} - 2 \times 10^{18} \text{ P}\left(\frac{\text{atoms}}{\text{cm}^3}\right)}{0.1 \text{ cm}}$$

$$= -1.995 \times 10^{19} \text{ P}\frac{\text{atoms}}{\text{cm}^3 \cdot \text{cm}}$$

5-6 Factors Affecting Diffusion

Temperature and the Diffusion Coefficient The kinetics of the process of diffusion are strongly dependent on temperature. The diffusion coefficient D is related to temperature by an Arrhenius-type equation,

$$D = D_0 \exp\left(\frac{-Q}{RT}\right) \tag{5-4}$$

where Q is the activation energy (in units of cal/mol) for diffusion of species under consideration (e.g., Al in Si), R is the gas constant $\left(1.987\frac{\text{cal}}{\text{mol} \cdot \text{K}}\right)$, and T is the absolute temperature (K). D_0 is the pre-exponential term, similar to c_0 in Equation 5-1. It is a constant for a given diffusion system and is equal to the value of the diffusion coefficient at $1/T = 0$ or $T = \infty$. Typical values for D_0 are given in Table 5-1, while the temperature dependence of D is shown in Figure 5-10 for some metals and ceramics. Covalently bonded materials, such as carbon and silicon (Table 5-1), have unusually high activation energies, consistent with the high strength of their atomic bonds.

In ionic materials, such as some of the oxide ceramics, a diffusing ion only enters a site having the same charge. In order to reach that site, the ion must physically squeeze past adjoining ions, pass by a region of opposite charge, and move a relatively long distance (Figure 5-11). Consequently, the activation energies are high and the rates of diffusion are lower for ionic materials than those for metals.

When the temperature of a material increases, the diffusion coefficient D increases (according to Equation 5-4) and, therefore, the flux of atoms increases as well. At higher temperatures, the thermal energy supplied to the diffusing atoms permits the atoms to overcome the activation energy barrier and more easily move to new sites in the atomic arrangements. At low temperatures—often below about 0.4 times the absolute melting temperature of the material—diffusion is very slow and may not be significant. For this reason, the heat treatments of metals and the processing of ceramics are done at high

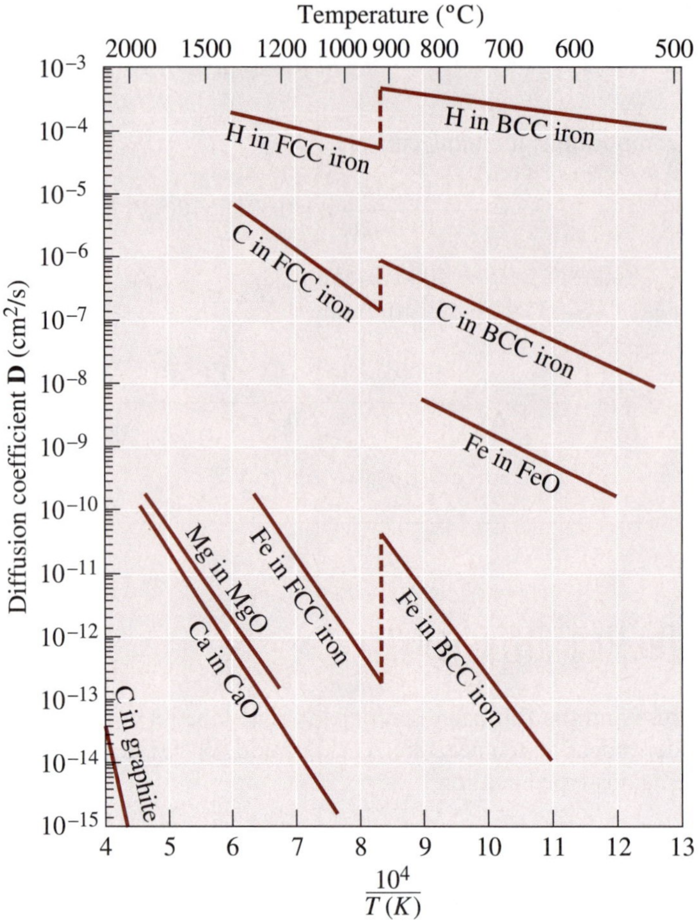

Figure 5-10 The diffusion coefficient D as a function of reciprocal temperature for some metals and ceramics. In this Arrhenius plot, D represents the rate of the diffusion process. A steep slope denotes a high activation energy.

temperatures, where atoms move rapidly to complete reactions or to reach equilibrium conditions. Because less thermal energy is required to overcome the smaller activation energy barrier, a small activation energy Q increases the diffusion coefficient and flux. The following example illustrates how Fick's first law and concepts related to temperature dependence of D can be applied to design an iron membrane.

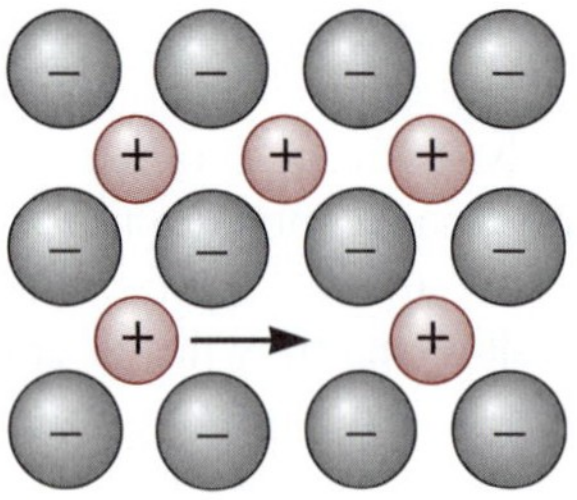

Figure 5-11
Diffusion in ionic compounds. Anions can only enter other anion sites. Smaller cations tend to diffuse faster.

EXAMPLE 5-4 *Diffusion in Yttrium in Chromia Film Formed by Oxidation of Ni-Cr Alloy*

The diffusion of yttrium ions in chromium oxide or chromia (Cr_2O_3) has been studied by Lesage and co-workers. This was done by forming a thin film of yttrium oxide on a nickel chromium alloy. The base alloy substrate formed a chromium oxide layer on the surface after oxidation and the yttrium ions from yttrium oxide layer subsequently diffused into the chromium oxide film formed by alloy oxidation. Their experiments show that the diffusion coefficient of yttrium ions while diffusing through the bulk chromia exhibited the following temperature dependence.

Temperature (°C)	Diffusion Coefficient (*D*) (cm²/s)
800	2.0×10^{-18}
850	1.8×10^{-18}
900	3.2×10^{-18}
950	7.9×10^{-18}
1000	2.4×10^{-17}

(Source: J. Li, M.K. Loudjani, B. Lesage, A.M. Huntz, Philosophical Magazine A, *1997,* ***76****[4], pp. 857–69).*

(a) What is the activation energy for this bulk diffusion of yttrium ions in chromia?

(b) What is value of the pre-exponential term D_0?

(c) What will be the equation of the straight line describing the relationship between $\ln(D)$ and $1/T$?

SOLUTION

(a) We assume that the Arrhenius relationship is applicable.

$$D = D_0 \exp\left[\frac{-Q}{RT}\right].$$

Note that the reported D value at 850°C is actually a bit lower than that for 800°C. This is unexpected, and the value may be within the experimental measurement error and hence not much different. There may have been some experimental artifact or some other unknown factor that caused this value to be lower. We will use these values, since the difference between the values of diffusion coefficients at 800 and 850°C is not extremely large.

Taking logarithms of both sides:

$$\ln D = \ln D_0 - \frac{Q}{RT}$$

Therefore, the diffusion data given are rewritten as

Temp (K)	1/*T* (K^{-1})	*D* cm^2/s	ln *D*
1073	0.000932	2E−18	−40.9159
1123	0.00089	1.7E−18	−40.7534
1173	0.000853	3.2E−18	−40.2834
1223	0.000818	7.9E−18	−39.3797
1273	0.000786	2.4E−17	−38.2685

We can then either plot the data or use a spreadsheet such as Excel™ to calculate the slope. In this case, we have used the SLOPE (known_y's,known_x's) function in Excel™ to calculate the slope of the expected straight line. The known *y*-axis values are those under the ln *D* column, and the known *x*-values are under the $1/T$ column. This calculation gives us a slope value of −17415.7. Note the negative sign for the slope value.

This value of slope is equal to $(-Q/R)$. If we use gas constant value of $R = 8.314$ Joules/K-mol, then the value of activation energy *Q* will be

$$Q = (-) \times (-17415.7) \times (8.314) \text{ Joules/mol}$$

$$= 144794.5 \text{ Joules/mol or } 144.8 \text{ kJ/mol}$$

If we use the value of $R = 1.987$ cal/K-mol, then the value of activation energy *Q* will be

$$Q = (-1) \times (-144794.5) \times (1.987) \text{ cal/mol}$$

$$= 34605.07 \text{ cal/mol or } 34.6 \text{ kcal/mol}$$

(b) We also calculate the slope of the straight line fitted to the ln(D) versus $1/T$ data. This can be done using graph paper or as is done here, using a spreadsheet. Here we used the INTERCEPT(known_y's,known_x's) function in Excel™ using the ln *D* values as known *y*-axis data and the $1/T$ values as the *x*-axis data. This gives us a value of the intercept as −25.01. Thus, the value of D_0 will be exp(intercept value) or exp(−25.01) = 1.36×10^{-11}, and the units will be cm^2/s.

(c) Thus, the relationship between *D* and *T* will be

$$D = \left(1.36 \times 10^{-11} \frac{\text{cm}^2}{\text{s}}\right) \exp\left(-\frac{34.6 \frac{\text{kcal}}{\text{mol}}}{\text{RT}}\right) \quad \text{or}$$

$$D = \left(1.36 \times 10^{-11} \frac{\text{cm}^2}{\text{s}}\right) \exp\left(-\frac{144.8 \frac{\text{kJ}}{\text{mol}}}{\text{RT}}\right)$$

EXAMPLE 5-5 *Design of an Iron Membrane*

An impermeable cylinder 3 cm in diameter and 10 cm long contains a gas that includes 0.5×10^{20} N atoms per cm^3 and 0.5×10^{20} H atoms per cm^3 on one side of an iron membrane (Figure 5-12). Gas is continuously introduced to the pipe to assure a constant concentration of nitrogen and hydrogen. The gas on the other side of the membrane includes a constant 1×10^{18} N atoms per cm^3 and 1×10^{18} H atoms per cm^3. The entire system is to operate at 700°C, where the iron has the BCC structure. Design an iron membrane that will allow no more than 1% of the nitrogen to be lost through the membrane each hour, while allowing 90% of the hydrogen to pass through the membrane per hour.

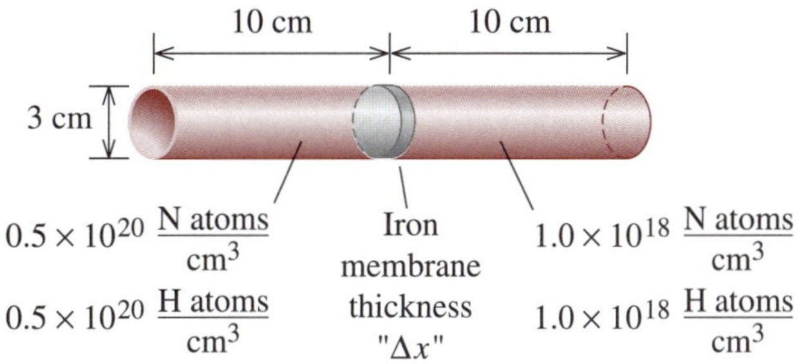

Figure 5-12 Design of an iron membrane (for Example 5-5).

SOLUTION

The total number of nitrogen atoms in the container is:

$$(0.5 \times 10^{20}\ \text{N/cm}^3)(\pi/4)(3\ \text{cm})^2(10\ \text{cm}) = 35.343 \times 10^{20}\ \text{N atoms}$$

The maximum number of atoms to be lost is 1% of this total, or:

$$\text{N atom loss per h} = (0.01)(35.34 \times 10^{20}) = 35.343 \times 10^{18}\ \text{N atoms/h}$$

$$\text{N atom loss per s} = (35.343 \times 10^{18}\ \text{N atoms/h})/(3600\ \text{s/h})$$

$$= 0.0098 \times 10^{18}\ \text{N atoms/s}$$

The flux is then:

$$J = \frac{(0.0098 \times 10^{18}(\text{N atoms/s})}{\left(\frac{\pi}{4}\right)(3\ \text{cm})^2}$$

$$= 0.00139 \times 10^{18}\ \text{N}\frac{\text{atoms}}{\text{cm}^2 \cdot \text{s}}$$

The diffusion coefficient of nitrogen in BCC iron at 700°C = 973 K is:

From Equation 5-4 and Table 5-1:

$$D = D_0 \exp\left(\frac{-Q}{RT}\right)$$

$$D_N = 0.0047\frac{\text{cm}^2}{\text{s}}\exp\left[\frac{-18{,}300\frac{\text{cal}}{\text{mol}}}{1.987\frac{\text{cal}}{\text{K}\cdot\text{mol}}(973)\text{K}}\right]$$

$$= 0.0047\exp(-9.4654) = (0.0047)(7.749\times10^{-5}) = 3.64\times10^{-7}\frac{\text{cm}^2}{\text{s}}$$

From Equation 5-3:

$$J = -D\left(\frac{\Delta c}{\Delta x}\right) = 0.00139\times10^{18}\frac{\text{N atoms}}{\text{cm}^2\cdot\text{s}}$$

$$\Delta x = -D\Delta c/J = \frac{\left[(-3.64\times10^{-7}\ \text{cm}^2/\text{s})\left(1\times10^{18} - 50\times10^{18}\frac{\text{N atoms}}{\text{cm}^3}\right)\right]}{0.00139\times10^{18}\frac{\text{N atoms}}{\text{cm}^2\cdot\text{s}}}$$

$\Delta x = 0.0128$ cm = minimum thickness of the membrane

In a similar manner, the maximum thickness of the membrane that will permit 90% of the hydrogen to pass can be calculated:

$$\text{H atom loss per h} = (0.90)(35.343\times10^{20}) = 31.80\times10^{20}$$

$$\text{H atom loss per s} = 0.0088\times10^{20}$$

$$J = 0.125\times10^{18}\frac{\text{H atoms}}{\text{cm}^2\cdot\text{s}}$$

From Equation 5-4:

$$D_H = 0.004\frac{\text{cm}^2}{\text{s}}\exp\left[\frac{-18{,}300\frac{\text{cal}}{\text{mol}}}{1.987\frac{\text{cal}}{\text{K}\cdot\text{mol}}(973)\text{K}}\right] = 1.86\times10^{-4}\ \text{cm}^2/\text{s}$$

Since

$$\Delta x = -D\,\Delta c/J$$

$$\Delta x = \frac{\left(1.86\times10^{-4}\frac{\text{cm}^2}{\text{s}}\right)\left(49\times10^{18}\frac{\text{H atoms}}{\text{cm}^3}\right)}{0.125\times10^{18}\frac{\text{H atoms}}{\text{cm}^2\cdot\text{s}}}$$

$$= 0.0729\ \text{cm} = \text{maximum thickness}$$

An iron membrane with a thickness between 0.0128 and 0.0729 cm will be satisfactory.

Types of Diffusion In **volume diffusion**, the atoms move through the crystal from one regular or interstitial site to another. Because of the surrounding atoms, the activation energy is large and the rate of diffusion is relatively slow.

However, atoms can also diffuse along boundaries, interfaces, and surfaces in the material. Atoms diffuse easily by **grain boundary diffusion** because the atom packing is poor at the grain boundaries. Because atoms can more easily squeeze their way through

TABLE 5-2 ■ *The effect of the type of diffusion for thorium in tungsten and for self-diffusion in silver**

	Diffusion Coefficient (*D*)			
Diffusion Type	**Thorium in Tungsten**		**Silver in Silver**	
	D_0 cm²/s	Q cal/mole	D_0 cm²/s	Q cal/mole
Surface	0.47	66,400	0.068	8,900
Grain boundary	0.74	90,000	0.24	22,750
Volume	1.00	120,000	0.99	45,700

** Given by parameters for Equation 5-4.*

the disordered grain boundary, the activation energy is low (Table 5-2). **Surface diffusion** is easier still because there is even less constraint on the diffusing atoms at the surface.

Time Diffusion requires time; the units for flux are $\frac{\text{atoms}}{\text{cm}^2 \cdot \text{s}}$! If a large number of atoms must diffuse to produce a uniform structure, long times may be required, even at high temperatures. Times for heat treatments may be reduced by using higher temperatures or by making the **diffusion distances** (related to Δx) as small as possible.

We find that some rather remarkable structures and properties are obtained if we prevent diffusion. Steels quenched rapidly from high temperatures to prevent diffusion form nonequilibrium structures and provide the basis for sophisticated heat treatments. Similarly, in forming metallic glasses (Chapter 4) we have to quench liquid metals at a very high cooling rate ($\sim 10^6$°C/second). This is to avoid diffusion of atoms by taking away their thermal energy and encouraging them to assemble into nonequilibrium amorphous and chemically homogeneous arrangements. Melts of silicate glasses, on the other hand, are viscous and diffusion of ions through these is slow. As a result, we do not have to cool these melts very rapidly.

EXAMPLE 5-6 ***Tungsten Thorium Diffusion Couple***

Consider a diffusion couple setup between pure tungsten and a tungsten alloy containing 1 at.% thorium. After several minutes of exposure at 2000°C, a transition zone of 0.01 cm thickness is established. What is the flux of thorium atoms at this time if diffusion is due to (a) volume diffusion, (b) grain boundary diffusion, and (c) surface diffusion? (See Table 5-2.)

SOLUTION

The lattice parameter of BCC tungsten is 3.165 Å. Thus, the number of tungsten atoms per cm³ is:

$$\frac{\text{W atoms}}{\text{cm}^3} = \frac{2 \text{ atoms/cell}}{(3.165 \times 10^{-8})^3 \text{ cm}^3\text{/cell}} = 6.3 \times 10^{22}$$

In the tungsten-1 at.% thorium alloy, the number of thorium atoms per cm³ is:

$$c_{\text{Th}} = (0.01)(6.3 \times 10^{22}) = 6.3 \times 10^{20} \text{ Th atoms/cm}^3$$

In pure tungsten, the number of thorium atoms is zero. Thus, the concentration gradient is:

$$\frac{\Delta c}{\Delta x} = \frac{0 - 6.3 \times 10^{20} \frac{\text{atoms}}{\text{cm}^3}}{0.01 \text{ cm}} = -6.3 \times 10^{22} \text{ Th} \frac{\text{atoms}}{\text{cm}^3 \cdot \text{cm}}$$

1. Volume diffusion

$$D = 1.0 \frac{\text{cm}^2}{\text{s}} \exp\left(\frac{-120{,}000 \frac{\text{cal}}{\text{mol}}}{\left(1.987 \frac{\text{cal}}{\text{deg} \cdot \text{mol}}\right)(2273 \text{ K})}\right) = 2.89 \times 10^{-12} \text{ cm}^2/\text{s}$$

$$J = -D\frac{\Delta c}{\Delta x} = -\left(2.89 \times 10^{-12} \frac{\text{cm}^2}{\text{s}}\right)\left(-6.3 \times 10^{22} \frac{\text{atoms}}{\text{cm}^3 \cdot \text{cm}}\right)$$

$$= 18.2 \times 10^{10} \frac{\text{Th atoms}}{\text{cm}^2 \cdot \text{s}}$$

2. Grain boundary diffusion

$$D = 0.74 \frac{\text{cm}^2}{\text{s}} \exp\left(\frac{-90{,}000 \frac{\text{cal}}{\text{mol}}}{\left(1.987 \frac{\text{cal}}{\text{deg} \cdot \text{mol}}\right)(2273 \text{ K})}\right) = 1.64 \times 10^{-9} \text{ cm}^2/\text{s}$$

$$J = -\left(1.64 \times 10^{-9} \frac{\text{cm}^2}{\text{s}}\right)\left(-6.3 \times 10^{22} \frac{\text{atoms}}{\text{cm}^3 \cdot \text{cm}}\right) = 10.3 \times 10^{13} \frac{\text{Th atoms}}{\text{cm}^2 \cdot \text{s}}$$

3. Surface diffusion

$$D = 0.47 \frac{\text{cm}^2}{\text{s}} \exp\left(\frac{-66{,}400 \frac{\text{cal}}{\text{mol}}}{\left(1.987 \frac{\text{cal}}{\text{mol} \cdot \text{deg}}\right)(2273 \text{ K})}\right) = 1.94 \times 10^{-7} \text{ cm}^2/\text{s}$$

$$J = -\left(1.94 \times 10^{-7} \frac{\text{cm}^2}{\text{s}}\right)\left(-6.3 \times 10^{22} \frac{\text{atoms}}{\text{cm}^3 \cdot \text{cm}}\right) = 12.2 \times 10^{15} \frac{\text{Th atoms}}{\text{cm}^2 \cdot \text{s}}$$

Dependence on Bonding and Crystal Structure A number of factors influence the activation energy for diffusion and, hence, the rate of diffusion. Interstitial diffusion, with a low-activation energy, usually occurs much faster than vacancy, or substitutional diffusion. Activation energies are usually lower for atoms diffusing through open crystal structures than for close-packed crystal structures. Because the activation energy depends on the strength of atomic bonding, it is higher for diffusion of atoms in materials with a high melting temperature (Figure 5-13).

We also find that, due to their smaller size, cations (with a positive charge) often have higher diffusion coefficients than those for anions (with a negative charge). In sodium chloride, for instance, the activation energy for diffusion of chloride ions (Cl^-) is about twice that for diffusion of sodium ions (Na^+).

Diffusion of ions also provides a transfer of electrical charge; in fact, the electrical conductivity of ionically bonded ceramic materials is related to temperature by an Arrhenius equation. As the temperature increases, the ions diffuse more rapidly, electrical charge is transferred more quickly, and the electrical conductivity is increased. As mentioned before, these are examples of ceramic materials that are good conductors of electricity.

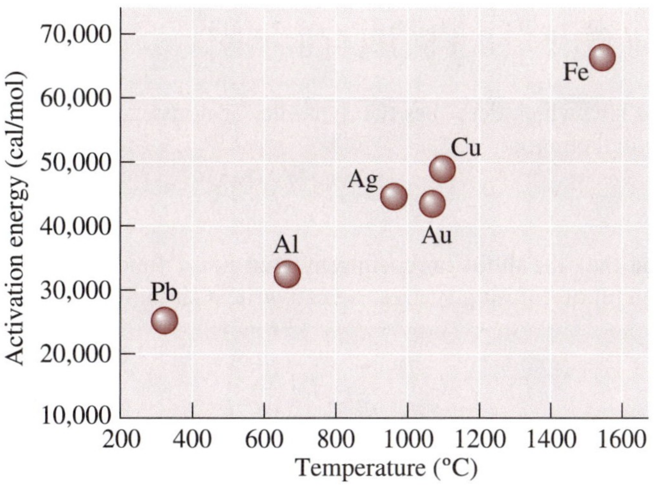

Figure 5-13
The activation energy for self-diffusion increases as the melting point of the metal increases.

Dependence on Concentration of Diffusing Species and Composition of Matrix The diffusion coefficient (D) depends not only on temperature, as given by Equation 5-4, but also on the concentration of diffusing species and composition of the matrix. These effects have not been included in our discussion so far. In many situations, the dependence of D on concentration of diffusing species can be ignored, for example, if the concentration of dopants is small.

5-7 Permeability of Polymers

In polymers, we are most concerned with the diffusion of atoms or small molecules between the long polymer chains. As engineers, we often cite the permeability of polymers and other materials, instead of the diffusion coefficients. The **permeability** is expressed in terms of the volume of gas or vapor that can permeate per unit area, per unit time, or per unit thickness at a specified temperature and relative humidity. Polymers that have a polar group (e.g., ethylene vinyl alcohol) have higher permeability for water vapor than that for oxygen gas. Polyethylene, on the other hand, has much higher permeability for oxygen than for water vapor. In general, the more compact the structure of polymers, the lesser the permeability. For example, low-density polyethylene has a higher permeability than high-density polyethylene. Polymers used for food and other applications need to have the appropriate barrier properties. For example, polymer films are typically used as packaging to store food. If air diffuses through the film, the food may spoil. Similarly, care has to be exercised in the storage of ceramic or metal powders that are sensitive to atmospheric water vapor, nitrogen, oxygen, or carbon dioxide. For example, zinc oxide powders used in rubbers, paints, and ceramics must be stored in polyethylene bags to avoid reactions with atmospheric water vapor. If air diffuses through the rubber inner tube of an automobile tire, the tire will deflate.

Diffusion of some molecules into a polymer can cause swelling problems. For example, in automotive applications, polymers used to make O-rings can absorb considerable amounts of oil, causing them to swell.

5-8 Composition Profile [Fick's Second Law]

Fick's second law, which describes the dynamic, or nonsteady state, diffusion of atoms, is the differential equation

$$\frac{\partial c}{\partial t} = \frac{\partial}{\partial x}\left(D\frac{\partial c}{\partial x}\right) \tag{5-5}$$

If we assume that the diffusion coefficient D is not a function of location x and the concentration (c) of diffusing species, we can write a simplified version of Fick's second law as follows

$$\frac{\partial c}{\partial t} = D\left(\frac{\partial^2 c}{\partial x^2}\right) \tag{5-6}$$

The solution to this equation depends on the boundary conditions for a particular situation. One solution is

$$\frac{c_s - c_x}{c_s - c_0} = \text{erf}\left(\frac{x}{2\sqrt{Dt}}\right) \tag{5-7}$$

where c_s is a constant concentration of the diffusing atoms at the surface of the material, c_0 is the initial uniform concentration of the diffusing atoms in the material, and c_x is the concentration of the diffusing atom at location x below the surface after time t. These concentrations are illustrated in Figure 5-14. In these equations we have assumed a one-dimensional model (i.e., we assume that atoms or other diffusing species are moving only in the direction x). The function "erf" is the error function and can be evaluated from Table 5-3 or Figure 5-15.

The mathematical definition of the error function is as follows:

$$\text{erf}(x) = \frac{2}{\sqrt{\pi}}\int_0^x \exp(-y^2)\,dy \tag{5-8}$$

In Equation 5-8, y is known as the argument of the error function. We also define the complementary error function as follows:

$$\text{erfc}(x) = 1 - \text{erf}(x) \tag{5-9}$$

This function is used in certain solution forms of Fick's second law.

As mentioned previously, depending upon the boundary conditions, different equations describe the solutions to Fick's second law. These solutions to Fick's second law permit us to calculate the concentration of one diffusing species as a function of time (t) and location (x). Equation 5-7 is *a possible solution* to Fick's law that describes

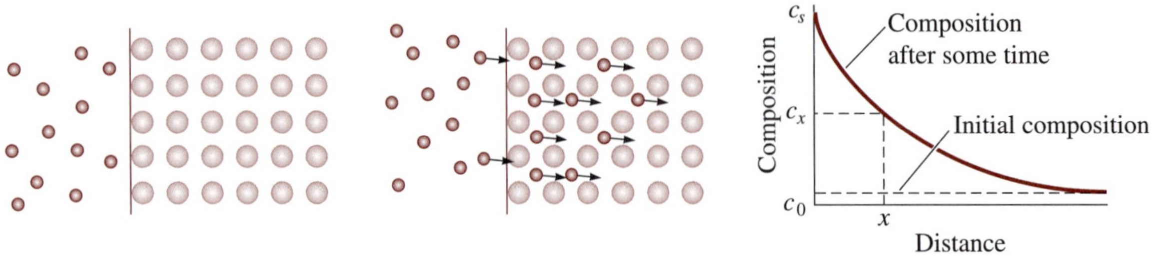

Figure 5-14 Diffusion of atoms into the surface of a material illustrating the use of Fick's second law.

TABLE 5-3 ■ ***Error function values for Fick's second law***

Argument of the error function $\frac{x}{2\sqrt{Dt}}$	Value of the error function $\text{erf}\frac{x}{2\sqrt{Dt}}$
0	0
0.10	0.1125
0.20	0.2227
0.30	0.3286
0.40	0.4284
0.50	0.5205
0.60	0.6039
0.70	0.6778
0.80	0.7421
0.90	0.7970
1.00	0.8427
1.50	0.9661
2.00	0.9953

Note that error function values are available on many software packages found on personal computers.

the variation in concentration of different species near the surface of the material as a function of time and distance, provided that the diffusion coefficient D remains constant and the concentrations of the diffusing atoms at the surface (c_s) and at large distance (x) within the material (c_0) remain unchanged. Fick's second law can also assist us in designing a variety of materials processing techniques, including the steel carburizing heat treatment and dopant diffusion in semiconductors as described in the following examples.

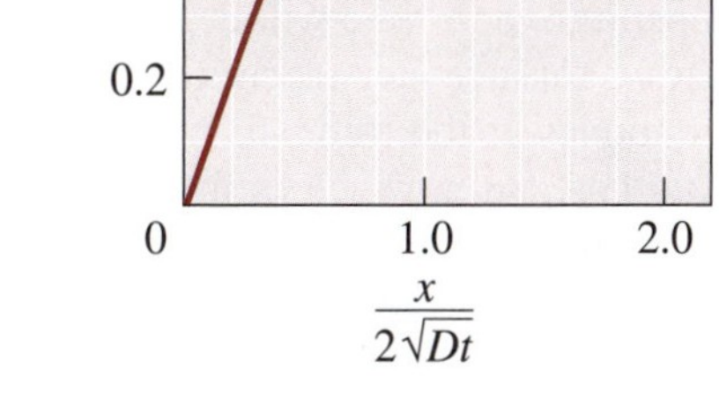

Figure 5-15
Graph showing the argument (on the *x*-axis) and error function value (on the *y*-axis) encountered in Fick's second law.

EXAMPLE 5-7 Design of a Carburizing Treatment

The surface of a 0.1% C steel gear is to be hardened by carburizing. In gas carburizing, the steel gears are placed in an atmosphere that provides 1.2% C at the surface of the steel at a high temperature (Figure 5-1). Carbon then diffuses from the surface into the steel. For optimum properties, the steel must contain 0.45% C at a depth of 0.2 cm below the surface. Design a carburizing heat treatment that will produce these optimum properties. Assume that the temperature is high enough (at least 900°C) so that the iron has the FCC structure.

SOLUTION

Since the boundary conditions for which Equation 5-7 was derived are assumed to be valid we can use this equation.

$$\frac{c_s - c_x}{c_s - c_0} = \text{erf}\left(\frac{x}{2\sqrt{Dt}}\right)$$

We know that $c_s = 1.2\%$ C, $c_0 = 0.1\%$ C, $c_x = 0.45\%$ C, and $x = 0.2$ cm. From Fick's second law,

$$\frac{c_s - c_x}{c_s - c_0} = \frac{1.2\%\text{C} - 0.45\%\text{C}}{1.2\%\text{C} - 0.1\%\text{C}} = 0.68 = \text{erf}\left(\frac{0.2 \text{ cm}}{2\sqrt{Dt}}\right) = \text{erf}\left(\frac{0.1 \text{ cm}}{\sqrt{Dt}}\right)$$

From Table 5-3, we find that:

$$\frac{0.1 \text{ cm}}{\sqrt{Dt}} = 0.71 \quad \text{or} \quad Dt = \left(\frac{0.1}{0.71}\right)^2 = 0.0198 \text{ cm}^2$$

Any combination of D and t whose product is 0.0198 cm^2 will work. For carbon diffusing in FCC iron, the diffusion coefficient is related to temperature by Equation 5-4

$$D = D_0 \exp\left(\frac{-Q}{RT}\right)$$

From Table 5-1,

$$D = 0.23 \exp\left(\frac{-32{,}900 \text{ cal/mol}}{1.987 \frac{\text{cal}}{\text{mol-K}} T(\text{K})}\right) = 0.23 \exp\left(\frac{-16{,}558}{T}\right)$$

Therefore, the temperature and time of the heat treatment are related by

$$t = \frac{0.0198 \text{ cm}^2}{D \frac{\text{cm}^2}{\text{s}}} = \frac{0.0198 \text{ cm}^2}{0.23 \exp(-16{,}558/T) \frac{\text{cm}^2}{\text{s}}} = \frac{0.0861}{\exp(-16{,}558/T)}$$

Some typical combinations of temperatures and times are:

If $T = 900°\text{C} = 1173$ K, then $t = 116{,}174$ s $= 32.3$ h

If $T = 1000°\text{C} = 1273$ K, then $t = 36{,}360$ s $= 10.7$ h

If $T = 1100°\text{C} = 1373$ K, then $t = 14{,}880$ s $= 4.13$ h

If $T = 1200°\text{C} = 1473$ K, then $t = 6{,}560$ s $= 1.82$ h

The exact combination of temperature and time will depend on the maximum temperature that the heat treating furnace can reach, the rate at which parts must be produced, and the economics of the tradeoffs between higher temperatures versus longer times. Another factor which is important is to consider changes in microstructure that occur in the rest of the material. For example, while carbon is diffusing into the surface, the rest of the microstructure can begin to experience grain growth or other changes.

Example 5-8 shows that one of the consequences of Fick's second law is that the same concentration profile can be obtained for different processing conditions, so long as the term Dt is constant. This permits us to determine the effect of temperature on the time required for a particular heat treatment to be accomplished.

EXAMPLE 5-8 Design of a More Economical Heat Treatment

We find that 10 h are required to successfully carburize a batch of 500 steel gears at 900°C, where the iron has the FCC structure. We find that it costs $1000 per hour to operate the carburizing furnace at 900°C and $1500 per hour to operate the furnace at 1000°C. Is it economical to increase the carburizing temperature to 1000°C? What other factors must be considered?

SOLUTION

Again, assuming we can use the solution to Ficks's second law given by Equation 5-7

$$\frac{c_s - c_x}{c_s - c_0} = \text{erf}\left(\frac{x}{2\sqrt{Dt}}\right)$$

Note that, since we are dealing with only changes in heat treatment time and temperature, the term Dt must be constant.

To achieve the same carburizing treatment at 1000°C as at 900°C:

$$D_{1273}t_{1273} = D_{1173}t_{1173}$$

The temperatures of interest are 900°C = 1173 K and 1000°C = 1273 K. For carbon diffusing in FCC iron, the activation energy is 32,900 cal/mol. Since we are dealing with the ratios of times, it does not matter whether we substitute for the time in hours or seconds. It is, however, always a good idea to use units that balance out. Therefore, we will show time in seconds. Note that temperatures must be converted into Kelvin.

$$D_{1273}t_{1273} = D_{1173}t_{1173}$$

$$D = D_0 \exp(-Q/RT)$$

$$t_{1273} = \frac{D_{1173}t_{1173}}{D_{1273}}$$

$$= \frac{D_0 \exp\left[-\dfrac{32{,}900\,\frac{\text{cal}}{\text{mol}}}{1.987\,\frac{\text{cal}}{\text{K}\cdot\text{mol}}\,1173\text{ K}}\right](10\text{ hours})(3600\text{ sec/hour})}{D_0 \exp\left[-\dfrac{32{,}900\,\frac{\text{cal}}{\text{mol}}}{1.987\,\frac{\text{cal}}{\text{K}\cdot\text{mol}}\,1273\text{ K}}\right]}$$

$$t_{1273} = \frac{\exp(-14.11562)(10)(3600)}{\exp(-13.00677)} = (10)\exp(-1.10885)$$

$$= (10)(0.3299)(3600)\text{ s}$$

$$t_{1273} = 3.299\text{ h} = 3\text{ h and }18\text{ min}$$

Notice, we really did not need the value of the pre-exponential term D_0, since it canceled out.
At 900°C, the cost per part is ($1000/h) (10 h)/500 parts = $20/part
At 1000°C, the cost per part is ($1500/h) (3.299 h)/500 parts = $9.90/part

Considering only the cost of operating the furnace, increasing the temperature reduces the heat-treating cost of the gears and increases the production rate. Another factor to consider is if the heat treatment at 1000°C could cause some other microstructural or other changes? For example, would increased temperature cause grains to grow significantly? If this is the case, we will be weakening the bulk of the material. How does the increased temperature affect the life of the other equipment such as the furnace itself and any accessories? How long would the cooling take? Will cooling from a higher temperature cause residual stresses? Would the product still meet all other specifications? These and other questions should be considered. The point is, as engineers, we need to ensure that the solution we propose is not only technically sound and economically sensible, it should recognize and make sense for the system as a whole (i.e., bigger picture). A good solution is often simple, solves problems for the system, and does not create new problems.

5-9 Diffusion and Materials Processing

We briefly discussed applications of diffusion in processing materials in Section 5-1. Many important examples related to solidification, phase transformations, heat treatments, etc., will be discussed in later chapters. In this section, we provide more information to highlight the importance of diffusion in the processing of engineered materials. Diffusional processes become very important when materials are used or processed at elevated temperatures.

Melting and Casting One of the most widely used methods to process metals, alloys, many plastics, and glasses involves melting and casting of materials into a desired shape. Diffusion plays a particularly important role in solidification of metals and alloys. In inorganic glasses, we rely on the fact that diffusion is slow and inorganic glasses do not crystallize easily.

Sintering Although casting and melting methods are very popular for many manufactured materials, the melting points of many ceramic and some metallic materials are too high for processing by melting and casting. These relatively refractory materials are manufactured into useful shapes by a process that requires the consolidation of small particles of a powder into a solid mass. **Sintering** is the high-temperature treatment that causes particles to join, gradually reducing the volume of pore space between them. When conducted properly, sintering will lead to densification of a powder compact.

Grain Growth A polycrystalline material contains a large number of grain boundaries, which represent a high-energy area because of the inefficient packing of the atoms. A lower overall energy is obtained in the material if the amount of grain boundary area is reduced by grain growth. **Grain growth** involves the movement of grain boundaries, permitting larger grains to grow at the expense of smaller grains.

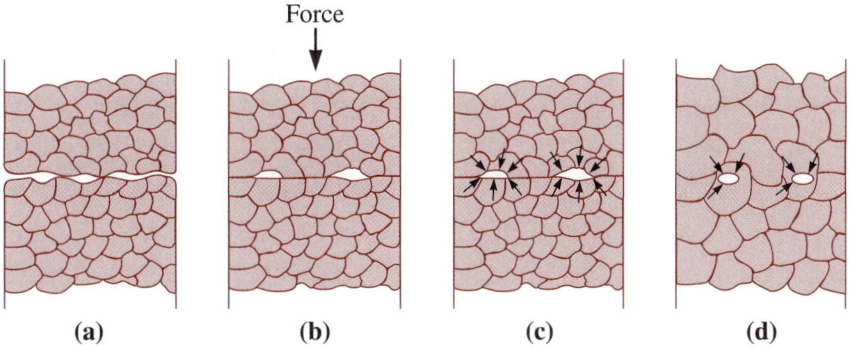

Figure 5-16 The steps in diffusion bonding: (a) Initially the contact area is small; (b) application of pressure deforms the surface, increasing the bonded area; (c) grain boundary diffusion permits voids to shrink; and (d) final elimination of the voids requires volume diffusion.

Diffusion Bonding A method used to join materials, **diffusion bonding** occurs in three steps (Figure 5-16). The first step forces the two surfaces together at a high temperature and pressure, flattening the surface, fragmenting impurities, and producing a high atom-to-atom contact area. As the surfaces remain pressed together at high temperatures, atoms diffuse along grain boundaries to the remaining voids; the atoms condense and reduce the size of any voids in the interface. Because grain boundary diffusion is rapid, this second step may occur very quickly. Eventually, however, grain growth isolates the remaining voids from the grain boundaries. For the third step—final elimination of the voids—volume diffusion, which is comparatively slow, must occur. The diffusion bonding process is often used for joining reactive metals such as titanium, for joining dissimilar metals and materials, and for joining ceramics.

SUMMARY

- The net flux of atoms, ions, etc. resulting from diffusion depends upon the initial concentration gradient.
- The kinetics of diffusion depend strongly on temperature. In general, diffusion is a thermally activated process and the dependence of the diffusion coefficient on temperature is given by the Arrhenius equation.
- The extent of diffusion depends on temperature, time, nature and concentration of diffusing species, crystal structure, and composition of the matrix, stoichiometry, and point defects.
- Encouraging or limiting the diffusion process forms the underpinning of many important technologies. Examples include the processing of semiconductors, heat treatments of metallic materials, sintering of ceramics and powdered metals, formation of amorphous materials, solidification of molten materials during a casting process, diffusion bonding, barrier plastics, films, and coatings.
- Fick's laws describe the diffusion process quantitatively. Fick's first law defines the relationship between chemical potential gradient and the flux of diffusing species. Fick's second law describes the variation of concentration of diffusing species under nonsteady state diffusion conditions.

- The activation energy Q describes the ease with which atoms diffuse, with rapid diffusion occurring for a low activation energy. A low activation energy and rapid diffusion rate are obtained for (1) interstitial diffusion compared with vacancy diffusion, (2) crystal structures with a smaller packing factor, (3) materials with a low melting temperature or weak atomic bonding, and (4) diffusion along grain boundaries or surfaces.

GLOSSARY

Activation energy The energy required to cause a particular reaction to occur. In diffusion, the activation energy is related to the energy required to move an atom from one lattice site to another.

Carburization A heat treatment for steels to harden the surface using a gaseous or solid source of carbon. The carbon diffusing into the surface makes the surface harder and more abrasion resistant.

Concentration gradient The rate of change of composition with distance in a nonuniform material, typically expressed as $\frac{\text{atoms}}{\text{cm}^3 \cdot \text{cm}}$ or $\frac{\text{at}\%}{\text{cm}}$.

Diffusion The net flux of atoms, ions, or other species within a material induced by concentration gradient and accelerated by increased temperature.

Diffusion bonding A joining technique in which two surfaces are pressed together at high pressures and temperatures. Diffusion of atoms to the interface fills in voids and produces a strong bond between the surfaces.

Diffusion coefficient (D) A temperature-dependent coefficient related to the rate at which atoms, ions, or other species diffuse. The diffusion coefficient depends on temperature, the composition and microstructure of the host material and also concentration of diffusing species.

Diffusion couple A combination of elements involved in diffusion studies (e.g., if we are considering diffusion of Al in Si, then Al-Si is a diffusion couple).

Diffusion distance The maximum or desired distance that atoms must diffuse; often, the distance between the locations of the maximum and minimum concentrations of the diffusing atom.

Diffusivity Another term for diffusion coefficient (D).

Drift Movement of electrons, holes, ions, or particles as a result of a gradient in temperature, electric, or magnetic field.

Driving force A cause that induces an effect. For example, an increased gradient in chemical potential enhances diffusion, similarly reduction in surface area of powder particles is the driving force for sintering.

Fick's first law The equation relating the flux of atoms by diffusion to the diffusion coefficient and the concentration gradient.

Fick's second law The partial differential equation that describes the rate at which atoms are redistributed in a material by diffusion. Many solutions exist to Fick's second law; Equation 5-7 is one possible solution.

Flux The number of atoms or other diffusing species passing through a plane of unit area per unit time. This is related to the rate at which mass is transported by diffusion in a solid.

Grain boundary diffusion Diffusion of atoms along grain boundaries. This is faster than volume diffusion, because the atoms are less closely packed in grain boundaries.

Grain growth Movement of grain boundaries by diffusion in order to reduce the amount of grain boundary area. As a result, small grains shrink and disappear and other grains become larger, similar to how some bubbles in soap froth become larger at the expense of smaller bubbles. In many situations, grain growth is not desirable.

Interstitial diffusion Diffusion of small atoms from one interstitial position to another in the crystal structure.

Permeability A relative measure of the diffusion rate in materials, often applied to plastics and coatings. It is often used as an engineering design parameter that describes the effectiveness of a particular material to serve as a barrier against diffusion.

Self-diffusion The random movement of atoms within an essentially pure material. No net change in composition results.

Sintering A high-temperature treatment used to join small particles. Diffusion of atoms to points of contact causes bridges to form between the particles. Further diffusion eventually fills in any remaining voids. The driving force for sintering is a reduction in total surface area of the powder particles.

Surface diffusion Diffusion of atoms along surfaces.

Vacancy diffusion Diffusion of atoms when an atom leaves a regular lattice position to fill a vacancy in the crystal. This process creates a new vacancy and the process continues.

Volume diffusion Diffusion of atoms through the interior of grains.

PROBLEMS

Section 5-1 Applications of Diffusion

5-1 What is the driving force for diffusion?

5-2 In the carburization treatment of steels, what are the diffusing species?

5-3 Why do we use PET plastic to make carbonated beverage bottles?

Section 5-2 Stability of Atoms and Ions

5-4 Atoms are found to move from one lattice position to another at the rate of $5 \times 10^5 \frac{\text{jumps}}{\text{s}}$ at 400°C when the activation energy for their movement is 30,000 cal/mol. Calculate the jump rate at 750°C.

5-5 The number of vacancies in a material is related to temperature by an Arrhenius equation. If the fraction of lattice points containing vacancies is 8×10^{-5} at 600°C, determine the fraction of lattice points at 1000°C.

Section 5-3 Mechanisms for Diffusion

Section 5-4 Activation Energy for Diffusion

5-6 The diffusion coefficient for Cr^{+3} in Cr_2O_3 is 6×10^{-15} cm^2/s at 727°C and is 1×10^{-9} cm^2/s at 1400°C. Calculate

(a) the activation energy; and
(b) the constant D_0.

5-7 The diffusion coefficient for O^{-2} in Cr_2O_3 is 4×10^{-15} cm^2/s at 1150°C, and 6×10^{-11} cm^2/s at 1715°C. Calculate

(a) the activation energy; and
(b) the constant D_0.

5-8 Without referring to the actual data, can you predict whether the activation energy for diffusion of carbon in FCC iron will be higher or lower than that in BCC iron? Explain.

Section 5-5 Rate of Diffusion (Fick's First Law)

5-9 Write down Fick's first law of diffusion. Clearly explain what each term means.

Section 5-6 Factors Affecting Diffusion

5-10 Write down the equation that describes the dependence of D on temperature.

5-11 Explain briefly the dependence of D on the concentration of diffusing species.

5-12 A 0.2-mm thick wafer of silicon is treated so that a uniform concentration gradient of antimony is produced. One surface contains 1 Sb atom per 10^8 Si atoms and the other surface contains 500 Sb atoms per 10^8 Si atoms. The lattice parameter for Si is 5.407 Å (Appendix A). Calculate the concentration gradient in

(a) atomic percent Sb per cm; and

(b) Sb $\frac{\text{atoms}}{\text{cm}^3 \cdot \text{cm}}$

5-13 When a Cu-Zn alloy solidifies, one portion of the structure contains 25 atomic percent zinc and another portion 0.025 mm away contains 20 atomic percent zinc. The lattice parameter for the FCC alloy is about 3.63×10^{-8} cm. Determine the concentration gradient in

(a) atomic percent Zn per cm;

(b) weight percent Zn per cm; and

(c) Zn $\frac{\text{atoms}}{\text{cm}^3 \cdot \text{cm}}$

5-14 A 0.0025-cm BCC iron foil is used to separate a high hydrogen gas from a low hydrogen gas at 650°C. 5×10^8 H atoms/cm^3 are in equilibrium on one side of the foil, and 2×10^3 H atoms/cm^3 are in equilibrium with the other side. Determine

(a) the concentration gradient of hydrogen; and

(b) the flux of hydrogen through the foil.

5-15 A 1-mm sheet of FCC iron is used to contain nitrogen in a heat exchanger at 1200°C. The concentration of N at one surface is 0.04 atomic percent and the concentration at the second surface is 0.005 atomic percent. Determine the flux of nitrogen through the foil in N atoms/cm$^2 \cdot$ s.

5-16 A 4-cm-diameter, 0.5-mm-thick spherical container made of BCC iron holds nitrogen at 700°C. The concentration at the inner surface is 0.05 atomic percent and at the outer surface is 0.002 atomic percent. Calculate the number of grams of nitrogen that are lost from the container per hour.

5-17 A BCC iron structure is to be manufactured that will allow no more than 50 g of hydrogen to be lost per year through each square centimeter of the iron at 400°C. If the concentration of hydrogen at one surface is 0.05 H atom per unit cell and is 0.001 H atom per unit cell at the second surface, determine the minimum thickness of the iron.

5-18 Determine the maximum allowable temperature that will produce a flux of less than 2000 H atoms/cm$^2 \cdot$ s through a BCC iron foil when the concentration gradient is $-5 \times 10^{16} \frac{\text{atoms}}{\text{cm}^3 \cdot \text{cm}}$. (Note the negative sign for the flux.)

5-19 As mentioned before in Example 5-4, the diffusion of yttrium ions in chromium oxide (Cr_2O_3) has been studied by Lesage and co-workers. In addition to the measurement of diffusion of yttrium ion in bulk chromia scale grown on a Ni-Cr alloy, these researchers also measured the diffusion of yttrium along the grain boundaries. These data are for grain-boundary diffusivities are shown here.

Temperature (°C)	Grain-Boundary Diffusion Coefficient (*D*) (cm^2/s)
800	1.2×10^{-13}
850	5.4×10^{-13}
900	6.7×10^{-13}
950	1.8×10^{-12}
1000	4.6×10^{-12}

(Source: J. Li, M.K. Loudjani, B. Lesage, A.M. Huntz, Philosophical Magazine A, *1997,* ***76****[4], pp. 857–69).*

(a) From these data, show that the activation energy for grain-boundary diffusion of yttrium in chromia oxide scale on nickel-chromium alloy is 190 kJ/cal.

(b) What is the value of the pre-exponential term D_0 in cm^2/s?

(c) What is the relationship between D and $1/T$ for the grain-boundary diffusivity in this temperature range?

(d) At any given temperature, the diffusivity of chromium along grain boundaries is several orders of magnitude higher than that for within the bulk (See Example 5-4). Is this to be expected? Explain.

5-20 Certain ceramic materials such as those based on oxides of yttrium, barium, and copper have been shown to be superconductors near liquid nitrogen temperature (~77 to 110 K). Since ceramics are brittle, it has been proposed to make long wires of these materials by encasing them in a silver tube. In this work, researchers investigated the diffusion of oxygen in a compound $YBa_2Cu_3O_7$. The data in the temperature range 500 to 650°C are shown below for undoped (i.e., silver free) samples.

Temperature (°C)	Diffusion Coefficient (*D*) (cm^2/s)
500	2.77×10^{-6}
600	5.2×10^{-6}
650	9.24×10^{-6}

(Source: D.K. Aswal, S.K. Gupta, P.K. Mishra, V.C. Sahni, Superconductor Science and Technology, *1998,* ***11****[7], pp. 631–6).*

Assume that these data are sufficient to make a straight line fit for the relationship between $\ln(D)$ and $1/T$ and calculate the values of the activation energy for diffusion of oxygen in $YBa_2Cu_3O_7$ containing no silver.

5-21 Diffusion of oxygen in $YBa_2Cu_3O_7$ doped with silver was measured. It was seen that the diffusion of oxygen was slowed down by silver doping, as shown in the data here.

Temperature (°C)	Diffusion Coefficient (D) (cm^2/s)
650	2.89×10^{-7}
700	8.03×10^{-7}
750	$3.07 \times 10^{2-6}$

(Source: D.K. Aswal, S.K. Gupta, P.K. Mishra, V.C. Sahni, Superconductor Science and Technology, *1998,* ***11****[7], pp. 631–6).*

Ideally, more data points would be better. However, assume that these data are sufficient to make a straight line fit for the relationship between $\ln(D)$ and $1/T$ and calculate the values of the activation energy for diffusion of oxygen in $YBa_2Cu_3O_7$ containing silver.

5-22 Zinc oxide (ZnO) ceramics are used in a variety of applications, such as surge-protection devices. The diffusion of oxygen in single crystals of ZnO was studied by Tomlins and co-workers. These data are shown in the table here.

Temperature (°C)	Diffusion Coefficient (D) (cm^2/s)
850	2.73×10^{-17}
925	8.20×10^{-17}
995	2.62×10^{-15}
1000	2.21×10^{-15}
1040	5.48×10^{-15}
1095	4.20×10^{-15}
1100	6.16×10^{-15}
1150	1.31×10^{-14}
1175	1.97×10^{-14}
1200	3.50×10^{-14}

(Source: G.W. Tomlins, J.L. Routbort, and T.O. Mason, Journal of the American Ceramic Society, *1998,* ***81****[4], pp. 869–76).*

Using these data, calculate the activation energy for the diffusion of oxygen in ZnO. What is the value of D_0 in cm^2/s?

Section 5-7 Permeability of Polymers

5-23 Amorphous PET is more permeable to CO_2 than PET that contains micro-crystallites. Explain why.

5-24 Explain why a rubber balloon filled with helium gas deflates over time.

Section 5-8 Composition Profile (Fick's Second Law)

5-25 Consider a 2-mm-thick silicon (Si) wafer to be doped using antimony (Sb). Assume that the dopant source (gas mixture of antimony chloride and other gases) provides a constant concentration of 10^{22} atoms/m^3. If we need a dopant profile such that the concentration of Sb at a depth of 1 micrometer is 5×10^{21} atoms/m^3. What will be the time for the diffusion heat treatment? Assume that the silicon wafer to begin with contains no impurities or dopants. Assume the activation energy for diffusion of Sb in silicon is 380 kJ/mole and D_0 for Sb diffusion in Si is 1.3×10^{-3} m^2/s.

5-26 Compare the diffusion coefficients of carbon in BCC and FCC iron at the allotropic transformation temperature of 912°C and explain the difference.

5-27 What is carburizing? Explain why this process is expected to cause an increase in the hardness of the surface of plain carbon steels?

5-28 A carburizing process is carried out on a 0.10% C steel by introducing 1.0% C at the surface at 980°C, where the iron is FCC. Calculate the carbon content at 0.01 cm, 0.05 cm, and 0.10 cm beneath the surface after 1 h.

5-29 Iron containing 0.05% C is heated to 912°C in an atmosphere that produces 1.20% C at the surface and is held for 24 h. Calculate the carbon content at 0.05 cm beneath the surface if

(a) the iron is BCC; and

(b) the iron is FCC. Explain the difference.

5-30 What temperature is required to obtain 0.50% C at a distance of 0.5 mm beneath the surface of a 0.20% C steel in 2 h, when 1.10% C is present at the surface? Assume that the iron is FCC.

5-31 A 0.15% C steel is to be carburized at 1100°C, giving 0.35% C at a distance of 1 mm beneath the surface. If the surface composition is maintained at 0.90% C, what time is required?

5-32 A 0.02% C steel is to be carburized at 1200°C in 4 h, with a point 0.6 mm beneath the surface reaching 0.45% C. Calculate the carbon content required at the surface of the steel.

5-33 A 1.2% C tool steel held at 1150°C is exposed to oxygen for 48 h. The carbon content at the steel surface is zero. To what depth will the steel be decarburized to less than 0.20% C?

5-34 A 0.80% C steel must operate at 950°C in an oxidizing environment where the carbon content at the steel surface is zero. Only the outermost 0.02 cm of the steel part can fall below 0.75% C. What is the maximum time that the steel part can operate?

5-35 A steel with BCC crystal structure containing 0.001% N is nitrided at 550°C for 5 h. If the nitrogen content at the steel surface is 0.08%, determine the nitrogen content at 0.25 mm from the surface.

5-36 What time is required to nitride a 0.002 N steel to obtain 0.12% N at a distance of 0.005 cm beneath the surface at 625°C? The nitrogen content at the surface is 0.15%.

5-37 We can successfully perform a carburizing heat treatment at 1200°C in 1 h. In an effort to reduce the cost of replacing the brick lining in our furnace, we propose to reduce the carburizing temperature to 950°C. What time will be required to give us a similar carburizing treatment?

Section 5-9 Diffusion and Materials Processing

5-38 Arrange the following materials in increasing order of self-diffusion coefficient: Ar gas, water, single crystal aluminum, and liquid aluminum at 700°C.

5-39 During freezing of a Cu-Zn alloy, we find that the composition is nonuniform. By heating the alloy to 600°C for 3 hours, diffusion of zinc helps to make the composition more uniform. What temperature would be required if we wished to perform this homogenization treatment in 30 minutes?

5-40 A ceramic part made of MgO is sintered successfully at 1700°C in 90 minutes. To minimize thermal stresses during the process, we plan to reduce the temperature to 1500°C. Which will limit the rate at which sintering can be done: diffusion of magnesium ions or diffusion of oxygen ions? What time will be required at the lower temperature?

5-41 A Cu-Zn alloy has an initial grain diameter of 0.01 mm. The alloy is then heated to various temperatures, permitting grain growth to occur. The times required for the grains to grow to a diameter of 0.30 mm are

Temperature (°C)	Time (minutes)
500	80,000
600	3,000
700	120
800	10
850	3

Determine the activation energy for grain growth. Does this correlate with the diffusion of zinc in copper? (*Hint*: Note that rate is the reciprocal of time.)

5-42 A sheet of gold is diffusion-bonded to a sheet of silver in 1 h at 700°C. At 500°C, 440 h are required to obtain the same degree of bonding, and at 300°C, bonding requires 1530 years. What is the activation energy for the diffusion bonding process? Does it appear that diffusion of gold or diffusion of silver controls the bonding rate? (*Hint*: Note that rate is the reciprocal of time.)

Design Problems

5-43 Design a spherical tank, with a wall thickness of 2 cm that will assure that no more than 50 kg of hydrogen will be lost per year. The tank, which will operate at 500°C, can be made of nickel, aluminum, copper, or iron. The diffusion coefficient of hydrogen and the cost per kg for each available material is listed here.

Diffusion Data

Material	D_0 (cm^2/s)	Q cal/mol	Cost ($/kg)
Nickel	0.0055	8,900	9.0
Aluminum	0.16	10,340	1.32
Copper	0.011	9,380	2.43
Iron (BCC)	0.0012	3,600	0.33

5-44 A steel gear initially containing 0.10% C is to be carburized so that the carbon content at a depth of 0.13 cm is 0.50% C. We can generate a carburizing gas at the surface that contains anywhere from 0.95% C to 1.15% C. Design an appropriate carburizing heat treatment.

5-45 When a valve casting containing copper and nickel solidifies under nonequilibrium conditions, we find that the composition of the alloy varies substantially over a distance of 0.005 cm. Usually we are able to eliminate this concentration difference by heating the alloy for 8 h at 1200°C; however, sometimes this treatment causes the alloy to begin to melt, destroying the part. Design a heat treatment that will permit elimination of the nonuniformity without danger of melting. Assume that the cost of operating the furnace per hour doubles for each 100°C increase in temperature.

6

Mechanical Properties: Fundamentals and Tensile, Hardness, and Impact Testing

Have You Ever Wondered?

- *Why Silly Putty® can be stretched a considerable amount when pulled slowly, but snaps when pulled fast?*
- *Why we can load the weight of a fire truck on four ceramic coffee cups, however, ceramic cups tend to break easily when we drop them on the floor?*
- *What materials related factors played an important role in the sinking of the* Titanic*?*
- *What factors played a major role in the 1986* Challenger *and the 2003* Columbia *space shuttle accidents?*
- *Why does Boeing's new Dreamliner airplane contain almost 50% composites?*

The mechanical properties of materials depend on their composition and microstructure. In Chapters 2, 3, and 4, we learned that a material's composition, nature of bonding, crystal structure, and defects such as dislocations, grain size, etc., have a profound influence on the strength and ductility of metallic materials. In this chapter, we will begin to evaluate other factors that affect the mechanical properties of materials, such as how lower temperatures can cause many metals and plastics to become brittle. Lower temperatures contributed to the brittleness of the plastic used for the O-rings, causing the 1986 *Challenger* accident. In 2003, the space shuttle *Columbia* was lost because of an impact of debris on the ceramic tiles and failure of carbon-carbon composites. Similarly, the special chemistry of the steel used on the *Titanic* and the stresses associated in the fabrication and embrittlement of this steel when subjected to lower temperatures have been identified as factors contributing to the failure of the ship's hull. Some researchers have shown that weaker rivets and design flaws also contributed to the failure.

The main goal of this chapter is to introduce the basic concepts associated with mechanical properties. We will learn basic terms such as hardness, stress, strain, elastic and plastic deformation, viscoelasticity, strain rate, etc. We will also review some of the basic testing procedures that engineers use to evaluate many of these properties. These concepts will be discussed using illustrations from real-world applications.

6-1 Technological Significance

With many of today's emerging technologies, the primary emphasis is on the mechanical properties of the materials used. For example, in aircraft manufacturing, aluminum alloys or carbon-reinforced composites used for aircraft components must be lightweight, strong, and able to withstand cyclic mechanical loading for a long and predictable period of time. The latest (2007) Dreamliner passenger aircraft designed by Boeing uses 50% composites and is 20% more fuel efficient. Steels used in the construction of structures such as buildings and bridges must have adequate strength so that these structures can be built without compromising safety. The plastics used for manufacturing pipes, valves, flooring, and the like also must have adequate mechanical strength. Materials such as pyrolytic graphite or cobalt chromium tungsten alloys, used for prosthetic heart valves, must not fail. Similarly, the performance of baseballs, cricket bats, tennis rackets, golf clubs, skis, and other sport equipment depends not only on the strength and weight of the materials used, but also on their ability to perform under an "impact" loading. The importance of mechanical properties is easy to appreciate in many of these "load-bearing" applications.

In many other applications, the mechanical properties of the material also play an important role. For example, an optical fiber must have a certain level of strength to withstand the stresses encountered in its application. A biocompatible titanium alloy used for a bone implant must have enough strength and toughness to survive in the human body for many years without failure. Coating on optical lenses must resist mechanical abrasion. An aluminum alloy or a glass-ceramic substrate used as a base for building magnetic hard drives must have sufficient mechanical strength so that it will not break or crack during operation that requires rotation at high speeds. Similarly, electronic packages used to house semiconductor chips and the thin-film structures created on the semiconductor chip must be able to withstand stresses encountered in various applications, as well as those encountered during the heating and cooling of electronic devices. The mechanical robustness of small devices prepared using micro-electro mechanical systems (*MEMS*) and nano-technology is also important. Float glass used in automotive and building applications must have sufficient strength and shatter resistance. Many components designed from plastics, metals, and ceramics must not only have adequate toughness and strength at room temperature but also at relatively high and low temperatures.

For load-bearing applications, engineered materials are selected by matching their mechanical properties to the design specifications and service conditions required of the component. The first step in the selection process requires an analysis of the material's application to determine its most important characteristics. Should it be strong, stiff, or ductile? Will it be subjected to an application involving high stress or sudden intense force, high stress at elevated temperature, cyclic stresses, corrosive or abrasive conditions? Once we know the required properties, we can make a preliminary selection of the appropriate material using various databases. We must, however, know how the properties listed in the handbook are obtained, know what the properties mean, and realize that the properties listed are obtained from idealized tests that may not apply exactly to real-life engineering applications. Materials with the same nominal chemical composition and other properties can show significantly different mechanical properties as dictated by microstructure. Furthermore, changes in temperature; the cyclical nature of stresses applied; the chemical changes due to oxidation, corrosion, or erosion; microstructural changes due to temperature; the effect of possible defects introduced during machining operations (e.g., grinding, welding, cutting, etc.); or other factors can also have a major effect on the mechanical behavior of materials (Chapter 7).

The mechanical properties of materials must also be understood so that we can process materials into useful shapes using materials processing techniques. **Materials processing**, such as the use of steels and plastics to fabricate car bodies, requires a detailed understanding of the mechanical properties of materials at different temperatures and conditions of loading, for example, the mechanical behavior of steels and plastics used to fabricate fuel-efficient cars and trucks.

In the sections that follow, we discuss mechanical properties of materials. We will define and discuss different terms that are used to describe the mechanical properties of engineered materials. Different tests used to determine mechanical properties of materials are discussed.

6-2 Terminology for Mechanical Properties

There are different types of forces or "stresses" that are encountered in dealing with mechanical properties of materials. In general, we define **stress** as force per unit area. Tensile, compressive, shear, and bending stresses are illustrated in Figure 6-1(a). **Strain** is defined as the change in length per unit length. Stress is typically expressed in Pa (Pascals). Strain has no dimensions and is often expressed as cm/cm.

When discussing stress and strain, it may be useful to think about stress as the *cause* and strain as the *effect*. Typically, tensile and shear stresses are designated by the symbols σ and τ, respectively. Tensile and shear strains are represented by the symbols ε and γ, respectively. Many load-bearing applications involve tensile or compressive stresses. Shear stresses are often encountered in the processing of materials using such techniques as polymer extrusion. Shear stresses are also found in structural applications. Note that even a simple tensile stress applied along one direction will cause a shear stress to components in other directions (similar to the situation discussed in Schmid's law, Chapter 4).

Elastic strain is defined as fully recoverable strain resulting from an applied stress. The strain is "elastic" if it develops instantaneously (i.e., the strain occurs as soon as the force is applied), remains as long as the stress is applied, and disappears as soon as the force is withdrawn. A material subjected to an elastic strain does not show any permanent deformation (i.e., it returns to its original shape after the force or stress is removed). Consider stretching a stiff metal spring by a small amount and letting go. If

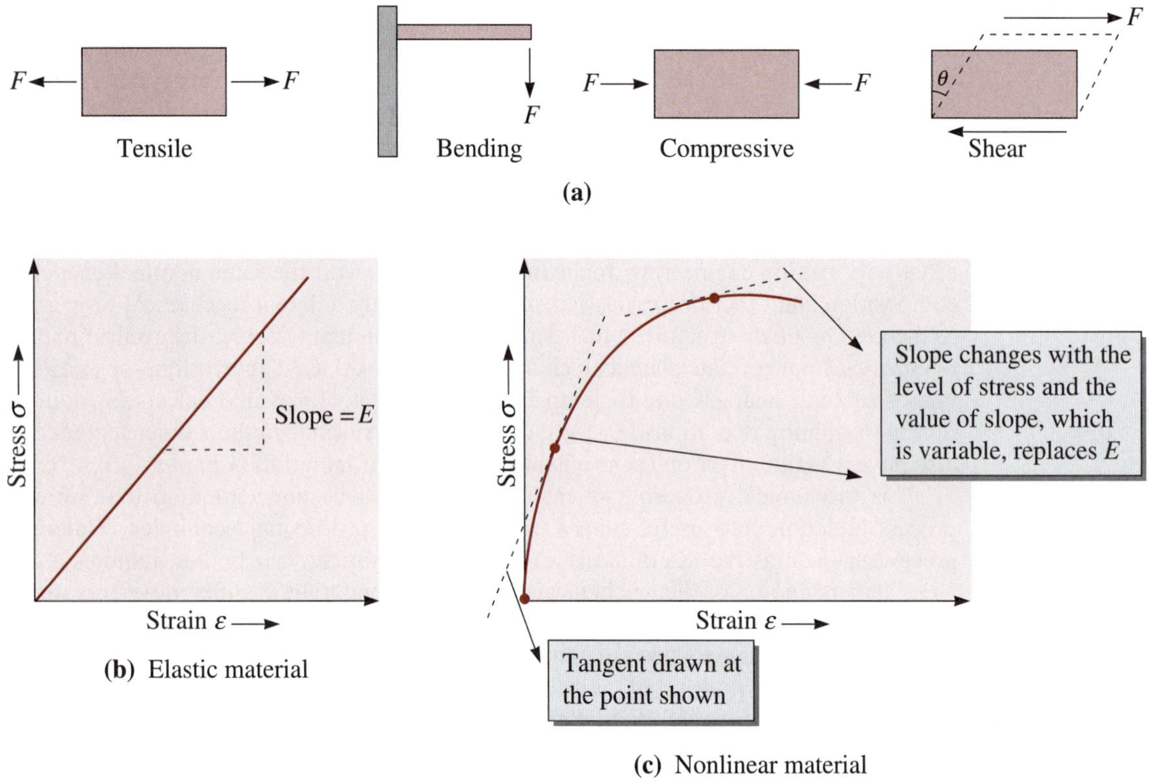

Figure 6-1 (a) Tensile, compressive, shear and bending stresses. (b) Illustration showing how Young's modulus is defined for an elastic material. (c) For nonlinear materials, we use the slope of a tangent as a variable quantity that replaces the Young's modulus constant.

the spring goes back quickly (within a few milliseconds or less) to its original dimensions, the strain developed in the spring was elastic.

In many materials, elastic stress and elastic strain are linearly related. The slope of a tensile stress-strain curve in the linear regime defines the **Young's modulus** or **modulus of elasticity** (E) of a material [Figure 6-1(b)]. The units of E are measured in Pascals (Pa) (same as those of stress). Large **elastic deformations** are observed in **elastomers** (e.g.,natural rubber, silicones), where the relationship between elastic strain and stress is non-linear. In elastomers, the large elastic strain is related to the coiling and uncoiling of spring-like molecules (Chapter 16). In dealing with such materials, we use the slope of the tangent at any given value of stress or strain and consider that as a variable quantity that replaces the Young's modulus [Figure 6-1(b)]. The inverse of Young's modulus is known as the **compliance** of the material. Similarly, we define **shear modulus** (G) as the slope of the linear part of the shear stress-shear strain curve.

Permanent or **plastic deformation** in a material is known as the **plastic strain**. In this case, when the stress is removed, the material does *not* go back to its original shape. A dent in a car is plastic deformation! Note that the word "plastic" here does not refer to strain in a plastic (polymeric) material, but rather to a type of strain in any material.

The rate at which strain develops in a material is defined as **strain rate** ($\dot{\varepsilon}$ or $\dot{\gamma}$ for tensile and shear strain rates, respectively). Units of strain rate are s^{-1}. You will learn later in this chapter that the rate at which a material is deformed is important from a mechanical properties perspective. Many materials considered to be ductile behave as brittle solids when the strain rates are high. Silly Putty® (a silicone polymer) is an example of such a material. When stretched slowly (smaller rate of strain), we can stretch this material by a large amount. However, when stretched rapidly (high strain rates), we do not allow the untangling and extension of the large polymer molecules and, hence, the material snaps. When the strain rates are low, Silly Putty® can show significant ductility. When materials are subjected to high strain rates we refer to this type of loading as **impact loading**.

A **viscous material** is one in which the strain develops over a period of time and the material does not go to its original shape after the stress is removed. The development of strain takes time and is not in phase with the applied stress. Also, the material will remain deformed when the applied stress is removed (i.e., the strain will be plastic). A **viscoelastic** (or **anelastic**) material can be thought of as a material whose response is between that of a viscous material and an elastic material. The term "anelastic" is typically used for metals, while the term "viscoelastic" is usually associated with polymeric materials. Many polymeric materials (solids and molten) are viscoelastic. A common example of a viscoelastic material is Silly Putty®.

In a viscoelastic material, the development of a permanent strain is similar to that in a viscous material. However, unlike a viscous material, when the applied stress is removed, part of the strain will recover over a period of time. Recovery of strain refers to a change in shape of a material after the stress causing deformation is removed. A qualitative description of development of strain as a function of time in relation to an applied force in elastic, viscous, and viscoelastic materials is shown in Figure 6-2. In viscoelastic materials held under constant strain, if we wait, the level of stress decreases over a period of time. This is known as **stress relaxation**. Recovery of strain and stress relaxation are different terms and should not be confused. A common example of stress relaxation is the nylon strings in a tennis racket. We know that the level of stress, or the "tension", as the tennis players call it, decreases with time.

While dealing with molten materials, liquids, and dispersions, such as paints or gels, a description of the resistance to flow under an applied stress is required. If the relationship between the applied stress (τ) and **shear strain rate** ($\dot{\gamma}$) is linear, we refer to that material as **Newtonian**. The slope of the shear stress versus the steady-state shear strain rate curve is defined as the **viscosity** (η) of the material. Water is an example of a Newtonian material. The following relationship defines viscosity:

$$\tau = \eta\dot{\gamma} \tag{6-1}$$

The units of η are Pa-s (in the SI system) or Poise (P) or $\dfrac{\text{g}}{\text{cm} - \text{s}}$ in the cgs system. Sometimes the term centipoise (cP) is used, 1 cP $= 10^{-2}$ P.

Conversion between these units is given by 1 Pa-s = 10 P = 1000 cP.

The **kinematic viscosity** (ν) is defined as:

$$\nu = \eta/\rho \tag{6-2}$$

where viscosity (η) is in Poise and density (ρ) is in g/cm^3. The kinematic viscosity unit is in Stokes (St). In this, St is cm^2/s. Sometimes the unit of centiStokes (cSt) is used, 1 cSt $= 10^{-2}$ St.

For many materials the relationship between shear stress and shear strain rate is nonlinear. These materials are **non-Newtonian**. The stress versus steady state shear strain rate relationship in these materials can be described as:

$$\tau = \eta\dot{\gamma}^m \tag{6-3}$$

where the exponent m is not equal to 1.

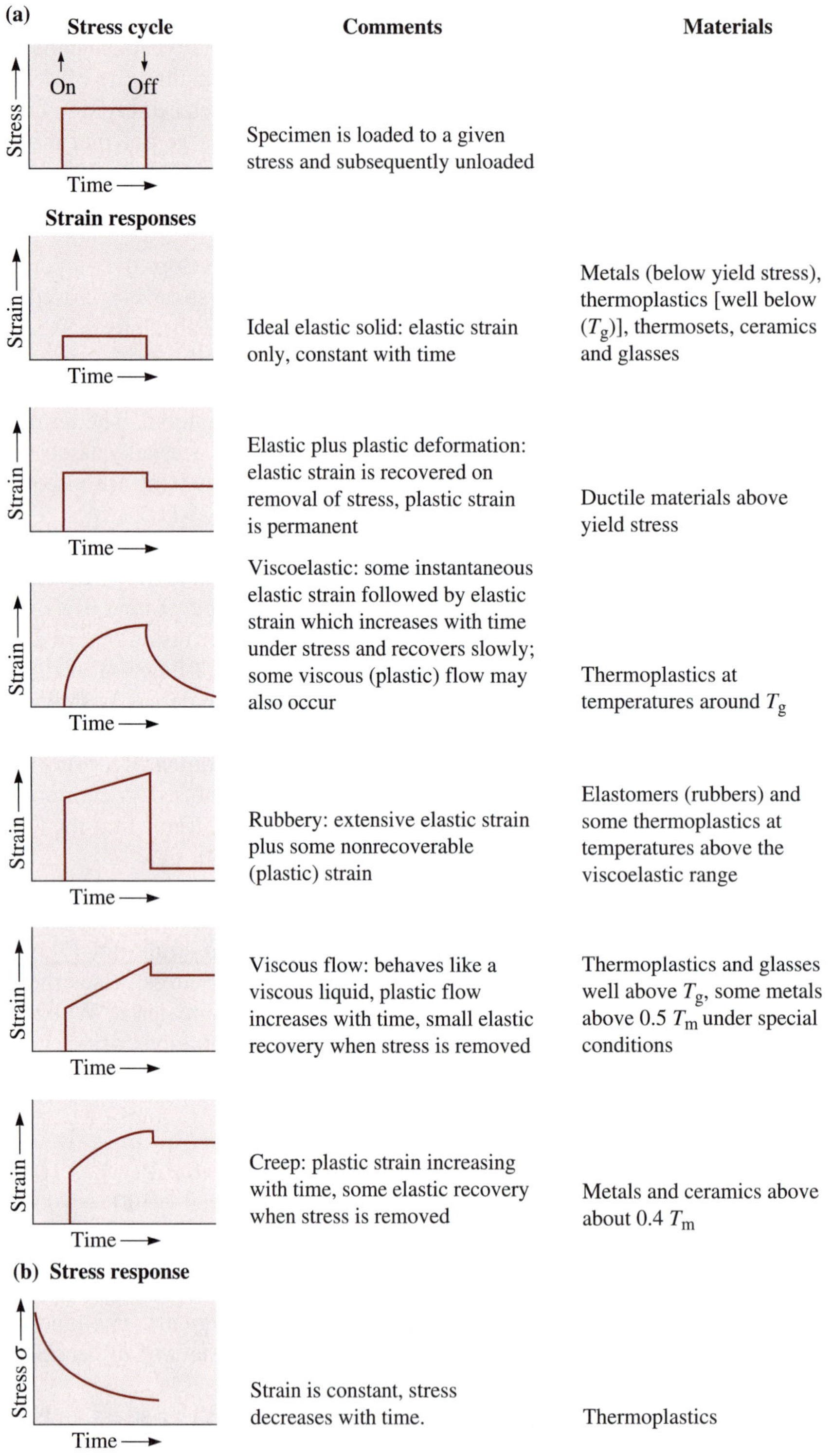

Figure 6-2 (a) Various types of strain response to an imposed stress. (*This article was published in* Materials: Principles and Practice, *C. Newey and G. Weaver (Eds.), Figure 6-9, p. 300. Copyright © Butterworth-Heinemann (1991).*) (b) Stress relaxation in a viscoelastic material. *Note the y-axis is stress.* Strain is constant.

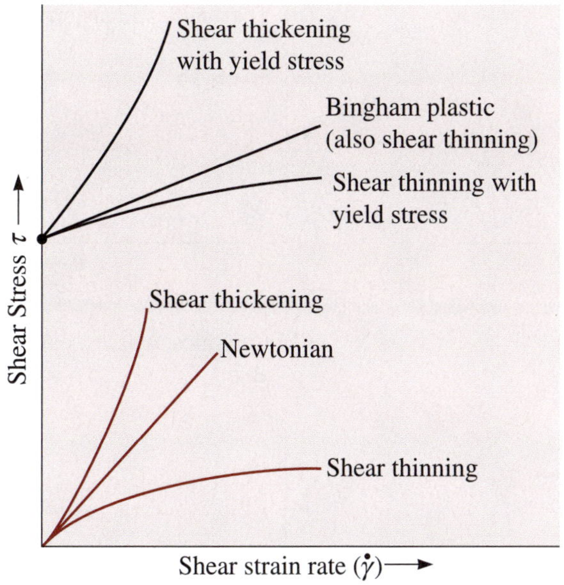

Figure 6-3
Shear stress-shear strain rate relationships for Newtonian and non-Newtonian materials.

Non-Newtonian materials are classified as **shear thinning** (or **pseudoplastic**) or **shear thickening** (or **dilatant**). The relationships between the shear stress and shear strain rate for different types of materials are shown in Figure 6-3.

In the sections that follow, we will discuss different mechanical properties of solid materials and some of their testing methods to evaluate these properties.

6-3 The Tensile Test: Use of the Stress-Strain Diagram

The **tensile test** is popular since the properties obtained could be applied to design different components. The tensile test measures the resistance of a material to a static or slowly applied force. The strain rates in a tensile test are very small ($\dot{\varepsilon} = 10^{-4}$ to $10^{-2} s^{-1}$). A test setup is shown in Figure 6-4; a typical specimen has a diameter of 1.263 cm and a

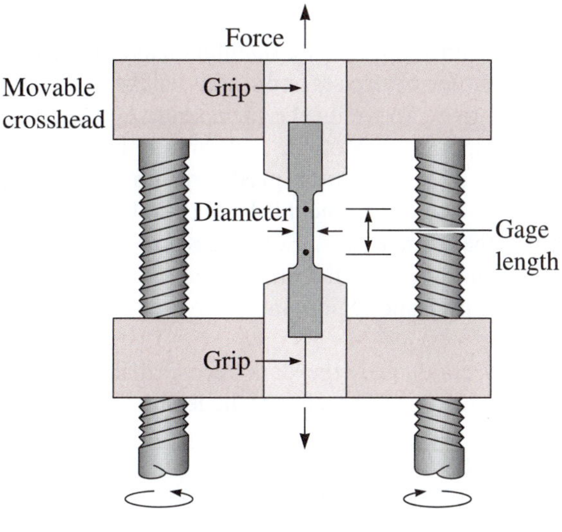

Figure 6-4
A unidirectional force is applied to a specimen in the tensile test by means of the moveable crosshead. The cross-head movement can be performed using screws or a hydraulic mechanism.

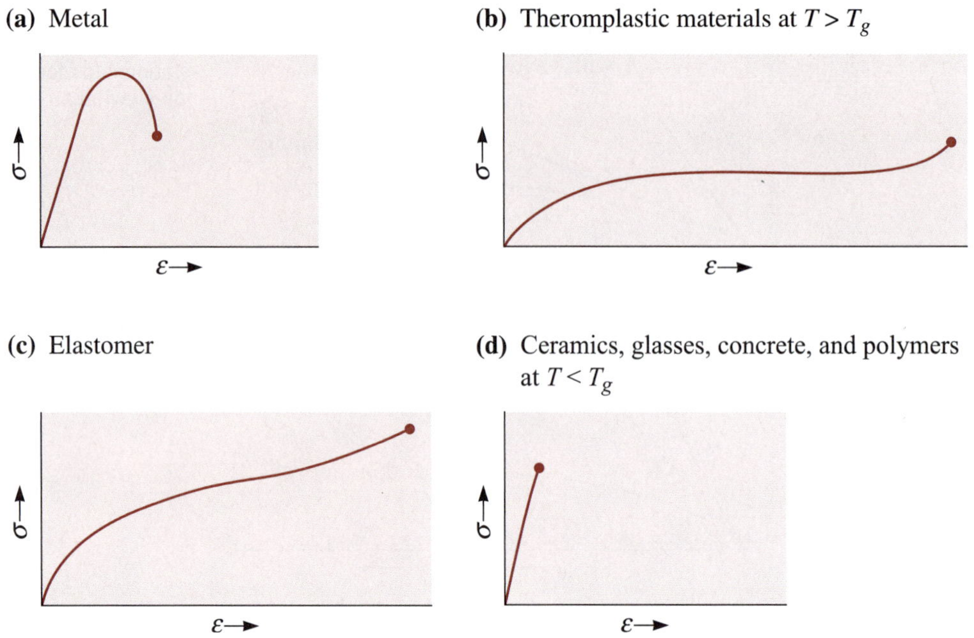

Figure 6-5 Tensile stress-strain curves for different materials. Note that these graphs are *qualitative.*

gage length of 5 cm. The specimen is placed in the testing machine and a force F, called the **load**, is applied. A **strain gage** or **extensometer** is used to measure the amount that the specimen stretches between the gage marks when the force is applied. Thus, what is measured is the change in length of the specimen (Δl) over a particular original length (l_0). Information concerning the strength, Young's modulus, and ductility of a material can be obtained from such a tensile test. Typically, a tensile test is conducted on metals, alloys, and plastics. Tensile tests can be used for ceramics, however, the test is not very useful for ceramics because the sample often easily fractures while it is being aligned. Civil engineers use a compression test to test materials such as concretes. The following discussion mainly applies to the tensile testing of metals and alloys. We will briefly discuss the stress-strain behavior of polymers as well.

Figure 6-5 shows *qualitatively* the stress-strain curves for a typical (a) metal, (b) thermoplastic material, (c) elastomer, and (d) ceramic (or glass), all under relatively small strain rates. The scales in this figure are qualitative and different for each material. In practice, the actual magnitude of stresses and strains will be very different. The thermoplastic material is assumed to be above its **glass temperature** (T_g). Metallic materials are assumed to be at room temperature. Metallic and thermoplastic materials show an initial elastic region followed by a non-linear plastic region. A separate curve for elastomers (e.g., rubber or silicones) is also included since the behavior of these materials is different from other polymeric materials. For elastomers, a large portion of the deformation is elastic and non-linear. On the other hand, ceramics, glasses, and polymers at $T < T_g$ show only a linear elastic region and almost no plastic deformation at room temperature.

When a tensile test is conducted, the data recorded includes load or force as a function of change in length (Δl). The change in length is typically measured using a strain gage. Table 6-1 shows the effect of the load on the changes in length of an aluminum alloy test bar. These data are then subsequently converted into stress and strain. The stress-strain curve is analyzed further to the extract properties of materials (e.g., Young's modulus, yield strength, etc.).

TABLE 6-1 ■ ***The results of a tensile test of a 1.263 cm diameter aluminum alloy test bar, initial length (l_0) = 5 cm***

		Calculated	
Load (N)	**Δl (cm)**	**Stress (MPa)**	**Strain (cm/cm)**
0	0.000	0	0
4450	0.0025	35.5	0.0005
13350	0.0075	106.5	0.0015
22240	0.0125	177.5	0.0025
31150	0.0175	248.6	0.0035
33360	0.075	266.2	0.0150
35140	0.2	280.4	0.0400
35580 (maximum load)	0.3	284	0.0600
35360	0.4	282.2	0.0800
33800 (fracture)	0.5125	269.8	0.1025

Engineering Stress and Strain The results of a single test apply to all sizes and cross-sections of specimens for a given material if we convert the force to stress and the distance between gage marks to strain. **Engineering stress** and **engineering strain** are defined by the following equations,

$$\text{Engineering stress} = \sigma = \frac{F}{A_0} \tag{6-4}$$

$$\text{Engineering strain} = \varepsilon = \frac{\Delta l}{l_0} \tag{6-5}$$

where A_0 is the *original* cross-sectional area of the specimen before the test begins, l_0 is the *original* distance between the gage marks, and Δl is the change in length after force F is applied. The conversions from load and sample length to stress and strain are included in Table 6-1. The stress-strain curve (Figure 6-6) is used to record the results of a tensile test.

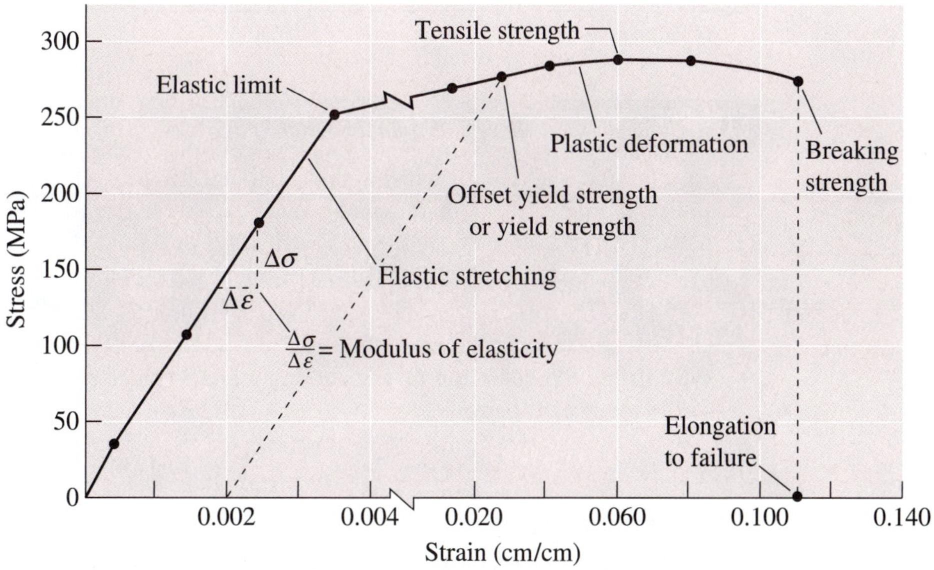

Figure 6-6 The stress-strain curve for an aluminum alloy from Table 6-1.

EXAMPLE 6-1 *Tensile Testing of Aluminum Alloy*

Convert the change in length data in Table 6-1 to engineering stress and strain and plot a stress-strain curve.

SOLUTION

For the 4450 N load:

$$\sigma = \frac{F}{A_0} = \frac{4450\text{ N}}{(\pi/4)(1.263\text{ cm})^2} = \frac{4450}{1.253\text{ cm}^2} = 35.5\text{ MPa}$$

$$\varepsilon = \frac{\Delta l}{l_0} = \frac{0.0025\text{ cm}}{5\text{ cm}} = 0.0005\text{ cm/cm}$$

The results of similar calculations for each of the remaining loads are given in Table 6-1 and are plotted in Figure 6-6.

Units Many different units are used to report the results of the tensile test. The most common units for stress are pounds per square inch (psi) and MegaPascals (MPa). The units for strain include inch/inch, centimeter/centimeter, and meter/meter. The conversion factors for stress are summarized in Table 6-2.

TABLE 6-2 ■ *Units and conversion factors*

1 pound (lb) = 4.448 Newtons (N)
1 psi = pounds per square inch
1 MPa = MegaPascal = MegaNewtons per square meter (MN/m^2)
= Newtons per square millimeter (N/mm^2) = 1,000,000 Pa
1 GPa = 1000 MPa = GigaPascal
1 ksi = 1000 psi = 6.895 MPa
1 psi = 0.006895 MPa
1 MPa = 0.145 ksi = 145 psi

EXAMPLE 6-2 *Design of a Suspension Rod*

An aluminum rod is to withstand an applied force of 200,200 N. To assure sufficient safety, the maximum allowable stress on the rod is limited to 172 MPa. The rod must be at least 375 cm long but must deform elastically no more than 0.625 cm when the force is applied. Design an appropriate rod.

SOLUTION

We can use the definition of engineering stress to calculate the required cross-sectional area of the rod:

$$A_0 = \frac{F}{\sigma} = \frac{200{,}200\text{ N}}{172\text{ MPa}} = 11.64\text{ cm}^2$$

The rod could be produced in various shapes, provided that the cross-sectional area is 11.64 cm^2. For a round cross-section, the minimum diameter to assure that the stress is not too high is:

$$A_0 = \frac{\pi d^2}{4} = 11.64 \text{ cm}^2 \quad \text{or} \quad d = 3.85 \text{ cm}$$

The maximum allowable elastic deformation is 0.625 cm. From the definition of engineering strain:

$$\varepsilon = \frac{\Delta l}{l_0} = \frac{0.625 \text{ cm}}{l_0}$$

From Figure 6-6, the strain expected for a stress of 172 MPa is 0.0024 cm/cm. If we use the cross-sectional area determined previously, the maximum length of the rod is:

$$0.0024 = \frac{\Delta l}{l_0} = \frac{0.625 \text{ cm}}{l_0} \quad \text{or} \quad l_0 = 260.42 \text{ cm}$$

However, the minimum length of the rod is specified as 375 cm. To produce a longer rod, we might make the cross-sectional area of the rod larger. The minimum strain allowed for the 375 cm rod is:

$$\varepsilon = \frac{\Delta l}{l_0} = \frac{0.625 \text{ cm}}{375 \text{ cm}} = 0.001667 \text{ cm/cm}$$

The stress, from Figure 6-6, is about 114.9 MPa, which is less than the maximum of 172 MPa. The minimum cross-sectional area then is:

$$A_0 = \frac{F}{\sigma} = \frac{200{,}200 \text{ N}}{114.9 \text{ MPa}} = 17.42 \text{ cm}^2$$

In order to satisfy both the maximum stress and the minimum elongation requirements, the cross-sectional area of the rod must be at least 17.42 cm^2, or a minimum diameter of 4.71 cm.

6-4 Properties Obtained from the Tensile Test

Yield Strength As we apply a low level of stress to a material, the material initially exhibits elastic deformation. In this region the strain that develops is completely and quickly recovered when the applied stress is removed. However, as we continue to increase the applied stress the material begins to exhibit both elastic and plastic deformation. The material eventually "yields" to the applied stress. The critical stress value needed to initiate plastic deformation is defined as the **elastic limit** of the material. In metallic materials, this is usually the stress required for dislocation motion, or slip to be initiated. In polymeric materials, this stress will correspond to disentanglement of polymer molecule chains or sliding of chains past each other. The **proportional limit** is defined as the level of stress above which the relationship between stress and strain is not linear.

In most materials the elastic limit and proportional limit are quite close. However, neither the elastic limit nor the proportional limit values can be determined precisely. Measured values depend on the sensitivity of the equipment used. We, therefore, define them at an **offset strain value** (typically, but not always, 0.002 or 0.2%). We then draw a

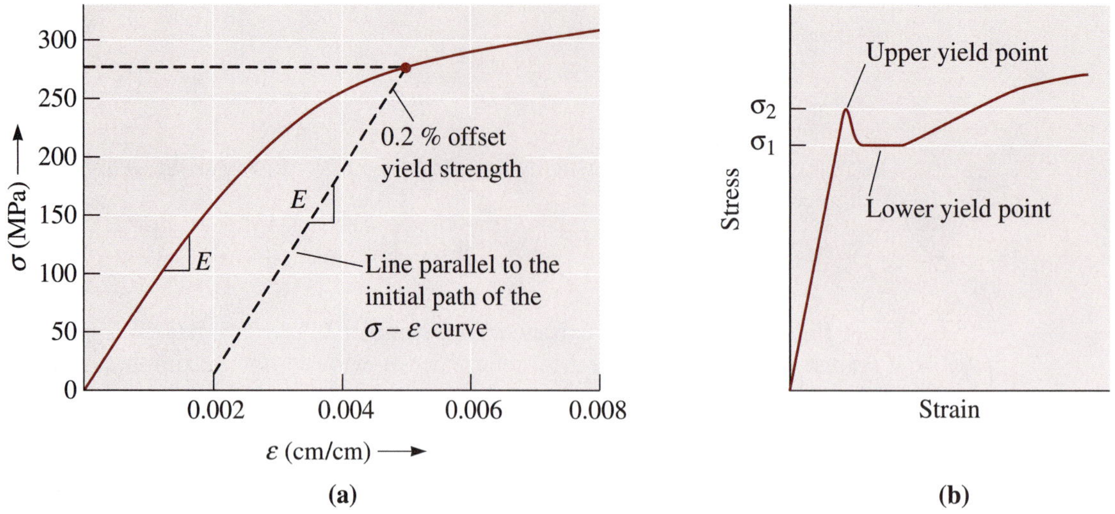

Figure 6-7 (a) Determining the 0.2% offset yield strength in gray cast iron, and (b) upper and lower yield point behavior in a low-carbon steel.

line starting with this offset value of strain and draw a line parallel to the linear portion of the engineering stress-strain curve. The stress value corresponding to the intersection of this line and the engineering stress-strain curve is defined as the **offset yield strength**, also often stated as the **yield strength**. The 0.2% offset yield strength for gray cast iron is 276 MPa as shown in Figure 6-7(a). Engineers normally prefer to use the offset yield strength for design purposes.

For some materials the transition from elastic deformation to plastic flow is rather abrupt. This transition is known as the **yield point phenomenon**. In these materials, as the plastic deformation begins the stress value drops first from the *upper yield point* (σ_2) [Figure 6-7(b)]. The stress value then decreases and oscillates around an average value defined as the *lower yield point* (σ_1). For these materials, the yield strength is usually defined from the 0.2% strain offset as shown in Figure 6-7(a).

The stress-strain curve for certain low-carbon steels displays a double yield point [Figure 6-7(b)]. The material is expected to plastically deform at stress σ_1. However, small interstitial atoms clustered around the dislocations interfere with slip and raise the yield point to σ_2. Only after we apply the higher stress σ_2 do the dislocations slip. After slip begins at σ_2, the dislocations move away from the clusters of small atoms and continue to move very rapidly at the lower stress σ_1.

When we design parts for load-bearing applications we prefer little or no plastic deformation. As a result we must select a material such that the design stress is considerably lower than the yield strength at the temperature at which the material will be used. We can also make the component cross-section larger so that the applied force produces a stress that is well below the yield strength. On the other hand, when we want to shape materials into components (e.g., take a sheet of steel and form a car chassis), we need to apply stresses that are well above the yield strength.

Tensile Strength The stress obtained at the highest applied force is the **tensile strength** (σ_{TS}), which is the maximum stress on the engineering stress-strain curve (Figure 6-6). In many ductile materials, deformation does not remain uniform. At some point, one region deforms more than others and a large, *local* decrease in the cross-sectional area occurs (Figure 6-8). This locally deformed region is called a "neck." This phenomenon

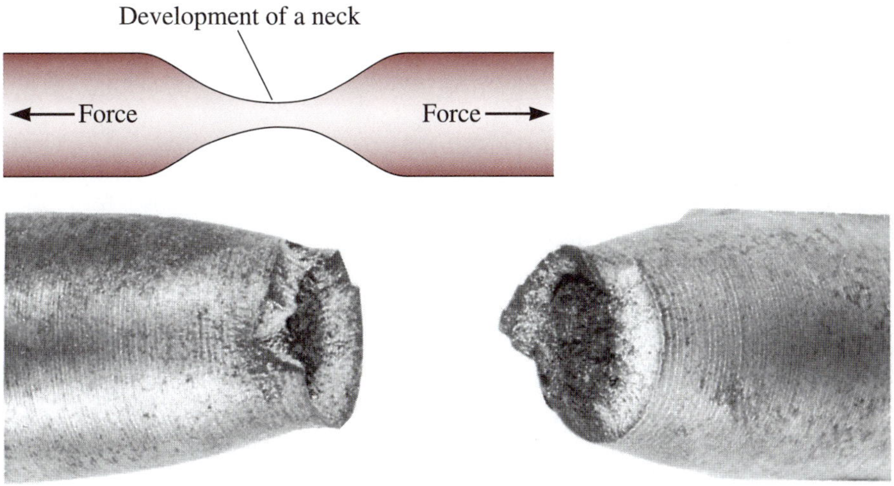

Figure 6-8 Localized deformation of a ductile material during a tensile test produces a necked region. The micrograph at the bottom shows necked region in a fractured sample.

is known as **necking**. Because the cross-sectional area becomes smaller at this point, a lower force is required to continue its deformation, and the engineering stress, calculated from the *original* area A_0, decreases. The tensile strength is the stress at which necking begins in ductile materials (Figure 6-6). With the reduced area, now less force is necessary for additional deformation. However, since engineering stress is based on A_0, the overall stress decreases after necking. Many ductile metals and polymers show the phenomenon of necking. In compression testing, the materials will bulge, thus necking is seen only in a tensile test.

Figure 6-9 shows typical yield strength values for different engineered materials. The yield strength of pure metals is smaller. For example, ultra-pure metals have a yield strength of ~(1–10 MPa). On the other hand, the yield strength of alloys is higher. Strengthening in alloys is achieved using different mechanisms described before (e.g., grain size, solid solution formation, strain hardening, etc.). The yield strength of plastics and elastomers is generally lower than metals and alloys, ranging up to about (10–100 MPa). The values for ceramics are for compressive strength (obtained using a hardness test). Tensile strength of most ceramics is much lower (~100–200 MPa). The tensile strength of glasses is about ~70 MPa and depends strongly on surface flaws.

Elastic Properties The modulus of elasticity, or *Young's modulus* (E), is the slope of the stress-strain curve in the elastic region. This relationship is **Hooke's Law**:

$$E = \frac{\sigma}{\varepsilon} \tag{6-6}$$

The modulus is closely related to the binding energies (Figure 2-21). A steep slope in the force-distance graph at the equilibrium spacing indicates that high forces are required to separate the atoms and cause the material to stretch elastically. Thus, the material has a high modulus of elasticity. Binding forces, and thus the modulus of elasticity, are typically higher for high melting-point materials (Table 6-3). In metallic materials, modulus of elasticity is considered a relative microstructure *insensitive* property since the value is dominated strongly by the strength of atomic bonds. Grain size or other microstructural features do not have a very large effect on the Young's modulus.

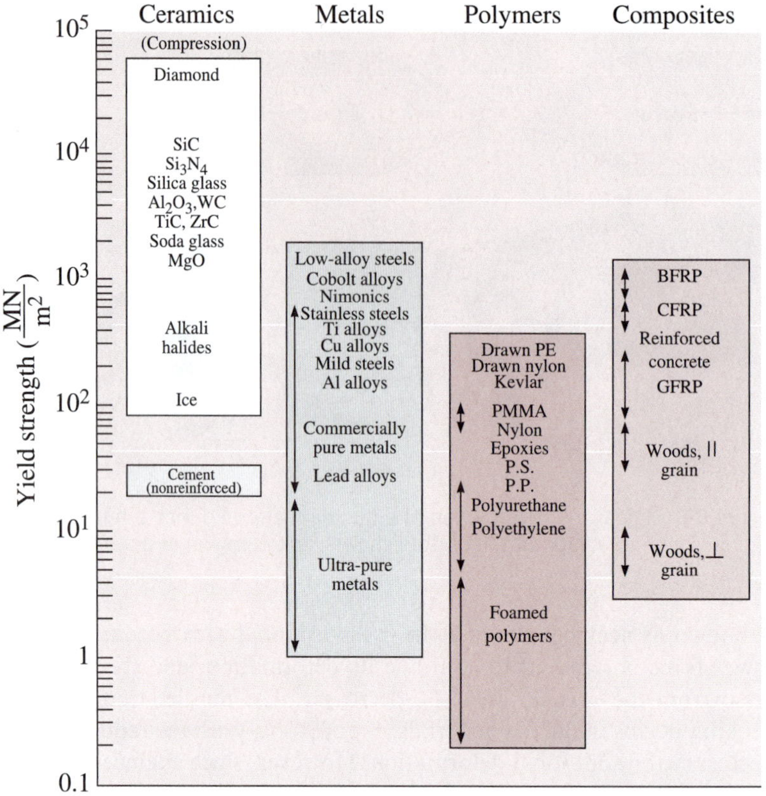

Figure 6-9 Typical yield strength values for different engineered materials. (*This article was published in* Engineering Materials I, *Second Edition, M.F. Ashby and D.R.H. Jones, Figure 8-12, p. 85. Copyright © Butterworth-Heinemann (1996).*)

Young's modulus is a measure of the stiffness of a component. A stiff component, with a high modulus of elasticity, will show much smaller changes in dimensions if the applied stress is relatively small and, therefore, causes only elastic deformation. Figure 6-10 compares the elastic behavior of steel and aluminum. If a stress of 207 MPa is applied to each material, the steel deforms elastically 0.001 cm/cm; at the same stress, aluminum deforms 0.003 cm/cm. In general, most engineers view **stiffness** as a function of both the Young's modulus and the geometry of a component.

TABLE 6-3 ■ ***Elastic properties and melting temperature (T_m) of selected materials***

Material	T_m (°C)	E (MPa)	Poisson's ratio (μ)
Pb	327	13.8×10^3	0.45
Mg	650	44.8×10^3	0.29
Al	660	69.0×10^3	0.33
Cu	1085	124.8×10^3	0.36
Fe	1538	206.9×10^3	0.27
W	3410	408.2×10^3	0.28
Al_2O_3	2020	379.2×10^3	0.26
Si_3N_4		303.4×10^3	0.24

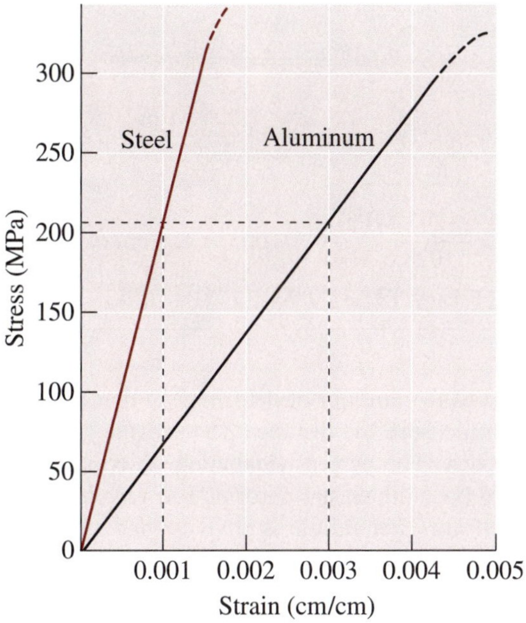

Figure 6-10
Comparison of the elastic behavior of steel and aluminum. For a given stress, aluminum deforms elastically three times as much as does steel.

Poisson's ratio, μ, relates the longitudinal elastic deformation produced by a simple tensile or compressive stress to the lateral deformation that occurs simultaneously:

$$\mu = \frac{-\varepsilon_{\text{lateral}}}{\varepsilon_{\text{longitudinal}}} \tag{6-7}$$

For many metals in the elastic region the Poisson's ratio is typically about 0.3 (Table 6-3).

The **modulus of resilience** (E_r), the area contained under the elastic portion of a stress-strain curve, is the elastic energy that a material absorbs during loading and subsequently releases when the load is removed. For linear elastic behavior:

$$E_r = \left(\tfrac{1}{2}\right)(\text{yield strength})(\text{strain at yielding}) \tag{6-8}$$

The ability of a spring or a golf ball to perform satisfactorily depends on a high modulus of resilience.

Tensile Toughness The energy absorbed by a material prior to fracturing is known as **tensile toughness** and is sometimes measured as the area under the true stress-strain curve (also known as **work of fracture**). We will define true stress and true strain in Section 6-5. Since it is easier to measure engineering stress-strain, engineers often equate tensile toughness to the area under the engineering stress-strain curve.

EXAMPLE 6-3 ***Young's Modulus of Aluminum Alloy***

From the data in Example 6-1, calculate the modulus of elasticity of the aluminum alloy. Use the modulus to determine the length after deformation of a bar of initial length of 125 cm. Assume that a level of stress of 207 MPa is applied.

SOLUTION

When a stress of 248.6 MPa is applied, a strain of 0.0035 is produced (Table 6-1). Thus:

$$\text{Modulus of elasticity} = E = \frac{\sigma}{\varepsilon} = \frac{248.6 \text{ MPa}}{0.0035} = 71028.6 \text{ MPa}$$

From Hooke's law:

$$\varepsilon = \frac{\sigma}{E} = \frac{207 \text{ MPa}}{71028.6 \text{ MPa}} = 0.003 = \text{cm/cm} = \frac{l - l_0}{l_0}$$

$$l = l_0 + \varepsilon l_0 = 125 + (0.003)(125) = 125.375 \text{ cm}$$

Ductility **Ductility** measures the amount of deformation that a material can withstand without breaking. We can measure the distance between the gage marks on our specimen before and after the test. The **percent elongation** describes the permanent plastic deformation before failure (i.e., the elastic deformation recovered after fracture is not included). Note that the strain after failure is smaller than the strain at the breaking point.

$$\% \text{ Elongation} = \frac{l_f - l_0}{l_0} \times 100 \quad (6\text{-}9)$$

where l_f is the distance between gage marks after the specimen breaks.

A second approach is to measure the percent change in the cross-sectional area at the point of fracture before and after the test. The **percent reduction in area** describes the amount of thinning undergone by the specimen during the test:

$$\% \text{ Reduction in area} = \frac{A_0 - A_f}{A_0} \times 100 \quad (6\text{-}10)$$

where A_f is the final cross-sectional area at the fracture surface.

Ductility is important to both designers of load-bearing components and manufacturers of components (bars, rods, wires, plates, I-beams, fibers, etc.) utilizing materials processing.

EXAMPLE 6-4 ***Ductility of an Aluminum Alloy***

The aluminum alloy in Example 6-1 has a final length after failure of 5.488 cm and a final diameter of 0.995 cm at the fractured surface. Calculate the ductility of this alloy.

SOLUTION

$$\% \text{ Elongation} = \frac{l_f - l_0}{l_0} \times 100 = \frac{5.488 - 5.000}{5.000} \times 100 = 9.76\%$$

$$\% \text{ Reduction in area} = \frac{A_0 - A_f}{A_0} \times 100$$

$$= \frac{(\pi/4)(1.263)^2 - (\pi/4)(0.995)^2}{(\pi/4)(1.263)^2} \times 100$$

$$= 37.9\%$$

The final length is less than 5.5125 cm (see Table 6-1) because, after fracture, the elastic strain is recovered.

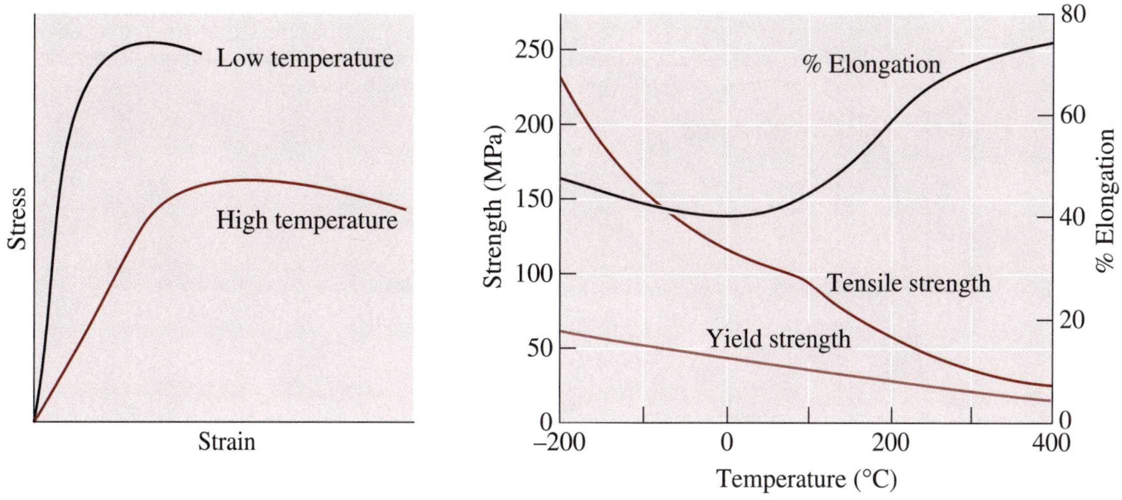

(a) (b)

Figure 6-11 The effect of temperature (a) on the stress-strain curve and (b) on the tensile properties of an aluminum alloy.

Effect of Temperature Mechanical properties of materials depend on temperature (Figure 6-11). Yield strength, tensile strength, and modulus of elasticity decrease at higher temperatures, whereas ductility commonly increases. A materials fabricator may wish to deform a material at a high temperature (known as *hot working*) to take advantage of the higher ductility and lower required stress. When temperatures are reduced, many, but not all, metals and alloys and polymers become brittle.

Increased temperatures also play an important role in forming polymeric materials and inorganic glasses. In many polymer-processing operations, such as extrusion, the increased ductility of polymers at higher temperatures is advantageous. Also, many polymeric materials will become harder and more brittle as they are exposed to temperatures that are below their glass temperatures (Figure 6-5). The loss of ductility played a role in failures of the *Titanic* in 1912 and the *Challenger* in 1986.

6-5 True Stress and True Strain

The decrease in engineering stress beyond the tensile strength point on an engineering stress-strain curve is related to the definition of engineering stress. We used the original area A_0 in our calculations, but this is not precise because the area continually changes. We define **true stress** and **true strain** by the following equations:

$$\text{True stress} = \sigma_t = \frac{F}{A} \tag{6-11}$$

$$\text{True strain} = \int \frac{dl}{l} = \ln\left(\frac{l}{l_0}\right) = \ln\left(\frac{A_0}{A}\right) \tag{6-12}$$

where A is the actual area at which the force F is applied. In plastic deformation we assume a constant volume $\left(\text{i.e., } \frac{A_0}{A} = \frac{l}{l_0}\right)$. The expression $\ln(A_0/A)$ should be used

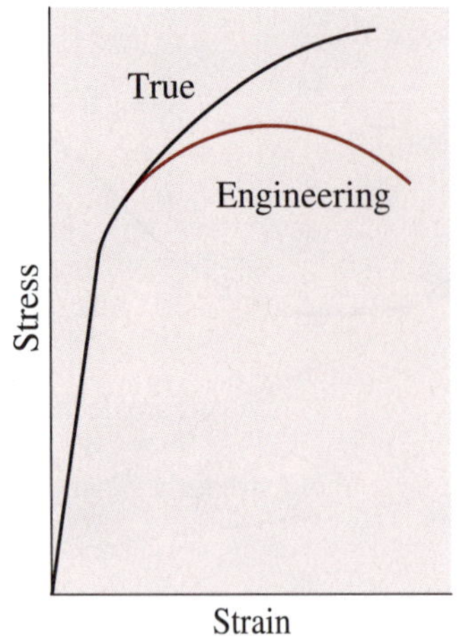

Figure 6-12
The relation between the true stress-true strain diagram and engineering stress–engineering strain diagram. The curves are identical to the yield point.

after necking begins. This is because beyond maximum load the distribution of strain along gage length is not uniform. The true stress-strain curve is compared with the engineering stress-strain curve in Figure 6-12. True stress continues to increase after necking because, although the load required decreases, the area decreases even more.

For structural applications we often do not require true stress and true strain. When we exceed the yield strength, the material deforms. The component would fail because it can no longer support the applied stress. Furthermore, a significant difference develops between the two curves only when necking begins. But when necking begins, our component is grossly deformed and no longer satisfies its intended use. Engineers dealing with materials processing require data related to true stress and strain.

EXAMPLE 6-5 *True Stress and True Strain Calculation*

Compare engineering stress and strain with true stress and strain for the aluminum alloy in Example 6-1 at (a) the maximum load and (b) fracture. The diameter at maximum load is 1.243 cm and at fracture is 0.995 cm.

SOLUTION

(a) At the tensile or maximum load:

$$\text{Engineering stress} = \frac{F}{A_0} = \frac{35580 \text{ N}}{(\pi/4)(1.263 \text{ cm})^2} = 283.9 \text{ MPa}$$

$$\text{True stress} = \frac{F}{A} = \frac{35580 \text{ N}}{(\pi/4)(1.243 \text{ cm})^2} = 293.1 \text{ MPa}$$

$$\text{Engineering strain} = \frac{l - l_0}{l_0} = \frac{5.3 - 5}{5} = 0.060 \text{ cm/cm}$$

$$\text{True strain} = \ln\left(\frac{l}{l_0}\right) = \ln\left(\frac{5.3}{5}\right) = 0.058 \text{ cm/cm}$$

(b) At fracture:

$$\text{Engineering stress} = \frac{F}{A_0} = \frac{33800\ \text{N}}{(\pi/4)(1.263\ \text{cm})^2} = 269.7\ \text{MPa}$$

$$\text{True stress} = \frac{F}{A} = \frac{33800\ \text{N}}{(\pi/4)(0.995\ \text{cm})^2} = 434.5\ \text{MPa}$$

$$\text{Engineering strain} = \frac{\Delta l}{l_0} = \frac{0.5125}{5} = 0.1025\ \text{cm/cm}$$

$$\text{True strain} = \ln\left(\frac{A_0}{A_f}\right) = \ln\left[\frac{(\pi/4)(1.263)^2}{(\pi/4)(0.995)^2}\right]$$

$$= \ln(1.610) = 0.476\ \text{cm/cm}$$

The true stress becomes much greater than the engineering stress only after necking begins. Note, the true strain was calculated using $\ln\left(\frac{A_0}{A_f}\right)$.

6-6 The Bend Test for Brittle Materials

In ductile metallic materials, the engineering stress-strain curve typically goes through a maximum; this maximum stress is the tensile strength of the material. Failure occurs at a lower stress after necking has reduced the cross-sectional area supporting the load. In more brittle materials, failure occurs at the maximum load, where the tensile strength and breaking strength are the same. In brittle materials, including many ceramics, yield strength, tensile strength, and breaking strength are all the same (Figure 6-13).

In many brittle materials, the normal tensile test cannot easily be performed because of the presence of flaws at the surface. Often, just placing a brittle material in the grips of the tensile testing machine causes cracking and fracture. These brittle materials may be tested using the **bend test** [Figure 6-14(a)]. By applying the load at three

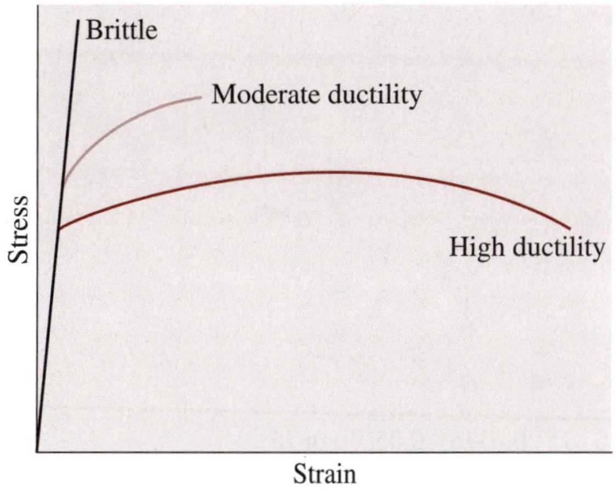

Figure 6-13
The stress-strain behavior of brittle materials compared with that of more ductile materials.

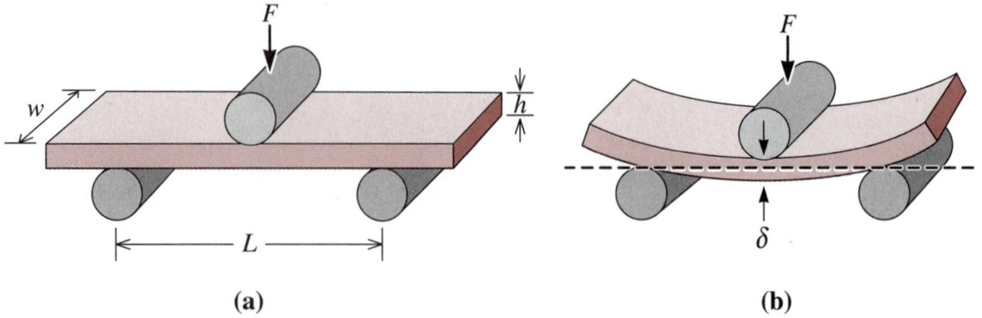

Figure 6-14 (a) The three-point bend test often used for measuring the strength of brittle materials, and (b) the deflection δ obtained by bending.

points and causing bending, a tensile force acts on the material opposite the midpoint. Fracture begins at this location. The **flexural strength**, or **modulus of rupture**, describes the material's strength:

$$\text{Flexural strength for three-point bend test} = \frac{3FL}{2wh^2} = \sigma_{\text{bend}} \tag{6-13}$$

where F is the fracture load, L is the distance between the two outer points, w is the width of the specimen, and h is the height of the specimen. The flexural strength has the units of stress and is designated by σ_{bend}. The results of the bend test are similar to the stress-strain curves; however, the stress is plotted versus deflection rather than versus strain (Figure 6-15).

The modulus of elasticity in bending, or the **flexural modulus** (E_{bend}), is calculated in the elastic region of Figure 6-15.

$$\text{Flexural modulus} = \frac{L^3F}{4wh^3\delta} = E_{\text{bend}} \tag{6-14}$$

where δ is the deflection of the beam when a force F is applied.

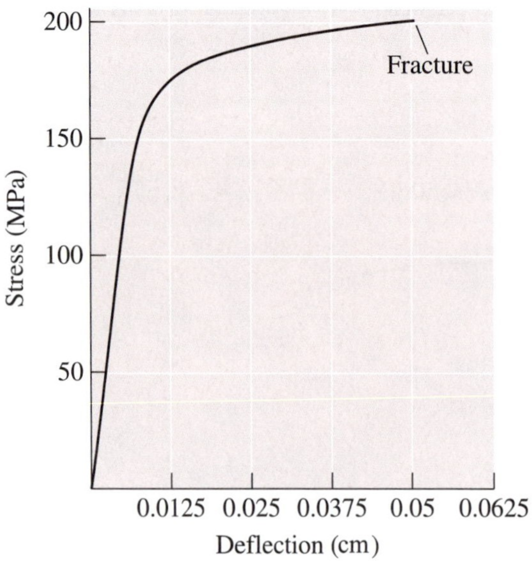

Figure 6-15 Stress-deflection curve for MgO obtained from a bend test.

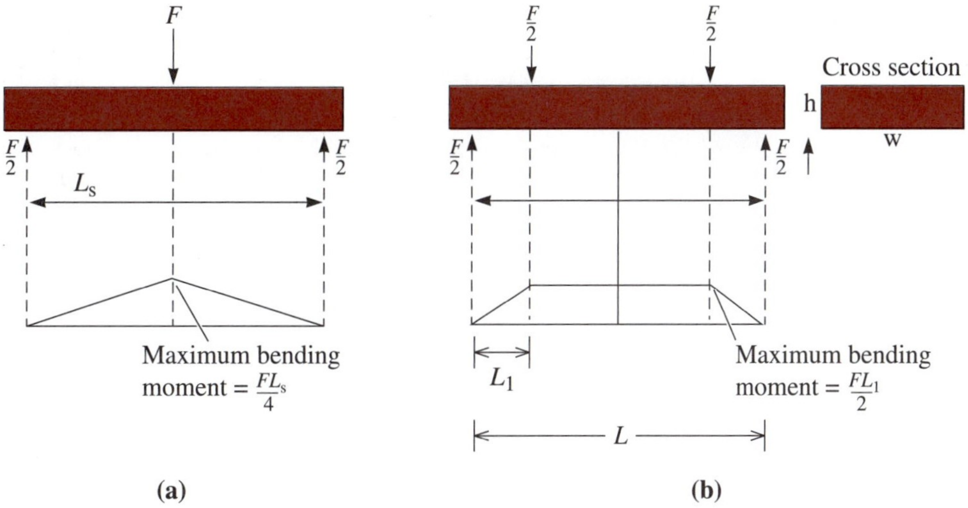

Figure 6-16 (a) Three-point and (b) four-point bend test setup.

This test can also be conducted using a setup known as the four-point bend test (Figure 6-16). The maximum stress or flexural stress for a four point bend test is given by:

$$\sigma_{\text{bend}} = \frac{3FL_1}{4wh^2} \tag{6-15}$$

Note that while deriving Equations 6-13 through 6-15, we assume a linear stress-strain response (thus cannot be correctly applied to many polymers). The four-point bend test is better suited for testing materials containing flaws. This is because the bending moment between inner platens is constant [Figure 6-16(b)], thus samples tend to break randomly unless there is a flaw that raises the stress concentration.

Since cracks and flaws tend to remain closed in compression, brittle materials such as concrete are often designed so that only compressive stresses act on the part. Often, we find that brittle materials fail at much higher compressive stresses than tensile stresses (Table 6-4). This is why it is possible to support the weight of a fire truck on four coffee cups. However, ceramics have very limited mechanical toughness and hence when we drop a ceramic coffee cup it can break easily.

TABLE 6-4 ■ ***Comparison of the tensile, compressive, and flexural strengths of selected ceramic and composite materials***

Material	Tensile Strength (MPa)	Compressive Strength (MPa)	Flexural Strength (MPa)
Polyester—50% glass fibers	159	221	310
Polyester—50% glass fiber fabric	255	186[a]	317
Al_2O_3 (99% pure)	207	2586	345
SiC (pressureless-sintered)	172	3861	552

[a] *A number of composite materials are quite poor in compression.*

EXAMPLE 6-6 *Flexural Strength of Composite Materials*

The flexural strength of a composite material reinforced with glass fibers is 310 MPa and the flexural modulus is 124.1×10^3 MPa. A sample, which is 1.25 cm wide, 0.938 cm high, and 20 cm long, is supported between two rods 12.5 cm apart. Determine the force required to fracture the material and the deflection of the sample at fracture, assuming that no plastic deformation occurs.

SOLUTION

Based on the description of the sample, $w = 1.25$ cm, $h = 0.938$ cm, and $L = 10$ cm. From Equation 6-15:

$$310 \text{ MPa} = \frac{3FL}{2wh^2} = \frac{(3)(F)(12.5)}{(2)(1.25 \text{ cm})(0.938 \text{ cm})^2} = 1.7 \times 10^5 F$$

$$F = \frac{310 \times 10^6}{1.7 \times 10^5} = 1823.5 \text{ N}$$

Therefore, the deflection, from Equation 6-14, is:

$$124.1 \times 10^3 \text{ MPa} = \frac{L^3 F}{4wh^3\delta} = \frac{(12.5 \text{ cm})^3(1823.5 \text{ N})}{(4)(1.25 \text{ cm})(0.938 \text{ cm})^3\delta}$$

$$\delta = 0.0696 \text{ cm}$$

In this calculation, we did assume that there is no viscoelastic behavior and a linear behavior of stress versus strain.

6-7 Hardness of Materials

The **hardness test** measures the resistance to penetration of the surface of a material by a hard object. Hardness can not be defined precisely. Hardness, depending upon the context, represents resistance to scratching or indentation and a qualitative measure of the strength of the material. In general, in **macrohardness** measurements the load applied is ~2 N. A variety of hardness tests have been devised, but the most commonly used are the Rockwell test and the Brinell test. Different indentors used in these tests are shown in Figure 6-17.

In the *Brinell hardness test*, the indentor is a hard steel sphere (usually 10 mm in diameter) that is forced into the surface of the material. The diameter of the impression,

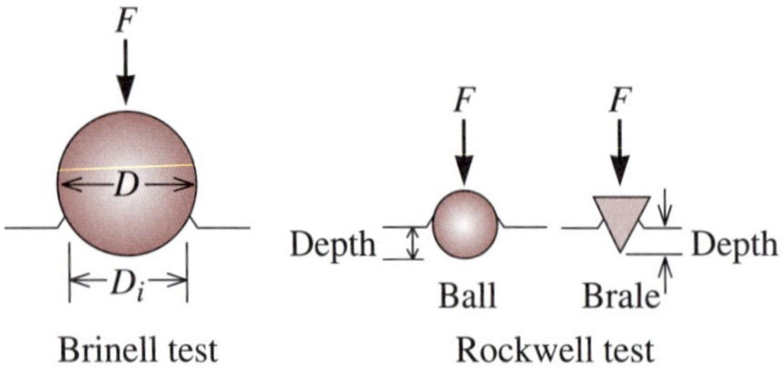

Figure 6-17
Indentors for the Brinell and Rockwell hardness tests.

TABLE 6-5 ■ *Comparison of typical hardness tests*

Test	Indentor	Load	Application
Brinell	10-mm ball	3000 kg	Cast iron and steel
Brinell	10-mm ball	500 kg	Nonferrous alloys
Rockwell *A*	Brale	60 kg	Very hard materials
Rockwell *B*	1/16-in. ball	100 kg	Brass, low-strength steel
Rockwell *C*	Brale	150 kg	High-strength steel
Rockwell *D*	Brale	100 kg	High-strength steel
Rockwell *E*	1/8-in. ball	100 kg	Very soft materials
Rockwell *F*	1/16-in. ball	60 kg	Aluminum, soft materials
Vickers	Diamond pyramid	10 g–1 kg (micro) 1 kg –30 kg (macro)	All materials
Knoop	Diamond pyramid	500 g	All materials

typically 2 to 6 mm, is measured and the Brinell hardness number (abbreviated as HB or BHN) is calculated from the following equation:

$$HB = \frac{2F}{\pi D\left[D - \sqrt{D^2 - D_i^2}\right]} \tag{6-16}$$

where F is the applied load in kilograms, D is the diameter of the indentor in millimeters, and D_i is the diameter of the impression in millimeters. The Brinell hardness has the units of stress (e.g., kg/mm^2).

The *Rockwell hardness test* uses a small-diameter steel ball for soft materials and a diamond cone, or Brale, for harder materials. The depth of penetration of the indentor is automatically measured by the testing machine and converted to a Rockwell hardness number: hardness Rockwell (HR). Since an optical measurement of the indention dimensions is not needed, the Rockwell test tends to be more popular than the Brinell test. Several variations of the Rockwell test are used, including those described in Table 6-5. A Rockwell *C* (HRC) test is used for hard steels, whereas a Rockwell *F* (HRF) test might be selected for aluminum. Rockwell tests provide a hardness number that has no units.

Hardness numbers are used primarily as a *qualitative* basis for comparison of materials, specifications for manufacturing and heat treatment, quality control, and correlation with other properties of materials. For example, Brinell hardness is closely related to the tensile strength of steel by the relationship:

$$\text{Tensile strength (psi)} = 500\ \text{HB} \tag{6-17}$$

where HB is in the units of kg/mm^2 and $1\ \text{psi} = 6.895 \times 10^3\ \text{Pa}$.

A Brinell hardness number can be obtained in just a few minutes with virtually no preparation of the specimen and without breaking the component (i.e., it is considered to be a nondestructive test), yet it provides a close approximation of the tensile strength. The Rockwell hardness number cannot be directly related to strength of metals and alloys; however, the test is rapid, easily performed, and therefore remains popular in industry.

Hardness correlates well with wear resistance. There is also a separate test available for measuring the wear resistance. A material used in crushing or grinding of ores should be very hard to assure that the material is not eroded or abraded by the hard feed materials. Similarly, gear teeth in the transmission or the drive system of a vehicle should be hard enough that the teeth do not wear out. Typically, we find that polymer materials are exceptionally soft, metals and alloys have intermediate hardness, and ceramics are exceptionally hard. We use materials such as tungsten carbide-cobalt composite (WC-Co), known as "carbide," for cutting tool applications. We also use microcrystalline diamond or diamond-like carbon (DLC) materials for cutting tools and other applications.

The Knoop (HK) test is a **microhardness test**, forming such small indentations that a microscope is required to obtain the measurement. In these tests, the load applied is less than 2 N. The Vickers test, which uses a diamond pyramid indentor, can be conducted either as a macro and microhardness test. Microhardness tests are suitable for materials that may have a surface that has a higher hardness than the bulk, materials in which different areas show different levels of hardness, or on samples that are not macroscopically flat.

In Chapter 2, we described nano-structured materials and devices. For some of the nano-technology applications, measurements of hardness at a nano-scale or **nano-hardness**, are important. Techniques for measuring hardness at very small length scales have become important for many applications. A nano-indentor is used for these applications.

6-8 Strain Rate Effects and Impact Behavior

When a material is subjected to a sudden, intense blow, in which the strain rate ($\dot{\gamma}$ or $\dot{\varepsilon}$) is extremely rapid, it may behave in much more brittle a manner than is observed in the tensile test. This, for example, can be seen with many plastics and materials such as Silly Putty®. If you stretch a plastic such as polyethylene or Silly Putty® very slowly, the polymer molecules have time to disentangle or the chains to slide past each other and cause large plastic deformations. If, however, we apply an impact loading, there is insufficient time for these mechanisms to play a role and the materials break in a brittle manner. An **impact test** is often used to evaluate the brittleness of a material under these conditions. In contrast to the tensile test, in this test the strain rates are much higher ($\dot{\varepsilon} \sim 10^3\ s^{-1}$).

Many test procedures have been devised, including the *Charpy* test and the *Izod* test (Figure 6-18). The Izod test is often used for plastic materials. The test specimen may be either notched or unnotched; V-notched specimens better measure the resistance of the material to crack propagation.

In this test, a heavy pendulum, starting at an elevation h_0, swings through its arc, strikes and breaks the specimen, and reaches a lower final elevation h_f. If we know the initial and final elevations of the pendulum, we can calculate the difference in potential energy. This difference is the **impact energy** absorbed by the specimen during failure. For the Charpy test, the energy is usually expressed in joules (J). The results of the Izod test are expressed in units of J/m. The ability of a material to withstand an impact blow is often referred to as the **impact toughness** of the material. As we mentioned before, in some situations, we consider the area under the true or engineering stress-strain curve as a measure of **tensile toughness**. In both cases, we are measuring the energy needed to fracture a material. The difference is that, in tensile tests, the strain rates are much smaller compared to those used in an impact test. Another difference is that in an impact test we usually deal with materials that have a notch. Fracture toughness of a material is defined as the ability of a material containing flaws to withstand an applied load. We will discuss fracture toughness in Section 7-1.

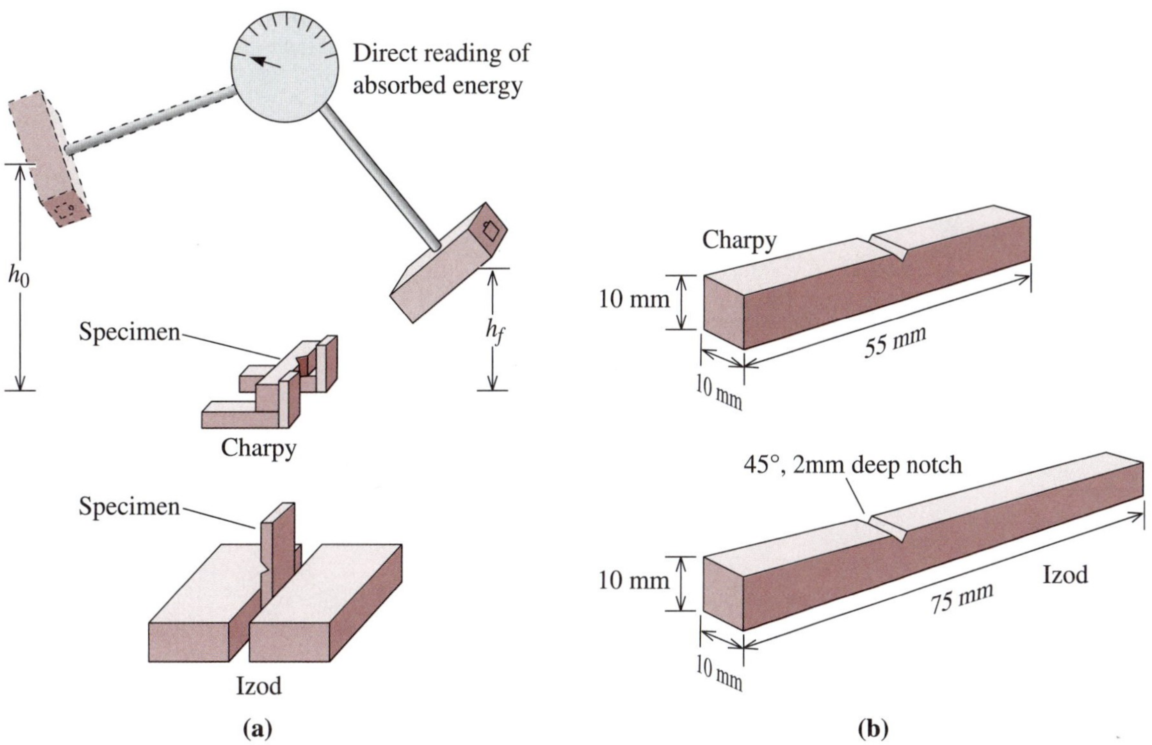

Figure 6-18 The impact test: (a) the Charpy and Izod tests, and (b) dimensions of typical specimens.

In another test, known as the Hopkinson bar test, a striker bar creates a stress wave in a sample. The absorption of this rapidly moving stress wave is measured to investigate mechanical properties at high strain rates.

6-9 Properties Obtained from the Impact Test

A curve showing the trends in the results of a series of impact tests performed on nylon at various temperatures is shown in Figure 6-19. In practice, the tests are conducted at a limited number of temperatures.

Ductile to Brittle Transition Temperature (DBTT) The **ductile to brittle transition temperature** is the temperature at which a material changes from ductile to brittle fracture. This temperature may be defined by the average energy between the ductile and brittle regions, at some specific absorbed energy, or by some characteristic fracture appearance. A material subjected to an impact blow during service should have a transition temperature *below* the temperature of the material's surroundings.

Not all materials have a distinct transition temperature (Figure 6-20). BCC metals have ductile to brittle transition temperatures, but most FCC metals do not. FCC metals have high absorbed energies, with the energy decreasing gradually and, sometimes, even increasing as the temperature decreases. As mentioned before, this transition may have contributed to the failure of the *Titanic*.

The ductile to brittle transition temperature is closely related to the glass temperature in polymers and for practical purposes is treated as the same. As mentioned before,

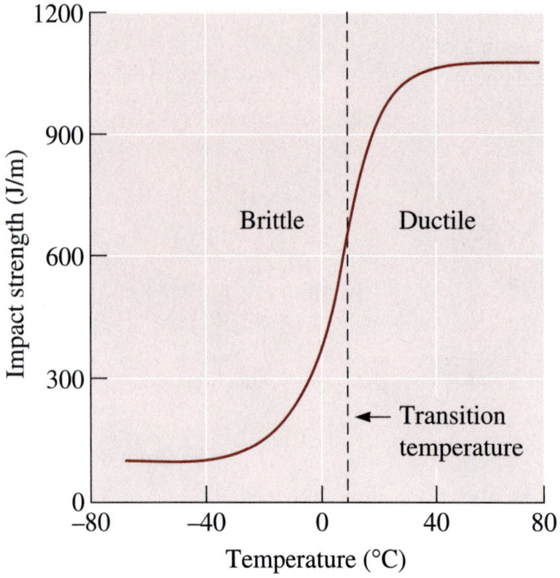

Figure 6-19
Results from a series of Izod impact tests for a super-tough nylon thermoplastic polymer.

the lower transition temperature of the polymers used in booster rocket O-rings and other factors led to the *Challenger* disaster.

Notch Sensitivity Notches caused by poor machining, fabrication, or design concentrate stresses and reduce the toughness of materials. The *notch sensitivity* of a material can be evaluated by comparing the absorbed energies of notched versus unnotched specimens. The absorbed energies are much lower in notched specimens if the material is notch-sensitive. We will discuss in Section 7-7 how the presence of notches affect the behavior of materials subjected to cyclical stress.

Relationship to the Stress-Strain Diagram The energy required to break a material during impact testing (i.e., the impact toughness) is not always related to the tensile toughness (i.e., the area contained within the true stress-true strain diagram (Figure 6-21)).

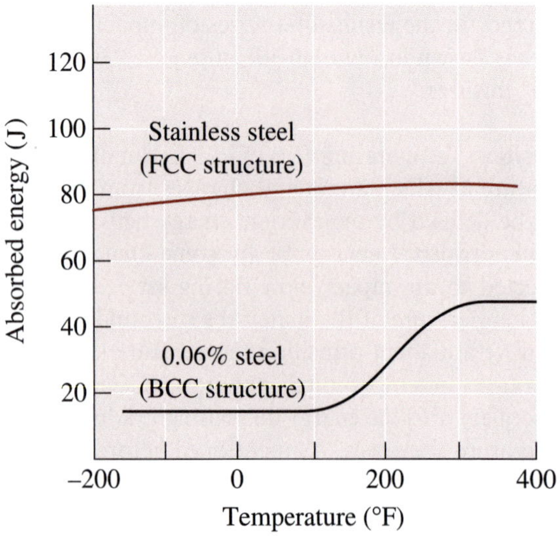

Figure 6-20
The Charpy V-notch properties for a BCC carbon steel and a FCC stainless steel. The FCC crystal structure typically leads to higher absorbed energies and no transition temperature.

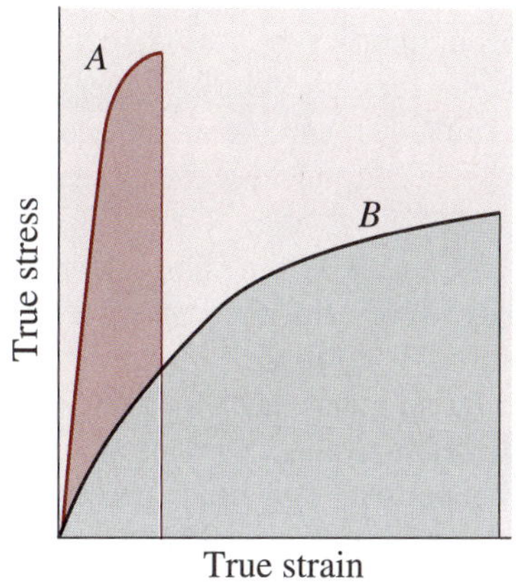

Figure 6-21
The area contained within the true stress-true strain curve is related to the tensile toughness. Although material *B* has a lower yield strength, it absorbs a greater energy than material *A*. The energies from these curves may not be the same as those obtained from impact test data.

As noted before, engineers often consider area under the engineering stress-strain curve as tensile toughness. In general, metals with both high strength and high ductility have good tensile toughness. However, this is not always the case when the strain rates are high. For example, metals that show excellent tensile toughness may show a brittle behavior under high strain rates (i.e., they may show poor impact toughness). Thus, strain rate can shift the ductile to brittle transition temperature (DBTT). Ceramics and many composites normally have poor toughness, even though they have high strength, because they display virtually no ductility. These materials show both poor tensile toughness and poor impact toughness.

Use of Impact Properties Absorbed energy and DBTT are very sensitive to loading conditions. For example, a higher rate of applying energy to the specimen reduces the absorbed energy and increases the DBTT. The size of the specimen also affects the results; because it is more difficult for a thick material to deform, smaller energies are required to break thicker materials. Finally, the configuration of the notch affects the behavior; a sharp, pointed surface crack permits lower absorbed energies than does a V-notch. Because we often cannot predict or control all of these conditions, the impact test is a quick, convenient, and inexpensive way to compare different materials.

EXAMPLE 6-7 *Design of a Sledgehammer*

Design an 3.6-kg sledgehammer for driving steel fence posts into the ground.

SOLUTION

First, we must consider the design requirements to be met by the sledgehammer. A partial list would include:

1. The handle should be light in weight, yet tough enough that it will not catastrophically break.
2. The head must not break or chip during use, even in subzero temperatures.
3. The head must not deform during continued use.
4. The head must be large enough to assure that the user doesn't miss the fence post, and it should not include sharp notches that might cause chipping.
5. The sledgehammer should be inexpensive.

Although the handle could be a lightweight, tough composite material (such as a polymer reinforced with Kevlar (a special polymer) fibers), a wood handle about 70 cm long would be much less expensive and would still provide sufficient toughness. As shown in a later chapter, wood can be categorized as a natural fiber-reinforced composite.

To produce the head, we prefer a material that has a low transition temperature, can absorb relatively high energy during impact, and yet also has enough hardness to avoid deformation. The toughness requirement would rule out most ceramics. A face-centered cubic metal, such as FCC stainless steel or copper, might provide superior toughness even at low temperatures; however, these metals are relatively soft and expensive. An appropriate choice might be a normal BCC steel. Ordinary steels are inexpensive, have good hardness and strength, and some have sufficient toughness at low temperatures.

In Appendix A, we find that the density of iron is 7.87 g/cm^3. We assume that the density of steel is about the same. The volume of steel required is $V = \dfrac{3.6 \times 10^3 \text{ g}}{7.87 \text{ g/cm}^3} = 0.46 \times 10^3 \text{ cm}^3$. To assure that we will hit our target, the head might have a cylindrical shape, with a diameter of 6.25 cm. The length of the head would then be 15 cm.

SUMMARY

- The mechanical behavior of materials is described by their mechanical properties, which are measured with idealized, simple tests. These tests are designed to represent different types of loading conditions. The properties of a material reported in various handbooks are the results of these tests. Consequently, we should always remember that handbook values are average results obtained from idealized tests and, therefore, must be used with care.
- The tensile test describes the resistance of a material to a slowly applied stress. Important properties include yield strength (the stress at which the material begins to permanently deform), tensile strength (the stress corresponding to the maximum applied load), modulus of elasticity (the slope of the elastic portion of the stress-strain curve), and % elongation and % reduction in area (both, measures of the ductility of the material).
- The bend test is used to determine the tensile properties of brittle materials. A modulus of elasticity and a flexural strength (similar to a tensile strength) can be obtained.
- The hardness test measures the resistance of a material to penetration and provides a measure of the wear and abrasion resistance of the material. A number of hardness tests, including the Rockwell and Brinell tests, are commonly used. Often the hardness can be correlated to other mechanical properties, particularly tensile strength.
- The impact test describes the response of a material to a rapidly applied load. The Charpy and Izod tests are typical. The energy required to fracture the specimen is measured and can be used as the basis for comparison of various materials tested under the same conditions. In addition, a transition temperature above which the material fails in a ductile, rather than a brittle, manner can be determined.

GLOSSARY

Anelastic (viscoelastic) material A material in which the total strain developed has elastic and viscous components. Part of the total strain recovers similar to elastic strain. Some part, though, recovers over a period of time. Examples of viscoelastic materials: polymer melts, many polymers including Silly Putty®. Typically, the term anelastic is used for metallic materials.

Bend test Application of a force to the center of a bar that is supported on each end to determine the resistance of the material to a static or slowly applied load. Typically used for brittle materials.

Compliance Inverse of Young's modulus or modulus of elasticity.

Dilatant (shear thickening) Materials in which the apparent viscosity increases with the increasing rate of shear.

Ductile to brittle transition temperature (DBTT) The temperature below which a material behaves in a brittle manner in an impact test. The ductile to brittle switchover also depends on the strain rate.

Ductility The ability of a material to be permanently deformed without breaking when a force is applied.

Elastic deformation Deformation of the material that is recovered instantaneously when the applied load is removed.

Elastic limit The magnitude of stress at which the relationship between stress and strain begins to depart from linearity.

Elastic strain Fully and instantaneously recoverable strain in a material.

Elastomers Natural or synthetic polymeric materials that are comprised of molecules with spring-like coils that lead to large elastic deformations (e.g., natural rubber, silicones).

Engineering strain The amount that a material deforms per unit length in a tensile test.

Engineering stress The applied load, or force, divided by the original cross-sectional area of the material.

Extensometer An instrument to measure change in length of a tensile specimen, thus allowing calculation of strain.

Flexural modulus The modulus of elasticity calculated from the results of a bend test, giving the slope of the stress-deflection curve.

Flexural strength The stress required to fracture a specimen in a bend test. Also called the modulus of rupture.

Glass temperature (T_g) A temperature below which an otherwise ductile material behaves as if it is brittle. Usually, this temperature is not fixed and is affected by processing of the material.

Hardness test Measures the resistance of a material to penetration by a sharp object. Common hardness tests include the Brinell test, Rockwell test, Knoop test, and Vickers test.

Hooke's law The relationship between stress and strain in the elastic portion of the stress-strain curve.

Impact energy The energy required to fracture a standard specimen when the load is applied suddenly.

Impact loading Application of stress at a very high strain rate ($\sim>100\ s^{-1}$).

Impact test Measures the ability of a material to absorb the sudden application of a load without breaking. The Charpy and Izod tests are the commonly used impact tests.

Impact toughness Energy absorbed by a material, usually notched, during fracture, under the conditions of an impact test.

Kinematic viscosity (μ) Ratio of viscosity and density, often expressed in centiStokes.

Load The force applied to a material during testing.

Materials processing Manufacturing or fabrication methods used for shaping of materials (e.g., extrusion, forging).

Macrohardness Overall bulk hardness of materials measured using loads >2 N.

Microhardness Hardness of materials typically measured using loads less than 2 N using such test as Knoop (HK).

Modulus of elasticity (E) Young's modulus, or the slope of the linear part of the stress-strain curve in the elastic region. It is a measure of the stiffness of a material, depends upon strength of interatomic bonds and composition, and is not strongly dependent upon microstructure.

Modulus of resilience (E_r) The maximum elastic energy absorbed by a material when a load is applied.

Modulus of rupture The stress required to fracture a specimen in a bend test. Also called the flexural strength.

Nano-hardness Hardness of materials measured at 1–10 nm length scale using extremely small ($\sim$100 μN) forces.

Necking Local deformation causing reduction in the cross-sectional area of a tensile specimen. Many ductile materials show this behavior. The engineering stress begins to decrease at the onset of necking.

Newtonian Materials in which the shear stress and shear strain rate are linearly related (e.g., light oil or water).

Non-Newtonian Materials in which shear stress and shear strain rate are not linearly related, these materials are shear thinning or shear thickening (e.g., polymer melts, slurries, paints, etc.).

Offset strain value A value of strain (e.g., 0.002) used to obtain the offset yield stress value.

Offset yield strength A stress value obtained graphically that describes the stress that gives no more than a specified amount of plastic deformation. Most useful for designing components. Also, simply stated as the yield strength.

Percent elongation The total percentage increase in the length of a specimen during a tensile test.

Percent reduction in area The total percentage decrease in the cross-sectional area of a specimen during the tensile test.

Plastic deformation or strain Permanent deformation of a material when a load is applied, then removed.

Poisson's ratio The ratio between the lateral and longitudinal strains in the elastic region.

Proportional limit A level of stress above which the relationship between stress and strain is not linear.

Pseudoplastics (shear thinning) Materials in which the apparent viscosity decreases with increasing rate of shear.

Shear modulus (G) The slope of the linear part of the shear stress-shear strain curve.

Shear-strain rate Time derivative of shear strain. *See* Strain rate.

Stiffness A qualitative measure of the elastic deformation produced in a material. A stiff material has a high modulus of elasticity. Stiffness also depends upon geometry.

Strain Elongation change in dimension per unit length.

Strain gage A device used for measuring change in length and hence strain.

Strain rate The rate at which strain develops in or is applied to a material indicated as $\dot{\varepsilon}$ or $\dot{\gamma}$ for tensile and shear-strain rates, respectively. Strain rate can have an effect on whether a material would behave in a ductile or brittle fashion.

Stress Force or load per unit area of cross-section over which the force or load is acting.

Stress relaxation Decrease in the stress for a material held under constant strain, as a function of time, observed in viscoelastic materials. Stress relaxation is different from time dependent recovery of strain.

Tensile strength The stress that corresponds to the maximum load in a tensile test.

Tensile test Measures the response of a material to a slowly applied uniaxial force. The yield strength, tensile strength, modulus of elasticity, and ductility are obtained.

Tensile toughness The area under the true stress-true strain tensile test curve. It is a measure of the energy required to cause fracture under tensile test conditions.

True strain The strain calculated using actual and not original dimensions, given by $\varepsilon_t = \ln(l/l_0)$.

True stress The load divided by the actual cross-sectional area of the specimen at that load.

Viscoelastic (or anelastic) material *See* Anelastic material.

Viscosity (η) Measure of resistance to flow, defined as the ratio of shear stress to shear strain rate (units Poise or Pa-s).

Viscous material A viscous material is one in which the strain develops over a period of time and the material does not go to its original shape after the stress is removed.

Work of fracture Area under the stress-strain curve, considered as a measure of tensile toughness.

Yield point phenomenon An abrupt transition, seen in some materials, from elastic deformation to plastic flow.

Yield strength A stress value obtained graphically that describes no more than a specified amount of deformation (usually 0.002). Also known as offset yield strength.

Young's modulus The slope of the elastic part of the stress-strain curve in the elastic region, same as modulus of elasticity. In some thermoplastic polymers, the Young's modulus depends on the level of stress.

PROBLEMS

Section 6-1 Technological Significance

Section 6-2 Terminology for Mechanical Properties

Section 6-3 The Tensile Test: Use of the Stress-Strain Diagram

6-1 Draw qualitative engineering stress-engineering strain curves for a ductile polymer, a ductile metal, a ceramic, a glass, and natural rubber. Label carefully. Rationalize your sketch for each material.

6-2 Why do some polymers get stronger as we stretch them beyond a region where necking occurs?

6-3 A 3780 N force is applied to a 0.375-cm-diameter nickel wire having a yield strength of 310 MPa and a tensile strength of 379 MPa. Determine

(a) whether the wire will plastically deform; and
(b) whether the wire will experience necking.

6-4 A force of 100,000 N is applied to a 10 mm × 20 mm iron bar having a yield strength of 400 MPa and a tensile strength of 480 MPa. Determine

(a) whether the bar will plastically deform; and
(b) whether the bar will experience necking.

6-5 Calculate the maximum force that a 0.5-cm-diameter rod of Al_2O_3, having a yield strength of 241 MPa, can withstand with no plastic deformation. Express your answer in Newtons.

6-6 A force of 20,000 N will cause a 1 cm × 1 cm bar of magnesium to stretch from 10 cm to 10.045 cm. Calculate the modulus of elasticity.

6-7 A polymer bar's dimensions are 2.5 cm × 5 cm × 37.5 cm. The polymer has a modulus of elasticity of 4137 MPa. What force is required to stretch the bar elastically to 38.13 cm?

6-8 An aluminum plate 0.5 cm thick is to withstand a force of 50,000 N with no permanent deformation. If the aluminum has a yield strength of 125 MPa, what is the minimum width of the plate?

6-9 A 7.5-cm-diameter rod of copper is to be reduced to a 5-cm-diameter rod by being pushed through an opening. To account for the elastic strain, what should be the diameter of the opening? The modulus of elasticity for the copper is 117.2×10^3 MPa and the yield strength is 276 MPa. A 0.15-cm-thick, 8-cm-wide sheet of magnesium that is originally 5 m long is to be stretched to a final length of 6.2 m. What should be the length of the sheet before the applied stress is released? The modulus of elasticity of magnesium is 45 GPa and the yield strength is 200 MPa.

6-10 A steel cable 3.13 cm in diameter and 1500 cm long is to lift a 18,140 kg load. What is the length of the cable during lifting? The modulus of elasticity of the steel is 207×10^3 MPa.

Section 6-4 Properties Obtained from the Tensile Test and Section 6-5 True Stress and True Strain

6-11 Define "true stress" and "true strain." Compare with engineering stress and engineering strain.

6-12 Write down the formulas for calculating the stress and strain for a sample subjected to a tensile test. Assume the sample shows necking.

6-13 The following data were collected from a standard 1.263-cm-diameter test specimen of a copper alloy (initial length (l_0) = 5 cm):

Load (N)	Δl (cm)
0	0.00000
13,340	0.00418
26,690	0.00833
33,360	0.01043
40,030	0.0225
46,700	0.1
53,380	0.65
55,160	1.25 (maximum load)
50,710	2.55 (fracture)

After fracture, the total length was 7.535 cm and the diameter was 0.935 cm. Plot the data and calculate the 0.2% offset yield strength along with

(a) the tensile strength;
(b) the modulus of elasticity;
(c) the % elongation;
(d) the % reduction in area;
(e) the engineering stress at fracture;
(f) the true stress at fracture; and
(g) the modulus of resilience.

6-14 The following data were collected from a 1-cm-diameter test specimen of polyvinyl chloride (l_0 = 5 cm):

Load (N)	Δl (cm)
0	0.00000
1334	0.01865
2669	0.0374
4003	0.05935
5338	0.08
6672	0.115
7384	0.175 (maximum load)
7117	0.235
6316	0.3 (fracture)

After fracture, the total length was 5.225 cm and the diameter was 0.983 cm. Plot the data and calculate

(a) the 0.2% offset yield strength;
(b) the tensile strength;
(c) the modulus of elasticity;
(d) the % elongation;
(e) the % reduction in area;
(f) the engineering stress at fracture;
(g) the true stress at fracture; and
(h) the modulus of resilience.

6-15 The following data were collected from a 12-mm-diameter test specimen of magnesium (l_0 = 30.00 mm):

Load (N)	Δl (mm)
0	0.0000
5,000	0.0296
10,000	0.0592
15,000	0.0888
20,000	0.15
25,000	0.51
26,500	0.90
27,000	1.50 (maximum load)
26,500	2.10
25,000	2.79 (fracture)

After fracture, the total length was 32.61 mm and the diameter was 11.74 mm. Plot the data and calculate

(a) the 0.2% offset yield strength;
(b) the tensile strength;
(c) the modulus of elasticity;
(d) the % elongation;
(e) the % reduction in area;
(f) the engineering stress at fracture;
(g) the true stress at fracture; and
(h) the modulus of resilience.

6-16 The following data were collected from a 20-mm-diameter test specimen of a ductile cast iron ($l_0 = 40.00$ mm):

Load (N)	Δl (mm)
0	0.0000
25,000	0.0185
50,000	0.0370
75,000	0.0555
90,000	0.20
105,000	0.60
120,000	1.56
131,000	4.00 (maximum load)
125,000	7.52 (fracture)

After fracture, the total length was 47.42 mm and the diameter was 18.35 mm. Plot the data and calculate

(a) the 0.2% offset yield strength;
(b) the tensile strength;
(c) the modulus of elasticity;
(d) the % elongation;
(e) the % reduction in area;
(f) the engineering stress at fracture;
(g) the true stress at fracture; and
(h) the modulus of resilience.

Section 6-6 The Bend Test for Brittle Materials

6-17 A bar of Al_2O_3 that is 0.625 cm thick, 1.25 cm wide, and 22.5 cm long is tested in a three-point bending apparatus, with the supports located 15 cm apart. The deflection of the center of the bar is measured as a function of the applied load. The data are shown below. Determine the flexural strength and the flexural modulus.

Force (N)	Deflection (cm)
64.5	0.00625
128.5	0.0125
193.0	0.01875
257.5	0.025
382.5	0.03725 (fracture)

6-18 A 1-cm-diameter, 30-cm-long titanium bar has a yield strength of 345 MPa, a modulus of elasticity of 110.3×10^3 MPa, and Poisson's ratio of 0.30. Determine the length and diameter of the bar when a 2224 N load is applied.

6-19 When a tensile load is applied to a 1.5-cm diameter copper bar, the diameter is reduced to 1.498-cm diameter. Determine the applied load, using the data in Table 6-3.

6-20 A three-point bend test is performed on a block of ZrO_2 that is 20 cm long, 1.25 cm wide, and 0.625 cm thick and is resting on two supports 10 cm apart. When a force of 1780 N is applied, the specimen deflects 0.093 cm and breaks. Calculate

(a) the flexural strength; and
(b) the flexural modulus, assuming that no plastic deformation occurs.

6-21 A three-point bend test is performed on a block of silicon carbide that is 10 cm long, 1.5 cm wide, and 0.6 cm thick and is resting on two supports 7.5 cm apart. The sample breaks when a deflection of 0.09 mm is recorded. The flexural modulus for silicon carbide is 480 GPa. Assume that no plastic deformation occurs. Calculate

(a) the force that caused the fracture; and
(b) the flexural strength.

6-22 A thermosetting polymer containing glass beads is required to deflect 0.5 mm when a force of 500 N is applied. The polymer part is 2 cm wide, 0.5 cm thick, and 10 cm long. If the flexural modulus is 6.9 GPa, determine the minimum distance between the supports. Will the polymer fracture if its flexural strength is 85 MPa? Assume that no plastic deformation occurs.

6-23 The flexural modulus of alumina is 310.3×10^3 MPa and its flexural strength is 317 MPa. A bar of alumina 0.75 cm thick, 2.5 cm wide, and 25 cm long is placed on supports 17.5 cm apart. Determine the amount of deflection at the moment the bar breaks, assuming that no plastic deformation occurs.

6-24 Ceramics are much stronger in compression than in tension. Explain why.

Section 6-7 Hardness of Materials

6-25 A Brinell hardness measurement, using a 10-mm-diameter indenter and a 500-kg load, produces an indentation of 4.5 mm on an aluminum plate. Determine the Brinell hardness number (HB) of the metal.

6-26 When a 3000-kg load is applied to a 10-mm-diameter ball in a Brinell test of a steel, an indentation of 3.1 mm is produced. Estimate the tensile strength of the steel.

Section 6-8 Strain Rate Effects and Impact Behavior and Section 6-9 Properties from the Impact Test

6-27 The following data were obtained from a series of Charpy impact tests performed on four steels, each having a different manganese content. Plot the data and determine

(a) the transition temperature (defined by the mean of the absorbed energies in the ductile and brittle regions); and

(b) the transition temperature (defined as the temperature that provides 50 J of absorbed energy).

Test Temperature °C	Impact Energy (J) 0.30% Mn	0.39% Mn	1.01% Mn	1.55% Mn
−100	2	5	5	15
−75	2	5	7	25
−50	2	12	20	45
−25	10	25	40	70
0	30	55	75	110
25	60	100	110	135
50	105	125	130	140
75	130	135	135	140
100	130	135	135	140

6-28 Plot the transition temperature versus manganese content and using the previous data shown discuss the effect of manganese on the toughness of steel. What would be the minimum manganese allowed in the steel if a part is to be used at 0°C?

6-29 The following data were obtained from a series of Charpy impact tests performed on four ductile cast irons, each having a different silicon content. Plot the data and determine

(a) the transition temperature (defined by the mean of the absorbed energies in the ductile and brittle regions); and

(b) the transition temperature (defined as the temperature that provides 10 J of absorbed energy).

Plot the transition temperature versus silicon content and discuss the effect of silicon on the toughness of the cast iron. What would be the maximum silicon allowed in the cast iron if a part is to be used at 25°C?

Test Temperature °C	Impact Energy (J) 2.55% Si	2.85% Si	3.25% Si	3.63% Si
−50	2.5	2.5	2	2
−5	3	2.5	2	2
0	6	5	3	2.5
25	13	10	7	4
50	17	14	12	8
75	19	16	16	13
100	19	16	16	16
125	19	16	16	16

6-30 FCC metals are often recommended for use at low temperatures, particularly when any sudden loading of the part is expected. Explain.

6-31 A steel part can be made by powder metallurgy (compacting iron powder particles and sintering to produce a solid) or by machining from a solid steel block. Which part is expected to have the higher toughness? Explain.

6-32 What is the difference between a tensile test and an impact test? Using this, explain why the toughness values measured using impact tests may not always correlate with tensile toughness measured using tensile tests.

6-33 A number of aluminum-silicon alloys have a structure that includes sharp-edged plates of brittle silicon in the softer, more ductile aluminum matrix. Would you expect these alloys to be notch-sensitive in an impact test? Would you expect these alloys to have good toughness? Explain your answers.

6-34 What caused NASA's *Challenger* 1986 accident?

6-35 How is tensile toughness defined in relation to the true stress-strain diagram? How is tensile toughness related to impact toughness?

6-36 What factors contributed to the NASA *Columbia* 2003 accident?

7

Fracture Mechanics, Fatigue, and Creep Behavior

Have You Ever Wondered?

- *Why is it that glass fibers of different lengths have different strengths?*
- *Why do some metals and plastics become brittle at low temperatures?*
- *Can a material or component ultimately fracture even if the overall stress does not exceed the yield strength?*
- *Why do aircrafts have a finite service life?*

One goal of this chapter is to introduce the basic concepts associated with the fracture toughness of materials. In this regard, we will examine what factors affect the strength of glasses and ceramics, and how the Weibull distribution quantitatively describes the variability in their strength. Another goal is to learn about time-dependent phenomena such as fatigue, creep, and stress corrosion. Materials ultimately fail because of excessive tensile load and/or corrosion. This chapter will review some of the basic testing procedures that engineers use to evaluate many of these properties and the failure of materials.

7-1 Fracture Mechanics

Fracture mechanics is the discipline concerned with the behavior of materials containing cracks or other small flaws. The term "flaw" refers to such features as small pores (holes), inclusions, or micro-cracks. The term "flaw" does *not* refer to atomic level defects such as vacancies, or dislocations. What we wish to know is the maximum stress that a material can withstand if it contains flaws of a certain size and geometry. **Fracture toughness** measures the ability of a material containing a flaw to withstand an applied load. Note that this does *not* require a high strain rate (impact).

A typical fracture toughness test may be performed by applying a tensile stress to a specimen prepared with a flaw of known size and geometry (Figure 7-1). The stress applied to the material is intensified at the flaw, which acts as a *stress raiser*. For a simple case, the *stress intensity factor*, K, is

$$K = f\sigma\sqrt{\pi a} \tag{7-1}$$

where f is a geometry factor for the specimen and flaw, σ is the applied stress, and a is the flaw size (as defined in Figure 7-1). Note that the analytical expression for K changes with the geometry of the flaw and specimen. If the specimen is assumed to have an "infinite" width, then $f \cong 1.0$. For a small single-edge notch [Figure 7-1(a)], $f = 1.12$.

By performing a test on a specimen with a known flaw size, we can determine the value of K at which a flaw would grow and cause failure. This critical stress intensity factor is defined as the *fracture toughness*, K_c,

$$K_c = K \text{ required for a crack to propagate} \tag{7-2}$$

Fracture toughness depends on the thickness of the sample: as thickness increases, fracture toughness K_c decreases to a constant value (Figure 7-2). This constant is called the *plane strain fracture toughness*, K_{Ic}. It is K_{Ic} that is normally reported as the property of a material. The value of K_{Ic} does not depend upon the thickness of the sample. Table 7-1 compares the value of K_{Ic} to the yield strength of several materials. Units for fracture toughness are 1.0989 MPa $\sqrt{\text{m}}$.

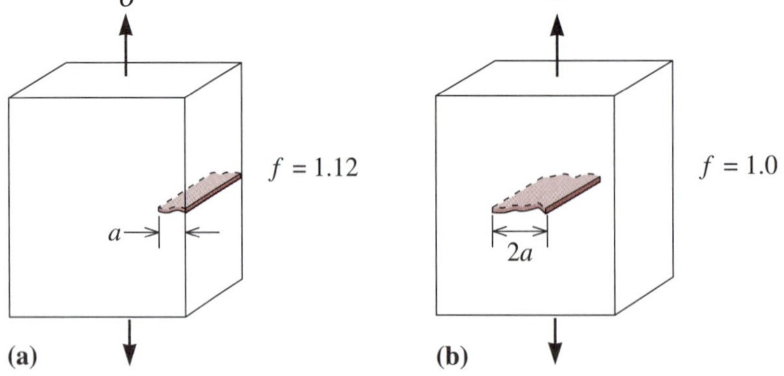

Figure 7-1 Schematic drawing of fracture toughness specimens with (a) edge and (b) internal flaws.

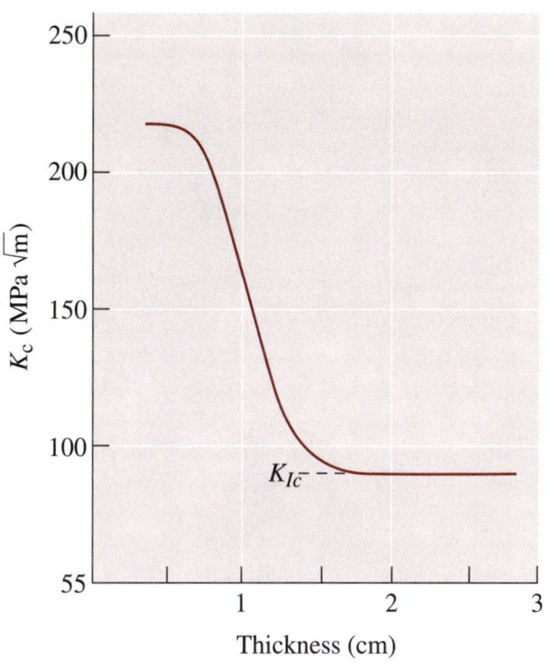

Figure 7-2
The fracture toughness K_c of a 2070 MPa yield strength steel decreases with increasing thickness, eventually leveling off at the plane strain fracture toughness K_{Ic}.

The ability of a material to resist the growth of a crack depends on a large number of factors:

1. Larger flaws reduce the permitted stress. Special manufacturing techniques, such as filtering impurities from liquid metals and hot pressing or hot isostatic pressing of powder particles to produce ceramic or superalloy components reduce flaw size and improve fracture toughness (Chapters 9 and 15).

TABLE 7-1 ■ ***The plane strain fracture toughness K_{Ic} of selected materials***

Material	Fracture Toughness K_{Ic} (MPa$\sqrt{m}$)	Yield Strength or Ultimate Strength (for Brittle Solids) (MPa)
Al-Cu alloy	24.2	455.1
	36.3	324.1
Ti-6% Al-4% V	54.9	896.4
	98.9	861.9
Ni-Cr steel	50.3	1641.0
	87.9	1420.4
Al_2O_3	1.8	206.9
Si_3N_4	4.9	551.6
Transformation toughened ZrO_2	11.0	413.7
Si_3N_4-SiC composite	56.0	827.4
Polymethyl methacrylate polymer	1.0	27.6
Polycarbonate polymer	3.3	57.9

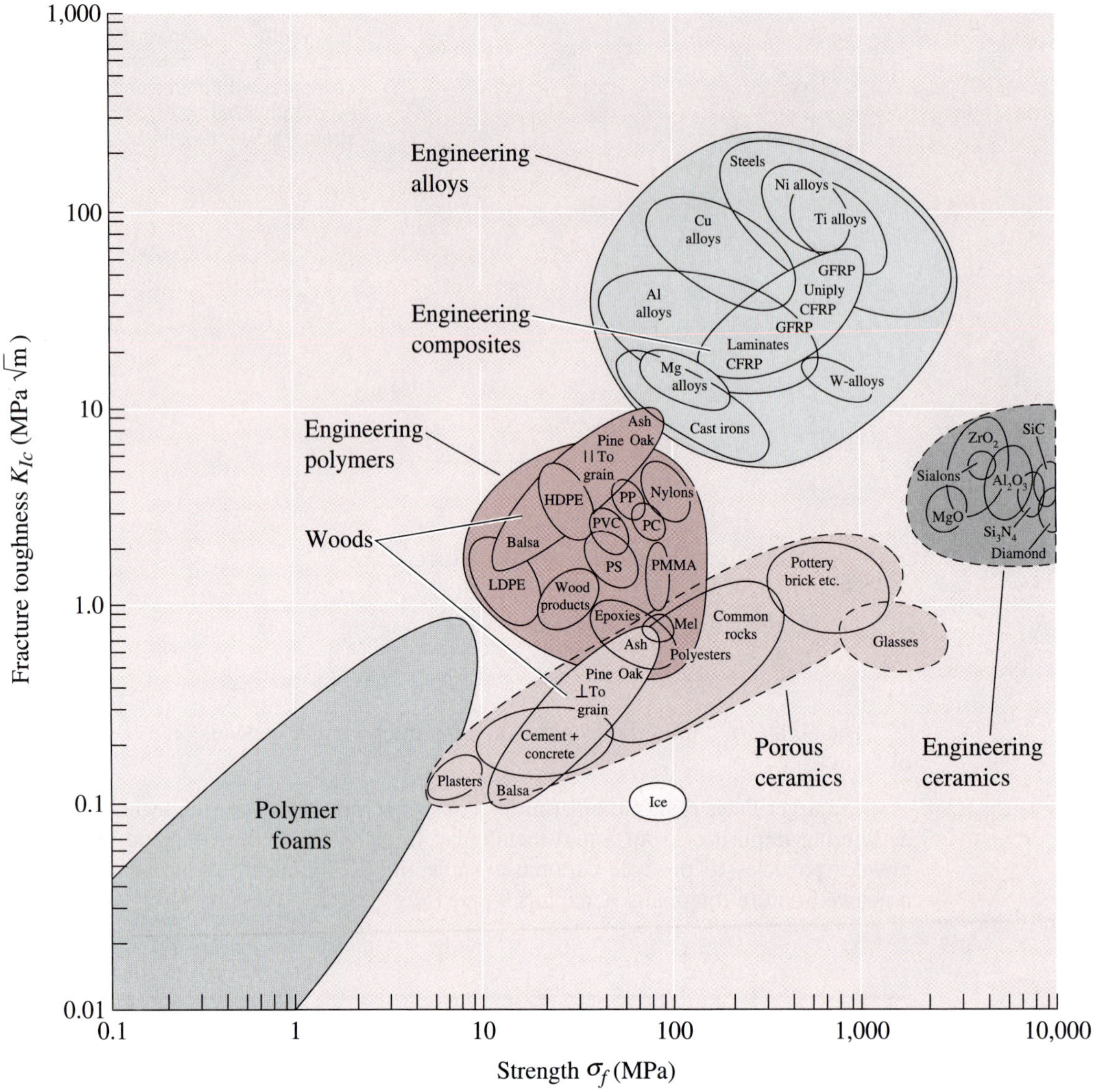

Figure 7-3 Fracture toughness versus strength of different engineered materials. (*Reprinted by permission of Waveland Press, Inc. from Courtney, T.,* Mechanical Behavior of Materials, *2/e, Long Grove, IL; Waveland Press, Inc., 2000 [reissued 2005]. All rights reserved.)*

2. The ability of a material to deform is critical. In ductile metals, the material near the tip of the flaw can deform, causing the tip of any crack to become blunt, reducing the stress intensity factor, and preventing growth of the crack. Increasing the strength of a given metal usually decreases ductility and gives a lower fracture toughness. (See Table 7-1.) Brittle materials such as ceramics and many polymers have much lower fracture toughness than metals (Figure 7-3).

3. Thicker, more rigid pieces of a given material have a lower fracture toughness than thinner materials.

4. Increasing the rate of application of the load, such as that encountered in an **impact test**, typically reduces the fracture toughness of the material.

5. Increasing the temperature normally increases the fracture toughness, just as in the impact test.

6. A small grain size normally improves fracture toughness, whereas more point defects and dislocations reduce fracture toughness. Thus, a fine-grained ceramic material may provide improved resistance to crack growth.

7. In certain ceramic materials we can also take advantage of stress-induced transformations that lead to compressive stresses that cause increased fracture toughness.

Fracture testing of ceramics cannot be performed easily using a sharp notch, since formation of such a notch often causes the samples to break. We can use hardness testing to gain a measure of the fracture toughness of many ceramics.

7-2 The Importance of Fracture Mechanics

The fracture mechanics approach allows us to design and select materials while taking into account the inevitable presence of flaws. There are three variables to consider: the property of the material (K_c or K_{Ic}), the stress σ that the material must withstand, and the size of the flaw a. If we know two of these variables, the third can be determined.

Selection of a Material If we know the maximum size a of flaws in the material and the magnitude of the applied stress, we can select a material that has a fracture toughness K_c or K_{Ic} large enough to prevent the flaw from growing.

Design of a Component If we know the maximum size of any flaw and the material (and therefore it's K_c or K_{Ic} has already been selected), we can calculate the maximum stress that the component can withstand. Then we can design the appropriate size of the part to assure that the maximum stress is not exceeded.

Design of a Manufacturing or Testing Method If the material has been selected, the applied stress is known, and the size of the component is fixed, we can calculate the maximum size of a flaw that can be tolerated. A nondestructive testing technique that detects any flaw greater than this critical size can help assure that the part will function safely. In addition, we find that, by selecting the correct manufacturing process, we can limit the size of the flaws to be smaller than this critical size.

EXAMPLE 7-1 *Design of a Nondestructive Test*

A large steel plate used in a nuclear reactor has a plane strain fracture toughness of 87.9 $MPa\sqrt{m}$ and is exposed to a stress of 310 MPa during service. Design a testing or inspection procedure capable of detecting a crack at the edge of the plate before the crack is likely to grow at a catastrophic rate.

SOLUTION

We need to determine the minimum size of a crack that will propagate in the steel under these conditions. From Equation 7-1, assuming that $f = 1.12$:

$$K_{Ic} = f\sigma\sqrt{a\pi}$$

$$87.9 = (1.12)(310)\sqrt{a\pi}$$

$$a = 2 \text{ cm}$$

A 2 cm deep crack on the edge should be relatively easy to detect. Often, cracks of this size can be observed visually. A variety of other tests, such as dye penetrant inspection, magnetic particle inspection, and eddy current inspection, also detect cracks much smaller than this. If the growth rate of a crack is slow and inspection is performed on a regular basis, a crack should be discovered long before reaching this critical size.

Brittle Fracture Any crack or imperfection limits the ability of a ceramic to withstand a tensile stress. This is because a crack (sometimes called a **Griffith flaw**) concentrates and magnifies the applied stress. Figure 7-4 shows a crack of length a at the surface of a brittle material. The radius r of the crack is also shown. When a tensile stress σ is applied, the actual stress at the crack tip is:

$$\sigma_{\text{actual}} \cong 2\sigma\sqrt{a/r} \tag{7-3}$$

For very thin cracks (small r) or long cracks (large a), the ratio $\sigma_{\text{actual}}/\sigma$ becomes large, or the stress is intensified. If the stress (σ_{actual}) at the crack tip exceeds the yield strength, the crack grows and eventually causes failure, even though the nominal applied stress σ is small.

In a different approach, we recognize that an applied stress causes an elastic strain, related to the modulus of elasticity, E, within the material. When a crack propagates, this strain energy is released, reducing the overall energy. At the same time, however,

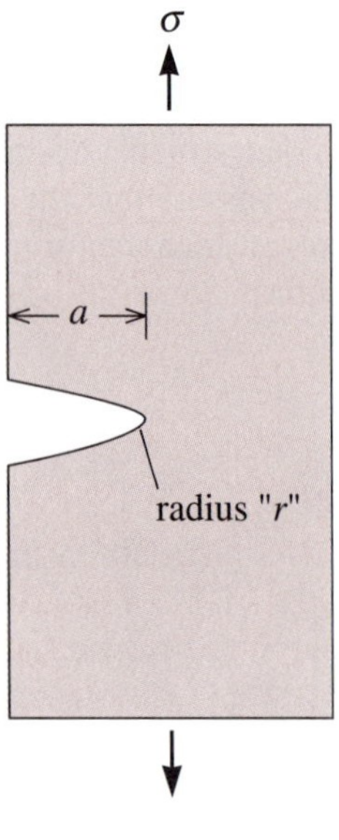

Figure 7-4
Schematic diagram of the Griffith flaw in a ceramic.

two new surfaces are created by the extension of the crack; this increases the energy associated with the surface. By balancing the strain energy and the surface energy, we find that the critical stress required to propagate the crack is given by the Griffith equation,

$$\sigma_{\text{critical}} \cong 2\sigma\sqrt{\frac{E\gamma}{\pi a}} \quad (7\text{-}4)$$

where a is the length of a surface crack (or one-half the length of an internal crack) and γ is the surface energy (per unit area). Again, this equation shows that even small flaws severely limit the strength of the ceramic.

We also note that if we rearrange Equation 7-1, which described the stress intensity factor K, we obtain:

$$\sigma = \frac{K}{f\sqrt{\pi a}} \quad (7\text{-}5)$$

This equation is similar in form to Equation 7-4. Each of these equations points out the dependence of the mechanical properties on the size of flaws present in the ceramic. Development of manufacturing processes (Chapter 15) to minimize the flaw size becomes crucial in improving the strength of ceramics.

The flaws are most important when tensile stresses act on the material. Compressive stresses try to close rather than open a crack; consequently, ceramics often have very good compressive strengths.

EXAMPLE 7-2 Properties of SiAlON Ceramics

Assume that an advanced ceramic, Sialon (acronym for silicon aluminum oxynitride), has a tensile strength of 414 MPa. Let us assume that this value is for a flaw-free ceramic. (In practice, it is almost impossible to produce flaw-free ceramics.) A thin crack 0.025 cm deep is observed before a Sialon part is tested. The part unexpectedly fails at a stress of 3.5 MPa by propagation of the crack. Estimate the radius of the crack tip.

SOLUTION

The failure occurred because the 3.5 MPa applied stress, magnified by the stress concentration at the tip of the crack, produced an actual stress equal to the tensile strength. From Equation 7-3:

$$\sigma_{\text{actual}} = 2\sigma\sqrt{a/r}$$

$$414 \text{ MPa} = (2)(3.5 \text{ MPa})\sqrt{0.025 \text{ cm}/r}$$

$$\sqrt{0.025/r} = 59.1 \quad \text{or} \quad 0.025/r = 3492.8$$

$$r = 7.1 \times 10^{-6} \text{ cm} = 710 \text{ Å}$$

The likelihood of our being able to measure a radius of curvature of this size by any method of nondestructive testing is virtually zero. Therefore, although Equation 7-3 may help illustrate the factors that influence how a crack propagates in a brittle material, it does not help in predicting the strength of actual ceramic parts.

EXAMPLE 7-3 *Design of a Ceramic Support*

Design a supporting 7.5-cm-wide plate made of Sialon, which has a fracture toughness of 9.9 MPa$\sqrt{m}$, that will withstand a tensile load of 177,920 N. The part is to be nondestructively tested to assure that no flaws are present that might cause failure.

SOLUTION

Let's assume that we have three nondestructive testing methods available to us: X-ray radiography can detect flaws larger than 0.05 cm, gamma-ray radiography can detect flaws larger than 0.02 cm, and ultrasonic inspection can detect flaws larger than 0.0125 cm. For these flaw sizes, we must now calculate the minimum thickness of the plate that will assure that flaws of these sizes will not propagate.

From our fracture toughness equation, assuming that $f = 1$:

$$\sigma_{\max} = \frac{K_{Ic}}{\sqrt{\pi a}} = \frac{F}{A}$$

$$A = \frac{F\sqrt{\pi a}}{K_{Ic}} = \frac{(177{,}920 \text{ N})(\sqrt{\pi})(\sqrt{a})}{9.9 \text{ MPa}\sqrt{\text{m}}}$$

$$A = 0.03\sqrt{a} \text{ m}^2 \quad \text{and} \quad \text{thickness} = \left(\frac{0.03 \text{ m}^2}{0.075 \text{ m}}\right)\sqrt{a} = 0.4\sqrt{a}$$

NDT Method	Smallest Detectable Crack (cm)	Minimum Area (cm^2)	Minimum Thickness (cm)	Maximum Stress (MPa)
X-ray radiography	0.05	6.7	0.89	270
γ-ray radiography	0.02	4.24	0.57	420
Ultrasonic inspection	0.0125	3.35	0.45	530

Our ability to detect flaws, coupled with our ability to produce a ceramic with flaws smaller than our detection limit, significantly affects the maximum stress than can be tolerated and, hence, the size of the part. In this example, the part can be smaller if ultrasonic inspection is available.

The fracture toughness is also important. Had we used Si_3N_4, with a fracture toughness of 3.3 MPa $\sqrt{m}$ instead of the Sialon, we could repeat the calculations and show that, for ultrasonic testing, the minimum thickness is 1.4 cm and the maximum stress is only 166 MPa.

7-3 Microstructural Features of Fracture in Metallic Materials

Ductile Fracture In metals that have good ductility and **toughness**, ductile fracture normally occurs in a **transgranular** manner (through the grains). Often, a considerable amount of deformation—including necking—is observed in the failed component. The deformation occurs before the final fracture. Ductile fractures are usually caused by simple overloads, or by applying too high a stress to the material.

In a simple tensile test, ductile fracture begins with the nucleation, growth, and coalescence of microvoids at the center of the test bar. **Microvoids** form when a high stress causes separation of the metal at grain boundaries or interfaces between the metal and small impurity particles (inclusions). As the local stress increases, the microvoids grow and coalesce into larger cavities. Eventually, the metal-to-metal contact area is too small to support the load and fracture occurs.

Deformation by slip also contributes to the ductile fracture of a metal. We know that slip occurs when the resolved shear stress reaches the critical resolved shear stress and that the resolved shear stresses are highest at a 45° angle to the applied tensile stress (Chapter 4, Schmid's law).

These two aspects of ductile fracture give the failed surface characteristic features. In thick metal sections, we expect to find evidence of necking, with a significant portion of the fracture surface having a flat face where microvoids first nucleated and coalesced, and a small shear lip, where the fracture surface is at a 45° angle to the applied stress. The shear lip, indicating that slip occurred, gives the fracture a cup and cone appearance (see Figure 6-8). Simple macroscopic observation of this fracture may be sufficient to identify the ductile fracture mode.

Examination of the fracture surface at a high magnification—perhaps using a scanning electron microscope—reveals a dimpled surface. The dimples are traces of the microvoids produced during fracture. Normally, these microvoids are round, or equiaxed, when a normal tensile stress produces the failure [Figure 7-5(a)]. However, on the shear lip, the dimples are oval-shaped, or elongated, with the ovals pointing toward the origin of the fracture [Figure 7-5(b)].

In a thin plate, less necking is observed and the entire fracture surface may be a shear face. Microscopic examination of the fracture surface shows elongated dimples rather than equiaxed dimples, indicating a greater proportion of 45° slip than in thicker metals.

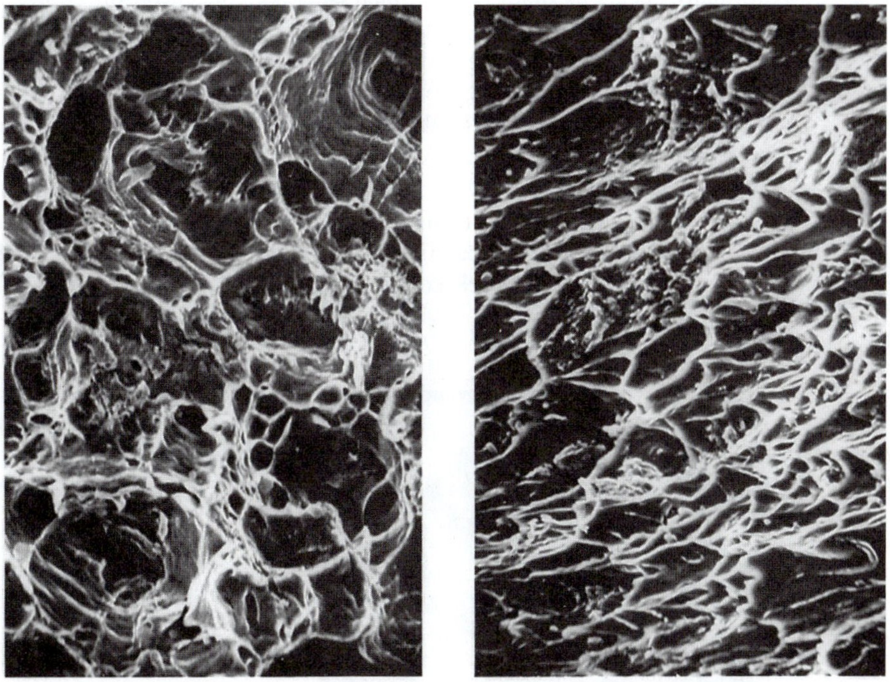

(a) (b)

Figure 7-5 Scanning electron micrographs of an annealed 1018 steel exhibiting ductile fracture in a tensile test. (a) Equiaxed dimples at the flat center of the cup and cone, and (b) elongated dimples at the shear lip (×1250).

EXAMPLE 7-4 *Hoist Chain Failure Analysis*

A chain used to hoist heavy loads failed. Examination of the failed link indicates considerable deformation and necking prior to failure. List some of the possible reasons for failure.

SOLUTION

This description suggests that the chain failed in a ductile manner by a simple tensile overload. Two factors could be responsible for this failure:

1. The load exceeded the hoisting capacity of the chain. Thus, the stress due to the load exceeded the yield strength of the chain, permitting failure. Comparison of the load to the manufacturer's specifications will indicate that the chain was not intended for such a heavy load. This is the fault of the user!

2. The chain material was of the wrong composition or was improperly heat-treated. Consequently, the yield strength was lower than intended by the manufacturer and could not support the load.

Brittle Fracture Brittle fracture occurs in metals and alloys of high strength or those with poor ductility and toughness. Furthermore, even metals that are normally ductile may fail in a brittle manner at low temperatures, in thick sections, at high strain rates (such as impact), or when flaws play an important role. Brittle fractures are frequently observed when impact, rather than overload, causes failure.

In brittle fracture, little or no plastic deformation is required. Initiation of the crack normally occurs at small flaws, which causes a concentration of stress. The crack may move at a rate approaching the velocity of sound in the metal. Normally, the crack propagates most easily along specific crystallographic planes, often the {100} planes, by cleavage. In some cases, however, the crack may take an **intergranular** (along the grain boundaries) path, particularly when segregation (preferential separation of different elements) or inclusions weaken the grain boundaries.

Brittle fracture can be identified by observing the features on the failed surface. Normally, the fracture surface is flat and perpendicular to the applied stress in a tensile test. If failure occurs by cleavage, each fractured grain is flat and differently oriented, giving a crystalline or "rock candy" appearance to the fracture surface (Figure 7-6).

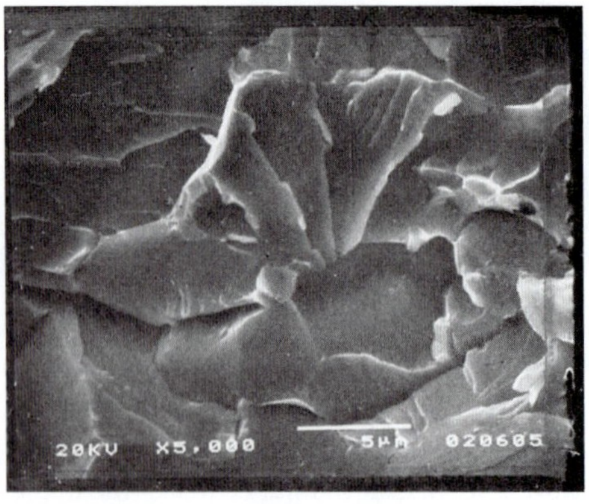

Figure 7-6
Scanning electron micrograph of a brittle fracture surface of a quenched 1010 steel (×5000). (*Courtesy of C.W. Ramsay.*)

Figure 7-7
The Chevron pattern in a 1.25-cm-diameter quenched 4340 steel. The steel failed in a brittle manner by an impact blow.

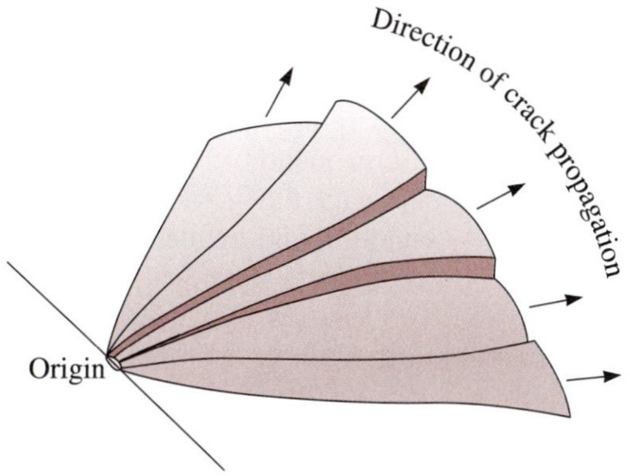

Figure 7-8
The Chevron pattern forms as the crack propagates from the origin at different levels. The pattern points back to the origin.

Often, the layman claims that the metal failed because it crystallized. Of course, we know that the metallic material was crystalline to begin with and the surface appearance is due to the cleavage faces.

Another common fracture feature is the **Chevron pattern** (Figure 7-7), produced by separate crack fronts propagating at different levels in the material. A radiating pattern of surface markings, or ridges, fans away from the origin of the crack (Figure 7-8). The Chevron pattern is visible with the naked eye or a magnifying glass and helps us identify both the brittle nature of the failure process as well as the origin of the failure.

EXAMPLE 7-5 *Automobile Axle Failure Analysis*

An engineer investigating the cause of an automobile accident finds that the right rear wheel has broken off at the axle. The axle is bent. The fracture surface reveals a Chevron pattern pointing toward the surface of the axle. Suggest a possible cause for the fracture.

SOLUTION

The evidence suggests that the axle did not break prior to the accident. The deformed axle means that the wheel was still attached when the load was applied. The Chevron pattern indicates that the wheel was subjected to an intense impact blow, which was transmitted to the axle, causing failure. The preliminary evidence suggests that the driver lost control and crashed, and the force of the crash caused the axle to break. Further examination of the fracture surface, microstructure, composition, and properties may verify that the axle was manufactured properly.

7-4 Microstructural Features of Fracture in Ceramics, Glasses, and Composites

In ceramic materials, the ionic or covalent bonds permit little or no slip (Chapter 4). Consequently, failure is a result of brittle fracture. Most crystalline ceramics fail by cleavage along widely spaced, closely packed planes. The fracture surface typically is smooth, and frequently no characteristic surface features point to the origin of the fracture [Figure 7-9(a)].

Glasses also fracture in a brittle manner. Frequently, a **conchoidal fracture** surface is observed. This surface contains a very smooth mirror zone near the origin of the fracture, with tear lines comprising the remainder of the surface [Figure 7-9(b)]. The tear lines point back to the mirror zone and the origin of the crack, much like the chevron pattern in metals.

Polymers can fail by either a ductile or a brittle mechanism. Below the glass temperature (T_g), thermoplastic polymers fail in a brittle manner—much like a glass. Likewise, the hard thermoset polymers, whose structure consists of inter-connected long chains of molecules, fail by a brittle mechanism. Some plastics whose structure consists of tangled but not chemically cross-linked chains, however, fail in a ductile manner above the glass temperature, giving evidence of extensive deformation and even necking prior to failure. The ductile behavior is a result of sliding of the polymer chains, which is not possible in glassy or thermosetting polymers. Thermosetting polymers have a rigid, three-dimensional cross-linked structure (Chapter 16).

Fracture in fiber-reinforced composite materials is more complex. Typically, these composites contain strong, brittle fibers surrounded by a soft, ductile matrix, as in boron-reinforced aluminum. When a tensile stress is applied along the fibers, the soft aluminum deforms in a ductile manner, with void formation and coalescence eventually producing a dimpled fracture surface. As the aluminum deforms, the load is no longer transmitted effectively to the fibers; the fibers break in a brittle manner until there are too few of them left intact to support the final load.

Fracturing is more common if the bonding between the fibers and matrix is poor. Voids can then form between the fibers and the matrix, causing pull-out. Voids can also form between layers of the matrix if composite tapes or sheets are not properly bonded, causing **delamination** (Figure 7-10). Delamination, in this context, means the layers of different materials in a composite begin to come apart.

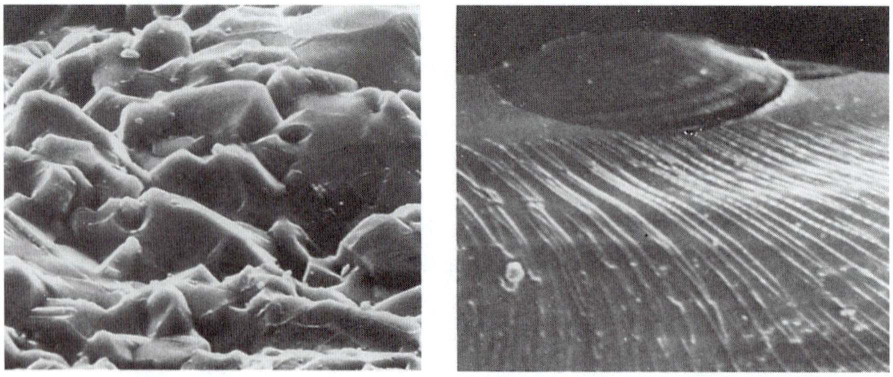

(a) (b)

Figure 7-9 Scanning electron micrographs of fracture surfaces in ceramics. (a) The fracture surface of Al_2O_3, showing the cleavage faces (×1250), and (b) the fracture surface of glass, showing the mirror zone (top) and tear lines characteristic of conchoidal fracture (×300).

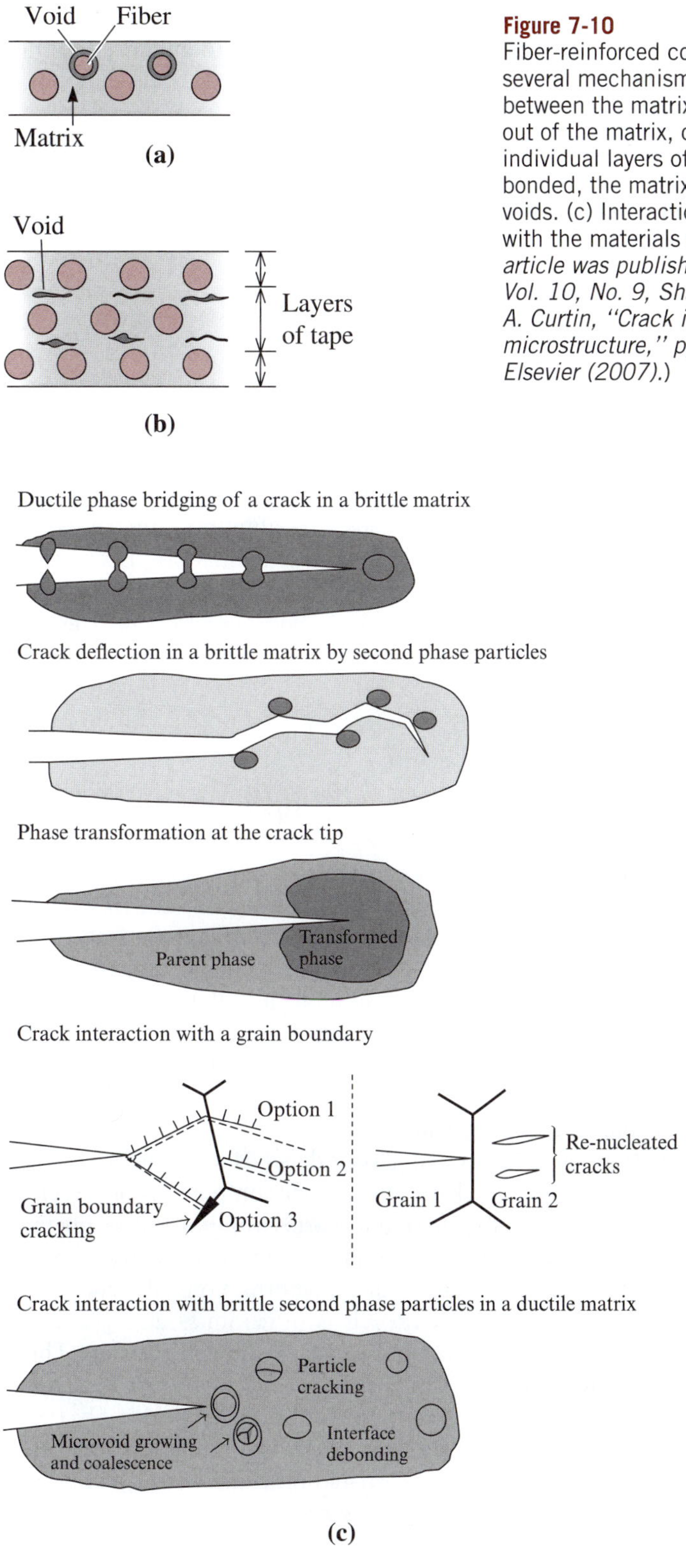

Figure 7-10
Fiber-reinforced composites can fail by several mechanisms. (a) Due to weak bonding between the matrix and fibers, fibers can pull out of the matrix, creating voids. (b) If the individual layers of the matrix are poorly bonded, the matrix may delaminate, creating voids. (c) Interactions of an advancing crack with the materials microstructure. (*This article was published in* Materials Today, *Vol. 10, No. 9, Sharvan Kumar and William A. Curtin, "Crack interaction with microstructure," pp. 34–43, Copyright Elsevier (2007).*)

There are several ways in which a crack will interact with a material's microstructure. Some of these interactions can be used to enhance the fracture toughness of materials. The interactions of an advancing crack include crack bridging by a ductile phase, deflection of the crack in a brittle matrix by another phase, and creating a stress-induced phase transformation at the crack tip [Figure 7-10(c)]. Other strategies include creating crack interactions with grain boundaries and crack interactions with brittle, second phase particles in which the second phase particles may crack or debond or create microvoids.

EXAMPLE 7-6 *Fracture in Composites*

Describe the difference in fracture mechanism between a boron-reinforced aluminum composite and a glass fiber-reinforced epoxy composite.

SOLUTION

In the boron-aluminum composite, the aluminum matrix is soft and ductile; thus we expect the matrix to fail in a ductile manner. Boron fibers, in contrast, fail in a brittle manner. Both glass fibers and epoxy are brittle; thus the composite as a whole should display little evidence of ductile fracture.

7-5 Weibull Statistics for Failure Strength Analysis

We need a statistical approach when evaluating the strength of ceramic materials. The strength of ceramics and glasses depends upon the size and distribution of sizes of flaws. In these materials, flaws originate from the ceramic manufacturing process. The flaws also can result during machining, grinding, etc. Glasses can also develop microcracks as a result of interaction with water vapor in air. If we test alumina or other ceramic components of different sizes and geometry, we often find a large scatter in the measured values—even if their nominal composition is the same. Similarly, if we are testing the strength of glass fibers of a given composition, we find that, on average, shorter fibers are stronger than longer fibers.

The strength of ceramics and glasses depends upon the probability of finding a flaw that exceeds a certain critical size. For larger components or larger fibers this probability increases. As a result, the strength of larger components or fibers is likely to be lower than that of smaller components or shorter fibers. In metallic or polymeric materials, which can exhibit relatively large plastic deformations, the effect of flaws and flaw size distribution is not felt to the extent it is in ceramics and glasses. In these materials, cracks initiating from flaws get blunted by plastic deformation. Thus, for ductile materials, the distribution of strength is narrow and close to a Gaussian distribution. The strength of ceramics and glasses, however, varies considerably (i.e., if we test a large number of identical samples of silica glass or alumina ceramic, the data will show a wide scatter owing to changes in distribution of flaw sizes). The strength of brittle materials, such as ceramics and glasses, is not Gaussian; it is given by the **Weibull distribution**. The Weibull distribution is an indicator of the variability of strength of materials resulting from a distribution of flaw sizes. This behavior results from critical sized flaws in materials with a distribution of flaw sizes (i.e., failure due to the weakest link of a chain).

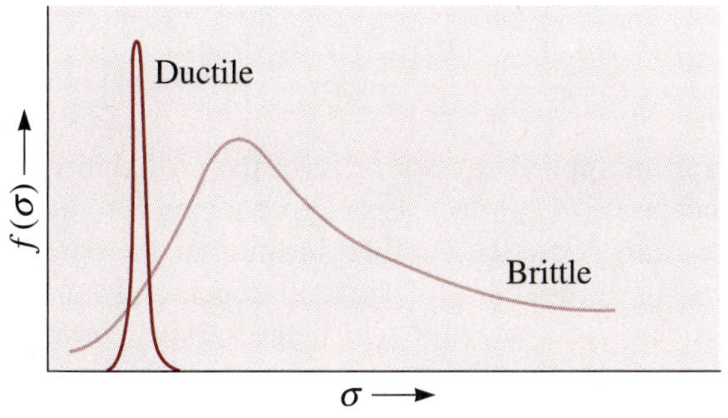

Figure 7-11 The Weibull distribution describes the fraction of the samples that fail at any given applied stress.

The Weibull distribution shown in Figure 7-11 describes the fraction of samples that fail at different applied stresses. At low stresses, a small fraction of samples contain flaws large enough to cause fracture; most fail at an intermediate applied stress, and a few contain only small flaws and do not fail until large stresses are applied. To provide predictability, we prefer a very narrow distribution.

Consider a body of volume V with a distribution of flaws and subjected to a stress σ. If we assumed that the volume, V, was made up of n elements with volume V_0 and each element had the same flaw-size distribution, it can be shown that the survival probability, $P(V_0)$, (i.e., the probability that a brittle material will not fracture under the applied stress σ) is given by:

$$P(V_0) = \exp\left[-\left(\frac{\sigma - \sigma_u}{\sigma_0}\right)^m\right] \tag{7-6}$$

The probability of failure, $F(V_0)$, can be written as:

$$F(V_0) = 1 - P(V_0) = 1 - \exp\left[-\left(\frac{\sigma - \sigma_u}{\sigma_0}\right)^m\right] \tag{7-7}$$

In Equations 7-6 and 7-7, σ is the applied stress, σ_0 is characteristic strength (often assumed to be—even though it is not—equal to the average strength), σ_u is the stress level below which the probability of failure is zero (i.e., the probability of survival is 1.0). In these equations, m is the Weibull modulus. In theory, Weibull modulus values can range from 0 to ∞. The **Weibull modulus** is a measure of the variability of the strength of the material.

The Weibull modulus m indicates the strength variability. For metals and alloys, the Weibull modulus is ~100. For traditional ceramics (e.g., bricks, pottery, etc.), the Weibull modulus is less than 3. Engineered ceramics, in which the processing is better controlled and hence the number of flaws is expected to be less, have a Weibull modulus of 5 to 10.

Note that for ceramics and other brittle solids, we can assume $\sigma_u = 0$. This is because there is no nonzero stress level for which we can claim a brittle material will not fail. For *brittle materials*, Equations 7-6 and 7-7 can be rewritten as follows:

$$P(V_0) = \exp\left[-\left(\frac{\sigma}{\sigma_0}\right)^m\right] \tag{7-8}$$

and

$$F(V_0) = 1 - P(V_0) = 1 - \exp\left[-\left(\frac{\sigma}{\sigma_0}\right)^m\right] \tag{7-9}$$

From Equation 7-8, for an applied stress σ of zero, the probability of survival is 1. As the applied stress σ increases, $P(V_0)$ decreases, approaching zero at very high values of applied stresses. We can also describe another meaning of the parameter σ_0. In Equation 7-8, when $\sigma = \sigma_0$, the probability of survival becomes $1/e \cong 0.37$. Therefore, in brittle materials, σ_0 is the stress level for which the survival probability is $\cong 0.37$ or 37%. We can also state that σ_0 is the stress level for which the failure probability is $\cong 0.63$ or 63%.

Equations 7-8 and 7-9 can be modified to account for samples with different volumes. It can be shown that, for an equal probability of survival, samples with larger volumes will have lower strengths. This is what we mentioned before (e.g., longer glass fibers will be weaker than shorter glass fibers).

The following examples illustrate how the Weibull plots can be used for the analyzing of mechanical properties of materials and designing of components.

EXAMPLE 7-7 *Weibull Modulus for Steel and Alumina Ceramics*

Figure 7-12 shows the log-log plots of the probability of failure and strength of a 0.2% plain carbon steel and an alumina ceramic prepared using conventional powder processing in which alumina powders are compacted in a press and sintered into a dense mass at high temperature. Also included is a plot for alumina ceramics prepared using special techniques that leads to much more

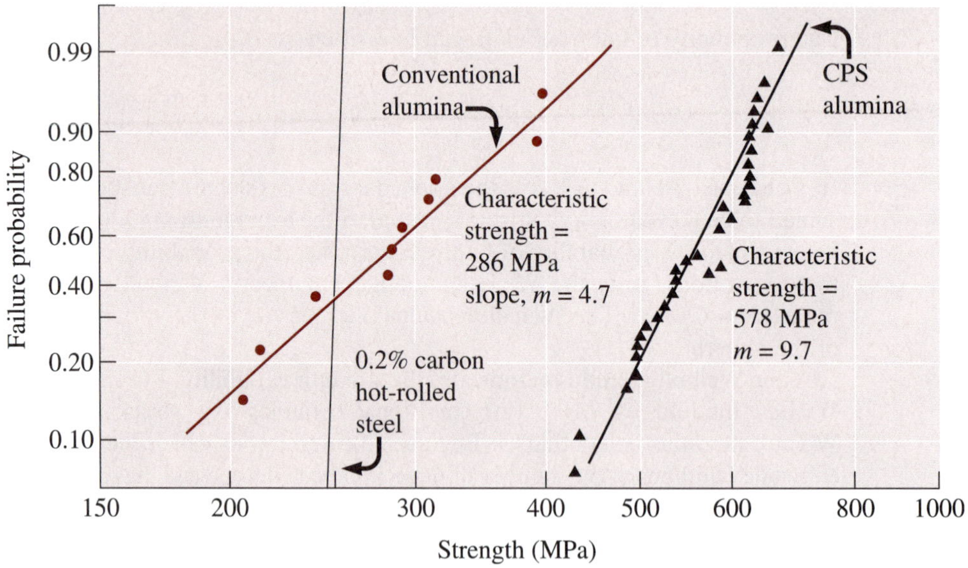

Figure 7-12 A cumulative plot (using special graph paper) of the probability that a sample will fail at any given stress yields the Weibull modulus or slope. Alumina produced by two different methods is compared with low carbon steel. Good reliability in design is obtained for a high Weibull modulus. (*Source: Used with permission from M.A. Meyers & K.K. Chawla*, Mechanical Behavior of Materials, *2nd ed., 2008, Cambridge University Press, UK.*)

uniform and controlled particle size. This in turn minimizes the flaws. The data for these samples are labeled as controlled particle size (CPS). Comment on the nature of these graphs.

SOLUTION

The failure probability and strength when plotted on a log-log scale result in data that can be fitted to a straight line. The slope of these lines provides us the measure of variability (i.e., the Weibull modulus).

For plain carbon steel the line is almost vertical (i.e., slope or m value is essentially approaching large values). This means that there is very little variation (5 to 10%) in the strength of different samples of the 0.2% C steel.

For alumina ceramics prepared using traditional processing, the variability is high (i.e., m is low $\sim$4.7).

For ceramics prepared using improved and controlled processing techniques the m is higher $\sim$9.7 indicating a more uniform distribution of flaws. The characteristic strength (σ_0) is also higher ($\sim$578 MPa) suggesting a lesser number of flaws that will lead to fracture.

EXAMPLE 7-8 *Strength of Ceramics and Probability of Failure*

An advanced engineered ceramic has a Weibull modulus $m = 9$. The flexural strength is 250 MPa at a probability of failure $F = 0.4$. What is the level of flexural strength if the probability of failure has to be 0.1?

SOLUTION

We assume all samples tested had the same volume, thus the size of the sample will not be a factor in this case. We can use the symbol V for sample volume instead of V_0. We are dealing with a brittle material, so we begin with Equation 7-9.

$$F(V) = 1 - P(V) = 1 - \exp\left[-\left(\frac{\sigma}{\sigma_0}\right)^m\right]$$

or

$$1 - F(V) = \exp\left[-\left(\frac{\sigma}{\sigma_0}\right)^m\right]$$

Take the logarithm of both sides to get

$$\ln[1 - F(V)] = \left[-\left(\frac{\sigma}{\sigma_0}\right)^m\right]$$

Take logarithms of both sides again,

$$\ln\{\ln[1 - F(V)]\} = -m(\ln \sigma - \ln \sigma_0) \qquad (7\text{-}10)$$

We can eliminate the minus sign on the right-hand side of Equation 7-10 by rewriting the equation as:

$$\ln\left[\ln\left(\frac{1}{1 - F(V)}\right)\right] = m(\ln \sigma - \ln \sigma_0) \qquad (7\text{-}11)$$

For $F = 0.4$, $\sigma = 250$ MPa, and $m = 9$, so from Equation 7-11, we have

$$\ln\left[\ln\left(\frac{1}{1-0.4}\right)\right] = 9(\ln 250 - \ln \sigma_0) \quad (7\text{-}12)$$

Therefore, $\ln\{\ln 1/0.6)\} = \ln\{(\ln 1.66667)\} = \ln\{0.510826\} = -0.67173 = 9(5.52146 - \ln \sigma_0)$.

Therefore, $\ln \sigma_0 = 5.52146 + 0.07464 = 5.5961$. This gives us a value of $\sigma_0 = 269.4$ MPa. This is the characteristic strength of the ceramic. For a stress level of 269.4 MPa, the probability of survival is 0.37 (or the probability of failure is 0.63). As the required probability of failure (F) goes down, the stress level to which the ceramic can be subjected (σ) also goes down.

Now, we want to determine the value of σ for $F = 0.1$. We know that $m = 9$ and $\sigma_0 = 269.4$ MPa, so we need to get the value of σ. We substitute these values into Equation 7-11:

$$\ln\left[\ln\left(\frac{1}{1-0.1}\right)\right] = 9(\ln \sigma - \ln 269.4)$$

$$\ln\left[\ln\left(\frac{1}{0.9}\right)\right] = 9(\ln \sigma - \ln 269.4)$$

$$\ln(\ln 1.11111) = \ln(0.105361) = -2.25037 = 9(\ln \sigma - 5.596097)$$

$$\therefore \quad -0.25004 = \ln \sigma - 5.596097 \quad \text{or}$$

$$\ln \sigma = 5.346056$$

or $\sigma = 209.8$ MPa. As expected, as we lowered the probability of failure to 0.1, we also decreased the level of stress that can be supported.

EXAMPLE 7-9 *Weibull Modulus Parameter Determination*

Seven silicon carbide specimens were tested and the following fracture strengths were obtained: 23, 49, 34, 30, 55, 43, and 40 MPa. Estimate the Weibull modulus for the data by fitting the data to Equation 7-11. Discuss the reliability of the ceramic.

SOLUTION

First, we point out that for any type of statistical analysis we need a large number of samples. Seven samples are not enough. The purpose of this example is to illustrate the calculation.

One simple, though not completely accurate, method for determining the behavior of the ceramic is to assign a numerical rank (1 to 7) to the specimens, with the specimen having the lowest fracture strength assigned the value 1. The total number of specimens is n (in our case, 7). The probability of failure F is then the numerical rank divided by $n + 1$ (in our case, 8). We can then plot

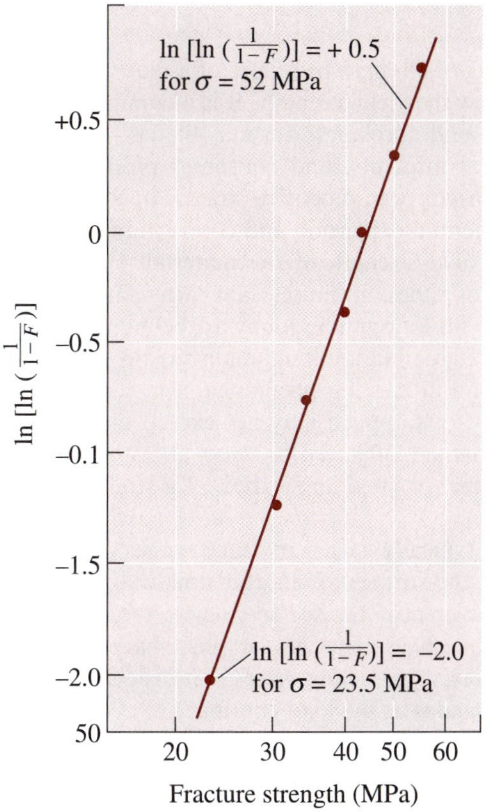

Figure 7-13 Plot of cumulative probability of failure versus fracture stress. Note the fracture strength is plotted on a log scale.

$\ln[\ln(1/1 - F(V_0))]$ versus $\ln \sigma$. The following table and Figure 7-13 show the results of these calculations. Note that σ is plotted on a log scale.

i^{th} Specimen	σ (MPa)	$F(V_0)$	$\ln\{\ln 1/[1 - F(V_0)]\}$
1	23	$1/8 = 0.125$	−2.013
2	30	$2/8 = 0.250$	−1.246
3	34	$3/8 = 0.375$	−0.755
4	40	$4/8 = 0.500$	−0.367
5	43	$5/8 = 0.625$	−0.019
6	49	$6/8 = 0.750$	+0.327
7	55	$7/8 = 0.875$	+0.732

The slope of the fitted line, or the Weibull modulus m, is (using the two points indicated on the curve):

$$m = \frac{0.5 - (-2.0)}{\ln(52) - \ln(23.5)} = \frac{2.5}{3.951 - 3.157} = 3.15$$

This low Weibull modulus of 3.15 suggests that the ceramic has a highly variable fracture strength, making it difficult to use reliably in high load-bearing applications.

7-6 Fatigue

Fatigue is the lowering of strength or failure of a material due to repetitive stress which may be above or below the yield strength. It is a common phenomenon in load-bearing components in cars and airplanes, turbine blades, springs, crankshafts and other machinery, biomedical implants, and consumer products, such as shoes or springs, that are subjected constantly to repetitive stresses in the form of tension, compression, bending, vibration, thermal expansion and contraction, or other stresses. These stresses are often *below* the yield strength of the material! However, when the stress occurs a sufficient number of times, it causes failure by fatigue! Quite a large fraction of components found in an automobile junkyard belongs to those that failed by fatigue. The possibility of a fatigue failure is the main reason why aircraft components have a finite life. Fatigue is an interesting phenomenon in that load-bearing components can fail while the overall stress applied may not exceed the yield stress! Fatigue can occur even if the components are subjected to stress above the yield strength. A component is often subjected to the repeated application of a stress *below* the yield strength of the material.

Fatigue failures typically occur in three stages. First, a tiny crack initiates or nucleates typically at the surface, often at a time well after loading begins. Normally, nucleation sites are at or near the surface, where the stress is at a maximum, and include surface defects such as scratches or pits, sharp corners due to poor design or manufacture, inclusions, grain boundaries, or dislocation concentrations. Next, the crack gradually propagates as the load continues to cycle. Finally, a sudden fracture of the material occurs when the remaining cross-section of the material is too small to support the applied load. Thus, components fail by fatigue because even though the overall applied stress may remain below the yield stress, at a local length scale the stress intensity exceeds the yield strength. For fatigue to occur, at least part of the stress in the material has to be tensile. We are normally concerned with fatigue of metallic and polymeric materials.

In ceramics, we normally do not consider fatigue since ceramics typically fail because of their low fracture toughness. Any fatigue cracks that may form will lower the useful life of the ceramic since it will cause the lowering of the fracture toughness. In general, we design ceramics for static (and not cyclic) loading and we factor in the Weibull modulus.

Polymeric materials also show fatigue failure. The mechanism of fatigue in polymers is different than that in metallic materials. In polymers, as the materials are subjected to repetitive stresses, considerable heating can occur near the crack tips and the inter-relationships between fatigue and another mechanism, known as **creep** (discussed in Section 7-9), affect the overall behavior.

Fatigue is also important in dealing with composites. As fibers or other reinforcing phases begin to degrade as a result of fatigue, the overall elastic modulus of the composite decreases and this weakening will be seen before the fracture due to fatigue.

Fatigue failures are often easy to identify. The fracture surface—particularly near the origin—is typically smooth. The surface becomes rougher as the original crack increases in size and may be fibrous during final crack propagation. Microscopic and macroscopic examinations reveal a fracture surface including a beach mark pattern and striations (Figure 7-14). **Beach** or **clamshell marks** (Figure 7-15) are normally formed when the load is changed during service or when the loading is intermittent, perhaps

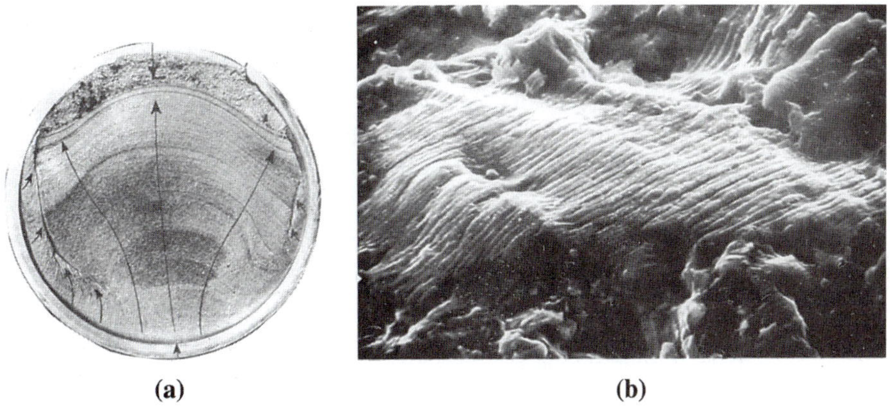

Figure 7-14 Fatigue fracture surface. (a) At low magnifications, the beach mark pattern indicates fatigue as the fracture mechanism. The arrows show the direction of growth of the crack front, whose origin is at the bottom of the photograph. (*Image* (a) is *from C.C. Cottell, "Fatigue Failures with Special Reference to Fracture Characteristics,"* Failure Analysis: The British Engine Technical Reports, *American Society for Metals, 1981, p. 318.*) (b) At very high magnifications, closely spaced striations formed during fatigue are observed (×1000).

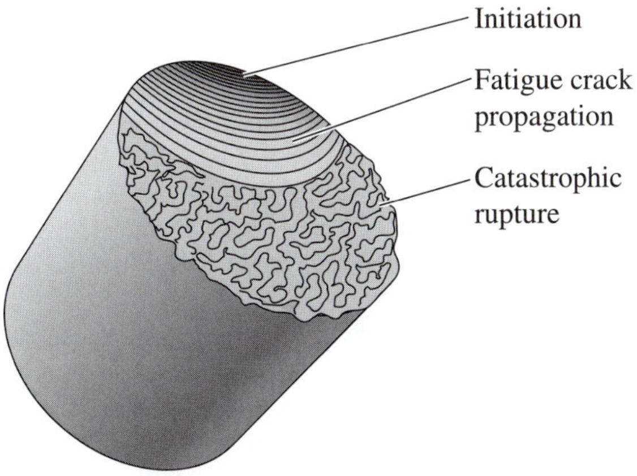

Figure 7-15 Schematic representation of a fatigue fracture surface in a steel shaft, showing the initiation region, the propagation of fatigue crack (with beach markings), and catastrophic rupture when the crack length exceeds a critical value at the applied stress.

permitting time for oxidation inside the crack. **Striations**, which are on a much finer scale, show the position of the crack tip after each cycle. Beach marks always suggest a fatigue failure, but—unfortunately—the absence of beach marks does not rule out fatigue failure.

EXAMPLE 7-10 *Fatigue Failure Analysis of a Crankshaft*

A crankshaft in a diesel engine fails. Examination of the crankshaft reveals no plastic deformation. The fracture surface is smooth. In addition, several other cracks appear at other locations in the crankshaft. What type of failure mechanism would you expect?

SOLUTION

Since the crankshaft is a rotating part, the surface experiences cyclical loading. We should immediately suspect fatigue. The absence of plastic deformation supports our suspicion. Furthermore, the presence of other cracks is consistent with fatigue; the other cracks didn't have time to grow to the size that produced catastrophic failure. Examination of the fracture surface will probably reveal beach marks or fatigue striations.

A conventional and older method used to measure a material's resistance to fatigue is the **rotating cantilever beam test** (Figure 7-16). One end of a machined, cylindrical specimen is mounted in a motor-driven chuck. A weight is suspended from the opposite end. The specimen initially has a tensile force acting on the top surface, while the bottom surface is compressed. After the specimen turns 90°, the locations that were originally in tension and compression have no stress acting on them. After a half revolution of 180°, the material that was originally in tension is now in compression. Thus, the stress at any one point goes through a complete sinusoidal cycle from maximum tensile stress to maximum compressive stress. The maximum stress acting on this type of specimen is given by

$$\pm\sigma = \frac{32\,M}{\pi d^3} \tag{7-13a}$$

In this equation M is the bending moment at the cross-section, and d is the specimen diameter. The bending moment $M = F \cdot (L/2)$, and therefore,

$$\pm\sigma = \frac{16\,FL}{\pi d^3} = 5.09\frac{FL}{d^3} \tag{7-13b}$$

where L is the distance between the bending force location and the support (Figure 7-16), F is the load, and d is the diameter.

Newer machines used for fatigue testing are known as direct-loading machines. In these machines, a servo-hydraulic system, an actuator, and a control system, driven by computers, applies a desired force, deflection, displacement or strain. In some of these machines, temperature and atmosphere (e.g., humidity level) also can be controlled.

After a sufficient number of cycles in a fatigue test, the specimen may fail. Generally, a series of specimens are tested at different applied stresses. The results are presented as an **S-N curve** (also known as the **W—hler curve**), with the stress (S) plotted versus the number of cycles (N) to failure (Figure 7-17).

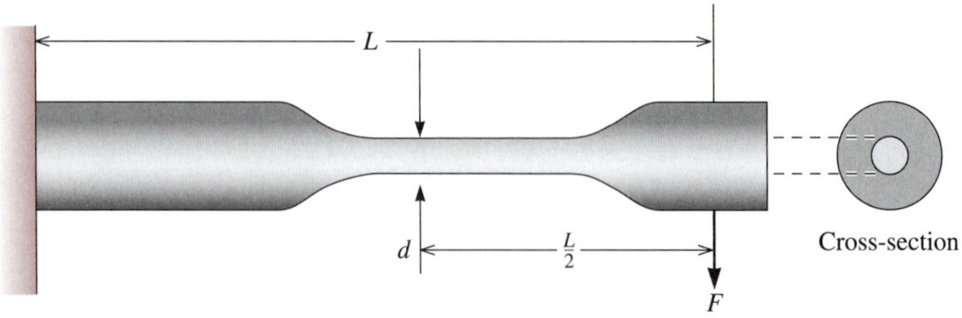

Figure 7-16 Geometry for the rotating cantilever beam specimen setup.

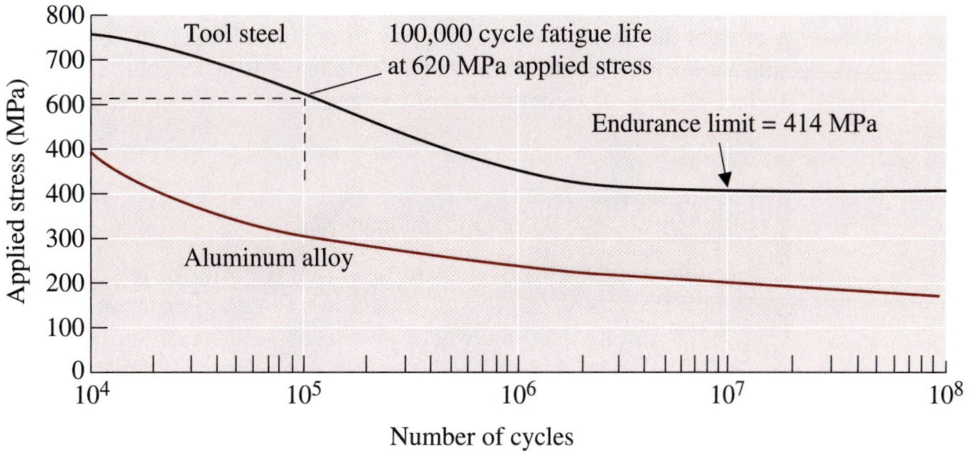

Figure 7-17 The stress-number of cycles to failure (S-N) curves for a tool steel and an aluminum alloy.

7-7 Results of the Fatigue Test

The **fatigue test** can tell us how long a part may survive or the maximum allowable loads that can be applied to prevent failure. The **endurance limit**, which is the stress below which there is a 50% probability that failure by fatigue will never occur, is our preferred design criterion. To prevent a tool steel part from failing (Figure 7-17), we must be sure that the applied stress is below 414 MPa. The assumption of existence of an endurance limit is a relatively older concept. Recent research on many metals has shown that probably an endurance limit does not exist. We also need to account for the presence of corrosion, occasional overloads, and other mechanisms that may cause the material to fail below the endurance limit. Thus, values for an endurance limit should be treated with caution.

Fatigue life tells us how long a component survives at a particular stress. For example, if the tool steel (Figure 7-17) is cyclically subjected to an applied stress of 620 MPa, the fatigue life will be 100,000 cycles. Knowing the time associated with each cycle, we can calculate a fatigue life value in years. **Fatigue strength** is the maximum stress for which fatigue will not occur within a particular number of cycles, such as 500,000,000. The fatigue strength is necessary for designing with aluminum and polymers, which have no endurance limit.

In some materials, including steels, the endurance limit is approximately half the tensile strength. The ratio is the **endurance ratio**:

$$\text{Endurance ratio} = \frac{\text{endurance limit}}{\text{tensile strength}} \approx 0.5 \tag{7-14}$$

The endurance ratio allows us to estimate fatigue properties from the tensile test. The endurance ratio values are $\sim$0.3 to 0.4 for metallic materials other than low and medium strength steels. *Again, note that research has shown that endurance limit does not exist for many materials.*

Most materials are **notch sensitive**, with the fatigue properties particularly sensitive to flaws at the surface. Design or manufacturing defects concentrate stresses and reduce the endurance limit, fatigue strength, or fatigue life. Sometimes highly polished surfaces are prepared in order to minimize the likelihood of a fatigue failure. **Shot peening** is also a process that is used very effectively to enhance fatigue life of materials. Small metal spheres are shot at the component. This leads to a residual compressive stress at the surface similar to strengthening of inorganic glasses by tempering (Section 7-9).

EXAMPLE 7-11 *Design of a Rotating Shaft*

A solid shaft for a cement kiln produced from tool steel must be 240 cm long and must survive continuous operation for one year with an applied load of 55,600 N. The shaft makes one revolution per minute during operation. Design a shaft that will satisfy these requirements. The S-N curve for the tool steel is shown in Figure 7-17.

SOLUTION

The fatigue life required for our design is the total number of cycles N that the shaft will experience in one year:

$$N = (1 \text{ cycle}/min)(60\ min/h)(24\ h/d)(365\ d/y)$$

$$N = 5.256 \times 10^5 \text{ cycles}/y$$

where $y =$ year, $d =$ day, and $h =$ hour.

From Figure 7-17, the applied stress therefore must be less than about 496 MPa. From Equation 7-13, the diameter of the shaft must be:

$$\pm\sigma = \frac{16FL}{\pi d^3} = 5.09\frac{FL}{d^3}$$

$$496 \text{ MPa} = \frac{(5.09)(240 \text{ cm})(55{,}600 \text{ N})}{d^3}$$

$$d = 11.1 \text{ cm}$$

A shaft with a diameter of 11.1 cm should operate for one year under these conditions. However, a significant margin of safety might be incorporated in the design. In addition, we might consider producing a shaft that would never fail.

Let us assume the factor of safety to be 2 (i.e., we will assume that the maximum allowed stress level will be 496/2 = 248 MPa). The minimum diameter required to prevent failure would now be:

$$248 \text{ MPa} = \frac{(5.09)(240 \text{ cm})(55{,}600 \text{ N})}{d^3}$$

$$d = 14 \text{ cm}$$

Selection of a larger shaft reduces the stress level and makes fatigue less likely to occur or delays the failure. Other considerations might, of course, be important. High temperatures and corrosive conditions are inherent in producing cement. If the shaft is heated or attacked by the corrosive environment, fatigue is accelerated. Thus, in the applications involving fatigue of components regular inspections of the components go a long way toward avoiding a catastrophic failure.

EXAMPLE 7-12 Do Materials have an Infinite Fatigue Life?

Typically, engineering materials such as alloys and composites are tested for fatigue up to about 10^7 cycles (known as low-cycle fatigue). Following are the data for an aluminum alloy designated as 7075T73 (the solid curve labeled as A), used for making helicopter propellers. The solid circles are the data for an aluminum matrix composite reinforced with SiC fibers (labeled as the B-dashed curve). (a) Based on these low-cycle fatigue data, which material appears to have better fatigue resistance? (b) What appears to be the fatigue strength of the aluminum matrix composites? (c) Can these data be used to predict the fatigue life of these materials for longer tests involving fatigue cycles up to 10^{10}?

SOLUTION

(a) From the low-cycle (i.e., up to 10^7 cycles) fatigue data, it is clear that the aluminum alloy reinforced with SiC fibers (material B) has a better fatigue behavior at a low number of cycles.

(b) The fatigue curve for material B appears to show a plateau at around 10^5 cycles. This would suggest a fatigue strength of ~300 MPa.

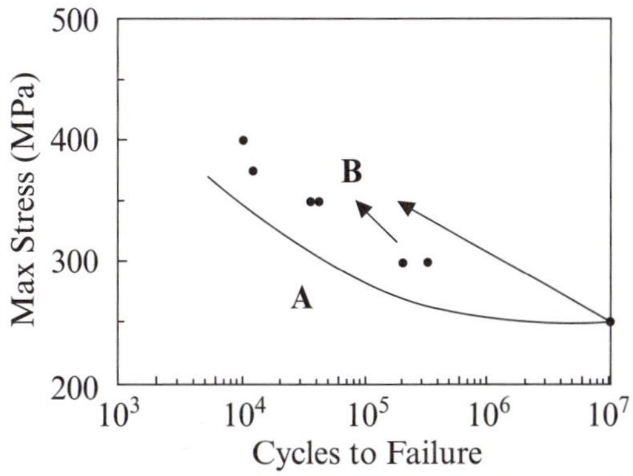

Figure 7-18 Low-cycle fatigue data for an aluminum alloy and an aluminum-alloy matrix composite reinforced with SiC. (*Source: Ref. Bathias, C.,* Fatigue and Fracture in Engineering Materials Structures, *Blackwell Publishing, page 559–565, Vol. 22 (1999).*)

(c) If we examine the data between 10^6 and 10^7 cycles, it appears that the stress to failure for the SiC reinforced composite (material B) now is lower and is approaching that for the alloy 7075T73 (material A). Thus, it probably would *not* be safe to assume that over a longer period of time (up to 10^{10} cycles) that the SiC reinforced material will have a better fatigue life than alloy 7075T73.

In fact, measurements conducted by Bathias and co-workers show that maximum stress that can be supported for the SiC reinforced alloy decreases even more to almost 200 MPa at ~10^{10} cycles.

Thus, an observation is that we must make fatigue measurements in the giga-cycle range to see the fatigue behavior and not just extend what is seen only in the low cycle fatigue regime. In fact, most materials may not have an infinite fatigue life, as is mostly suggested by the low-cycle fatigue data. These giga-cycle fatigue tests can be done on modern fatigue machines that are driven by piezoelectric materials at a frequency of ~20 kHz, compared to 100 Hz.

7-8 Application of Fatigue Testing

Material components are often subjected to loading conditions that do not give equal stresses in tension and compression (Figure 7-19). For example, the maximum stress during compression may be less than the maximum tensile stress. In other cases, the loading may be between a maximum and a minimum tensile stress; here the S-N curve is presented as the stress amplitude versus number of cycles to failure. *Stress amplitude* (σ_a) is defined as half of the difference between the maximum and minimum stresses;

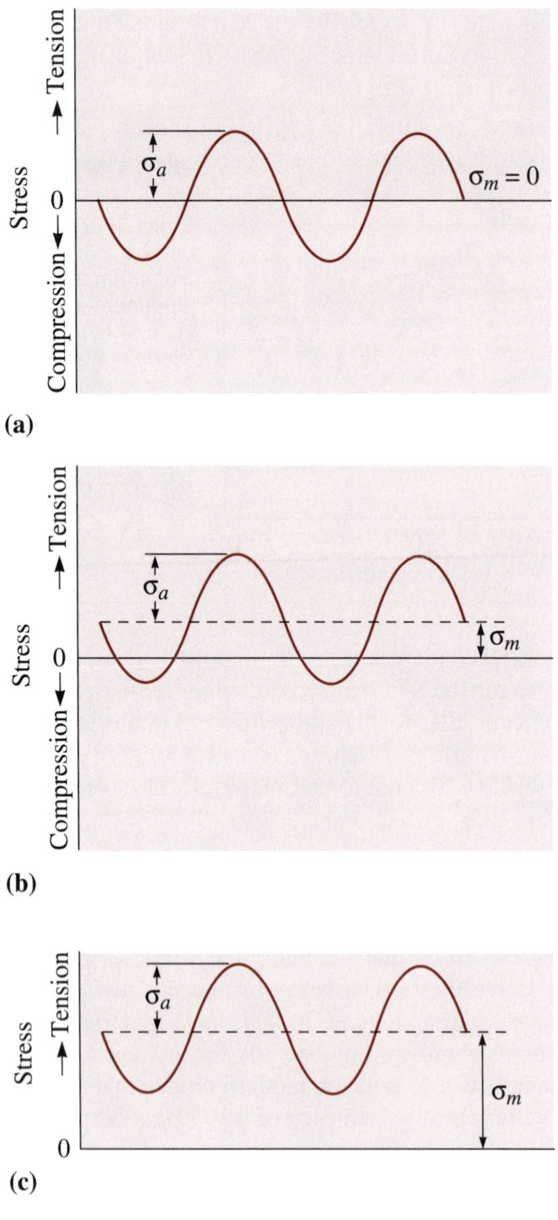

Figure 7-19
Examples of stress cycles. (a) Equal stress in tension and compression, (b) greater tensile stress than compressive stress, and (c) all of the stress is tensile.

mean stress (σ_m) is defined as the average between the maximum and minimum stresses:

$$\sigma_a = \frac{\sigma_{max} - \sigma_{min}}{2} \tag{7-15}$$

$$\sigma_m = \frac{\sigma_{max} + \sigma_{min}}{2} \tag{7-16}$$

A compressive stress is considered a "negative" stress. Thus, if the maximum tensile stress is 345 MPa and the minimum stress is a 69 MPa compressive stress, using Equations 7-15 and 7-16 the stress amplitude is 138 MPa and the mean stress is 207 MPa.

As the mean stress increases, the stress amplitude must decrease in order for the material to withstand the applied stresses. This condition can be summarized by the Goodman relationship:

$$\sigma_a = \sigma_{fs}\left[1 - \left(\frac{\sigma_m}{\sigma_{TS}}\right)\right] \tag{7-17}$$

where σ_{fs} is the desired fatigue strength for zero mean stress and σ_{TS} is the tensile strength of the material. Therefore, in a typical rotating cantilever beam fatigue test, where the mean stress is zero, a relatively large stress amplitude can be tolerated without fatigue. If, however, an airplane wing is loaded near its yield strength, vibrations of even a small amplitude may cause a fatigue crack to initiate and grow.

Crack Growth Rate In many cases, a component may not be in danger of failure even when a crack is present. To estimate when failure might occur, the rate of propagation of a crack becomes important. Figure 7-20 shows the crack growth rate versus the range of the stress-intensity factor ΔK, which characterizes crack geometry and the stress amplitude. Below a threshold ΔK, a crack does not grow; for somewhat higher

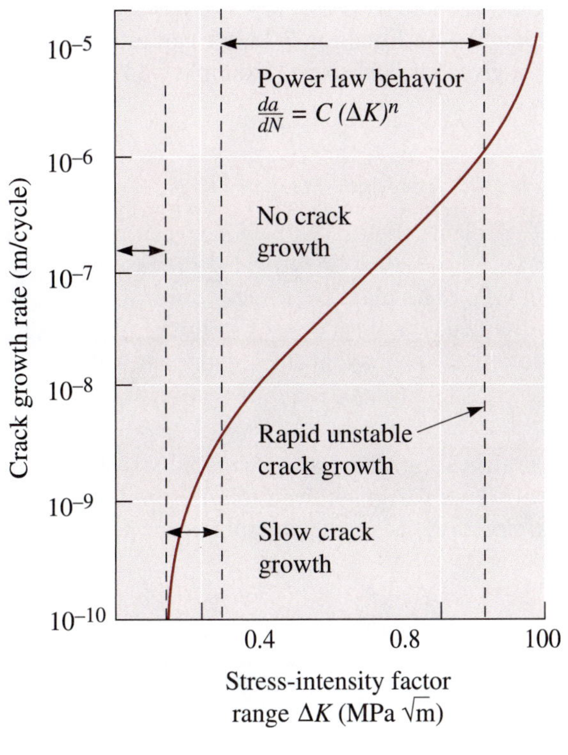

Figure 7-20
Crack growth rate versus stress-intensity factor range for a high-strength steel. For this steel, $C = 1.62 \times 10^{-12}$ and $n = 3.2$ for the units shown.

stress-intensities, cracks grow slowly; and at still higher stress-intensities, a crack grows at a rate given by:

$$\frac{da}{dN} = C(\Delta K)^n \tag{7-18}$$

In this equation, C and n are empirical constants that depend upon the material. Finally, when ΔK is still higher, cracks grow in a rapid and unstable manner until fracture occurs.

The rate of crack growth increases as a crack increases in size, as predicted from the stress intensity factor (Equation 7-1):

$$\Delta K = K_{\text{max}} - K_{\text{min}} = f\sigma_{\text{max}}\sqrt{\pi a} - f\sigma_{\text{min}}\sqrt{\pi a} = f\Delta\sigma\sqrt{\pi a} \tag{7-19}$$

If the cyclical stress $\Delta\sigma(\sigma_{\text{max}} - \sigma_{\text{min}})$ is not changed, then as crack length a increases, ΔK and the crack growth rate da/dN increase. In using this expression, however, one should note that a crack will not propagate during compression. Therefore, if σ_{min} is compressive, or less than zero, then σ_{min} should be set equal to zero.

Knowledge of crack growth rate is of assistance in designing components and in nondestructive evaluation to determine if a crack poses imminent danger to the structure. One approach to this problem is to estimate the number of cycles required before failure occurs. By rearranging Equation 7-18 and substituting for ΔK:

$$dN = \frac{1}{Cf^n\Delta\sigma^n\pi^{n/2}}\frac{da}{a^{n/2}}$$

If we integrate this expression between the initial size of a crack and the crack size required for fracture to occur, we find that

$$N = \frac{2[(a_c)^{(2-n)/2} - (a_i)^{(2-n)/2}]}{(2-n)Cf^n\Delta\sigma^n\pi^{n/2}} \tag{7-20}$$

where a_i is the initial flaw size and a_c is the flaw size required for fracture. If we know the material constants n and C in Equation 7-18, we can estimate the number of cycles required for failure for a given cyclical stress (Example 7-13).

EXAMPLE 7-13 *Design of a Fatigue-Resistant Plate*

A high-strength steel plate (Figure 7-20), which has a plane strain fracture toughness of 80 MPa $\sqrt{\text{m}}$, is alternately loaded in tension to 500 MPa and in compression to 60 MPa. The plate is to survive for 10 years, with the stress being applied at a frequency of once every 5 minutes. Design a manufacturing and testing procedure that assures that the component will serve as intended.

SOLUTION

To design our manufacturing and testing capability, we must determine the maximum size of any flaws that might lead to failure within the 10-year period. The critical crack size (a_c), using the fracture toughness and the maximum stress, is:

$$K_{Ic} = f\sigma\sqrt{\pi a_c}$$

$$80 \text{ MPa}\sqrt{\text{m}} = (1)(500 \text{ MPa})\sqrt{\pi a_c}$$

$$a_c = 0.0081 \text{ m} = 8.1 \text{ mm}$$

The maximum stress is 500 MPa; however, the minimum stress is zero, not 60 MPa in compression, because cracks do not propagate in compression. Thus, $\Delta\sigma$ is

$$\Delta\sigma = \sigma_{\text{max}} - \sigma_{\text{min}} = 500 - 0 = 500 \text{ MPa}$$

We need to determine the minimum number of cycles that the plate must withstand:

$$N = (1 \text{ cycle}/5 \text{ min})(60 \text{ min/h})(24 \text{ h/d})(365 \text{ d/y})(10 \text{ y})$$

$$N = 1{,}051{,}200 \text{ cycles}$$

If we assume that $f = 1$ for all crack lengths and note that $C = 1.62 \times 10^{-12}$ and $n = 3.2$ in Equation 7-20, then

$$1{,}051{,}200 = \frac{2[(0.008)^{(2-3.2)/2} - (a_i)^{(2-3.2)/2}]}{(2-3.2)(1.62 \times 10^{-12})(1)^{3.2}(500)^{3.2}\pi^{3.2/2}}$$

$$1{,}051{,}200 = \frac{2[18 - a_i^{-0.6}]}{(-1.2)(1.62 \times 10^{-12})(1)(4.332 \times 10^{8})(6.244)}$$

$$a_i^{-0.6} = 18 + 2764 = 2782$$

$$a_i = 1.82 \times 10^{-6} \text{ m} = 0.00182 \text{ mm for surface flaws}$$

$$2a_i = 0.00364 \text{ mm for internal flaws}$$

The manufacturing process must produce surface flaws smaller than 0.00182 mm in length. We can conduct a similar calculation for specifying a limit on edge cracks. In addition, nondestructive tests must be available to assure that cracks exceeding this length are not present.

Effect of Temperature As the material's temperature increases, both fatigue life and endurance limit decrease. Furthermore, a cyclical temperature change encourages failure by thermal fatigue; when the material heats in a nonuniform manner, some parts of the structure expand more than others. This nonuniform expansion introduces a stress within the material, and when the structure later cools and contracts, stresses of the opposite sign are imposed. As a consequence of the thermally induced stresses and strains, fatigue may eventually occur. The frequency with which the stress is applied also influences fatigue behavior. In particular, high-frequency stresses may cause polymer materials to heat; at increased temperature, polymers fail more quickly. Chemical effects of temperature (e.g., oxidation) must also be considered.

7-9 Creep, Stress Rupture, and Stress Corrosion

If we apply stress to a material at an elevated temperature, the material may stretch and eventually fail, even though the applied stress is *less* than the yield strength at that temperature. A time dependent permanent deformation under a constant load or constant stress and at high temperatures is known as **creep**. A large number of failures occurring in components used at high temperatures can be attributed to creep or a combination of creep and fatigue. Essentially, in creep the material begins to flow slowly. Diffusion, dislocation glide or climb, or grain boundary sliding can contribute

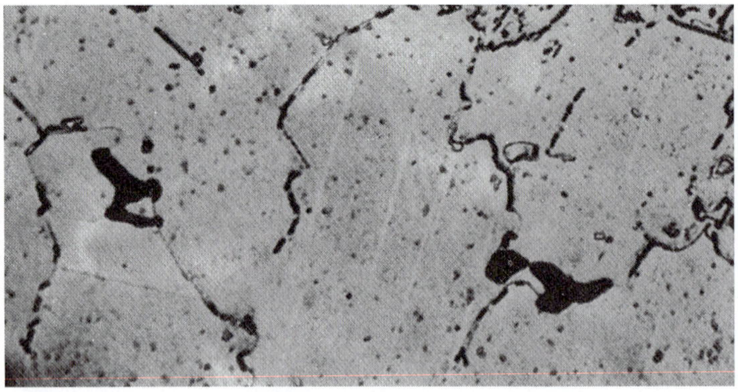

Figure 7-21 Creep cavities formed at grain boundaries in an austentic stainless steel (×500). (*From* ASM Handbook, *Vol. 7, (1972) ASM International, Materials Park, OH 44073.*)

to the creep of metallic materials. Polymeric materials also show creep. In ductile metals and alloys subjected to creep, fracture is accompanied by necking, void nucleation and coalescence, or grain boundary sliding.

A material is considered failed by creep even if it has *not* actually fractured. When a material does actually creep and then ultimately break the fracture is defined as *stress rupture*. Normally, ductile stress-rupture fractures include necking and the presence of many cracks that did not have an opportunity to produce final fracture. Furthermore, grains near the fracture surface tend to be elongated. Ductile stress-rupture failures generally occur at high creep rates and relatively low exposure temperatures and have short rupture times. Brittle stress-rupture failures usually show little necking and occur more often at smaller creep rates and high temperatures. Equiaxed grains are observed near the fracture surface. Brittle failure typically occurs by formation of voids at the intersection of three grain boundaries and precipitation of additional voids along grain boundaries by diffusion processes (Figure 7-21).

Stress-Corrosion **Stress-corrosion** is a phenomenon in which materials react with corrosive chemicals in the environment. This leads to formation of cracks and lowering of strength. Stress-corrosion can occur at stresses well below the yield strength of the metallic, ceramic, or glassy material due to attack by a corrosive medium. In metallic materials, deep, fine corrosion cracks are produced, even though the metal as a whole shows little uniform attack. The stresses can be either externally applied or stored residual stresses. Stress-corrosion failures are often identified by microstructural examination of the nearby metal. Ordinarily, extensive branching of the cracks along grain boundaries is observed (Figure 7-22). The location at which cracks initiated may be identified by the presence of a corrosion product.

Inorganic silicate glasses are especially prone to failure by reaction with water vapor. It is well known that the strength of silica fibers or silica glass products is very high when these materials are protected from water vapor. As the fibers or silica glass components get exposed to water vapor, corrosion reactions begin leading to formation of surface flaws, which ultimately cause the cracks to grow when stress is applied. Polymeric coatings are applied to optical fibers to prevent them from reacting with water vapor. For bulk glasses, special heat treatments such as tempering are used. Tempering produces an overall compressive stress on the surface of glass. Thus, even if the glass surface reacts with water vapor the cracks do not grow since the overall stress at the surface is compressive. If we create a flaw that will penetrate the compressive

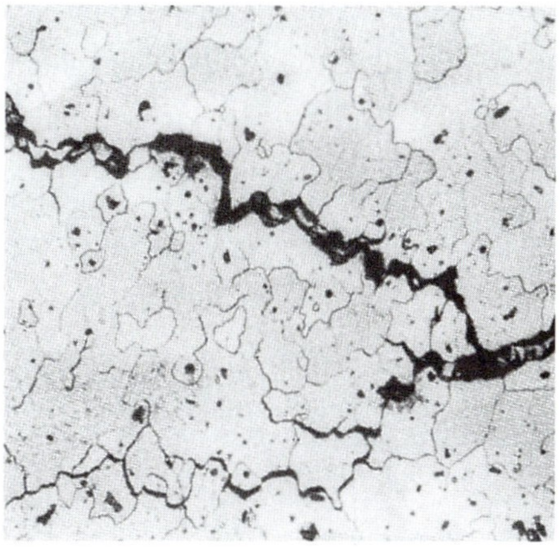

Figure 7-22
Photomicrograph of a metal near a stress-corrosion fracture, showing the many intergranular cracks formed as a result of the corrosion process (×200). (*From* ASM Handbook, *Vol. 7, (1972) ASM International, Materials Park, OH 44073.*)

stress region on the surface, tempered glass will shatter. Tempered glass is used widely in building and automotive applications.

EXAMPLE 7-14 *Failure Analysis of a Pipe*

A titanium pipe used to transport a corrosive material at 400°C is found to fail after several months. How would you determine the cause for the failure?

SOLUTION

Since a period of time at a high temperature was required before failure occurred, we might first suspect a creep or stress-corrosion mechanism for failure. Microscopic examination of the material near the fracture surface would be advisable. If many tiny, branched cracks leading away from the surface are noted, stress-corrosion is a strong possibility. However, if the grains near the fracture surface are elongated, with many voids between the grains, creep is a more likely culprit.

7-10 Evaluation of Creep Behavior

To determine the creep characteristics of a material, a constant stress is applied to a heated specimen in a **creep test**. As soon as the stress is applied, the specimen stretches elastically a small amount ε_0 (Figure 7-23), depending on the applied stress and the modulus of elasticity of the material at the high temperature. Creep testing can also be conducted under a constant load and is important from an engineering design viewpoint.

Dislocation Climb High temperatures permit dislocations in a metallic material to **climb**. In climb, atoms move either to or from the dislocation line by diffusion, causing the dislocation to move in a direction that is perpendicular, not parallel, to the slip plane (Figure 7-24). The dislocation escapes from lattice imperfections, continues to slip, and causes additional deformation of the specimen even at low applied stresses.

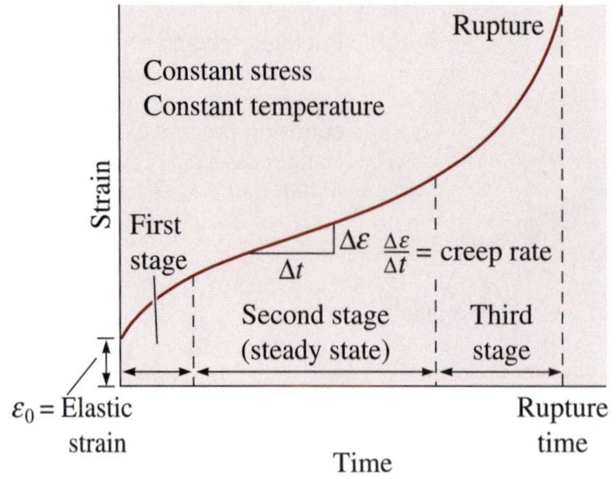

Figure 7-23
A typical creep curve showing the strain produced as a function of time for a constant stress and temperature.

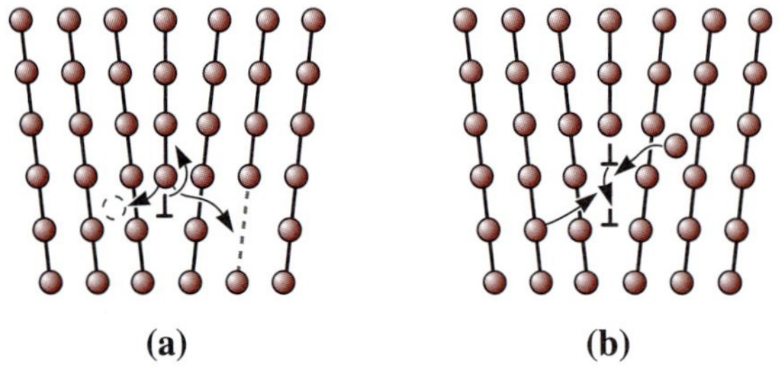

Figure 7-24
Dislocations can climb (a) when atoms leave the dislocation line to create interstitials or to fill vacancies or (b) when atoms are attached to the dislocation line by creating vacancies or eliminating interstitials.

Creep Rate and Rupture Times During the creep test, strain or elongation is measured as a function of time and plotted to give the creep curve (Figure 7-23). In the first stage of creep of metals, many dislocations climb away from obstacles, slip, and contribute to deformation. Eventually, the rate at which dislocations climb away from obstacles equals the rate at which dislocations are blocked by other imperfections. This leads to second-stage, or steady-state, creep. The slope of the steady-state portion of the creep curve is the **creep rate**:

$$\text{Creep rate} = \frac{\Delta \text{ strain}}{\Delta \text{ time}} \tag{7-21}$$

Eventually, during third-stage creep, necking begins, the stress increases, and the specimen deforms at an accelerated rate until failure occurs. The time required for failure to occur is the **rupture time**. Either a higher stress or a higher temperature reduces the rupture time and increases the creep rate (Figure 7-25).

The combined influence of applied stress and temperature on the creep rate and rupture time (t_r) follows an Arrhenius relationship:

$$\text{Creep rate} = \mathrm{C}\sigma^{\mathrm{n}} \exp\left(-\frac{Q_c}{\mathrm{R}T}\right) \tag{7-22}$$

$$t_r = \mathrm{K}\sigma^{\mathrm{m}} \exp\left(\frac{Q_r}{\mathrm{R}T}\right) \tag{7-23}$$

where R is the gas constant, T is the temperature in Kelvin, C, K, n, and m are constants for the material, Q_c is the activation energy for creep, and Q_r is the activation energy for rupture. In particular, Q_c is related to the activation energy for self-diffusion

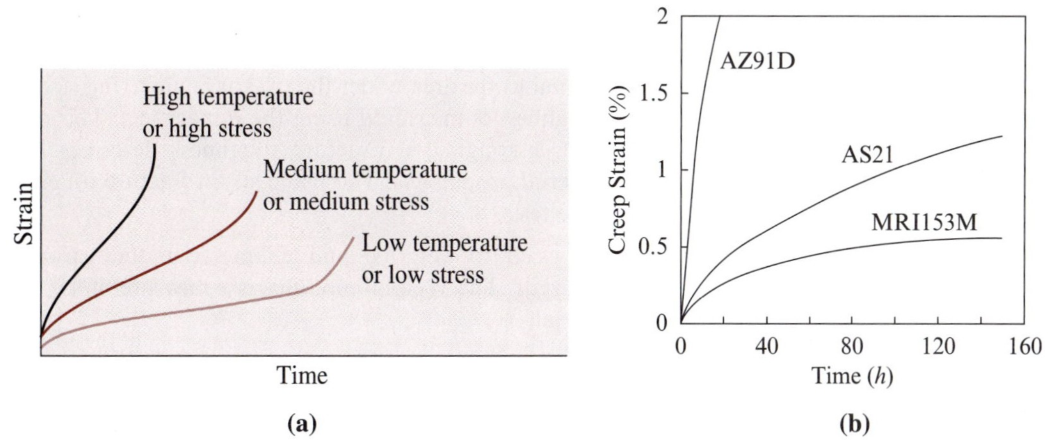

Figure 7-25 (a) The effect of temperature or applied stress on the creep curve. (b) Relative creep resistance of Mg alloys AZ91D, AS21, and MRI153M. (*Ref. A.M. Russell and K.L. Lee in* Structure-Property Relations in Nonferrous Metals, *Publ. Wiley, (2005), page 176.*)

when dislocation climb is important. Relative creep resistance of three magnesium alloys, namely AZ91D, AS21, and MRI153M is shown in Figure 7-25(b).

In creep of polycrystalline ceramics, other factors—including grain boundary sliding and nucleation of microcracks—are particularly important. Often, a noncrystalline or glassy material is present at the grain boundaries; the activation energy required for the glass to deform is low, leading to high creep rates compared with completely crystalline ceramics. For the same reason, creep occurs at a rapid rate in ceramic glasses and amorphous polymers.

The stress exponent (n) and the creep activation energy (Q_c) encountered in Equation 7-22 for Mg alloys MRI 151, MRI 153, and As21 are shown in Table 7-2.

In Mg alloys containing Al, an intermetallic compound $Mg_{12}Al_{12}$ can form at grain boundaries, and this causes the creep resistance to be lowered. The creep resistance is enhanced by adding small concentrations of alkaline earth metals, such as Ca or Mg. These metals react with Al preferentially and form other intermetallics (i.e., Al_2Ca and Al_2Mg). These intermetallics have melting temperatures greater than 1000°C. This translates into enhanced creep resistance, as grain boundary sliding is suppressed.

TABLE 7-2 ■ *Creep exponent (n) and activation energy (Q_c) for some Mg Alloys*

Alloy	Stress Exponent (n) T = 135°C, stress 85–110 MPa	Activation energy (Q_c) kJ/mol (90 MPa, 130–150°C)
MRI 151	7.0	175
MRI 153	7.6	181
AS21	19.5	166

(Source: A.M. Russell and K.L. Lee in Structure-Property Relations in Nonferrous Metals, Publ. Wiley, (2005), page 176.)

SUMMARY

- Toughness refers to the ability of materials to absorb energy before they fracture. Tensile toughness is equal to the area under the true or engineering stress-true strain curve. The impact toughness is measured using the impact test. This could be very different from the tensile toughness. Fracture toughness describes how easily a crack or flaw in a material propagates. The plane strain fracture toughness K_{Ic} is a common result of these tests.
- Weibull statistics are used to describe and characterize the variability in the strength of brittle materials. The Weibull modulus is a measure of the variability of the strength of a material.
- The fatigue test permits us to understand how a material performs when a cyclical stress is applied. Knowledge of the rate of crack growth can help determine fatigue life.
- Microstructural analysis of fractured surfaces can lead to better insights into the origin and cause of fracture. Different microstructural features are associated with ductile and brittle fracture as well as fatigue failure.
- The creep test provides information on the load-carrying ability of a material at high temperatures. Creep rate and rupture time are important properties obtained from these tests.

GLOSSARY

Beach or clamshell marks Patterns often seen on a component subjected to fatigue. Normally formed when the load is changed during service or when the loading is intermittent, perhaps permitting time for oxidation inside the crack.

Chevron pattern A common fracture feature produced by separate crack fronts propagating at different levels in the material.

Climb Movement of a dislocation perpendicular to its slip plane by the diffusion of atoms to or from the dislocation line.

Conchoidal fracture Fracture surface containing a very smooth mirror zone near the origin of the fracture, with tear lines comprising the remainder of the surface. This is typical of amorphous materials.

Creep A time dependent, permanent deformation at high temperatures, occurring at constant load or constant stress.

Creep rate The rate at which a material deforms when a stress is applied at a high temperature.

Creep test Measures the resistance of a material to deformation and failure when subjected to a static load below the yield strength at an elevated temperature.

Delamination The process by which different layers in a composite will begin to debond.

Endurance limit An older concept that defined a stress below which a material will not fail in a fatigue test. Factors as corrosion or occasional overloading can cause materials to fail at stresses below the assumed endurance limit.

Endurance ratio The endurance limit divided by the tensile strength of the material. The ratio is about 0.5 for many ferrous metals. See the cautionary note on endurance limit.

Fatigue life The number of cycles permitted at a particular stress before a material fails by fatigue.

Fatigue strength The stress required to cause failure by fatigue in a given number of cycles, such as 500 million cycles.

Fatigue test Measures the resistance of a material to failure when a stress below the yield strength is repeatedly applied.

Fracture mechanics The study of a material's ability to withstand stress in the presence of a flaw.

Fracture toughness The resistance of a material to failure in the presence of a flaw.

Griffith flaw A crack or flaw in a material that concentrates and magnifies the applied stress.

Impact test Measures the ability of a material to absorb the sudden application of a load without breaking. The Charpy and Izod tests are commonly used impact tests.

Intergranular In between grains or along the grain boundaries.

Microvoids Development of small holes in a material. These form when a high stress causes separation of the metal at grain boundaries or interfaces between the metal and inclusions.

Notch sensitivity Measures the effect of a notch, scratch, or other imperfection on a material's properties, such as toughness or fatigue life.

Rotating cantilever beam test An older test for fatigue testing.

Rupture time The time required for a specimen to fail by creep at a particular temperature and stress.

S-N curve (also known as the Wöhler curve) A graph showing stress as a function of number of cycles in fatigue.

Shot peening A process in which metal spheres are shot at a component. This leads to a residual compressive stress at the surface of a component and this enhances fatigue life.

Stress-corrosion A phenomenon in which materials react with corrosive chemicals in the environment, leading to the formation of cracks and lowering of strength.

Striations Patterns seen on a fractured surface of a fatigued sample. These are on a much finer scale than beach marks and show the position of the crack tip after each cycle.

Tempering A glass heat treatment that makes the glass safer; it does so by creating a compressive stress layer at the surface.

Toughness A qualitative measure of the energy required to cause fracture of a material. A material that resists failure by impact is said to be tough. One measure of toughness is the area under the true stress-strain curve (tensile toughness), another is the impact energy measured during an impact test (impact toughness). The ability of materials containing flaws to withstand load is known as fracture toughness.

Transgranular Meaning across the grains (e.g., a transgranular fracture would be fracture in which cracks would go through the grains).

Weibull distribution A mathematical distribution showing the probability of failure or survival of a material as a function of the stress.

Weibull modulus (*m*) A parameter related to the Weibull distribution. It is an indicator of the variability of the strength of materials resulting from a distribution of flaw sizes.

Wöhler curve Graph showing fatigue stress as a function of number of cycles (also known as the S-N curve).

PROBLEMS

Section 7-1 Fracture Mechanics

Section 7-2 The Importance of Fracture Mechanics

7-1 Alumina Al_2O_3 is a brittle ceramic with low toughness. Suppose that fibers of silicon carbide SiC, another brittle ceramic with low toughness, could be embedded within the alumina. Would doing this affect the toughness of the ceramic matrix composite? Explain.

7-2 A ceramic matrix composite contains internal flaws as large as 0.001 cm in length. The plane strain fracture toughness of the composite is 45 $MPa\sqrt{m}$ and the tensile strength is 550 MPa. Will the stress cause the composite to fail before the tensile strength is reached? Assume that $f = 1$.

7-3 An aluminum alloy that has a plane strain fracture toughness of 27.5 $MPa\sqrt{m}$ fails when a stress of 290 MPa is applied. Observation of the fracture surface indicates that fracture began at the surface of the part. Estimate the size of the flaw that initiated fracture. Assume that $f = 1.1$.

7-4 A polymer that contains internal flaws 1 mm in length fails at a stress of 25 MPa. Determine the plane strain fracture toughness of the polymer. Assume that $f = 1$.

7-5 A ceramic part for a jet engine has a yield strength of 517 MPa and a plane strain fracture toughness of 5.5 $MPa\sqrt{m}$. To be sure that the part does not fail, we plan to assure that the maximum applied stress is only one-third the yield strength. We use a nondestructive test that will detect any internal flaws greater than 0.125 cm long. Assuming that $f = 1.4$, does our nondestructive test have the required sensitivity? Explain.

7-6 Assume that the critical stress intensity factor or fracture toughness (K_{Ic}) for a partially stabilized zirconia is 10 $MPa\text{-}m^{1/2}$. If there is a plate of this ceramic with a sharp edge notch 100 μm deep and subjected to a stress of 300 MPa, will this plate be able to withstand this stress?

7-7 Assume that the critical stress intensity factor or fracture toughness (K_{Ic}) for a partially stabilized zirconia is 10 $MPa\text{-}m^{1/2}$. If there is a plate of this ceramic with an internal notch 100 μm deep and subjected to a stress of 300 MPa, will this plate be able to withstand this stress?

Section 7-3 Microstructural Features of Fracture in Metallic Materials

Section 7-4 Microstructural Features of Fracture in Ceramics, Glasses, and Composites

7-8 Concrete has exceptional strength in compression but it fails rather easily in tension. Explain why.

7-9 What controls the strength of glasses? What can be done to enhance the strength of silicate glasses?

Section 7-5 Weibull Statistics for Failure Strength Analysis

7-10 Sketch a schematic of the strength of ceramics and that of metals and alloys as a function of probability of failure. Explain the differences you anticipate.

7-11 Why does the strength of ceramics vary considerably with the size of ceramic components?

7-12 Explain the significance of the Weibull distribution.

Section 7-6 Fatigue

Section 7-7 Results of the Fatigue Test

Section 7-8 Application of Fatigue Testing

7-13 A cylindrical tool steel specimen that is 15 cm long and 0.625 cm in diameter rotates as a cantilever beam and is to be designed so that failure never occurs. Assuming that the maximum tensile and compressive stresses are equal, determine the maximum load that can be applied to the end of the beam. (See Figure 7-17.)

7-14 A 2-cm-diameter, 20-cm-long bar of an acetal polymer (Figure 7-26) is loaded on one end and

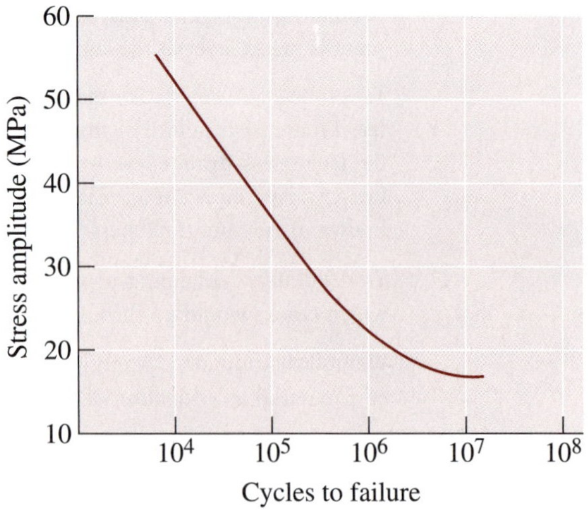

Figure 7-26 The S-N fatigue curve for an acetal polymer (for Problems 7-14, 7-16, and 7-17).

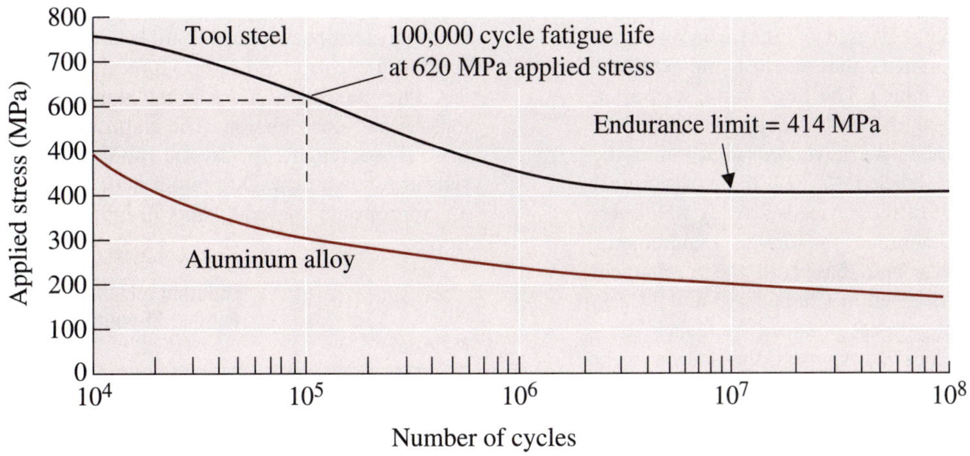

Figure 7-17 (Repeated for Problems 7-13 and 7-15) The stress-number of cycles to failure (S-N) curves for a tool steel and an aluminum alloy.

is expected to survive one million cycles of loading, with equal maximum tensile and compressive stresses, during its lifetime. What is the maximum permissible load that can be applied?

7-15 A cyclical load of 6672 N is to be exerted at the end of a 25-cm-long aluminum beam (Figure 7-17). The bar must survive for at least 10^6 cycles. What is the minimum diameter of the bar?

7-16 A cylindrical acetal polymer bar 20 cm long and 1.5 cm in diameter is subjected to a vibrational load at a frequency of 500 vibrations per minute, with a load of 50 N. How many hours will the part survive before breaking? (See Figure 7-26.)

7-17 Suppose that we would like a part produced from the acetal polymer shown in Figure 7-26 to survive for one million cycles under conditions that provide for equal compressive and tensile stresses. What is the fatigue strength, or maximum stress amplitude, required? What are the maximum stress, the minimum stress, and the mean stress on the part during its use? What effect would the frequency of the stress application have on your answers? Explain.

7-18 Explain how fatigue failure occurs even if the material does not see overall stress levels higher than the yield strength.

7-19 A fatigue test is conducted on an aluminum alloy at a frequency of 100 Hz. If the number of cycles is 10^7, how much time will this fatigue test take? How much will be the time if this test were conducted for 10^8 cycles?

7-20 How much time will a piezoelectric fatigue testing machine take to conduct a fatigue test on a titanium alloy for 10^{10} cycles? Assume that the frequency of this test is 10 kHz.

Section 7-9 Creep, Stress Rupture, and Stress Corrosion

7-21 Define the term "creep" and differentiate creep from stress relaxation.

7-22 What is meant by the terms "stress rupture" and "stress corrosion?"

7-23 What is the difference between failure of a material by creep and that by stress rupture?

Design Problems

7-24 A hook (Figure 7-27) for hoisting containers of ore in a mine is to be designed using a non-

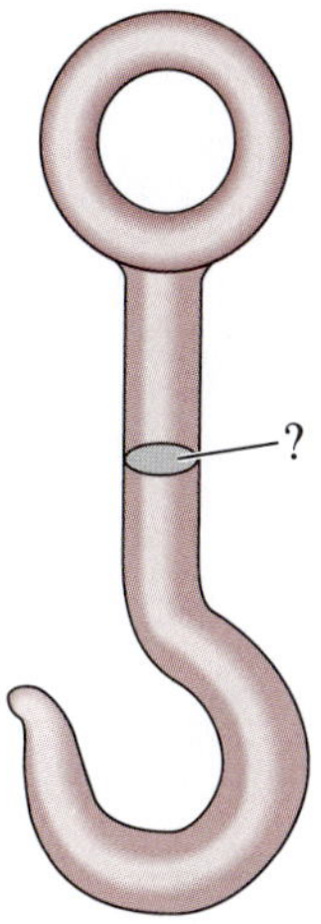

Figure 7-27 Schematic of a hook (for Problem 7-24).

ferrous (not based on iron) material. (A nonferrous material is used because iron and steel could cause a spark that would ignite explosive gases in the mine.) The hook must support a load of 111,200 N, and a factor of safety of 2 should be used. We have determined that the cross-section labeled "?" is the most critical area; the rest of the device is already well overdesigned. Determine the design requirements for this device and, based on the mechanical property data given in Chapters 13 and 14 and the metal/alloy prices obtained from such sources as your local newspapers, the internet website of London Metal Exchange or *The Wall Street Journal*, design the hook and select an economical material for the hook.

7-25 A support rod for the landing gear of a private airplane is subjected to a tensile load during landing. The loads are predicted to be as high as 177,920 N. Because the rod is crucial and failure could lead to a loss of life, the rod is to be designed with a factor of safety of 4 (that is, designed so that the rod is capable of supporting loads four times as great as expected). Operation of the system also produces loads that may induce cracks in the rod. Our nondestructive testing equipment can detect any crack greater than 0.05 cm deep. Based on the materials given in Table 7-1, design the support rod and the material, and justify your answer.

7-26 A lightweight rotating shaft for a pump on the national aerospace plane is to be designed to support a cyclical load of 66,720 N during service. The maximum stress is the same in both tension and compression. The endurance limits or fatigue strengths for several candidate materials are shown here. Design the shaft, including an appropriate material, and justify your solution.

Material	Endurance Limit/ Fatigue Strength (MPa)
Al-Mn alloy	110
Al-Mg-Zn alloy	225
Cu-Be alloy	295
Mg-Mn alloy	80
Be alloy	180
Tungsten alloy	320

7-27 A ductile cast-iron bar is to support a load of 177,920 N in a heat-treating furnace used to make malleable cast iron. The bar is located in a spot that is continuously exposed to 500°C. Design the bar so that it can operate for at least 10 years without failing.

8

Strain Hardening and Annealing

Have You Ever Wondered?

- *Why does bending a copper wire make it stronger?*
- *What type of steel improves the crashworthiness of cars?*
- *How are aluminum beverage cans made?*
- *Why do thermoplastics get stronger when strained?*
- *Why is it that the strength of the metallic material around a weld could be lower than that of the surrounding material?*

In this chapter, we will learn how the strength of metals and alloys is influenced by mechanical processing and heat treatments. In Chapter 4, we learned about the different techniques that can strengthen metals and alloys (e.g., enhancing dislocation density, decreasing grain size, alloying, etc.). In this chapter, we will learn how to enhance the strength of metals and alloys using cold working, a process by which a metallic material is simultaneously deformed and strengthened. We will also see how hot working can be used to shape metals and alloys by deformation at high temperatures without strengthening. We will learn how the annealing heat treatment can be used to enhance ductility and counter the increase in hardness caused by cold working. The strengthening we obtain during cold working, which is brought about by increasing the dislocation density, is called **strain hardening** or **work hardening**. By controlling the **thermo-mechanical processing** (i.e., combinations of mechanical processing and heat treatment), we are able to process metallic materials into a usable shape yet still improve and control their mechanical properties.

The topics discussed in this chapter pertain particularly to metals and alloys. Strain hardening (obtained by multiplication of dislocations) requires that the materials have ductility. We use strain hardening as a tool to enhance strength of a material. We have to counter the effects of strain hardening in manufacturing processes. For example, when we draw a wire or extrude a tube, strain hardening can occur and we have to ensure that the product still has acceptable ductility. Cars and trucks are made by stamping out a material known as sheet steel. This process leads to aerodynamic and aesthetically pleasing car chassis. The sheet steel used must exhibit an ability to stretch and bend easily during stamping. However, we must ultimately produce a strong steel that can withstand minor bumps and major impacts. The increase in the strength of steel as a result of strain hardening helps us in this regard. Furthermore, for better crashworthiness we must use steels that exhibit rapid strain hardening during impact loading.

What about polymers, glasses, and ceramics? Do they also exhibit strain hardening? We will show that the deformation of thermoplastic polymers often produces a strengthening effect. However, the mechanism of deformation strengthening is completely different in polymers than that in metallic materials. The strength of most brittle materials such as ceramics and glasses depends upon the flaws and flaw-size distribution (Chapters 6 and 7). Therefore, inorganic glasses and ceramics do not respond well to strain hardening. We, therefore, consider different strategies to strengthen these materials. In this context, we will learn the principles of tempering and annealing of glasses. These processes make glass stronger and safer. We will also examine conditions under which ceramic materials can show large (several hundred percent) plastic deformations. Thus, all ceramic materials are not intrinsically brittle! There are conditions under which many ceramics can exhibit considerable ductility.

We begin by discussing strain hardening in metallic materials in the context of stress-strain curves.

8-1 Relationship of Cold Working to the Stress-Strain Curve

A stress-strain curve for a ductile metallic material is shown in Figure 8-1(a). If we apply a stress σ_1 that is greater than the yield strength (σ_y), it causes a permanent deformation or strain. When the stress is removed, it leaves behind a strain of ε_1. If we make a tensile test sample from the metallic material that had been previously stressed to σ_1 and retest that material, we obtain the stress-strain curve shown in Figure 8-1(b). Our new test specimen would begin to deform plastically or flow at stress level σ_1. We define *flow stress* as the stress that is needed to initiate plastic flow in a previously deformed material. Thus, σ_1 is now the flow stress of the material. If we continue to apply a stress until we reach σ_2, then release the stress and again retest the metallic material, the new flow stress is σ_2. Each time we apply a higher stress, the flow stress and tensile strength increase and the ductility decreases. We eventually strengthen the metallic material until

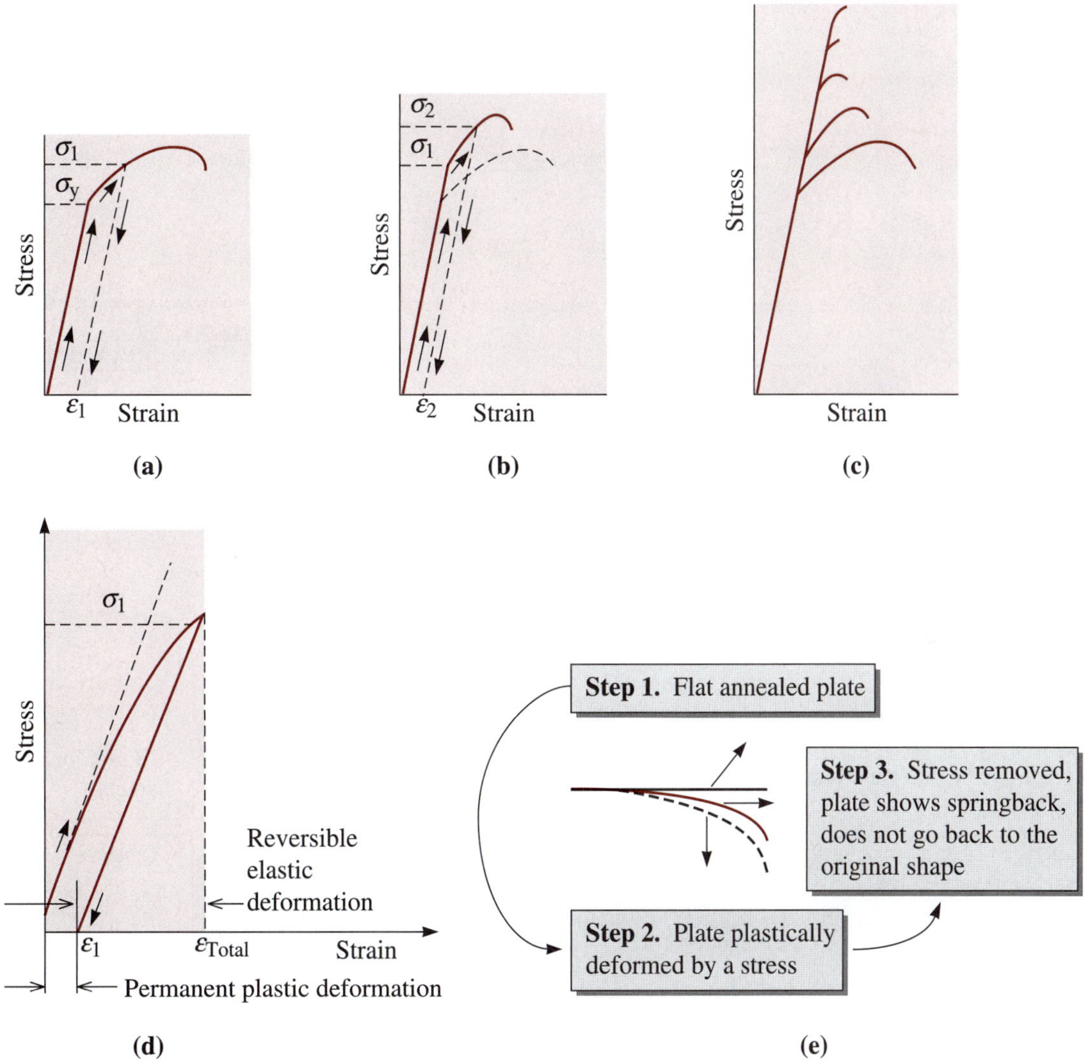

Figure 8-1 Development of strain hardening from the stress-strain diagram. (a) A specimen is stressed beyond the yield strength before the stress is removed. (b) Now the specimen has a higher yield strength and tensile strength, but lower ductility. (c) By repeating the procedure, the strength continues to increase and the ductility continues to decrease until the alloy becomes very brittle. (d) Note the total strain and the elastic strain recovery lead to remnant plastic strain and (e) illustration of springback. (*This article was published in* Engineering Materials I, *Second Edition, M.F. Ashby and D.R.H. Jones. Copyright © Butterworth-Heinemann (1996).*)

the flow stress, tensile, and breaking strengths are equal and there is no ductility [Figure 8-1(c)]. At this point, the metallic material can be plastically deformed no further. Figures 8-1(d) and (e) are related to springback, a concept that is discussed a bit later in this section.

By applying a stress that exceeds the original yield strength of the metallic material, we have **strain hardened** or **cold worked** the metallic material, while simultaneously deforming it. This is the basis for many manufacturing techniques, such as wire drawing. Figure 8-2 illustrates several manufacturing processes that make use of both

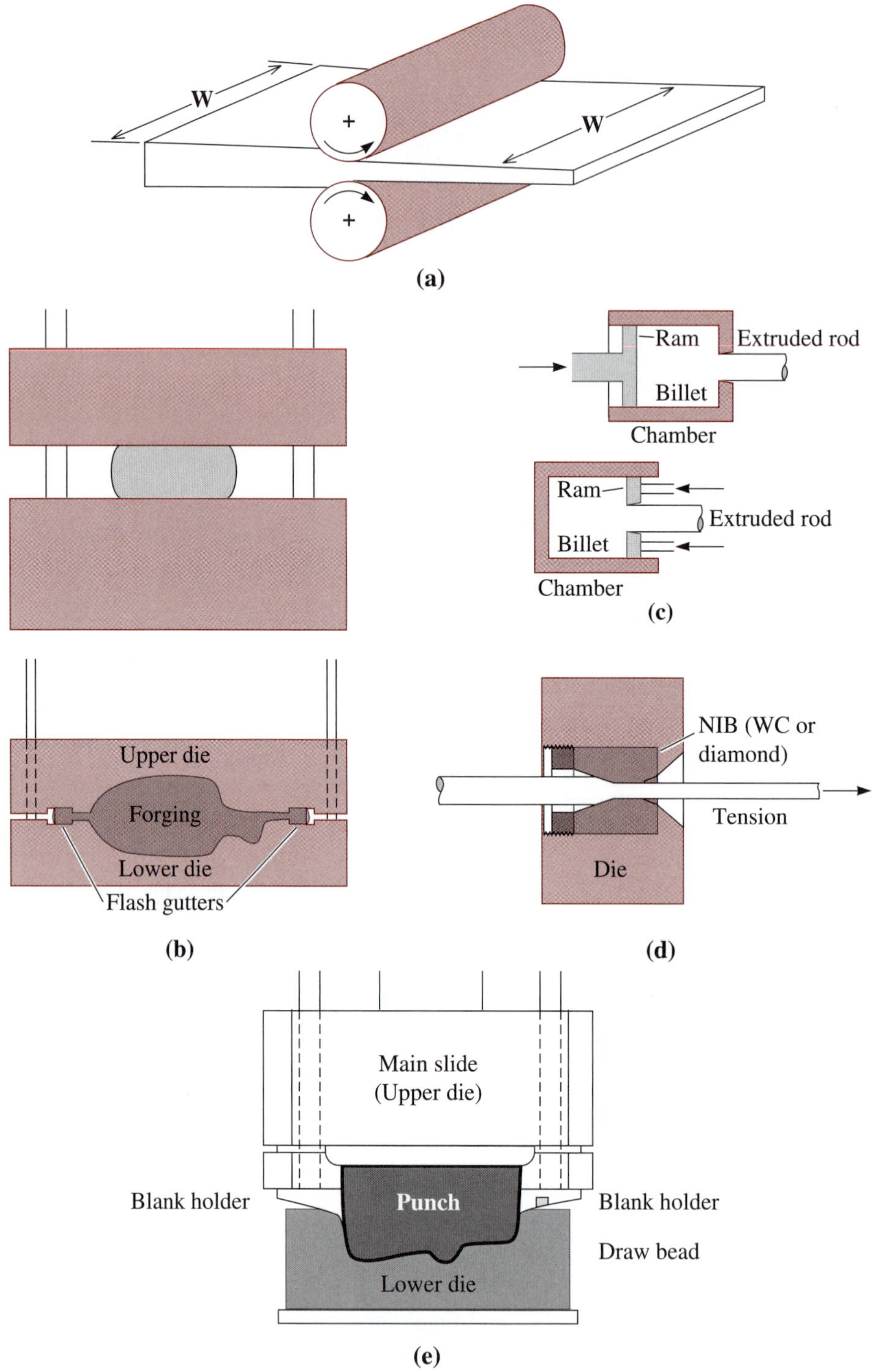

Figure 8-2 Manufacturing processes that make use of cold working as well as hot working. Common metalworking methods. (a) Rolling. (b) Forging (open and closed die). (c) Extrusion (direct and indirect). (d) Wire drawing. (e) Stamping. (Adapted from Mechanical Behavior of Materials by Meyers, M.A. & Chawla, K.K. 1999, Prentice Hall.) (*Source: Adapted from* Mechanical Behavior of Materials, *by M.A. Meyers and K.K. Chawla, p. 292, Fig. 6-1. Copyright © 1999 Prentice Hall. Adapted with permission of Pearson Education, Inc., Upper Saddle River, NJ.*)

cold-working and hot-working processes. We will discuss the difference between **hot working** and **cold working** later in this chapter (Section 8-7). Many techniques are used to simultaneously shape and strengthen a material by cold working (Figure 8-2). For example, **rolling** is used to produce metal plate, sheet, or foil. **Forging** deforms the metal into a die cavity, producing relatively complex shapes such as automotive crankshafts or connecting rods. In **drawing**, a metallic rod is pulled through a die to produce a wire or fiber. In **extrusion**, a material is pushed through a die to form products of uniform cross-sections, including rods, tubes, or aluminum trims for doors or windows. *Deep drawing* is used to form the body of aluminum beverage cans. *Stretch forming* and *bending* are used to shape sheet material. Thus, cold working is an effective way of shaping metallic materials while simultaneously increasing their strength. The down side of this process is the loss of ductility. If you take a metal wire and bend it repeatedly it will harden and eventually break because of strain hardening. Strain hardening is used in many products, especially those that are not going to be exposed to very high temperatures. For example, an aluminum beverage can derives almost 70% of its strength as a result of strain hardening that occurs during its fabrication. Some of the strength of aluminum cans also comes from the alloying elements (e.g., Mg) added. Note that many of the processes such as rolling, can be conducted using both cold and hot working. The pros and cons of using each will be discussed later in this chapter.

Strain-Hardening Exponent (*n*) The response of a metallic material to cold working is given by the **strain-hardening exponent**, *n*, which is the slope of the plastic portion of the *true stress-true strain* curve in Figure 8-3 when a logarithmic scale is used:

$$\sigma_t = K\varepsilon_t^n \tag{8-1a}$$

or

$$\ln \sigma_t = \ln K + n \ln \varepsilon_t \tag{8-1b}$$

The constant K (strength coefficient) is equal to the stress when $\varepsilon_t = 1$. The strain-hardening exponent is relatively low for HCP metals, but is higher for BCC and, particularly, for FCC metals (Table 8-1). Metals with a low strain-hardening exponent respond poorly to cold working. If we take a copper wire and bend, the bent wire is stronger as a result of strain hardening.

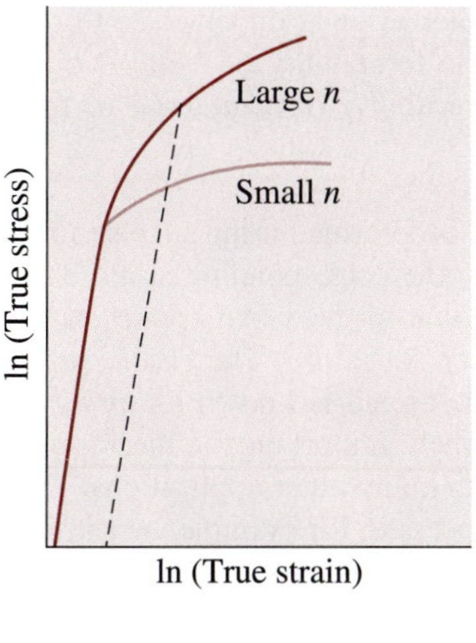

Figure 8-3
The true stress-true strain curves for metals with large and small strain-hardening exponents. Larger degrees of strengthening are obtained for a given strain for the metal with the larger *n*.

TABLE 8-1 ■ *Strain-hardening exponents and strength coefficients of typical metals and alloys*

Metal	Crystal Structure	*n*	K (MPa)
Titanium	HCP	0.05	1207
Annealed alloy steel	BCC	0.15	641
Quenched and tempered medium-carbon steel	BCC	0.10	1572
Molybdenum	BCC	0.13	724
Copper	FCC	0.54	317
Cu-30% Zn	FCC	0.50	896
Austenitic stainless steel	FCC	0.52	1517

Adapted from G. Dieter, Mechanical Metallurgy, *McGraw-Hill, 1961, and other sources.*

Strain-Rate Sensitivity (*m*) The **strain-rate sensitivity** (m) of stress is defined as:

$$m = \left[\frac{\partial(\ln \sigma)}{\partial(\ln \dot{\varepsilon})}\right] \tag{8-2}$$

This describes how fast strain hardening occurs in response to plastic deformation. As mentioned before, the mechanical behavior of sheet steels under high **strain rates** ($\dot{\varepsilon}$) is important not only for shaping, but also for how well the steel will perform under high-impact loading. The crashworthiness of sheet steels is an important consideration for the automotive industry. Steels that harden rapidly under impact loading are useful in absorbing mechanical energy.

A positive value of m implies that material will resist necking (Chapter 6). High values of m and n mean the material can exhibit a better formability in stretching. However, these values do not affect the deep drawing characteristics. For deep drawing, the *plastic strain ratio* (r) is important. We define plastic strain ratio as:

$$r = \frac{\varepsilon_w}{\varepsilon_t} = \frac{\ln\left(\frac{w}{w_0}\right)}{\ln\left(\frac{h}{h_0}\right)} \tag{8-3}$$

In this equation, w and h correspond to the width and thickness of the material being processed and the subscript zero indicates original dimensions. Forming limit diagrams are often used to better understand the **formability** of metallic materials. Overall, we define formability of a material as the ability of a material to maintain its integrity while being shaped.

Springback When a metallic material is deformed using a stress above its yield strength to a higher level (σ_1 in Figure 8-1(d)), the corresponding strain existing at stress σ_1 is obtained by dropping a perpendicular line to the x-axis (point $\varepsilon_{\text{total}}$). A strain equal to ($\varepsilon_{\text{total}} - \varepsilon_1$) is recovered since it is elastic in nature. The *elastic strain* that is recovered after a material has been *plastically* deformed is known as *springback* [Figure 8-1(e)]. The occurrence of springback is extremely important for the formation of automotive body panels from sheet steels along with many other applications. This effect is also seen in processing of polymeric materials processed, for example, by extrusion. This is because many polymers are viscoelastic (Chapter 6) and recovery of elastic strain does occur.

It is possible to account for springback in designing components; however, variability in springback makes this very difficult. For example, an automotive supplier will receive coils of sheet steel from different steel manufacturers, and even though the specifications for the steel are identical, the springback variation in steels received from each manufacturer (or even for different lots from the same manufacturer) will make it harder to obtain cold worked components that have precisely the same shape and dimensions.

Bauschinger Effect Consider a material that has been subjected to tensile plastic deformation. Then, consider two separate samples (A and B) of this material that have been previously deformed. Test sample A in tension, and sample B under compression. We notice that for the deformed material the flow stress in tension ($\sigma_{\text{flow, tension}}$) for sample A is greater than the compressive yield strength ($\sigma_{\text{flow, compression}}$) for sample B. This effect, in which a material subjected to tension shows reduction in compressive strength, is known as the **Bauschinger effect**. Note that we are comparing the yield strength of a material under compression and tension after the material has been subjected to plastic deformation under a tensile stress. The Bauschinger effect is also seen on stress reversal. Consider a sample deformed under compression. We can then evaluate two separate samples C and D of this material. The sample subjected to another compressive test (C) now shows a higher flow stress than that for the sample D subjected to tensile stress. The Bauschinger effect plays an important role in mechanical processing of steels and other alloys.

8-2 Strain-Hardening Mechanisms

We obtain strengthening during deformation of a metallic material by increasing the number of dislocations. Before deformation, the dislocation density is about 10^6 cm of dislocation line per cubic centimeter of metal—a relatively small concentration of dislocations.

When we apply a stress greater than the yield strength, dislocations begin to slip (Schmid's Law, Chapter 4). Eventually, a dislocation moving on its slip plane encounters obstacles that pin the ends of the dislocation line. As we continue to apply the stress, the dislocation attempts to move by bowing in the center. The dislocation may move so far that a loop is produced. When the dislocation loop finally touches itself, a new dislocation is created. The original dislocation is still pinned and can create additional dislocation loops. This mechanism for generating dislocations is called a **Frank-Read source**.

The dislocation density may increase to about 10^{12} cm of dislocation line per cubic centimeter of metal during strain hardening. As discussed in Chapter 4, dislocation motion is the cause for the plastic flow that occurs in metallic materials; however, when we have too many dislocations, they interfere with their own motions. An analogy for this is when we have too many people in a room it is difficult for them to move around. The result of increased dislocation density is an increased strength, but reduced ductility, for metallic materials that have undergone **cold working** or **work hardening**.

Ceramics contain dislocations and can even be strain-hardened to a small degree. However, dislocations in ceramics are normally not very mobile. Polycrystalline ceramics also contain porosity. As a result, ceramics behave as brittle materials and significant deformation and strengthening by cold working are not possible. Likewise, covalently bonded materials such as silicon (Si) are too brittle to workharden appreciably. Glasses are amorphous and do not contain dislocations and therefore cannot be strain hardened.

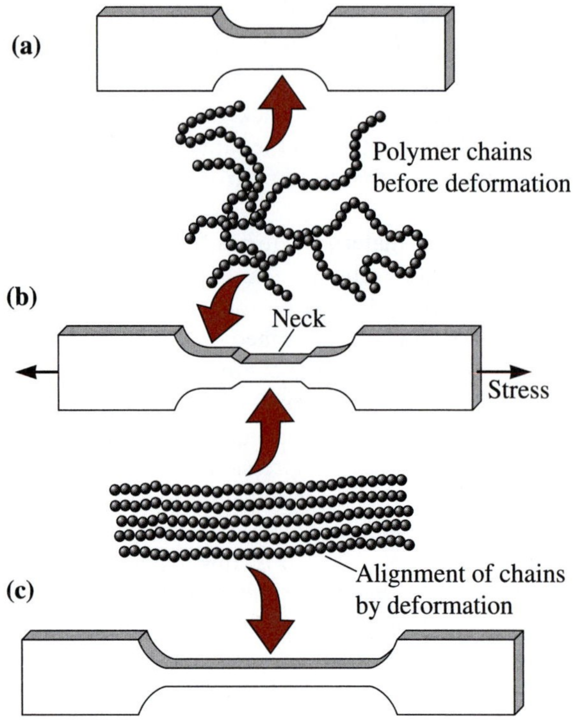

Figure 8-4
In an undeformed thermoplastic polymer tensile bar, (a) the polymer chains are randomly oriented. (b) When a stress is applied, a neck develops as chains become aligned locally. The neck continues to grow until the chains in the entire gage length have aligned. (c) The strength of the polymer is increased.

Thermoplastics are polymers such as polyethylene, polystyrene, and nylon. These materials consist of molecules that are long spaghetti-like chains (Chapter 16). Thermoplastics will strengthen when they are deformed. However, this is ***not*** strain hardening due to dislocation multiplication but, instead, strengthening of these materials involves alignment and possibly localized crystallization of the long, chainlike molecules. When a stress greater than the yield strength is applied to thermoplastic polymers such as polyethylene, the van der Waals bonds (Chapter 3) between the molecules in different chains are broken. The chains straighten and become aligned in the direction of the applied stress (Figure 8-4). The strength of the polymer, particularly in the direction of the applied stress, increases as a result of the alignment of polymeric chains in the direction of the applied stress. The processing of polyethylene terephthalate (PET) bottles made by the blow-stretch process involves such stress-induced crystallization. Thermoplastic polymers get stronger as a result of local alignments of polymer chains occurring as a result of applied stress. This strength increase is seen in the stress-strain curve of typical thermoplastics (Chapter 6).

8-3 Properties versus Percent Cold Work

By controlling the amount of plastic deformation, we control strain hardening. We normally measure the amount of deformation by defining the *percent cold work*:

$$\text{Percent cold work} = \left[\frac{A_0 - A_f}{A_0}\right] \times 100 \tag{8-4}$$

where A_0 is the original cross-sectional area of the metal and A_f is the final cross-sectional area after deformation.

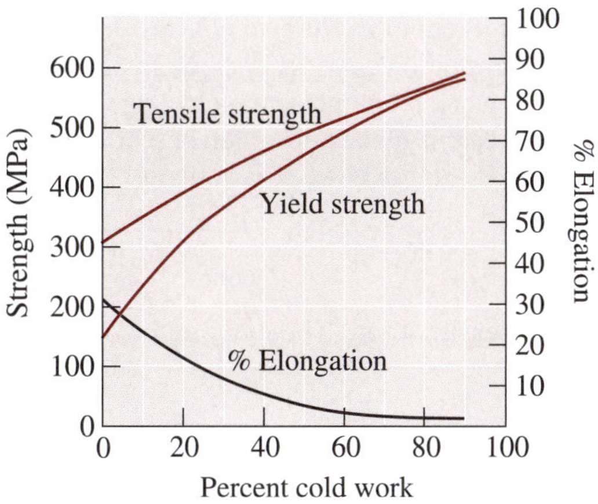

Figure 8-5
The effect of cold work on the mechanical properties of copper.

The effect of cold work on the mechanical properties of commercially pure copper is shown in Figure 8-5. As the cold work increases, both the yield and the tensile strength increase; however, the ductility decreases and approaches zero. The metal breaks if more cold work is attempted. Therefore, there is a maximum amount of cold work or deformation that we can perform on a metallic material before it becomes too brittle and breaks.

EXAMPLE 8-1 *Cold Working a Copper Plate*

A 1-cm-thick copper plate is cold-reduced to 0.50 cm, and later further reduced to 0.16 cm. Determine the total percent cold work and the tensile strength of the 0.16-cm plate. (See Figure 8-6.)

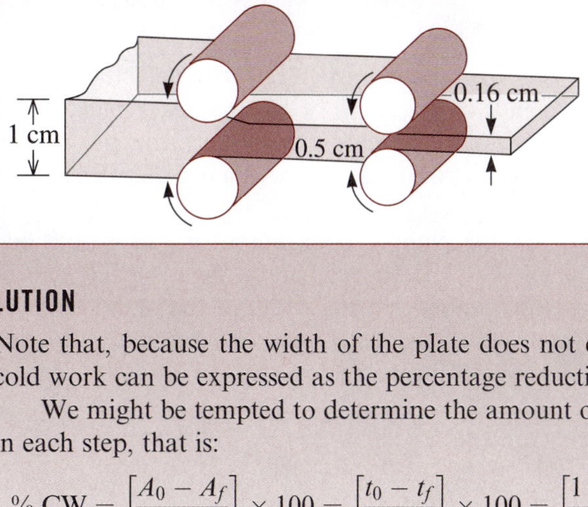

Figure 8-6
Diagram showing the rolling of a 1-cm plate to a 0.16-cm plate (for Example 8-1).

SOLUTION

Note that, because the width of the plate does not change during rolling, the cold work can be expressed as the percentage reduction in the thickness t.

We might be tempted to determine the amount of cold work accomplished in each step, that is:

$$\% \text{ CW} = \left[\frac{A_0 - A_f}{A_0}\right] \times 100 = \left[\frac{t_0 - t_f}{t_0}\right] \times 100 = \left[\frac{1 \text{ cm} - 0.50 \text{ cm}}{1 \text{ cm}}\right] \times 100$$

$$= 50\%$$

$$\% \text{ CW} = \left[\frac{A_0 - A_f}{A_0}\right] \times 100 = \left[\frac{t_0 - t_f}{t_0}\right] \times 100 = \left[\frac{0.50 \text{ cm} - 0.16 \text{ cm}}{0.50 \text{ cm}}\right] \times 100$$

$$= 68\%$$

We might then be tempted to combine the two cold work percentages (50% + 68% = 118%) to obtain the total cold work. *This would be incorrect.*

Our definition of cold work is the percentage change between the original and final cross-sectional areas; it makes no difference how many intermediate steps are involved. Thus, the total cold work is actually

$$\% \text{ CW} = \left[\frac{t_0 - t_f}{t_0}\right] \times 100 = \left[\frac{1 \text{ cm} - 0.16 \text{ cm}}{1 \text{ cm}}\right] \times 100 = 84\%$$

and, from Figure 8-5, the tensile strength is about 565 MPa.

We can predict the properties of a metal or an alloy if we know the amount of cold work during processing. We can then decide whether the component has adequate strength at critical locations.

When we wish to select a material for a component that requires certain minimum mechanical properties, we can design the *deformation process*. We first determine the necessary percent cold work and then, using the final dimensions we desire, calculate the original metal dimensions from the cold work equation.

EXAMPLE 8-2 *Design of a Cold Working Process*

Design a manufacturing process to produce a 0.1-cm-thick copper plate having at least 448 MPa tensile strength, 414 MPa yield strength, and 5% elongation.

SOLUTION

From Figure 8-5, we need at least 35% cold work to produce a tensile strength of 448 MPa and 40% cold work to produce a yield strength of 414 MPa, but we need less than 45% cold work to meet the 5% elongation requirement. Therefore, any cold work between 40% and 45% gives the required mechanical properties.

To produce the plate, a cold-rolling process would be appropriate. The original thickness of the copper plate prior to rolling can be calculated from Equation 8-4, assuming that the width of the plate does not change. Because there is a range of allowable cold work—between 40% and 45%—there is a range of initial plate thicknesses:

$$\% \text{ CW}_{\text{min}} = 40 = \left[\frac{t_{\text{min}} \text{ cm} - 0.1 \text{ cm}}{t_{\text{min}} \text{ cm}}\right] \times 100, \quad \therefore t_{\text{min}} = 0.167 \text{ cm}$$

$$\% \text{ CW}_{\text{max}} = 45 = \left[\frac{t_{\text{max}} \text{ cm} - 0.1 \text{ cm}}{t_{\text{max}} \text{ cm}}\right] \times 100, \quad \therefore t_{\text{max}} = 0.182 \text{ cm}$$

To produce the 0.1-cm copper plate, we begin with a 0.167- to 0.182-cm copper plate in the softest possible condition, then cold roll the plate 40% to 45% to achieve the 0.1 cm thickness.

8-4 Microstructure, Texture Strengthening, and Residual Stresses

During plastic deformation using cold or hot working, a microstructure consisting of grains that are elongated in the direction of the stress applied is often produced (Figure 8-7).

Anisotropic Behavior During deformation, the grains rotate as well as elongate, causing certain crystallographic directions and planes to become aligned with the direction in which stress is applied. Consequently, preferred orientations, or textures, develop and cause anisotropic behavior.

In processes such as wire drawing and extrusion, a **fiber texture** is produced. The term "fiber" refers to the way grains in the metallic material which become elongated in a direction parallel to the axis of the wire or an extruded product. In BCC metals, $\langle 110 \rangle$ directions line up with the axis of the wire. In FCC metals, $\langle 111 \rangle$ or $\langle 100 \rangle$ directions are aligned. This gives the highest strength along the axis of the wire or the extrudate (product being extruded, such as a tube), which is what we desire.

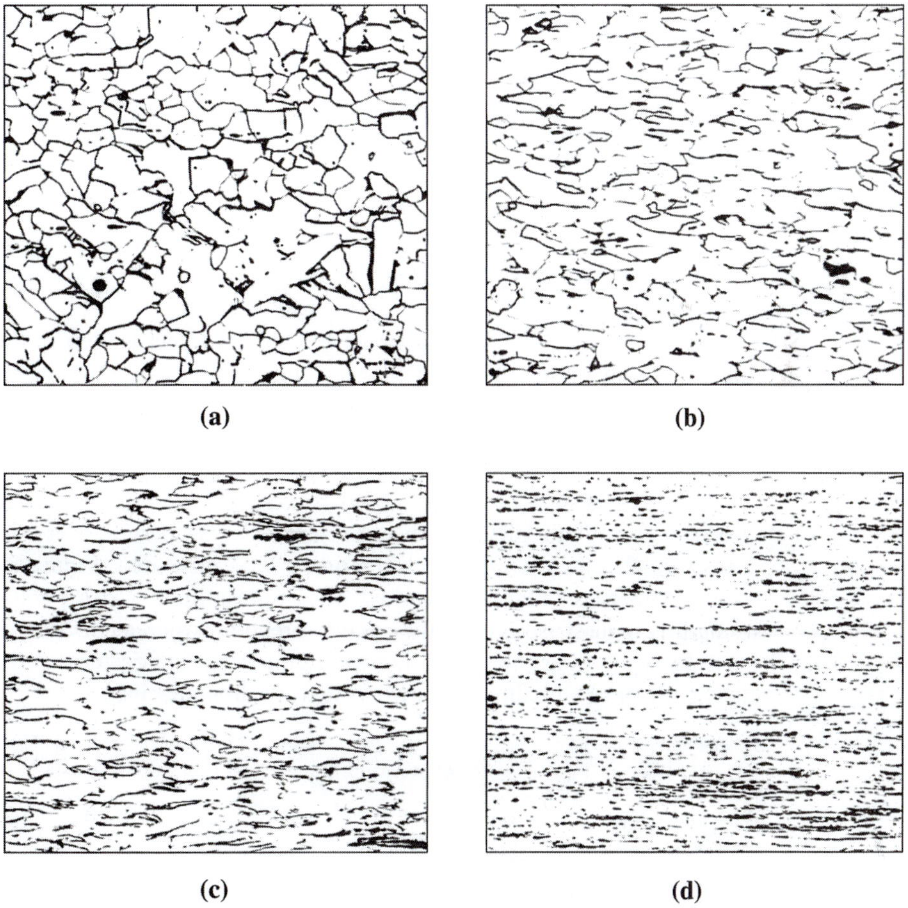

Figure 8-7 The fibrous grain structure of a low carbon steel produced by cold working: (a) 10% cold work, (b) 30% cold work, (c) 60% cold work, and (d) 90% cold work ($\times 250$). (*Source: From* ASM Handbook Vol. 9, Metallography and Microstructure, *(1985) ASM International, Materials Park, OH 44073. Used with permission.*)

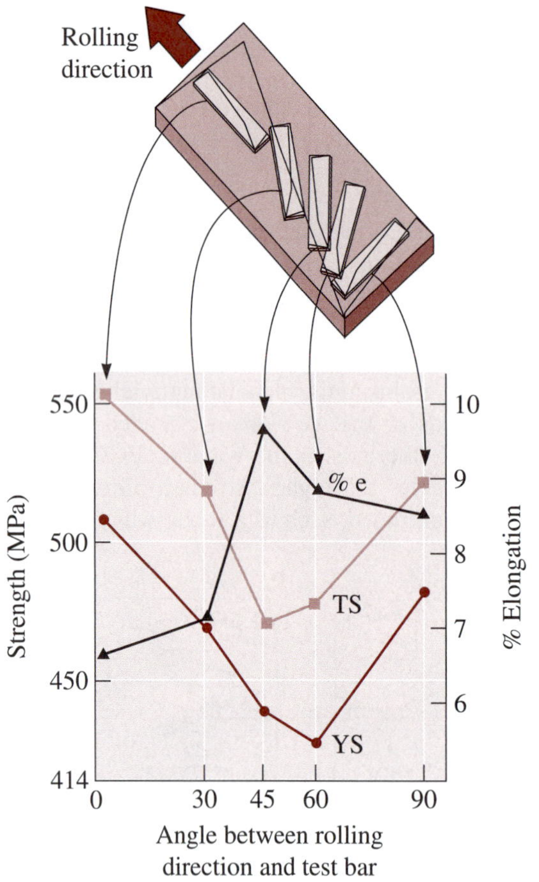

Figure 8-8
Anisotropic behavior in a rolled aluminum-lithium sheet material used in aerospace applications. The sketch relates the position of tensile bars to the mechanical properties that are obtained.

As mentioned previously, a somewhat similar effect is seen in thermoplastic materials when they are drawn into fibers or other shapes. The cause, as discussed before, is that polymer chains line up side-by-side (i.e., form crystalline regions) along the length of the fiber. As in metallic materials, the strength is greatest along the axis of the polymer fiber.

In processes such as rolling, grains become oriented in a preferred crystallographic direction and plane, giving a **sheet texture**. The properties of a rolled sheet or plate depend on the direction in which the property is measured. Figure 8-8 summarizes the tensile properties of a cold worked aluminum-lithium (Al-Li) alloy. For this alloy, strength is highest parallel to the rolling direction, whereas ductility is highest at a 45° angle to the rolling direction. The strengthening that occurs by the development of anisotropy or of a texture, is known as **texture strengthening**.

EXAMPLE 8-3 *Design of a Stamping Process*

One method for producing fans for cooling automotive and truck engines is to stamp the blades from cold-rolled steel sheet, then attach the blades to a "spider" that holds the blades in the proper position. A number of fan blades, all produced at the same time, have failed by the initiation and propagation of a fatigue crack transverse to the axis of the blade (Figure 8-9). All other fan blades perform satisfactorily. Provide an explanation for the failure of the blades and redesign the manufacturing process to prevent these failures.

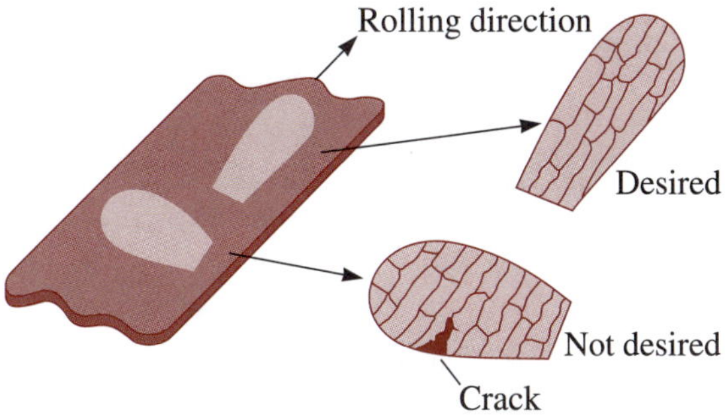

Figure 8-9
Orientations of samples (for Example 8-3).

SOLUTION

There may be several explanations for the failure of the blades—for example, the wrong type of steel may have been selected, the dies used to stamp the blades from the sheet may be worn, or the clearance between the parts of the dies may be incorrect, producing defects that initiate fatigue failure.

The failures could also be related to the anisotropic behavior of the steel sheet caused by rolling. To achieve the best performance from the blade, the axis of the blade should be aligned with the rolling direction of the steel sheet. This procedure produces high strength along the axis of the blade and, by assuring that the grains are aligned with the blade axis, reduces the number of grain boundaries along the leading edge of the blade that might help initiate a fatigue crack. Suppose your examination of the blade using, for example, pole figure analysis or metallographic analysis, indicates that the steel sheet was aligned 90° from its usual position during stamping. Now the blade has a low strength in the critical direction and, in addition, fatigue cracks will more easily initiate and grow. This mistake in manufacturing can cause failures and injuries to mechanics performing maintenance on automobiles.

You might recommend that the manufacturing process be redesigned to assure that the blades cannot be stamped from misaligned sheet. Perhaps special guides or locking devices on the die will assure that the die is properly aligned with the sheet.

Residual Stresses A small portion of the applied stress—perhaps about 10%—is stored in the form of **residual stresses** within the structure as a tangled network of dislocations. The residual stresses increase the total energy of the structure. The presence of dislocations increases the total internal energy of the structure. The higher the extent of cold working, the higher would be the level of total internal energy of the material. Residual stresses generated by cold working may not always be desirable and can be relieved by a heat treatment known as **stress-relief anneal** (Section 8-6). As will be discussed shortly, in some instances we deliberately create residual compressive stresses at the surface of materials to enhance their mechanical properties.

The residual stresses are not uniform throughout the deformed metallic material. For example, high compressive residual stresses may be present at the surface of a rolled plate and high tensile stresses may be stored in the center. If we machine a small amount of metal from one surface of a cold-worked part, we remove metal that contains only compressive residual stresses. To restore the balance, the plate must distort. If there is a net compressive residual stress at the surface of a component, this may be beneficial from a viewpoint of mechanical properties since any crack or flaw on the

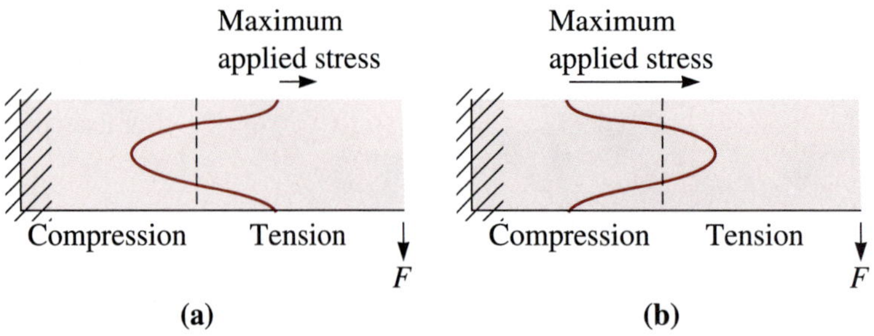

Figure 8-10 The compressive residual stresses can be harmful or beneficial. (a) A bending force applies a tensile stress on the top of the beam. Since there are already tensile residual stresses at the top, the load-carrying characteristics are poor. (b) The top contains compressive residual stresses. Now the load-carrying characteristics are very good.

surface will not likely grow. This is why any residual stresses, originating from cold work or any other source, affect the ability of the part to carry a load (Figure 8-10). If a tensile stress is applied to a material that already contains tensile residual stresses, the total stress acting on the part is the sum of the applied and residual stresses. If, however, compressive stresses are stored at the surface of a metal part, an applied tensile stress must first balance the compressive residual stresses. Now the part may be capable of withstanding a larger load. In Chapters 6 and 7, we learned that fatigue is a common mechanism of failure for load-bearing components. Sometimes, components that are subject to fatigue failure can be strengthened by **shot peening**. Bombarding the surface with steel shot propelled at a high velocity introduces compressive residual stresses at the surface that increase the resistance of the metal surface to fatigue failure (Chapter 7). The following example explains the use of shot peening.

EXAMPLE 8-4 *Design of a Fatigue-Resistant Shaft*

Your company has produced several thousand shafts that have a fatigue strength of 138 MPa. The shafts are subjected to high-bending loads during rotation. Your sales engineers report that the first few shafts placed into service failed in a short period of time by fatigue. Design a process by which the remaining shafts can be salvaged by improving their fatigue properties.

SOLUTION

Fatigue failures typically begin at the surface of a rotating part; thus, increasing the strength at the surface improves the fatigue life of the shaft. A variety of methods might be used to accomplish this.

If the shaft is made of steel, we could carburize the surface of the part (Chapter 5). In carburizing, carbon is diffused into the surface of the shaft. After an appropriate heat treatment, the higher carbon at the surface increases the strength of the surface and, perhaps more importantly, introduces *compressive* residual stresses at the surface.

We might consider cold working the shaft; cold working increases the yield strength of the metal and, if done properly, introduces compressive residual stresses. However, the cold work also reduces the diameter of the shaft and, because of the dimensional change, the shaft may not be able to perform its function.

Another alternative would be to shot peen the shaft. Shot peening introduces local compressive residual stresses at the surface without changing the dimensions of the part. This process, which is also inexpensive, might be sufficient to salvage the remaining shafts.

8-5 Characteristics of Cold Working

There are a number of advantages and limitations to strengthening a metallic material by cold working or strain hardening:

- We can simultaneously strengthen the metallic material and produce the desired final shape.
- We can obtain excellent dimensional tolerances and surface finishes by the cold-working process.
- The cold-working process is an inexpensive method for producing large numbers of small parts, since high forces and expensive forming equipment are not needed. Also, no alloying elements are needed, which means lower-cost raw materials can be used.
- Some metals, such as HCP magnesium, have a limited number of slip systems and are rather brittle at room temperature; thus, only a small degree of cold working can be accomplished.
- Ductility, electrical conductivity, and corrosion resistance are impaired by cold working. However, since the extent to which electrical conductivity is reduced by cold working is less than that for other strengthening processes, such as introducing alloying elements (Figure 8-11), cold working is a satisfactory way to strengthen conductor materials, such as the copper wires used for transmission of electrical power.

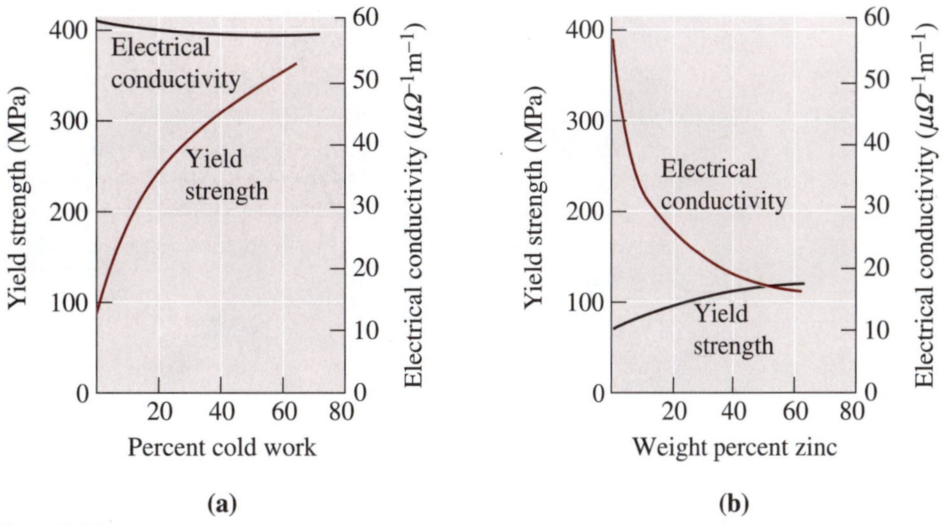

Figure 8-11 A comparison of strengthening copper by (a) cold working and (b) alloying with zinc. Note that cold working produces greater strengthening, yet has little effect on electrical conductivity.

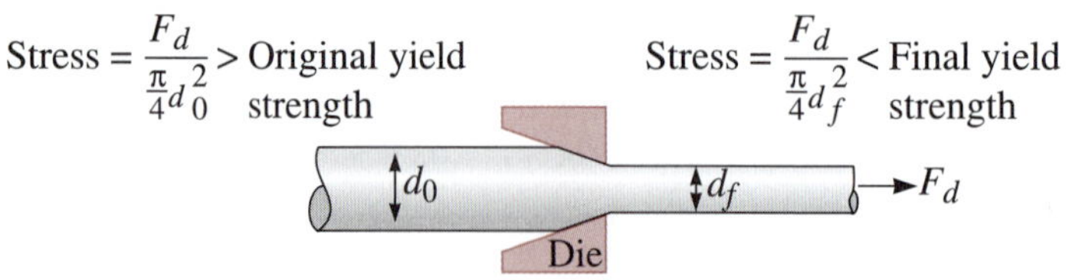

Figure 8-12 The wire-drawing process. The force F_d acts on both the original and final diameters. Thus, the stress produced in the final wire is greater than that in the original. If the wire did not strain harden during drawing, the final wire would break before the original wire was drawn through the die.

- Properly controlled residual stresses and anisotropic behavior may be beneficial. However, if residual stresses are not property controlled, the materials properties are greatly impaired.
- As will be seen in Section 8-6, since the effect of cold working is decreased or eliminated at higher temperatures, we cannot use cold working as a strengthening mechanism for components that will be subjected to high temperatures during application or service.
- Some **deformation processing** techniques can be accomplished only if cold working occurs. For example, wire drawing requires that a rod be pulled through a die to produce a smaller cross-sectional area (Figure 8-12). For a given draw force F_d, a different stress is produced in the original and final wire. The stress on the initial wire must exceed the yield strength of the metal to cause deformation. The stress on the final wire must be less than its yield strength to prevent failure. This is accomplished only if the wire strain hardens during drawing.

EXAMPLE 8-5 *Design of a Wire Drawing Process*

Design a process to produce 0.5 cm diameter copper wire. The mechanical properties of copper are to be assumed as those shown in Figure 8-5.

SOLUTION

Wire drawing manufacturing technique will be suitable for this application. To produce the copper wire as efficiently as possible, we make the largest reduction in the diameter possible. Our design must assure that the wire strain hardens sufficiently during drawing to prevent the drawn wire from breaking.

As an example calculation, let's assume that the starting diameter of the copper wire is 1 cm and that the wire is in the softest possible condition. The cold work is:

$$\% \text{ CW} = \left[\frac{A_0 - A_f}{A_0}\right] \times 100 = \left[\frac{(\pi/4)d_0^2 - (\pi/4)d_f^2}{(\pi/4)d_0^2}\right] \times 100$$

$$= \left[\frac{(1 \text{ cm})^2 - (0.5 \text{ cm})^2}{(1 \text{ cm})^2}\right] \times 100 = 75\%$$

From Figure 8-5, the initial yield strength with 0% cold work is 152 MPa. The final yield strength with 75% cold work is about 534 MPa (with very little ductility). The draw force required to deform the initial wire is:

$$F = \sigma_y A_0 = (152 \text{ MPa})(\pi/4)(1 \text{ cm})^2 = 11,943 \text{ N}$$

The stress acting on the wire after passing through the die is:

$$\sigma = \frac{F_d}{A_f} = \frac{11,943 \text{ N}}{(\pi/4)(0.5 \text{ cm})^2} = 608 \text{ MPa}$$

The applied stress of 608 MPa is greater than the 534 MPa yield strength of the drawn wire. Therefore, the wire breaks since the percent elongation is almost zero.

We can perform the same set of calculations for other initial diameters, with the results shown in Table 8-2 and Figure 8-13.

The graph shows that the draw stress exceeds the yield strength of the drawn wire when the original diameter is about 0.925 cm. To produce the wire as efficiently as possible, the original diameter should be just under 0.925 cm.

TABLE 8-2 ■ ***Mechanical properties of copper wire (see Example 8-5)***

d_0 (cm)	% CW	Yield Strength of Drawn Wire (MPa)	Force (N)	Draw Stress on Drawn Wire (MPa)
0.625	36	400	4665	238
0.75	56	469	6718	342
0.875	67	510	9144	466
1	75	534	11,943	608

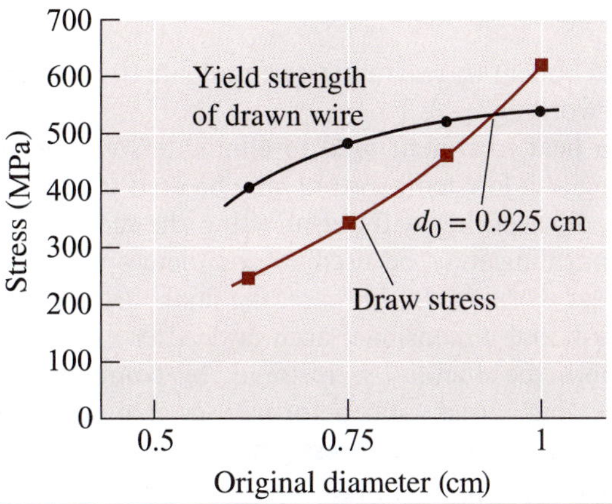

Figure 8-13 Yield strength and draw stress of wire (for Example 8-5).

8-6 The Three Stages of Annealing

Cold working is a useful strengthening mechanism. It is also a very effective tool for shaping materials using wire drawing, rolling, extrusion, etc. However, cold working leads to some effects that are sometimes undesirable. For example, the loss of ductility or development of residual stresses may not be desirable for certain applications. Since cold working or strain hardening results from increased dislocation density we can

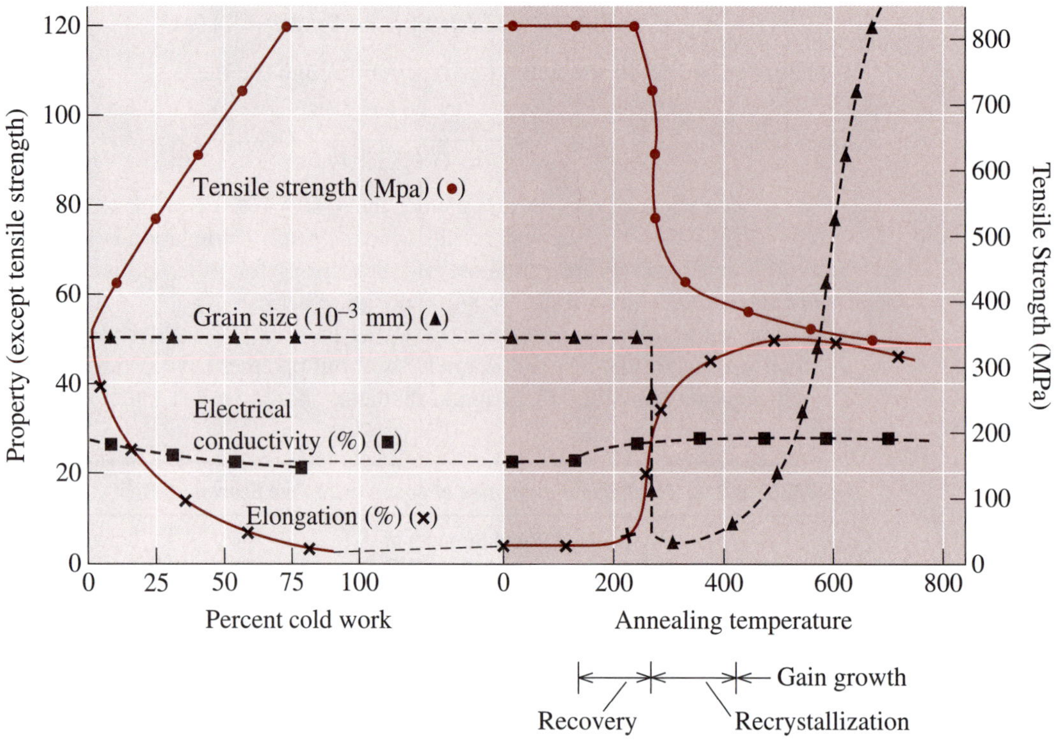

Figure 8-14 The effect of cold work on the properties of a Cu-35% Zn alloy and the effect of annealing temperature on the properties of a Cu-35% Zn alloy that is cold-worked 75%.

assume that any treatment to rearrange or annihilate dislocations would begin to undo the effects of cold working.

Annealing is a heat treatment used to eliminate some or all of the effects of cold working. Annealing at a low temperature may be used to eliminate the residual stresses produced during cold working without affecting the mechanical properties of the finished part. Or, annealing may be used to completely eliminate the strain hardening achieved during cold working. In this case, the final part is soft and ductile but still has a good surface finish and dimensional accuracy. After annealing, additional cold work could be done, since the ductility is restored; by combining repeated cycles of cold working and annealing, large total deformations may be achieved. There are three possible stages in the annealing process; their effects on the properties of brass are shown in Figure 8-14.

Note that the term "annealing" is also used to describe other thermal treatments. For example, glasses may be annealed, or heat treated, to eliminate residual stresses. Cast irons and steels may be annealed to produce the maximum ductility, even though no prior cold work was done to the mateiral. These annealing heat treatments will be discussed in later chapters.

Recovery The original cold-worked microstructure is composed of deformed grains containing a large number of tangled dislocations. When we first heat the metal, the additional thermal energy permits the dislocations to move and form the boundaries of a **polygonized subgrain structure** (Figure 8-15). The dislocation density, however, is virtually unchanged. This low-temperature treatment removes the residual stresses due to cold working without causing a change in dislocation density and is called **recovery**.

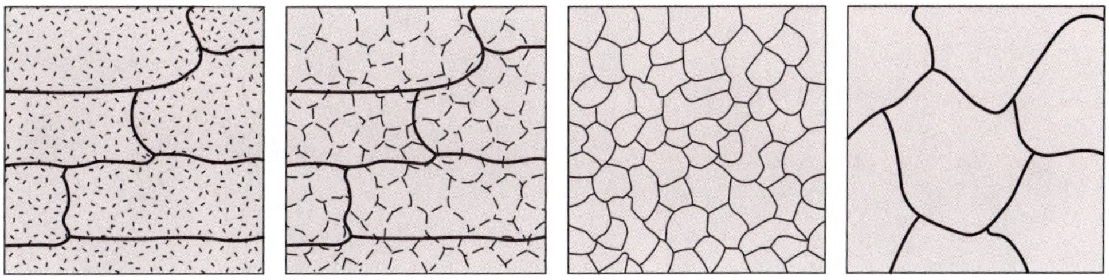

Figure 8-15 The effect of annealing temperature on the microstructure of cold-worked metals. (a) cold-worked, (b) after recovery, (c) after recrystallization, and (d) after grain growth.

The mechanical properties of the metallic material are relatively unchanged because the number of dislocations is not reduced during recovery. However, since residual stresses are reduced or even eliminated when the dislocations are rearranged, recovery is often called a **stress-relief anneal**. In addition, recovery restores high electrical conductivity to the metal, permitting us to manufacture copper or aluminum wire for transmission of electrical power that is strong yet still has high conductivity. Finally, recovery often improves the corrosion resistance of the material.

Recrystallization When a cold worked metallic material is heated above a certain temperature, rapid recovery eliminates residual stresses and produces the polygonized dislocation structure. New small grains then nucleate at the cell boundaries of the polygonized structure, eliminating most of the dislocations (Figure 8-15). Because the number of dislocations is greatly reduced, the recrystallized metal has low strength but high ductility. The temperature at which a microstructure of new grains that have very low dislocation density appears is known as the **recrystallization temperature**. The process of formation of new grains by heat treating a cold-worked material is known as **recrystallization**. As will be seen in Section 8-7, recrystallization temperature depends on many variables and is not a fixed temperature similar to melting temperature of elements and compounds.

Grain Growth At still higher annealing temperatures, both recovery and recrystallization occur rapidly, producing a fine, recrystallized grain structure. If the temperature is high enough, the grains begin to grow, with favored grains consuming the smaller grains (Figure 8-16). This phenomenon, called **grain growth**, is driven by the reduction in grain boundary area and was described in Chapter 5. Illustrated for a copper-zinc alloy in Figure 8-14, grain growth is almost always undesirable. Remember that grain growth will occur in most materials if they are subjected to a high enough temperature and, as such, is not related to cold working. Thus, recrystallization or recovery are not needed for grain growth to occur.

You may be aware that incandescent light bulbs contain filaments that are made from tungsten (W). High temperature causing grain growth and the subsequent reduction in strength is one of the factors that causes the filament to fail.

Ceramic materials, which normally do not show any significant strain hardening, show a considerable amount of grain growth. Also, abnormal grain growth can occur in some materials as a result of a formation of liquid phase during their sintering (Chapter 15). An example of where grain growth is useful is the application of alumina ceramics for making optical materials used in lighting. In this application, we want very

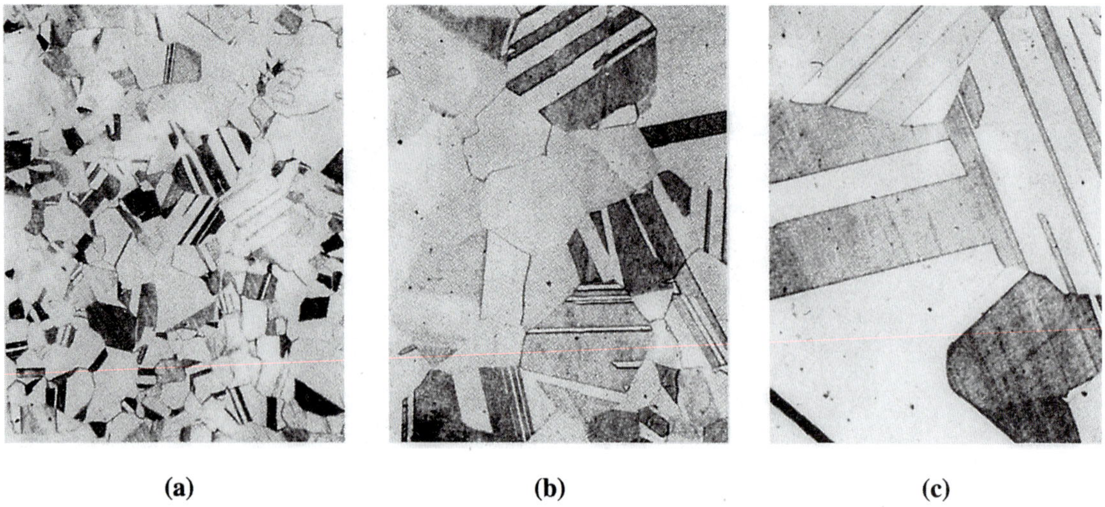

Figure 8-16 Photomicrographs showing the effect of annealing temperature on grain size in brass. Twin boundaries can also be observed in the structures. (a) Annealed at 400°C, (b) annealed at 650°C, and (c) annealed at 800°C (×75). (*Adapted from Brick, R. and Phillips, A.,* The Structure and Properties of Alloys, *1949: McGraw-Hill.*)

large grains since the scattering of light from grain boundaries has to be minimized. Some researchers have also developed methods for growing single crystals of ceramic materials using grain growth.

8-7 Control of Annealing

In many metallic materials applications, we need a combination of strength and toughness. Therefore, we need to design processes that involve shaping via cold working. We then need to control the annealing process to obtain a level of ductility. To design an appropriate annealing heat treatment, we need to know the recrystallization temperature and the size of the recrystallized grains.

Recrystallization Temperature This is the temperature at which grains in the cold-worked microstructure begin to transform into new, equiaxed, and dislocation-free grains. The driving force for recrystallization is the difference between the internal energy between a cold worked material and that of a recrystallized material. It is important for us to emphasize that the recrystallization temperature is *not a* fixed temperature and is influenced by a variety of processing variables:

- Recrystallization temperature decreases when the amount of cold work increases. Greater amounts of cold work make the metallic material less stable and encourage nucleation of recrystallized grains. There is a minimum amount of cold work, about 30 to 40%, below which recrystallization will not occur.
- A smaller original cold-worked grain size reduces the recrystallization temperature by providing more sites—the former grain boundaries—at which new grains can nucleate.
- Pure metals recrystallize at lower temperatures than alloys.

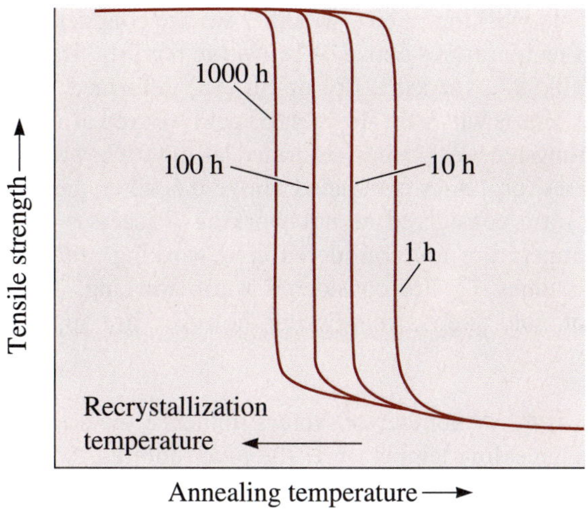

Figure 8-17
Longer annealing times reduce the recrystallization temperature. Note that the recrystallization temperature is not a fixed temperature.

- Increasing the annealing time reduces the recrystallization temperature (Figure 8-17), since more time is available for nucleation and growth of the new recrystallized grains.
- Higher melting-point alloys have a higher recrystallization temperature. Since recrystallization is a diffusion-controlled process, the recrystallization temperature is roughly proportional to $0.4T_m$ Kelvin. Typical recrystallization temperatures for selected metals are shown in Table 8-3.

The concept of recrystallization temperature is very important since it also defines the boundary between cold working and hot working of a metallic material. If we conduct deformation (shaping) of a material above the recrystallization temperature, we refer to it as hot working. If we conduct the shaping or deformation at a temperature below the recrystallization temperature, we refer to this as cold working. Therefore, the

TABLE 8-3 ■ ***Typical recrystallization temperatures for selected metals***

Metal	Melting Temperature (°C)	Recrystallization Temperature (°C)
Sn	232	−4
Pb	327	−4
Zn	420	10
Al	660	150
Mg	650	200
Ag	962	200
Cu	1085	200
Fe	1538	450
Ni	1453	600
Mo	2610	900
W	3410	1200

(*From R. Brick, A. Pense and R. Gordon,* Structure and Properties of Engineering Materials, *McGraw-Hill, 1977.*)

terms "hot" and "cold" working refer to whether we are conducting the shaping process or deformation at temperatures above or below the recrystallization temperature. As can be seen from Table 8-3, for lead (Pb) or tin (Sn) deformed at 25°C, we are conducting hot working! This is why iron (Fe) can be cold worked at room temperature but not lead (Pb). For tungsten (W) being deformed at 1000°C, we are conducting cold working! In some cases, processes conducted above 0.6 times the melting temperature (T_m) of metal (in K) are considered as hot working. Processes conducted below 0.3 times the melting temperature are considered cold working and processes conducted between 0.3 and 0.6 times T_m are considered **warm working**. These descriptions of ranges that define hot, cold and warm working, however, are approximate and should be used with caution.

Recrystallized Grain Size A number of factors influence the size of the recrystallized grains. Reducing the annealing temperature, the time required to heat to the annealing temperature, or the annealing time reduces grain size by minimizing the opportunity for grain growth. Increasing the initial cold work also reduces final grain size by providing a greater number of nucleation sites for new grains. Finally, the presence of a second phase in the microstructure helps prevent grain growth and keeps the recrystallized grain size small.

8-8 Annealing and Materials Processing

The effects of recovery, recrystallization, and grain growth are important in the processing and eventual use of a metal or an alloy.

Deformation Processing By taking advantage of the annealing heat treatment, we can increase the total amount of deformation we can accomplish. If we are required to reduce a 12.5 cm thick plate to a 0.125 cm thick sheet, we can do the maximum permissible cold work, anneal to restore the metal to its soft, ductile condition, then cold work again. We can repeat the cold work-anneal cycle until we approach the proper thickness. The final cold-working step can be designed to produce the final dimensions and properties required (Example 8-6).

EXAMPLE 8-6 *Design of a Process to Produce Copper Strip*

We wish to produce a 0.1-cm-thick, 6-cm-wide copper strip having at least 414 MPa yield strength and at least 5% elongation. We are able to purchase 6-cm-wide strip only in thicknesses of 5 cm. Design a process to produce the product we need.

SOLUTION

In Example 8-2, we found that the required properties can be obtained with a cold work of 40 to 45%. Therefore, the starting thickness must be between 0.167 cm and 0.182 cm, and this starting material must be as soft as possible—that is, in the annealed condition. Since we are able to purchase only 5-cm-thick stock, we must reduce the thickness of the 5-cm strip to between 0.167

and 0.182 cm, then anneal the strip prior to final cold working. But can we successfully cold work from 5 cm to 0.182 cm?

$$\% \text{ CW} = \left[\frac{t_o - t_f}{t_o}\right] \times 100 = \left[\frac{5 \text{ cm} - 0.182 \text{ cm}}{5 \text{ cm}}\right] \times 100 = 96.4\%$$

Based on Figure 8-5, a maximum of about 90% cold work is permitted. Therefore, we must do a series of cold work and anneal cycles. Although there are many possible combinations, one is as follows:

1. Cold work the 5-cm strip 80% to 1 cm:

$$80 = \left[\frac{t_o - t_f}{t_o}\right] \times 100 = \left[\frac{5 \text{ cm} - t_i \text{ cm}}{5 \text{ cm}}\right] \times 100 \quad \text{or} \quad t_f = 1 \text{ cm}$$

2. Anneal the 1-cm strip to restore the ductility. If we don't know the recrystallization temperature, we can use the $0.4T_m$ relationship to provide an estimate. The melting point of copper is 1085°C:

$$T_r \cong (0.4)(1085 + 273) = 543 \text{ K} = 270°\text{C}$$

3. Cold work the 1-cm-thick strip to 0.182 cm:

$$\% \text{ CW} = \left[\frac{1 \text{ cm} - 0.182 \text{ cm}}{1 \text{ cm}}\right] \times 100 = 81.8\%$$

4. Again anneal the copper at 270°C to restore ductility.
5. Finally cold work 45%, from 0.182 cm to the final dimension of 0.1 cm. This process gives the correct final dimensions and properties.

High-Temperature Service As mentioned previously, neither strain hardening nor grain size strengthening (Hall-Petch equation, Chapter 4) are appropriate for an alloy that is to be used at elevated temperatures, as in creep-resistant applications. When the cold-worked metal is placed into service at a high temperature, recrystallization immediately causes a catastrophic decrease in strength. In addition, if the temperature is high enough, the strength continues to decrease because of growth of the newly recrystallized grains.

Joining Processes Metallic materials can be joined using processes such as welding. When we join a cold-worked metal using a welding process, the metal adjacent to the weld heats above the recrystallization and grain growth temperatures and subsequently cools slowly. This region is called the **heat-affected zone (HAZ)**. The structure and properties in the heat-affected zone of a weld are shown in Figure 8-18. The mechanical properties are reduced catastrophically by the heat of the welding process.

Welding processes, such as electron-beam welding or laser welding, which provide high rates of heat input for brief times, and, thus, subsequent fast cooling, minimize the exposure of the metallic materials to temperatures above recrystallization and minimize this type of damage. Similarly a process known as friction stir welding provides almost no HAZ and is being commercially used for welding of aluminum alloys (Chapter 9).

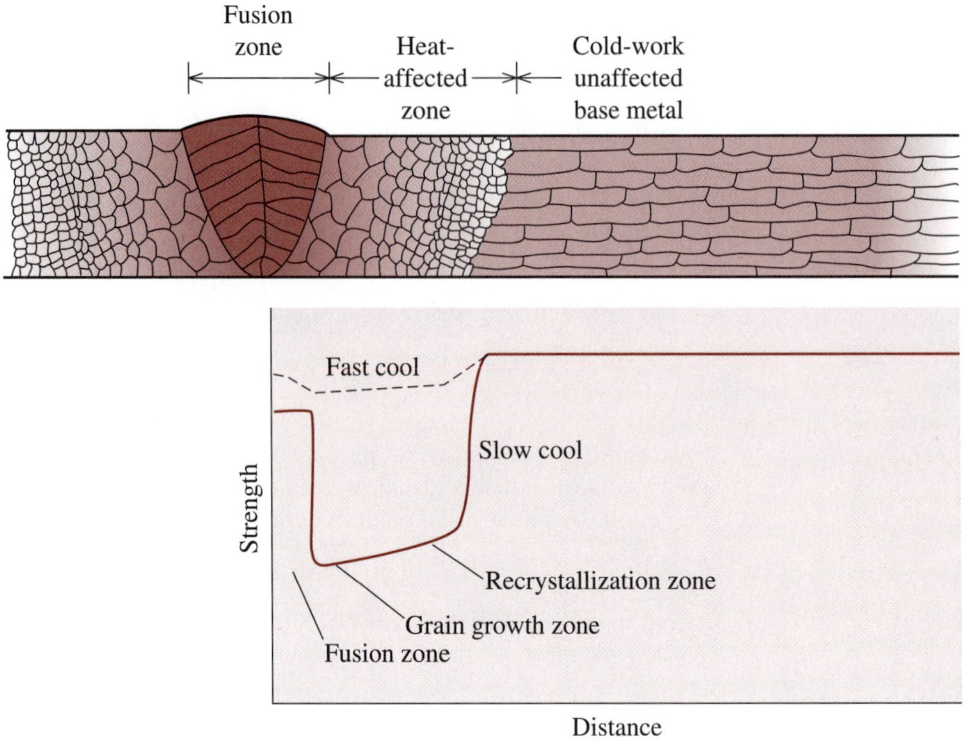

Figure 8-18 The structure and properties surrounding a fusion weld in a cold-worked metal. *Note*: only the right-hand side of the heat-affected zone is marked on the diagram. Note the loss in strength caused by recrystallization and grain growth in the heat-affected zone.

8-9 Hot Working

We can deform a metal into a useful shape by hot working rather than cold working. As described previously, hot working is defined as plastically deforming the metallic material at a temperature above the recrystallization temperature. During hot working, the metallic material is continually recrystallized (Figure 8-19).

Figure 8-19
During hot working, the elongated, anisotropic grains immediately recrystallize. If the hot-working temperature is properly controlled, the final hot-worked grain size can be very fine.

Lack of Strengthening No strengthening occurs during deformation by hot working; consequently, the amount of plastic deformation is almost unlimited. A very thick plate can be reduced to a thin sheet in a continuous series of operations. The first steps in the process are carried out well above the recrystallization temperature to take advantage of the lower strength of the metallic material. The last step is performed just above the recrystallization temperature, using a large percent deformation in order to produce the finest possible grain size.

Hot working is well-suited for forming large parts, since the metallic material has a low yield strength and high ductility at elevated temperatures. In addition, HCP metals such as magnesium have more active slip systems at hot-working temperatures; the higher ductility permits larger deformations than are possible by cold working. The following example illustrates design of a hot-working process.

EXAMPLE 8-7 *Design of a Process to Produce a Copper Strip*

We want to produce a 0.1-cm-thick, 6-cm-wide copper strip having at least 414 MPa yield strength and at least 5% elongation. We are able to purchase 6-cm-wide strip only in thicknesses of 5 cm. Design a process to produce the product we need, but in fewer steps than were required in Example 8-6.

SOLUTION

In Example 8-6, we relied on a series of cold work-anneal cycles to obtain the required thickness. We could reduce the steps by hot rolling to the required intermediate thickness:

$$\%\ \text{HW} = \left[\frac{t_o - t_f}{t_o}\right] \times 100 = \left[\frac{5\ \text{cm} - 0.182\ \text{cm}}{5\ \text{cm}}\right] \times 100 = 96.4\%$$

$$\%\ \text{HW} = \left[\frac{t_o - t_f}{t_o}\right] \times 100 = \left[\frac{5\ \text{cm} - 0.167\ \text{cm}}{5\ \text{cm}}\right] \times 100 = 96.7\%$$

Note that the formulas for hot and cold work are the same.

Because recrystallization occurs simultaneously with hot working, we can obtain these large deformations and a separate annealing treatment is not required. Thus our design might be:

1. Hot work the 5-cm strip 96.4% to the intermediate thickness of 0.182 cm.
2. Cold work 45% from 0.182 cm to the final dimension of 0.1 cm. This design gives the correct dimensions and properties.

Elimination of Imperfections Some imperfections in the original metallic material may be eliminated or their effects minimized. Gas pores can be closed and welded shut during hot working—the internal lap formed when the pore is closed is eliminated by diffusion during the forming and cooling process. Composition differences in the metal can be reduced as hot working brings the surface and center of the plate closer together, thereby reducing diffusion distances.

Anisotropic Behavior The final properties in hot-worked parts are *not* isotropic. The forming rolls or dies, which are normally at a lower temperature than the metal, cool the surface more rapidly than the center of the part. The surface then has a finer grain size than the center. In addition, a fibrous structure is produced because inclusions and second-phase particles are elongated in the working direction.

Surface Finish and Dimensional Accuracy The surface finish is usually poorer than that obtained by cold working. Oxygen often reacts with the metallic material at the surface to form oxides, which are forced into the surface during forming. Hot worked steels and other metals are often subjected to a "pickling" treatment in which acids are used to dissolve the oxide scale. In some metals, such as tungsten (W) and beryllium (Be), hot-working must be done in a protective atmosphere to prevent oxidation.

The dimensional accuracy is also more difficult to control during hot working. A greater elastic strain must be considered, since the modulus of elasticity is low at hot-working temperatures. In addition, the metal contracts as it cools from the hot-working temperature. The combination of elastic strain and thermal contraction requires that the part be made oversize during deformation; forming dies must be carefully designed, and precise temperature control is necessary if accurate dimensions are to be obtained.

SUMMARY

- When a metallic material is deformed by cold working, strain hardening occurs as additional dislocations are introduced into the structure. Very large increases in strength may be obtained in this manner. The ductility of the strain hardened metallic material is reduced.
- Wire drawing, stamping, rolling, and extrusion are some examples of manufacturing methods for shaping metallic materials. Some of the underlying principles for these processes can also be used for the manufacturing of polymeric materials.
- Annealing of metallic materials is a heat treatment intended to eliminate all, or a portion of, the effects of strain hardening. The annealing process may involve as many as three steps.
- Recovery occurs at low temperatures, eliminating residual stresses and restoring electrical conductivity without reducing the strength. A "stress relief anneal" refers to recovery.
- Recrystallization occurs at higher temperatures and eliminates almost all of the effects of strain hardening. The dislocation density decreases dramatically during recrystallization as new grains nucleate and grow.
- Grain growth, which normally should be avoided, occurs at still higher temperatures. In cold-worked metallic materials, grain growth follows recovery and recrystallization.
- Hot working combines plastic deformation and annealing in a single step, permitting large amounts of plastic deformation without embrittling the material.
- Residual stresses in materials need to be controlled. In cold-worked metallic materials, residual stresses can be eliminated using a stress-relief anneal.

GLOSSARY

Annealing In the context of metallic material, annealing is a heat treatment used to eliminate part or all of the effects of cold working. For glasses, annealing is a heat treatment that removes thermally induced stresses.

Bauschinger effect A material previously plastically deformed under tension shows decreased flow stress when tested again under compression or vice versa.

Cold working Deformation of a metal below the recrystallization temperature. During cold working, the number of dislocations increases, causing the metal to be strengthened as its shape is changed.

Deformation processing Techniques for the manufacturing of metallic and other materials using such processes as rolling, extrusion, drawing, etc.

Drawing A deformation processing technique in which a material is pulled through an opening in a die (e.g., wire drawing).

Extrusion A deformation processing technique in which a material is pushed through an opening in a die. Used for metallic and polymeric materials.

Fiber texture A preferred orientation of grains obtained during the wire drawing process. Certain crystallographic directions in each elongated grain line up with the drawing direction, causing anisotropic behavior.

Formability The ability of a material to stretch and bend without breaking. Forming diagrams describe the ability to stretch and bend materials.

Frank-Read source A pinned dislocation that, under an applied stress, produces additional dislocations. This mechanism is at least partly responsible for strain hardening.

Heat-affected zone (HAZ) The volume of material adjacent to a weld that is heated during the welding process above some critical temperature at which a change in the structure, such as grain growth or recrystallization, occurs.

Hot working Deformation of a metal above the recrystallization temperature. During hot working, only the shape of the metal changes; the strength remains relatively unchanged because no strain hardening occurs.

Polygonized subgrain structure A subgrain structure produced in the early stages of annealing. The subgrain boundaries are a network of dislocations rearranged during heating.

Recovery A low-temperature annealing heat treatment designed to eliminate residual stresses introduced during deformation without reducing the strength of the cold-worked material. This is the same as a stress-relief anneal.

Recrystallization A medium-temperature annealing heat treatment designed to eliminate all of the effects of the strain hardening produced during cold working.

Recrystallization temperature A temperature above which essentially dislocation-free and new grains emerge from a material that was previously cold worked. This depends upon the extent of cold work, time of heat treatment, etc., and is not a fixed temperature.

Residual stresses Stresses introduced in a material during processing. These can originate as a result of cold working or differential thermal expansion and contraction. A stress-relief anneal in metallic materials and the annealing of glasses minimize residual stresses. Compressive residual stresses deliberately introduced on the surface by the tempering of glasses or shot peening of metallic materials improve their mechanical properties.

Sheet texture A preferred orientation of grains obtained during the rolling process. Certain crystallographic directions line up with the rolling direction, and certain preferred crystallographic planes become parallel to the sheet surface.

Shot peening Introducing compressive residual stresses at the surface of a part by bombarding that surface with steel shot. The residual stresses may improve the overall performance of the material.

Strain hardening Strengthening of a material by increasing the number of dislocations by deformation, or cold working. Also known as "work hardening."

Strain-hardening exponent (*n*) A parameter that describes susceptibility of a material to cold working. It describes the effect that strain has on the resulting strength of the material. A material with a high strain-hardening coefficient obtains high strength with only small amounts of deformation or strain.

Strain rate The rate at which a material is deformed.

Strain-rate sensitivity (*m*) The rate at which stress develops changes as a function of strain rate. A material may behave much differently if it is slowly pressed into a shape rather than transformed rapidly into a shape by an impact loading.

Stress-relief anneal The recovery stage of the annealing heat treatment during which residual stresses are relieved without reducing the mechanical properties of the material.

Texture strengthening Increase in the yield strength of a material as a result of preferred crystallographic texture.

Thermomechanical processing Processes involved in the manufacturing of metallic components using mechanical deformation and various heat treatments.

Thermoplastics A class of polymers that consist of large, long spaghetti-like molecules that are intertwined (e.g., polyethylene, nylon, *PET*, etc.).

Warm working A term used to indicate the processing of metallic materials in a temperature range that is between those that define cold and hot working (usually a temperature between 0.3 to 0.6 of melting temperature in K).

Work hardening A term sometimes used instead of strain hardening or cold working to describe the effect of deformation on the strengthening of metallic materials.

PROBLEMS

Section 8-1 Relationship of Cold Working to the Stress-Strain Curve

8-1 A 1.263-cm-diameter metal bar with a 5-cm gage length l_0 is subjected to a tensile test. The following measurements are made in the plastic region:

Force (N)	Change in Gage length (cm) (Δl)	Diameter (cm)
122,320	0.5258	1.2
120,100	1.107	1.1415
114,310	1.7493	1.0858

Determine the strain-hardening exponent for the metal. Is the metal most likely to be FCC, BCC, or HCP? Explain.

8-2 Define the following terms: strain-hardening exponent (n), strain-rate sensitivity (m), and plastic strain ratio (r). Use appropriate equations.

8-3 A 1.5-cm-diameter metal bar with a 3-cm gage length (l_0) is subjected to a tensile test. The following measurements are made:

Force (N)	Change in Gage Length (cm) (Δl)	Diameter (cm)
16,240	0.6642	1.2028
19,066	1.4754	1.0884
19,273	2.4663	0.9848

Determine the strain-hardening coefficient for the metal. Is the metal most likely to be FCC, BCC, or HCP? Explain.

8-4 A true stress-true strain curve is shown in Figure 8-20. Determine the strain-hardening exponent for the metal.

8-5 A Cu-30% Zn alloy tensile bar has a strain-hardening coefficient of 0.50. The bar, which has an initial diameter of 1 cm and an initial gage

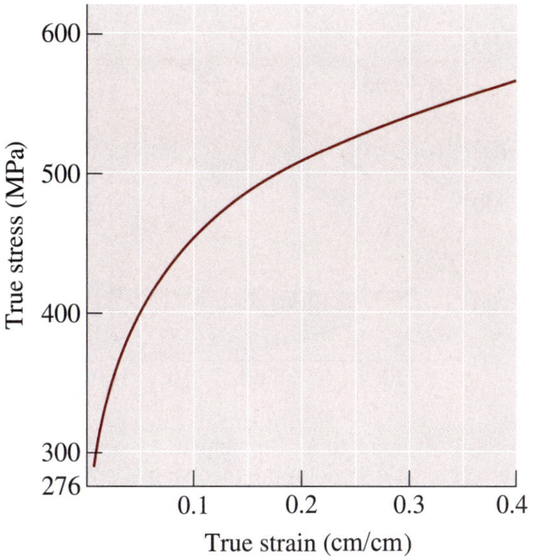

Figure 8-20 True stress-true strain curve (for Problem 8-4).

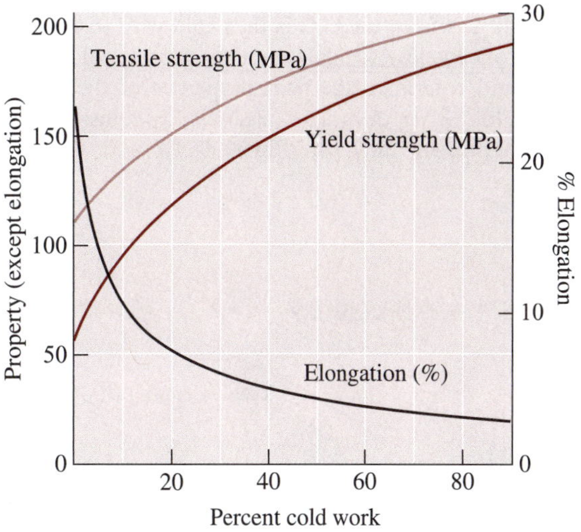

Figure 8-21 The effect of percent cold work on the properties of a 3105 aluminum alloy (for Problems 8-9 and 8-11).

length of 3 cm, fails at an engineering stress of 120 MPa. After fracture, the gage length is 3.5 cm and the diameter is 0.926 cm. No necking occurred. Calculate the true stress when the true strain is 0.05 cm/cm.

Section 8-2 Strain-Hardening Mechanisms

Section 8-3 Properties versus Percent Cold Work

8-6 A 0.625-cm-thick copper plate is to be cold worked 63%. Find the final thickness.

8-7 A 0.625-cm-diameter copper bar is to be cold worked 63%. Find the final diameter.

8-8 A 5-cm-diameter copper rod is reduced to a 3.75-cm diameter, then reduced again to a final diameter of 2.5 cm. In a second case, the 5-cm-diameter rod is reduced in one step from a 5 cm to a 2.5-cm diameter. Calculate the % CW for both cases.

8-9 A 3105 aluminum plate is reduced from 4.38 cm to 2.88 cm. Determine the final properties of the plate. Note 3105 designates a special composition of aluminum alloy. (See Figure 8-21.)

8-10 A Cu-30% Zn brass bar is reduced from a 2.5-cm diameter to a 1.13-cm diameter. Determine the final properties of the bar. (See Figure 8-22.)

8-11 A 3105 aluminum bar is reduced from a 2.5-cm diameter, to a 2-cm diameter, to a 1.5-cm diameter, to a final 1-cm diameter. Determine the % CW and the properties after each step of the process. Calculate the total percent cold work. Note 3105 designates a special composition of aluminum alloy. (See Figure 8-21.)

8-12 We want a Cu-30% Zn brass plate originally 3-cm thick to have a yield strength greater than 345 MPa and a % elongation of at least 10%. What range of final thicknesses must be obtained? (See Figure 8-22.)

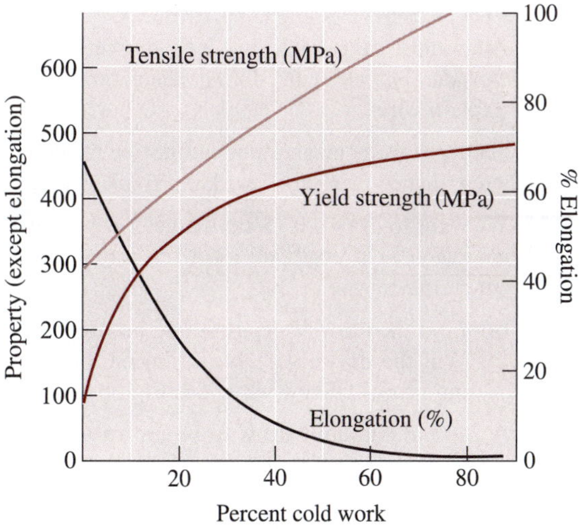

Figure 8-22 The effect of percent cold work on the properties of a Cu-30% Zn brass (for Problems 8-10 and 8-12).

8-13 A 3105 aluminum plate previously cold worked 20% is 5-cm thick. It is then cold worked further to 3.25 cm. Calculate the total percent cold work and determine the final properties of the plate. Note 3105 designates a special composition of aluminum alloy. (See Figure 8-21.)

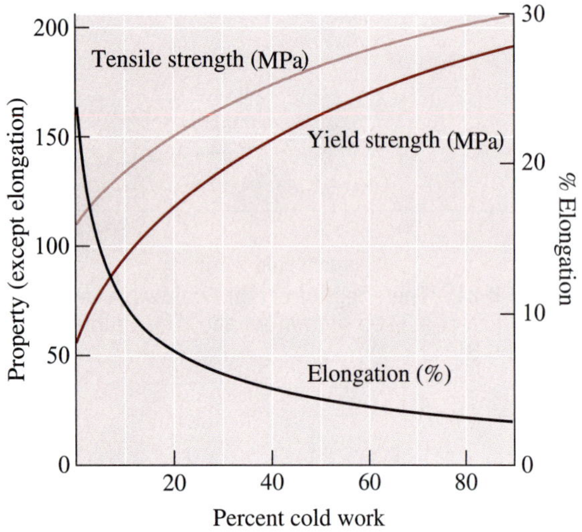

Figure 8-21 (Repeated for Problems 8-13, 8-17 and 8-22.) The effect of percent cold work on the properties of a 3105 aluminum alloy.

Section 8-4 Microstructure, Texture Strengthening, and Residual Stresses

Section 8-5 Characteristics of Cold Working

8-14 Aluminum cans made by deep drawing derive considerable strength during their fabrication. Explain why.

8-15 Such metals as magnesium can not be effectively strengthened using cold working. Explain why.

8-16 We want to draw a 0.75-cm-diameter copper wire having yield strength of 138 MPa into a 0.625-cm-diameter wire.

- **(a)** Find the draw force, assuming no friction;
- **(b)** Will the drawn wire break during the drawing process? Show why. (See Figure 8-5.)

8-17 A 3105 aluminum wire is to be drawn to give a 1-mm-diameter wire having yield strength of 138 MPa. Note 3105 designates a special composition of aluminum alloy.

- **(a)** Find the original diameter of the wire;
- **(b)** Calculate the draw force required; and

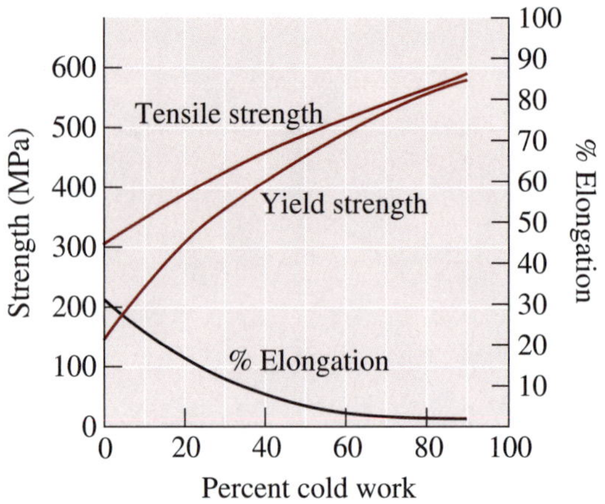

Figure 8-5 (Repeated for Problems 8-16 and 8-23.) The effect of cold work on the mechanical properties of copper.

- **(c)** Determine whether the as-drawn wire will break during the process. (See Figure 8-21.)

Section 8-6 The Three Stages of Annealing

8-18 The following data were obtained when a cold-worked metal was annealed.

- **(a)** Estimate the recovery, recrystallization, and grain growth temperatures;
- **(b)** Recommend a suitable temperature for a stress-relief heat treatment;
- **(c)** Recommend a suitable temperature for a hot-working process; and
- **(d)** Estimate the melting temperature of the alloy.

Annealing Temperature (°C)	Electrical Conductivity ($ohm^{-1} \cdot cm^{-1}$)	Yield Strength (MPa)	Grain Size (mm)
400	3.04×10^5	86	0.10
500	3.05×10^5	85	0.10
600	3.36×10^5	84	0.10
700	3.45×10^5	83	0.098
800	3.46×10^5	52	0.030
900	3.46×10^5	47	0.031
1000	3.47×10^5	44	0.070
1100	3.47×10^5	42	0.120

8-19 The following data were obtained when a cold-worked metallic material was annealed.

- **(a)** Estimate the recovery, recrystallization, and grain growth temperatures;
- **(b)** Recommend a suitable temperature for obtaining a high-strength, high-electrical-conductivity wire;

(c) Recommend a suitable temperature for a hot working process; and
(d) Estimate the melting temperature of the alloy.

Annealing Temperature (°C)	Residual Stresses (MPa)	Tensile Strength (MPa)	Grain Size (cm)
250	145	359	0.0075
275	145	359	0.0075
300	35	359	0.0075
325	0	359	0.0075
350	0	234	0.0025
375	0	207	0.0025
400	0	186	0.0088
425	0	172	0.0018

Section 8-7 Control of Annealing

8-20 Two sheets of steel have cold works of 20% and 80%, respectively. Which one would likely have a lower recrystallization temperature? Why?

Section 8-8 Annealing and Materials Processing

8-21 Using the data in Table 8-3, plot the recrystallization temperature versus the melting temperature of each metal, using absolute temperatures (Kelvin). Measure the slope and compare with the expected relationship between these two temperatures. Is our approximation a good one?

TABLE 8-3 ■ ***Typical recrystallization temperatures for selected metals (Repeated for Problem 8-21)***

Metal	Melting Temperature (°C)	Recrystallization Temperature (°C)
Sn	232	−4
Pb	327	−4
Zn	420	10
Al	660	150
Mg	650	200
Ag	962	200
Cu	1085	200
Fe	1538	450
Ni	1453	600
Mo	2610	900
W	3410	1200

(Source: Adapted from Structure and Properties of Engineering Materials, *by R. Brick, A. Pense, and R. Gordon, 1977. Copyright © 1977 The McGraw-Hill Companies. Adapted by permission.)*

8-22 We wish to produce a 0.75-cm-thick plate of 3105 aluminum having a tensile strength of at least 172 MPa and a % elongation of at least 5%. The original thickness of the plate is 7.5 cm. The maximum cold work in each step is 80%. Describe the cold working and annealing steps required to make this product. Compare this process with what you would recommend if you could do the initial deformation by hot working. (See Figure 8-21.)

8-23 We wish to produce a 0.5-cm-diameter wire of copper having a minimum yield strength of 414 MPa and a minimum % elongation of 5%. The original diameter of the rod is 5 cm and the maximum cold work in each step is 80%. Describe the cold working and annealing steps required to make this product. Compare this process with what you would recommend if you could do the initial deformation by hot working. (See Figure 8-5.)

8-24 What is a heat-affected zone? Why do some welding processes result in a joint where the material in the heat-affected zone is weaker than the base metal?

8-25 What welding techniques can be used to avoid loss of strength in the material in the heat-affected zone? Explain why these techniques are effective.

Section 8-9 Hot Working

8-26 The amount of plastic deformation that can be performed during hot working is almost unlimited. Justify this statement.

8-27 Compare and contrast hot working and cold working.

Design Problems

8-28 Design, using one of the processes discussed in this chapter, a method to produce each of the following products. Should the process include hot working, cold working, annealing, or some combination of these? Explain your decisions.

(a) paper clips;
(b) I-beams that will be welded to produce a portion of a bridge;
(c) copper tubing that will connect a water faucet to the main copper plumbing;
(d) the steel tape in a tape measure; and
(e) a head for a carpenter's hammer formed from a round rod.

8-29 We plan to join two sheets of cold-worked copper by soldering. Soldering involves heating the metal

9

Principles and Applications of Solidification

Have You Ever Wondered?

- *Whether water really does "freeze" at 0° C and "boil" at 100° C?*
- *What is the process used to produce several million kilograms of steels and other alloys?*
- *What factors determine the strength of a cast product?*
- *What is a Liquidmetal™?*

Of all the processing techniques used in the manufacturing of materials, solidification is probably the most important. All metallic materials, as well as many ceramics, inorganic glasses, and thermoplastic polymers, are liquid or molten at some point during processing. Like water freezes to ice, molten materials solidify as they cool below their freezing temperature. In Chapter 2, we learned how materials are classified based on their atomic, ionic, or molecular order. During the solidification of materials that crystallize, the atomic arrangement changes from a short-range order (SRO) in a liquid to a long-range order (LRO) in the crystalline solid. The solidification of materials that crystallize requires two steps: In the first step, ultra-fine crystallites, known as the nuclei of a solid phase, form from the liquid. In the second step, which can overlap with the first, the ultra-fine solid crystallites begin to grow as atoms from the liquid are attached to the nuclei until no liquid remains. Some materials, such as inorganic silicate glasses, will turn into a solid without developing a long-range order (i.e., they remain amorphous). Many polymeric materials may develop partial crystallinity during solidification or processing.

The solidification of metallic, polymeric, and ceramic materials is an important process to study because of its effect on the properties of the materials involved. In Chapter 8, we examined how strain hardening can be used to strengthen and shape metallic materials. We learned in Chapter 4 that grain size plays an important role in determining the strength of metallic materials. In this chapter, we will study the principles of solidification as they apply to pure metals. We will discuss solidification of alloys and more complex materials in subsequent chapters. We will first discuss the technological significance of solidification, and then examine the mechanisms by which solidification occurs. This will be followed by an examination of the microstructure of cast metallic materials and its effect on the material's mechanical properties. We will also examine the role of casting as a materials shaping process. We will examine how techniques such as welding, brazing, and soldering are used for joining metallic materials. Applications of the solidification process in single crystal growth and the solidification of glasses and polymers also will be discussed.

9-1 Technological Significance

The ability to use fire to produce, melt, and cast metals such as copper, bronze, and steel indeed is regarded as an important hallmark in the development of mankind. The use of fire for reducing naturally occurring ores into metals and alloys led to the production of useful tools and other products. Today, thousands of years later, **solidification** is still considered one of the most important manufacturing processes. Several million pounds of steel, aluminum alloys, copper, and zinc are produced through the casting process. The solidification process is also used to manufacture specific components (e.g., aluminum alloy for automotive wheels) (Figure 9-1). Industry also uses the solidification process as a **primary processing** step to produce metallic slabs or ingots (a simple, and often large casting that is processed later into useful shapes). The ingots or slabs are then hot and cold worked through **secondary processing** steps into more useful shapes (i.e., sheets, wires, rods, plates, etc.). Solidification also is applied when joining metallic materials using techniques such as welding, brazing, and soldering.

We also use solidification for processing inorganic glasses; silicate glass, for example, is processed using the float-glass process. High-quality optical fibers and other materials, such as fiberglass fibers, also are produced from the solidification of molten glasses. During the solidification of inorganic glasses, amorphous rather than crystalline, materials are produced. In the manufacture of **glass-ceramics**, we first shape the materials by casting amorphous glasses, and then crystallize them using a heat treatment to enhance their strength. Many thermoplastic materials such as polyethylene, polyvinyl chloride (PVC), polypropylene, and the like are processed into useful shapes (i.e., fibers, tubes, bottles, toys, utensils, etc.) using a process that involves melting and

Figure 9-1
Aluminum alloy wheels for automobiles. (*Courtesy of Simon Askham/Stockphoto.*)

solidification. Therefore, solidification is an extremely important technology used to control the properties of many melt-derived products as well as a tool for the manufacturing of modern engineered materials. In the sections that follow, we first discuss the nucleation and **growth** processes.

9-2 Nucleation

In the context of solidification, the term **nucleation** refers to the formation of the first nano-sized crystallites from molten material. For example, as water begins to freeze, nano-sized ice crystals, known as nuclei, form first. In a broader sense, the term nucleation refers to the initial stage of formation of one phase from another phase. When a vapor condenses into liquid, the nanoscale sized drops of liquid that appear when the condensation begins are referred to as **nuclei**. Later, we will also see that there are many systems in which the nuclei of a solid (β) will form a second solid material (α) (i.e., α- to β-phase transformation). What is interesting about these transformations is that, in most engineered materials, many of them occur while the material is in the solid state (i.e., there is no melting involved). Therefore, although we discuss nucleation from a solidification perspective, it is important to note that the phenomenon of nucleation is general and is associated with phase transformations.

We expect a material to solidify when the liquid cools to just below its freezing (or melting) temperature, because the energy associated with the crystalline structure of the solid is then less than the energy of the liquid. This energy difference between the liquid and the solid is the free energy per unit volume (ΔG_v) and is the driving force for solidification.

When the solid forms a solid-liquid interface is created (Figure 9-2). A surface free energy σ_{sl} is associated with this interface; the larger the solid, the greater the increase in surface energy. Thus, the total change in energy ΔG, shown in Figure 9-3, is:

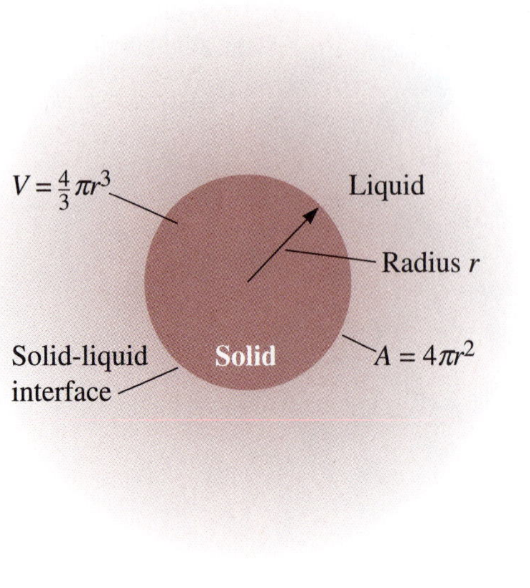

Figure 9-2
An interface is created when a solid forms from the liquid.

$$\Delta G = \tfrac{4}{3}\pi r^3 \Delta G_v + 4\pi r^2 \sigma_{sl} \tag{9-1}$$

where $\frac{4}{3}\pi r^3$ is the volume of a spherical solid of radius r, $4\pi r^2$ is the surface area of a spherical solid, σ_{sl} is the surface free energy of the solid-liquid interface (in this case), and ΔG_v is the free energy change for the solidification process per unit volume, which is a negative since the phase transformation is assumed to be thermodynamically feasible. Note that σ_{sl} is not a strong function of r and is assumed constant. The ΔG_v also does not depend on r.

An **embryo** is a tiny particle of solid that forms from the liquid as atoms cluster together. The embryo is unstable, and may either grow into a stable nucleus or redissolve back into the liquid.

In Figure 9-3, the top curve shows the parabolic variation of the total surface energy $(4\pi r^2 \cdot \sigma_{sl})$. The bottom most curve shows the total volume free-energy change term $(\frac{4}{3}\pi r^3 \cdot \Delta G_v)$. The curve in the middle shows the variation of ΔG. At the temperature where the solid and liquid phases are predicted to be in thermodynamic equilibrium (i.e., at the freezing temperature), the free energy of the solid phase and that of the

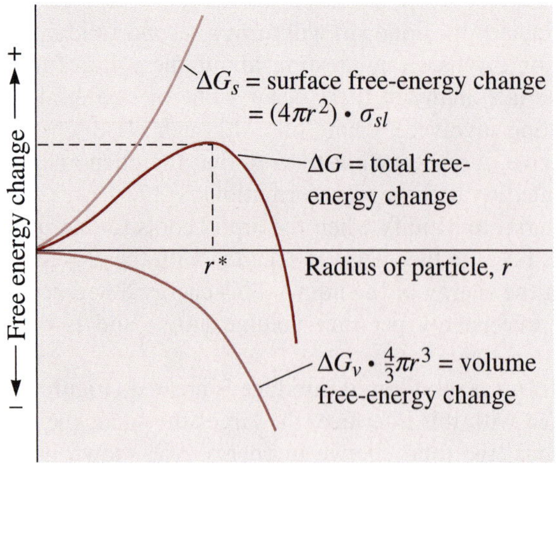

Figure 9-3
The total free energy of the solid-liquid system changes with the size of the solid. The solid is an embryo if its radius is less than the critical radius, and is a nucleus if its radius is greater than the critical radius.

liquid phase are equal ($\Delta G_v = 0$), so the total free energy change (ΔG) will be positive. When the solid is very small, with a radius less than the **critical radius** for nucleation (r^*) (Figure 9-3), further *growth* causes the total free energy to increase. The critical radius (r^*) is the minimum size of a crystal that must be formed by atoms clustering together in the liquid before the solid particle is stable and begins to grow. But instead of growing, the solid has a tendency to remelt, causing the free energy to decrease; thus, the bulk of the material remains liquid leaving just a small crystal solid. At freezing temperatures, embryos are thermodynamically unstable. So how can they grow?

The formation of embryos is a statistical process. Many embryos form and redissolve. If by chance, an embryo forms which has a radius that is larger than r^*, further growth causes the total free energy to decrease. The new solid is then stable and sustainable since nucleation has occurred, and growth of the solid particle—which is now called a nucleus—begins. At the thermodynamic freezing temperature, the probability of forming stable, sustainable nuclei is extremely small. Therefore, solidification does not begin at the thermodynamic melting or freezing temperature. If the temperature continues to decrease below the equilibrium freezing temperature, the liquid phase that should have transformed into a solid becomes increasingly unstable thermodynamically speaking. Because the liquid is below the equilibrium freezing temperature, the liquid is considered undercooled. The **undercooling** of ΔT is the equilibrium freezing temperature minus the actual temperature of the liquid. As the extent of undercooling increases, the thermodynamic driving force for the formation of a solid phase from liquid overtakes the resistance to create a solid-liquid interface.

This phenomenon can be seen in many other phase transformations. When one solid phase (α) transforms into another solid phase (β), the system has to be cooled to a temperature that is below the thermodynamic phase transformation temperature (at which free energies of the α and β phases are equal). When a liquid is transformed into a vapor, a bubble of vapor is created in the liquid. In order to create the transformation, though, we need to **superheat** the liquid above its boiling temperature! Therefore, we can see that liquids do not really freeze at their freezing temperature and do not really boil at their boiling point! We need to undercool the liquid for it to solidify and superheat it for it to boil!

Homogeneous Nucleation As liquid cools to temperatures below the equilibrium freezing temperature, two factors combine to favor nucleation. First, since atoms are losing their thermal energy the probability of forming clusters to form larger embryos increases. Second, the larger volume free energy difference between the liquid and the solid reduces the critical size (r^*) of the nucleus. **Homogeneous nucleation** occurs when the undercooling becomes large enough to cause the formation of a stable nucleus.

The critical radius r^* is given by

$$r^* = \frac{2\sigma_{sl}\, T_m}{\Delta H_f\, \Delta T} \tag{9-2}$$

where ΔH_f is the **latent heat of fusion**, T_m is the equilibrium solidification temperature in Kelvin, and $\Delta T = (T_m - T)$ is the undercooling when the liquid temperature is T. The latent heat of fusion represents the heat given off during the liquid-to-solid transformation. As the undercooling increases, the critical radius required for nucleation decreases. Table 9-1 presents values for σ_{sl}, ΔH_f, and typical undercoolings observed experimentally for homogeneous nucleation.

The following example shows how we can calculate the critical radius of the nucleus for the solidification of copper.

TABLE 9-1 ■ ***Values for freezing temperature, latent heat of fusion, surface energy, and maximum undercooling for selected materials***

Metal	Freezing Temperature (T_m) (°C)	Heat of Fusion (ΔH_f) (J/cm³)	Solid-Liquid Interfacial Energy (σ_{sl}) (J/cm²)	Typical Undercooling for Homogeneous Nucleation (ΔT) (°C)
Ga	30	488	56×10^{-7}	76
Bi	271	543	54×10^{-7}	90
Pb	327	237	33×10^{-7}	80
Ag	962	965	126×10^{-7}	250
Cu	1085	1628	177×10^{-7}	236
Ni	1453	2756	255×10^{-7}	480
Fe	1538	1737	204×10^{-7}	420
NaCl	801			169
CsCl	645			152
H_2O	0			40

EXAMPLE 9-1 *Calculation of Critical Radius for the Solidification of Copper*

Calculate the size of the critical radius and the number of atoms in the critical nucleus when solid copper forms by homogeneous nucleation. Comment on the size of the nucleus and assumptions we made while deriving the equation for the radius of the nucleus.

SOLUTION

From Table 9-1:

$$\Delta T = 236°\text{C} \quad T_m = 1085 + 273 = 1358 \text{ K}$$

$$\Delta H_f = 1628 \text{ J/cm}^3$$

$$\sigma_{sl} = 177 \times 10^{-7} \text{ J/cm}^2$$

$$r^* = \frac{2\sigma_{sl} T_m}{\Delta H_f \Delta T} = \frac{(2)(177 \times 10^{-7})(1358)}{(1628)(236)} = 12.51 \times 10^{-8} \text{ cm}$$

The lattice parameter for FCC copper is $a_0 = 0.3615$ nm $= 3.615 \times 10^{-8}$ cm

$$V_{\text{unit cell}} = (a_0)^3 = (3.615 \times 10^{-8})^3 = 47.24 \times 10^{-24} \text{ cm}^3$$

$$V_{r^*} = \tfrac{4}{3}\pi r^3 = \left(\tfrac{4}{3}\pi\right)(12.51 \times 10^{-8})^3 = 8200 \times 10^{-24} \text{ cm}^3$$

The number of unit cells in the critical nucleus is

$$\frac{8200 \times 10^{-24}}{47.24 \times 10^{-24}} = 174 \text{ unit cells}$$

Since there are four atoms in each unit cell of FCC metals, the number of atoms in the critical nucleus must be:

$$(4 \text{ atoms/cell})(174 \text{ cells/nucleus}) = 696 \text{ atoms/nucleus}$$

In these types of calculations, we assume that a nucleus that is made from only a few hundred atoms still exhibits properties similar to those of bulk materials. This is not strictly correct and as such considered to be a weakness of the classical theory of nucleation.

Heterogeneous Nucleation From Table 9-1, we can see that water will not solidify into ice via homogeneous nucleation until we reach a temperature of –40°C (undercooling of 40°C)! Except in controlled laboratory experiments, homogeneous nucleation almost never occurs in liquids. Instead, impurities in contact with the liquid, either suspended in the liquid or on the walls of the container that holds the liquid, provide a surface on which the solid can form (Figure 9-4). Now, a radius of curvature greater than the critical radius is achieved with very little total surface between the solid and liquid. Only a few atoms must cluster together to produce a solid particle that has the required radius of curvature. Much less undercooling is required to achieve the critical size, so nucleation occurs more readily. Nucleation on preexisting surfaces is known as **heterogeneous nucleation**. This process is dependent on the contact angle (θ) for the nucleating phase and the surface on which nucleation occurs. The same type of phenomenon occurs in solid-state transformations.

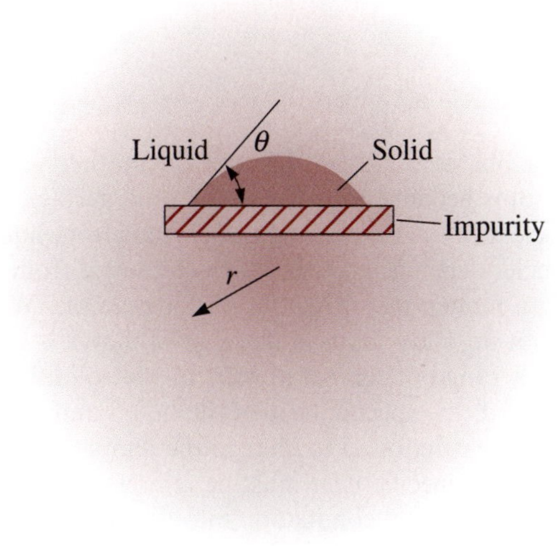

Figure 9-4
A solid forming on an impurity can assume the critical radius with a smaller increase in the surface energy. Thus, heterogeneous nucleation can occur with relatively low undercoolings.

Rate of Nucleation The *rate of nucleation* (the number of nuclei formed per unit time) is a function of temperature. Prior to solidification, of course, there is no nucleation and, at temperatures above the freezing point, the rate of nucleation is zero. As the temperature drops, the driving force for nucleation increases. However, as the temperature becomes lower, atomic diffusion becomes slower, hence slowing the nucleation process. Thus, a typical rate of nucleation (I) reaches a maximum at some temperature below the transformation temperature (Figure 9-5). In heterogeneous nucleation, the rate of nucleation is dictated by the concentration of the nucleating agents introduced. By considering the rates of nucleation and growth, we can predict the overall rate of a phase transformation.

Control of nucleation is important in the processing of metals, alloys, inorganic glasses, and other engineered materials.

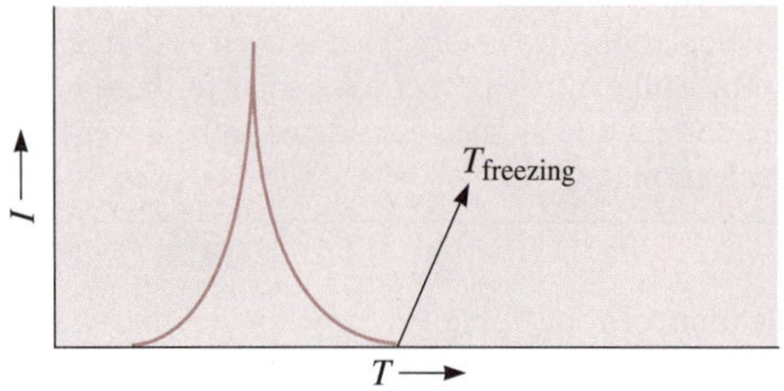

Figure 9-5
Rate of nucleation (I) as a function of temperature of the liquid (T).

Grain Size Strengthening When a metal casting freezes, impurities in the melt and walls of the mold in which solidification occurs serve as heterogeneous nucleation sites. Sometimes we intentionally introduce nucleating particles into the liquid. Such practices are called **grain refinement** or **inoculation**. Chemicals added to molten metals to promote nucleation and, hence, a finer grain size, are known as grain refiners or **inoculants**. For example, a combination of 0.03% titanium (Ti) and 0.01% boron (B) is added to many liquid-aluminum alloys. Tiny particles of an aluminum titanium compound (Al_3Ti) or titanium diboride (TiB_2) form and serve as sites for heterogeneous nucleation. Grain refining or inoculation produces a large number of grains, each beginning to grow from one nucleus. The greater grain boundary area provides grain size strengthening in metallic materials. This was discussed using the Hall-Petch equation in Chapter 4.

9-3 Growth Mechanisms

Once the solid nuclei of a phase forms (in a liquid or another solid phase), growth begins to occur as more atoms become attached to the solid surface. In this discussion, we will concentrate on nucleation and the growth of crystals from liquid. The nature of the growth of the solid nuclei depends on how heat is removed from the molten material. Let's consider casting a molten metal in a mold, for example. We assume we have a nearly pure metal and not an alloy (as solidification of alloys is different in that, in most cases, it occurs over a range of temperatures). In the solidification process, two types of heat must be removed: the specific heat of the liquid and the latent heat of fusion. The **specific heat** is the heat required to change the temperature of a unit mass of the material by one degree. The specific heat must be removed first, either by radiation into the surrounding atmosphere or by conduction into the surrounding mold, until the liquid cools to its freezing temperature. This is simply the cooling of the liquid from one temperature to a temperature at which nucleation begins.

We know that to melt a solid we need to supply heat. Therefore, when solid crystals form from a liquid, heat is generated! This heat is the latent heat of fusion (ΔH_f) and must be removed from the solid-liquid interface before solidification is completed. The manner in which we remove the latent heat of fusion determines the material's growth mechanism and final structure of a casting.

Planar Growth When a well-inoculated liquid (i.e., a liquid containing nucleating agents) cools under equilibrium conditions, there is no need for undercooling since heterogeneous nucleation can occur readily. Therefore, the temperature of the liquid ahead of the **solidification front** (i.e., solid–liquid interface) is greater than the freezing temperature. The temperature of the solid is at or below the freezing temperature. During solidification, the latent heat of fusion is removed by conduction from the solid-liquid interface through the solid. Any small protuberance that begins to grow on the interface

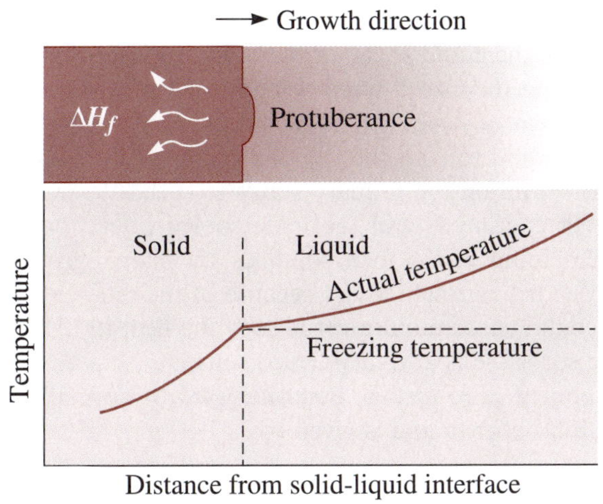

Figure 9-6
When the temperature of the liquid is above the freezing temperature, a protuberance on the solid-liquid interface will not grow, leading to maintenance of a planar interface. Latent heat is removed from the interface through the solid.

is surrounded by liquid above the freezing temperature (Figure 9-6). The growth of the protuberance then stops until the remainder of the interface catches up. This growth mechanism, known as **planar growth**, occurs by the movement of a smooth or planar solid-liquid front into the liquid.

Dendritic Growth When the liquid is not inoculated and the nucleation is poor, the liquid has to be undercooled before the solid forms (Figure 9-7). Under these condi-

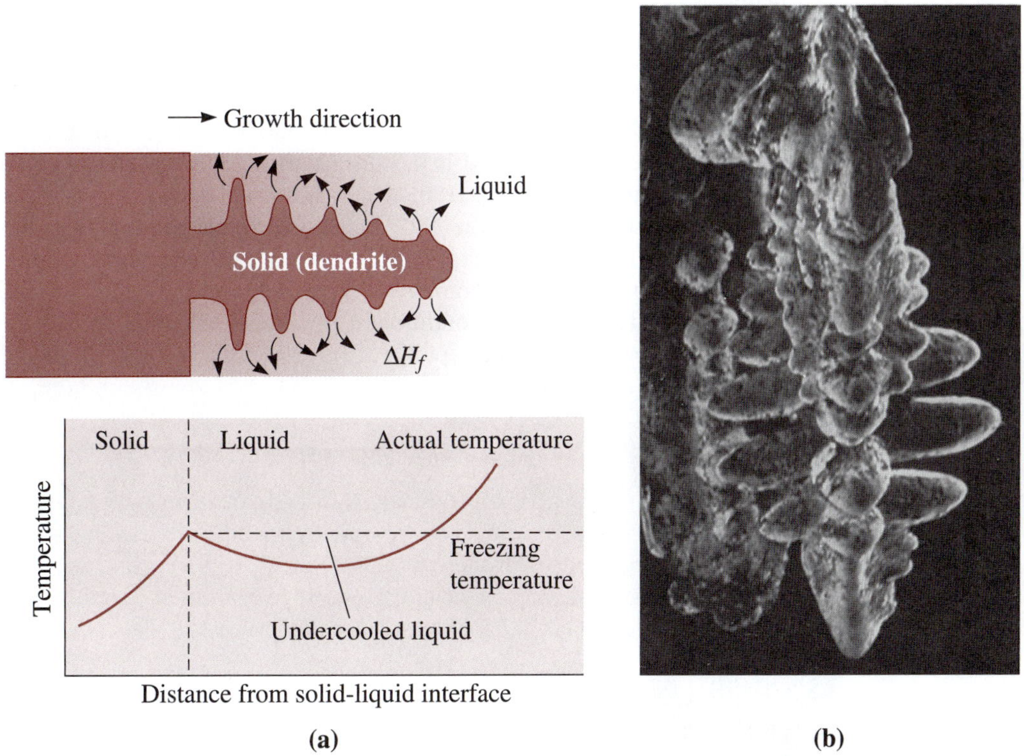

Figure 9-7 (a) If the liquid is undercooled, a protuberance on the solid-liquid interface can grow rapidly as a dendrite. The latent heat of fusion is removed by raising the temperature of the liquid back to the freezing temperature. (b) Scanning electron micrograph of dendrites in steel (×15).

tions, a small solid protuberance called a **dendrite**, which forms at the interface, is encouraged to grow since the liquid ahead of the solidification front is undercooled. The word dendrite comes from the Greek word *dendron* that means tree. As the solid dendrite grows, the latent heat of fusion is conducted into the undercooled liquid, raising the temperature of the liquid toward the freezing temperature. Secondary and tertiary dendrite arms can also form on the primary stalks to speed the evolution of the latent heat. Dendritic growth continues until the undercooled liquid warms to the freezing temperature. Any remaining liquid then solidifies by planar growth. The difference between planar and dendritic growth arises because of the different sinks for the latent heat of fusion. The container or mold must absorb the heat in planar growth, but the undercooled liquid absorbs the heat in dendritic growth.

In the solidification of pure metals, dendritic growth normally represents only a small fraction of the total growth and is given by:

$$\text{Dendritic fraction} = f = \frac{c\Delta T}{\Delta H_f} \tag{9-3}$$

where c is the specific heat of the liquid. The numerator represents the heat that the undercooled liquid can absorb, and the latent heat in the denominator represents the total heat that must be given up during solidification. As the undercooling ΔT increases, more dendritic growth occurs. If the liquid is well inoculated, undercooling is almost zero and growth would be mainly via the planar front solidification mechanism.

The rate at which growth of the solid occurs depends on the cooling rate, or the rate of heat extraction. A higher cooling rate produces rapid solidification, or short solidification times. The time t_s required for a simple casting to solidify completely can be calculated using ***Chvorinov's rule***:

$$t_s = \mathrm{B}\left(\frac{V}{A}\right)^n \tag{9-4}$$

where V is the volume of the casting and represents the amount of heat that must be removed before freezing occurs, A is the surface area of the casting in contact with the mold and represents the surface from which heat can be transferred away from the casting, n is a constant (usually about 2), and B is the **mold constant**. The mold constant depends on the properties and initial temperatures of both the metal and the mold. This rule basically accounts for the geometry of a casting and the heat transfer conditions. The rule states that for the same conditions a casting with a small volume and relatively large surface area will cool more rapidly.

EXAMPLE 9-2 *Redesign of a Casting for Improved Strength*

Your company currently is producing a disk-shaped brass casting 5 cm thick and 45 cm in diameter. You believe that by making the casting solidify 25% faster, the improvement in the tensile properties of the casting will permit the casting to be made lighter in weight. Design the casting to permit this. Assume that the mold constant is 3.5 min/cm^2 for this process.

SOLUTION

One approach would be to use the same casting process, but reduce the thickness of the casting. The thinner casting would solidify more quickly and, because of the faster cooling, should have improved mechanical properties.

Chvorinov's rule helps us calculate the required thickness. If d is the diameter and x is the thickness of the casting, then the volume, surface area, and solidification time of the 5 cm thick casting are:

$$V = (\pi/4)d^2x = (\pi/4)(45)^2(5) = 7955.4 \text{ cm}^3$$

$$A = 2(\pi/4)d^2 + \pi dx = 2(\pi/4)(45)^2 + \pi(45)(5) = 3889.2 \text{ cm}^2$$

$$t = \mathrm{B}\left(\frac{V}{A}\right)^2 = (3.5)\left(\frac{7955.4}{3889.2}\right)^2 = 14.71 \text{ min}$$

The solidification time of the redesigned casting should be 25% shorter than the current time, or $t_r = 0.75t$, where:

$$t_r = 0.75t = (0.75)(14.71) = 11.03 \text{ min}$$

Since the casting conditions have not changed, the mold constant B is unchanged. The V/A ratio of the new casting is:

$$t_r = \mathrm{B}\left(\frac{V}{A}\right)^2 = (3.5)\left(\frac{V}{A}\right)^2 = 11.03 \text{ min}$$

$$\left(\frac{V}{A}\right)^2 = 3.1514 \text{ cm}^2 \quad \text{or} \quad \frac{V}{A} = 1.775 \text{ cm}$$

If x is the required thickness for our redesigned casting, then:

$$\frac{V_r}{A_r} = \frac{(\pi/4)d^2x}{2(\pi/4)d^2 + \pi dx} = \frac{(\pi/4)(45)^2(x)}{2(\pi/4)(45)^2 + \pi(45)(x)} = 1.775 \text{ cm}$$

Therefore, $x = 4.22$ cm

This thickness provides the required solidification time, while reducing the overall weight of the casting by nearly 15%.

Solidification begins at the surface, where heat is dissipated into the surrounding mold material. The rate of solidification of a casting can be described by how rapidly the thickness d of the solidified skin grows:

$$d = \mathrm{k}_{\text{solidification}}\sqrt{t} - \mathrm{c}_1 \tag{9-5}$$

where t is the time after pouring, $\mathrm{k}_{\text{solidification}}$ is a constant for a given casting material and mold, and c_1 is a constant related to the **pouring temperature**.

Effect on Structure and Properties The solidification time affects the size of the dendrites. Normally, dendrite size is characterized by measuring the distance between the secondary dendrite arms (Figure 9-8). The **secondary dendrite arm spacing (SDAS)** is reduced when the casting freezes more rapidly. The finer, more extensive dendritic network serves as a more efficient conductor of the latent heat to the undercooled liquid. The SDAS is related to the solidification time by

$$\text{SDAS} = \mathrm{k}t_s^{\mathrm{m}} \tag{9-6}$$

where m and k are constants depending on the composition of the metal. This relationship

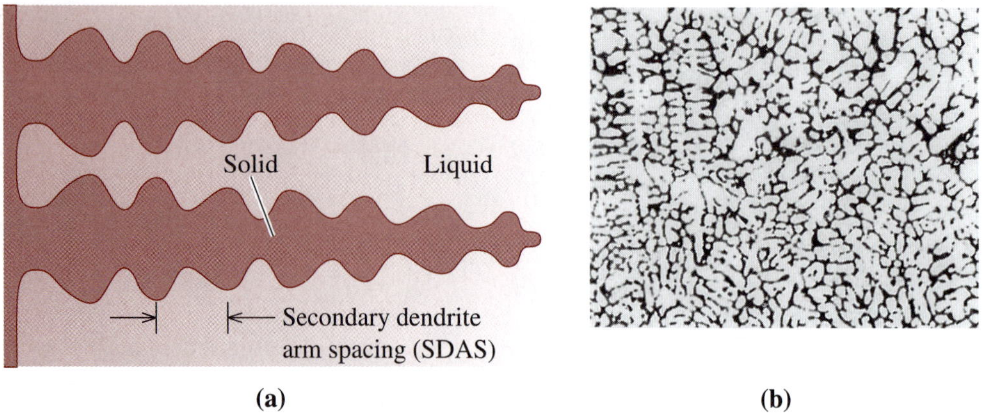

Figure 9-8 (a) The secondary dendrite arm spacing (SDAS). (b) Dendrites in an aluminum alloy (×50). (*From* ASM Handbook, *Vol. 9, Metallography and Microstructure (1985), ASM International, Materials Park, OH 44073-0002.*)

is shown in Figure 9-9 for several alloys. Small secondary dendrite arm spacings are associated with higher strengths and improved ductility (Figure 9-10).

Rapid solidification processing is used to produce exceptionally fine secondary dendrite arm spacings; a common method is to produce very fine liquid droplets that freeze

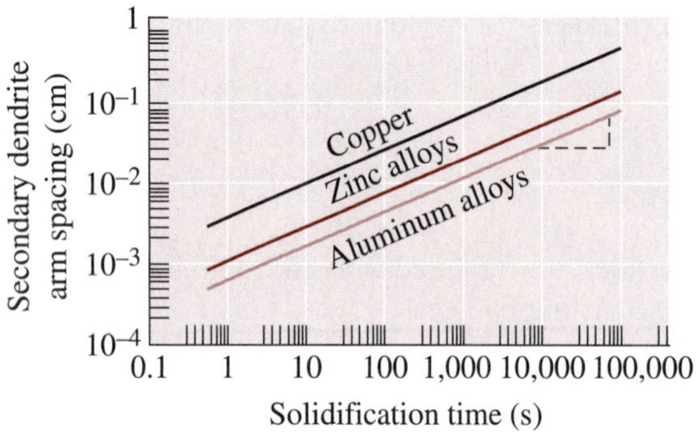

Figure 9-9 The effect of solidification time on the secondary dendrite arm spacings of copper, zinc, and aluminum.

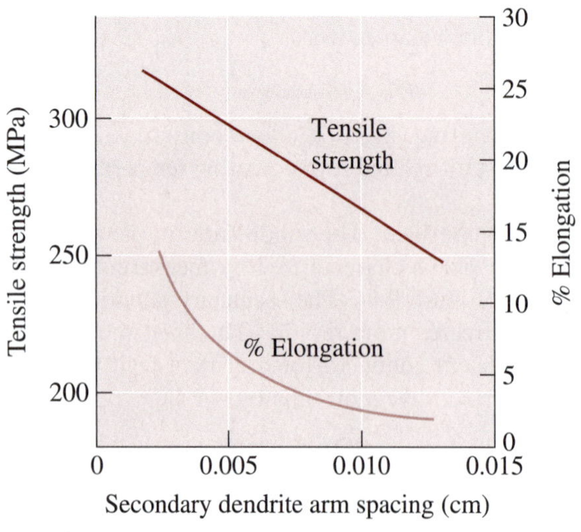

Figure 9-10 The effect of the secondary dendrite arm spacing on the properties of an aluminum casting alloy.

into solid particles. This process is known as spray atomization. The tiny droplets freeze at a rate of about 10^6C/s, producing powder particles that range from ~5–100 μm. This cooling rate is not rapid enough to form a metallic glass, but does produce a fine dendritic structure. By carefully consolidating the solid droplets by powder metallurgy processes, improved properties in the material can be obtained. Since the particles are derived from melt, many complex alloy compositions can be produced in the form of chemically homogenous powders.

The following example discusses how Chvorinov's rule, the relationship between SDAS and the time of solidification, and the SDAS and mechanical properties can be used to design casting processes.

EXAMPLE 9-3 *Secondary Dendrite Arm Spacing for Aluminum Alloys*

Determine the constants in the equation that describe the relationship between secondary dendrite arm spacing and solidification time for aluminum alloys (Figure 9-9).

SOLUTION

We could obtain the value of SDAS at two times from the graph and calculate k and m using simultaneous equations. However, if the scales on the ordinate and abscissa are equal for powers of ten (as in Figure 9-9), we can obtain the slope m from the log-log plot by directly measuring the slope of the graph. In Figure 9-9, we can mark five equal units on the vertical scale and 12 equal units on the horizontal scale. The slope is:

$$\mathrm{m} = \frac{5}{12} = 0.42$$

The constant k is the value of SDAS when $t_s = 1$ s, since:

$$\log \mathrm{SDAS} = \log \mathrm{k} + \mathrm{m} \log t_s$$

If $t_s = 1$ s, m log $t_s = 0$, and SDAS = k, from Figure 9-9:

$$\mathrm{k} = 8 \times 10^{-4} \text{ cm}$$

9-4 Cooling Curves

A cooling curve shows how the temperature of a material (in this case, a pure metal) changes with time [Figure 9-11(a) and (b)]. The liquid is poured into a mold at pouring temperature, point *A*. The difference between the pouring temperature and the freezing temperature is the superheat. The specific heat is extracted by the mold until the liquid reaches the freezing temperature (point *B*). If the liquid is not well inoculated it must be undercooled (point *B* to *C*). The slope of the cooling curve before solidification begins is the cooling rate $\frac{\Delta T}{\Delta t}$. As nucleation begins (point *C*) latent heat of fusion is given off and the temperature rises. This increase in temperature of the undercooled liquid as a result of nucleation is known as **recalescence** (point *C* to *D*). Solidification proceeds

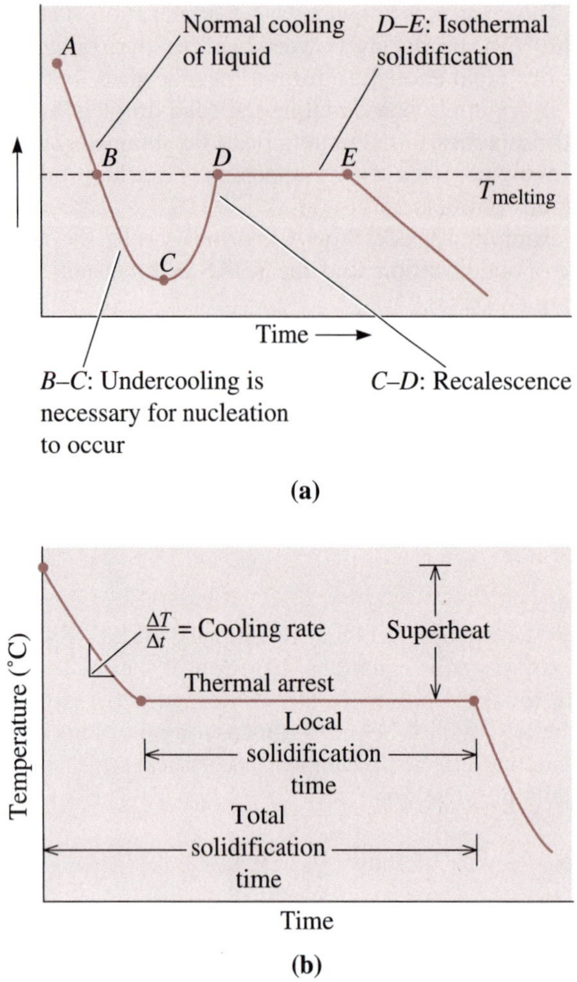

Figure 9-11
(a) Cooling curve for a pure metal that has not been well inoculated. Liquid cools as specific heat is removed (between points *A* and *B*). Undercooling is thus necessary (between points *B* and *C*). As the nucleation begins (point *C*), latent heat of fusion is released causing an increase in the temperature of the liquid. This process is known as recalescence (point *C* to point *D*). Metal continues to solidify at a constant temperature ($T_{melting}$). At point *E*, solidification is complete. Solid casting continues to cool from the point. (b) Cooling curve for a well inoculated, but otherwise pure, metal. No undercooling is needed. Recalescence is not observed. Solidification begins at the melting temperature.

isothermally at the melting temperature (point *D* to *E*) as the latent heat given off from continued solidification is balanced by the heat lost by cooling. This region between points *D* and *E*, where the temperature is constant, is known as the **thermal arrest**. A thermal arrest, or plateau, is produced because of the evolution of the latent heat of fusion balances the heat being lost because of cooling. At point *E*, solidification is complete and the solid casting cools from point *E* to room temperature.

If the liquid is well-inoculated, the extent of undercooling is usually very small. The undercooling and recalescence are very small and can be observed in cooling curves only by very careful measurements. If effective heterogeneous nuclei are present in the liquid, solidification begins at the freezing temperature [Figure 9-11(b)]. The latent heat keeps the remaining liquid at the freezing temperature until all of the liquid has solidified and no more heat can be evolved. Growth under these conditions is planar. The **total solidification time** of the casting is the time required to remove both the specific heat of the liquid and the latent heat of fusion. Measured from the time of pouring until solidification is complete, this time is given by Chvorinov's rule. The **local solidification time** is the time required to remove only the latent heat of fusion at a particular location in the casting; it is measured from when solidification begins until solidification is completed. The local solidification times (and the total solidification times) for liquids solidified via undercooled and inoculated liquids will be slightly different.

9-5 Cast Structure

In manufacturing components by casting, molten metals are often poured into molds and permitted to solidify. The mold produces a finished shape, known as a **casting**. In other cases, the mold produces a simple shape, called an **ingot**. An ingot usually requires extensive plastic deformation before a finished product is created. A *macrostructure*, sometimes referred to as the **ingot structure**, consists of as many as three parts (Figure 9-12). (Recall that in Chapter 2 we had used the term "macrostructure" to describe the structure of a material at a macroscopic scale. Hence, the term "ingot structure" may be more appropriate.)

Chill Zone The **chill zone** is a narrow band of randomly oriented grains at the surface of the casting. The metal at the mold wall is the first to cool to the freezing temperature. The mold wall also provides many surface sites at which heterogeneous nucleation takes place.

Columnar Zone The **columnar zone** contains elongated grains oriented in a particular crystallographic direction. As heat is removed from the casting by the mold material, the grains in the chill zone grow in the direction opposite the heat flow, or from the coldest toward the hottest areas of the casting. This tendency usually means that the grains grow perpendicular to the mold wall.

Equiaxed Zone Although the solid may continue to grow in a columnar manner until all of the liquid has solidified, an equiaxed zone frequently forms in the center of the casting or ingot. The **equiaxed zone** contains new, randomly oriented grains, often caused by a low pouring temperature, alloying elements, or grain refining or inoculating

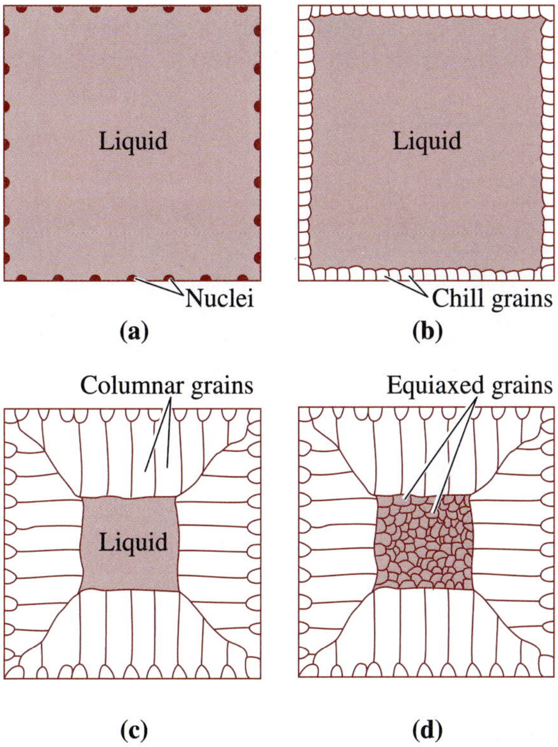

Figure 9-12 Development of the ingot structure of a casting during solidification: (a) Nucleation begins, (b) the chill zone forms, (c) preferred growth produces the columnar zone, and (d) additional nucleation creates the equiaxed zone.

agents. Small grains or dendrites in the chill zone may also be torn off by strong convection currents that are set up as the casting begins to freeze. These also provide the heterogeneous nucleation sites for what ultimately become equiaxed grains. These grains grow as relatively round, or equiaxed, grains with a random orientation, and they stop the growth of the columnar grains. The formation of the equiaxed zone is a nucleation-controlled process and causes that portion of the casting to display isotropic behavior.

9-6 Solidification Defects

Although there are many defects that potentially can be introduced during solidification, shrinkage and the porosity deserve special mention. If a casting contains pores (small holes), the cast component can fail catastrophically when used for load-bearing applications (e.g., turbine blades).

Shrinkage Almost all materials are more dense in the solid state than in the liquid state. During solidification, the material contracts, or shrinks, as much as 7% (Table 9-2). Often, the bulk of the **shrinkage** occurs as **cavities**, if solidification begins at all surfaces of the casting, or *pipes*, if one surface solidifies more slowly than the others (Figure 9-13). The presence of such pipes can pose problems. For example, if in the production of zinc ingots a shrinkage pipe remains, water vapor can condense in it. This water can lead to an explosion if the ingot gets introduced in a furnace in which zinc is being remelted for such applications as hot-dip galvanizing.

A common technique for controlling **cavity** and **pipe shrinkage** is to place a **riser**, or an extra reservoir of metal, adjacent and connected to the casting. As the casting solidifies and shrinks, liquid metal flows from the riser into the casting to fill the shrinkage void. We need only assure that the riser solidifies after the casting and that there is an internal liquid channel that connects the liquid in the riser to the last liquid to solidify in the casting. Chvorinov's rule can be used to help design the size of the riser. The following example illustrates how risers can be designed to compensate for shrinkage.

TABLE 9-2 ■ *Shrinkage during solidification for selected materials*

Material	Shrinkage (%)
Al	7.0
Cu	5.1
Mg	4.0
Zn	3.7
Fe	3.4
Pb	2.7
Ga	+3.2 (expansion)
H_2O	+8.3 (expansion)
Low-carbon steel	2.5–3.0
High-carbon steel	4.0
White Cast Iron	4.0–5.5
Gray Cast Iron	+1.9 (expansion)

Note: Some data from DeGarmo, E.P., Black, J.T., and Koshe, R.A., Materials and Processes in Manufacturing *(1997), Prentice Hall.*

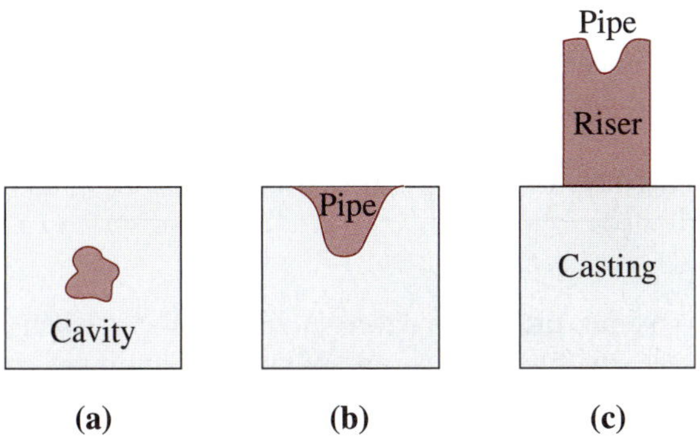

Figure 9-13
Several types of macroshrinkage can occur, including cavities and pipes. Risers can be used to help compensate for shrinkage.

EXAMPLE 9-4 *Design of a Riser for a Casting*

Design a cylindrical riser, with a height equal to twice its diameter, that will compensate for shrinkage in a 2 cm × 8 cm × 16 cm casting (Figure 9-14).

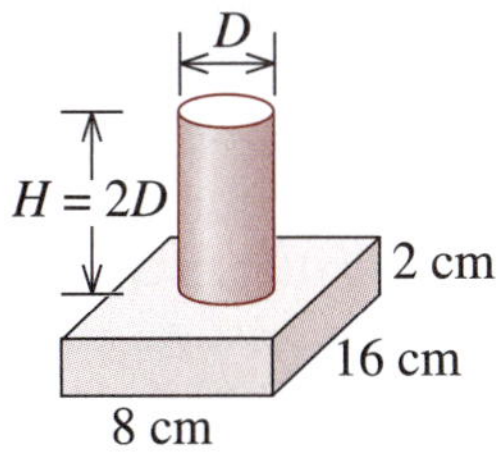

Figure 9-14
The geometry of the casting and riser (for Example 9-4).

SOLUTION

We know that the riser must freeze after the casting. To be conservative, however, we typically require that the riser take 25% longer to solidify than the casting. Therefore:

$$t_{\text{riser}} = 1.25 t_{\text{casting}} \quad \text{or} \quad B\left(\frac{V}{A}\right)_r^2 = 1.25B\left(\frac{V}{A}\right)_c^2$$

Subscripts r and c stand for riser and casting, respectively.

The mold constant B is the same for both casting and riser, so:

$$(V/A)_r = \sqrt{1.25(V/A)_c}$$

$$V_c = (2)(8)(16) = 256 \text{ cm}^3$$

$$A_c = (2)(2)(8) + (2)(2)(16) + (2)(8)(16) = 352 \text{ cm}^2$$

We can write equations for the volume and area of a cylindrical riser, noting that $H = 2D$:

$$V_r = (\pi/4)D^2H = (\pi/4)D^2(2D) = (\pi/2)D^3$$

$$A_r = 2(\pi/4)D^2 + \pi DH = 2(\pi/4)D^2 + \pi D(2D) = (5\pi/2)D^2$$

$$\frac{V_r}{A_r} = \frac{(\pi/2)(D)^3}{(5\pi/2)(D)^2} = \frac{D}{5} > \sqrt{\frac{(1.25)(256)}{352}}$$

$$D = 4.77 \text{ cm} \quad H = 2D = 9.54 \text{ cm} \quad V_r = 170.5 \text{ cm}^3$$

Although the volume of the riser is less than that of the casting, the riser solidifies more slowly because of its compact shape.

Interdendritic Shrinkage This consists of small shrinkage pores between dendrites. This defect, also called **microshrinkage** or **shrinkage porosity**, is difficult to prevent by the use of risers. Fast cooling rates may reduce problems with interdendritic shrinkage; the dendrites may be shorter, permitting liquid to flow through the dendritic network to the solidifying solid interface. In addition, any shrinkage that remains may be finer and more uniformly distributed.

Gas Porosity Many metals dissolve a large quantity of gas when they are molten. Aluminum, for example, dissolves hydrogen. When the aluminum solidifies, however, only a small fraction of the hydrogen is retained, since the solubility is remarkably lower. The excess hydrogen that cannot be incorporated in the solid metal or alloy crystal structure forms bubbles that may be trapped in the solid metal, producing **gas porosity**. The amount of gas that can be dissolved in molten metal is given by **Sievert's law**:

$$\text{Percent of gas} = \mathrm{K}\sqrt{p_{\text{gas}}} \tag{9-7}$$

where p_{gas} is the partial pressure of the gas in contact with the metal and K is a constant which, for a particular metal-gas system, increases with increasing temperature. We can minimize gas porosity in castings by keeping the liquid temperature low, by adding materials to the liquid to combine with the gas and form a solid, or by assuring that the partial pressure of the gas remains low. The latter may be achieved by placing the molten metal in a vacuum chamber or bubbling an inert gas through the metal. Because p_{gas} is low in the vacuum, the gas leaves the metal, enters the vacuum, and is carried away. **Gas flushing** is a process in which bubbles of a gas, inert or reactive, are injected into a molten metal to remove undesirable elements from molten metals and alloys. For example, hydrogen in aluminum can be removed using nitrogen or chlorine. Degassing a molten metal is important in the production of steels and many other alloys.

9-7 Casting Processes for Manufacturing Components

Figure 9-15 summarizes four of the dozens of commercial casting processes. In some processes the same mold can be used; in others the mold is expendable. **Sand casting** processes include green sand molding, for which silica (SiO_2) sand grains bonded with wet clay are packed around a removable pattern. Ceramic casting processes use a fine-grained ceramic material as the mold; a slurry containing the ceramic may be poured around a reusable pattern, which is removed after the ceramic hardens. In **investment casting**, the ceramic slurry of a material such as colloidal silica (consisting of nano-sized ceramic particles) coats a wax pattern. After the ceramic hardens (i.e., the colloidal silica dispersion gels), the wax is melted and drained from the ceramic shell, leaving behind a cavity that is filled with molten metal. The investment casting process, also known as the **lost wax process**, is best suited for generating most complex shapes. Dentists and jewelers originally used the precision investment casting process. Currently, this process is used to produce such components as turbine blades, titanium heads of golf clubs, and parts for knee and hip prosthesis. In another process known as the **lost foam process**, polystyrene beads, similar to those used to make coffee cups or packaging materials, are used to produce a foam pattern. Loose sand is compacted around the

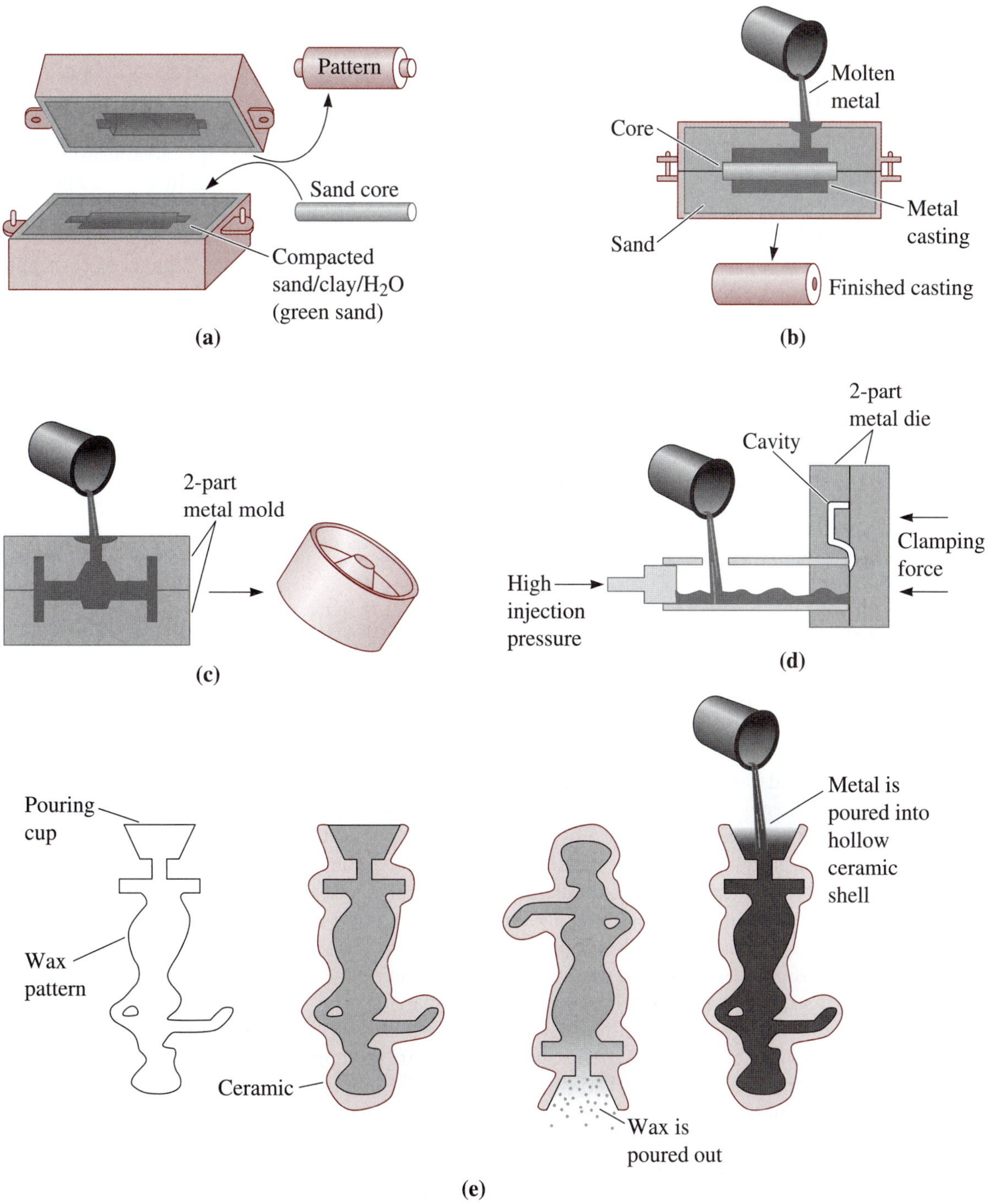

Figure 9-15 Four typical casting processes: (a) and (b) Green sand molding, in which clay-bonded sand is packed around a pattern. Sand cores can produce internal cavities in the casting. (c) The permanent mold process, in which metal is poured into an iron or steel mold. (d) Die casting, in which metal is injected at high pressure into a steel die. (e) Investment casting, in which a wax pattern is surrounded by a ceramic; after the wax is melted and drained, metal is poured into the mold.

pattern to produce a mold. When molten metal is poured into the mold, the polymer foam pattern melts and decomposes, with the metal taking the place of the pattern.

In the permanent mold and pressure die casting processes, a cavity is machined from metallic material. After the liquid poured into the cavity solidifies, the mold is opened, the casting is removed, and the mold is reused. The processes using metallic molds tend to give the highest strength castings because of the rapid solidification. Ceramic molds, including those used in investment casting, are good insulators and give the slowest-cooling and lowest-strength castings. Millions of truck and car pistons are made in foundries using permanent mold casting. Good surface finish and dimensional accuracy are the advantages of **permanent mold casting** technique. High mold costs and limited complexity in shape are the limitations of this technique.

In **pressure die casting**, molten metallic material is forced into the mold under high pressures and is held under pressure during solidification. Many zinc, aluminum, and magnesium-based alloys are processed using pressure die casting. Extremely smooth surface finishes, very good dimensional accuracy, the ability to cast intricate shapes, and high production rates are the advantages of the pressure die casting process. Since the mold is metallic and must withstand high pressures, the dies used are expensive and the technique is limited to smaller sized components.

9-8 Continuous Casting, Ingot Casting, and Single Crystal Growth

As discussed in the prior section, casting is a tool used for the manufacturing of components. It is also a process for producing ingots or slabs that can be further processed into different shapes (e.g., rods, bars, wires, etc.). In the steel industry, millions of pounds of steels are produced using a blast furnace, an electric arc furnace and other processes. Although the details change, most metals and alloys (e.g., copper and zinc) are extracted from their ores using similar processes. Certain metals, such as aluminum, are produced using an electrolytic process since aluminum oxide is too stable and can not be readily reduced to aluminum metal using coke or other reducing agents.

In many cases, we begin with scrap metals and recyclable alloys. In this case, the scrap metal is melted and processed, removing the impurities and adjusting the composition. Considerable amounts of steels, aluminum, zinc, stainless steels, titanium, and many other materials are recycled every year.

In **ingot casting**, molten steels or alloys obtained from a furnace are cast into large molds. The resultant castings, called ingots, are then processed for conversion into useful shapes via thermomechanical processing, often at another location. In the **continuous casting** process, the idea is to go from molten metallic material to some more useful "semi-finished" shape such as a plate, slab, etc. The liquid metal is fed from a holding vessel (a tundish) into a water-cooled oscillating copper mold, which rapidly cools the surface of the steel. The partially solidified steel is withdrawn from the mold at the same rate that additional liquid steel is introduced. The center of the steel casting finally solidifies well after the casting exits the mold. The continuously cast material is then cut into appropriate lengths by special cutting machines.

The secondary processing steps in the processing of steels and other alloys are shown in Figure 9-16.

There are some applications for which a small equiaxed grain structure in the casting is not desired. Castings used for blades and vanes in turbine engines are an example

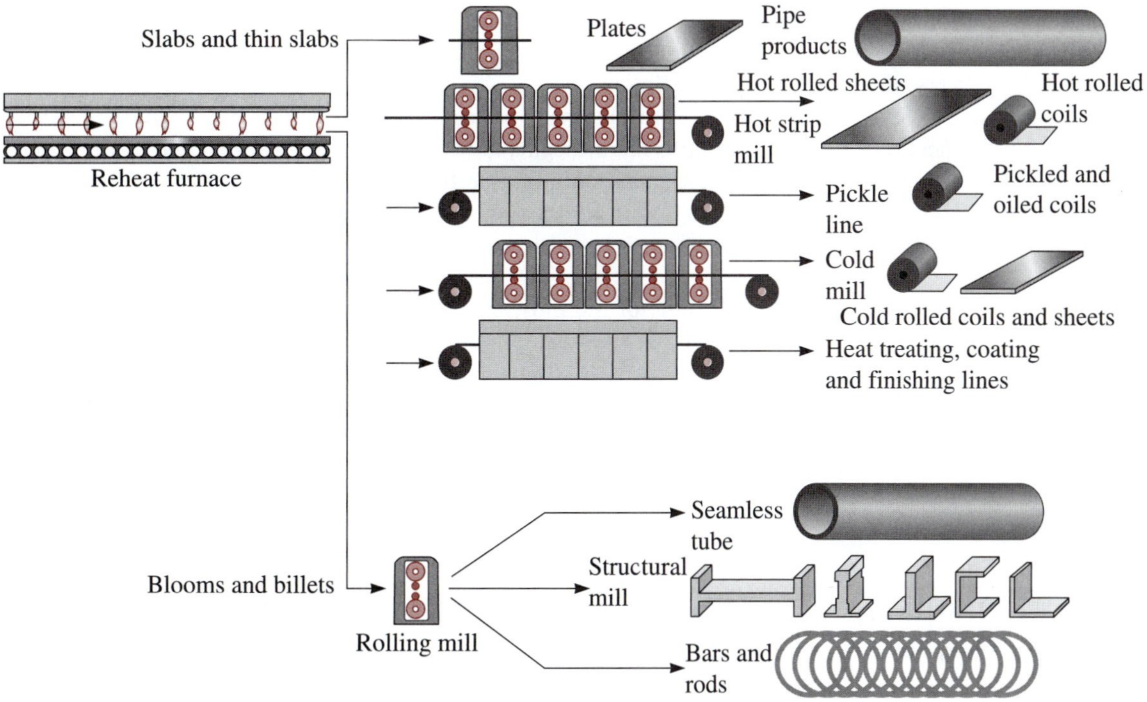

Figure 9-16 Secondary processing steps in processing of steel and alloys. (*Source: www.steel.org. Used with permission of the American Iron and Steel Institute.*)

(Figure 9-17). These castings are often made of titanium, cobalt or nickel-based superalloys using precision investment casting.

In conventionally cast parts, an equiaxed grain structure is often produced. However, blades and vanes for turbine and jet engines fail along transverse grain boundaries. Better creep and fracture resistance are obtained using the **directional solidification** (DS) growth technique. In the DS process, the mold is heated from one end and cooled from

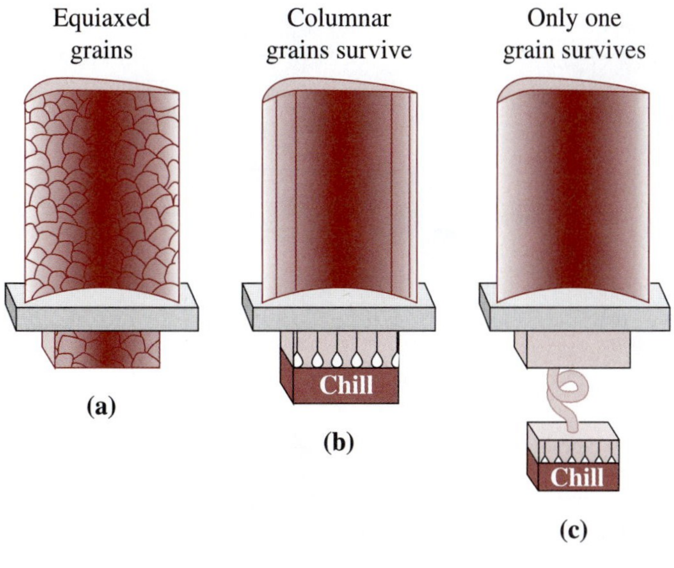

Figure 9-17 Controlling grain structure in turbine blades: (a) conventional equiaxed grains, (b) directionally solidified columnar grains, and (c) single crystal.

the other, producing a columnar microstructure with all of the grain boundaries running in the longitudinal direction of the part. No grain boundaries are present in the transverse direction [Figure 9-17(b)].

Still better properties are obtained by using a *single crystal* (SC) technique. Solidification of columnar grains again begins at a cold surface; however, due to the helical connection, only one columnar grain is able to grow to the main body of the casting [Figure 9-17(c)]. The single crystal casting has no grain boundaries, so its crystallographic planes and directions can be directed in an optimum orientation.

Single Crystal Growth One of the most important applications of solidification is the growth of single crystals. Polycrystalline materials cannot be used effectively in many electronic and optical applications. Grain boundaries and other defects interfere with the mechanisms that provide useful electrical or optical functions. For example, in order to utilize the semiconducting behavior of doped silicon, high-purity single crystals must be used. The current technology for silicon makes use of large (up to 12-in. diameter) single crystals.

9-9 Solidification of Polymers and Inorganic Glasses

Similar to the processing of metals and alloys, the processing of thermoplastics depends critically on our ability to melt and process them via extrusion and other processes. We will discuss these processes in later chapters.

Many polymers do not crystallize, but solidify, when cooled. In these materials, the thermodynamic driving force may exist; however, the rate of nucleation of the solid may be too slow, or the complexity of the polymer chains may be so great that a crystalline solid does not form. Crystallization in polymers is almost never complete and is significantly different from that of metallic materials, requiring long polymer chains to become closely aligned over relatively large distances. By doing so, the polymer grows as **lamellar**, or plate-like, crystals (Figure 9-18). Amorphous regions are present between

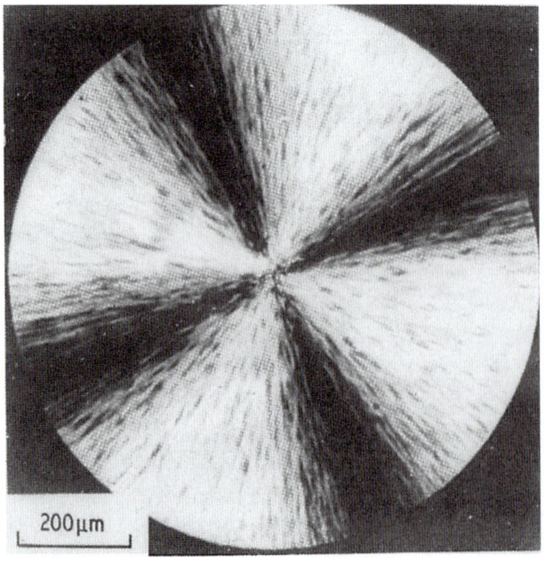

Figure 9-18
A spherulite in polystyrene (×8000). (*From R. Young and P. Lovell,* Introduction to Polymers, *2nd Ed., Chapman & Hall, 1991*).

the individual lamella, bundles of lamellae, and individual spherulites. In addition, bundles of lamellae grow from a common nucleus, but the crystallographic orientation of the lamellae within any one bundle is different from that in another. As the bundles grow, they may produce a spheroidal shape called a **spherulite**. The spherulite is composed of many individual bundles of differently oriented lamellae.

Many polymers of commercial interest develop crystallinity during their processing. Crystallinity can originate from cooling as discussed previously, or from the application of stress. For example, PET plastic bottles are prepared using the blow-stretch process, and they can develop considerable crystallinity during formation. This crystallization is a result of the application of stress, and thus, is different from that encountered in the solidification of metals and alloys. In general, such polymers as nylon and polyethylene crystallize more easily when compared to many other thermoplastics. Increased crystallinity due to fine crystals adds to the strength of thermoplastics.

Inorganic glasses, such as silicate glasses, also do not crystallize easily for kinetic reasons. While the thermodynamic driving force exists, similar to the solidification of metals and alloys; however, the melts are often too viscous and the diffusion is too slow for crystallization to proceed during solidification. The float-glass process is used to melt and cast large flat pieces of glasses. In this process, molten glass is made to float on molten tin. As discussed in Chapters 6 and 7, since the strength of inorganic glasses depends critically on surface flaws produced by the manufacturing process or the reaction with atmospheric moisture, most glasses are strengthened using tempering. When safety is not a primary concern, annealing is used to reduce stresses. Long lengths of glass fibers, such as those used with fiber optics, are produced by melting a high-purity glass rod known as a *preform*. As mentioned earlier, careful control of nucleation in glasses can lead to glass-ceramics, colored glasses, and photochromic glasses (glasses that can change their color or tint upon exposure to sunlight).

9-10 Joining of Metallic Materials

In **brazing**, an alloy, known as a filler, is used to join one metal to itself or to another metal. The brazing filler metal has a melting temperature above about 450°C. **Soldering** is a brazing process in which the filler has a melting temperature below 450°C. Lead-tin and antimony-tin alloys are the most common materials used for soldering. New lead-free soldering materials have been developed, since lead is toxic. Alloys being developed include those that are based on Sn-Cu-Ag. In brazing and soldering, the metallic materials being joined do not melt; only the filler material melts. In both brazing and soldering, the composition of the filler material is different from that of the base material being joined. Various aluminum-silicon, copper, magnesium and precious metals are used for brazing.

Solidification is also important in the joining of metals through **fusion welding**. In the fusion-welding processes, a portion of the metals to be joined is melted and, in many instances, additional molten filler metal is added. The pool of liquid metal is called the **fusion zone** (Figures 9-19 and 9-20). When the fusion zone subsequently solidifies, the original pieces of metal are joined together. During solidification of the fusion zone, nucleation is not required. The solid simply begins to grow from existing grains, frequently in a columnar manner. Growth of the solid grains in the fusion zone from the pre-existing grains is via epitaxial growth.

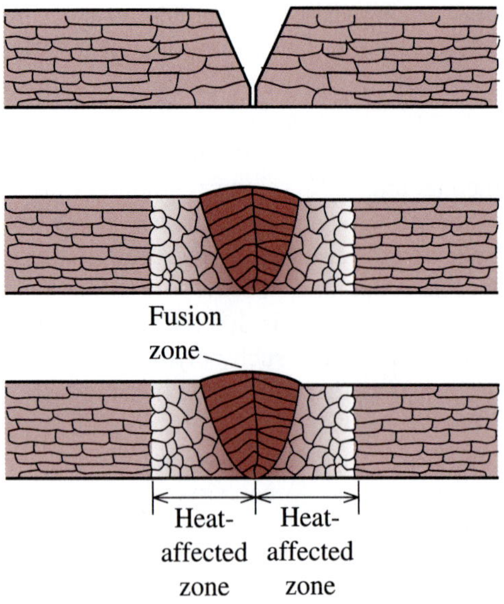

Figure 9-19
A schematic diagram of the fusion zone and solidification of the weld during fusion welding: (a) initial prepared joint, (b) weld at the maximum temperature, with joint filled with filler metal, and (c) weld after solidification.

The structure and properties of the fusion zone depend on many of the same variables as in a metal casting. Addition of inoculating agents to the fusion zone reduces the grain size. Fast cooling rates or short solidification times promote a finer microstructure and improved properties. Factors that increase the cooling rate include increased thickness of the metal, smaller fusion zones, low original metal temperatures, and certain types of welding processes. Oxyacetylene welding, for example, uses a relatively low-intensity heat source; consequently, welding times are long and the surrounding solid metal, which becomes very hot, is not an effective heat sink. Arc-welding processes provide a more intense heat source, thus reducing heating of the surrounding metal and providing faster cooling. Laser welding and electron-beam welding are exceptionally intense heat sources and produce very rapid cooling rates and potentially strong welds. The friction stir welding process has been developed for Al and Al-Li alloys for aerospace applications.

9-11 Bulk Metallic Glasses [BMG]

Typically, metals and alloys crystallize easily during solidification. However, in certain complex alloys, it is possible to conduct the solidification process in such a way that **bulk metallic glasses (BMG)** are formed. In ceramic systems, such as those based on silica, the formation of amorphous glasses is relatively easy because the kinetics of crystallization is quite slow (Chapter 15). However, in metallic materials, special formulations need to be developed so as to form bulk metallic glasses. Formation of metallic glasses as such was observed in the 1960s with a Au-Si alloy. What is new with bulk metallic glasses is that relatively large (>10 mm) pieces of materials can be prepared using conventional casting process with cooling rates of as low as 1 to 100°C/s. This is very different from the cooling rates used in rapid solidification of metallic alloys (Section 9-3).

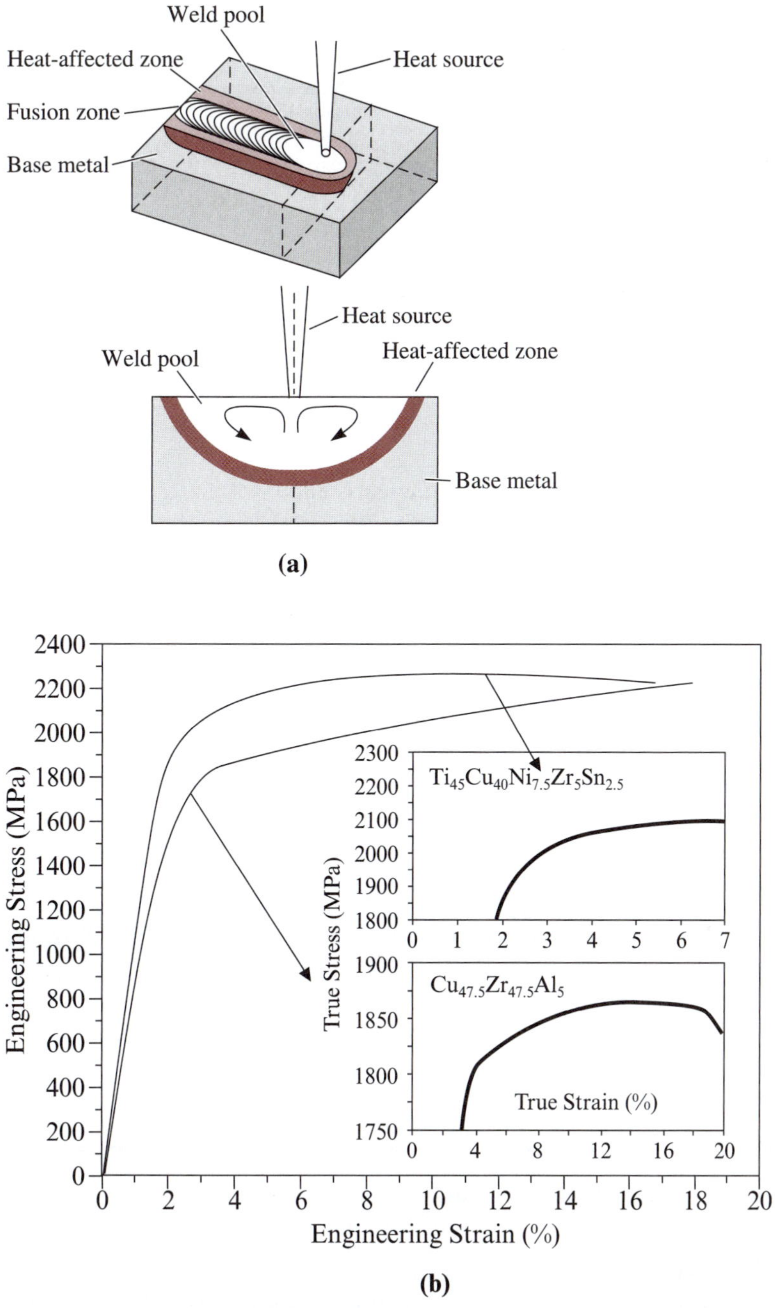

Figure 9-20 (a) Schematic diagram showing interaction between the heat source and the base metal. Three distinct regions in the weldment are the fusion zone, the heat-affected zone, and the base metal. (*Source: From Fig 2 from David & DebRoy,* Science, *257: 497–502, (1992). Reprinted with permission from AAAS.*) (b) Room temperature engineering compression stress-strain curves for titanium- and copper-based bulk metallic glasses. The inset curves show the true stress-true strain relationships. (*Source: Ref. Das, J., Kim, K.B., Xu, W., Wei, B.C., Zhang, Z.F., Wang, W.H., Wi, S., and Eckert,* J. Mater. Trans., *Vol. 47, p. 2606 (2006). Adapted from Yavari, A.R., Lewandowski, J.J., and Eckert, J. in* MRS Bulletin, *Vol. 32, pages 635–638, August 2007.*)

TABLE 9-3 ■ *Properties of V1 bulk metallic glass alloy compared to other materials*

Properties	V1 Bulk Metallic Glass	Al-Alloys	Ti-Alloys	Steels
Density g/cm^3	6.1	2.6–2.9	4.3–5.1	7.8
Yield Strength (Tensile)	1.9	0.1–0.63	0.18–1.32	0.5–1.6
Elastic Strain Limit	2%	~0.5%	~0.5%	0.5%
Fracture Toughness K_{Ic} (MPam$^{1/2}$)	20–140	23–45	55–115	50–154
Specific Strength (GPag^{-1}cm^{-3})	0.32	<0.24	<0.31	<0.21

(Source: Telford, M., Materials Today, *March 2004, pp. 36–44.)*

An example of a bulk metallic-glass alloy is a material known as Vitreoly 1 or V1. Its composition is $Zr_{41.2}Ti_{13.8}Cu_{12.5}Ni_{10.0}Be_{22.5}$. Such compositions of materials that form BMG show very deep eutectics (Chapter 11), unusually high viscosities, and also lower thermodynamic driving force for crystallization.

As can be expected, the mechanical properties of metallic glasses are different from those of polycrystalline metals and alloys. Since there are no dislocations and slip systems (Chapter 4), slip is not possible in amorphous materials. As a result, metallic glasses exhibit much higher elastic strains (~2%), as opposed to ~0.2% in polycrystalline metals and alloys (Chapter 6). This large elastic deformation can be seen in the stress-strain compression test curves for Cu- and Ti-based metallic glasses [Figure 9-20(b)]. Under tensile loading, the ductility of metallic glasses is limited, however, it is better than that for many ceramics (Chapter 15).

The combination of to develop high elastic strain (like polymers) and high yield stress provides for high "spring back" for these materials (Table 9-3). Bulk metallic glasses have been used to make golf club heads. Liquidmetal^{TM} technology also makes use of bulk metallic glasses for sports equipment applications, including tennis rackets. Some recent research has shown it may be possible to enhance the apparent ductility of bulk metallic glasses by using shot peening to introduce compressive stresses at the surface.

SUMMARY

- Transformation of a liquid to a solid is probably the most important phase transformation in applications of materials science and engineering.

- Nucleation produces a critical-size solid particle from the liquid melt. Formation of nuclei is determined by the thermodynamic driving force for solidification, and is opposed by the energy needed to create a solid-liquid interface.

- Homogeneous nucleation requires large undercoolings of the liquid and is not observed in the normal solidification processing. By introducing foreign particles or external surfaces into the liquid, sites are provided for heterogeneous nucleation. This is done in practice by inoculation or grain refining. This process permits the grain size of the casting to be controlled.

- Rapid cooling of the liquid can prevent nucleation and growth, giving amorphous solids, or glasses, with unusual mechanical and physical properties. Polymeric, metallic, and inorganic materials can be made in the form of amorphous glasses.

- In solidification from melts, growth occurs as the nuclei grow into the liquid melt. Either planar or dendritic modes of growth may be observed.
- Cooling curves can be used to determine pouring temperature, observe any undercooling, recalescence, and time for solidification.
- Porosity and shrinkage are major defects that can be present in cast products. If present, it can cause cast products to fail catastrophically.
- Sand casting, investment casting, and pressure die casting are some of the processes for casting components. Ingot casting and continuous casting are employed in the production and recycling of metals and alloys.
- Many metal joining processes (e.g., welding) involve solidification.

GLOSSARY

Brazing An alloy, known as a filler, is used to join two materials to one another. The composition of the filler, which has a melting temperature above about 450°C, is quite different from the metal being joined.

Bulk metallic glasses (BMG) Amorphous metallic glasses made from compositions that show unusually high liquid viscosities and resistance to crystallization. They exhibit high (~2% elastic strain limit) and strength and have found applications in sporting goods.

Cavities Small holes present in a casting.

Cavity shrinkage A large void within a casting caused by the volume contraction that occurs during solidification.

Chill zone A region of small, randomly oriented grains that forms at the surface of a casting as a result of heterogeneous nucleation.

Chvorinov's rule The solidification time of a casting is directly proportional to the square of the volume-to-surface area ratio of the casting.

Columnar zone A region of elongated grains having a preferred orientation that forms as a result of competitive growth during the solidification of a casting.

Continuous casting A process to convert molten metal or an alloy into a semi-finished product such as a slab.

Critical radius (r^*) The minimum size that must be formed by atoms clustering together in the liquid before the solid particle is stable and begins to grow.

Dendrite The treelike structure of the solid that grows when an undercooled liquid solidifies.

Directional solidification (DS) A solidification technique in which cooling in a given direction leads to preferential growth of grains in the opposite direction, leading to an anisotropic-oriented microstructure.

Embryo A tiny particle of solid that forms from the liquid as atoms cluster together. The embryo may grow into a stable nucleus or redissolve.

Equiaxed zone A region of randomly oriented grains in the center of a casting produced as a result of widespread nucleation.

Fusion welding Joining processes in which a portion of the materials must melt in order to achieve good bonding.

Fusion zone The portion of a weld heated to produce all liquid during the welding process. Solidification of the fusion zone provides joining.

Gas flushing A process in which a stream of gas is injected into a molten metal in order to eliminate a dissolved gas that might produce porosity.

Gas porosity Bubbles of gas trapped within a casting during solidification, caused by the lower solubility of the gas in the solid compared with that in the liquid.

Glass-ceramics Polycrystalline, ultrafine grained ceramic materials obtained by controlled crystallization of amorphous glasses.

Grain refinement The addition of heterogeneous nuclei in a controlled manner to increase the number of grains in a casting.

Growth The physical process by which a new phase increases in size. In the case of solidification, this refers to the formation of a stable solid as the liquid freezes.

Heterogeneous nucleation Formation of a critically-sized solid from the liquid on an impurity or an external surface.

Homogeneous nucleation Formation of a critically sized solid from the liquid by the clustering together of a large number of atoms at a high undercooling (without an external interface).

Ingot A simple casting that is usually remelted or reprocessed by another user to produce a more useful shape.

Ingot casting The process of casting ingots. This is different from the continuous casting route.

Ingot structure The macrostructure of a casting, including the chill zone, columnar zone, and equiaxed zone.

Inoculants Materials that provide heterogeneous nucleation sites during the solidification of a material.

Inoculation The addition of heterogeneous nuclei in a controlled manner to increase the number of grains in a casting.

Interdendritic shrinkage Small pores between the dendrite arms formed by the shrinkage that accompanies solidification. Also known as microshrinkage or shrinkage porosity.

Investment casting A casting process that is used for making complex shapes such as turbine blades, also known as the lost wax process.

Lamellar A plate-like arrangement of crystals within a material.

Latent heat of fusion (ΔH_f) The heat evolved when a liquid solidifies. The latent heat of fusion is related to the energy difference between the solid and the liquid.

Local solidification time The time required for a particular location in a casting to solidify once nucleation has begun.

Lost foam process A process in which a polymer foam is used as a pattern to produce a casting.

Lost wax process A casting process in which a wax pattern is used to cast a metal—same as investment casting.

Microshrinkage Small, frequently isolated pores between the dendrite arms formed by the shrinkage that accompanies solidification. Also known as microshrinkage or shrinkage porosity.

Mold constant (B) A characteristic constant in Chvorinov's rule.

Nucleation The physical process by which a new phase is produced in a material. In the case of solidification, this refers to the formation of a tiny, stable solid particles in the liquid.

Nuclei Tiny particles of solid that form from the liquid as atoms cluster together. Because these particles are large enough to be stable, nucleation has occurred and growth of the solid can begin.

Permanent mold casting A casting process in which a mold can be used many times.

Pipe shrinkage A large conical-shaped void at the surface of a casting caused by the volume contraction that occurs during solidification.

Planar growth The growth of a smooth solid-liquid interface during solidification, when no undercooling of the liquid is present.

Pouring temperature The temperature of a metal or an alloy when it is poured into a mold during the casting process.

Pressure die casting A casting process in which molten metal/alloys is forced into a die under pressure.

Primary processing Processes involving casting of molten metals into ingots or semi-finished useful shapes such as slabs.

Rapid solidification processing Producing unique material structures by promoting unusually high cooling rates during solidification.

Recalescence The increase in temperature of an undercooled liquid metal as a result of the liberation of heat during nucleation.

Riser An extra reservoir of liquid metal connected to a casting. If the riser freezes after the casting, the riser can provide liquid metal to compensate for shrinkage.

Sand casting A casting process using sand molds.

Secondary dendrite arm spacing (SDAS) The distance between the centers of two adjacent secondary dendrite arms.

Secondary processing Processes such as rolling, extrusion, etc. used to process ingots or slabs and other semi-finished shapes.

Shrinkage Contraction of a casting during solidification.

Shrinkage porosity Small pores between the dendrite arms formed by the shrinkage that accompanies solidification. Also known as microshrinkage or interdendritic porosity.

Sievert's law The amount of a gas that dissolves in a metal is proportional to the partial pressure of the gas in the surroundings.

Soldering Soldering is a joining process in which the filler has a melting temperature below 450°C, no melting of the base materials occurs.

Solidification front Interface between a solid and liquid.

Solidification process Processing of materials involving solidification (e.g., single crystal growth, continuous casting, etc.).

Specific heat The heat required to change the temperature of a unit mass of the material one degree.

Spherulite Spherical-shaped crystals produced when certain polymers solidify.

Superheat The pouring temperature minus the freezing temperature.

Thermal arrest A plateau on the cooling curve during the solidification of a material caused by the evolution of the latent heat of fusion during solidification. This heat generation balances the heat being lost as a result of cooling.

Total solidification time The time required for the casting to solidify completely after the casting has been poured.

Undercooling The temperature to which the liquid metal must cool below the equilibrium freezing temperature before nucleation occurs.

PROBLEMS

Section 9-1 Technological Significance

Section 9-2 Nucleation

9-1 Define the following terms: nucleation, embryo, heterogeneous nucleation, and homogeneous nucleation.

9-2 Suppose that liquid nickel is undercooled until homogeneous nucleation occurs. Calculate

(a) the critical radius of the nucleus required; and
(b) the number of nickel atoms in the nucleus.

Assume that the lattice parameter of the solid FCC nickel is 0.356 nm.

9-3 Suppose that liquid iron is undercooled until homogeneous nucleation occurs. Calculate

(a) the critical radius of the nucleus required; and
(b) the number of iron atoms in the nucleus.

Assume that the lattice parameter of the solid BCC iron is 2.92 Å.

9-4 Suppose that solid nickel was able to nucleate homogeneously with an undercooling of only 22°C. How many atoms would have to group together spontaneously for this to occur? Assume that the lattice parameter of the solid FCC nickel is 0.356 nm.

9-5 Suppose that solid iron was able to nucleate homogeneously with an undercooling of only 15°C. How many atoms would have to group together spontaneously for this to occur? Assume that the lattice parameter of the solid BCC iron is 2.92 Å.

9-6 Calculate the fraction of solidification that occurs dendritically when iron nucleates

(a) at 10°C undercooling;
(b) at 100°C undercooling; and
(c) homogeneously.

The specific heat of iron is 5.78 $J/cm^3 \cdot °C$.

Section 9-3 Growth Mechanisms

9-7 Calculate the fraction of solidification that occurs dendritically when silver nucleates

(a) at 10°C undercooling;
(b) at 100°C undercooling; and
(c) homogeneously.

The specific heat of silver is 3.25 $J/cm^3 \cdot °C$.

9-8 Analysis of a nickel casting suggests that 28% of the solidification process occurred in a dendritic manner. Calculate the temperature at which nucleation occurred. The specific heat of nickel is 4.1 $J/cm^3 \cdot °C$.

9-9 Write down Chvorinov's rule and explain the meaning of each term.

9-10 A 5-cm cube solidifies in 4.6 min. Calculate

(a) the mold constant in Chvorinov's rule; and
(b) the solidification time for a 1.25 cm × 1.25 cm × 15 cm bar cast under the same conditions.

Assume that $n = 2$.

9-11 A 5-cm-diameter sphere solidifies in 1050 s. Calculate the solidification time for a 0.3 cm × 10 cm × 20 cm plate cast under the same conditions. Assume that $n = 2$.

9-12 Find the constants B and n in Chvorinov's rule by plotting the following data on a log-log plot:

Casting Dimensions (cm)	Solidification Time (min)
1.25 × 20 × 30	3.48
5 × 7.5 × 25	15.78
6.25 cube	10.17
2.5 × 10 × 22.5	8.13

9-13 Find the constants B and n in Chvorinov's rule by plotting the following data on a log-log plot:

Casting Dimensions (cm)	Solidification Time (s)
1 × 1 × 6	28.58
2 × 4 × 4	98.30
4 × 4 × 4	155.89
8 × 6 × 5	306.15

9-14 A 7.5-cm-diameter casting was produced. The times required for the solid-liquid interface to reach different distances beneath the casting surface were measured and are shown in the following table:

Distance from Surface (cm)	Time (s)
0.25	32.6
0.75	73.5
1.25	130.6
1.875	225.0
2.5	334.9

Determine

(a) the time at which solidification begins at the surface; and

(b) the time at which the entire casting is expected to be solid.

(c) Suppose the center of the casting actually solidified in 720 s. Explain why this time might differ from the time calculated in part (b).

9-15 A 10-cm-diameter aluminum bar solidifies to a depth of 1.25 cm beneath the surface in 5 minutes. After 20 minutes, the bar has solidified to a depth of 3.75 cm. How much time is required for the bar to solidify completely?

9-16 Design the thickness of an aluminum alloy casting whose length is 30 cm and width is 20 cm, in order to produce a tensile strength of 276 MPa. The mold constant in Chvorinov's rule for aluminum alloys cast in a sand mold is 7.2 min/cm^2. Assume that data shown in Figures 9-9 and 9-10 can be used.

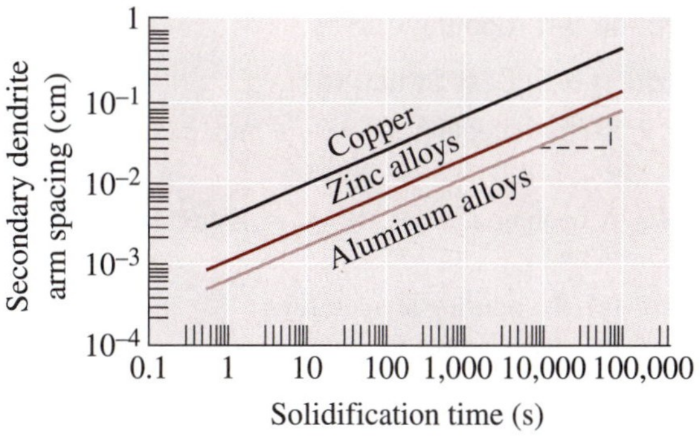

Figure 9-9 (Repeated for Problem 9-16) The effect of solidification time on the secondary dendrite arm spacings of copper, zinc, and aluminum.

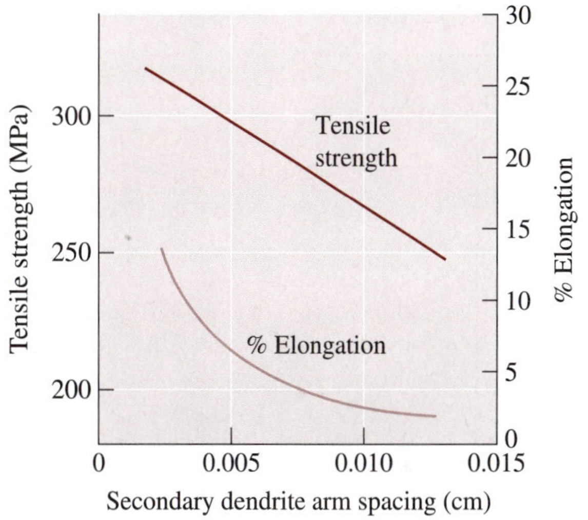

Figure 9-10 (Repeated for Problem 9-16) The effect of the secondary dendrite arm spacing on the properties of an aluminum casting alloy.

9-17 Find the constants c and m relating the secondary dendrite arm spacing to the local solidification time by plotting the following data on a log-log plot:

Solidification Time (s)	SDAS (cm)
156	0.0176
282	0.0216
606	0.0282
1356	0.0374

Section 9-4 Cooling Curves

Section 9-5 Cast Structure

9-18 Sketch a cooling curve for a pure metal and label different regions carefully.

9-19 A cooling curve is shown in Figure 9-21. Determine

(a) the pouring temperature;
(b) the solidification temperature;
(c) the superheat;
(d) the cooling rate, just before solidification begins;
(e) the total solidification time;
(f) the local solidification time; and
(g) the probable identity of the metal.
(h) If the cooling curve was obtained at the center of the casting sketched in the figure, determine the mold constant, assuming that $n = 2$.

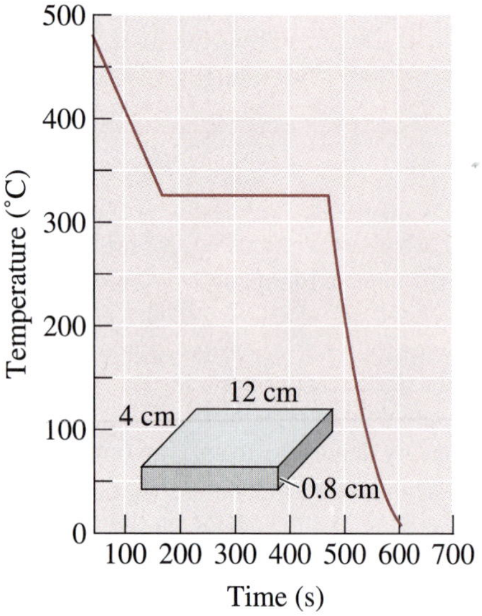

Figure 9-21 Cooling curve (for Problem 9-19).

9-20 A cooling curve is shown in Figure 9-22. Determine

(a) the pouring temperature;
(b) the solidification temperature;
(c) the superheat;
(d) the cooling rate, just before solidification begins;
(e) the total solidification time;
(f) the local solidification time;
(g) the undercooling; and
(h) the probable identity of the metal.

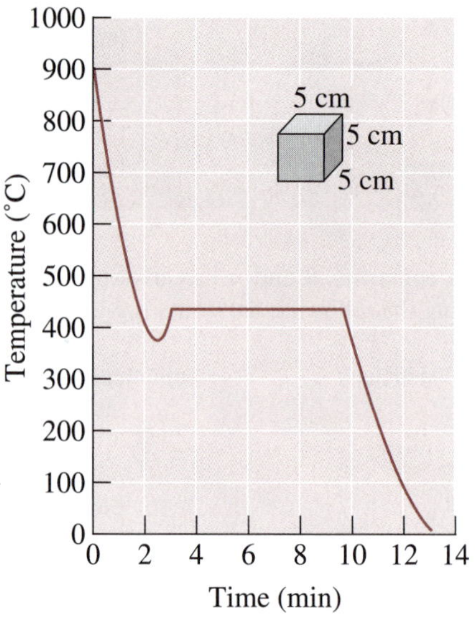

Figure 9-22 Cooling curve (for Problem 9-20).

(i) If the cooling curve was obtained at the center of the casting sketched in the figure, determine the mold constant, assuming that $n = 2$.

9-21 Figure 9-23 shows the cooling curves obtained from several locations within a cylindrical aluminum casting. Determine the local solidification times and the SDAS at each location, then plot the tensile strength versus distance from the casting surface. Would you recommend that the casting be designed so that a large or small amount of material must be machined from the surface during finishing? Explain.

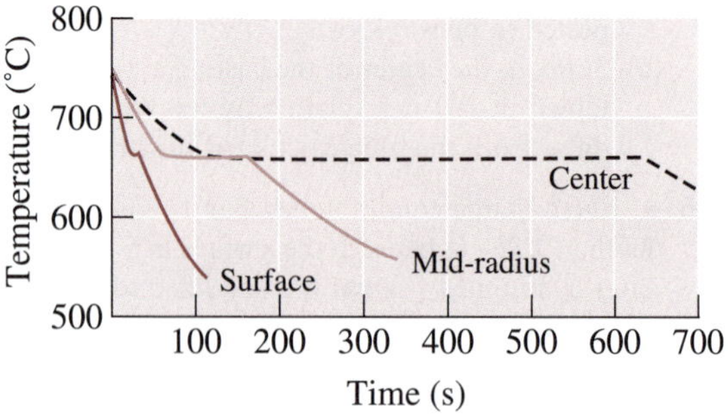

Figure 9-23 Cooling curves (for Problem 9-21).

9-22 Calculate the volume, diameter, and height of the cylindrical riser required to prevent shrinkage in a 10 cm × 25 cm × 50 cm casting if the H/D of the riser is 1.5.

Section 9-6 Solidification Defects

9-23 In general, compared to components prepared using forging, rolling, extrusion, etc., cast products tend to have lower fracture toughness. Explain why this may be the case.

9-24 What is a riser? Why should it be designed so as to freeze after the casting?

9-25 Calculate the volume, diameter, and height of the cylindrical riser required to prevent shrinkage in a 2.5 cm × 15 cm × 15 cm casting if the H/D of the riser is 1.0.

9-26 Figure 9-24 shows a cylindrical riser attached to a casting. Compare the solidification times for each casting section and the riser and determine whether the riser will be effective.

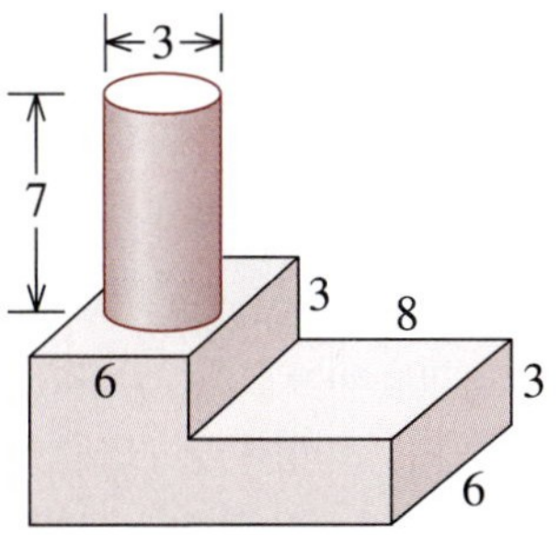

Figure 9-24 Step-block casting (for Problem 9-26).

9-27 Figure 9-25 shows a cylindrical riser attached to a casting. Compare the solidification times for each casting section and the riser and determine whether the riser will be effective.

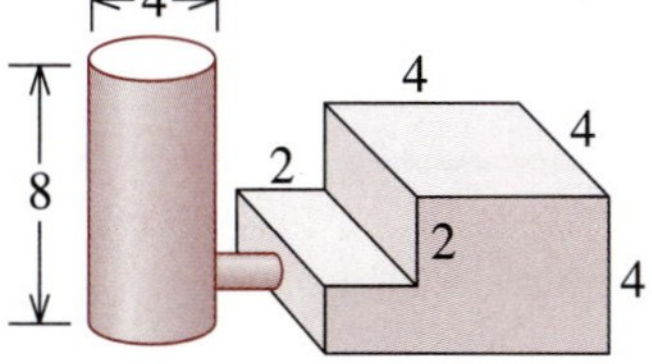

Figure 9-25 Step-block casting (for Problem 9-27).

9-28 A 10-cm-diameter sphere of liquid copper is allowed to solidify, producing a spherical shrinkage cavity in the center of the casting. Compare the volume and diameter of the shrinkage cavity in the copper casting to that obtained when a 10-cm sphere of liquid iron is allowed to solidify.

9-29 A 10-cm cube of a liquid metal is allowed to solidify. A spherical shrinkage cavity with a diameter of 3.725 cm is observed in the solid casting. Determine the percent volume change that occurs during solidification.

9-30 A 2 cm × 4 cm × 6 cm magnesium casting is produced. After cooling to room temperature, the casting is found to weigh 80 g. Determine

(a) the volume of the shrinkage cavity at the center of the casting; and
(b) the percent shrinkage that must have occurred during solidification.

9-31 A 5 cm × 20 cm × 25 cm iron casting is produced and, after cooling to room temperature, is found to weigh 19 kg. Determine

(a) the percent of shrinkage that must have occurred during solidification; and
(b) the number of shrinkage pores in the casting if all of the shrinkage occurs as pores with a diameter of 0.125 cm.

9-32 Liquid magnesium is poured into a 2 cm × 2 cm × 24 cm mold and, as a result of directional solidification, all of the solidification shrinkage occurs along the length of the casting. Determine the length of the casting immediately after solidification is completed.

9-33 A liquid cast iron has a density of 7.65 g/cm^3. Immediately after solidification, the density of the solid cast iron is found to be 7.71 g/cm^3. Determine the percent volume change that occurs during solidification. Does the cast iron expand or contract during solidification?

Section 9-7 Casting Processes for Manufacturing Components

9-34 An alloy is cast into a shape using a sand mold and a metallic mold. Which casting will be expected to be stronger and why?

Section 9-8 Continuous Casting, Ingot Casting, and Single Crystal Growth

Section 9-9 Solidification of Polymers and Inorganic Glasses

9-35 Why do most plastics contain amorphous and crystalline regions?

9-36 Explain why silicate glasses tend to form amorphous glasses, however, metallic melts typically crystallize easily.

Section 9-10 Joining of Metallic Materials

9-37 Define the terms brazing and soldering.

9-38 What is the difference between fusion welding, brazing, and soldering?

9-39 What is a heat affected by zone?

9-40 Explain why, while using low intensity heat sources, the strength of the material in a weld region can be reduced.

Design Problems

9-41 When two 1.25-cm-thick copper plates are joined using an arc-welding process, the fusion zone contains dendrites having a SDAS of 0.006 cm. However, this process produces large residual stresses in the weld. We have found that residual stresses are low when the welding conditions produce a SDAS of more than 0.02 cm. Design a process by which we can accomplish low residual stresses. Justify your design.

9-42 Design an efficient risering system for the casting shown in Figure 9-26. Be sure to include a sketch of the system, along with appropriate dimensions.

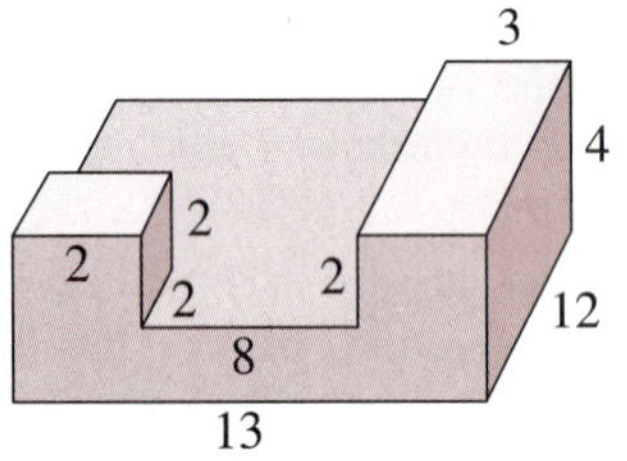

Figure 9-26 Casting to be risered (for Problem 9-42).

9-43 An aluminum casting is to be injected into a steel mold under pressure (die casting). The casting is essentially a 30-cm-long, 5-cm-diameter cylinder with a uniform wall thickness, and it must have a minimum tensile strength of 276 MPa. Based on the properties given in Figure 9-10, design the casting and process.

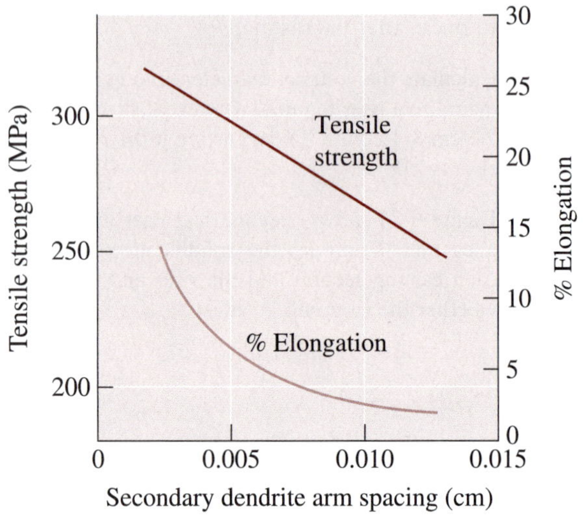

Figure 9-10 (Repeated for Problem 9-43) The effect of the secondary dendrite arm spacing on the properties of an aluminum casting alloy.

9-44 *Baseball Bat Materials.* What sort of new materials can you suggest for making baseball bats? Think about the specific strength of the materials, their ability to withstand impact, and corrosion. Also, assume that the goal is to be able to allow a player hitting the ball as far as possible, and this means that springback or the trampoline effect will be important.

10

Solid Solutions and Phase Equilibrium

Have You Ever Wondered?

- *Is it possible for the solid, liquid, and gaseous forms of a material to coexist?*
- *How ice skaters move so easily on ice?*
- *When an alloy known as brass solidifies, which element solidifies first—copper or zinc?*

The yield strength of metallic materials can increase if more obstacles for dislocation motion are created. In this chapter, we begin to explore the formation of a solid solution in metallic and ceramic systems. A solid solution is a solid material in which atoms or ions of elements constituting it are dispersed uniformly. The mechanical and other properties of materials can be controlled by creating point defects, such as substitutional and interstitial atoms. In metallic materials, the point defects disturb the atomic arrangement in the crystalline material and interfere with the movement of dislocations or slip. The point defects, therefore, cause a metallic material to be solid-solution strengthened.

The introduction of alloying elements or impurities, during processing, changes the composition of the material and influences its solidification behavior. In this chapter, we examine this effect by introducing the concept of an equilibrium phase diagram. For now, we consider a "phase" as a unique form in which a material exists. We will define the term "phase" more precisely later in this chapter. A phase diagram depicts the stability of different phases for a set of elements (e.g., Al and Si). From the phase diagram, we can predict how a material will solidify under equilibrium conditions. We can also predict what phases will be expected to be thermodynamically stable and in what concentrations such phases should be present.

Therefore, the major objectives of this chapter are to explore:

1. the formation of solid solutions;
2. the effects of solid-solution formation on the mechanical properties on metallic materials;
3. the conditions under which solid solutions can form;
4. the development of some basic ideas concerning phase diagrams; and
5. the solidification process in simple alloys.

10-1 Phases and the Phase Diagram

In most applications of engineered metallic materials, we use **alloys** rather than pure elements. We define an "alloy" as a material that exhibits properties of a metallic material and is made from multiple elements. A *plain carbon steel* is an alloy of iron (Fe) and carbon (C). Corrosion-resistant **stainless steels** are alloys that usually contain iron (Fe), carbon (C), chromium (Cr), nickel (Ni), and some other elements. Similarly, there are alloys based on aluminum (Al), copper (Cu), cobalt (Co), nickel (Ni), titanium (Ti), zinc (Zn), and zirconium (Zr). There are two types of alloys: **single-phase alloys** and **multiple-phase alloys**. In this chapter, we will examine the behavior of single-phase alloys. As a first step, let's define a "phase" and determine how the **phase rule** helps us to determine the state—solid, liquid, or gas—in which a pure material exists.

A **phase** can be defined as any portion, including the whole, of a system which is physically homogeneous within itself and bounded by a surface so that it is mechanically separable from any other portions. For example, water has three phases—liquid water, solid ice, and steam. A phase has the following characteristics:

1. the same structure or atomic arrangement throughout;
2. roughly the same composition and properties throughout; and
3. a definite interface between the phase and any surrounding or adjoining phases.

For example, if we enclose a block of ice in a vacuum chamber [(Figure 10-1(a)], the ice begins to melt and some of the water vaporizes. Under these conditions, we have three phases coexisting: solid H_2O, liquid H_2O, and gaseous H_2O. Each of these forms of H_2O is a distinct phase; each has a unique atomic arrangement, unique properties, and a definite boundary between each form. Although, in this case the phases have identical compositions.

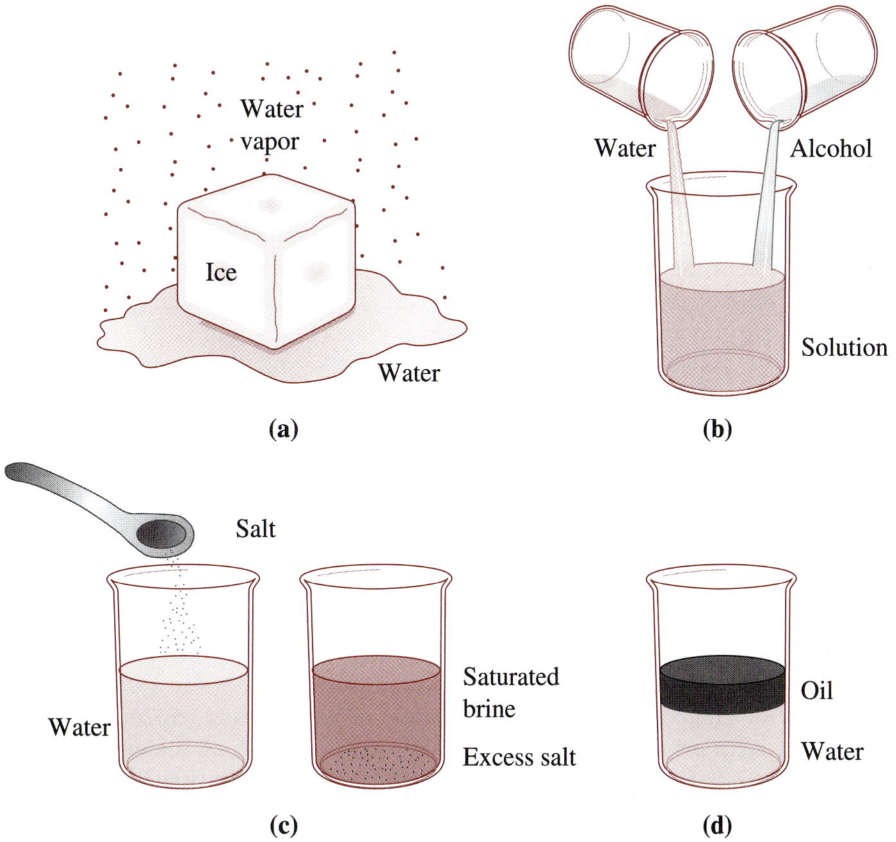

Figure 10-1 Illustration of phases and solubility: (a) The three forms of water—gas, liquid, and solid—are each a phase. (b) Water and alcohol have unlimited solubility. (c) Salt and water have limited solubility. (d) Oil and water have virtually no solubility in each other.

Josiah Willard Gibbs developed the **phase rule** in 1875–1876. It describes the relationship between the number of components and the number of phases for a given system and the conditions that may be allowed to change (e.g., temperature, pressure, etc.). It has the general form:

$$2 + C = F + P \quad \text{(when temperature and pressure both can vary)} \qquad (10\text{-}1)$$

A useful mnemonic (something that will allow you to remember) for the Gibbs phase rule is to start with a numeric and follow with the rest of the terms alphabetically (i.e., C, F, and P) using all positive signs. In the phase rule, C is the number of chemically independent components, usually elements or compounds, in the system; F is the number of degrees of freedom, or the number of variables (such as temperature, pressure, or composition), that are allowed to change independently without changing the number of phases in equilibrium; and P is the number of phases present (please do not confuse P with "pressure.") The constant "2" in Equation 10-1 implies that both the temperature and pressure are allowed to change. The term "chemically independent" refers to the number of different elements or compounds needed to specify a system. For example, water (H_2O) is considered as a one component system, since the concentrations of H and O in H_2O cannot be independently varied.

It is important to note that the Gibbs phase rule assumes thermodynamic equilibrium and, more often than not in materials processing, we encounter conditions in which equilibrium is *not* maintained. Therefore, you should not be surprised to see that the number and compositions of phases seen in practice are dramatically different from those predicted by the Gibbs phase rule.

Note that phases do not always have to be solid, liquid, and gaseous forms of a material. An element, such as iron (Fe), can exist in FCC and BCC crystal structures. These two solid forms of iron are two different phases of iron that will be stable at different temperatures and pressure conditions. Similarly, ice, itself, can exist in several crystal structures. Carbon can exist in many forms (e.g., graphite or diamond). These are only two of the many possible phases of carbon.

As an example of the use of the phase rule, let's consider the case of pure magnesium (Mg). Figure 10-2 shows a **unary** ($C = 1$) **phase diagram** in which the lines divide the liquid, solid, and vapor phases. This unary phase diagram is also called a pressure-temperature or **P-T diagram**. In the unary phase diagram, there is only one component; in this case, magnesium (Mg). Depending on the temperature and pressure, however, there may be one, two, or even three *phases* present simultaneously: solid magnesium, liquid magnesium, and magnesium vapor. Note that at atmospheric pressure (one atmosphere, given by the dashed line), the intersection of the lines in the phase diagram give the usual melting and boiling temperatures for magnesium. At very low pressures, a solid such as magnesium (Mg) can *sublime*, or go directly to a vapor form without melting when it is heated.

Suppose we have a combination of pressure and temperature that put us at point A in the phase diagram (Figure 10-2). At this point, all magnesium is liquid. The number of phases is one (liquid). The phase rule tells us that there are two degrees of freedom. From Equation 10-1:

$$2 + C = F + P, \quad \text{therefore, } 2 + 1 = F + 1 \text{ (i.e., } F = 2)$$

What does this mean? Within limits, as seen in Figure 10-2, we can change the pressure, the temperature, or both, and still be in an all-liquid portion of the diagram. Put another way, we must fix both the temperature and the pressure to know precisely where we are in the liquid portion of the diagram.

Consider point B, the boundary between the solid and liquid portions of the diagram. The number of components, C, is still one, but at point B, the solid and liquid coexist, or the number of phases P is two. From the phase rule Equation 10-1,

$$2 + C = F + P, \quad \text{therefore, } 2 + 1 = F + 2 \text{ (i.e., } F = 1)$$

or there is only one degree of freedom. For example, if we change the temperature, the pressure must also be adjusted if we are to stay on the boundary where the liquid and solid coexist. On the other hand, if we fix the pressure, the phase diagram tells us the temperature that we must have if solid and liquid are to coexist.

Finally, at point X, solid, liquid, and vapor coexist. While the number of components is still one, there are three phases. The number of degrees of freedom is

$$2 + C = F + P, \quad \text{therefore, } 2 + 1 = F + 3 \text{ (i.e., } F = 0)$$

Now we have no degrees of freedom; all three phases coexist only if both the temperature and the pressure are fixed. A point on the phase diagram at which the solid, liquid, and gaseous phases coexist under equilibrium conditions is the **triple point**. In the following example, we see how some of these ideas underlying the Gibbs phase rule can be applied.

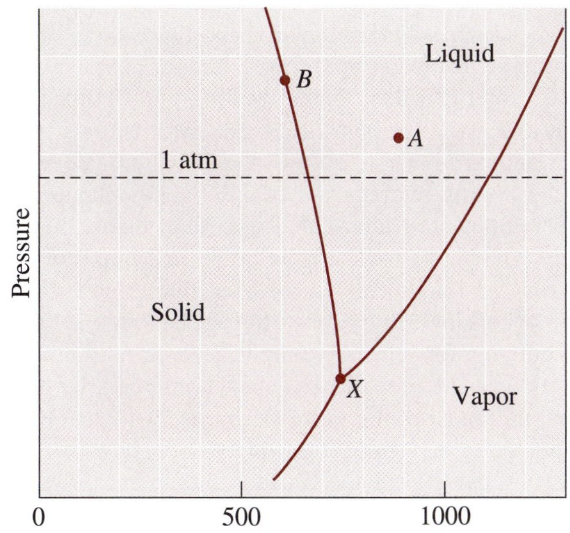

(a)

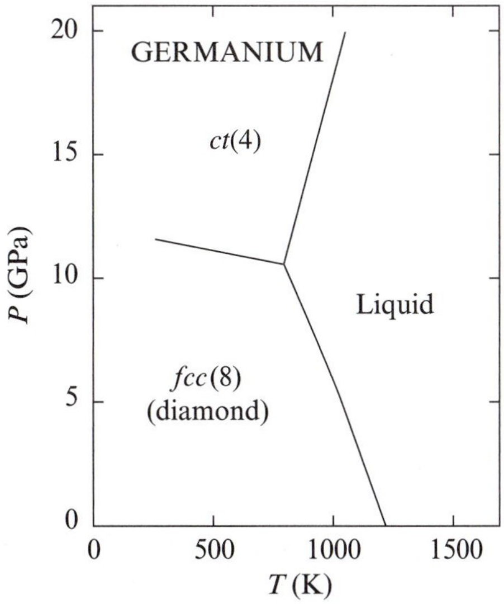

(b)

Figure 10-2
(a) Schematic unary phase diagram for magnesium, showing the melting and boiling temperatures at one atmosphere pressure. (b) Pressure-temperature phase diagram for germanium (Ge), ct (4): body-centered tetragonal lattice with four atoms per unit cell, fcc (8): face centered diamond cubic with eight atoms per unit cell (i.e., diamond cubic structure). *(Source: D. Young,* Phase Diagrams of the Elements. *1991, Publisher: University of California Press, Berkeley, p. 106.)*

EXAMPLE 10-1 *Design of an Aerospace Component*

Because magnesium (Mg) is a low-density material ($\rho_{Mg} = 1.738$ g/cm^3), it has been suggested for use in an aerospace vehicle intended to enter the outer space environment. Is this a good design?

SOLUTION

In space the pressure is very low. Even at relatively low temperatures, solid magnesium can begin to change to a vapor, causing metal loss that could damage a space vehicle. In addition, solar radiation could cause the vehicle to heat, increasing the rate of magnesium loss.

A low-density material with a higher boiling point (and, therefore, lower vapor pressure at any given temperature) might be a better choice. At atmospheric pressure, aluminum boils at 2494°C and beryllium (Be) boils at 2770°C, compared with the boiling temperature of 1107°C for magnesium. Although aluminum and beryllium are somewhat denser than magnesium, either might be a better design. Given the toxic effects of Be and many of its compounds when in powder form, we may want to consider aluminum first.

Other factors to consider: In load-bearing applications, we should not only look for density but also for relative strength. Therefore, the ratio of Young's modulus to density or yield strength to density could be a better parameter to compare different materials. In this comparison, we will have to be aware that yield strength, for example, depends strongly on microstructure. Also, the strength of aluminum can be enhanced using aluminum alloys, while keeping the density about the same.

These materials and some composites may make a good choice. Other factors such as oxidation during reentry into Earth's atmosphere may be applicable and will also have to be considered.

If you examine the P-T diagram for water, you will notice how the melting temperature of ice decreases with increasing pressure. This is rather unusual. It has been suggested that one of the reasons why skaters can skate on ice is that the pressure from their skates actually melts the ice, thus maintaining a layer of water. This, however, has been shown *not* to be an important factor. Instead, a phenomenon known as "*surface melting*" of ice has been shown to be the main factor in providing a lubricating film of water allowing the skater to move easily across the ice.

10-2 Solubility and Solid Solutions

Often, it is beneficial to know how much of each material or component we can combine without producing an additional phase. When we begin to combine different components or materials, as when we add alloying elements to a metal, solid or liquid, solutions can form. For example, when we add sugar to water, we form a sugar solution. When we diffuse a small number of phosphorus (P) atoms into single crystal silicon (Si), we produce a solid solution of P in Si. In other words, we are interested in the **solubility** of one material into another (e.g., sugar in water, copper in nickel, phosphorus in silicon, etc.).

Unlimited Solubility Suppose we begin with a glass of water and a glass of alcohol. The water is one phase, and the alcohol is a second phase. If we pour the water into the alcohol and stir, only one phase is produced [Figure 10-1(b)]. The glass contains a solution of water and alcohol that has unique properties and composition. Water and alcohol are soluble into each other. Furthermore, they display **unlimited solubility**. Regardless of the ratio of water and alcohol, only one phase is produced when they are mixed together.

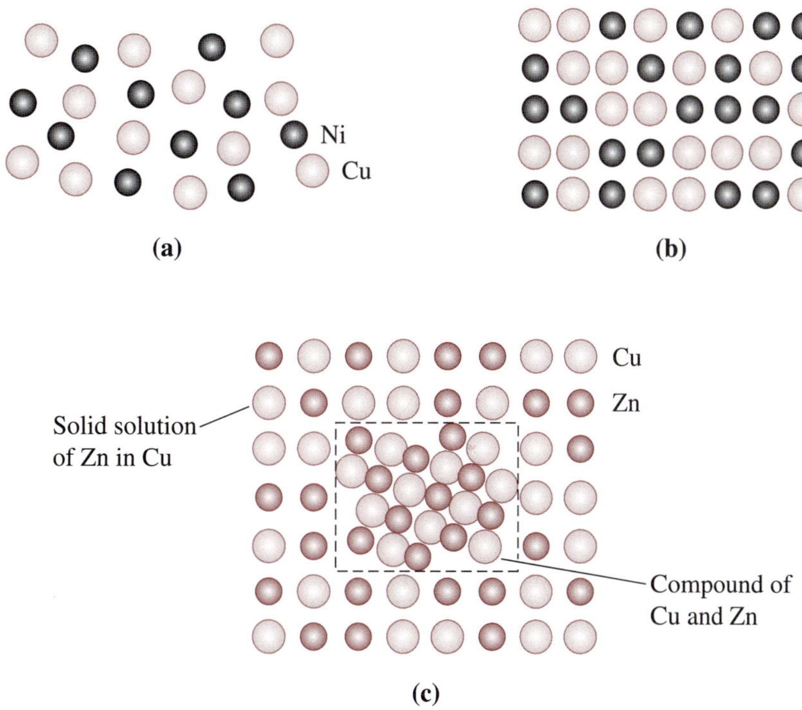

Figure 10-3 (a) Liquid copper and liquid nickel are completely soluble in each other. (b) Solid copper-nickel alloys display complete solid solubility, with copper and nickel atoms occupying random lattice sites. (c) In copper-zinc alloys containing more than 30% Zn, a second phase forms because of the limited solubility of zinc in copper.

Similarly, if we were to mix any amounts of liquid copper and liquid nickel, only one liquid phase would be produced. This liquid alloy has the same composition and properties everywhere [Figure 10-3(a)], because nickel and copper have unlimited liquid solubility.

If the liquid copper-nickel alloy solidifies and cools to room temperature while maintaining thermal equilibrium, only one solid phase is produced. After solidification, the copper and nickel atoms do not separate but, instead, are randomly located within the FCC crystal structure. Within the solid phase, the structure, properties, and composition are uniform and no interface exists between the copper and nickel atoms. Therefore, copper and nickel also have unlimited solid solubility. The solid phase is a solid solution of copper and nickel [Figure 10-3(b)].

A solid solution is *not* a mixture. A mixture contains more than one type of phase whose characteristics are retained when the mixture is formed. In contrast to this, the components of a solid solution completely dissolve in one another and do not retain their individual characteristics.

An example of a ceramic system forming a solid solution is that of barium titanate ($BaTiO_3$) and strontium titanate ($SrTiO_3$), which are compounds found in the BaO-TiO_2-SrO ternary system.

Limited Solubility When we add a small quantity of salt (one phase) to a glass of water (a second phase) and stir, the salt dissolves completely in the water. Only one phase—salty water or brine—is found. However, if we add too much salt to the water, the excess salt sinks to the bottom of the glass [Figure 10-1(c)]. Now we have two phases

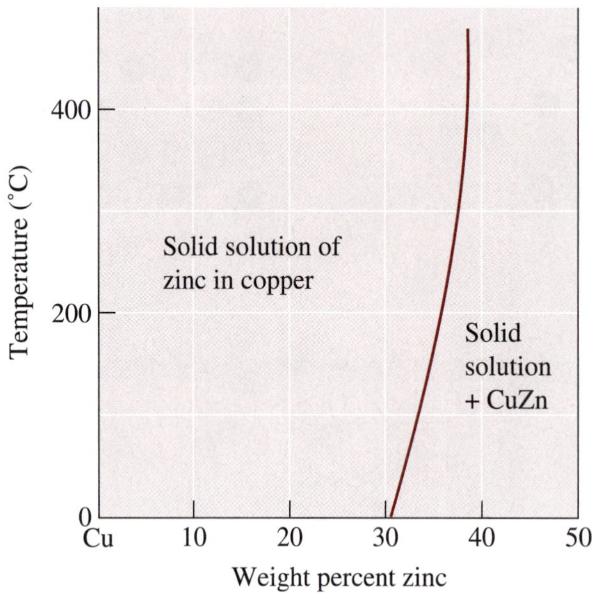

Figure 10-4
The solubility of zinc in copper. The solid line represents the solubility limit; when excess zinc is added, the solubility limit is exceeded and two phases coexist.

in equilibrium with one another—water that is saturated with salt plus excess solid salt. We find that salt or sugar have a **limited solubility** in water.

If we add a small amount of liquid zinc to liquid copper, a single liquid solution is produced. When that copper-zinc solution cools and solidifies, a single solid solution having an FCC structure results, with copper and zinc atoms randomly located at the normal lattice points. However, if the liquid solution contains more than about 30% Zn freezes, some of the excess zinc atoms combine with some of the copper atoms to form a CuZn compound [Figure 10-3(c)]. Two solid phases now coexist: a solid solution of copper saturated with about 30% Zn plus a CuZn compound. The solubility of zinc in copper is limited. Figure 10-4 shows a portion of the Cu-Zn phase diagram illustrating the solubility of zinc in copper at low temperatures. The solubility increases with increasing temperature. This is similar to how we can dissolve more sugar or salt in water by increasing the temperature.

In the extreme case, there may be almost no solubility of one material in another. This is true for oil and water [Figure 10-1(d)] or for copper-lead (Cu-Pb) alloys. Note that even though materials do not dissolve into one another they can be dispersed into one another. For example, oil-like phases and aqueous liquids can be mixed, often using surfactants (soap-like molecules), to form emulsions. Immiscibility, or lack of solubility, is seen in many molten, and solid ceramic and metallic materials.

Polymeric Systems We can process polymeric materials to enhance their usefulness using a concept similar to the formation of solid solutions in metallic and ceramic systems. We can form materials that are known as **copolymers** that consist of different monomers. For example, acrylonitrile (A), butadiene (B), and styrene (S) monomers can be made to react to form a copolymer known as ABS. This resultant copolymer is similar to a solid solution in that it has the functionalities of the three monomers from which it is derived, blending their properties. Similar to the Cu-Ni or $BaTiO_3$-$SrTiO_3$ solid solutions, we will not be able to separate out the acrylonitrile, butadiene, or styrene from an ABS plastic. Injection molding is used to convert ABS into telephones, helmets, steering wheels, small appliance cases, and other products. Figure 10-5 illustrates the properties of different copolymers in the ABS system. Note that this is *not* a phase diagram. This means the phases formed may not be thermodynamically stable.

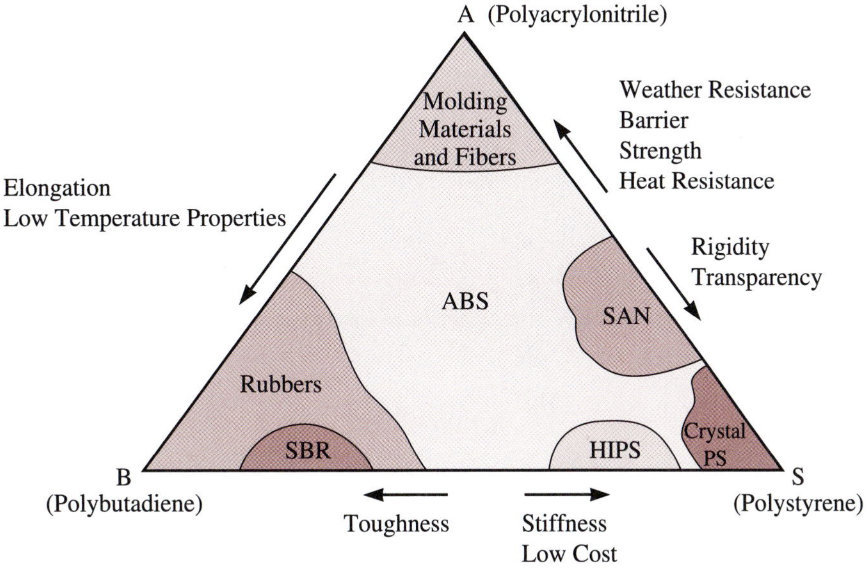

Figure 10-5 Diagram showing how the properties of copolymers formed in the ABS system vary. This is not a phase diagram. (*Source: Strong, A. Brent,* Plastics: Materials and Processing, *2nd, © 2000. Electronically reproduced by permission of Pearson Education, Inc., Upper Saddle River, New Jersey.*)

Dylark^{TM} is another example of a copolymer. It is formed using maleic anhydride and a styrene monomer. The Dylark^{TM} copolymer, with carbon black for UV protection, reinforced with fiberglass, and toughened with rubber, is used for instrument panels in many automobiles (Chapter 16).

10-3 Conditions for Unlimited Solid Solubility

In order for an alloy system (e.g., copper-nickel) to have unlimited solid solubility, certain conditions must be satisfied. These conditions, the **Hume-Rothery rules**, are as follows:

1. *Size factor*: The atoms or ions must be of similar size, with no more than a 15% difference in atomic radius, in order to minimize the lattice strain (i.e., to minimize, at an atomic level, the deviations caused in interatomic spacing).

2. *Crystal structure*: The materials must have the same crystal structure; otherwise, there is some point at which a transition occurs from one phase to a second phase with a different structure.

3. *Valence*: The ions must have the same valence; otherwise, the valence electron difference encourages the formation of compounds rather than solutions.

4. *Electronegativity*: The atoms must have approximately the same electronegativity. Electronegativity is the affinity for electrons (Chapter 2). If the electronegativities differ significantly, compounds form—as when sodium and chlorine atoms combine to form sodium chloride.

Hume-Rothery's conditions must be met, but they are not necessarily sufficient, for two metals (e.g., Cu and Ni) or compounds (e.g., $BaTiO_3$-$SrTiO_3$) to have unlimited solid solubility.

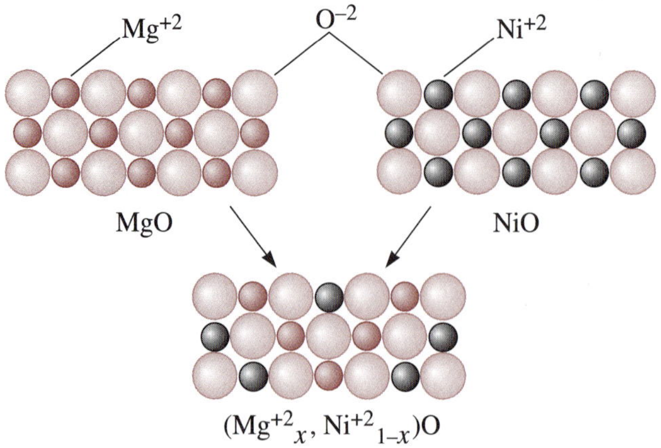

Figure 10-6
MgO and NiO have similar crystal structures, ionic radii, and valences; thus the two ceramic materials can form solid solutions.

Figure 10-6 shows schematically the two-dimensional structures of MgO and NiO. The Mg^{+2} and Ni^{+2} ions are similar in size and valence and, consequently, can replace one another in a sodium chloride (NaCl) crystal structure (Chapter 3), forming a complete series of solid solutions of the form $(Mg^{+2}{}_x Ni^{+2}{}_{1-x})O$, where $x =$ the mole fraction of Mg^{+2} or MgO.

The solubility of interstitial atoms is always limited. Interstitial atoms are much smaller than the atoms of the host element, thereby violating the first of Hume-Rothery's conditions.

EXAMPLE 10-2 *Ceramic Solid Solutions of MgO*

NiO can be added to MgO to produce a solid solution. What other ceramic systems are likely to exhibit 100% solid solubility with MgO?

SOLUTION

In this case, we must consider oxide additives that have metal cations with the same valence and ionic radius as the magnesium cations. The valence of the magnesium ion is +2 and its ionic radius is 0.66 Å. From Appendix B, some other possibilities in which the cation has a valence of +2 include the following:

	r (A)	$\left[\frac{r_{ion} - r_{Mg^{+2}}}{r_{Mg^{+2}}}\right] \times 100\%$	Crystal Structure
Cd^{+2} in CdO	$r_{Cd^{+2}} = 0.97$	47	NaCl
Ca^{+2} in CaO	$r_{Ca^{+2}} = 0.99$	50	NaCl
Co^{+2} in CoO	$r_{Co^{+2}} = 0.72$	9	NaCl
Fe^{+2} in FeO	$r_{Fe^{+2}} = 0.74$	12	NaCl
Sr^{+2} in SrO	$r_{Sr^{+2}} = 1.12$	70	NaCl
Zn^{+2} in ZnO	$r_{Zn^{+2}} = 0.74$	12	NaCl

The percent difference in ionic radii and the crystal structures are also shown and suggest that the FeO-MgO system will probably display unlimited solid solubility. The CoO and ZnO systems also have appropriate radius ratios and crystal structures.

10-4 Solid-Solution Strengthening

In metallic materials, one of the important effects of solid-solution formation is the resultant **solid-solution strengthening** (Figure 10-7). This strengthening, via solid-solution formation, is caused by increased resistance to dislocation motion. This is one of the important reasons why brass (Cu-Zn alloy) is stronger than pure copper. Similarly small levels of carbon strengthen iron. We will learn later that carbon also plays another role in additional strengthening of steels by forming iron carbide (Fe_3C) and other phases (Chapter 12). Jewelry could be made out of pure gold or silver. However, pure gold and pure silver are extremely soft and malleable and the jewelry pieces made will not retain their shape. This is also why jewelers add copper to gold or silver.

In the copper-nickel (Cu-Ni) system, we intentionally introduce a solid substitutional atom (nickel) into the original crystal structure (copper). The copper-nickel alloy is stronger than pure copper. Similarly, if less than 30% Zn is added to copper, the zinc behaves as a substitutional atom that strengthens the copper-zinc alloy, as compared with pure copper.

Recall from Chapter 7 that the strength of ceramics is mainly dictated by the distribution of flaws; solid-solution formation does not have a strong effect on the mechanical properties. This is similar to why strain hardening was not much of a factor in enhancing the strength of ceramics or semiconductors such as silicon (Chapter 8). As discussed before, solid-solution formation in ceramics and semiconductors (such as Si, GaAs, etc.) has considerable influence on their magnetic, optical, and dielectric prop-

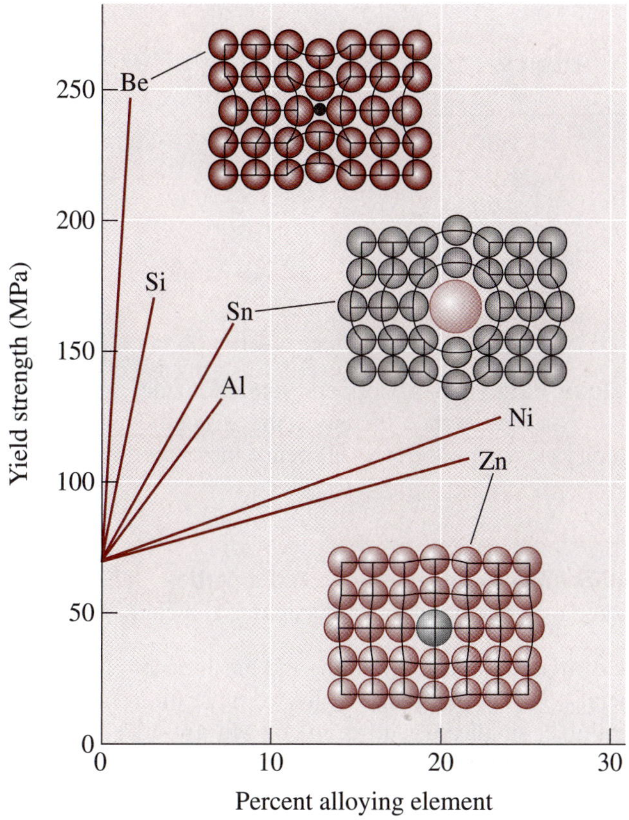

Figure 10-7
The effects of several alloying elements on the yield strength of copper. Nickel and zinc atoms are about the same size as copper atoms, but beryllium and tin atoms are much different from copper atoms. Increasing both atomic size difference and amount of alloying element increases solid-solution strengthening.

erties. The following discussion related to mechanical properties, therefore, applies mainly to metallic materials.

Degree of Solid-Solution Strengthening The degree of solid-solution strengthening depends on two factors. First, a large difference in atomic size between the original (host or solvent) atom and the added (guest or solute) atom increases the strengthening effect. A larger size difference produces a greater disruption of the initial crystal structure, making slip more difficult (Figure 10-7).

Second, the greater the amount of alloying element added, the greater the strengthening effect (Figure 10-7). A Cu-20% Ni alloy is stronger than a Cu-10% Ni alloy. Of course, if too much of a large or small atom is added, the solubility limit may be exceeded and a different strengthening mechanism, *dispersion strengthening*, may come in to play. In dispersion strengthening, the interface between the host phase and guest phase resists dislocation motion and contributes to strengthening. This mechanism is discussed further in Chapter 11.

EXAMPLE 10-3 *Solid-Solution Strengthening*

From the atomic radii, show whether the size difference between copper atoms and alloying atoms accurately predicts the amount of strengthening found in Figure 10-7.

SOLUTION

The atomic radii and percent size difference are shown below:

Metal	Atomic Radius (A)	$\left[\frac{r - r_{Cu}}{r_{Cu}}\right] \times 100\%$
Cu	1.278	0
Zn	1.332	+4.2
Al	1.432	+12.1
Sn	1.509	+18.1
Ni	1.243	−2.7
Si	1.176	−8.0
Be	1.143	−10.6

For atoms larger than copper—namely, zinc, aluminum, and tin—increasing the size difference increases the strengthening effect. Likewise for smaller atoms, increasing the size difference increases strengthening.

Effect of Solid-Solution Strengthening on Properties The effects of solid-solution strengthening on the properties of a metallic material include the following (Figure 10-8):

1. The yield strength, tensile strength, and hardness of alloys are greater than those of the pure metals. This is one reason why we most often use alloys rather than pure metals. For example, small concentrations of Mg are added to aluminum to provide higher strength to the aluminum alloys used in making aluminum beverage cans.

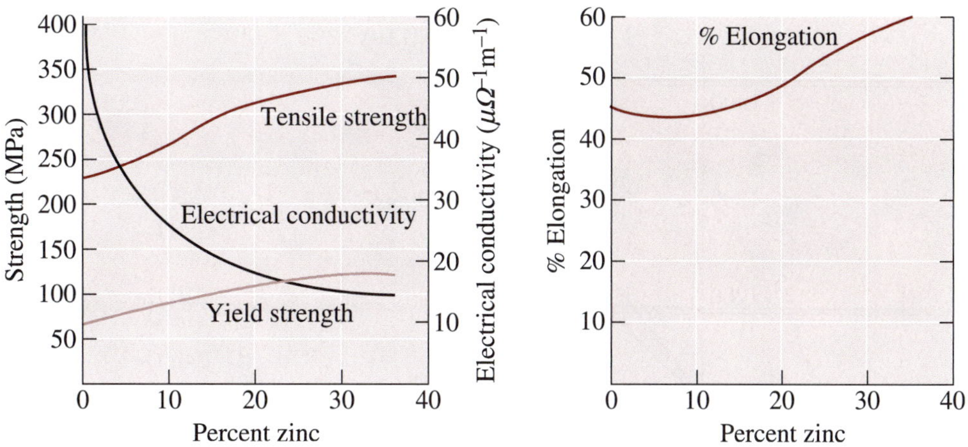

Figure 10-8 The effect of additions of zinc to copper on the properties of the solid-solution-strengthened alloy. The increase in % elongation with increasing zinc content is *not* typical of solid-solution strengthening.

2. Almost always, the ductility of the alloy is less than that of the pure metal. Only rarely, as in copper-zinc alloys (Figure 10-8), does solid-solution strengthening increase both strength and ductility.

3. Electrical conductivity of the alloy is much lower than that of the pure metal. This is because electrons get more scattered off the atoms of the alloying elements. Solid-solution strengthening of copper or aluminum wires used for transmission of electrical power is not recommended because of this pronounced effect. Electrical conductivity of many alloys, although lower than that of pure metals, is often more stable as a function of temperature.

4. The resistance to creep, or loss of strength at elevated temperatures, is improved by solid-solution strengthening. High temperatures do not cause a catastrophic change in the properties of solid-solution-strengthened alloys. Many high-temperature alloys, such as those used for jet engines, rely partly on extensive solid-solution strengthening.

10-5 Isomorphous Phase Diagrams

A **phase diagram** shows the phases and their compositions at any combination of temperature and alloy composition. When only two elements or two compounds are present in a material, a **binary phase diagram** can be constructed. **Isomorphous binary phase diagrams** are found in a number of metallic and ceramic systems. In the isomorphous systems, which include the copper-nickel and NiO-MgO systems [Figure 10-9(a) and (b)], only one solid phase forms; the two components in the system display complete solid solubility. As shown in the phase diagrams for $CaO \cdot SiO_2$-$SrO \cdot SiO_2$, and thallium-lead (Tl-Pb) systems, it is possible to have phase diagrams show a minimum or maximum point, respectively [Figure 10-9(c) and (d)]. Notice the *x*-axis scale can represent either mole% or weight% of one of the components. We can also plot atomic% or mole fraction of one of the components. Also, notice that the CaO-SiO_2 and SrO-SiO_2 diagram could be plotted as a *ternary phase diagram*. A ternary phase diagram is a phase

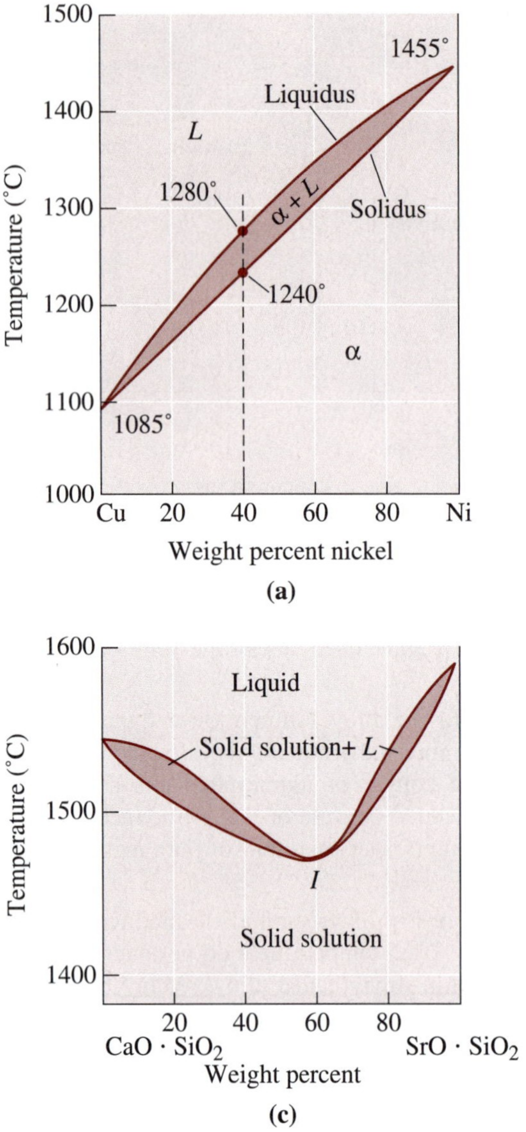

Figure 10-9 (a) The equilibrium phase diagrams for the Cu-Ni and NiO-MgO systems. (b) The liquidus and solidus temperatures are shown for a Cu-40% Ni alloy. (c) and (d) Systems with solid solution maxima and minima. (*Source: Adapted from* Introduction to Phase Equilibria, *by C.G. Bergeron, and S.H. Risbud. Copyright © 1984 American Ceramic Society. Adapted by permission.*)

diagram for systems consisting of three components. Here, we represent it as a *pseudo-binary diagram* (i.e., we assume that this is a diagram that represents phase equilibria between $CaO \cdot SiO_2$ and $SrO \cdot SiO_2$). In a pseudo-binary diagram, we represent equilibria between three or more components using two compounds. Ternary phase diagrams are often encountered in ceramic and metallic systems.

Recently, considerable developments also have occurred in the development of phase diagrams using computer databases containing thermodynamic properties of different elements and compounds.

Liquidus and Solidus Temperatures We define **liquidus temperature** as the temperature above which a material is completely liquid. The upper curve in Figure 10-9(a) represents the liquidus temperatures for copper-nickel alloys of different compositions. We must heat a copper-nickel alloy above the liquidus temperature to produce a completely liquid alloy that can then be cast into a useful shape. The liquid alloy begins to solidify when the temperature decreases to the liquidus temperature. For the Cu-40% Ni alloy in Figure 10-9(a), the liquidus temperature is 1280°C.

The **solidus temperature** for the copper-nickel alloys is the temperature below which the alloy is 100% solid. The lower curve in Figure 10-9(a) represents the solidus temperatures for Cu-Ni alloys of different compositions. A copper-nickel alloy is not completely solid until the material cools below the solidus temperature. If we use a copper-nickel alloy at high temperatures, we must be sure that the service temperature is below the solidus so that no melting occurs. For the Cu-40% Ni alloy in Figure 10-9(a), the solidus temperature is 1240°C.

Copper-nickel alloys melt and freeze over a range of temperatures between the liquidus and the solidus. The temperature difference between the liquidus and the solidus is the **freezing range** of the alloy. Within the freezing range, two phases coexist: a liquid and a solid. The solid is a solution of copper and nickel atoms and is designated as the α phase. For the Cu-40% Ni alloy in Figure 10-9(a), the freezing range is $1280 - 1240 = 40$°C. Note that pure metals solidify at a fixed temperature (i.e., the freezing range is zero degrees).

Phases Present Often we are interested in which phases are present in an alloy at a particular temperature. If we plan to make a casting, we must be sure that the metal is initially all liquid; if we plan to heat treat an alloy component, we must be sure that no liquid forms during the process. Different solid phases have different properties. For example, BCC Fe (indicated as α phase on the iron carbon phase diagram) is magnetic. However, FCC iron (indicated as γ phase on the Fe-C diagram) is not.

The phase diagram can be treated as a road map; if we know the coordinates—temperature and alloy composition—we can determine the phases present, assuming we know that thermodynamic equilibrium exists. There are many examples of technologically important situations where we do not want equilibrium phases to form.

The following two examples illustrate the applications of some of these concepts.

EXAMPLE 10-4 *NiO-MgO Isomorphous System*

From the phase diagram for the NiO-MgO binary system [Figure 10-9(b)], describe a composition that can melt at 2600°C but will not melt when placed into service at 2300°C.

SOLUTION

The material must have a liquidus temperature below 2600°C, but a solidus temperature above 2300°C. The NiO-MgO phase diagram [Figure 10-9(b)] permits us to design an appropriate composition.

To identify a composition with a liquidus temperature below 2600°C, there must be less than 65 mol% MgO in the refractory. To identify a composition with a solidus temperature above 2300°C, there must be at least 50 mol% MgO present. Consequently, we can use any composition between 50 mol% MgO and 65 mol% MgO.

EXAMPLE 10-5 *Design of a Composite Material*

One method to improve the fracture toughness of a ceramic material (Chapter 7) is to reinforce the ceramic matrix with ceramic fibers. A materials designer has suggested that Al_2O_3 could be reinforced with 25% Cr_2O_3 fibers, which would interfere with the propagation of any cracks in the alumina. The resulting composite is expected to operate under load at 2000°C for several months. Criticize the appropriateness of this design.

SOLUTION

Since the composite will operate at high temperatures for a substantial period of time, the two phases—the Cr_2O_3 fibers and the Al_2O_3 matrix—must not react with one another. In addition, the composite must remain solid to at least 2000°C. The phase diagram in Figure 10-10 permits us to consider this choice for a composite.

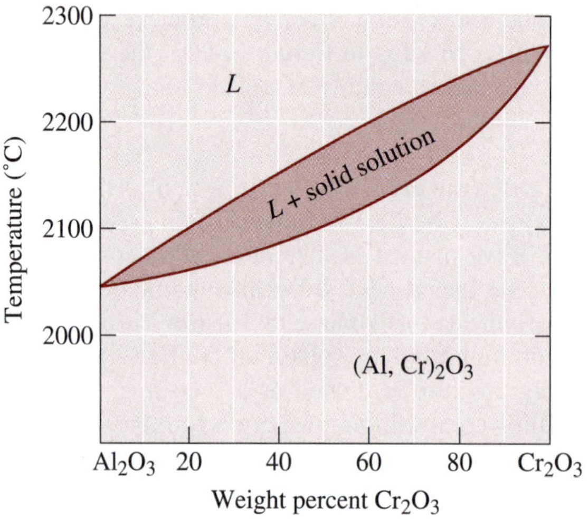

Figure 10-10 The Al_2O_3-Cr_2O_3 phase diagram (for Example 10-5).

Pure Cr_2O_3, pure Al_2O_3, and Al_2O_3-25% Cr_2O_3 have solidus temperatures above 2000°C; consequently, there is no danger of melting any of the constituents. However, Cr_2O_3 and Al_2O_3 display unlimited solid solubility. At the high service temperature, 2000°C, Al^{3+} ions will diffuse from the matrix into the fiber, replacing Cr^{3+} ions in the fibers. Simultaneously, Cr^{3+} ions will replace Al^{3+} ions in the matrix. Long before several months have elapsed, these diffusion processes cause the fibers to completely dissolve into the matrix. With no fibers remaining, the fracture toughness will again be poor.

Composition of Each Phase For each phase we can specify a composition, expressed as the percentage of each element in the phase. Usually the composition is expressed in weight percent (wt%). When only one phase is present in the alloy or a ceramic solid solution, the composition of the phase equals the overall composition of the material. If the original composition of a single phase alloy or ceramic material changes, then the composition of the phase must also change.

However, when two phases, such as liquid and solid, coexist, their compositions differ from one another and also differ from the original overall composition. In this case, if the original composition changes slightly, the composition of the two phases is unaffected, provided that the temperature remains constant.

This difference is explained by the Gibbs phase rule. In this case, unlike the example of pure magnesium (Mg) described earlier, we keep the pressure fixed (e.g., one atmosphere), which is normal for binary phase diagrams. The phase rule given by Equation 10-1 can be rewritten as:

$$1 + C = F + P \quad \text{(for constant pressure)} \tag{10-2}$$

where, again, C is the number of independent chemical components, P is the number of phases (*not pressure*), and F is the number of degrees of freedom. We now use number 1 instead of number 2 because we are holding the pressure constant. This reduces the number of degrees of freedom by 1. The pressure is typically, although not necessarily, one atmosphere. In a binary system, the number of components C is two; the degrees of freedom that we have include changing the temperature and changing the composition of the phases present. We can apply this form of the phase rule to the Cu-Ni system, as shown in Example 10-6.

EXAMPLE 10-6 *Gibbs Rule for Isomorphous Phase Diagram*

Determine the degrees of freedom in a Cu-40% Ni alloy at (a) 1300°C, (b) 1250°C, and (c) 1200°C. Use Figure 10-9(a).

SOLUTION

This is a binary system ($C = 2$) with components Cu and Ni. We will assume constant pressure. Therefore, Equation 10-2 ($1 + C = F + P$) can be used as follows:

(a) At 1300°C, $P = 1$, since only one phase (liquid) is present; $C = 2$, since both copper and nickel atoms are present. Thus:

$$1 + C = F + P \quad \therefore\ 1 + 2 = F + 1 \text{ or } F = 2$$

We must fix both the temperature and the composition of the liquid phase to completely describe the state of the copper-nickel alloy in the liquid region.

(b) At 1250°C, $P = 2$, since both liquid and solid are present; $C = 2$, since copper and nickel atoms are present. Now:

$$1 + C = F + P \quad \therefore\ 1 + 2 = F + 2 \text{ or } F = 1$$

If we fix the temperature in the two-phase region, the compositions of the two phases are also fixed. Or, if the composition of one phase is fixed, the temperature and composition of the second phase are automatically fixed.

(c) At 1200°C, $P = 1$, since only one phase, solid, is present; $C = 2$, since both copper and nickel atoms are present. Again,

$$1 + C = F + P \quad \therefore\ 1 + 2 = F + 1 \text{ or } F = 2$$

and we must fix both temperature and composition to completely describe the state of the solid.

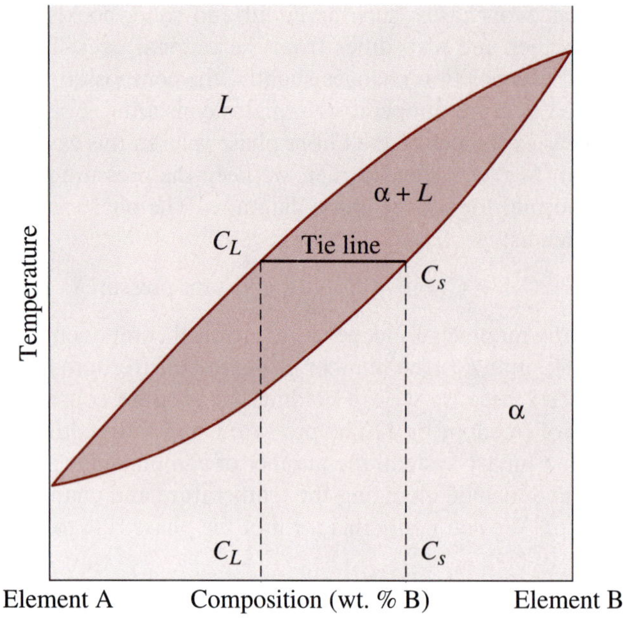

Figure 10-11
A hypothetical binary phase diagram between two elements A and B. When an alloy is present in a two-phase region, a tie line at the temperature of interest fixes the composition of the two phases. This is a consequence of the Gibbs phase rule, which provides only one degree of freedom.

Because there is only one degree of freedom in a two-phase region of a binary phase diagram, the compositions of the two phases are always fixed when we specify the temperature. This is true even if the overall composition of the alloy changes. Therefore, we can use a tie line to determine the composition of the two phases. A **tie line** is a horizontal line within a two-phase region drawn at the temperature of interest (see Figure 10-11). In an isomorphous system, the tie line connects the liquidus and solidus points at the specified temperature. The ends of the tie line represent the compositions of the two phases in equilibrium. Tie lines are not used in single-phase regions because we do not have two phases to "tie" in.

For any alloy with overall or bulk composition lying between c_L and c_S, the composition of the liquid is c_L and the composition of the solid α is c_S.

The following example illustrates how the concept of the tie line is used to determine the composition of different phases in equilibrium.

EXAMPLE 10-7 *Compositions of Phases in Cu-Ni Phase Diagram*

Determine the composition of each phase in a Cu-40% Ni alloy at 1300°C, 1270°C, 1250°C, and 1200°C. (See Figure 10-12.)

SOLUTION

The vertical line at 40% Ni represents the overall composition of the alloy:

- **1300°C**: Only liquid is present. The liquid must contain 40% Ni, the overall composition of the alloy.
- **1270°C**: Two phases are present. A horizontal line within the $\alpha + L$ field is drawn. The endpoint at the liquidus, which is in contact with the liquid region, is at 37% Ni. The endpoint at the solidus, which is in contact with the

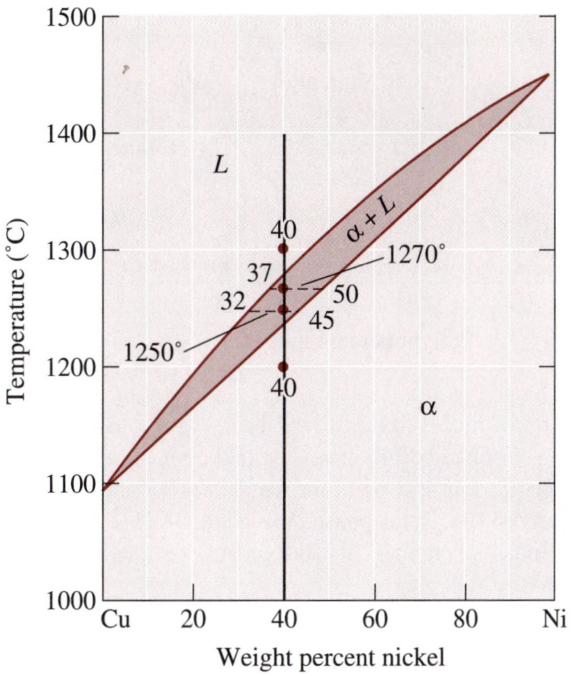

Figure 10-12 Tie lines and phase compositions for a Cu-40% Ni alloy at several temperatures (for Example 10-7).

α region, is at 50% Ni. Therefore, the liquid contains 37% Ni and the solid contains 50% Ni.

- **1250°C**: Again two phases are present. The tie line drawn at this temperature shows that the liquid contains 32% Ni and the solid contains 45% Ni.
- **1200°C**: Only solid α is present, so the solid must contain 40% Ni.

In Example 10-7, we find that the solid α contains more nickel than the overall alloy and the liquid *L* contains more copper than the original alloy. Generally, the higher melting point element (in this case, nickel) is concentrated in the first solid that forms.

Amount of Each Phase (the Lever Rule) Lastly, we are interested in the relative amounts of each phase present in the alloy. These amounts are normally expressed as weight percent (wt%). We express absolute amounts of different phases in units of mass or weight (grams, kilograms, pounds, etc.). The following example illustrates the rationale for the **lever rule**.

EXAMPLE 10-8 *Application of Lever Rule*

Calculate the amounts of α and *L* at 1250°C in the Cu-40% Ni alloy shown in Figure 10-13.

SOLUTION

Let's say that x = mass fraction of the alloy that is solid α. Since we have only two phases, the balance of nickel must be in the liquid phase (*L*). Thus, the

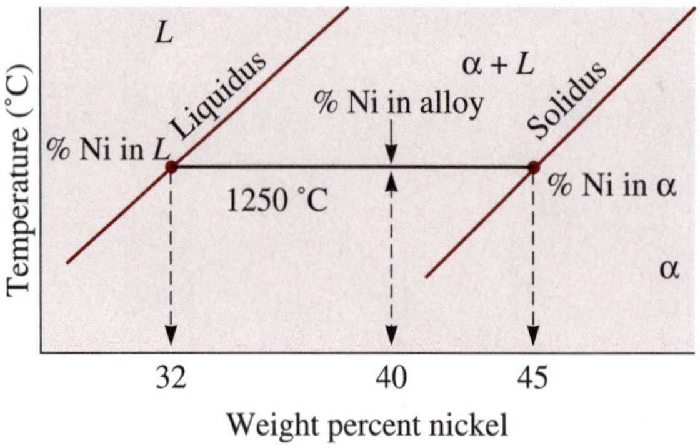

Figure 10-13 A tie line at 1250°C in the copper-nickel system that is used in Example 10-8 to find the amount of each phase.

mass fraction of nickel in liquid will be $1 - x$. Consider 100 grams of the alloy. This alloy will consist of 40 grams of nickel at all temperatures. At 1250°C, let us write an equation that will represent mass balance for nickel. At 1250°C, we have x grams of the alpha phase. We have $100(1 - x)$ grams of liquid.

Total mass of nickel in 100 grams of the alloy = mass of nickel in liquid + mass of nickel in α

$$\therefore\ 100 \times (\%\text{ Ni in alloy}) = [(100)(1 - x)](\%\text{ Ni in L}) + (100)[x](\%\text{ Ni in }\alpha)$$

$$\therefore\ (\%\text{ Ni in alloy}) = (\%\text{ Ni in }L)(1 - x) + (\%\text{ Ni in }\alpha)(x)$$

By multiplying and rearranging,

$$x = \frac{(\%\text{ Ni in alloy}) - (\%\text{ Ni in }L)}{(\%\text{ Ni in }\alpha) - (\%\text{ Ni in }L)}$$

From the phase diagram at 1250°C:

$$x = \frac{40 - 32}{45 - 32} = \frac{8}{13} = 0.62$$

If we convert from mass fraction to mass percent, the alloy at 1250°C contains 62% α and 38% L. Note that the concentration of Ni in alpha phase (at 1250°C) is 45% and the concentration of nickel in liquid phase (at 1250°C) is 32%.

To calculate the amounts of liquid and solid, we construct a lever on our tie line, with the fulcrum of our lever being the original composition of the alloy. The leg of the lever *opposite* to the composition of the phase, whose amount we are calculating, is divided by the total length of the lever to give the amount of that phase. In Example 10-8, note that the denominator represents the total length of the tie line and the numerator is the portion of the lever that is *opposite* the composition of the solid we are trying to calculate.

The lever rule in general can be written as:

$$\text{Phase percent} = \frac{\text{opposite arm of lever}}{\text{total length of tie line}} \times 100 \qquad (10\text{-}3)$$

We can work the lever rule in any two-phase region of a binary phase diagram. The lever rule calculation is not used in single-phase regions because the answer is trivial (there is 100% of that phase present). The lever rule is used to calculate the relative fraction or % of a phase in a two-phase mixture. The end points of the tie line we use give us the composition (i.e., chemical concentration of different components) of each phase.

The following example reinforces the application of the lever rule for calculating the amounts of phases for an alloy at different temperatures. This is one way to track the solidification behavior of alloys, something we had not seen previously.

EXAMPLE 10-9 *Solidification of a Cu-40% Ni Alloy*

Determine the amount of each phase in the Cu-40% Ni alloy shown in Figure 10-12 at 1300°C, 1270°C, 1250°C, and 1200°C.

SOLUTION

- **1300°C**: There is only one phase, so 100% L.
- **1270°C**: $\%L = \dfrac{50 - 40}{50 - 37} \times 100 = 77\%$

  $\%\alpha = \dfrac{40 - 37}{50 - 37} \times 100 = 23\%$
- **1250°C**: $\%L = \dfrac{45 - 40}{45 - 32} \times 100 = 38\%$

  $\%\alpha = \dfrac{40 - 32}{45 - 32} \times 100 = 62\%$
- **1200°C**: There is only one phase, so 100% α.

Note that at each temperature, we can determine the composition of the phases in equilibrium from the ends of the tie line drawn at that temperature.

This may seem a little odd at first. How does the α phase change its composition? The liquid phase also changes its composition and the amounts of each phase change with temperature as the alloy cools from liquidus to solidus.

Sometimes we wish to express composition as atomic percent (at%) rather than weight percent (wt%). For a Cu-Ni alloy, where M_{Cu} and M_{Ni} are the molecular weights, the following equations provide examples for making these conversions:

$$\text{at\% Ni} = \left(\frac{\left[\dfrac{\text{wt\% Ni}}{M_{Ni}}\right]}{\left[\dfrac{(\text{wt\% Ni})}{M_{Ni}} + \dfrac{(\text{wt\% Cu})}{M_{Cu}}\right]} \right) \times 100 \tag{10-4}$$

$$\text{wt\% Ni} = \left(\frac{(\text{at\% Ni})/(M_{Ni})}{(\text{at\% Ni})/(M_{Ni}) + (\text{at\% Cu})/(M_{Cu})} \right) \times 100 \tag{10-5}$$

10-6 Relationship Between Properties and the Phase Diagram

We have previously mentioned that a copper-nickel alloy will be stronger than either pure copper or pure nickel because of solid solution strengthening. The mechanical properties of a series of copper-nickel alloys can be related to the phase diagram as shown in Figure 10-14.

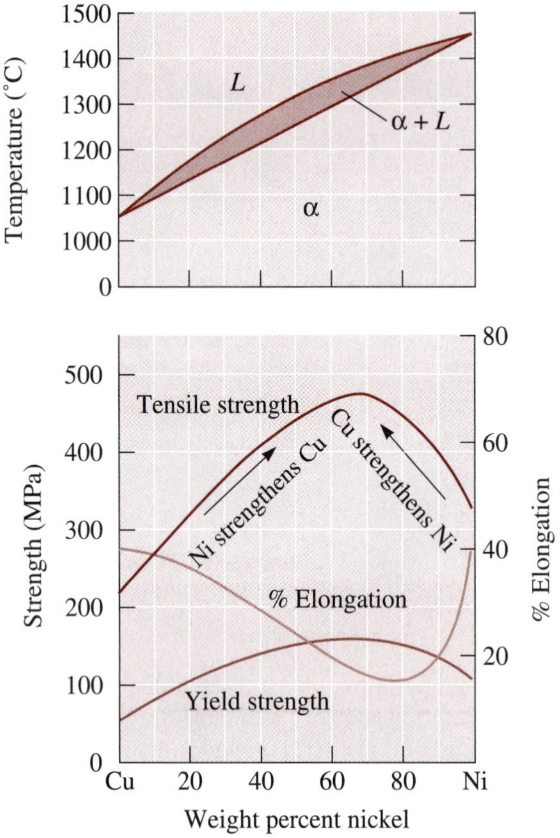

Figure 10-14
The mechanical properties of copper-nickel alloys. Copper is strengthened by up to 60% Ni and nickel is strengthened by up to 40% Cu.

The strength of copper increases by solid-solution strengthening until about 60% Ni is added. Pure nickel is solid-solution strengthened by the addition of copper until 40% Cu is added. The maximum strength is obtained for a Cu-60% Ni alloy, known as *Monel*. The maximum is closer to the pure nickel side of the phase diagram because pure nickel is stronger than pure copper.

EXAMPLE 10-10 *Design of a Melting Procedure for a Casting*

You need to produce a Cu-Ni alloy having minimum yield strength of 138 MPa, a minimum tensile strength of 414 MPa, and a minimum % elongation of 20%. You have in your inventory a Cu-20% Ni alloy and pure nickel. Design a method for producing castings having the required properties.

SOLUTION

From Figure 10-14, we determine the required composition of the alloy. To meet the required yield strength, the alloy must contain between 30 and 90% Ni; for the tensile strength, 33 to 90% Ni is required. The required % elongation can be obtained for alloys containing less than 60% Ni or more than 90% Ni. To satisfy all of these conditions, we could use:

Cu-90% Ni or Cu-33% to 60% Ni

We prefer to select a low nickel content, since nickel tends to be more expensive than copper. In addition, the lower nickel alloys have a lower liquidus, permitting castings to be made with less energy being expended. Therefore, a reasonable alloy might be Cu-35% Ni.

To produce this composition from the available melting stock, we must blend some of the pure nickel with the Cu-20% Ni ingot. Assume we wish to produce 10 kg of the alloy. Let x be the mass of Cu-20% Ni alloy we will need. The mass of pure Ni needed will be $10 - x$.

Since the final alloy consists of 35% Ni, the total mass of Ni needed will be:

$$(10 \text{ Kg})\left(\frac{35\% \text{ Ni}}{100\%}\right) = 3.5 \text{ kg Ni}$$

Now let's write a mass balance for nickel. Nickel from the Cu-20% alloy + pure nickel added = total nickel in the 35% alloy being produced.

$$(x \text{ kg})\left(\frac{20\%}{100\%}\right) + (10 - x \text{ kg Ni})\left(\frac{100\%}{100\%}\right) = 3.5 \text{ kg Ni}$$

$$0.2x + 10 - x = 3.5$$

$$6.5 = 0.8x$$

$$x = 8.125 \text{ kg}$$

Therefore, we need to melt 8.125 kg of Cu-20% Ni with 1.875 kg of pure nickel to produce the required alloy. We would then heat the alloy above the liquidus temperature, which is 1250°C for the Cu-35% Ni alloy, before pouring the liquid into an appropriate mold.

We need to conduct such calculations for many practical situations dealing with the processing of alloys because when we make these materials, we use new and recycled materials.

The following example illustrates how mass balance calculations are used to calculate quantities of materials needed for ceramics manufacturing.

EXAMPLE 10-11 ***Yttria Stabilized Zirconia (YSZ) Ceramics***

Essentially pure zirconia ceramics are not used, because the phase transformation involved with the change in crystal structure from monoclinic to tetragonal causes significant stress (due to changes in the volume) and consequent failure of the ceramics. It is well known that such oxides as yttria (Y_2O_3) will stabilize the high-temperature cubic form of zirconia and thus avoid this problem. An engineer working on manufacturing oxygen sensors wants to make 2.0 kilograms of such yttria stabilized zirconia (YSZ) powder containing 9 mol.% Y_2O_3. How much yttrium oxide and zirconium oxide powders will she need? The atomic masses of yttrium, zirconium, and oxygen are 88.9, 91.2, and 16, respectively.

SOLUTION

The molecular weight of zirconia (ZrO_2) is $= 91.2 + 2 \times 16 = 123.2$. Similarly, the molecular weight of yttria will be 285.8.

Let us start with 1 mole of the 7% YSZ material containing 7 mol.% yttria. This has 0.91 moles of zirconia and 0.09 moles of yttria.

$$\text{The mass of 0.91 moles of zirconia will be} = 0.91 \text{ moles} \times 123.2 \text{ grams/mole}$$
$$= 112.1 \text{ grams}$$

Similarly, the mass of 0.09 moles of yttria will be 20.3 grams.

The weight fraction of zirconia in this 9 mol.% YSZ material will be

$$\text{YSZ} = \frac{112.1}{(112.1 + 20.3)} = 0.846$$

The weight fraction of yttria will be $= 1 - 0.846 = 0.154$.

For 2000 grams of YSZ (with 9 mol.% zirconia), the amount of zirconia will be

$$ZrO_2 = 2000 \times 0.846 = 1693.1 \text{ grams}$$

The balance of 306.9 grams of yttrium oxide will be required. This combination of materials will be sufficient to produce 2 kilograms of powder of 9 mol.% yttria-based YSZ. In this calculation, we did not account for any loss of material, incorporation of impurities, or even yttrium oxide or zirconium oxide from other sources (such as grinding media used for breaking the agglomerates in the powders). These grinding media also are often made using YSZ ceramic spheres.

10-7 Solidification of a Solid-Solution Alloy

When an alloy such as Cu-40% Ni is melted and cooled, solidification requires both nucleation and growth. Heterogeneous nucleation permits little or no undercooling, so solidification begins when the liquid reaches the liquidus temperature (Chapter 9). The phase diagram (Figure 10-15), with a tie line drawn at the liquidus temperature, tells that the *first solid to form* has a composition of Cu-52% Ni.

Two conditions are required for growth of the solid α. First, growth requires that the latent heat of fusion (ΔH_f), which evolves as the liquid solidifies, be removed from the solid–liquid interface. Second, unlike the case of pure metals, diffusion must occur so that the compositions of the solid and liquid phases follow the solidus and liquidus curves during cooling. The latent heat of fusion (ΔH_f) is removed over a range of temperatures so that the cooling curve shows a change in slope, rather than a flat plateau

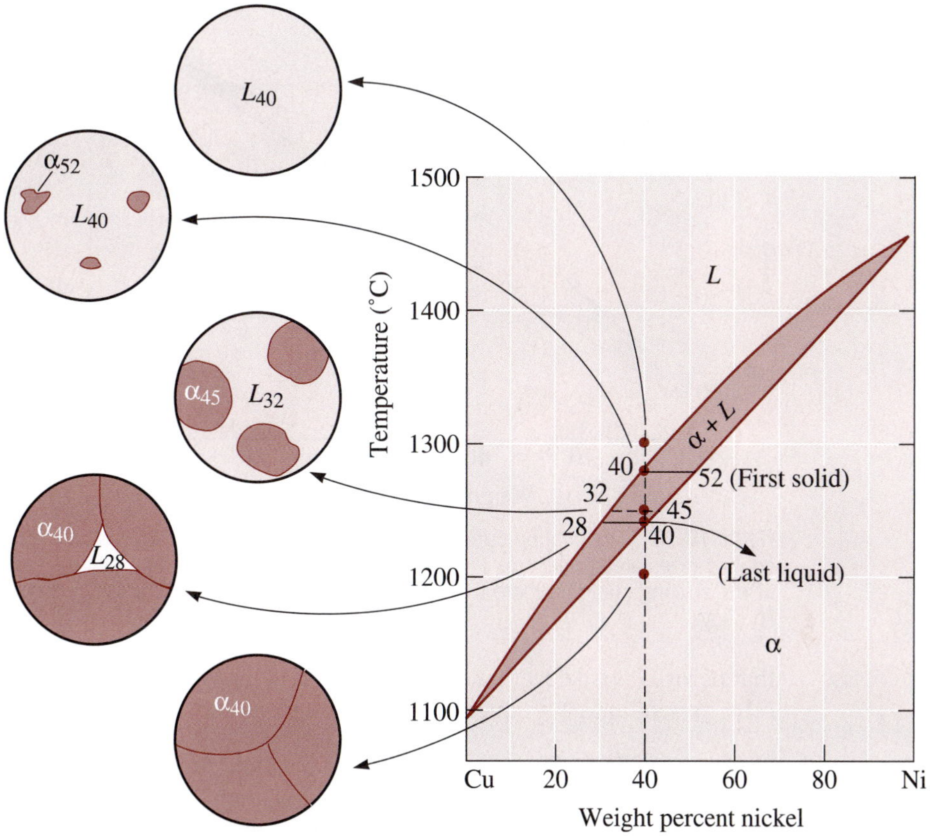

Figure 10-15 The change in structure of a Cu-40% Ni alloy during equilibrium solidification. The nickel and copper atoms must diffuse during cooling in order to satisfy the phase diagram and produce a uniform equilibrium structure.

(Figure 10-16). Thus, as we mentioned before in Chapter 9, the solidification of alloys is different from that of pure metals.

At the start of freezing, the liquid contains Cu-40% Ni and the first solid contains Cu-52% Ni. Nickel atoms must have diffused to and concentrated at the first solid to form. But after cooling to 1250°C, solidification has advanced and the phase diagram tells us that now all of the liquid must contain 32% Ni and all of the solid must contain 45% Ni. On cooling from the liquidus to 1250°C, some nickel atoms must diffuse from the first solid to the new solid, reducing the nickel concentration in the first solid. Additional nickel atoms diffuse from the solidifying liquid to the new solid. Meanwhile, copper atoms have concentrated—by diffusion—into the remaining liquid. This process must continue until we reach the solidus temperature, where the last liquid to freeze, which contains Cu-28% Ni, solidifies and forms a solid containing Cu-40% Ni. Just below the solidus, all of the solid must contain a uniform concentration of 40% Ni throughout.

In order to achieve this equilibrium final structure, the cooling rate must be extremely slow. Sufficient time must be permitted for the copper and nickel atoms to diffuse and produce the compositions given by the phase diagram. In many practical casting situations, the cooling rate is too rapid to permit equilibrium. Therefore, in most castings made from alloys we expect chemical segregation. We had seen in Chapter 9

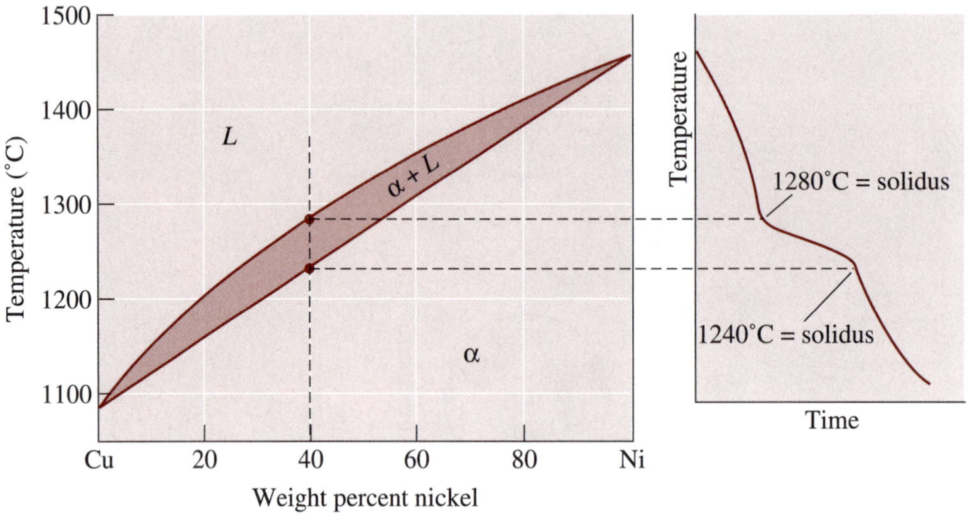

Figure 10-16 The cooling curve for an isomorphous alloy during solidification. We assume that cooling rates are small so as to allow thermal equilibrium to take place. The changes in slope of the cooling curve indicate the liquidus and solidus temperatures, in this case for a Cu-40% Ni alloy.

that porosity is a defect that could be present in many cast products. Another such defect often present in cast products is chemical **segregation**.

Microsegregation The nonuniform composition produced by nonequilibrium solidification is known as segregation. **Microsegregation**, also known as **interdendritic segregation** and **coring**, occurs over short distances, often between small dendrite arms. The centers of the dendrites, which represent the first solid to freeze, are rich in the higher melting point element in the alloy. The regions between the dendrites are rich in the lower melting point element, since these regions represent the last liquid to freeze. The composition and properties of the phase α (in the case of Cu-Ni alloys) differ from one region to the next, and we expect the casting to have poorer properties as a result.

Microsegregation can cause **hot shortness**, or melting of the lower melting point interdendritic material at temperatures below the equilibrium solidus. When we heat the Cu-40% Ni alloy to 1225°C, below the equilibrium solidus but above the nonequilibrium solidus, the low-nickel regions between the dendrites melt.

Homogenization We can reduce the interdendritic segregation and problems with hot shortness by means of a **homogenization heat treatment**. If we heat the casting to a temperature below the nonequilibrium solidus, the nickel atoms in the centers of the dendrites diffuse to the interdendritic regions; copper atoms diffuse in the opposite direction.

Macrosegregation There exists another type of segregation, known as **macrosegregation**, which occurs over a larger distance, between the surface and the center of the casting, with the surface (which freezes first) containing slightly more than the average amount of the higher melting point metal. We cannot eliminate macrosegregation by a homogenization treatment, because the diffusion distances are too great. Macrosegregation can be reduced by hot working. This is because in hot working we basically are breaking down the cast macrostructure.

Rapidly Solidified Powders In applications where porosity, microsegregation, and macrosegregation must be minimized, powders of complex alloys are prepared using

spray atomization. In spray atomization, homogeneous melts of complex compositions are prepared and sprayed through a ceramic nozzle. Since the solidification in droplets occurs very quickly, there is very little time for diffusion. Therefore, chemical segregation does not occur. Many complex nickel- and cobalt-based superalloys and stainless steel powders are examples of materials prepared using this technique.

SUMMARY

- A phase is any portion, including the whole, of a system that is physically homogeneous within it and bounded by a surface so that it is mechanically separable from any other portions.
- A phase diagram typically shows phases that are expected to be present in a system under thermodynamic equilibrium conditions. Sometimes metastable phases may also be shown.
- Solid solutions in metallic or ceramic materials form when elements or compounds with similar crystal structure can form a phase that is chemically homogeneous.
- In plastics, we can form copolymers by reacting different type of monomers. Forming copolymers is similar to forming solid solutions of metallic or ceramic systems.
- Solid-solution strengthening is accomplished in metallic materials by formation of solid solutions. The point defects created restrict dislocation motion and cause strengthening.
- The degree of solid-solution strengthening increases when (1) the amount of alloying element increases and (2) the atomic size difference between the host material and the alloying element increases.
- The amount of alloying element (or compound) that we can add to produce solid-solution strengthening is limited by the solubility of the alloying element or compound in the host material. The solubility is limited when (1) the atomic size difference is more than about 15%, (2) the alloying element (or compound) has a different crystal structure than the host element (or compound), and (3) the valence and electronegativity of the alloying element or constituent ions are different from those of the host element (or compound).
- In addition to increasing strength and hardness, solid-solution strengthening typically decreases ductility and electrical conductivity of metallic materials. An important function of solid-solution strengthening is to provide good high-temperature properties to the alloy.
- The addition of alloying elements to provide solid-solution strengthening also changes the physical properties, including the melting temperature, of the alloy. The phase diagram helps explain these changes.
- A phase diagram in which constituents exhibit complete solid solubility is known as an isomorphous phase diagram.
- As a result of solid-solution formation, solidification begins at the liquidus temperature and is completed at the solidus temperature; the temperature difference over which solidification occurs is the freezing range.
- In two-phase regions of the phase diagram, the ends of a tie line fix the composition of each phase and the lever rule permits the amount of each phase to be calculated.

- Microsegregation and macrosegregation can occur during solidification.
- Homogenization can reduce microsegregation. Macrosegregation describes differences in composition over long distances, such as between the surface and center of a casting. Hot working may reduce macrosegregation.

GLOSSARY

Alloy A material that exhibits properties of a metallic material and is made from multiple elements.

Binary phase diagram A phase diagram for a system with two components.

Copolymer A polymer that is formed by combining two or more different types of monomers, usually with the idea of blending the properties affiliated with individual polymers. For example, Dylark™ is a copolymer of maleic anhydride and styrene.

Coring Chemical segregation in cast products, also known as microsegregation or interdendritic segregation. The centers of the dendrites are rich in the higher melting point element, whereas interdendritic regions, which solidify last, are rich in the lower melting point element.

Freezing range The temperature difference between the liquidus and solidus temperatures.

Gibbs phase rule Describes the number of degrees of freedom, or the number of variables that must be fixed to specify the temperature and composition of a phase ($2 + C = F + P$, where pressure and temperature can change, $1 + C = F + P$, where pressure or temperature is constant).

Homogenization heat treatment The heat treatment used to reduce the microsegregation caused during nonequilibrium solidification. This heat treatment cannot eliminate macrosegregation.

Hot shortness Melting of the lower melting point nonequilibrium material that forms due to segregation, even though the temperature is below the equilibrium solidus temperature.

Hume-Rothery rules The conditions that an alloy or ceramic system must meet if the system is to display unlimited solid solubility. Hume-Rothery's rules are necessary but are not sufficient for materials to show unlimited solid solubility.

Interdendritic segregation *See* Coring.

Isomorphous phase diagram A phase diagram in which components display unlimited solid solubility.

Lever rule A technique for determining the amount of each phase in a two-phase system.

Limited solubility When only a maximum amount of a solute material can be dissolved in a solvent material.

Liquidus Curves on phase diagrams that describe the liquidus temperatures of all possible alloys.

Liquidus temperature The temperature at which the first solid begins to form during solidification.

Macrosegregation The presence of composition differences in a material over large distances caused by nonequilibrium solidification. The only way to remove this type of segregation is to break down the cast structure by hot working.

Microsegregation *See* Coring.

Multiple-phase alloy An alloy that consists of two or more phases.

Phase Any portion, including the whole of a system, which is physically homogeneous within it and bounded by a surface so that it is mechanically separable from any other portions.

Phase diagrams Diagrams showing phases present under equilibrium conditions and the phase compositions at each combination of temperature and overall composition. Sometimes phase diagrams also indicate metastable phases.

Phase rule *See* Gibbs phase rule.

P-T diagram A diagram describing thermodynamic stability of phases under different temperature and pressure conditions (same as a unary phase diagram).

Segregation The presence of composition differences in a material, often caused by insufficient time for diffusion during solidification.

Single-phase alloy An alloy consisting of one phase.

Solid solution A solid phase formed by combining multiple elements or compounds such that the overall phase has uniform composition and properties that are different from those of the elements or compounds forming it.

Solid-solution strengthening Increasing the strength of a metallic material via the formation of a solid solution.

Solidus Curves on phase diagrams that trace the solidus temperatures for all possible alloys.

Solidus temperature The temperature below which all liquid has completely solidified.

Solubility The amount of one material that will completely dissolve in a second material without creating a second phase.

Spray atomization A process in which molten alloys or metals are sprayed using a ceramic nozzle. The molten material stream is broken using a gas (e.g., Ar, N_2) or water. This leads to finer droplets that solidify rapidly, forming metal or alloy powders with ~10–100 μm particle size range.

Stainless steels Corrosion-resistant alloys that usually contain iron (Fe), carbon (C), chromium (Cr), nickel (Ni), and some other elements.

Tie line A horizontal line drawn in a two-phase region of a phase diagram to assist in determining the compositions of the two phases.

Triple point A pressure and temperature at which three phases of a single material are in equilibrium.

Unary phase diagram A phase diagram in which there is only one component.

Unlimited solubility When the amount of one material that will dissolve in a second material without creating a second phase is unlimited.

PROBLEMS

Section 10-1 Phases and the Phase Diagram

10-1 Write down the Gibbs phase rule, assuming temperature and pressure are allowed to change. Explain clearly the meaning of each term.

10-2 **(a)** What is a phase diagram? **(b)** Explain why the P-T diagram for H_2O is considered to be a unary diagram.

Section 10-2 Solubility and Solid Solutions

10-3 What is a solid solution?

10-4 How can solid solutions form in ceramic systems?

10-5 Do we need 100% solid solubility to form a solid solution of one material in another?

10-6 What is a copolymer? What is the advantage to forming copolymers?

10-7 What is the ABS copolymer? State some of the applications of this material.

Section 10-3 Conditions for Unlimited Solid Solubility

10-8 Briefly state the Hume-Rothery rules and explain the rationale.

10-9 Based on Hume-Rothery's conditions, which of the following systems would be expected to display unlimited solid solubility? Explain. **(a)** Au-Ag; **(b)** Al-Cu; **(c)** Al-Au; **(d)** U-W; **(e)** Mo-Ta; **(f)** Nb-W; **(g)** Mg-Zn; and **(h)** Mg-Cd.

Section 10-4 Solid-Solution Strengthening

10-10 Suppose 1 at% of the following elements is added to copper (forming a separate alloy with each element) without exceeding the solubility limit. Which one would be expected to give the higher strength alloy? Are any of the alloying elements expected to have unlimited solid solubility in copper? **(a)** Au; **(b)** Mn; **(c)** Sr; **(d)** Si; and **(e)** Co.

10-11 Suppose 1 at% of the following elements is added to aluminum (forming a separate alloy with each element) without exceeding the solubility limit. Which one would be expected to give the least reduction in electrical conductivity? Are any of the alloy elements expected to have unlimited solid solubility in aluminum? **(a)** Li; **(b)** Ba; **(c)** Be; **(d)** Cd; and **(e)** Ga.

10-12 Which of the following oxides is expected to have the largest solid solubility in Al_2O_3? **(a)** Y_2O_3; **(b)** Cr_2O_3; and **(c)** Fe_2O_3.

10-13 What is the role of small concentrations of Mg in aluminum alloys used to make beverage cans?

10-14 Why do jewelers add small amounts of copper to gold and silver?

10-15 Why is it not a good idea to use solid solution strengthening as a mechanism to increase the strength of copper for electrical applications?

Section 10-5 Isomorphous Phase Diagrams

10-16 Determine the liquidus temperature, solidus temperature, and freezing range for the following MgO-FeO ceramic compositions. (See Figure 10-17.)

(a) MgO-25 wt% FeO; **(b)** MgO-45 wt% FeO; **(c)** MgO-65 wt% FeO; **(d)** MgO-80 wt% FeO.

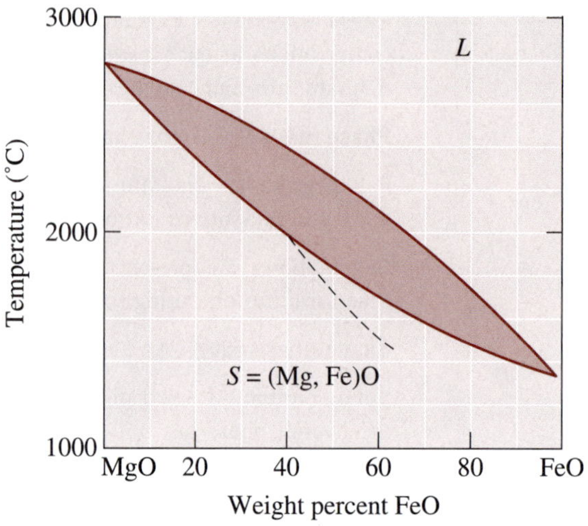

Figure 10-17 The equilibrium phase diagram for the MgO-FeO system (for Problems 10-16, 10-17, 10-24, 10-25, 10-32 and 10-36). The dashed curve represents the solidus for non-equilibrium cooling.

10-17 **(a)** Determine the phases present, the compositions of each phase, and the amount of each phase in wt% for the following MgO-FeO ceramics at 2000°C. (See Figure 10-17.) **(i)** MgO-25 wt% FeO; **(ii)** MgO-45 wt% FeO; **(iii)** MgO-60 wt% FeO; and **(iv)** MgO-80 wt% FeO. **(b)** Consider an alloy of 65 wt% Cu and 35 wt% Al. Calculate the composition of the alloy in at%.

10-18 Consider a ceramic composed of 30 mol% MgO and 70 mol% FeO. Calculate the composition of the ceramic in wt%.

10-19 A Nb-60 wt% W alloy is heated to 2800°C. Determine **(a)** the composition of the solid and liquid phases in both wt% and at%; **(b)** the amount of each phase in both wt% and at%; and **(c)** assuming that the density of the solid is 16.05 g/cm^3 and that of the liquid is 13.91 g/cm^3, determine the amount of each phase in vol%. (See Figure 10-18.)

10-20 How many grams of nickel must be added to 500 grams of copper to produce an alloy that has a liquidus temperature of 1350°C? What is the ratio of the number of nickel atoms to copper atoms in this alloy?

10-21 How many grams of nickel must be added to 500 grams of copper to produce an alloy that contains 50 wt% α at 1300°C?

10-22 How many grams of MgO must be added to 1 kg of NiO to produce a ceramic that has a solidus temperature of 2200°C?

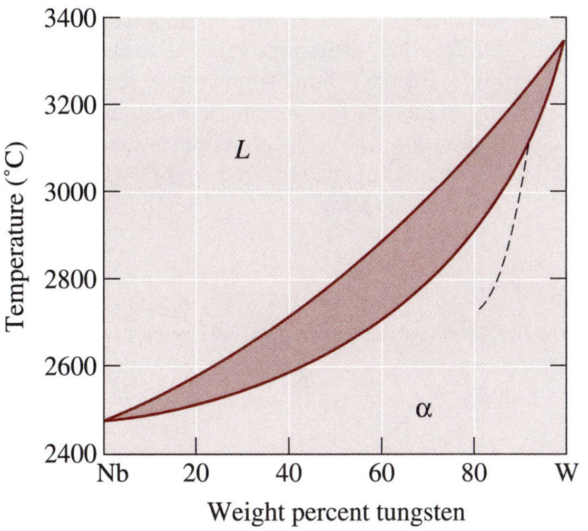

Figure 10-18 The equilibrium phase diagram for the Nb-W system (for Problems 10-19, 10-26, 10-27, 10-28, 10-29, 10-33 and 10-35). The dashed curve represents the solidus for non-equilibrium cooling.

10-23 How many grams of MgO must be added to 1 kg of NiO to produce a ceramic that contains 25 mol% solid at 2400°C?

10-24 We would like to produce a solid MgO-FeO ceramic that contains equal mol percentages of MgO and FeO at 1200°C. Determine the wt% FeO in the ceramic. (See Figure 10-17.)

10-25 We would like to produce a MgO-FeO ceramic that is 30 wt% solid at 2000°C. Determine the original composition of the ceramic in wt%. (See Figure 10-17.)

10-26 A Nb-W alloy held at 2800°C is partly liquid and partly solid. **(a)** If possible, determine the composition of each phase in the alloy; and **(b)** if possible, determine the amount of each phase in the alloy. (See Figure 10-18.)

10-27 A Nb-W alloy contains 55% α at 2600°C. Determine **(a)** the composition of each phase; and **(b)** the original composition of the alloy. (See Figure 10-18.)

10-28 Suppose a 544-kg bath of a Nb-40 wt% W alloy is held at 2800°C. How many kilograms of tungsten can be added to the bath before any solid forms? How many pounds of tungsten must be added to cause the entire bath to be solid? (See Figure 10-18.)

10-29 A fiber-reinforced composite material is produced, in which tungsten fibers are embedded in a Nb matrix. The composite is composed of 70 vol% tungsten. **(a)** Calculate the wt% of tungsten fibers in the composite; and **(b)** suppose the composite is heated to 2600°C and held for several years. What happens to the fibers? Explain. (See Figure 10-18.)

Section 10-6 Relationship between Properties and the Phase Diagram

10-30 What is brass? Explain which element strengthens the matrix for this alloy.

10-31 What is the composition of Monel alloy?

Section 10-7 Solidification of a Solid-Solution Alloy

10-32 Equal moles of MgO and FeO are combined and melted. Determine **(a)** the liquidus temperature, the solidus temperature, and the freezing range of the ceramic; and **(b)** determine the phase(s) present, their composition(s), and their amount(s) at 1800°C. (See Figure 10-17.)

10-33 Suppose 75 cm^3 of Nb and 45 cm^3 of W are combined and melted. Determine **(a)** the liquidus temperature, the solidus temperature, and the freezing range of the alloy; and **(b)** determine the phase(s) present, their composition(s), and their amount(s) at 2800°C. (See Figure 10-18.)

10-34 A NiO-60 mol% MgO ceramic is allowed to solidify. Determine **(a)** the composition of the first solid to form; and **(b)** the composition of the last liquid to solidify under equilibrium conditions.

10-35 A Nb-35% W alloy is allowed to solidify. Determine **(a)** the composition of the first solid to form; and **(b)** the composition of the last liquid to solidify under equilibrium conditions. (See Figure 10-18.)

10-36 For equilibrium conditions and a MgO-65 wt% FeO ceramic, determine **(a)** the liquidus temperature; **(b)** the solidus temperature; **(c)** the freezing range; **(d)** the composition of the first solid to form during solidification; **(e)** the composition of the last liquid to solidify; **(f)** the phase(s) present, the composition of the phase (s), and the amount of the phase(s) at 1800°C; and **(g)** the phase(s) present, the composition of the phase(s), and the amount of the phase(s) at 1600°C. (See Figure 10-17.)

10-37 Figure 10-19 on the next page shows the cooling curve for a NiO-MgO ceramic. Determine **(a)** the liquidus temperature; **(b)** the solidus temperature; **(c)** the freezing range; **(d)** the pouring temperature; **(e)** the superheat; **(f)** the local solidification time; **(g)** the total solidification time; and **(h)** the composition of the ceramic.

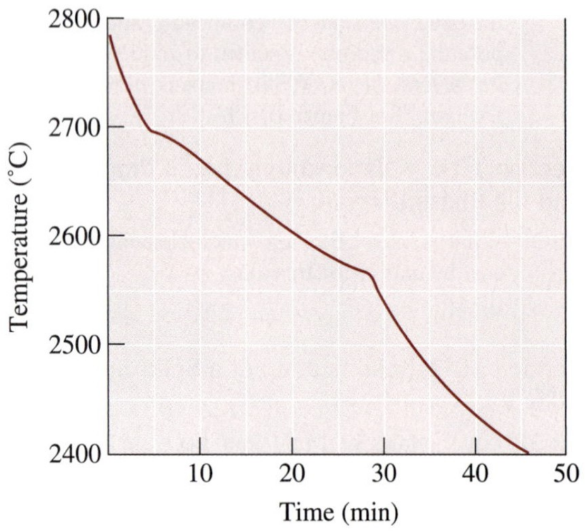

Figure 10-19 Cooling curve for a NiO-MgO ceramic (for Problem 10-37).

10-38 For equilibrium conditions and a Nb-80 wt% W alloy, determine **(a)** the liquidus temperature; **(b)** the solidus temperature; **(c)** the freezing range; **(d)** the composition of the first solid to form during solidification; **(e)** the composition of the last liquid to solidify; **(f)** the phase(s) present, the composition of the phase(s), and the amount of the phase(s) at 3000°C; and **(g)** the phases(s) present, the composition of the phase(s), and the amount of the phase(s) at 2800°C. (See Figure 10-18.)

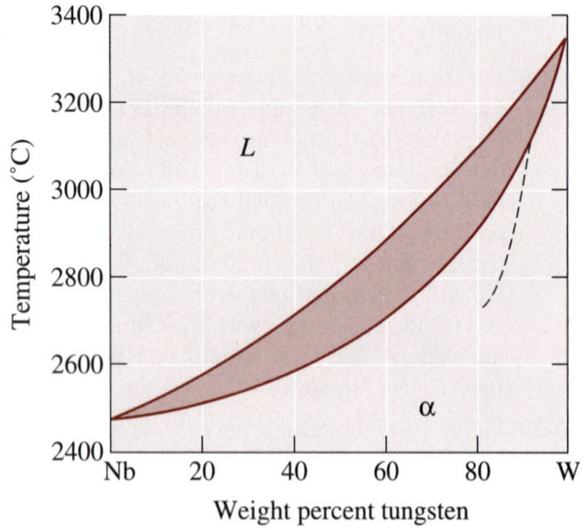

Figure 10-18 (Repeated for Problem 10-38). The equilibrium phase diagram for the Nb-W system. The dashed curve represents the solidus for non-equilibrium cooling.

10-39 Figure 10-20 shows the cooling curve for a Nb-W alloy. Determine **(a)** the liquidus temperature; **(b)** the solidus temperature; **(c)** the freezing range; **(d)** the pouring temperature; **(e)** the superheat; **(f)** the local solidification time; **(g)** the total solidification time; and **(h)** the composition of the alloy.

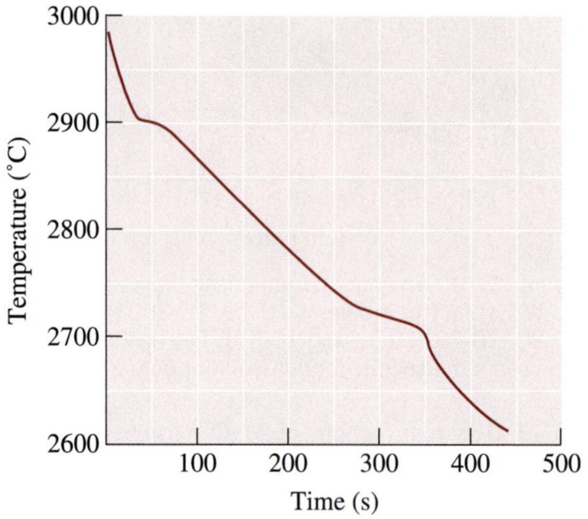

Figure 10-20 Cooling curve for a Nb-W alloy (for Problem 10-39).

10-40 Cooling curves are shown in Figure 10-21 for several Mo-V alloys. Based on these curves, construct the Mo-V phase diagram.

10-41 What are the origins of chemical segregation in cast products?

10-42 How can microsegregation be removed?

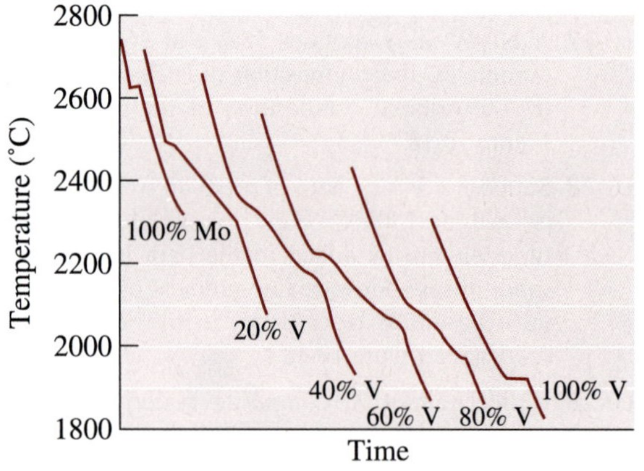

Figure 10-21 Cooling curves for a series of Mo-V alloys (for Problem 10-40).

10-43 What is macrosegregation? Is there a way to remove it without breaking up the cast structure?

10-44 What is homogenization? What type of segregation can it remove?

10-45 What is spray atomization?

Design Problems

10-46 Homogenization of a slowly cooled Cu-Ni alloy having a secondary dendrite arm spacing (SDAS) of 0.025 cm requires 8 hours at 1000°C. Design a process to produce a homogeneous structure in a more rapidly cooled Cu-Ni alloy having a SDAS of 0.005 cm.

10-47 Design a process to produce a NiO-60% MgO refractory whose structure is 40% glassy phase at room temperature. Include all relevant temperatures.

10-48 Design a method by which glass beads (having a density of 2.3 g/cm^3) can be uniformly mixed and distributed in a Cu-20% Ni alloy (density of 8.91 g/cm^3).

11

Dispersion Strengthening and Eutectic Phase Diagrams

Have You Ever Wondered?

- *Why did some of the earliest glassmakers use plant ash to make glass?*
- *What alloys are most commonly used for soldering?*
- *What is a fiberglass?*
- *Is there an alloy that freezes at a constant temperature?*
- *What is Pyrex®?*

When the solubility of a material is exceeded by adding too much of an alloying element or compound, a second phase forms and a two-phase material is produced. The boundary between the two phases, known as the **interphase interface**, is a surface where the atomic arrangement is not perfect. In metallic materials, this boundary interferes with the slip or movement of dislocations, causing their strengthening. The general term for such strengthening by the introduction of a second phase is known as dispersion strengthening. In this chapter, we first discuss the fundamentals of dispersion strengthening to determine the microstructure we should aim to produce. Next, we examine the types of reactions that produce multiple-phase alloys. Finally, we examine, in some detail, methods to achieve dispersion strengthening by controlling the solidification process. We will concentrate on eutectic phase diagrams that involve the formation of multiple phases. In Chapter 12, we will learn about a special way of producing a dispersion of a second phase via a solid-state phase transformation sequence known as **precipitation hardening** or **age hardening**. Precipitation hardening is a sub-set of the overall class of dispersion-strengthened materials.

11-1 Principles and Examples of Dispersion Strengthening

Most engineered materials are multi-phase, and many of these materials are designed to provide a certain level of strength. In simple **dispersion-strengthened alloys**, tiny particles of one phase, usually very strong and hard, are created within a second phase, which is weaker but more ductile. The soft phase, usually continuous and present in larger amounts, is called the **matrix**. The hard-strengthening phase may be called the **dispersed phase** or the **precipitate**, depending on how the alloy is formed. In some cases, a phase or a mixture of phases may have a very characteristic appearance – in these cases this phase or phase mixture may be called a **microconstituent**. For dispersion strengthening to occur, the dispersed phase or precipitate must be small enough to provide effective obstacles to dislocation movement, thus providing the strengthening mechanism.

In most alloys, dispersion strengthening is produced by phase transformations. In this chapter, we will concentrate on a solidification transformation by which a liquid freezes to simultaneously form two solid phases. This is called the **eutectic reaction** and is of particular importance in cast irons and many aluminum alloys. In the next chapter, we will discuss the *eutectoid reaction*, by which one solid phase leads to simultaneously form two different solid phases; this reaction is key in the control of properties in steels. In Chapter 12, we will also discuss precipitation, or age, hardening which produces precipitates by a sophisticated heat treatment.

There are some general considerations for determining how the characteristics of the matrix and the dispersed phase affect the overall properties of an alloy (Figure 11-1).

1. The matrix should be soft and ductile, however, the dispersed phase should be hard and strong. The dispersed phase particles interfere with slip, while the matrix provides at least some ductility to the overall alloy.

2. The hard dispersed phase should be discontinuous, while the soft, ductile matrix should be continuous. If the hard and brittle dispersed phase were continuous, cracks could propagate through the entire structure.

3. The dispersed phase particles should be small and numerous, increasing the likelihood that they interfere with the slip process since the area of the interphase interface is increased significantly.

4. The dispersed phase particles should be round, rather than needlelike or sharp edged, because the rounded shape is less likely to initiate a crack or to act as a notch.

5. Higher concentrations of the dispersed phase increase the strength of the alloy.

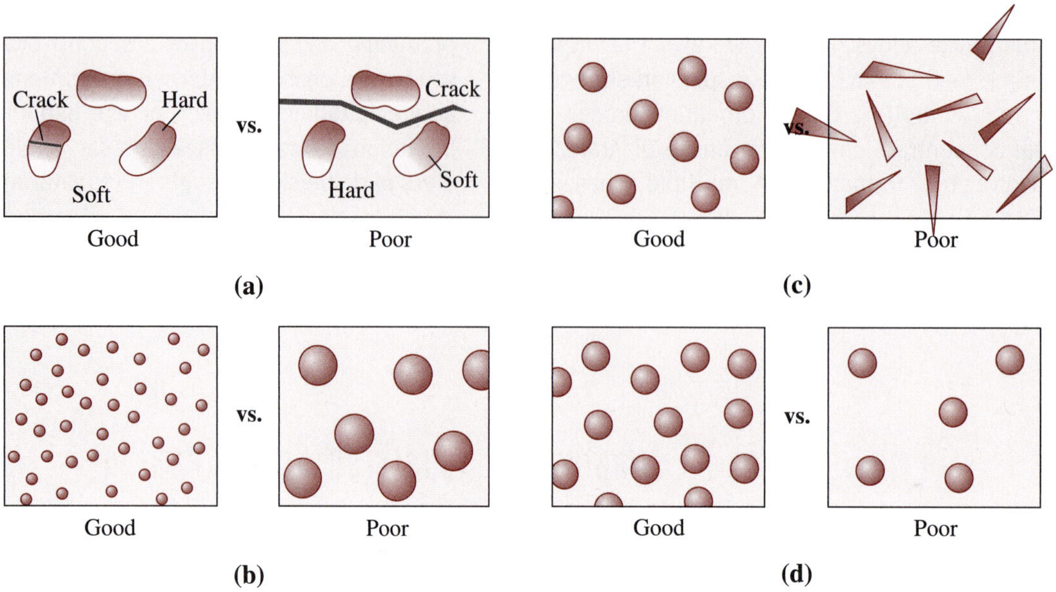

Figure 11-1 Considerations for effective dispersion strengthening: (a) The precipitate phase should be hard and discontinuous, (b) the dispersed phase particles should be small and numerous, (c) the dispersed phase particles should be round rather than needlelike, and (d) larger amounts of dispersed phase increase strengthening.

11-2 Intermetallic Compounds

An **intermetallic compound** is made up of two or more metallic elements, producing a new phase with its own composition, crystal structure, and properties. Intermetallic compounds are almost always very hard and brittle. Intermetallics or intermetallic compounds are similar to ceramic materials in terms of their mechanical properties, however, in that they are formed by combining two or more metallic elements. Our interest in intermetallics is two-fold. First, often dispersion-strengthened alloys contain an intermetallic compound as the dispersed phase. Secondly, many intermetallic compounds, on their own (and not as a second phase) are being investigated and developed for high-temperature applications. In this section, we will discuss properties of intermetallics as stand-alone materials. In the sections that follow, we will discuss how intermetallic phases help strengthen metallic materials. Table 11-1 summarizes the properties of some intermetallic compounds.

Stoichiometric intermetallic compounds have a fixed composition. Steels are often strengthened by a stoichiometric compound, iron carbide (Fe_3C), which has a fixed ratio of three iron atoms to one carbon atom. Stoichiometric intermetallic compounds are represented in the phase diagram by a vertical line [Figure 11-2(a)]. An example of a useful intermetallic compound is molybdenum disilicide ($MoSi_2$). This material is used for making heating elements of high temperature furnaces. At high temperatures (~1000–1600°C), $MoSi_2$ shows outstanding oxidation resistance. At low temperatures (~500°C and below), $MoSi_2$ is brittle and shows catastrophic oxidation known as pesting.

Nonstoichiometric intermetallic compounds have a range of compositions and are sometimes called **intermediate solid solutions**. In the molybdenum-rhodium system, the γ phase is a nonstoichiometric intermetallic compound [Figure 11-2(b)]. Because the

TABLE 11-1 ■ ***Physical and mechanical properties of important intermetallic compounds.***

	Density (g/cm^3)	Crystal Structure (ordered)	Young's Modulus (GPa)	Coefficient of thermal expansion (10^{-6}/°C)	Tensile yield stress (MPa)	Melting point (°C)
Al_3Ti	3.4–4.0	DO_{22}(tetr.)	215	12–15	120–425	1350
TiAl	3.8–4.0	Ll_0(tetr)	160–175	11.7	400–775	1480
Ti_3Al	4.1–4.7	DO_{19}(CPH)	120	12	700–900	1680
$MoSi_2$	6.1	Tetragonal	380–440	8.1–8.5	200–400	2020
Ni_3Al	7.4–7.7	Ll_2(FCC)	180–200	14–16	200–900	1397
NiAl	5.9	B2(FCC)	177–190	14–16	175–300	1638
Ni_5Si_3	7.2		340	N/A	550	N/A
Fe_3Al	6.7	DO_3	140–170	19	600–1350	1540
FeAl	5.6–5.8	B2	160–250	21.5	500–700	N/A

(*Source: Used with permission from M.A. Meyers & K.K. Chawla,* Mechanical Behavior of Materials, *2nd ed., 2008, Cambridge University Press, UK.*)

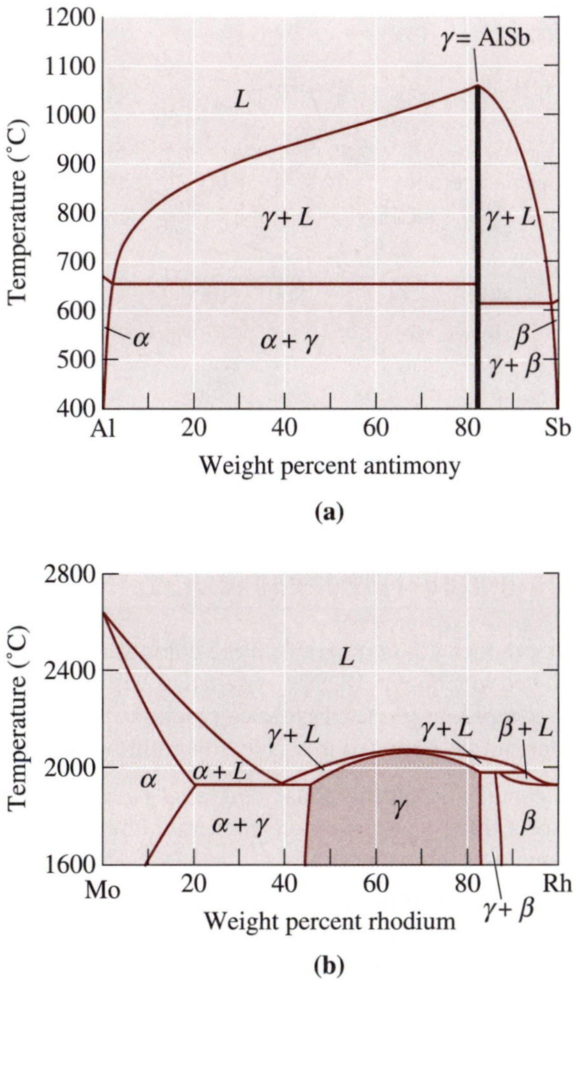

Figure 11-2
(a) The aluminum-antimony phase diagram includes a stoichiometric intermetallic compound γ.
(b) The molybdenum-rhodium phase diagram includes a nonstoichiometric intermetallic compound γ.

molybdenum-rhodium atom ratio is not fixed, the γ phase can contain from 45 wt% to 83 wt% Rh at 1600°C. Precipitation of the nonstoichiometric intermetallic copper aluminide $CuAl_2$ causes strengthening in a number of important aluminum alloys.

Properties and Applications of Intermetallics Intermetallics such as Ti_3Al and Ni_3Al maintain their strength and even develop usable ductility at elevated temperatures (Figure 11-3). Lower ductility, though, has impeded further development of these materials. It has been shown that the addition of small levels of boron (B) (up to 0.2%) can enhance the ductility of polycrystalline Ni_3Al. Environmental effects also probably play a role in limiting the ductility levels in intermetallics. Enhanced ductility levels could make it possible for intermetallics to be used in many high temperature and load-bearing applications. Ordered compounds of NiAl and Ni_3Al are also candidates for supersonic aircraft, jet engines, and high-speed commercial aircraft. Not all applications of intermetallics are structural. Intermetallics based on silicon (e.g., platinum silicide) play a useful role in microelectronics and certain intermetallics such as Nb_3Sn are useful as superconductors.

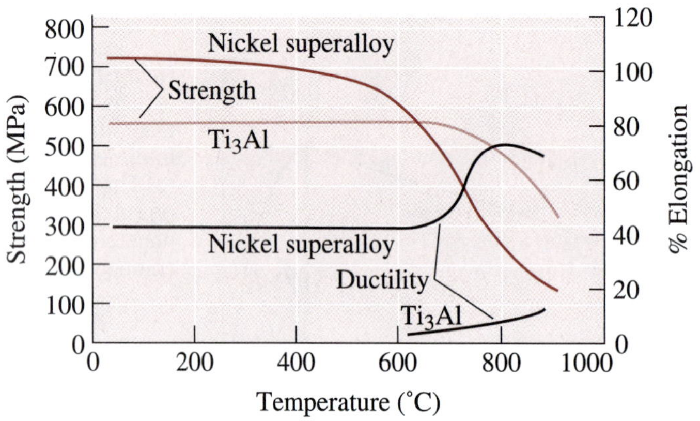

Figure 11-3 The strength and ductility of the intermetallic compound Ti_3Al compared with that of a conventional nickel superalloy. The Ti_3Al maintains its strength to higher temperatures than does the nickel superalloy.

The titanium aluminides, TiAl (also called the gamma (γ) alloy) and Ti_3Al (the α_2 alloy) are being considered for a variety of applications, including gas turbine engines.

11-3 Phase Diagrams Containing Three-Phase Reactions

Many binary systems produce phase diagrams more complicated than the isomorphous phase diagrams discussed in Chapter 10. The systems we will discuss here contain reactions that involve three separate phases. Five such reactions are defined in Figure 11-4. Each of these reactions can be identified in a phase diagram by the following procedure:

1. Locate a horizontal line on the phase diagram. The horizontal line, which indicates the presence of a three-phase reaction, represents the temperature at which the reaction occurs under equilibrium conditions.

2. Locate three distinct points on the horizontal line: the two endpoints plus a third point, in between the two endpoints of the horizontal line. This third point represents the composition at which the three-phase reaction occurs. In Figure 11-4 the point in

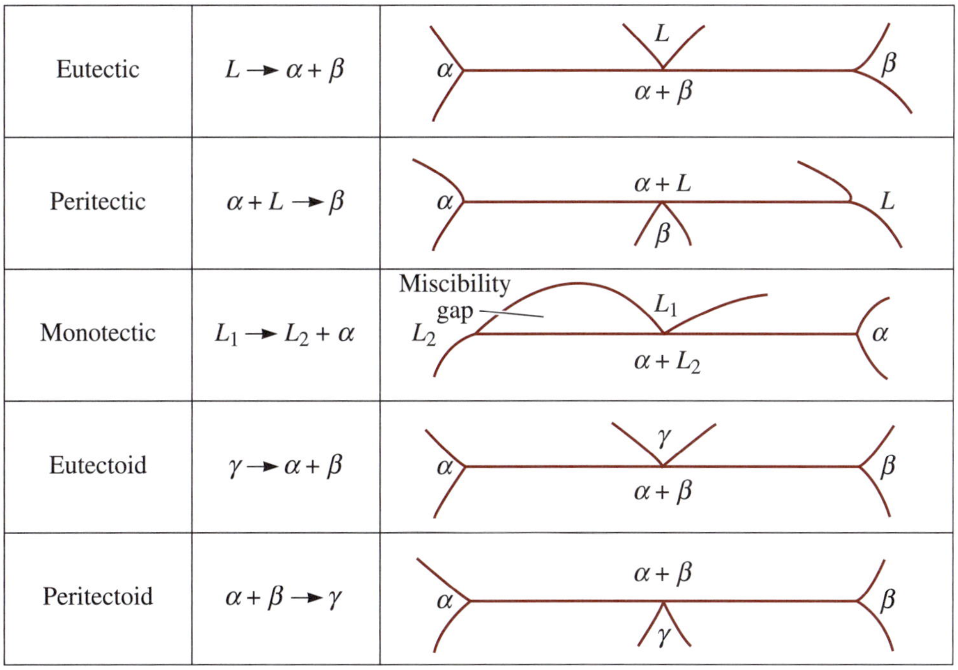

Figure 11-4 The five most important three-phase reactions in binary phase diagrams.

between has been shown at the center. However, on a real phase diagram this point is not necessarily at the center.

3. Look immediately above the in-between point and identify the phase or phases present; look immediately below the point in between the end points, and identify the phase or phases present. Then write in the reaction form the phase(s) above the point which are transforming to the phase(s) below the point. Compare this reaction with those in Figure 11-4 to identify the reaction.

EXAMPLE 11-1 *Identifying Three-Phase Reactions*

Consider the binary phase diagram in Figure 11-5. Identify the three-phase reactions that occur.

SOLUTION

We find horizontal lines at 1150°C, 920°C, 750°C, 450°C, and 300°C:

1150°C: The in-betwen point is at 15% B. $\delta + L$ are present above the point, γ is present below. The reaction is:

$$\delta + L \rightarrow \gamma, \quad \text{a } \textit{peritectic}$$

920°C: This reaction occurs at 40% B:

$$L_1 \rightarrow \gamma + L_2, \quad \text{a } \textit{monotectic}$$

750°C: This reaction occurs at 70% B:

$$L \rightarrow \gamma + \beta, \quad \text{a } \textit{eutectic}$$

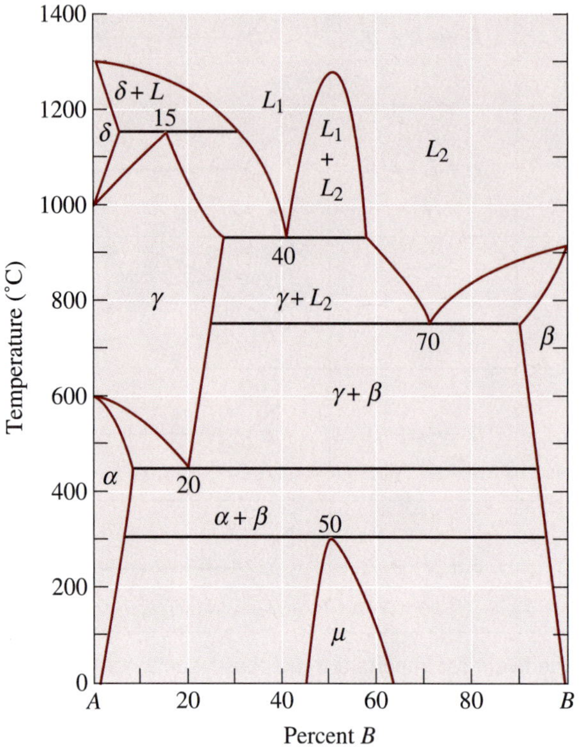

Figure 11-5 A hypothetical phase diagram (for Example 11-1).

450°C: This reaction occurs at 20% B:

$$\gamma \rightarrow \alpha + \beta, \quad \text{a } eutectoid$$

300°C: This reaction occurs at 50% B:

$$\alpha + \beta \rightarrow \mu \quad \text{or a } peritectoid$$

The **eutectic**, **peritectic**, and **monotectic** reactions are part of the solidification process. Alloys used for casting or soldering often take advantage of the low melting point and no freezing range of the eutectic reaction. The phase diagram of monotectic alloys contains a dome, or a **miscibility gap**, in which two liquid phases coexist. In the copper-lead system, the monotectic reaction produces tiny globules of dispersed lead, which improve the machinability of the copper alloy. Peritectic reactions lead to non-equilibrium solidification and segregation.

In many systems, there is **metastable miscibility gap**. In this case, the immiscibility dome extends into the sub-liquidus region. In some cases, the entire miscibility gap is metastable (i.e., the immiscibility dome is completely under the liquidus). These systems form such materials as Vycor™ and Pyrex® glasses, also known as phase-separated glasses. R. Roy was the first scientist to describe the underlying science for the formation of these glasses using the concept of a metastable miscibility gap existing below the liquidus.

The eutectoid and **peritectoid** reactions are completely solid-state reactions. The eutectoid reaction forms the basis for the heat treatment of several alloy systems, including steel (Chapter 12). The peritectoid reaction is extremely slow, often producing undesirable, nonequilibrium structures in alloys. As noted in Chapter 5, the rate of diffusion of atoms in solids is much smaller than that in liquids.

Each of these three-phase reactions occurs at a fixed temperature and composition. The Gibbs phase rule for a three-phase reaction is (at a constant pressure),

$$1 + C = F + P$$
$$F = 1 + C - P = 1 + 2 - 3 = 0, \tag{11-1}$$

since there are two components C in a binary phase diagram and three phases P are involved in the reaction. When the three phases are in equilibrium during the reaction, there are no degrees of freedom. As a result, these three phase reactions are known as invariant. The temperature and the composition of each phase involved in the three-phase reaction are fixed. Note that of these five reactions discussed here only eutectic and eutectoid reactions can lead to dispersion strengthening.

11-4 The Eutectic Phase Diagram

The lead-tin (Pb-Sn) system contains only a simple eutectic reaction (Figure 11-6). This alloy system is the basis for the most common alloys used for soldering. As mentioned before, because of the toxicity of Pb, there is an intense effort underway to replace lead in Pb-Sn solders with other alloys. We will continue to use a Pb-Sn system, though, as a convenient way to discuss the eutectic phase diagram. Let's examine four classes of alloys in this system.

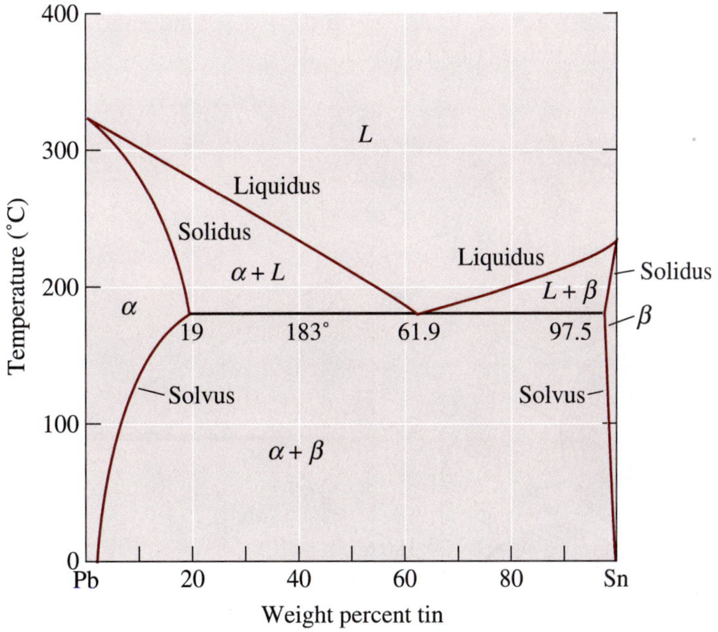

Figure 11-6 The lead-tin equilibrium phase diagram.

Solid Solution Alloys Alloys that contain 0 to 2% Sn behave exactly like the copper-nickel alloys; a single-phase solid solution α forms during solidification (Figure 11-7). These alloys are strengthened by solid-solution strengthening, by strain hardening, and by controlling the solidification process to refine the grain structure.

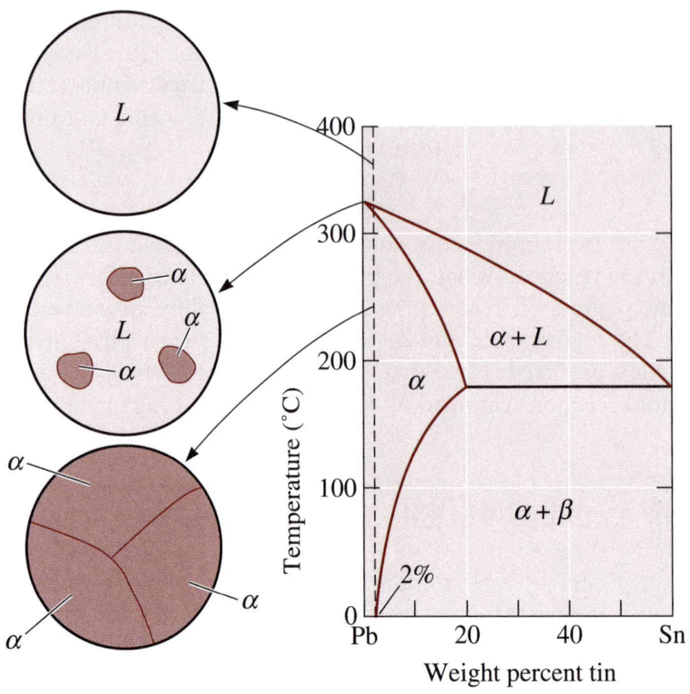

Figure 11-7 Solidification and micro-structure of a Pb-2% Sn alloy. The alloy is a single-phase solid solution.

Alloys That Exceed the Solubility Limit Alloys containing between 2% and 19% Sn also solidify to produce a single solid solution α. However, as the alloy continues to cool, a solid-state reaction occurs, permitting a second solid phase (β) to precipitate from the original α phase (Figure 11-8).

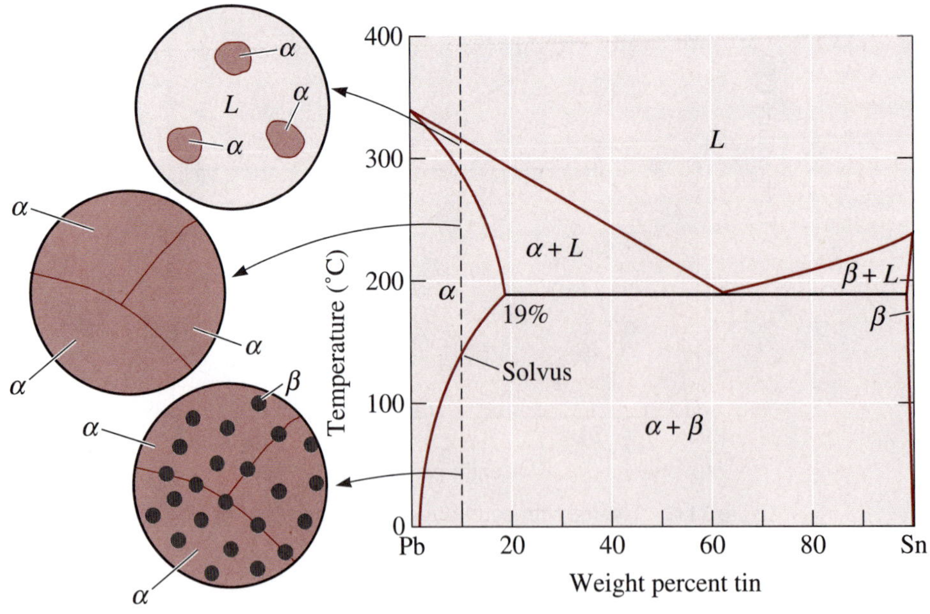

Figure 11-8 Solidification, precipitation, and microstructure of a Pb-10% Sn alloy. Some dispersion strengthening occurs as the β solid precipitates.

On this phase diagram, the α is a solid solution of tin in lead. However, the solubility of tin in the α solid solution is limited. At 0°C, only 2% Sn can dissolve in α. As the temperature increases, more tin dissolves into the lead until, at 183°C, the solubility of tin in lead has increased to 19% Sn. This is the maximum solubility of tin in lead. The solubility of tin in solid lead at any temperature is given by the **solvus** curve. Any alloy containing between 2% and 19% Sn cools past the solvus, the solubility limit is exceeded, and a small amount of β forms.

We control the properties of this type of alloy by several techniques, including solid-solution strengthening of the α portion of the structure, controlling the microstructure produced during solidification, and controlling the amount and characteristics of the β phase. These types of compositions, which form a single solid phase at high temperatures and two solid phases at lower temperatures, are suitable for age or precipitate hardening. In Chapter 12, we will learn how nonequilibrium processes are needed to make precipitation hardened alloys. A phase diagram (e.g., Figure 11-8) that shows a specific composition is known as an **isopleth**. Determination of reactions that occur upon the cooling of a particular composition is known as an **isoplethal study**. The following example illustrates how certain calculations related to the composition of phases and their relative concentrations can be performed.

EXAMPLE 11-2 *Phases in the Lead–Tin (Pb-Sn) Phase Diagram*

Determine (a) the solubility of tin in solid lead at 100°C, (b) the maximum solubility of lead in solid tin, (c) the amount of β that forms if a Pb-10% Sn alloy is cooled to 0°C, (d) the masses of tin contained in the α and β phases, and (e) the mass of lead contained in the α and β phases. Assume that the total mass of the Pb-10% Sn alloy is 100 grams. The phase diagram we need is shown in Figure 11-8. All percentages shown are weight %.

SOLUTION

(a) The 100°C temperature intersects the solvus curve at 5% Sn. The solubility of tin (Sn) in lead (Pb) at 100°C therefore is 5%.

(b) The maximum solubility of lead (Pb) in tin (Sn), which is found from the tin-rich side of the phase diagram, occurs at the eutectic temperature of 183°C and is 97.5% Sn.

(c) At 0°C, the 10% Sn alloy is in a $\alpha + \beta$ region of the phase diagram. By drawing a tie line at 0°C and applying the lever rule, we find that:

$$\%\,\beta = \frac{10 - 2}{100 - 2} \times 100 = 8.2\%$$

Note that the tie line intersects the solvus curve for solubility of Pb in Sn (on the right-hand side of the β-phase field) at a non-zero concentration of Sn. However, we can not read this accurately from the diagram; hence, we assume that the right-hand point for the tie line is 100% Sn. The % of α would be $(100 - \%\,\beta) = 91.8\%$. This means if we have 100 g of the 10% Sn alloy, it will consist of 8.2 g of the β phase and will consist of 91.8 g of the α phase.

(d) Note that 100 g of the alloy will consist of 10 g of Sn and 90 g of Pb. The Pb and Sn are distributed in two phases (i.e., α and β). The mass of Sn in the α phase = 2% Sn × 91.8 g of α phase = 0.02 × 91.8 g = 1.836 g. Since tin

(Sn) appears in both the α and β phases, the mass of Sn in the β phase will be $=(11 - 1.836)$ g $= 8.164$ g. Note that in this case the β phase at 0°C is nearly pure Sn.

(e) Let's now calculate the mass of lead in the two phases. The mass of Pb in the α phase will be equal to the mass of the α phase minus the mass of Sn in the α phase $= 91.8$ g $- 1.836$ g $= 89.964$ g. We could have also calculated this as:

$$\text{Mass of Pb in the } \alpha \text{ phase} = 98\% \text{ Sn} \times 91.8 \text{ g of } \alpha \text{ phase} = 0.98 \times 91.8 \text{ g}$$
$$= 89.964 \text{ g}$$

We know the total mass of the lead (90 g) and we also know the mass of lead in the α phase, therefore, the mass of Pb in the β phase $= 90 - 89.964 = 0.036$ g. This is consistent with what we said earlier (i.e., the β phase, in this case, is almost pure tin).

Eutectic Alloys The alloy containing 61.9% Sn has the eutectic composition (Figure 11-9). The word eutectic comes from the Greek word *eutectos* that means easily fused. Indeed, in a binary system showing one eutectic reaction, an alloy with a eutectic composition has the lowest melting temperature. This is the composition for which there is no freezing range (i.e., solidification of this alloy occurs at one temperature, 183°C in the Pb-Sn system). Above 183°C the alloy is all liquid and, therefore, must contain 61.9% Sn. After the liquid cools to 183°C, the eutectic reaction begins:

$$L_{61.9\%\text{ Sn}} \rightarrow \alpha_{19\%\text{ Sn}} + \beta_{97.5\%\text{ Sn}}$$

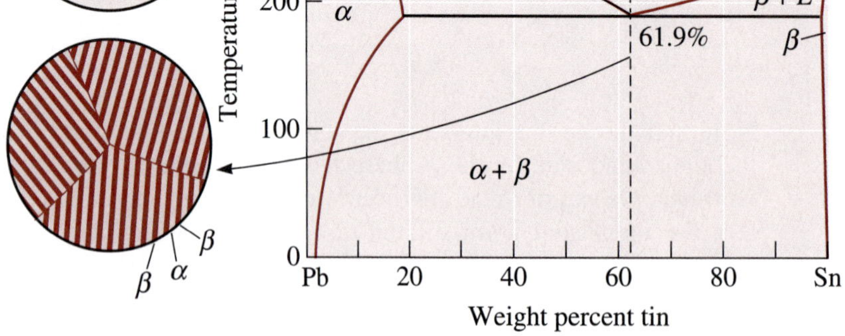

Figure 11-9 Solidification and microstructure of the eutectic alloy Pb-61.9% Sn.

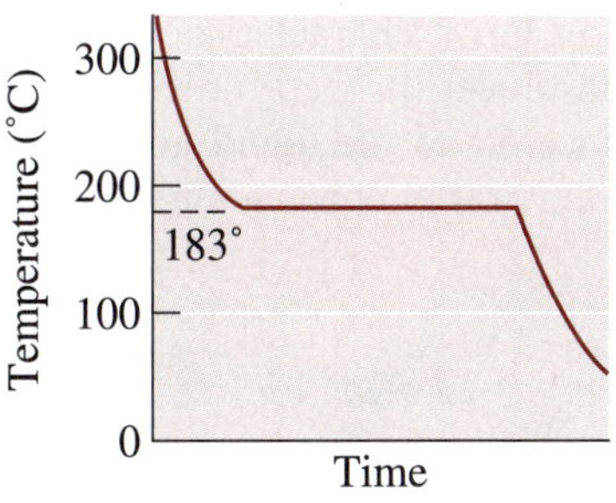

Figure 11-10
The cooling curve for an eutectic alloy is a simple thermal arrest, since eutectics freeze or melt at a single temperature.

Two solid solutions—α and β—are formed during the eutectic reaction. The compositions of the two solid solutions are given by the ends of the eutectic line.

During solidification, growth of the eutectic requires both removal of the latent heat of fusion and redistribution of the two different atom species by diffusion. Since solidification occurs completely at 183°C, the cooling curve (Figure 11-10) is similar to that of a pure metal; that is, a thermal arrest or plateau occurs at the eutectic temperature. In Chapter 9, we had stated that alloys solidify over a range of temperatures (between the liquidus and solidus) known as the freezing range. Eutectic compositions are an exception to this rule since they transform from a liquid to a solid at a constant temperature (i.e., the eutectic temperature).

As atoms are redistributed during eutectic solidification, a characteristic microstructure develops. In the lead-tin system, the solid α and β phases grow from the liquid in a **lamellar**, or plate-like, arrangement (Figure 11-11). The lamellar structure permits the lead and tin atoms to move through the liquid, in which diffusion is rapid, without having to move an appreciable distance. This lamellar structure is characteristic of numerous other eutectic systems.

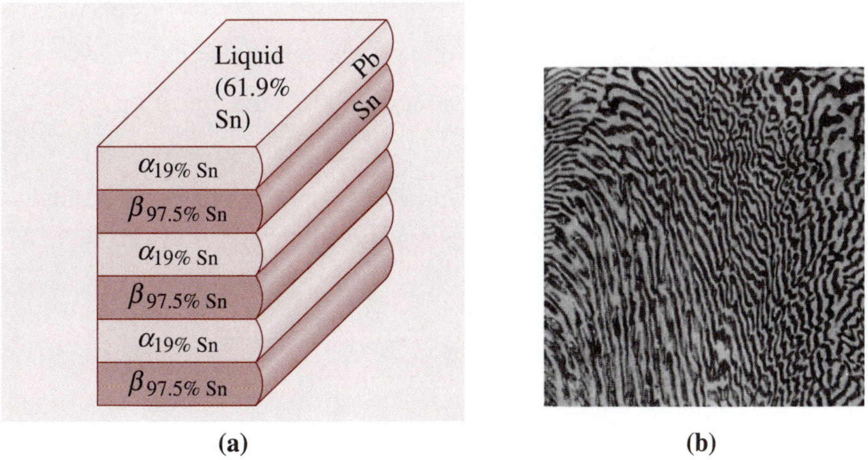

Figure 11-11 (a) Atom redistribution during lamellar growth of a lead-tin eutectic. Tin atoms from the liquid preferentially diffuse to the β plates, and lead atoms diffuse to the α plates. (b) Photomicrograph of the lead-tin eutectic microconstituent ($\times$400).

The product of the eutectic reaction has a characteristic arrangement of the two solid phases called the **eutectic microconstituent**. In the Pb-61.9% Sn alloy, 100% of the eutectic microconstituent is formed, since all of the liquid goes through the reaction. The following example shows how the amount of eutectic alloy can be calculated.

EXAMPLE 11-3 *Amount of Phases in the Eutectic Alloy*

(a) Determine the amount and composition of each phase in a lead-tin alloy of eutectic composition. (b) Calculate the mass of phases present. (c) Calculate the amount of lead and tin in each phase, assuming you have 200 g of the alloy.

SOLUTION

(a) The eutectic alloy contains 61.9% Sn. We work the lever law at a temperature just below the eutectic—say, at 182°C, since that is the temperature at which the eutectic reaction is just completed. The fulcrum of our lever is at 61.9% Sn. The ends of the tie line coincide approximately with the ends of the eutectic line.

$$\alpha\text{: (Pb} - 19\% \text{ Sn)} \quad \%\,\alpha = \frac{97.5 - 61.9}{97.5 - 19.0} \times 100 = 45.35\%$$

$$\beta\text{: (Pb} - 97.5\% \text{ Sn)} \quad \%\,\beta = \frac{61.9 - 19.0}{97.5 - 19.0} \times 100 = 54.65\%$$

Or we could state that the weight fraction of the α phase = 0.4535, that of the β phase is 0.5565.

A 200 g sample of the alloy would contain a total of $200 \times 0.6190 =$ 123.8 g Sn and a balance of 76.2 g lead. The total mass of lead and tin cannot change as a result of conservation of mass. What changes is the mass of lead and tin in the different phases.

(b) At a temperature of 182°C, just below the eutectic:

The mass of the α phase in 200 g of the alloy

= mass of the alloy × fraction of the α phase

$= 200 \text{ g} \times 0.4535 = 90.7 \text{ g}$

The amount of the β phase in 200 g of the alloy

= (mass of the alloy – mass of the α phase)

$= 200.0 \text{ g} - 90.7 \text{ g} = 109.3 \text{ g}$

We could have also written this as:

Amount of β phase in 200 g of the alloy

= mass of the alloy × fraction of the β phase

$= 200 \text{ g} \times 0.5465 = 109.3 \text{ g}$

Thus, at a temperature just below the eutectic (i.e., at 182°C), the alloy contains 109.3 g of the β phase and 90.7 g of the α phase.

(c) Now let's calculate the masses of lead and tin in the α and β phases:

$$\text{Mass of Pb in the } \alpha \text{ phase} = \text{mass of the } \alpha \text{ phase in 200 g} \times (\text{concentration of Pb in } \alpha)$$

$$\text{Mass of Pb in the } \alpha \text{ phase} = (90.7\text{ g}) \times (1 - 0.190) = 73.467\text{ g}$$

$$\text{Mass of Sn in the } \alpha \text{ phase} = \text{mass of the } \alpha \text{ phase} - \text{mass of Pb in the } \alpha \text{ phase}$$

$$\text{Mass of Sn in the } \alpha \text{ phase} = (90.7 - 73.467\text{ g}) = 17.233\text{ g}$$

$$\text{Mass of Pb in } \beta \text{ phase} = \text{mass of the } \beta \text{ phase in 200 g} \times (\text{wt. fraction Pb in } \beta)$$

$$\text{Mass of Pb in the } \beta \text{ phase} = (109.3\text{ g}) \times (1 - 0.975) = 2.73\text{ g}$$

$$\text{Mass of Sn in the } \beta \text{ phase} = \text{total mass of Sn} - \text{mass of Sn in the } \alpha \text{ phase}$$

$$= 123.8\text{ g} - 17.233\text{ g} = 106.57\text{ g}$$

Notice, that we could have obtained the same result by considering the total lead mass balance as follows:

$$\text{Total mass of lead in the alloy} = \text{mass of lead in the } \alpha \text{ phase} + \text{mass of lead in the } \beta \text{ phase}$$

$$76.2\text{ g} = 73.467\text{ g} + \text{mass of lead in the } \beta \text{ phase}$$

$$\text{Mass of lead in the } \beta \text{ phase} = 76.2 - 73.467\text{ g} = 2.73\text{ g}$$

This is same as what we calculated before. Figure 11-12 summarizes the various concentrations and masses.

This analysis confirms that most of the lead in the eutectic alloy gets concentrated in the α phase. Most of the tin gets concentrated in the β phase.

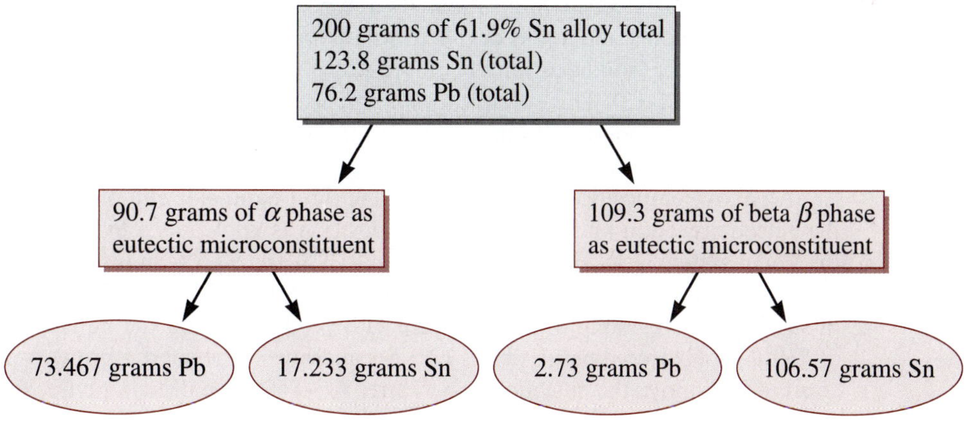

Figure 11-12 Summary of calculations (for Example 11-3).

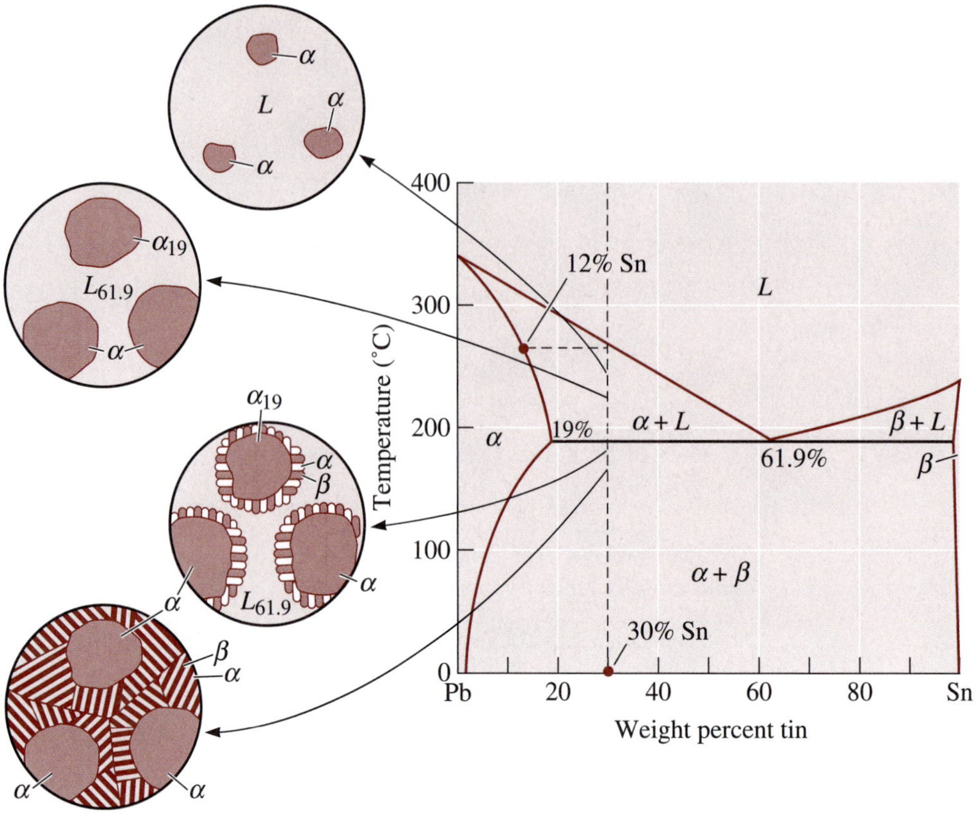

Figure 11-13 The solidification and microstructure of a hypoeutectic alloy (Pb-30% Sn).

Hypoeutectic and Hypereutectic Alloys A **hypoeutectic alloy** is an alloy whose composition will be between that of the left-hand-side end of the tie line defining the eutectic reaction and the eutectic composition. As a hypoeutectic alloy containing between 19% and 61.9% Sn cools, the liquid begins to solidify at the liquidus temperature, producing solid α. However, solidification is completed only after going through the eutectic reaction (Figure 11-13). This solidification sequence occurs for compositions in which the vertical line corresponding to the original composition of the alloy crosses both the liquidus and the eutectic.

An alloy composition between that of the right-hand-side end of the tie line defining the eutectic reaction and the eutectic composition is known as a **hypereutectic alloy**. In the Pb-Sn system, any composition between 61.9% and 97.5% Sn is hypereutectic.

Let's consider a hypoeutectic alloy containing Pb-30% Sn and follow the changes in structure during solidification (Figure 11-13). On reaching the liquidus temperature of 260°C, solid α containing about 12% Sn nucleates. The solid α grows until the alloy cools to just above the eutectic temperature. At 184°C, we draw a tie line and find that the solid α contains 19% Sn and the remaining liquid contains 61.9% Sn. We note that at 184°C, the liquid contains the eutectic composition! When the alloy is cooled below 183°C, all of the remaining liquid goes through the eutectic reaction and transforms to a lamellar mixture of α and β. The microstructure shown in Figure 11-14(a) results. Notice that the eutectic microconstituent surrounds the solid α that formed between the liquidus and eutectic temperatures. The eutectic microconstituent is continuous and the primary phase is dispersed between the colonies of the eutectic microconstituent.

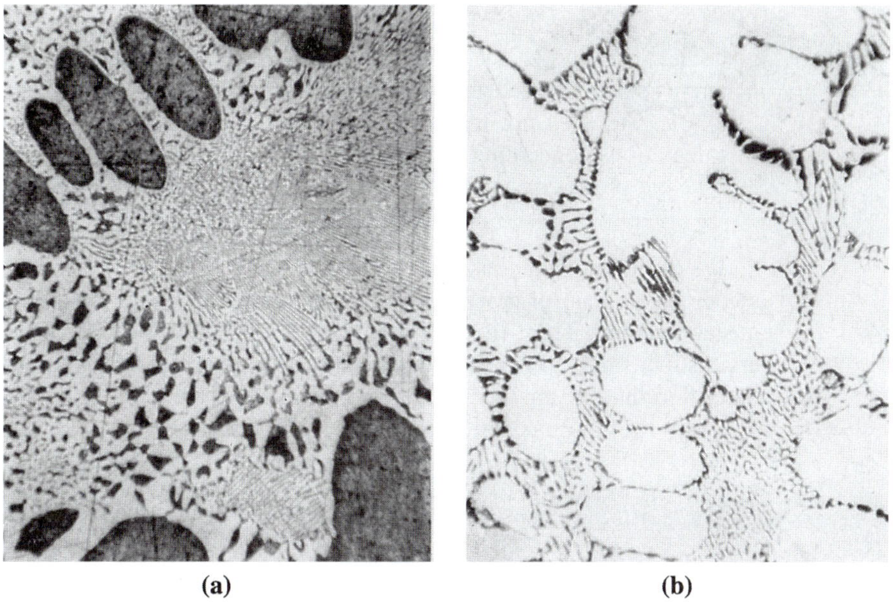

Figure 11-14 (a) A hypoeutectic lead-tin alloy. (b) A hypereutectic lead-tin alloy. The dark constituent is the lead-rich solid α, the light constituent is the tin-rich solid β, and the fine plate structure is the eutectic (×400).

EXAMPLE 11-4 *Determination of Phases and Amounts in a Pb-30% Sn Hypoeutectic Alloy*

For a Pb-30% Sn alloy, determine the phases present, their amounts, and their compositions at 300°C, 200°C, 184°C, 182°C, and 0°C.

SOLUTION

Temperature (°C)	Phases	Compositions	Amounts
300	L	L: 30% Sn	$L = 100\%$
200	$\alpha + L$	L: 55% Sn	$L = \frac{30-18}{55-18} \times 100 = 32\%$
		α: 18% Sn	$\alpha = \frac{55-30}{55-18} \times 100 = 68\%$
184	$\alpha + L$	L: 61.9% Sn	$L = \frac{30-19}{61.9-19} \times 100 = 26\%$
		α: 19% Sn	$\alpha = \frac{61.9-30}{61.9-19} \times 100 = 74\%$
182	$\alpha + \beta$	α: 19% Sn	$\alpha = \frac{97.5-30}{97.5-19} \times 100 = 86\%$
		β: 97.5% Sn	$\beta = \frac{30-19}{97.5-19} \times 100 = 14\%$
0	$\alpha + \beta$	α: 2% Sn	$\alpha = \frac{100-30}{100-2} \times 100 = 71\%$
		β: 100% Sn	$\beta = \frac{30-2}{100-2} \times 100 = 29\%$

Note that in these calculations, the fractions have been rounded off to the nearest percent. This can pose problems if we were to calculate masses of different phases, in that you may not be able to preserve mass balance. It is usually a good idea not to round these percentages if you are going to perform calculations concerning amounts of different phases or masses of elements in different phases.

We call the solid α phase that forms when the liquid cooled from the liquidus to the eutectic the **primary** or **proeutectic microconstituent**. This solid α did not take part in the eutectic reaction. Thus, the morphology and appearance of this proeutectic α phase is very distinct from that of the α phase that appears in the eutectic microconstituent. Often we find that the amounts and compositions of the microconstituents are of more use to us than the amounts and compositions of the phases.

EXAMPLE 11-5 *Microconstituent Amount and Composition for a Hypoeutectic Alloy*

Determine the amounts and compositions of each microconstituent in a Pb-30% Sn alloy immediately after the eutectic reaction has been completed.

SOLUTION

This is a hypoeutectic composition. Therefore, the microconstituents expected are primary α and eutectic. Note that we still have only two phases (α and β). We can determine the amounts and compositions of the microconstituents if we look at how they form. The *primary* α microconstituent is all of the solid α that forms before the alloy cools to the eutectic temperature; the eutectic microconstituent is all of the liquid that goes through the eutectic reaction. At a temperature just above the eutectic—say, 184°C—the amounts and compositions of the two phases are:

$$\alpha\text{: 19\% Sn} \quad \%\,\alpha = \frac{61.9 - 30}{61.9 - 19} \times 100 = 74\% = \%\text{ primary }\alpha$$

$$L\text{: 61.9\% Sn} \quad \%\,L = \frac{30 - 19}{61.9 - 19} \times 100 = 26\% = \%\text{ eutectic at 182°C}$$

Thus, the primary alpha microconstituent is obtained by determining the amount of α present at the temperature just above the eutectic. The amount of eutectic microconstituent at a temperature *just below* the eutectic (e.g., 182°C) is determined by calculating the amount of liquid *just above* the eutectic temperature (e.g., at 184°C), since all of this liquid of eutectic composition is transformed into the eutectic microconstituent. Note that at the eutectic temperature (183°C) the eutectic reaction is in progress (formation of the proeutectic α is complete); hence, the amount of the eutectic microconstituent at 183°C will change with time (starting at 0% and ending at 26% eutectic, in this case). Please be certain that you understand this example since many students tend to miss how this calculation is performed.

When the alloy cools below the eutectic to 182°C, all of the liquid at 184°C transforms to eutectic and the composition of the eutectic microconstituent is 61.9% Sn. The solid α phase present at 184°C remains unchanged after cooling to 182°C and is the primary microconstituent.

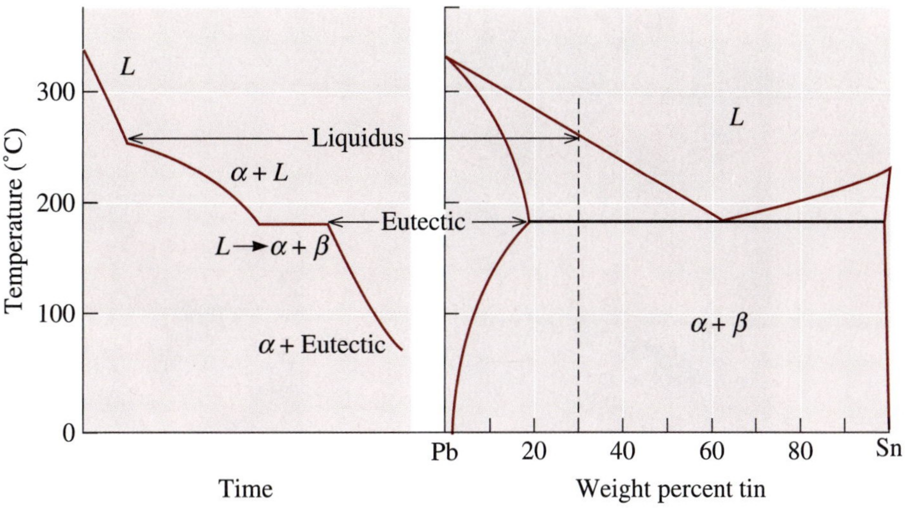

Figure 11-15 The cooling curve for a hypoeutectic Pb-30% Sn alloy.

The cooling curve for a hypoeutectic alloy is a composite of those for solid-solution alloys and "straight" eutectic alloys (Figure 11-15). A change in slope occurs at the liquidus as primary α begins to form. Evolution of the latent heat of fusion slows the cooling rate as the solid α grows. When the alloy cools to the eutectic temperature, a thermal arrest is produced as the eutectic reaction proceeds at 183°C. The solidification sequence is similar in a hypereutectic alloy, giving the microstructure shown in Figure 11-14(b).

11-5 Strength of Eutectic Alloys

Each phase in the eutectic alloy is, to some degree, solid-solution strengthened. In the lead-tin system, α, which is a solid solution of tin in lead, is stronger than pure lead (Chapter 10). Some eutectic alloys can be strengthened by cold working. We also control grain size by adding appropriate inoculants or grain refiners during solidification. Finally, we can influence the properties by controlling the amount and microstructure of the eutectic.

Eutectic Colony Size Eutectic colonies, or grains, each nucleate and grow independently. Within each colony, the orientation of the **lamellae** in the eutectic microconstituent is identical. The orientation changes on crossing a colony boundary [Figure 11-16(a)]. We can refine the eutectic colonies and improve the strength of the eutectic alloy by inoculation (Chapter 9).

Interlamellar Spacing The **interlamellar spacing** of a eutectic is the distance from the center of one α lamella to the center of the next α lamella [Figure 11-16(b)]. A small interlamellar spacing indicates that the amount of α–β interface area is large. A small interlamellar spacing therefore increases the strength of the eutectic.

The interlamellar spacing (λ) is determined primarily by the growth rate of the eutectic,

$$\lambda = cR^{-1/2} \tag{11-2}$$

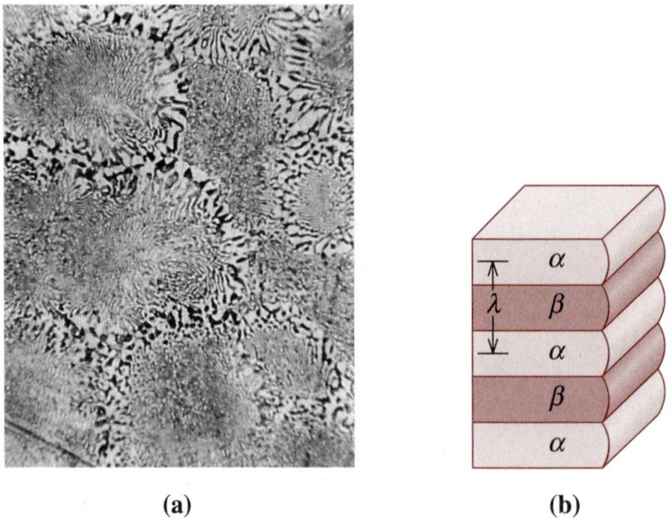

Figure 11-16
(a) Colonies in the lead-tin eutectic (×300).
(b) The interlamellar spacing in a eutectic microstructure.

where R is the growth rate (cm/s) and c is a constant. The interlamellar spacing for the lead-tin eutectic is shown in Figure 11-17. We can increase the growth rate R, and consequently reduce the interlamellar spacing, by increasing the cooling rate or reducing the solidification time.

Amount of Eutectic We also control the properties by the relative amounts of the **primary microconstituent** and the eutectic. In the lead-tin system, the amount of the eutectic microconstituent changes from 0% to 100% when the tin content increases from 19% to 61.9%. With increasing amounts of the stronger eutectic microconstituent, the strength of the alloy increases (Figure 11-18). Similarly, when we increase the lead added to tin from 2.5% to 38.1% Pb, the amount of primary β in the hypereutectic alloy decreases, the amount of the strong eutectic increases, and the strength increases. When both individual phases have about the same strength, the eutectic alloy is expected to have the highest strength due to effective dispersion strengthening.

Microstructure of the Eutectic Not all eutectics give a lamellar structure. The shapes of the two phases in the microconstituent are influenced by the cooling rate, the presence of impurity elements, and the nature of the alloy.

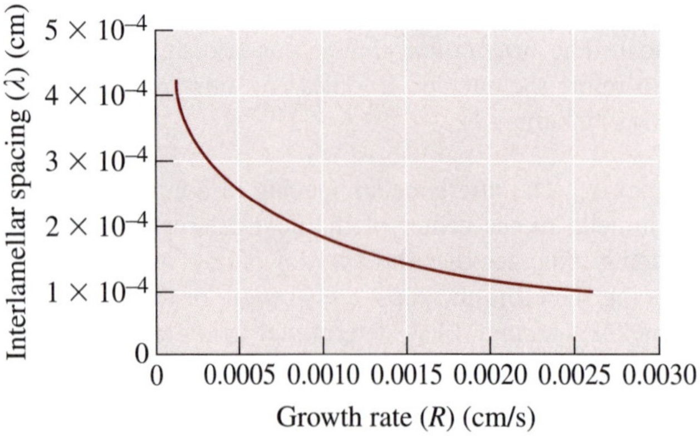

Figure 11-17
The effect of growth rate (R) on the interlamellar spacing (λ) in the lead-tin eutectic.

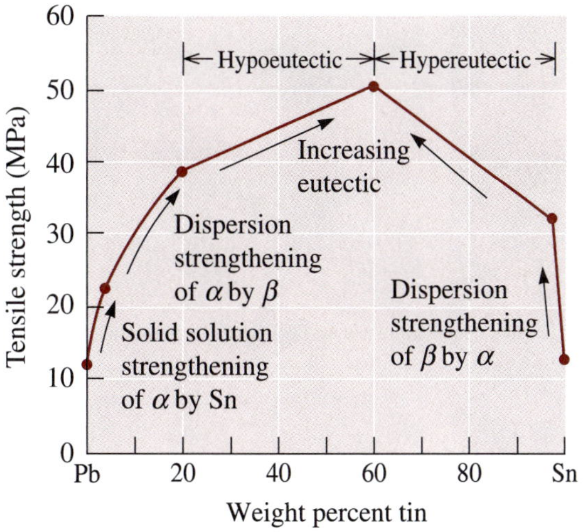

Figure 11-18
The effect of the composition and strengthening mechanism on the tensile strength of lead-tin alloys.

The aluminum-silicon eutectic phase diagram (Figure 11-19) forms the basis for a number of important commercial alloys. However, the silicon portion of the eutectic grows as thin, flat plates that appear needle-like in a photomicrograph [Figure 11-20(a)]. The brittle silicon platelets concentrate stresses and reduce ductility and toughness.

The eutectic microstructure in aluminum-silicon alloys is altered by a process known as modification. **Modification** causes the silicon phase to grow as thin, interconnected rods between aluminum dendrites [Figure 11-20(b)], improving both tensile strength and % elongation. In two dimensions, the modified silicon appears to be composed of tiny, round particles. Rapidly cooled alloys, such as those in die casting, are naturally modified during solidification. At slower cooling rates, however, about 0.02% Na or 0.01% Sr must be added to cause modification.

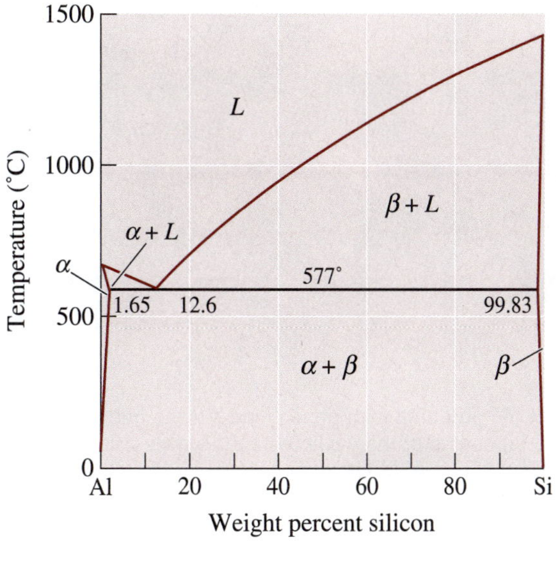

Figure 11-19
The aluminum-silicon phase diagram.

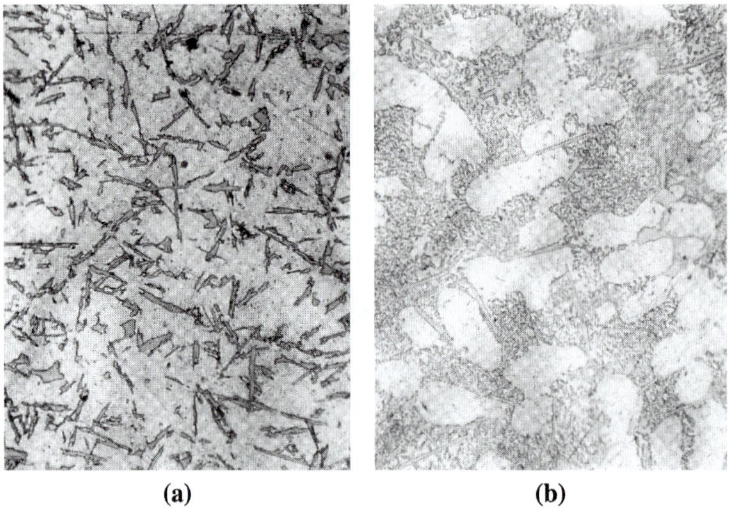

Figure 11-20 Typical eutectic microstructures: (a) needle-like silicon plates in the aluminum-silicon eutectic (×100), and (b) rounded silicon rods in the modified aluminum-silicon eutectic (×100).

The shape of the primary phase is also important. Often the primary phase grows in a dendritic manner; decreasing the secondary dendrite arm spacing of the primary phase may improve the properties of the alloy. However, in hypereutectic aluminum-silicon alloys, coarse β is the primary phase [Figure 11-21(a)]. Because β phase is hard, the hypereutectic alloys are wear-resistant and are used to produce automotive engine parts. However, the coarse β causes poor machinability and gravity segregation (where the primary β floats to the surface of the casting during freezing). Addition of 0.05% phosphorus (P) encourages nucleation of primary silicon, refines its size, and minimizes its deleterious qualities [Figure 11-21(b)]. The two examples following show how eutectic compositions can be designed to achieve certain levels of mechanical properties.

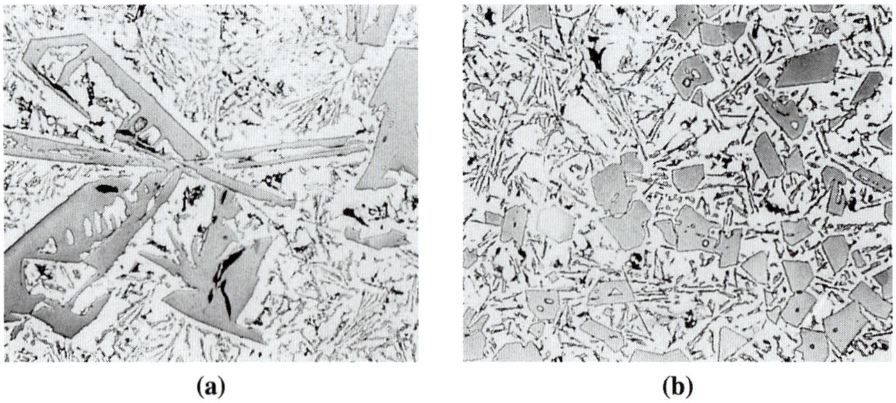

Figure 11-21 The effect of hardening with phosphorus on the microstructure of hypereutectic aluminum-silicon alloys: (a) coarse primary silicon, and (b) fine primary silicon, as refined by phosphorus addition (×75). (*From* ASM Handbook, *Vol. 7, (1972), ASM International, Materials Park, OH 44073.*)

EXAMPLE 11-6 Design of Materials for a Wiping Solder

One way to repair dents in a metal is to wipe a partly liquid-partly solid material into the dent, then allow this filler material to solidify. For our application, the wiping material should have the following specifications: (1) a melting temperature below 230°C, (2) a tensile strength in excess of 41 MPa, (3) be 60% to 70% liquid during application, and (4) the lowest possible cost. Design an alloy and repair procedure that will meet these specifications.

SOLUTION

Let's see if one of the Pb-Sn alloys will satisfy these conditions. First, the alloy must contain more than 40% Sn in order to have a melting temperature below 230°C (Figure 11-6). This low temperature will make it easier for the person doing the repairs to apply the filler.

Second, Figure 11-18 indicates that the tin content must lie between 23% and 80% to achieve the required 41 MPa tensile strength. In combination with the first requirement, any alloy containing between 40 and 80% Sn will be satisfactory.

Third, the cost of tin is about \$5500/ton whereas that of lead is \$550/ton. Thus, an alloy of Pb-40% Sn might be the most economical choice. There are other considerations, as well, such as: What is the geometry? Can the alloy flow well under that geometry (i.e., the viscosity of the molten metal)?

Finally, the filler material must be at the correct temperature in order to be 60% to 70% liquid. As the calculations below show, the temperature must be between 200°C and 210°C:

$$\% L_{200} = \frac{40 - 18}{55 - 18} \times 100 = 60\%$$

$$\% L_{210} = \frac{40 - 17}{50 - 17} \times 100 = 70\%$$

Our recommendation, therefore, is to use a Pb-40% Sn alloy applied at 205°C, a temperature at which there will be 65% liquid and 35% primary α. As mentioned before, we should also pay attention to the toxicity of lead and any legal liabilities the use of such materials may cause. A number of new lead free solders have been developed.

EXAMPLE 11-7 Design of a Wear-Resistant Part

Design a lightweight, cylindrical component that will provide excellent wear-resistance at the inner wall, yet still have reasonable ductility and toughness overall. Such a product might be used as a cylinder liner in an automotive engine.

SOLUTION

Many wear-resistant parts are produced from steels, which have a relatively high density, but the hypereutectic Al-Si alloys containing primary β may provide the wear-resistance that we wish at one-third the weight of the steel.

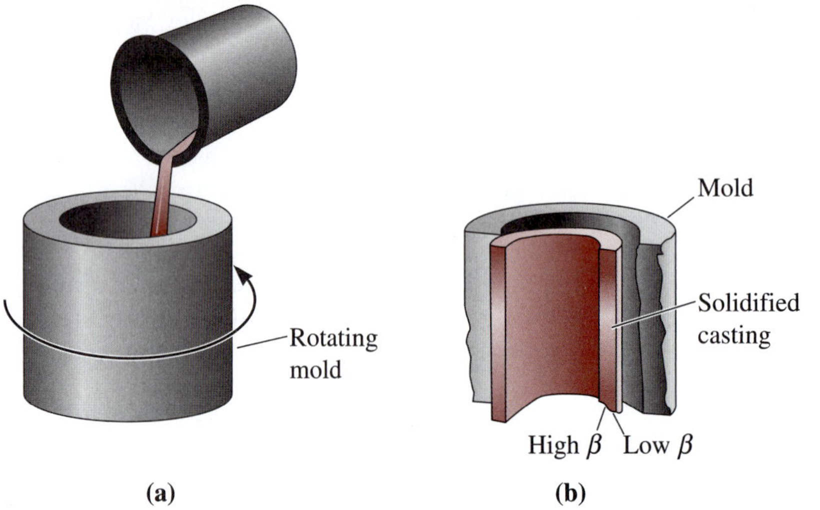

Figure 11-22 Centrifugal casting of a hypereutectic Al-Si alloy: (a) Liquid alloy is poured into a rotating mold, and (b) the solidified casting is hypereutectic at the inner diameter and eutectic at the outer diameter (for Example 11-7).

Since the part to be produced is cylindrical in shape, centrifugal casting (Figure 11-22) might be a unique method for producing it. In centrifugal casting, liquid metal is poured into a rotating mold and the centrifugal force produces a hollow shape. In addition, material that has a high density is spun to the outside wall of the casting, while material that has a lower density than the liquid migrates to the inner wall.

When we centrifugally cast a hypereutectic Al-Si alloy, primary β nucleates and grows. The density of the β phase (if we assume it to be same as that of pure Si) is, according to Appendix A, 2.33 g/cm^3, compared with a density near 2.7 g/cm^3 for aluminum. As the primary β particles precipitate from the liquid, they are spun to the inner surface. The result is a casting that is composed of eutectic microconstituent (with reasonable ductility) at the outer wall and a hypereutectic composition, containing large amounts of primary β, at the inner wall.

A typical alloy used to produce aluminum engine components is Al-17% Si. From Figure 11-19, the total amount of primary β that can form is calculated at 578°C, just above the eutectic temperature:

$$\%\text{ Primary } \beta = \frac{17 - 12.6}{99.83 - 12.6} \times 100 = 5.0\%$$

Although only 5.0% primary β is expected to form, the centrifugal action can double or triple the amount of β at the inner wall of the casting.

11-6 Eutectics and Materials Processing

Several manufacturing processes take advantage of the low melting temperature associated with the eutectic reaction. The Pb-Sn alloys are the basis for a series of alloys used to produce filler materials for soldering (Chapter 9). If, for example, we wish to join copper pipe, individual segments can be joined by introducing the low-melting-point eutectic Pb-Sn alloy into the joint [Figure 11-23(a)]. The copper is heated just above the eutectic temperature. The heated copper melts the Pb-Sn alloy, which is then drawn into the thin gap by capillary action. When the Pb-Sn alloy cools and solidifies, the copper is joined. The prospects of corrosion of such pipes and the introduction of lead (Pb) into water must also be factored in.

Many casting alloys are also based on eutectic alloys. Liquid can be melted and solidified into a mold at low and at constant temperatures, reducing energy costs involved in melting, minimizing casting defects such as gas porosity, and preventing liquid metal-mold reactions. Cast iron and many aluminum alloys are based on eutectic compositions.

Although most of this discussion has been centered around metallic materials, it is important to recognize that the eutectics are very important in many ceramic systems as well (Chapter 15). Formation of eutectics played a role in the successful formation of glass-like materials known as the Egyptian faience. The sands of the Nile River Valley contained appreciable amounts of limestone ($CaCO_3$). Plant ash contains considerable amounts of potassium oxide and sodium oxide and is used to cause the sand to melt at lower temperatures by the formation of the eutectics.

Silica and alumina are the most widely used ceramic materials. Figure 11-23(b) shows a phase diagram for the Al_2O_3-SiO_2. Notice the eutectic at ~1587°C. The dashed lines on this diagram show metastable extensions of the liquidus and **metastable miscibility gaps**. As mentioned before, the existence of these gaps makes it possible to make technologically useful products such as the Vycor™ and the Pyrex® glasses. A Vycor™ glass is made by first melting (approximately at 1500°C) silica (63%), boron oxide (27%), sodium oxide (7%), and alumina (3%). The glass is then formed into desired shapes. During glass formation, the glass has phase separated (because of the metastable miscibility gap) into boron oxide rich and silica rich regions. The boron oxide rich regions are dissolved using an acid. The porous object is sintered to form Vycor™ glass that contains 95% silica, 4% boron oxide, and 1% sodium oxide. It would be very difficult to achieve a high silica glass such as this without resorting to the technique described above. Pyrex® contains about 80% silica, 13% boron oxide, 4% sodium oxide, 2% alumina. These are used widely in making laboratory ware (i.e., beakers, etc.) and household products.

Figure 11-23(c) shows a binary phase diagram for the CaO-SiO_2 system. Compositions known as the E-glass or S-glass are used to make the fibers that go into fiber-reinforced plastics. These glasses are made by melting silica sand, limestone, boric acid at about 1260°C. The glass is then drawn into fibers. The E-glass (the letter "E" stands for "electrical", as the glass was originally made for electrical insulation) contains approximately 52–56 wt.% silica, 12–16% Al_2O_3, 5–10% B_2O_3, 0–5% MgO, 0–2% Na_2O, 0–2% K_2O. The "S"-glass (the letter "S" represents "strength") contains approximately 65 wt.% silica, 12–25% Al_2O_3, 10% MgO, 0–2% Na_2O, 0–2% K_2O.

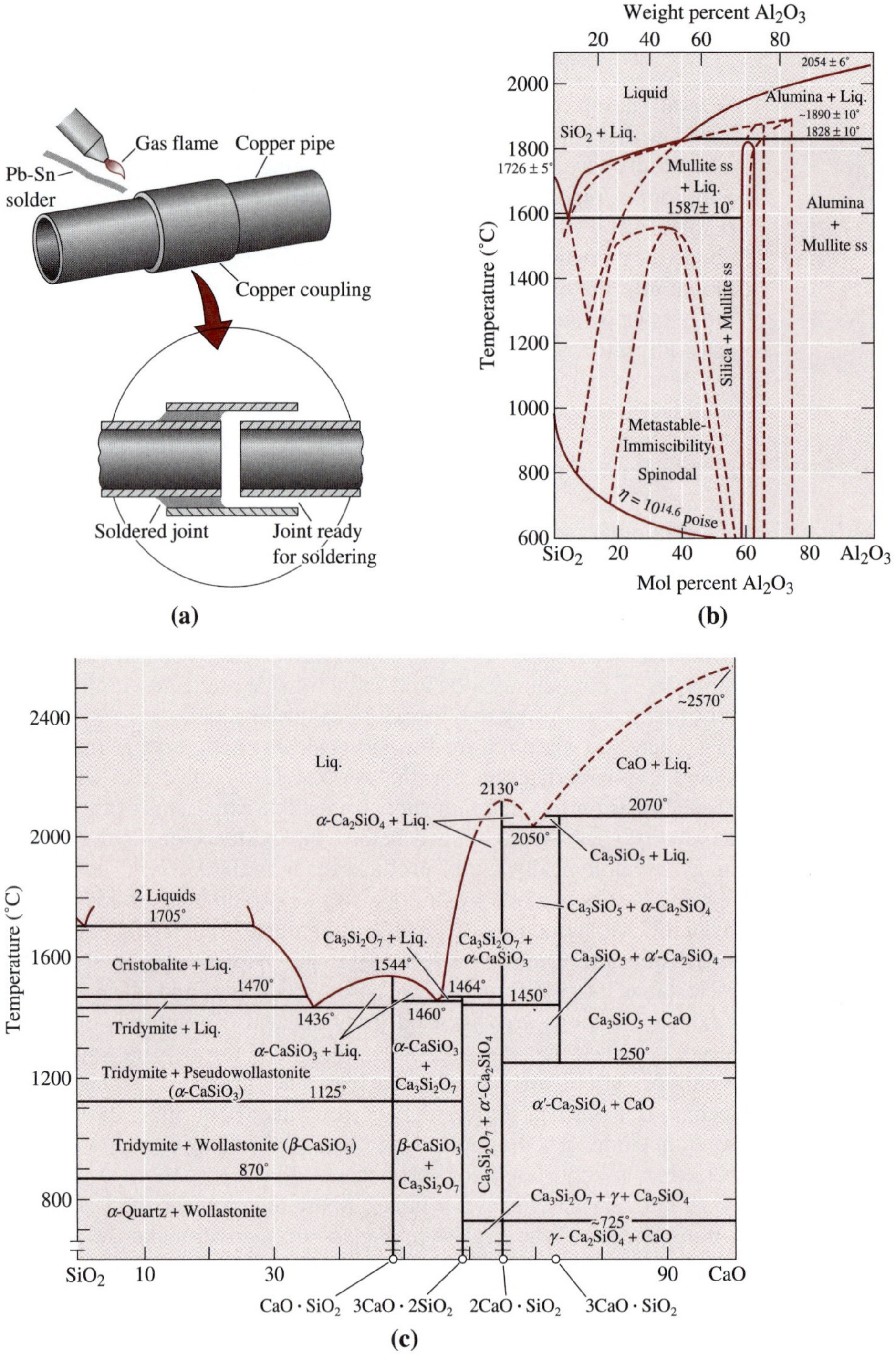

Figure 11-23 (a) A Pb-Sn eutectic alloy is often used during soldering to assemble parts. A heat source, such as a gas flame, heats both the parts and the filler material. The filler is drawn into the joint and solidifies. (b) A phase diagram for Al_2O_3-SiO_2. (Adapted from *Introduction to Phase Equilibria in Ceramics*, by Bergeron, C.G. and Risbud, S.H., The American Ceramic Society, Inc., 1984, page 44.) (c) A phase diagram for the CaO-SiO_2 system. (*Source: Adapted from* Introduction to Phase Equilibria, *by C.G. Bergeron and S.H. Risbud, pp. 44 and 45, Figs. 3-36 and 3-37. Copyright © 1984 American Ceramic Society.*)

11-7 Nonequilibrium Freezing in the Eutectic System

Suppose we have an alloy, such as Pb-15% Sn, that ordinarily solidifies as a solid solution alloy. The last liquid should freeze near 230°C, well above the eutectic. However, if the alloy cools too quickly, a nonequilibrium solidus curve is produced (Figure 11-24). The primary α continues to grow until, just above 183°C, the remaining nonequilibrium liquid contains 61.9% Sn. This liquid then transforms to the eutectic microconstituent, surrounding the primary α. For the conditions shown in Figure 11-24, the amount of nonequilibrium eutectic is:

$$\% \text{ eutectic} = \frac{15 - 10}{61.9 - 10} \times 100 = 9.6\%$$

When heat treating an alloy such as Pb-15% Sn, we must keep the maximum temperature below the eutectic temperature of 183°C to prevent hot shortness or partial melting. This concept is very important in the precipitation, or age, hardening of metallic alloys such as those in the Al-Cu system.

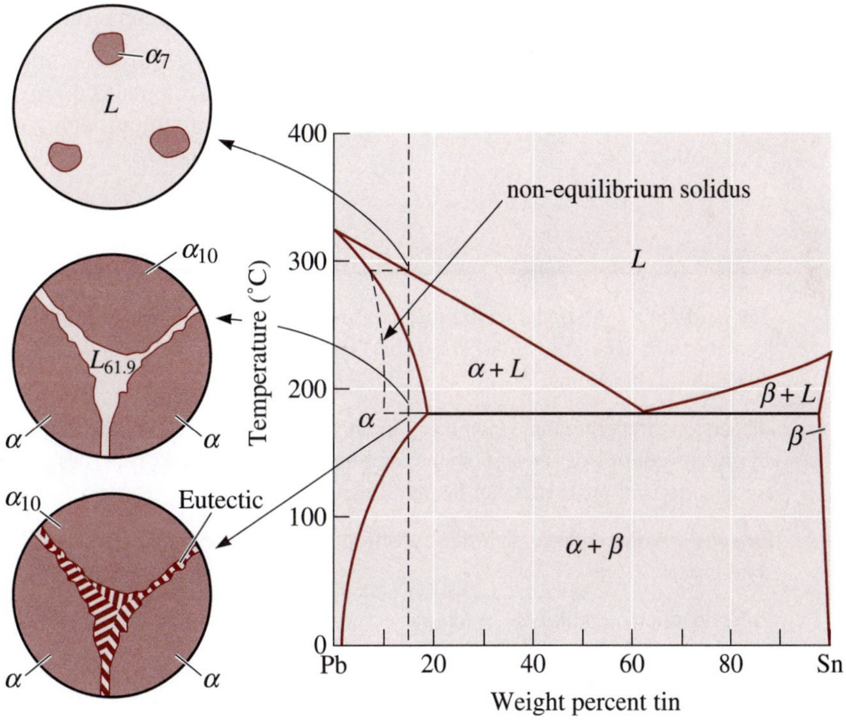

Figure 11-24 Nonequilibrium solidification and microstructure of a Pb-15% Sn alloy. A nonequilibrium eutectic microconstituent can form if the solidification is too rapid.

SUMMARY

- By producing a material containing two or more phases, dispersion strengthening is obtained. In metallic materials, the boundary between the phases impedes the movement of dislocations and improves strength. Introduction of multiple phases may provide other benefits, including improvement of the fracture toughness of ceramics and polymers.
- For optimum dispersion strengthening, particularly in metallic materials, a large number of small, hard, discontinuous dispersed phase particles should form in a soft, ductile matrix to provide the most effective obstacles to dislocations. Round dispersed phase particles minimize stress concentrations, and the final properties of the alloy can be controlled by the relative amounts of these and the matrix.
- Intermetallic compounds, which normally are strong but brittle, are frequently introduced as dispersed phases in the microstructure.
- Phase diagrams for materials containing multiple phases normally contain one or more three-phase reactions:
    - The eutectic reaction permits liquid to solidify as an intimate mixture of two phases. By controlling the solidification process, we can achieve a wide range of properties. Some of the factors that can be controlled include: the grain size or secondary dendrite arm spacings of primary microconstituents, the colony size of the eutectic microconstituent, the interlamellar spacing within the eutectic microconstituent, the microstructure, or shape, of the phases within the eutectic microconstituent, and the amount of the eutectic microconstituent that forms.
    - The eutectoid reaction causes a solid to transform to a mixture of two other solids. As shown in the next chapter, heat treatments to control the eutectoid reaction provide an excellent basis for dispersion strengthening of many steels and other alloys.

GLOSSARY

Age hardening A strengthening mechanism that relies on a sequence of solid-state phase transformations in generating a dispersion of ultrafine particles of a second phase. This is the same as precipitation hardening.

Dispersion strengthening Increasing the strength of a material by forming more than one phase. By proper control of the size, shape, amount, and individual properties of the phases, excellent combinations of properties can be obtained.

Eutectic A three-phase, invariant reaction in which one liquid phase solidifies to produce two solid phases.

Eutectic microconstituent A characteristic mixture of two phases formed as a result of the eutectic reaction.

Eutectoid A three-phase, invariant reaction in which one solid phase transforms to two different solid phases.

Hyper- A prefix indicating that the composition of an alloy is more than the composition at which a three-phase reaction occurs.

Hypereutectic alloys An alloy composition between that of the right-hand-side end of the tie line defining the eutectic reaction and the eutectic composition.

Hypo- A prefix indicating that the composition of an alloy is less than the composition at which a three-phase reaction occurs.

Hypoeutectic alloy An alloy composition between that of the left-hand-side end of the tie line defining the eutectic reaction and the eutectic composition.

Interlamellar spacing The distance between the center of a lamella or plate of one phase and the center of the adjoining lamella or plate of the same phase.

Intermediate solid solution A nonstoichiometric intermetallic compound displaying a range of compositions.

Intermetallic compound A compound formed of two or more metals that has its own unique composition, structure, and properties.

Interphase interface The boundary between two phases in a microstructure. In metallic materials, this boundary resists dislocation motion and provides dispersion strengthening and precipitation hardening.

Isopleth A line on a phase diagram that shows constant chemical composition.

Isoplethal study Determination of reactions and microstructural changes that are expected while studying a particular chemical composition in a system.

Lamella A thin plate of a phase that forms during certain three-phase reactions, such as the eutectic or eutectoid.

Matrix The continuous solid phase in a complex microstructure. Solid dispersed phase particles may form within the matrix.

Metastable miscibility gap A miscibility gap that extends below the liquidus or exists completely below the liquidus. Two liquids that are immiscible continue to exist as liquids and remain unmixed. These systems form the basis for Vycor™ and Pyrex® glasses.

Microconstituent A phase or mixture of phases in an alloy that has a distinct appearance. Frequently, we describe a microstructure in terms of the microconstituents rather than the actual phases.

Miscibility gap A region in a phase diagram in which two phases, with essentially the same structure, do not mix, or have no solubility in one another.

Modification Addition of alloying elements, such as sodium or strontium, which change the microstructure of the eutectic microconstituent in aluminum-silicon alloys.

Monotectic A three-phase reaction in which one liquid transforms to a solid and a second liquid on cooling.

Nonstoichiometric intermetallic compound A phase formed by the combination of two components into a compound having a structure and properties different from either component. The nonstoichiometric compound has a variable ratio of the components present in the compound (see also Intermediate solid solution).

Peritectic A three-phase reaction in which a solid and a liquid combine to produce a second solid on cooling.

Peritectoid A three-phase reaction in which two solids combine to form a third solid on cooling.

Precipitate A solid phase that forms from the original matrix phase when the solubility limit is exceeded.

Precipitation hardening A strengthening mechanism that relies on a sequence of solid-state phase transformations in generating a dispersion of ultrafine precipitates of a second phase (Chapter 12). This is the same as age hardening. It is a form of dispersion strengthening.

Primary microconstituent The microconstituent that forms before the start of a three-phase reaction.

Solvus A solubility curve that separates a single-solid phase region from a two-solid phase region in the phase diagram.

Stoichiometric intermetallic compound A phase formed by the combination of two components into a compound having a structure and properties different from either component. The stoichiometric intermetallic compound has a fixed ratio of the components present in the compound.

PROBLEMS

Section 11-1 Principles and Examples of Dispersion Strengthening

11-1 What are the requirements of a matrix and precipitate for dispersion strengthening to be effective?

Section 11-2 Intermetallic Compounds

11-2 What is an intermetallic compound? How is it different from other compounds? For example, other than the obvious difference in composition how is TiAl different from, for example, Al_2O_3?

11-3 Explain clearly the two different ways in which intermetallic compounds can be used.

11-4 What are some of the major problems in the utilization of intermetallics for high-temperature applications?

Section 11-3 Phase Diagrams Containing Three-Phase Reactions

11-5 Define the terms eutectic, eutectoid, peritectic, peritectoid, and monotectic reactions.

11-6 What is an invariant reaction? Show that for a two-component system the number of degrees of freedom for an invariant reaction is zero.

11-7 A hypothetical phase diagram is shown in Figure 11-25.

(a) Are there any intermetallic compounds present? If so, identify them and determine whether they are stoichiometric or nonstoichiometric.

(b) Identify the solid solutions present in the system. Is either material A or B allotropic? Explain.

(c) Identify the three-phase reactions by writing down the temperature, the reaction in equation form, the composition of each phase in the reaction, and the name of the reaction.

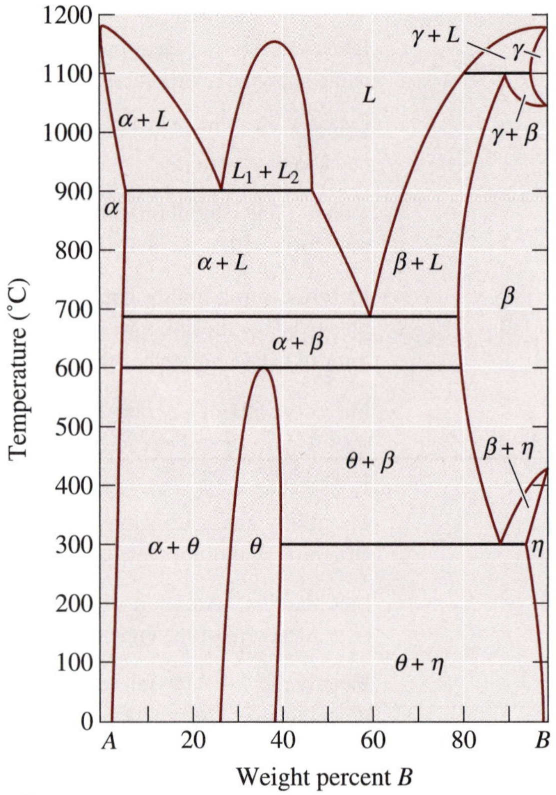

Figure 11-25 Hypothetical phase diagram (for Problem 11-7).

11-8 The Cu-Zn phase diagram is shown in Figure 11-26.

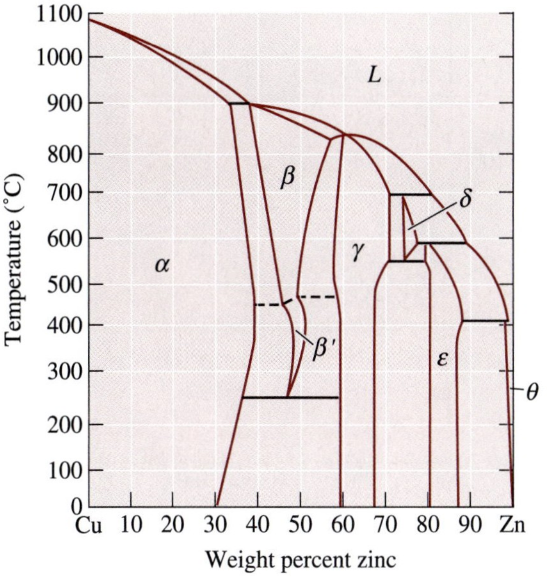

Figure 11-26 Binary phase diagram for the copper-zinc system (for Problem 11-8).

(a) Are any intermetallic compounds present? If so, identify them and determine whether they are stoichiometric or nonstoichiometric.

(b) Identify the solid solutions present in the system.

(c) Identify the three-phase reactions by writing down the temperature, the reaction in equation form, and the name of the reaction.

11-9 The Al-Li phase diagram is shown in Figure 11-27.

(a) Are any intermetallic compounds present? If so, identify them and determine whether they are stoichiometric or nonstoichiometric. Determine the formula for each compound.

(b) Identify the three-phase reactions by writing down the temperature, the reaction in equation form, the composition of each phase in the reaction, and the name of the reaction.

11-10 An intermetallic compound is found for 38 wt% Sn in the Cu-Sn phase diagram. Determine the formula for the compound.

11-11 An intermetallic compound is found for 10 wt% Si in the Cu-Si phase diagram. Determine the formula for the compound.

Section 11-4 The Eutectic Phase Diagram

11-12 Consider a Pb-15% Sn alloy. During solidification, determine

(a) the composition of the first solid to form,

(b) the liquidus temperature, solidus temperature, solvus temperature, and freezing range of the alloy,

(c) the amounts and compositions of each phase at 260°C,

(d) the amounts and compositions of each phase at 183°C, and

(e) the amounts and compositions of each phase at 25°C.

11-13 Consider an Al-12% Mg alloy (Figure 11-28). During solidification, determine

(a) the composition of the first solid to form,

(b) the liquidus temperature, solidus temperature, solvus temperature, and freezing range of the alloy,

(c) the amounts and compositions of each phase at 525°C,

(d) the amounts and compositions of each phase at 450°C, and

(e) the amounts and compositions of each phase at 25°C.

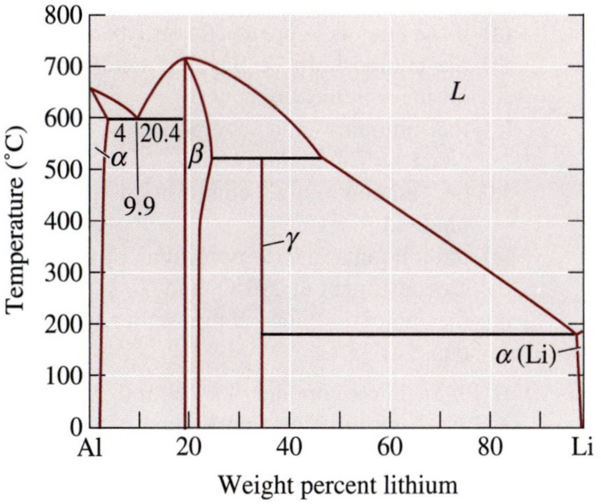

Figure 11-27 The aluminum-lithium phase diagram (for Problem 11-9).

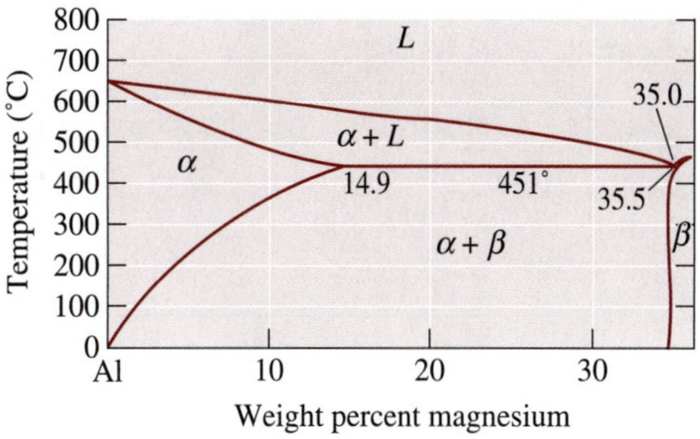

Figure 11-28 Portion of the aluminum-magnesium phase diagram (for Problems 11-13).

11-14 Consider a Pb-35% Sn alloy. Determine

(a) if the alloy is hypoeutectic or hypereutectic,
(b) the composition of the first solid to form during solidification,
(c) the amounts and compositions of each phase at 184°C,
(d) the amounts and compositions of each phase at 182°C,
(e) the amounts and compositions of each microconstituent at 182°C, and
(f) the amounts and compositions of each phase at 25°C.

11-15 Consider a Pb-70% Sn alloy. Determine

(a) if the alloy is hypoeutectic or hypereutectic,
(b) the composition of the first solid to form during solidification,
(c) the amounts and compositions of each phase at 184°C,
(d) the amounts and compositions of each phase at 182°C,
(e) the amounts and compositions of each microconstituent at 182°C, and
(f) the amounts and compositions of each phase at 25°.

11-16 Calculate the total % β and the % eutectic microconstituent at room temperature for the following lead-tin alloys: 10% Sn, 20% Sn, 50% Sn, 60% Sn, 80% Sn, and 95% Sn. Using Figure 11-18, plot the strength of the alloys versus the % β and the % eutectic and explain your graphs.

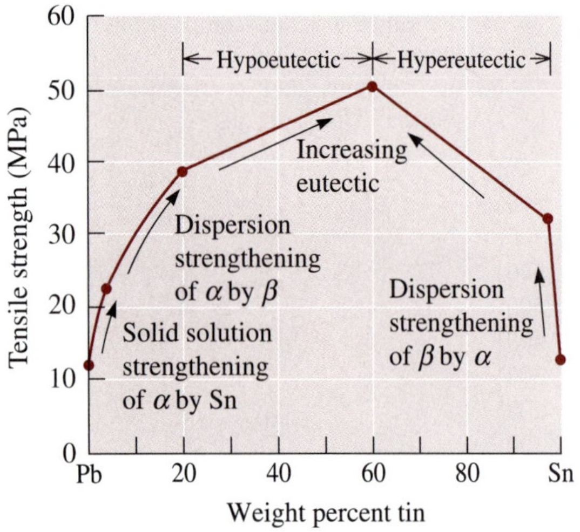

Figure 11-18 (Repeated for Problem 11-16) The effect of the composition and strengthening mechanism on the tensile strength of lead-tin alloys.

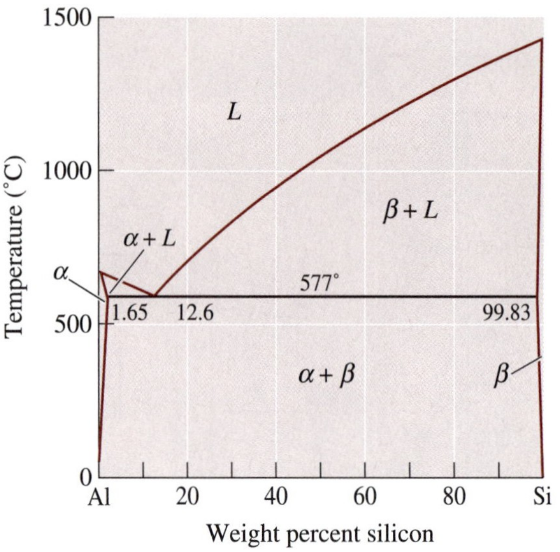

Figure 11-19 (Repeated for Problems 11-17 and 11-18) The aluminum-silicon phase diagram.

11-17 Consider an Al-4% Si alloy. (See Figure 11-19.) Determine

(a) if the alloy is hypoeutectic or hypereutectic,
(b) the composition of the first solid to form during solidification,
(c) the amounts and compositions of each phase at 578°C,
(d) the amounts and compositions of each phase at 576°C, the amounts and compositions of each microconstituent at 576°C, and
(e) the amounts and compositions of each phase at 25°C.

11-18 Consider an Al-25% Si alloy. (See Figure 11-19.) Determine

(a) if the alloy is hypoeutectic or hypereutectic,
(b) the composition of the first solid to form during solidification,
(c) the amounts and compositions of each phase at 578°C,
(d) the amounts and compositions of each phase at 576°C,
(e) the amounts and compositions of each microconstituent at 576°C, and
(f) the amounts and compositions of each phase at 25°C.

11-19 A Pb-Sn alloy contains 45% α and 55% β at 100°C. Determine the composition of the alloy. Is the alloy hypoeutectic or hypereutectic?

11-20 An Al-Si alloy contains 85% α and 15% β at 500°C. Determine the composition of the alloy. Is the alloy hypoeutectic or hypereutectic?

11-21 A Pb-Sn alloy contains 23% primary α and 77% eutectic microconstituent. Determine the composition of the alloy.

11-22 An Al-Si alloy contains 15% primary β and 85% eutectic microconstituent. Determine the composition of the alloy.

11-23 Observation of a microstructure shows that there is 28% eutectic and 72% primary β in an Al-Li alloy (Figure 11-27).

(a) Determine the composition of the alloy and whether it is hypoeutectic or hypereutectic.

(b) How much α and β are in the eutectic microconstituent?

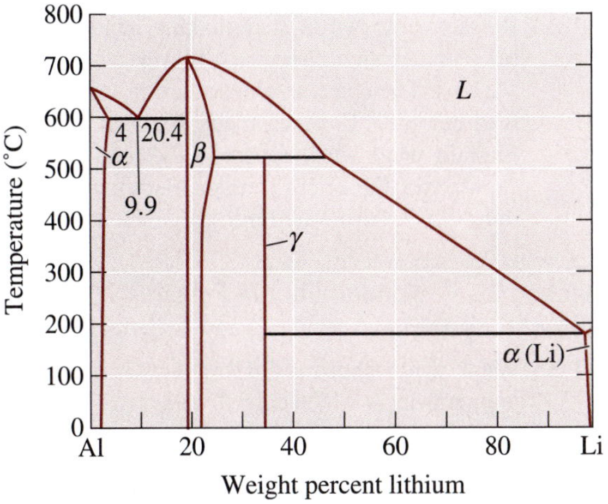

Figure 11-27 (Repeated for Problem 11-23) The aluminum-lithium phase diagram.

11-24 Write the eutectic reaction that occurs, including the compositions of the three phases in equilibrium, and calculate the amount of α and β in the eutectic microconstituent in the Mg-Al system (Figure 11-28).

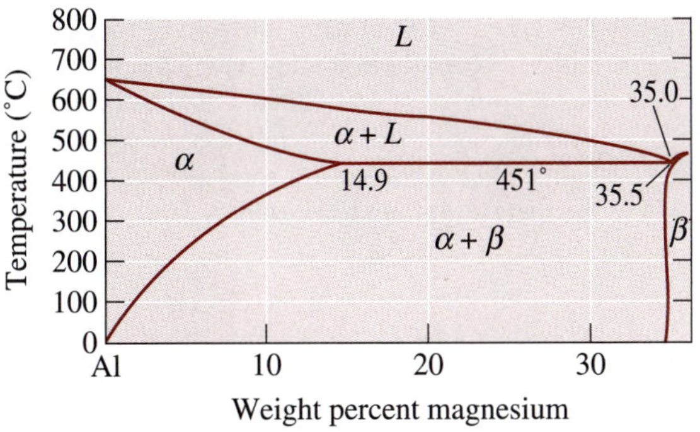

Figure 11-28 Portion of the aluminum-magnesium phase diagram (for Problems 11-24).

11-25 Calculate the total amount of α and β and the amount of each microconstituent in a Pb-50% Sn alloy at 182°C. What fraction of the total α in the alloy is contained in the eutectic microconstituent?

11-26 Figure 11-29 shows a cooling curve for a Pb-Sn alloy. Determine

(a) the pouring temperature,
(b) the superheat,
(c) the liquidus temperature,
(d) the eutectic temperature,
(e) the freezing range,
(f) the local solidification time,
(g) the total solidification time, and
(h) the composition of the alloy.

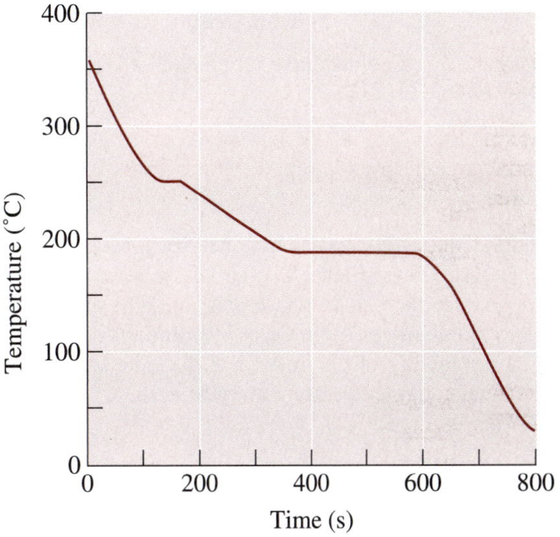

Figure 11-29 Cooling curve for a Pb-Sn alloy (for Problem 11-26).

11-27 Figure 11-30 shows a cooling curve for an Al-Si alloy. Determine

(a) the pouring temperature,
(b) the superheat,
(c) the liquidus temperature,
(d) the eutectic temperature,
(e) the freezing range,
(f) the local solidification time,
(g) the total solidification time, and
(h) the composition of the alloy.

11-28 Draw the cooling curves, including appropriate temperatures, expected for the following Al-Si alloys:

(a) Al-4% Si,
(b) Al-12.6% Si,

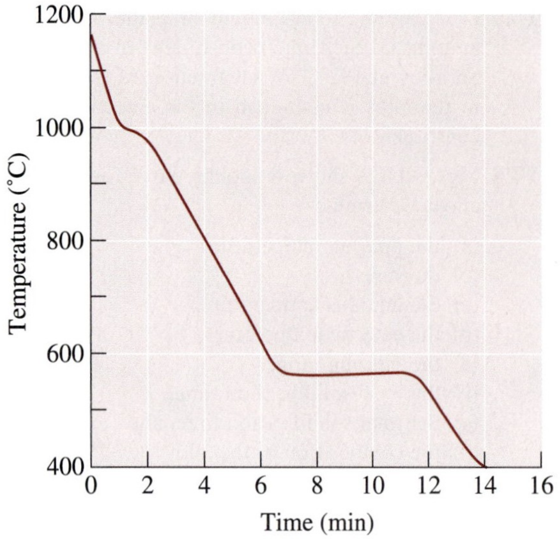

Figure 11-30 Cooling curve for an Al-Si alloy (for Problem 11-27).

(c) Al-25% Si, and
(d) Al-65% Si.

11-29 Cooling curves are obtained for a series of Cu-Ag alloys (Figure 11-31). Use this data to produce the Cu-Ag phase diagram. The maximum solubility of Ag in Cu is 7.9% and the maximum solubility of Cu in Ag is 8.8%. The solubilities at room temperature are near zero.

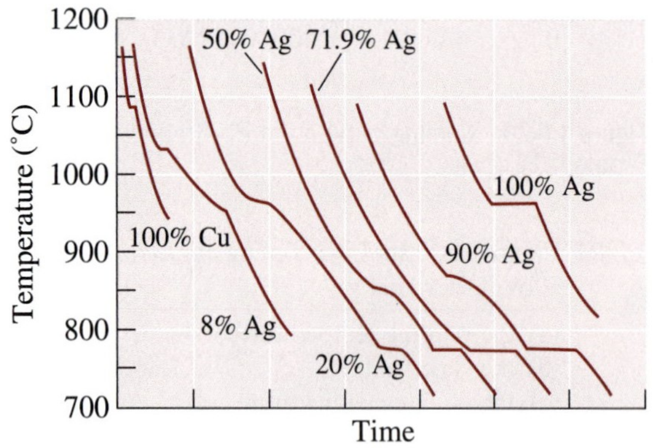

Figure 11-31 Cooling curves for a series of Cu-Ag alloys (for Problem 11-29).

Section 11-5 Strength of Eutectic Alloys

11-30 In regards to eutectic alloys, what does the term "modification" mean? How does it help properties of the alloy?

11-31 For the Pb-Sn system, explain why the tensile strength is maximum at the eutectic composition.

11-32 Does the shape of the proeutectic phase have an effect on the strength of eutectic alloys? Explain.

Section 11-6 Eutectics and Materials Processing

11-33 Explain why Pb-Sn alloys are used for soldering.

11-34 Refractories used in steel making include silica brick that contain very small levels of alumina (Al_2O_3). The eutectic temperature in this system is about 1587°C. Silica melts at about 1725°C. Explain what will happen to the load bearing capacity of the bricks if a small amount of alumina gets incorporated into the silica bricks.

Section 11-7 Nonequilibrium Freezing in the Eutectic System

11-35 What is hot shortness? How does it affect the temperature at which eutectic alloys can be used?

Design Problems

11-36 Design a processing method that permits a Pb-15% Sn alloy solidified under nonequilibrium conditions to be hot worked.

11-37 Design a eutectic diffusion bonding process to join aluminum to silicon. Describe the changes in microstructure at the interface during the bonding process.

11-38 Design an Al-Si brazing alloy and process that will be successful in joining an Al-Mn alloy that has a liquidus of 659°C and a solidus of 656°C. Brazing, like soldering, involves introducing a liquid filler metal into a joint without melting the metals that are to be joined.

12

Dispersion Strengthening by Phase Transformations and Heat Treatment

Have You Ever Wondered?

- *Who invented and flew the first controllable airplane?*
- *How do engineers strengthen aluminum alloys used in aircrafts?*
- *Why do some steels become very hard upon quenching from high temperatures?*
- *What alloys are used to make orthodontic braces?*
- *Would it be possible to further enhance the strength and, hence, the dent resistance of sheet steels after the car chassis is made?*

In Chapter 11, we examined in detail how second-phase particles can increase the strength of metallic materials. In this chapter, we further discuss dispersion strengthening as we describe a variety of solid-state transformation processes including precipitation or age hardening and the eutectoid reaction. We also examine how nonequilibrium phase transformations in particular, the martensitic reaction can provide strengthening.

As we discuss these strengthening mechanisms, keep in mind the characteristics that produce the most desirable dispersion strengthening as discussed in Chapter 11:

- The matrix should be relatively soft and ductile and the precipitate, or second phase, should be strong;
- the precipitate particles should be round and discontinuous;
- the second-phase particles should be small and numerous; and
- in general, the more precipitate phase we have, the stronger the alloy will be.

As in Chapter 11, we will concentrate on how phase transformations influence the strength of the materials and how heat treatments can influence other properties. Since we will be dealing with solid-state phase transformations, we will begin with a discussion on the nucleation and growth of second-phase particles in solid-state phase transformations.

12-1 Nucleation and Growth in Solid-State Reactions

In Chapter 9, we discussed nucleation of a solid nucleus from a melt. We also discussed the concepts of supersaturation, undercooling, and homogeneous and heterogeneous nucleation. Let's now see how these concepts apply to solid-state phase transformations such as the eutectoid reaction. In order for a precipitate of phase β to form from a solid matrix of phase α, both nucleation and growth must occur. The total change in free energy required for nucleation of a spherical solid precipitate from the matrix is:

$$\Delta G = \tfrac{4}{3}\pi r^3 \Delta G_{v(\alpha\rightarrow\beta)} + 4\pi r^2 \sigma_{\alpha\beta} + \tfrac{4}{3}\pi r^3 \varepsilon \tag{12-1}$$

The first two terms include the free energy change per unit volume (ΔG_v), and the energy change needed to create the unit area of the interface ($\sigma_{\alpha\beta}$), just as in solidification. However, the third term takes into account the **strain energy** per unit volume (ε), the energy required to permit a precipitate to fit into the surrounding matrix during the nucleation and growth of the precipitate, introduced when the precipitate forms in a solid, rigid matrix. The precipitate does not occupy the same volume that is displaced, so additional energy is required to accommodate the precipitate in the matrix.

Nucleation As in solidification, nucleation occurs most easily on surfaces already present in the structure, thereby minimizing the surface energy term. Thus, the precipitates heterogeneously nucleate most easily at grain boundaries and other defects.

Growth Growth of the precipitates normally occurs by long-range diffusion and redistribution of atoms. Diffusing atoms must be detached from their original locations (perhaps at lattice points in a solid solution), move through the surrounding material to the nucleus, and be incorporated into the crystal structure of the precipitate. In some cases, the diffusing atoms might be so tightly bonded within an existing phase that the detachment process limits the rate of growth. In other cases, attaching the diffusing atoms to the precipitate—perhaps because of the lattice strain—limits growth. This result sometimes leads to the formation of precipitates that have a special relationship to the matrix structure that minimizes the strain at the interface between the parent phase and the precipitate particles. In most cases, however, the controlling factor is the diffusion step.

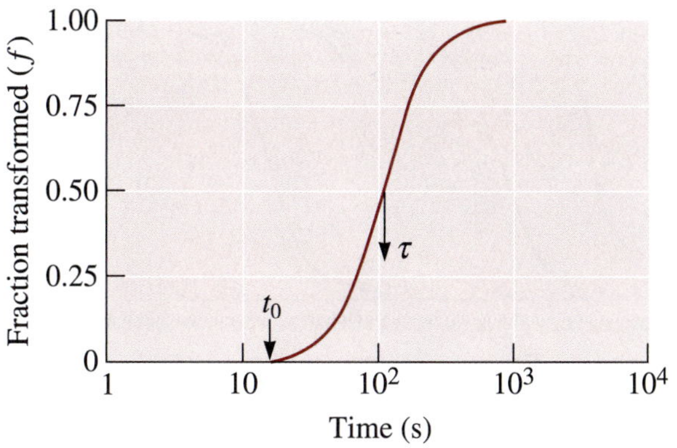

Figure 12-1
Sigmoidal curve showing the rate of transformation of FCC iron at a constant temperature. The incubation time t_0 and the time τ for 50% transformation are also shown.

Kinetics The overall rate, or *kinetics*, of a transformation depends on both nucleation and growth. If more nuclei are present at a particular temperature, growth occurs from a larger number of sites and the phase transformation is completed in a shorter period of time. At higher temperatures, the diffusion coefficient is higher, growth rates are higher, and again we expect the transformation to be completed in a shorter time, assuming an equal number of nuclei.

The rate of transformation is given by the Avrami equation (Equation 12-2), with the fraction of the transformation, f, related to time, t, by

$$f = 1 - \exp(-ct^{n}) \tag{12-2}$$

where c and n are constants for a particular temperature. This **Avrami relationship**, shown in Figure 12-1, produces a sigmoidal, or S-shaped, curve. This equation can describe most solid-state phase transformations. An incubation time, t_0, during which no observable transformation occurs, is the time required for nucleation to occur. Initially, the transformation occurs slowly as nuclei form.

The incubation period is followed by rapid growth as atoms diffuse to the growing precipitate. Near the end of the transformation, the rate again slows as the source of atoms available to diffuse to the growing precipitate is depleted. The transformation is 50% complete in time τ; the rate of transformation is often given by the reciprocal of τ:

$$\text{Rate} = 1/\tau \tag{12-3}$$

Effect of Temperature In many phase transformations, the material undercools below the temperature at which the phase transformation occurs under equilibrium conditions. Recall from Chapter 9 the undercooling of water and other liquids and other supersaturation phenomena. Because both nucleation and growth are temperature-dependent, the rate of phase transformation depends on the undercooling (ΔT). The rate of nucleation is low for small undercoolings (since the thermodynamic driving force is low) and increases for larger undercoolings as the thermodynamic driving force increases at least up to a certain point (since diffusion becomes slower as temperature decreases). At the same time, the growth rate of the new phase decreases continuously, because of slower diffusion, as the undercooling increases. The growth rate follows an Arrhenius relationship (recall, Equation 5-1):

$$\text{Growth rate} = A \exp\left(\frac{-Q}{RT}\right) \tag{12-4}$$

where Q is the activation energy (in this case for the phase transformation), R is the gas constant, T is the temperature, and A is a constant.

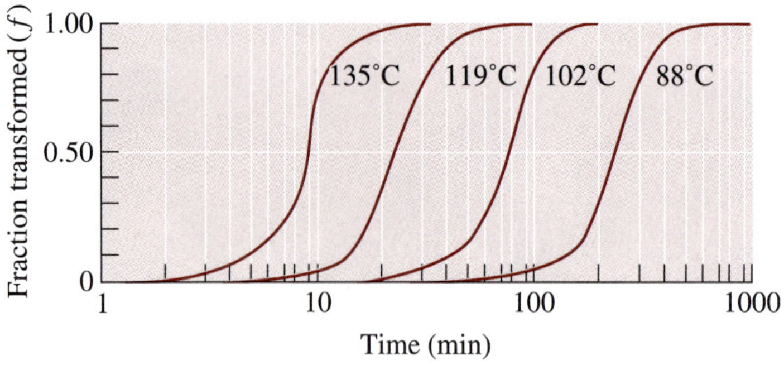

Figure 12-2 The effect of temperature on recrystallization of cold-worked copper.

Figure 12-2 shows sigmoidal curves at different temperatures for the recrystallization of copper; as the temperature increases, the rate of recrystallization of copper *increases*, because growth is the most important factor for copper.

At any particular temperature, the overall rate of transformation is the product of the nucleation and growth rates. In Figure 12-3(a), the combined effect of the nucleation and growth rates is shown. A maximum transformation rate may be observed at a critical undercooling. The time required for transformation is inversely related to the rate of transformation; Figure 12-3(b) describes the time (on a log scale) required for the transformation. This *C*-shaped curve is common for many transformations in metals, ceramics, glasses, and polymers. Notice that the time required at a temperature corresponding to the equilibrium phase transformation would be ∞ (i.e., the phase transformation will not occur). This is because there is no undercooling and, hence, the rate of homogenous nucleation is zero.

In some processes, such as the recrystallization of a cold-worked metal, we find that the transformation rate continually decreases with decreasing temperature. In this case, nucleation occurs easily, and diffusion—or growth—predominates (i.e., the growth is the rate limiting step for the transformation). The following example illustrates how the activation energy for a solid-state phase transformation such as recrystallization can be obtained from data related to the kinetics of the process.

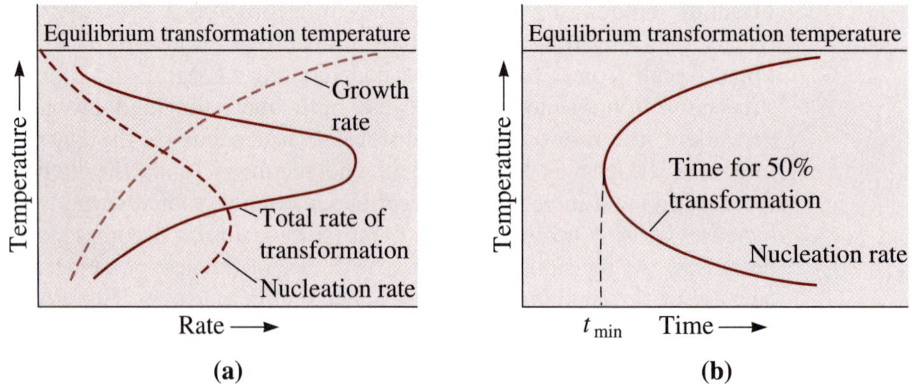

Figure 12-3 (a) The effect of temperature on the rate of a phase transformation is the product of the growth rate and nucleation rate contributions, giving a maximum transformation rate at a critical temperature. (b) Consequently, there is a minimum time (t_{min}) required for the transformation, given by the "*C*-curve".

EXAMPLE 12-1 Activation Energy for the Recrystallization of Copper

Determine the activation energy for the recrystallization of copper from the sigmoidal curves in Figure 12-2.

SOLUTION

The rate of transformation is the reciprocal of the time τ required for half of the transformation to occur. From Figure 12-2, the times required for 50% transformation at several different temperatures can be calculated:

T(°C)	T (K)	τ (min)	Rate (min^{-1})
135	408	9	0.111
119	392	22	0.045
102	375	80	0.0125
88	361	250	0.0040

The rate of transformation is an Arrhenius equation, so a plot of ln (rate) versus $1/T$ (Figure 12-4 and Equation 12-4) allows us to calculate the constants in the equation. Taking natural log of both sides of Equation 12-4:

$$\ln(\text{Growth rate}) = \ln A - \frac{Q}{RT}$$

Thus, if we plot ln(Growth rate) as a function of $1/T$, we expect a straight line that has a slope of $-Q/R$.

$$\text{Slope} = \frac{-Q}{R} = \left[\frac{\Delta \ln(\text{rate})}{\Delta\left(\frac{1}{T}\right)}\right] = \frac{[(\ln(0.111) - \ln(0.004))]}{\left[\frac{1}{408} - \frac{1}{361}\right]}$$

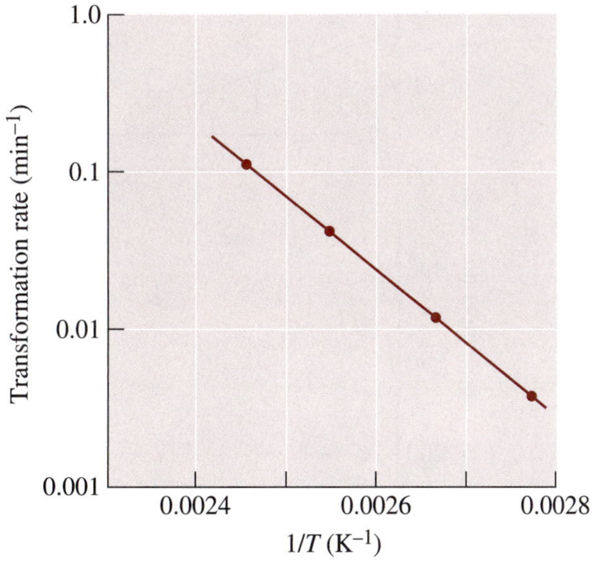

Figure 12-4 Arrhenius plot of transformation rate versus reciprocal temperature for recrystallization of copper (for Example 12-1).

$$Q/R = 10{,}414$$

$$Q = 20{,}693 \frac{\text{cal}}{\text{mol}}$$

$$\therefore 0.111 = A \exp\left(\frac{-20{,}693 \text{ cal/mol}}{\left(1.987 \frac{\text{cal}}{\text{deg} - \text{mol}}\right) \times (408 \text{ deg})}\right)$$

$$A = 0.111/8.21 \times 10^{-12} = 1.351 \times 10^{10} \text{ s}^{-1}$$

$$\therefore \text{rate} = 1.351 \times 10^{10} \exp\left(\frac{-20{,}693}{RT}\right)$$

In this particular example, the rate at which the reaction occurs *increases* as the temperature increases, indicating that the reaction may be dominated by diffusion.

12-2 Alloys Strengthened by Exceeding the Solubility Limit

In Chapter 11, we learned that lead-tin (Pb-Sn) alloys containing about 2 to 19% Sn can be dispersion-strengthened because the solubility of tin in lead is exceeded.

A similar situation occurs in aluminum-copper alloys. For example, the Al-4% Cu alloy (shown in Figure 12-5) is 100% α above 500°C. The α phase is a solid solution of aluminum containing copper up to 5.65 wt%. On cooling below the solvus tempera-

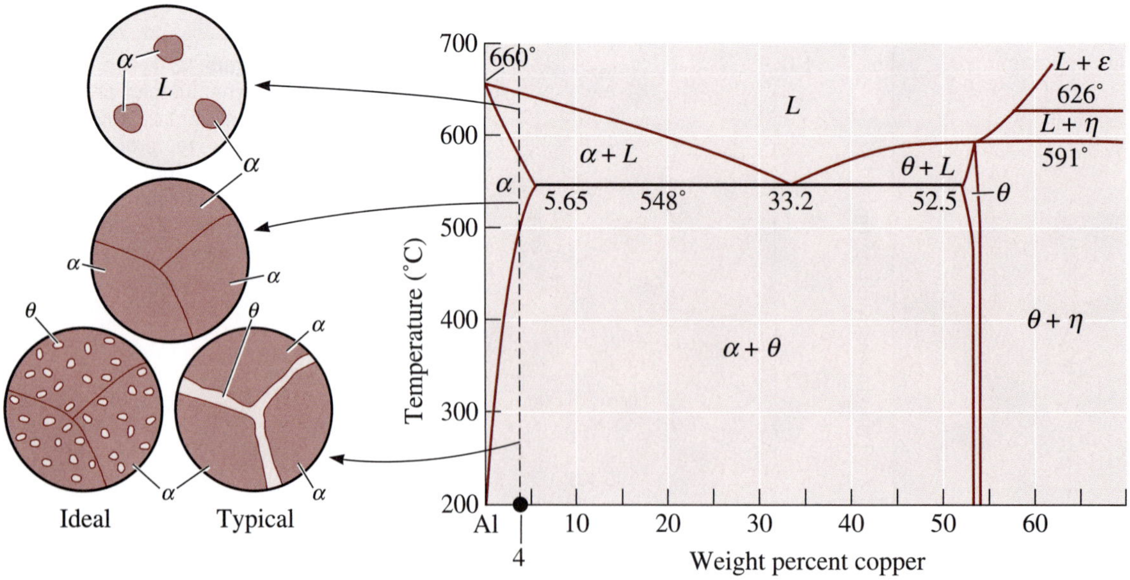

Figure 12-5 The aluminum-copper phase diagram and the microstructures that may develop during cooling of an Al-4% Cu alloy.

ture, a second phase, θ, precipitates. The θ phase, which is the hard, brittle intermetallic compound $CuAl_2$, provides dispersion strengthening. Applying the lever rule to the phase diagram shown in Figure 12-5, we can show that at 200°C and below, in a 4% Cu alloy, only about 7.5% of the final structure is θ. We must control the precipitation of the second phase to satisfy the requirements of good dispersion strengthening.

Widmanstðtten Structure The second phase may grow so that certain crystallographic planes and directions in the precipitate are parallel to preferred planes and directions in the matrix, creating a basket-weave pattern known as the **Widmanstðtten structure**. This growth mechanism minimizes strain and surface energies and permits faster growth rates. Widmanstðtten growth produces a characteristic appearance for the precipitate. When a needle-like shape is produced [Figure 12-6(a)], the Widmanstðtten precipitate may encourage the nucleation of cracks, thus reducing the ductility of the material. However, some of these structures make it more difficult for cracks, once formed, to propagate, therefore providing good fracture toughness. Certain titanium alloys and ceramics obtain toughness in this way.

Interfacial Energy Relationships We expect the precipitate to have a spherical shape in order to minimize surface energy. However, when the precipitate forms at an interface, the precipitate shape is also influenced by the **interfacial energy** (γ_{pm}) of the boundary between the matrix (m) grains and the precipitate (p).

If the precipitate phase completely wets the grain (similar to how water wets glass) then the dihedral angle is zero and the second phase would grow as a continuous layer along the grain boundaries of the matrix phase. If the dihedral angle is small, the precipitate may be continuous. If the precipitate is also hard and brittle, the thin film that surrounds the matrix grains causes the alloy to be very brittle [Figure 12-6(b)].

On the other hand, discontinuous and even spherical precipitates form when the dihedral angle is large [Figure 12-6(c)]. This occurs if the precipitate phase does not wet the matrix.

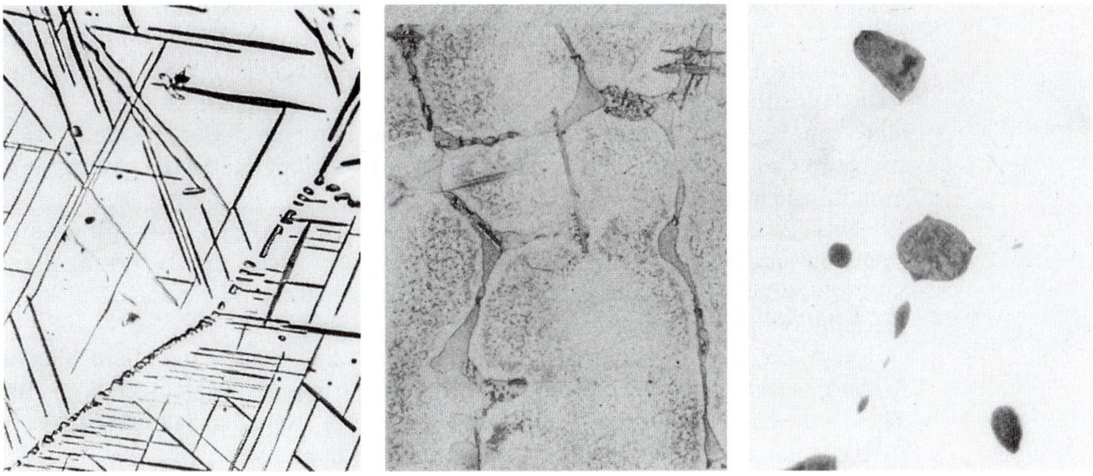

(a) (b) (c)

Figure 12-6 (a) Widmanstðtten needles in a Cu-Ti alloy (×420). (*From* ASM Handbook, *Vol. 9, Metallography and Microstructure (1985), ASM International, Materials Park, OH 44073.*) (b) Continuous θ precipitate in an Al-4% Cu alloy, caused by slow cooling (×500). (c) Precipitates of lead at grain boundaries in copper (×500).

Coherent Precipitate Even if we produce a uniform distribution of discontinuous precipitate, the precipitate may not significantly disrupt the surrounding matrix structure. Consequently, the precipitate blocks slip only if it lies directly in the path of the dislocation.

But when a **coherent precipitate** forms, the planes of atoms in the crystal structure of the precipitate are related to—or even continuous with—the planes in the crystal structure of the matrix. Now a widespread disruption of the matrix crystal structure is created, and the movement of a dislocation is impeded even if the dislocation merely passes near the coherent precipitate. A special heat treatment, such as age hardening, may produce the coherent precipitate.

12-3 Age or Precipitation Hardening

Age hardening, or **precipitation hardening**, is produced by a sequence of phase transformations that leads to a uniform dispersion of nano-sized, coherent precipitates in a softer, more ductile matrix. The inadvertent occurrence of this process may have helped the Wright brothers, who, on December 17, 1903, made the first controllable airplane flight that changed the world forever. Recently, Dr. Frank Gayle and co-workers at the National Institute of Standards and Technology (NIST) have shown that the aluminum alloy used by the Wright brothers for making the engine of the first airplane ever flown picked up copper from the casting mold. The age hardening occurred inadvertently as the mold remained hot during the casting process.

12-4 Applications of Age-Hardened Alloys

Before we examine the details of the mechanisms of phase transformations that are needed for age hardening to occur, let's examine some of the applications of this technique. A major advantage of precipitation hardening is that it can be used to increase the yield strength of many metallic materials via relatively simple heat treatments and without creating significant changes in density. Thus, the strength-to-density ratio of an alloy can be improved substantially using age hardening. For example, the yield strength of an aluminum alloy can be increased from about 138 MPa to 414 MPa as a result of age hardening.

Nickel-based super alloys (alloys based on Ni, Cr, Al, Ti, Mo, and C) are precipitation hardened by precipitation of a Ni_3Al-like γ' phase that is rich in Al and Ti. Similarly, titanium alloys (e.g., Ti-6Al-4V), stainless steels, Be-Cu and many steels are precipitation hardened and used for a variety of applications.

New sheet-steel formulations are designed such that precipitation hardening occurs in the material while the paint on the chassis is being "baked" or cured (~100°C). These bake-hardenable steels are just one example of steels that take advantage of the strengthening effect provided by age-hardening mechanisms.

A limitation associated with this mechanism is that age-hardened alloys can be used over a limited range of temperatures. At higher temperatures, the precipitates formed initially begin to grow and eventually dissolve if the temperatures are high enough (Section 12-8). This is where alloys in which dispersion strengthening is achieved by using a second phase that is insoluble are more effective than age-hardened alloys.

12-5 Microstructural Evolution in Age or Precipitation Hardening

How do precipitates form in precipitation hardening? How do they grow or age? Can the precipitates grow too much, or overage, so that they can not provide maximum dispersion strengthening? Answers to these questions can be found by following the microstructural evolution in the sequence of phase transformations that are necessary for age hardening.

Let's use Al-Cu as an archetypal system to illustrate these ideas. The Al-4% Cu alloy is a classical example of an age-hardenable alloy. There are three steps in the age-hardening heat treatment (Figure 12-7).

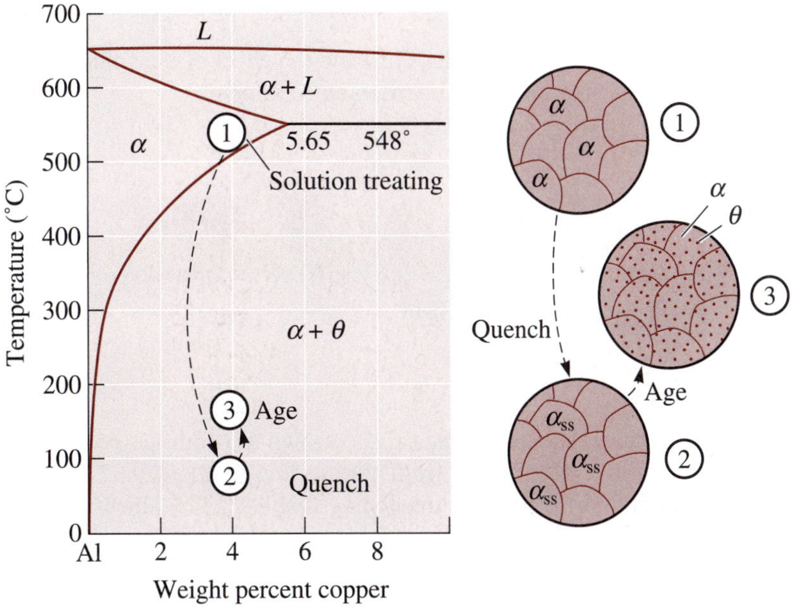

Figure 12-7 The aluminum-rich end of the aluminum-copper phase diagram showing the three steps in the age-hardening heat treatment and the microstructures that are produced.

Step 1: Solution Treatment In the **solution treatment**, the alloy is first heated above the solvus temperature and held until a homogeneous solid solution α is produced. This step dissolves the θ phase precipitate and reduces any microchemical segregation present in the original alloy.

We could heat the alloy to just below the solidus temperature and increase the rate of homogenization. However, the presence of a nonequilibrium eutectic microconstituent may cause melting (hot shortness, Chapter 10). Thus, the Al-4% Cu alloy is solution-treated between 500°C and 548°C, that is, between the solvus and the eutectic temperatures.

Step 2: Quench After solution treatment, the alloy, which contains only α in its structure, is rapidly cooled, or quenched. The atoms do not have time to diffuse to potential nucleation sites, so the θ does not form. After the quench, the structure is a **supersaturated solid solution** ($α_{ss}$) containing excess copper, and it is not an equilibrium structure. It is a metastable structure. This situation is effectively the same as undercooling of water, molten metals, and silicate glasses (Chapter 9). The only difference is we are dealing with materials in their solid state.

Step 3: Age Finally, the supersaturated solid solution (α_{ss}) is heated at a temperature below the solvus temperature. At this *aging* temperature, atoms diffuse only short distances. Because the α_{ss} phase is metastable, the extra copper atoms diffuse to numerous nucleation sites and precipitates grow. Eventually, if we hold the alloy for a sufficient time at the aging temperature, the equilibrium $\alpha + \theta$ structure is produced. Note that even though the structure that is formed has two equilibrium phases (i.e., $\alpha + \theta$), the morphology of the phases is different from the structure that would have been obtained by the slow cooling of this alloy (Figure 12-5). When we go through the three steps described previously, we produce the θ phase in the form of ultrafine uniformly dispersed second-phase precipitate particles. This is what we need for effective precipitation strengthening.

The following two examples illustrate the effect of quenching on the composition of phases and a design for an age-hardening treatment.

EXAMPLE 12-2 *Composition of Al-4% Cu Alloy Phases*

Compare the composition of the α solid solution in the Al-4% Cu alloy at room temperature when the alloy cools under equilibrium conditions with that when the alloy is quenched.

SOLUTION

From Figure 12-5, a tie line can be drawn at room temperature. The composition of the α determined from the tie line is about 0.02% Cu. However, the composition of the α after quenching is still 4% Cu. Since α contains more than the equilibrium copper content, the α is supersaturated with copper.

EXAMPLE 12-3 *Design of an Age-Hardening Treatment*

A portion of a magnesium-aluminum phase diagram is shown in Figure 12-8. Suppose a Mg-8% Al alloy is responsive to an age-hardening heat treatment. Design a heat treatment for the alloy.

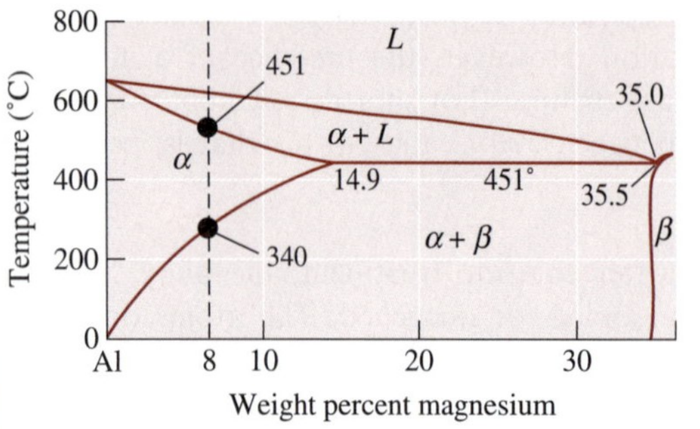

Figure 12-8 Portion of the aluminum-magnesium phase diagram.

SOLUTION

Step 1: Solution-treat at a temperature between the solvus and the eutectic to avoid hot shortness. Thus, heat between 340°C and 451°C (Figure 12-8).

Step 2: Quench to room temperature fast enough to prevent the precipitate phase β from forming.

Step 3: Age at a temperature below the solvus, that is, below 340°C, to form a fine dispersion of β phase.

Nonequilibrium Precipitates during Aging During aging of aluminum-copper alloys, a continuous series of other precursor precipitate phases forms prior to the formation of the equilibrium θ phase. This is fairly common in precipitation-hardened alloys. The simplified diagram in Figure 12-7 does not show these intermediate phases. At the start of aging, the copper atoms concentrate on $\{100\}$ planes in the α matrix and produce very thin precipitates called **Guinier-Preston** (GP) **zones**. As aging continues, more copper atoms diffuse to the precipitate and the GP-I zones thicken into thin disks, or GP-II zones. With continued diffusion, the precipitates develop a greater degree of order and are called θ'. Finally, the stable θ precipitate is produced.

The nonequilibrium precipitates—GP-I, GP-II, and θ'—are coherent precipitates. The strength of the alloy increases with aging time as these coherent phases grow in size during the initial stages of the heat treatment. When these coherent precipitates are present, the alloy is in the aged condition. This important development in the microstructure evolution of precipitation-hardened alloys is the reason the time for heat treatment during aging is very important. Since the changes in the structure occur at a nano-scale, no change is seen in the structure of the alloy at a micro-scale. This is one reason why the mechanisms of precipitation hardening were established more firmly only in the 1950s, (when electron microscopes were starting to be used), even though the phenomenon was discovered earlier.

When the stable noncoherent θ phase precipitates, the strength of the alloy begins to decrease. Now the alloy is in the overaged condition. The θ phase still provides some dispersion strengthening, but with increasing time, the θ phase grows larger and even the simple dispersion-strengthening effect diminishes.

12-6 Effects of Aging Temperature and Time

The properties of an age-hardenable alloy depend on both aging temperature and aging time (Figure 12-9). At 260°C, diffusion in the Al-4% Cu alloy is rapid, and precipitates quickly form. The strength reaches a maximum after less than 0.1 h (6 minutes) exposure. Overaging occurs if the alloy is held for longer than 0.1 h.

At 190°C, which is a typical aging temperature for many aluminum alloys, a longer time is required to produce the optimum strength. However, there are benefits to using the lower temperature. First, the maximum strength increases as the aging temperature decreases. Second, the alloy maintains its maximum strength over a longer period of time. Third, the properties are more uniform. If the alloy is aged for only 10 min at 260°C, the surface of the part reaches the proper temperature and strengthens, but the center remains cool and ages only slightly. The example that follows illustrates the effect of aging heat treatment time on the strength of aluminum alloys.

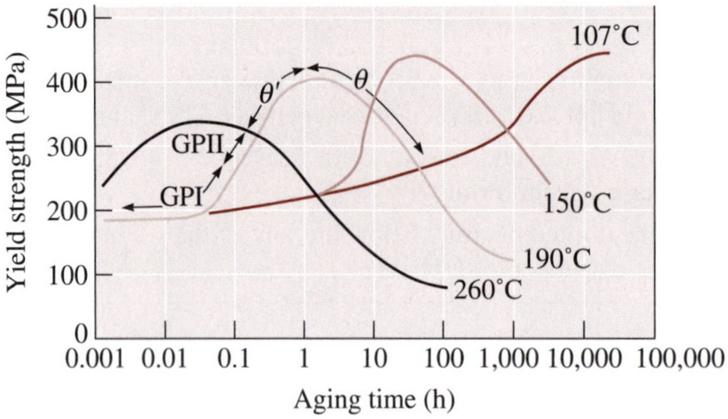

Figure 12-9 The effect of aging temperature and time on the yield strength of an Al-4% Cu alloy.

EXAMPLE 12-4 *Effect of Aging Heat Treatment Time on the Strength of Aluminum Alloys*

The operator of a furnace left for his hour lunch break without removing the Al-4% Cu alloy from the furnace used for the aging treatment. Compare the effect on the yield strength of the extra hour of aging for the aging temperatures of 190°C and 260°C.

SOLUTION

At 190°C, the peak strength of 400 MPa occurs at 2 h (Figure 12-9). After 3 h, the strength is essentially the same.

At 260°C, the peak strength of 340 MPa occurs at 0.06 h. However, after 1 h, the strength decreases to 250 MPa.

Thus, the higher aging temperature gives lower peak strength and makes the strength more sensitive to aging time.

Aging at either 190°C or 260°C is called **artificial aging**, because the alloy is heated to produce precipitation. Some solution-treated and quenched alloys age at room temperature; this is called **natural aging**. Natural aging requires long times—often several days—to reach maximum strength. However, the peak strength is higher than that obtained in artificial aging, and no overaging occurs.

An interesting observation made by Dr. Gayle and coworkers of NIST is a striking example of the difference between natural aging and artificial aging. Dr. Gayle and coworkers analyzed the aluminum alloy of the engine used in the Wright brothers' airplane. They found two interesting things. First, they found that the original alloy had undergone precipitation hardening as a result of being held in the mold for a period of time and at a temperature that was sufficient to cause precipitation hardening. Second, since when the research was done, from 1903 until about 1993 (almost ninety years), the alloy had continued to age naturally! This could be seen from two different size distributions for the precipitate particles using transmission electron microscopy. In some aluminum alloys (designated as T4) used to make tapered poles or fasteners, it may be necessary to refrigerate the alloy to avoid natural aging at room temperature. If not, the alloy would age at room temperature, become harder and not be workable!

12-7 Requirements for Age Hardening

Not all alloys are age hardenable. Four conditions must be satisfied for an alloy to have an age-hardening response during heat treatment:

1. The alloy system must display decreasing solid solubility with decreasing temperature. In other words, the alloy must form a single phase on heating above the solvus line, then enter a two-phase region on cooling.

2. The matrix should be relatively soft and ductile, and the precipitate should be hard and brittle. In most age hardenable alloys, the precipitate is a hard, brittle intermetallic compound.

3. The alloy must be quenchable. Some alloys cannot be cooled rapidly enough to suppress the formation of the precipitate. Quenching may, however, introduce residual stresses that cause distortion of the part (Chapter 8). To minimize residual stresses, aluminum alloys are quenched in hot water, at about 80°C.

4. A coherent precipitate must form.

As mentioned before in Section 12-4, a number of important alloys, including certain stainless steels and alloys based on aluminum, magnesium, titanium, nickel, chromium, iron, and copper, meet these conditions and are age-hardenable.

12-8 Use of Age-Hardenable Alloys at High Temperatures

Based on our previous discussion, we would not select an age-hardened Al-4% Cu alloy for use at high temperatures. At service temperatures ranging from 100°C to 500°C, the alloy overages and loses its strength. Above 500°C, the second phase redissolves in the matrix and we do not even obtain dispersion strengthening. In general, the aluminum age-hardenable alloys are best suited for service near room temperature. However, some magnesium alloys may maintain their strength to about 250°C and certain nickel superalloys resist overaging at 1000°C.

We may also have problems when welding age-hardenable alloys (Figure 12-10). During welding, the metal adjacent to the weld is heated. The heat-affected zone (HAZ) contains two principle regions. The lower temperature zone near the unaffected base metal is exposed to temperatures just below the solvus and may overage. The higher temperature zone is solution-treated, eliminating the effects of age hardening. If the solution-treated zone cools slowly, stable θ may form at the grain boundaries, embrittling the weld area. Very fast welding processes such as electron-beam welding, complete reheat treatment of the area after welding, or welding the alloy in the solution-treated condition improve the quality of the weld (Chapter 9). Welding of nickel-based superalloys strengthened by precipitation hardening does not pose such problems since the precipitation process is sluggish and the welding process simply acts as a solution and quenching treatment. The process of friction stir welding has also been recently applied to welding of Al and Al-Li alloys for aerospace and aircraft applications.

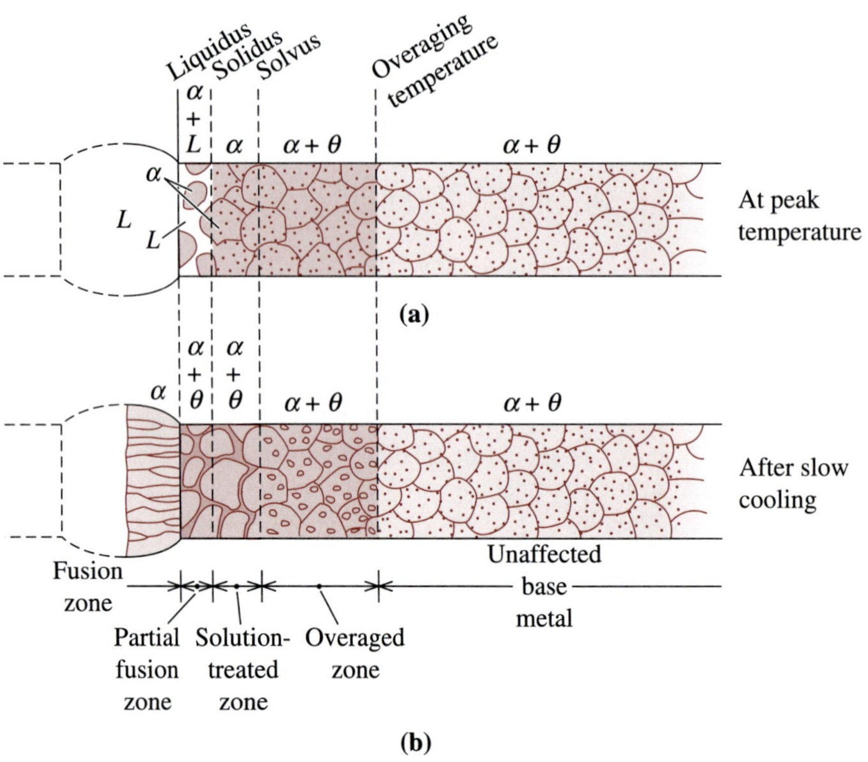

Figure 12-10 Microstructural changes that occur in age-hardened alloys during fusion welding: (a) microstructure in the weld at the peak temperature, and (b) microstructure in the weld after slowly cooling to room temperature.

12-9 The Eutectoid Reaction

In Chapter 11, we defined the eutectoid as a solid-state reaction in which one solid phase (S_1) transforms to two other solid phases (S_2 and S_3):

$$S_1 \rightarrow S_2 + S_3 \tag{12-5}$$

As an example of how we can use the eutectoid reaction to control the microstructure and properties of an alloy, let's examine the technologically important portion of the iron-iron carbide (Fe-Fe_3C) phase diagram (Figure 12-11), which is the basis for steels and cast irons. The formation of the two solid phases (α and Fe_3C) permits us to obtain dispersion strengthening. The ability to control the occurrence of the eutectoid reaction (this includes either making it happen, slowing it down, or avoiding it all together) is probably the most important step in the thermomechanical processing of steels. On the Fe-Fe_3C diagram, the eutectoid temperature is known as the A_1 temperature. The boundary between austenite (γ) and the two-phase field consisting of ferrite (α) and austenite is known as the A_3. The boundary between austenite (γ) and the two-phase field consisting of cementite (Fe_3C) and austenite is known as the A_{cm}.

We are normally not interested in the carbon-rich end of the Fe-C phase diagram and this is why we examine the Fe-Fe_3C diagram as part of the Fe-C binary phase diagram.

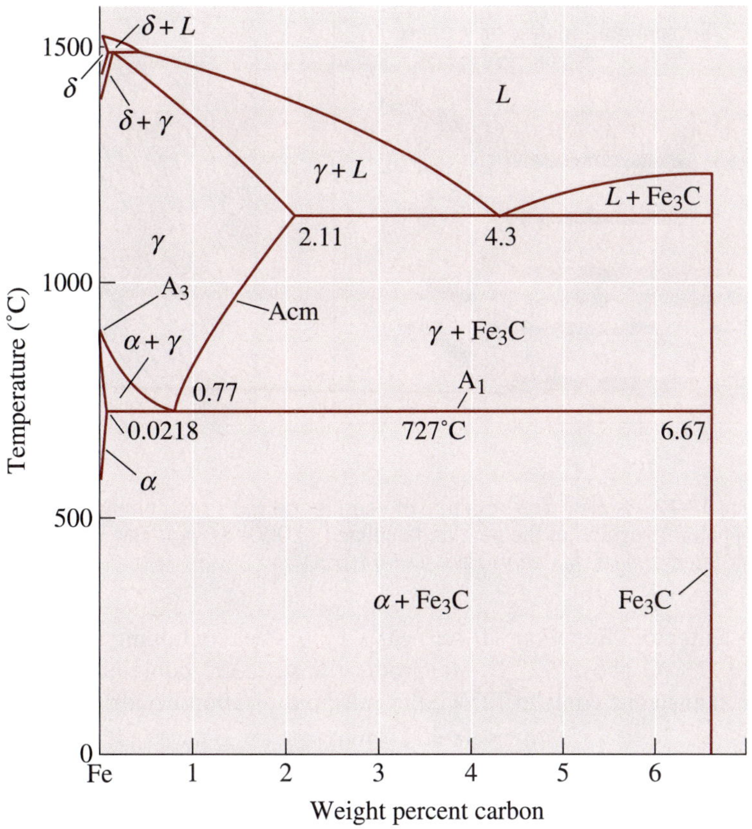

Figure 12-11 The Fe-Fe_3C phase diagram (a portion of the Fe-C diagram). The vertical line at 6.67% C is the stoichiometric compound Fe_3C.

Solid Solutions Iron goes through two allotropic transformations (Chapter 3) during heating or cooling. Immediately after solidification, iron forms a BCC structure called δ-ferrite. On further cooling, the iron transforms to a FCC structure called gamma (γ) or **austenite**. Finally, iron transforms back to the BCC structure at lower temperatures; this structure is called alpha (α) or **ferrite**. Both of the ferrites (α and δ) and the austenite are solid solutions of interstitial carbon atoms in iron (Example 4-3). Normally, when no specific reference is made, the term ferrite refers to the α ferrite, since this is the phase we encounter more often during the heat treatment of steels. Note that certain ceramic materials used in magnetic applications are also known as ferrites but are not related to the ferrite phase in the Fe-Fe_3C system.

Because interstitial holes in the FCC crystal structure are somewhat larger than the holes in the BCC crystal structure, a greater number of carbon atoms can be accommodated in FCC iron. Thus, the maximum solubility of carbon in austenite is 2.11% C, whereas the maximum solubility of carbon in BCC iron is much lower (i.e., ~0.0218% C in α and 0.09% C in δ). The solid solutions of carbon in iron are relatively soft and ductile, but are stronger than pure iron due to solid-solution strengthening by the carbon.

Compounds A stoichiometric compound Fe_3C, or **cementite**, forms when the solubility of carbon in solid iron is exceeded. The Fe_3C contains 6.67% C, is extremely hard and brittle (like a ceramic material), and is present in all commercial steels. By properly controlling the amount, size, and shape of Fe_3C, we control the degree of dispersion strengthening and the properties of the steel.

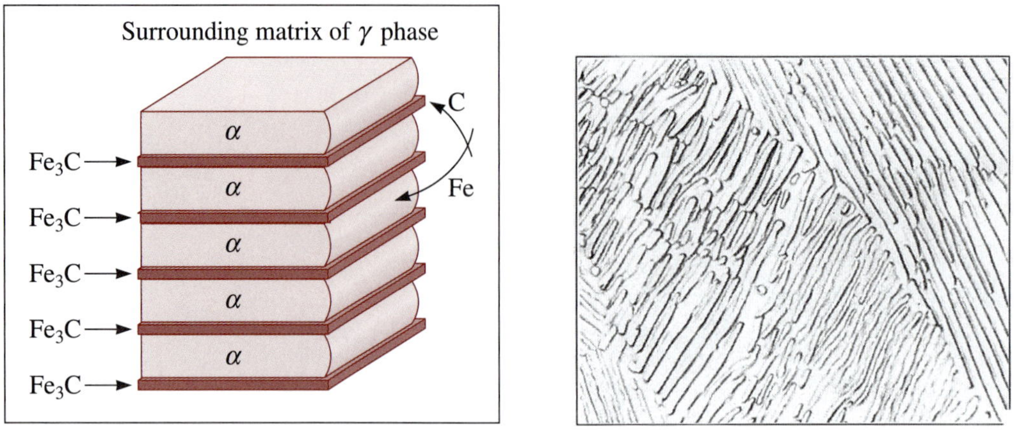

(a) (b)

Figure 12-12 Growth and structure of pearlite: (a) redistribution of carbon and iron, and (b) photomicrograph of the pearlite lamellae (×2000). (*From* ASM Handbook, *Vol. 7, (1972), ASM International, Materials Park, OH 44073.*)

The Eutectoid Reaction If we heat a Fe-C steel containing the eutectoid composition of 0.77% C above 727°C, we produce a structure containing only austenite grains. When austenite cools to 727°C, the eutectoid reaction begins:

$$\gamma(0.77\%\ \text{C}) \rightarrow \alpha(0.0218\%\ \text{C}) + \text{Fe}_3\text{C}(6.67\%\ \text{C}) \tag{12-6}$$

As in the eutectic reaction, the two phases that form have different compositions, so atoms must diffuse during the reaction (Figure 12-12). Most of the carbon in the austenite diffuses to the Fe_3C, and most of the iron atoms diffuse to α. This redistribution of atoms is easiest if the diffusion distances are short, which is the case when the α and Fe_3C grow as thin lamellae, or plates.

Pearlite The lamellar structure of α and Fe_3C that develops in the iron-carbon system is called **pearlite**, which is a microconstituent in steel. This was so named because a polished and etched pearlite shows the colorfulness of mother-of-pearl. The lamellae in pearlite are much finer than the lamellae in the lead-tin eutectic because the iron and carbon atoms must diffuse through solid austenite rather than through liquid. One way to think about pearlite is to consider it as a metal-ceramic nano-composite. The following example shows the calculation of the amounts of the phases in the pearlite microconstituent.

EXAMPLE 12-5 *Phases and Composition of Pearlite*

Calculate the amounts of ferrite and cementite present in pearlite.

SOLUTION

Since pearlite must contain 0.77% C, using the lever rule:

$$\%\ \alpha = \frac{6.67 - 0.77}{6.67 - 0.0218} \times 100 = 88.7\%$$

$$\%\ \text{Fe}_3\text{C} = \frac{0.77 - 0.0218}{6.67 - 0.0218} \times 100 = 11.3\%$$

In Example 12-5, we saw that most of the pearlite is composed of ferrite. In fact, if we examine the pearlite closely, we find that the Fe_3C lamellae are surrounded by α. The pearlite structure therefore provides dispersion strengthening—the continuous ferrite phase is relatively soft and ductile and the hard, brittle cementite is dispersed. The next example illustrates the similarity and differences between composites and pearlites.

EXAMPLE 12-6 *Tungsten Carbide (WC)-Cobalt (Co) Composite and Pearlite*

Tungsten carbide-cobalt composites, known as cemented carbides or carbides, are used as bits for cutting tools and drills (Chapter 1). What features are similar between these "cemented carbides" and pearlite, a microconstituent in steels? What are some of the major differences?

SOLUTION

Pearlite is very similar to the tungsten carbide-cobalt (WC-Co) composites known in industry as carbides. You may recall from earlier chapters that WC-Co are ceramic-metal composites (known as cermets) and used as cutting tools, drill bits, etc. In both materials, we take advantage of the toughness of one phase (ferrite or cobalt metal, in the case of pearlite in steel and WC-Co, respectively) and the hard ceramic-like phase (WC and Fe_3C, in WC-Co and steel, respectively). The metallic phase helps with ductility and the hard phase helps with strength. The difference is, WC and Co are two separate compounds that are sintered together using the powder metallurgy route. Pearlite is a microconstituent made up of two phases derived from same two elements (Fe-C). Another difference is in pearlite, the phases are formed via a eutectoid reaction. No such reaction occurs in the formation of WC-Co composites. Typically, WC-Co microstructure consists mainly of WC grains that are "glued" by cobalt grains. In pearlite, the metal-like ferrite phase dominates.

Primary Microconstituents Hypoeutectoid steels contain less than 0.77% C, and hypereutectoid steels contain more than 0.77% C. Ferrite is the primary or proeutectoid microconstituent in hypoeutectoid alloys, and cementite is the primary or proeutectoid microconstituent in hypereutectoid alloys. If we heat a hypoeutectoid alloy containing 0.60% C above 750°C, only austenite remains in the microstructure. Figure 12-13 shows what happens when the austenite cools. Just below 750°C, ferrite nucleates and grows, usually at the austenite grain boundaries. Primary ferrite continues to grow until the temperature falls to 727°C. The remaining austenite at that temperature is now surrounded by ferrite and has changed in composition from 0.60% C to 0.77% C. Subsequent cooling to below 727°C causes all of the remaining austenite to transform to pearlite by the eutectoid reaction. The structure contains two phases—ferrite and cementite—arranged as two microconstituents—primary ferrite and pearlite. The final microstructure contains islands of pearlite surrounded by the primary ferrite [Figure 12-14(a)]. This structure permits the alloy to be strong, due to the dispersion-strengthened pearlite, yet ductile, due to the continuous primary ferrite.

In hypereutectoid alloys, the primary phase is Fe_3C, which forms at the austenite grain boundaries. After the austenite cools through the eutectoid reaction, the steel contains hard, brittle cementite surrounding islands of pearlite [Figure 12-14(b)]. Now, because the hard, brittle microconstituent is continuous, the steel is also brittle.

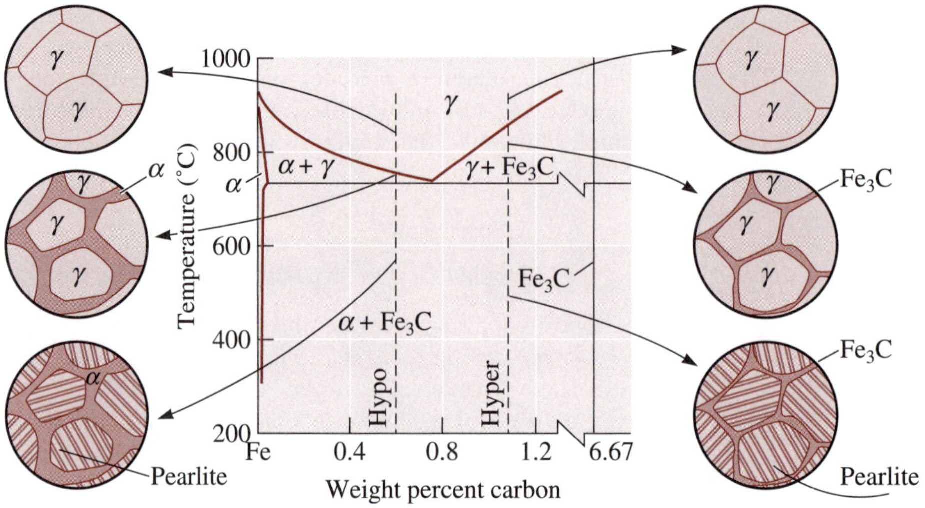

Figure 12-13 The evolution of the microstructure of hypoeutectoid and hypereutectoid steels during cooling, in relationship to the Fe-Fe_3C phase diagram.

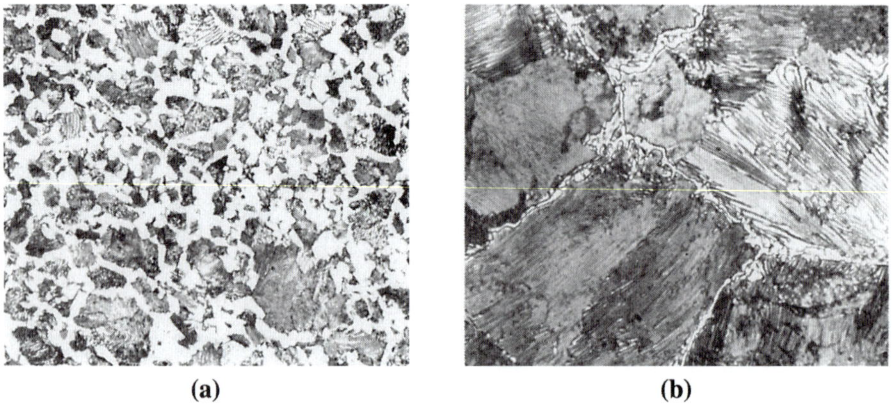

Figure 12-14 (a) A hypoeutectoid steel showing primary α (white) and pearlite ($\times$400). (b) A hypereutectoid steel showing primary Fe_3C surrounding pearlite ($\times$800). (*From* ASM Handbook, *Vol. 7, (1972), ASM International, Materials Park, OH 44073.*)

Fortunately, we can improve the microstructure and properties of the hypereutectoid steels by heat treatment. The following example shows the calculation for the composition of phases and microconstituent in a plain carbon steel.

EXAMPLE 12-7 *Phases in Hypoeutectoid Plain Carbon Steel*

Calculate the amounts and compositions of phases and microconstituents in a Fe-0.60% C alloy at 726°C.

SOLUTION

The phases are ferrite and cementite. Using a tie line and working the lever law at 726°C, we find:

$$\alpha\ (0.0218\%\ C) \quad \%\ \alpha = \left[\frac{6.67 - 0.60}{6.67 - 0.0218}\right] \times 100 = 91.3\%$$

$$Fe_3C\ (6.67\%\ C) \quad \%\ Fe_3C = \left[\frac{0.60 - 0.0218}{6.67 - 0.0218}\right] \times 100 = 8.7\%$$

The microconstituents are primary ferrite and pearlite. If we construct a tie line just above 727°C, we can calculate the amounts and compositions of ferrite and austenite just before the eutectoid reaction starts. All of the austenite at that temperature will have the eutectoid composition (i.e., it will contain 0.77% C) and will transform to pearlite; all of the proeutectoid ferrite will remain as primary ferrite.

$$\text{Primary}\ \alpha : 0.0218\%\ C \quad \%\ \text{Primary}\ \alpha = \left[\frac{0.77 - 0.60}{0.77 - 0.0218}\right] \times 100 = 22.7\%$$

$$\text{Austentite just above } 727°C = \text{Pearlite} : 0.77\%\ C$$

$$\%\ \text{Pearlite} = \left[\frac{0.60 - 0.0218}{0.77 - 0.0218}\right] \times 100 = 77.3\%$$

12-10 Controlling the Eutectoid Reaction

We control dispersion strengthening in the eutectoid alloys in much the same way that we did in eutectic alloys (Chapter 11).

Controlling the Amount of the Eutectoid By changing the composition of the alloy, we change the amount of the hard second phase. As the carbon content of steel increases towards the eutectoid composition of 0.77% C, the amounts of Fe_3C and pearlite increase, thus increasing the strength. However, this strengthening effect eventually peaks and the properties level out or even decrease when the carbon content is too high (Table 12-1).

Controlling the Austenite Grain Size Pearlite grows as grains or *colonies*. Within each colony, the orientation of the lamellae is identical. The colonies nucleate most easily

TABLE 12-1 ■ ***The effect of carbon on the strength of steels***

	Slow Cooling (Coarse Pearlite)			Fast Cooling (Fine Pearlite)		
Carbon %	Yield Strength (MPa)	Tensile Strength (MPa)	% Elongation	Yield Strength (MPa)	Tensile Strength (MPa)	% Elongation
0.20	295	394	36.5	347	441	36.0
0.40	353	519	30.0	374	590	28.0
0.60	372	626	23.0	420	776	18.0
0.80	376	615	25.0	524	1010	11.0
0.95	379	657	13.0	500	1014	9.5

After Metals Progress Materials and Processing Databook, *1981.*

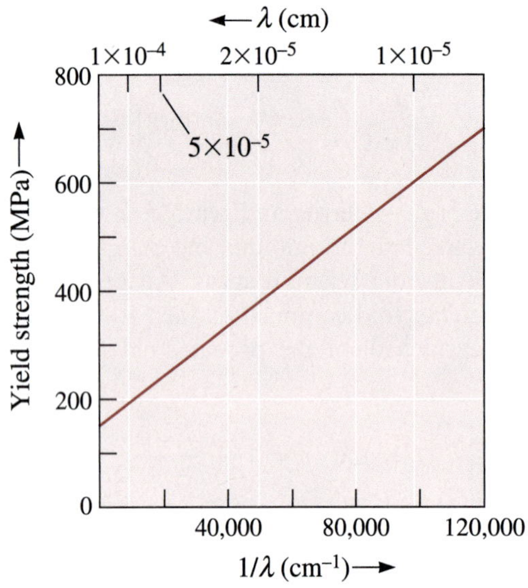

Figure 12-15
The effect of interlamellar spacing (λ) on the yield strength of pearlite.

at the grain boundaries of the original austenite grains. We can increase the number of pearlite colonies by reducing the prior austenite grain size, usually by using low temperatures to produce the austenite. Typically, we can increase the strength of the alloy by reducing the initial austenite grain size, thus increasing the number of colonies.

Controlling the Cooling Rate By increasing the cooling rate during the eutectoid reaction, we reduce the distance that the atoms are able to diffuse. Consequently, the lamellae produced during the reaction are finer or more closely spaced. By producing fine pearlite, we increase the strength of the alloy (Table 12-1 and Figure 12-15).

Controlling the Transformation Temperature The solid-state eutectoid reaction is rather slow, and the steel may cool below the equilibrium eutectoid temperature before the transformation begins (i.e., the austenite phase can be undercooled). Lower transformation temperatures give a finer, stronger structure (Figure 12-16), influence the time required for transformation, and even alter the arrangement of the two phases. This information is contained in the **time-temperature-transformation** (TTT) **diagram** (Figure 12-17). This diagram, also called the **isothermal transformation** (IT) diagram or

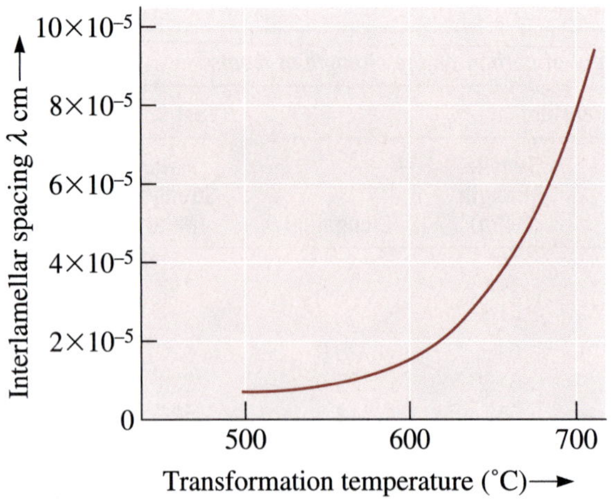

Figure 12-16
The effect of the austenite transformation temperature on the interlamellar spacing in pearlite.

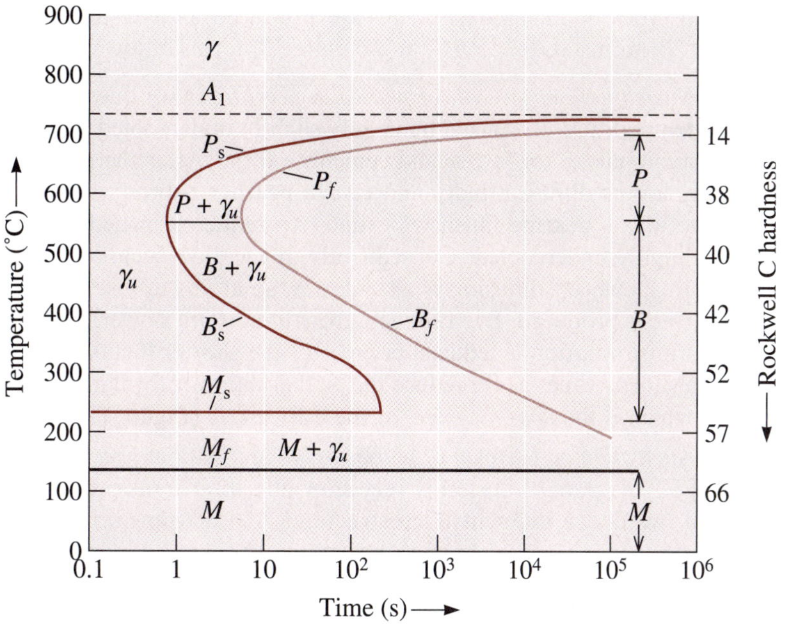

Figure 12-17 The time-temperature-transformation (TTT) diagram for an eutectoid steel.

the C-curve, permits us to predict the structure, properties, and heat treatment required in steels.

The shape of the TTT diagram is a consequence of the kinetics of the eutectoid reaction and is similar to the diagram shown by the Avrami relationship (Figure 12-3). At any particular temperature, a sigmoidal curve describes the rate at which the austenite transforms to a mixture of ferrite and cementite (Figure 12-18). An incubation time is required for nucleation. The P_s (pearlite start) curve represents the time at which austenite starts to transform to ferrite and cementite via the eutectoid transformation. The sigmoidal curve also gives the time at which the transformation is completed; this time is given by the P_f (pearlite finish) curve. When the temperature decreases from 727°C, the rate of nucleation increases, while the rate of growth of the eutectoid decreases. As in Figure 12-3, a maximum transformation rate, or minimum transformation time, is found; the maximum rate of transformation occurs near 550°C for an eutectoid steel (Figure 12-17).

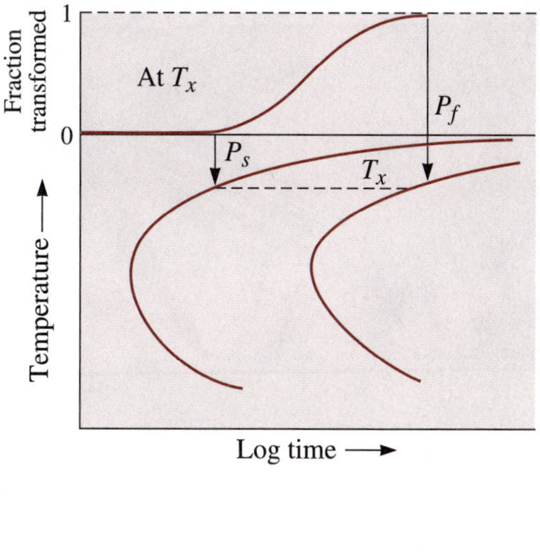

Figure 12-18
The sigmoidal curve is related to the start and finish times on the TTT diagram for steel. In this case, austenite is transforming to pearlite.

Two types of microconstituents are produced as a result of the transformation. Pearlite (P) forms above 550°C, and bainite (B) forms at lower temperatures:

1. *Nucleation and growth of phases in pearlite*: If we quench to just below the eutectoid temperature, the austenite is only slightly undercooled. Long times are required before stable nuclei for ferrite and cementite form. After the phases that form pearlite nucleate, atoms diffuse rapidly and *coarse* pearlite is produced; the transformation is complete at the pearlite finish (P_f) time. Austenite quenched to a lower temperature is more highly undercooled. Consequently, nucleation occurs more rapidly and the P_s is shorter. However, diffusion is also slower, so atoms diffuse only short distances and *fine* pearlite is produced. Even though growth rates are slower, the overall time required for the transformation is reduced because of the shorter incubation time. Finer pearlite forms in shorter times as we reduce the isothermal transformation temperature to about 550°C, which is the *nose*, or *knee*, of the TTT curve (Figure 12-17).

2. *Nucleation and growth of phases in bainite*: At a temperature just below the nose of the TTT diagram, diffusion is very slow and total transformation times increase. In addition, we find a different microstructure! At low transformation temperatures, the lamellae in pearlite would have to be extremely thin and, consequently, the boundary area between the ferrite and Fe_3C lamellae would be very large. Because of the energy associated with the ferrite-cementite interface, the total energy of the steel would have to be very high. Instead the steel reduces its internal energy by permitting the cementite to precipitate as discrete, rounded particles in a ferrite matrix. This new microconstituent, or arrangement of ferrite and cementite, is called **bainite**. Transformation begins at a bainite start (B_s) time and ends at a bainite finish (B_f) time.

The times required for austenite to begin and finish its transformation to bainite increase and the bainite becomes finer as the transformation temperature continues to decrease. The bainite that forms just below the nose of the curve is called coarse bainite, upper bainite, or feathery bainite. The bainite that forms at lower temperatures is called fine bainite, lower bainite, or acicular bainite. Figure 12-19 shows typical microstructures of bainite. Note that the morphology of bainite depends on the heat treatment used.

Figure 12-20 shows the effect of transformation temperature on the properties of eutectoid (0.77% C) steel. As the temperature decreases, there is a general trend toward higher strength and lower ductility due to the finer microstructure that is produced. The following two examples illustrate how we can design heat treatments of steels to produce desired microstructures and properties.

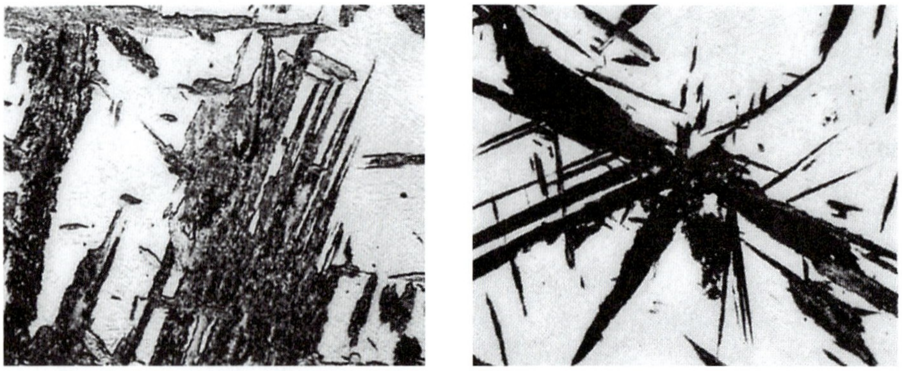

(a) **(b)**

Figure 12-19 (a) Upper bainite (gray, feathery plates) (×600). (b) Lower bainite (dark needles) (×400). (*From* ASM Handbook, *Vol. 8, (1973), ASM International, Materials Park, OH 44073.*)

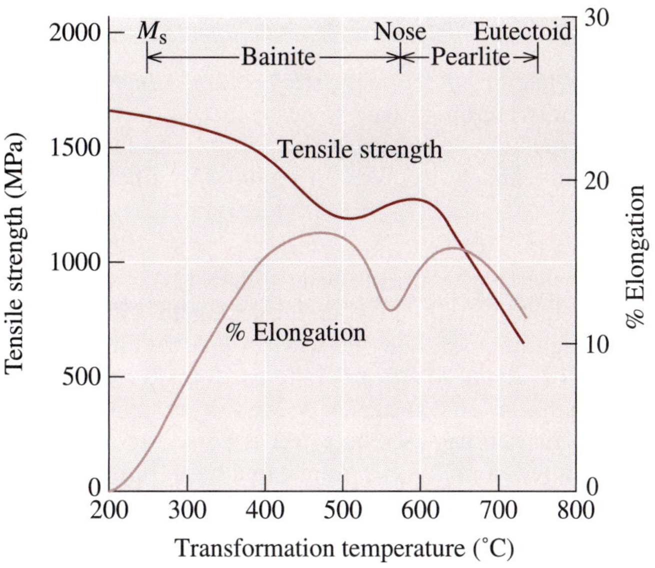

Figure 12-20 The effect of transformation temperature on the properties of an eutectoid steel (0.77% C).

EXAMPLE 12-8 *Design of a Heat Treatment to Generate Pearlite Microstructure*

Design a heat treatment to produce the pearlite structure shown in Figure 12-12(b).

SOLUTION

First, we need to determine the interlamellar spacing of the pearlite. If we count the number of lamellar spacings in the upper right of Figure 12-12(b), remembering that the interlamellar spacing is measured from one α plate to the next α plate, we find 14 spacings over a 2-cm distance. Due to the $\times$ 2000 magnification, this 2-cm distance is actually 0.001 cm. Thus:

$$\lambda = \left[\frac{0.001 \text{ cm}}{14 \text{ spacings}}\right] = 7.14 \times 10^{-5} \text{ cm}$$

If we assume that the pearlite is formed by an isothermal transformation, we find from Figure 12-16 that the transformation temperature must have been approximately 700°C. From the TTT diagram (Figure 12-17), our heat treatment must have been:

1. Heat the steel to about 750°C and hold—perhaps for 1 h—to produce all austenite. A higher temperature may cause excessive growth of austenite grains (Chapter 5).
2. Quench to 700°C and hold for at least 10^5 s (the P_f time). We assume here that the steel cools instantly to 700°C. In practice, this does not happen and thus the transformation does not occur at one temperature. We may need to use the continuous cooling transformation diagrams to be more precise (Chapter 13).
3. Cool to room temperature.

The steel should have a hardness of HRC 14 (Figure 12-17) and a yield strength of about 200 MPa, as shown in Figure 12-15.

EXAMPLE 12-9 Heat Treatment to Generate Bainite Microstructure

Excellent combinations of hardness, strength, and toughness are obtained from bainite. One heat treatment facility austenitized an eutectoid steel at 750°C, quenched and held the steel at 250°C for 15 min, and finally permitted the steel to cool to room temperature. Was the required bainitic structure produced?

SOLUTION

Let's examine the heat treatment using Figure 12-17. After heating at 750°C, the microstructure is 100% γ. After quenching to 250°C, unstable austenite remains for slightly more than 100 s, when fine bainite begins to grow. After 15 min, or 900 s, about 50% fine bainite has formed and the remainder of the steel still contains unstable austenite. As we will see later, the unstable austenite transforms to martensite when the steel is cooled to room temperature and the final structure is a mixture of bainite and hard, brittle martensite. The heat treatment was not successful! The heat treatment facility should have held the steel at 250°C for at least 10^4 s, or about 3 h.

12-11 The Martensitic Reaction and Tempering

Martensite is a phase that forms as the result of a diffusionless solid-state transformation. In this transformation there is no diffusion and, hence, it does not follow the Avrami transformation kinetics. The growth rate in **martensitic transformations** (also known as **displacive** or **athermal transformations**) is so high that nucleation becomes the controlling step. The phase that forms upon the quenching of steels was named "martensite" by Floris Osmond in 1895 in honor of German metallurgist Adolf Martens. Similar martensitic phase transformations occur in other systems as well.

Cobalt, for example, transforms from a FCC to a HCP crystal structure by a slight shift in the atom locations that alters the stacking sequence of close-packed planes. Because the reaction does not depend on diffusion, the martensite reaction is an athermal transformation—that is, the reaction depends only on the temperature, not on the time. The martensite reaction often proceeds rapidly, at speeds approaching the velocity of sound in the material.

Many other alloys such as Cu-Zn-Al and Cu-Al-Ni and Ni-Ti show martensitic phase transformations. These transformations can also be driven by the application of mechanical stress. Other than martensite that forms in certain type of steels, the Ni-Ti alloy, known as nitinol (which stands for Nickel Titanium Naval Ordinance Laboratory, developed by the U.S. Naval Ordinance Laboratory in the 1940s) is perhaps the best-known example of alloys that make use of martensitic phase transformations. These materials can remember their shape (i.e. shape memory effect) and are known as **shape-memory alloys** (SMAs).

Martensite in Steels In steels with less than about 0.2% C, the FCC austenite transforms to a supersaturated BCC martensite structure. In higher carbon steels, the martensite reaction occurs as FCC austenite transforms to BCT (body centered tetragonal) martensite. The relationship between the FCC austenite and the BCT martensite [Figure 12-21(a)] shows that carbon atoms in the 1/2, 0, 0 type of interstitial sites in the

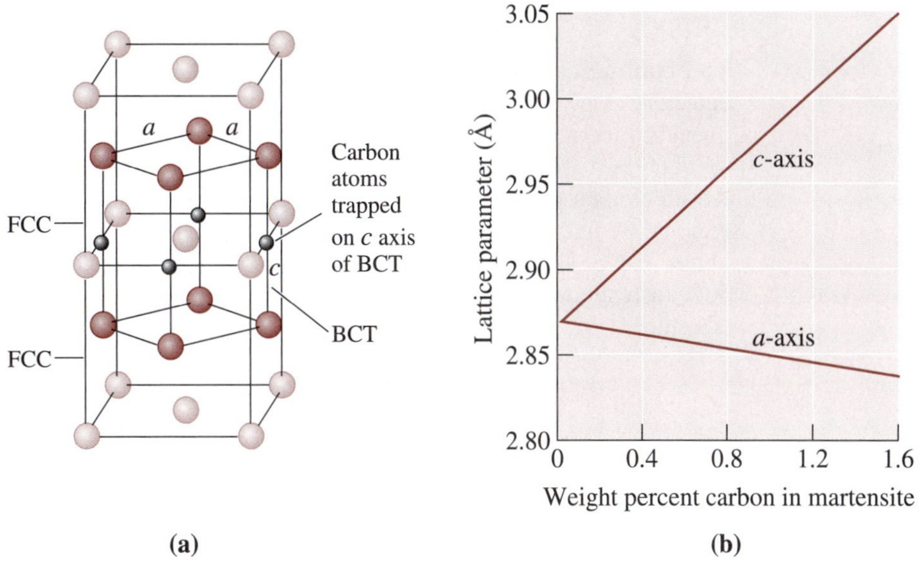

Figure 12-21 (a) The unit cell of BCT martensite is related to the FCC austenite unit cell. (b) As the percentage of carbon increases, more interstitial sites are filled by the carbon atoms and the tetragonal structure of the martensite becomes more pronounced.

FCC cell can be trapped during the transformation to the body-centered structure, causing the tetragonal structure to be produced. As the carbon content of the steel increases, a greater number of carbon atoms are trapped in these sites, thereby increasing the difference between the *a*- and *c*-axes of the martensite [Figure 12-21(b)].

The steel must be quenched, or rapidly cooled, from the stable austenite region to prevent the formation of pearlite, bainite, or primary microconstituents. The martensite reaction begins in an eutectoid steel when austenite cools below 220°C, the martensite start (M_s) temperature (Figure 12-17). The amount of martensite increases as the temperature decreases. When the temperature passes below the martensite finish temperature (M_f), the steel should contain 100% martensite. At any intermediate temperature, the amount of martensite does not change as the time at that temperature increases.

Owing to the conservation of mass, the composition of martensite must be the same as that of the austenite from which it forms. There is no long-range diffusion during the transformation that can change the composition. Thus, in iron-carbon alloys, the initial austenite composition and the final martensite composition are the same. The following example illustrates how heat treatment is used to produce a dual phase steel.

EXAMPLE 12-10 *Design of a Heat Treatment for a Dual Phase Steel*

Unusual combinations of properties can be obtained by producing a steel whose microstructure contains 50% ferrite and 50% martensite; the martensite provides strength and the ferrite provides ductility and toughness. Design a heat treatment to produce a dual phase steel in which the composition of the martensite is 0.60% C.

SOLUTION

To obtain a mixture of ferrite and martensite, we need to heat-treat a hypoeutectoid steel into the $\alpha + \gamma$ region of the phase diagram. The steel is then quenched, permitting the γ portion of the structure to transform to martensite.

The heat treatment temperature is fixed by the requirement that the martensite contain 0.60% C. From the solubility line between the γ and the $\alpha + \gamma$ regions, we find that 0.60% C is obtained in austenite when the temperature is about 750°C. To produce 50% martensite, we need to select a steel that gives 50% austenite when the steel is held at 750°C. If the carbon content of the steel is x, then:

$$\%\,\gamma = \left[\frac{(x - 0.02)}{(0.60 - 0.02)}\right] \times 100 = 50 \quad \text{or} \quad x = 0.31\%\ \text{C}$$

Our final design is:

1. Select a hypoeutectoid steel containing 0.31% C.
2. Heat the steel to 750°C and hold (perhaps for 1 h, depending on the thickness of the part) to produce a structure containing 50% ferrite and 50% austenite, with 0.60% C in the austenite.
3. Quench the steel to room temperature. The austenite transforms to martensite, also containing 0.60% C.

Properties of Steel Martensite Martensite in steels is very hard and brittle, just like ceramics. The BCT crystal structure has no close-packed slip planes in which dislocations can easily move. The martensite is highly supersaturated with carbon, since iron normally contains less than 0.0218% C at room temperature, and martensite contains the amount of carbon present in the steel. Finally, martensite has a fine grain size and an even finer substructure within the grains.

The structure and properties of steel martensites depend on the carbon content of the alloy (Figure 12-22). When the carbon content is low, the martensite grows in a "lath" shape, composed of bundles of flat, narrow plates that grow side by side

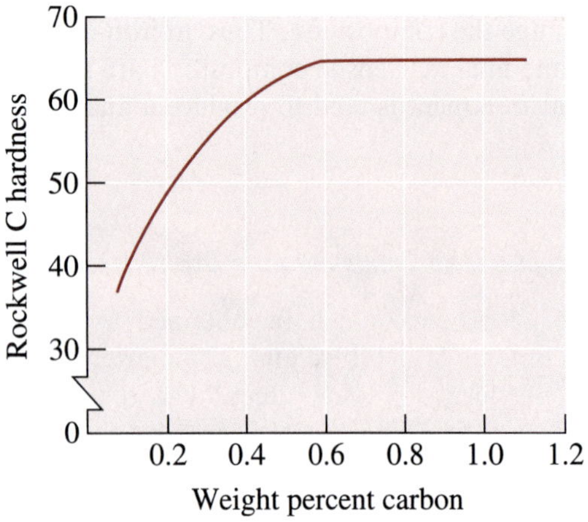

Figure 12-22
The effect of carbon content on the hardness of martensite in steels.

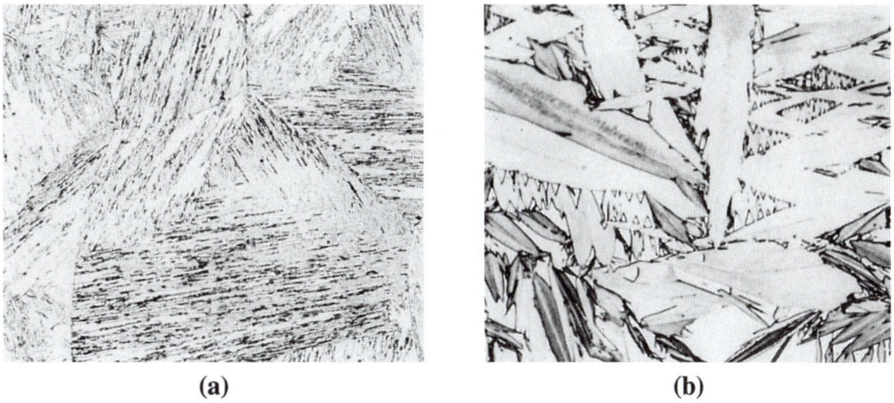

Figure 12-23 (a) Lath martensite in low-carbon steel (×80). (b) Plate martensite in high-carbon steel (×400). (*From* ASM Handbook, *Vol. 8, (1973), ASM International, Materials Park, OH 44073.*)

[Figure 12-23(a)]. This martensite is not very hard. At a higher carbon content, plate martensite grows, in which flat, narrow plates grow individually rather than as bundles [Figure 12-23(b)]. The hardness is much greater in the higher carbon, plate martensite structure, partly due to the greater distortion, or large c/a ratio, of the crystal structure.

Tempering of Steel Martensite Martensite is not an equilibrium phase. This is why it does not appear on the Fe-Fe_3C phase diagram (Figure 12-11). When martensite in a steel is heated below the eutectoid temperature, the thermodynamically stable α and Fe_3C phases precipitate. This process is called **tempering**. The decomposition of martensite in steels causes the strength and hardness of the steel to decrease while the ductility and impact properties are improved (Figure 12-24). Note that the term tempering here is different from the term we used for tempering of silicate glasses. In both

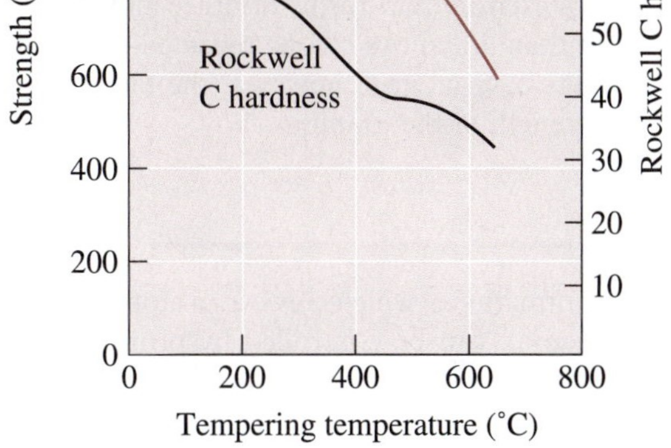

Figure 12-24 Effect of tempering temperature on the properties of an eutectoid steel.

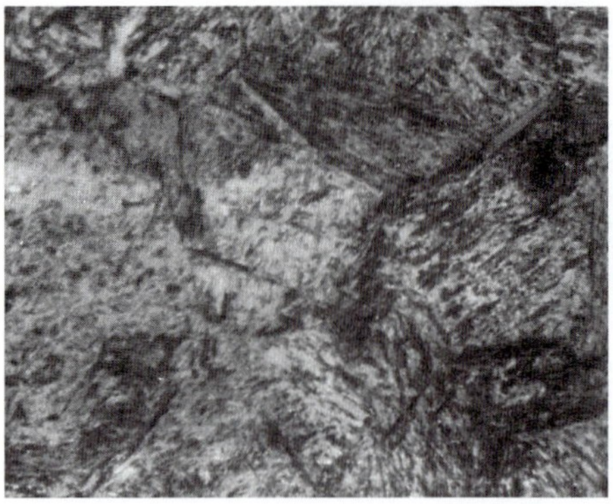

Figure 12-25
Tempered martensite in steel (×500). (*From* ASM Handbook, *Vol. 9, Metallography and Microstructure (1985), ASM International Materials Park, OH 44073.*)

tempering of glasses and tempering of steels, however, the key result is an increase in the toughness of the material.

At low tempering temperatures, the martensite may form two transition phases—a lower carbon martensite and a very fine nonequilibrium ε-carbide, or $Fe_{2.4}C$. The steel is still strong, brittle, and perhaps even harder than before tempering. At higher temperatures, the stable α and Fe_3C form and the steel becomes softer and more ductile. If the steel is tempered just below the eutectoid temperature, the Fe_3C becomes very coarse and the dispersion-strengthening effect is greatly reduced. By selecting the appropriate tempering temperature, a wide range of properties can be obtained. The product of the tempering process is a microconstituent called tempered martensite (Figure 12-25).

Martensite in Other Systems The characteristics of the martensite reaction are different in other alloy systems. For example, martensite can form in iron-based alloys that contain little or no carbon by a transformation of the FCC crystal structure to a BCC crystal structure. In certain high-manganese steels and stainless steels, the FCC structure changes to a HCP crystal structure during the martensite transformation. In addition, the martensitic reaction occurs during the transformation of many polymorphic ceramic materials, including ZrO_2, and even in some crystalline polymers. Thus the terms martensitic reaction and martensite are rather generic. In the context of steel properties, microstructure, and heat treatment, the term "martensite" refers to the hard and brittle bct phase obtained upon the quenching of steels.

The properties of martensite in other alloys are also different from the properties of steel martensite. In titanium alloys, BCC titanium transforms to a HCP martensite structure during quenching. However, the titanium martensite is softer and weaker than the original structure. The martensite that forms in other alloys can also be tempered. The martensite produced in titanium alloys can be reheated to permit the precipitation of a second phase. Unlike the case of steel, however, the tempering process *increases*, rather than decreases, the strength of the titanium alloy.

SUMMARY

- Solid-state phase transformations, which have a profound effect on the structure and properties of a material, can be controlled by proper heat treatments. These heat treatments are designed to provide an optimum distribution of two or more phases in the microstructure. Dispersion strengthening permits a wide variety of structures and properties to be obtained.

- Age hardening, or precipitation hardening, is one powerful method for controlling the optimum dispersion strengthening in many metallic alloys. In age hardening, a very fine widely dispersed coherent precipitate is allowed to precipitate by a heat treatment that includes (a) solution treating to produce a single-phase solid solution, (b) quenching to retain that single phase, and (c) aging to permit a precipitate to form. In order for age hardening to occur, the phase diagram must show decreasing solubility of the solute in the solvent as the temperature decreases.
- The most widely used eutectoid reaction occurs in producing steels from iron-carbon alloys: Either pearlite or bainite can be produced as a result of the eutectoid reaction in steel. In addition, primary ferrite or primary cementite may be present, depending on the carbon content of the alloy. The trick is to formulate a microstructure that consists of a right mix of metal-like phases that are tough and ceramic-like phases that are hard and brittle.
- Factors that influence the mechanical properties of the microconstituent produced by the eutectoid reaction include (a) the composition of the alloy (amount of eutectoid microconstituent), (b) the grain size of the original solid, the eutectoid microconstituent, and any primary microconstituents, (c) the fineness of the structure within the eutectoid microconstituent (interlamellar spacing), (d) the cooling rate during the phase transformation, and (e) the temperature at which the transformation occurs (the amount of undercooling).
- A martensitic reaction occurs with no long-range diffusion. Again, the best known example occurs in steels:
    - The amount of martensite that forms depends on the temperature of the transformation (an athermal reaction).
    - Martensite is very hard and brittle, with the hardness determined primarily by the carbon content.
    - The amount and composition of the martensite are the same as the austenite from which it forms.
- Martensite can be tempered. During tempering, a dispersion-strengthened structure is produced. In steels, tempering reduces the strength and hardness but improves the ductility and toughness.

GLOSSARY

Age hardening A special dispersion-strengthening heat treatment. By solution treatment, quenching, and aging, a coherent precipitate forms that provides a substantial strengthening effect. Also known as precipitation hardening.

Artificial aging Reheating a solution-treated and quenched alloy to a temperature below the solvus in order to provide the thermal energy required for a precipitate to form.

Athermal transformation When the amount of the transformation depends only on the temperature, not on the time (same as martensitic transformation or displacive transformation).

Austenite The name given to the FCC crystal structure of iron.

Avrami relationship Describes the fraction of a transformation that occurs as a function of time. This describes most solid-state transformations that involve diffusion, thus martensitic transformations are not described.

Bainite A two-phase microconstituent, containing ferrite and cementite, that forms in steels that are isothermally transformed at relatively low temperatures.

Cementite The hard, brittle ceramic-like compound Fe_3C that, when properly dispersed, provides the strengthening in steels.

Coherent precipitate A precipitate whose crystal structure and atomic arrangement have a continuous relationship with the matrix from which the precipitate is formed. The formation of coherent precipitate provides excellent disruption of the atomic arrangement in the matrix and provides excellent strengthening.

Displacive transformation A phase transformation that occurs via small displacements of atoms or ions and without diffusion. Same as athermal or martensitic transformation.

Ferrite The name given to the BCC crystal structure of iron that can occur as α or δ. This is not to be confused with ceramic ferrites which are magnetic materials.

Guinier-Preston (GP) zones Tiny clusters of atoms that precipitate from the matrix in the early stages of the age-hardening process. Although the GP zones are coherent with the matrix, they are too small to provide optimum strengthening.

Interfacial energy The energy associated with the boundary between two phases.

Isothermal transformation When the amount of a transformation at a particular temperature depends on the time permitted for the transformation.

Martensite A metastable phase formed in steel and other materials by a diffusionless, athermal transformation.

Martensitic transformation A phase transformation that occurs without diffusion. Same as athermal or displacive transformation. These occur in steels, Ni-Ti and many ceramic materials.

Natural aging When a coherent precipitate forms from a solution-treated and quenched age-hardenable alloy at room temperature, providing optimum strengthening.

Pearlite A two-phase lamellar microconstituent, containing ferrite and cementite, that forms in steels cooled in a normal fashion or isothermally transformed at relatively high temperatures.

Shape-memory alloys (SMAs) Certain materials which develop microstructures that, after being deformed, can return the material to its initial shape when heated (e.g. Ni-Ti alloys).

Solution treatment The first step in the age-hardening heat treatment. The alloy is heated above the solvus temperature to dissolve any second phase and to produce a homogeneous single-phase structure.

Strain energy The energy required to permit a precipitate to fit into the surrounding matrix during nucleation and growth of the precipitate.

Supersaturated solid solution The solid solution formed when a material is rapidly cooled from a high-temperature single-phase region to a low-temperature two-phase region without the second phase precipitating. Because the quenched phase contains more alloying element than the solubility limit, it is supersaturated in that element.

Tempering A low-temperature heat treatment used to reduce the hardness of martensite by permitting the martensite to begin to decompose to the equilibrium phases. This leads to increased toughness.

TTT diagram The time-temperature-transformation diagram describes the time required at any temperature for a phase transformation to begin and end. The TTT diagram assumes that the temperature is constant during the transformation.

Widmanstðtten structure The precipitation of a second phase from the matrix when there is a fixed crystallographic relationship between the precipitate and matrix crystal structures. Often needle-like or plate-like structures form in the Widmanstðtten structure.

PROBLEMS

Section 12-1 Nucleation and Growth in Solid-State Reactions

12-1 How is the equation for nucleation of a phase in the solid state different from that for a liquid to solid transformation?

12-2 Determine the constants c and n in Equation 12-2 that describe the rate of crystallization of polypropylene at 140°C. (See Figure 12-26.)

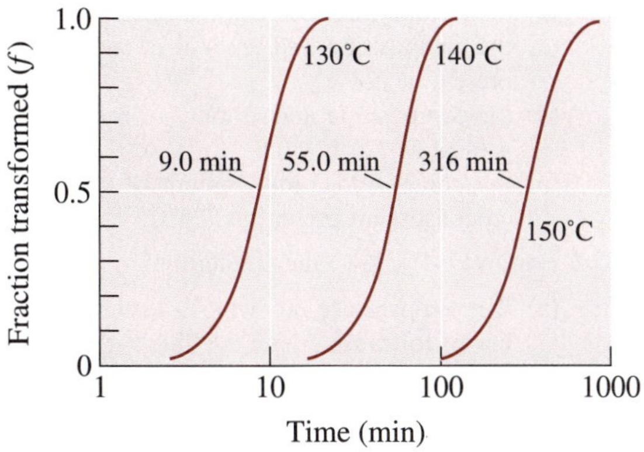

Figure 12-26 The effect of temperature on the crystallization of polypropylene (for Problem 12-2).

Section 12-2 Alloys Strengthened By Exceeding the Solubility Limit

12-3 What are the different ways by which a second phase can be made to precipitate in a two-phase microstructure?

12-4 Explain, when cooled slowly, why it is that the second phase in Al-4% Cu alloys nucleates and grows along the grain boundaries. Is this usually desirable?

12-5 What properties of the precipitate phase are needed for precipitation hardening? Why?

Section 12-3 Age or Precipitation Hardening

12-6 What is the principle of precipitation hardening?

12-7 What is a supersaturated solution? How do we obtain supersaturated solutions during precipitation hardening? Why is the formation of a supersaturated solution necessary?

12-8 Suppose that age hardening is possible in the Al-Mg system. (See Figure 12-8.)

(a) Recommend an artificial age-hardening heat treatment for each of the following alloys, and

(b) compare the amount of the β precipitate that forms from your treatment of each alloy.
(i) Al-4% Mg **(ii)** Al-6% Mg
(iii) Al-12% Mg

(c) Testing of the alloys after the heat treatment reveals that little strengthening occurs as a result of the heat treatment. Which of the requirements for age hardening is likely not satisfied?

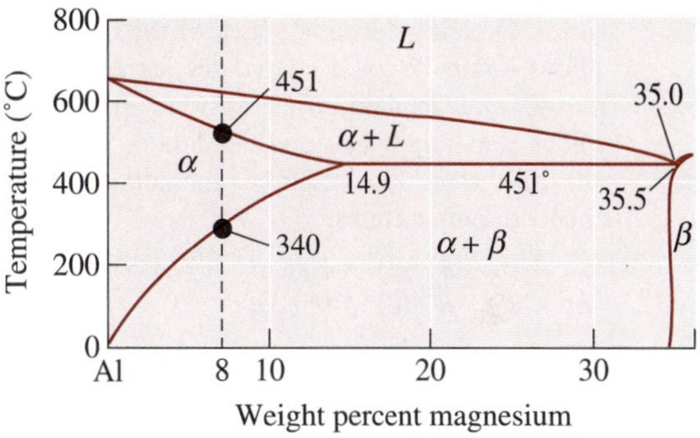

Figure 12-8 (Repeated for Problem 12-8) Portion of the aluminum-magnesium phase diagram.

12-9 An Al-2.5% Cu alloy is solution-treated, quenched, and overaged at 230°C to produce a stable microstructure. If the θ precipitates as spheres with a diameter of 9000 A and a density of 4.26 g/cm^3, determine the number of precipitate particles per cm^3.

Section 12-4 Applications of Age-Hardened Alloys

12-10 Why is precipitation hardening an attractive mechanism of strengthening for aircraft materials?

12-11 Why are most precipitation-hardened alloys suitable only for relatively low-temperature applications?

Section 12-5 Microstructural Evolution in Age or Precipitation Hardening

12-12 Explain the three basic steps encountered during precipitation hardening.

Section 12-6 Effects of Aging Temperature and Time

12-13 What is aging? Why is this step needed in precipitation hardening?

12-14 What do the terms "natural aging" and "artificial aging" mean?

12-15 In the plane flown by the Wright brothers, how was the alloy precipitation strengthened?

Section 12-7 Requirements for Age Hardening

12-16 Can all alloy compositions be strengthened using precipitation hardening? Can we use this mechanism for the strengthening of ceramics, glasses, or polymers?

12-17 A conductive copper wire is to be made. Would you choose precipitation hardening as a way of strengthening this wire? Explain.

12-18 Figure 12-27 shows a hypothetical phase diagram. Determine whether each of the following alloys might be good candidates for age hardening, and explain your answer. For those alloys that might be good candidates, describe the heat treatment required, including recommended temperatures.

(a) *A*-10% *B* **(b)** *A*-20% *B* **(c)** *A*-55% *B*
(d) *A*-87% *B* **(e)** *A*-95% *B*

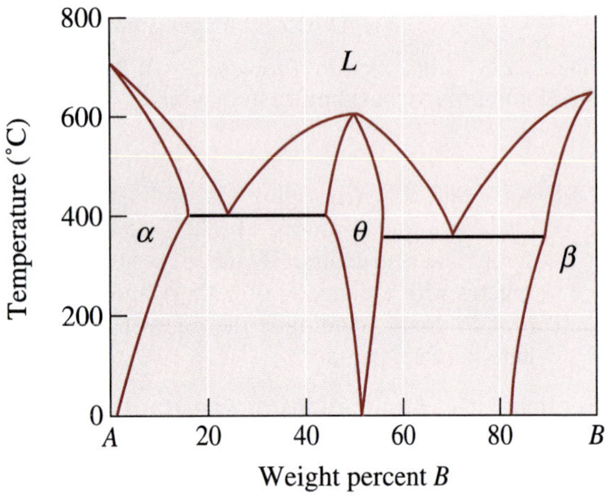

Figure 12-27 Hypothetical phase diagram (for Problem 12-18).

Section 12-8 Use of Age-Hardenable Alloys at High Temperatures

12-19 Why is it that certain aluminum (not nickel-based) alloys strengthened using age hardening can lose their strength on welding?

12-20 Would you choose a precipitation-hardened alloy to make an aluminum alloy baseball bat?

Section 12-9 The Eutectoid Reaction

12-21 Write down the eutectoid reaction in Fe-Fe_3C system.

12-22 Compare and contrast eutectic and eutectoid reactions.

12-23 Define the following terms: ferrite, austenite, pearlite, and cementite.

12-24 The pearlite microstructure is similar to a ceramic-metal nanocomposite. True or False. Comment.

12-25 What is the difference between a microconstituent and a phase?

12-26 For an Fe-0.35% C alloy, determine

(a) the temperature at which austenite first begins to transform on cooling,
(b) the primary microconstituent that forms,
(c) the composition and amount of each phase present at 728°C,
(d) the composition and amount of each phase present at 726°C, and
(e) the composition and amount of each microconstituent present at 726°C.

12-27 For an Fe-1.15% C alloy, determine

(a) the temperature at which austenite first begins to transform on cooling,
(b) the primary microconstituent that forms,
(c) the composition and amount of each phase present at 728°C,
(d) the composition and amount of each phase present at 726°C, and
(e) the composition and amount of each microconstituent present at 726°C.

12-28 A steel contains 8% cementite and 92% ferrite at room temperature. Estimate the carbon content of the steel. Is the steel hypoeutectoid or hypereutectoid?

12-29 A steel contains 18% cementite and 82% ferrite at room temperature. Estimate the carbon content of the steel. Is the steel hypoeutectoid or hypereutectoid?

12-30 A steel contains 18% pearlite and 82% primary ferrite at room temperature. Estimate the carbon content of the steel. Is the steel hypoeutectoid or hypereutectoid?

12-31 A steel contains 94% pearlite and 6% primary cementite at room temperature. Estimate the carbon content of the steel. Is the steel hypoeutectoid or hypereutectoid?

12-32 A steel contains 55% α and 45% γ at 750°C. Estimate the carbon content of the steel.

12-33 A steel contains 96% γ and 4% Fe_3C at 800°C. Estimate the carbon content of the steel.

12-34 A steel is heated until 40% austenite, with a carbon content of 0.5%, forms. Estimate the temperature and the overall carbon content of the steel.

12-35 A steel is heated until 85% austenite, with a carbon content of 1.05%, forms. Estimate the temperature and the overall carbon content of the steel.

Section 12-10 Controlling the Eutectoid Reaction

12-36 Why are the distances between lamellae formed in an eutectoid reaction typically separated by distances smaller than those formed in eutectic reactions?

12-37 Compare the interlamellar spacing and the yield strength when an eutectoid steel is isothermally transformed to pearlite at

(a) 700°C, and
(b) 600°C.

12-38 Why is it that a eutectoid steel exhibits different yield strengths and % elongations, depending upon if it was cooled slowly or relatively fast?

12-39 What is a TTT diagram?

12-40 Sketch and label clearly different parts of a TTT diagram for a plain-carbon steel with 0.77% carbon.

12-41 On the TTT diagram what is the difference between the γ and γ_u phases?

12-42 How is it that bainite and pearlite do not appear in the Fe-Fe_3C diagram? Are these phases or microconstituents?

12-43 Why is it that we cannot make use of TTT diagrams for describing heat treatment profiles in which samples are getting cooled over a period of time (i.e., why are TTT diagrams suitable for only following isothermal transformations)?

12-44 What is bainite? Why do steels containing bainite exhibit higher levels of toughness?

12-45 An isothermally transformed eutectoid steel is found to have a yield strength of 410 MPa. Estimate

(a) the transformation temperature, and
(b) the interlamellar spacing in the pearlite.

12-46 Determine the required transformation temperature and microconstituent if an eutectoid steel is to have the following hardness values:

(a) HRC 38 (b) HRC 42
(c) HRC 48 (d) HRC 52

12-47 Describe the hardness and microstructure in an eutectoid steel that has been heated to 800°C for 1 h, quenched to 350°C and held for 750 s, and finally quenched to room temperature.

12-48 Describe the hardness and microstructure in an eutectoid steel that has been heated to 800°C, quenched to 650°C, held for 500 s, and finally quenched to room temperature.

12-49 Describe the hardness and microstructure in an eutectoid steel that has been heated to 800°C, quenched to 300°C and held for 10 s, and finally quenched to room temperature.

12-50 Describe the hardness and microstructure in an eutectoid steel that has been heated to 800°C, quenched to 300°C and held for 10 s, quenched to room temperature, and then reheated to 400°C before finally cooling to room temperature again.

12-51 A steel containing 0.3% C is heated to various temperatures above the eutectoid temperature, held for 1 h, and then quenched to room temperature. Using Figure 12-28, determine the amount, composition, and hardness of any martensite that forms when the heating temperature is:

(a) 728°C (b) 750°C
(c) 790°C (d) 850°C

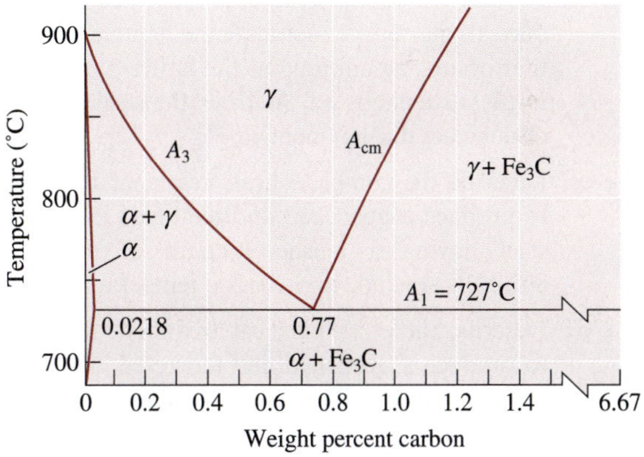

Figure 12-28 The eutectoid portion of the Fe-Fe_3C phase diagram (for Problems 12-51, 12-56, 12-57, and 12-58).

Section 12-11 The Martensitic Reaction and Tempering

12-52 What is the difference between solid-state phase transformations such as the eutectoid reaction and the martensitic phase transformation?

12-53 What is the difference between isothermal and athermal transformations?

12-54 Why does the martensite phase not appear on the Fe-Fe_3C phase diagram?

12-55 Compare the mechanical properties of martensite, pearlite, and bainite formed from eutectoid steel composition.

12-56 A steel containing 0.95% C is heated to various temperatures above the eutectoid temperature, held for 1 h, and then quenched to room temperature. Using Figure 12-28 determine the amount and composition of any martensite that forms when the heating temperature is:

(a) 728°C **(b)** 750°C
(c) 780°C **(d)** 850°C

12-57 A steel microstructure contains 75% martensite and 25% ferrite; the composition of the martensite is 0.6% C. Using Figure 12-28, determine

(a) the temperature from which the steel was quenched, and
(b) the carbon content of the steel.

12-58 A steel microstructure contains 92% martensite and 8% Fe_3C; the composition of the martensite is 1.10% C. Using Figure 12-28, determine

(a) the temperature from which the steel was quenched, and
(b) the carbon content of the steel.

12-59 A steel containing 0.8% C is quenched to produce all martensite. Estimate the volume change that occurs, assuming that the lattice parameter of the austenite is 3.6 A. Does the steel expand or contract during quenching?

12-60 Describe the complete heat treatment required to produce a quenched and tempered eutectoid steel having a tensile strength of at least 862 MPa. Include appropriate temperatures.

12-61 Describe the complete heat treatment required to produce a quenched and tempered eutectoid steel having a HRC hardness of less than 50. Include appropriate temperatures.

12-62 In eutectic alloys, the eutectic microconstituent is generally the continuous one, but in the eutectoid structures, the primary microconstituent is normally continuous. By describing the changes that occur with decreasing temperature in each reaction, explain why this difference is expected.

12-63 What is the tempering of steels? Why is tempering necessary?

12-64 What phases are formed by the decomposition of martensite?

12-65 What is tempered martensite?

12-66 If tempering results in the decomposition of martensite, why should we form martensite in the first place?

12-67 Describe the changes in properties that occur upon the tempering of an eutectoid steel.

Design Problems

12-68 You wish to attach aluminum sheets to the frame of the twenty-fourth floor of a skyscraper. You plan to use rivets made of an age-hardenable aluminum, but the rivets must be soft and ductile in order to close. After the sheets are attached, the rivets must be very strong. Design a method for producing, using, and strengthening the rivets.

12-69 An age-hardened, Al-Cu bracket is used to hold a heavy electrical-sensing device on the outside of a steel-making furnace. Temperatures may exceed 200°C. Is this a good design? Explain. If it is not, design an appropriate bracket and explain why your choice is acceptable.

12-70 You use an arc-welding process to join an eutectoid steel. Cooling rates may be very high following the joining process. Describe what happens in the heat-affected zone of the weld and discuss the problems that might occur. Design a joining process that may minimize these problems.

13

Heat Treatment of Steels and Cast Irons

Have You Ever Wondered?

- *What is the most widely used engineered material?*
- *What makes stainless steels "stainless"?*
- *What is the difference between cast iron and steel?*
- *Are stainless steels magnetic?*
- *Is a tin can made out of tin?*

Ferrous alloys, which are based on iron-carbon alloys, include plain-carbon steels, alloy and tool steels, stainless steels, and cast irons. These are the most widely used materials in the world. In the history of civilization, these materials made their mark by defining the *Iron Age*. Steels typically are produced in two ways: by refining iron ore or by recycling scrap steel.

In producing primary steel, iron ore (processed to contain 50 to 70% iron oxide, Fe_2O_3 or Fe_3O_4) is heated in a *blast furnace* in the presence of coke (a form of carbon) and oxygen. The iron oxide is reduced into a crude molten iron known as **hot metal** or **pig iron**. At about $\sim$1600°C, this material contains about 95% iron; 4% carbon; 0.3 to 0.9% silicon; 0.5% Mn;

and 0.025 to 0.05% of sulfur, phosphorus, and titanium. *Slag* is a byproduct of the blast furnace process. It contains silica, CaO, and other impurities in the form of a silicate melt.

Because the liquid pig iron contains a large amount of carbon, oxygen is blown into it in the *basic oxygen furnace* (BOF) to eliminate the excess carbon and produce liquid steel. Steel has a carbon content up to a maximum of ~2%. In the second method, steel scrap is often melted in an **electric arc furnace**. Many alloy and specialty steels, such as stainless steels, are produced using electric melting. Steel is one of the few materials that is nearly 100% recycled. Molten steels (including stainless steels) often undergo further refining. The goal here is to reduce the levels of impurities such as phosphorus, sulfur, etc., and to bring the carbon to a desired level.

Molten steel made by either route is poured into molds to produce finished steel castings, or cast continuously into shapes that are later processed through metal-forming techniques such as rolling or forging. In the latter case, the steel is either poured into large ingot molds or continuously cast into regular shapes.

All of the strengthening mechanisms discussed in the previous chapters apply to at least some of the ferrous alloys. In this chapter, we will discuss how to use the eutectoid reaction to control the structure and properties of steels through heat treatment and alloying. We will also examine two special classes of ferrous alloys: stainless steels and cast irons.

13-1 Designations and Classification of Steels

The dividing point between "steels" and "cast irons" is 2.11% C, where the eutectic reaction becomes possible. For steels, we concentrate on the eutectoid portion of the diagram (Figure 13-1) in which the solubility lines and the eutectoid isotherm are specially identified. The A_3 shows the temperature at which ferrite starts to form on cooling; the A_{cm} shows the temperature at which cementite starts to form; and the A_1 is the eutectoid temperature.

Almost all of the heat treatments of steel are directed toward producing the mixture of ferrite and cementite that gives the proper combination of properties. Figure 13-2 shows the three important microconstituents, or arrangements of ferrite and cementite, that are usually sought. Pearlite is a microconstituent consisting of a lamellar mixture of ferrite and cementite. In bainite, which is obtained by transformation of austenite at a large undercooling, the cementite is more rounded than in pearlite. Tempered martensite, a mixture of very fine and nearly round cementite in a matrix of ferrite, forms when martensite is reheated following its formation.

Designations The American Iron and Steel Institute (AISI) and Society of Automotive Engineers (SAE) provide designation systems (Table 13-1) that use a four- or five-digit number. The first two numbers refer to the major alloying elements present, and the last two or three numbers refer to the percentage of carbon. An AISI 1040 steel is a plain-carbon steel with 0.40% C. An SAE 10120 steel is a plain-carbon steel containing 1.20% C. An AISI 4340 steel is an alloy steel containing 0.40% C. Note that the American Society for Testing of Materials (ASTM) has a different way of classifying steels. The ASTM has a list of specifications that describe steels suitable for different applications. The following example illustrates the use of AISI numbers.

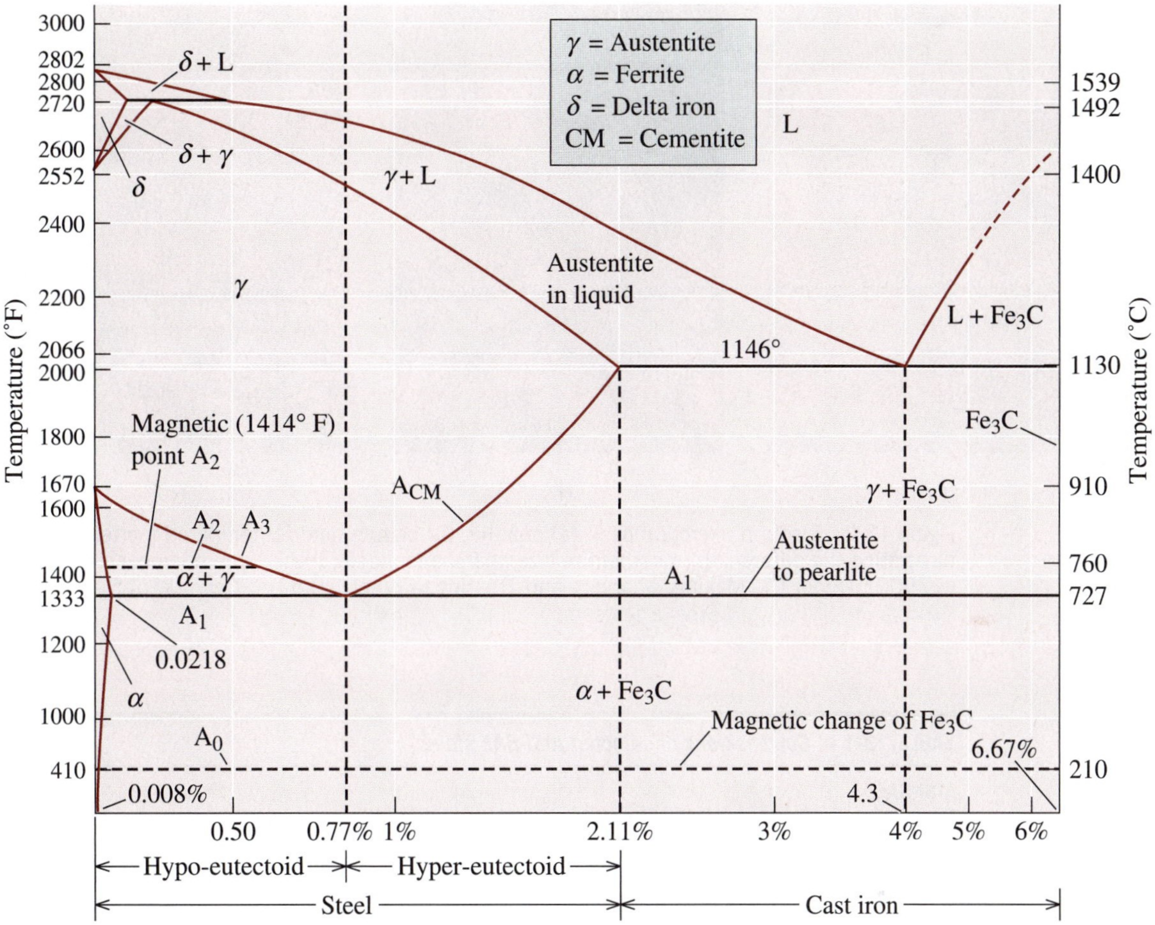

(a)

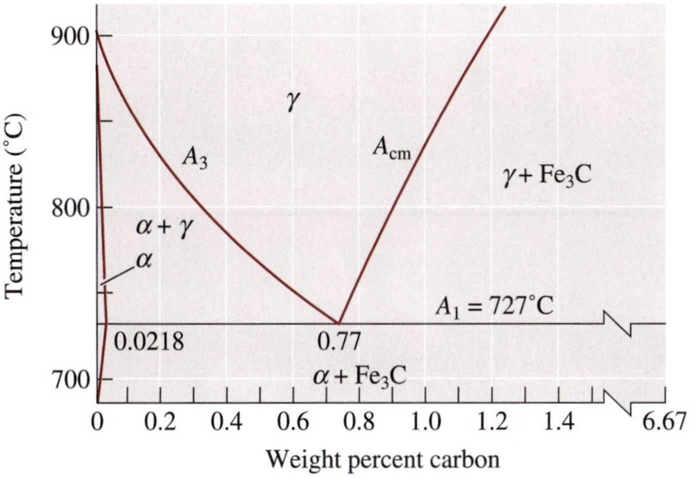

(b)

Figure 13-1 (a) An expanded version of the Fe-C diagram, adapted from several sources. (b) The eutectoid portion of the Fe-Fe_3C phase diagram.

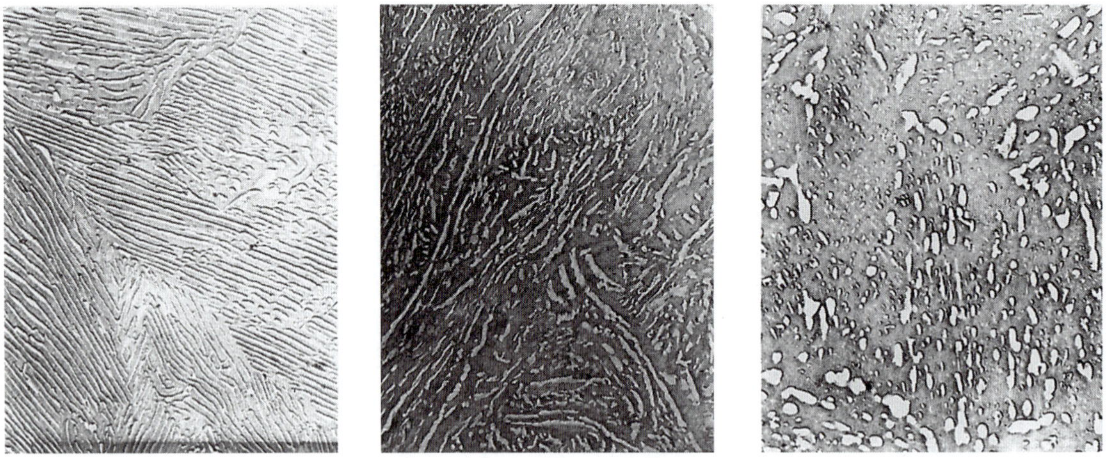

(a) (b) (c)

Figure 13-2 Electron micrographs of (a) pearlite, (b) bainite, and (c) tempered martensite, illustrating the differences in cementite size and shape among these three microconstituents (×7500). (*From* The Making, Shaping, and Treating of Steel, 10th Ed. *Courtesy of the Association of Iron and Steel Engineers.*)

TABLE 13-1 ■ *Compositions of selected AISI-SAE steels*

AISI-SAE Number	% C	% Mn	% Si	% Ni	% Cr	Others
1020	0.18–0.23	0.30–0.60				
1040	0.37–0.44	0.60–0.90				
1060	0.55–0.65	0.60–0.90				
1080	0.75–0.88	0.60–0.90				
1095	0.90–1.03	0.30–0.50				
1140	0.37–0.44	0.70–1.00				0.08–0.13% S
4140	0.38–0.43	0.75–1.00	0.15–0.30		0.80–1.10	0.15–0.25% Mo
4340	0.38–0.43	0.60–0.80	0.15–0.30	1.65–2.00	0.70–0.90	0.20–0.300% Mo
4620	0.17–0.22	0.45–0.65	0.15–0.30	1.65–2.00		0.20–0.30% Mo
52100	0.98–1.10	0.25–0.45	0.15–0.30		1.30–1.60	
8620	0.18–0.23	0.70–0.90	0.15–0.30	0.40–0.70	0.40–0.60	0.15–0.25% Y
9260	0.56–0.64	0.75–1.00	1.80–2.20			

EXAMPLE 13-1 ***Design of a Method to Determine AISI Number***

An unalloyed steel tool used for machining aluminum automobile wheels has been found to work well, but the purchase records have been lost and you do not know the steel's composition. The microstructure of the steel is tempered martensite, and assume that you cannot estimate the composition of the steel from the structure. Design a treatment that may help determine the steel's carbon content.

SOLUTION

Assume that there is no access to equipment that would permit you to analyze the chemical composition directly. Since the entire structure of the steel is a very fine tempered martensite, we can do a simple heat treatment to produce a structure that can be analyzed more easily. This can be done in two different ways.

The first way is to heat the steel to a temperature just below the A_1 temperature and hold for a long time. The steel overtempers and large Fe_3C spheres form in a ferrite matrix. We then estimate the amount of ferrite and cementite and calculate the carbon content using the lever law. If we measure 16% Fe_3C using this method, the carbon content is:

$$\%\ Fe_3C = \left[\frac{(x - 0.0218)}{(6.67 - 0.0218)}\right] \times 100 = 16 \quad \text{or} \quad x = 1.086\%\ C$$

A better approach, however, is to heat the steel above the A_{cm} to produce all austenite. If the steel then cools slowly, it transforms to pearlite and a primary microconstituent. If, when we do this, we estimate that the structure contains 95% pearlite and 5% primary Fe_3C, then:

$$\%\ \text{Pearlite} = \left[\frac{(6.67 - x)}{(6.67 - 0.77)}\right] \times 100 = 95 \quad \text{or} \quad x = 1.065\%\ C$$

The carbon content is on the order of 1.065 to 1.086%, consistent with a 10110 steel. In this procedure, we assumed that the weight and volume percentages of the microconstituents are the same, which is nearly the case in steels.

Classifications Steels can be classified based on their composition or the way they have been processed. Carbon steels contain up to ~2% carbon. These steels may also contain other elements, such as Si (maximum 0.6%), copper (up to 0.6%), and Mn (up to 1.65%). Decarburized steels contain less than 0.005% C. Ultra-low carbon steels contain a maximum of 0.03% carbon. They also contain very low levels of other elements such as Si and Mn. Low-carbon steels contain 0.04 to 0.15% carbon. These low-carbon steels are used for making car bodies and hundreds of other applications. Mild steel contains 0.15 to 0.3% carbon. This steel is used in buildings, bridges, piping, etc. Medium-carbon steels contain 0.3 to 0.6% carbon. These are used in making machinery, tractors, mining equipment, etc. High-carbon steels contain above 0.6% carbon. These are used in making springs, railroad car wheels, and the like. Note that cast irons are Fe-C alloys containing about 2 to 4% carbon.

Alloy steels are compositions that contain more significant levels of alloying elements. We will discuss the effect of alloying elements later in this chapter. They improve the hardenability of steels. The AISI defines alloy steels as steels that exceed in one or more of these elements: ≥1.65% Mn, 0.6% Si, 0.6% Cu. The total carbon content is up to 1% and the total alloying elements content is below 5%. A material is also an alloy steel if a definite concentration of alloying elements, such as Ni, Cr, Mo, Ti, etc., is specified. These steels are used for making tools (hammers, chisels, etc.) and also in making parts such as axles, shafts, and gears.

Certain specialty steels may consist of higher levels of sulfur (>0.1%) or lead (~0.15–0.35%) to provide machinability. These, however, can not be welded easily. Recently, researchers have developed "green steel" in which lead, an environmental toxin, was replaced with tin (Sn) and/or antimony (Sb). Steels can also be classified based on their processing. For example, the term "concast steels" refers to continuously cast steels. Galvanized steels have a zinc coating for corrosion resistance. Similarly, tin-plated steel is used to make corrosion-resistant tin cans and other products. Tin is deposited using electroplating—a process known as "continuous web electrodeposition." "E-steels" are steels that are melted using an electric furnace, while "B-steels" contain a small (0.0005 to 0.003%), yet significant, concentration of boron. Recently, a "germ-resistant" coated stainless steel has been developed. Some biomedical products, such as cardiovascular stents, also make use of stainless steel coated with heprin, a chemical that makes the stents more biocompatible.

13-2 Simple Heat Treatments

Four simple heat treatments—process annealing, annealing, normalizing, and spheroidizing—are commonly used for steels (Figure 13-3). These heat treatments are used to accomplish one of three purposes: (1) eliminating the effects of cold work, (2) controlling dispersion strengthening, or (3) improving machinability.

Process Annealing—Eliminating Cold Work The recrystallization heat treatment used to eliminate the effect of cold working in steels with less than about 0.25% C is called a **process anneal**. The process anneal is done 80 to 170°C below the A_1 temperature. The goal of the process anneal treatment for steels is similar to the annealing of inorganic glasses in that the main idea is to significantly reduce or eliminate residual stresses.

Annealing and Normalizing—Dispersion Strengthening Steels can be dispersion strengthened by controlling the fineness of pearlite. The steel is initially heated to produce homogeneous austenite (FCC γ phase), a step called **austenitizing**. **Annealing**, or a full anneal, allows the steel to cool slowly in a furnace, producing coarse pearlite. In **normalizing**, the steel is cooled more rapidly in air, producing fine pearlite. Figure 13-4 shows the typical properties obtained by annealing and normalizing plain-carbon steels.

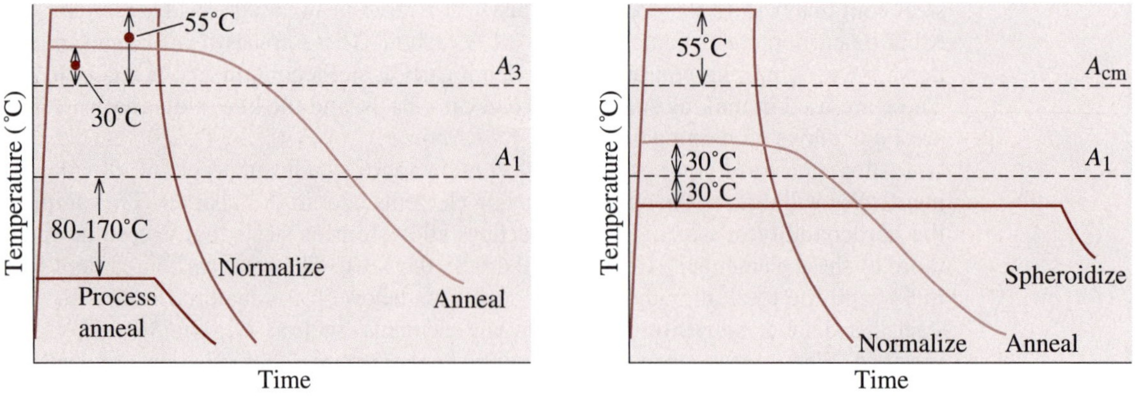

(a) Hypoeutectoid **(b) Hypereutectoid**

Figure 13-3 Schematic summary of the simple heat treatments for (a) hypoeutectoid steels and (b) hypereutectoid steels.

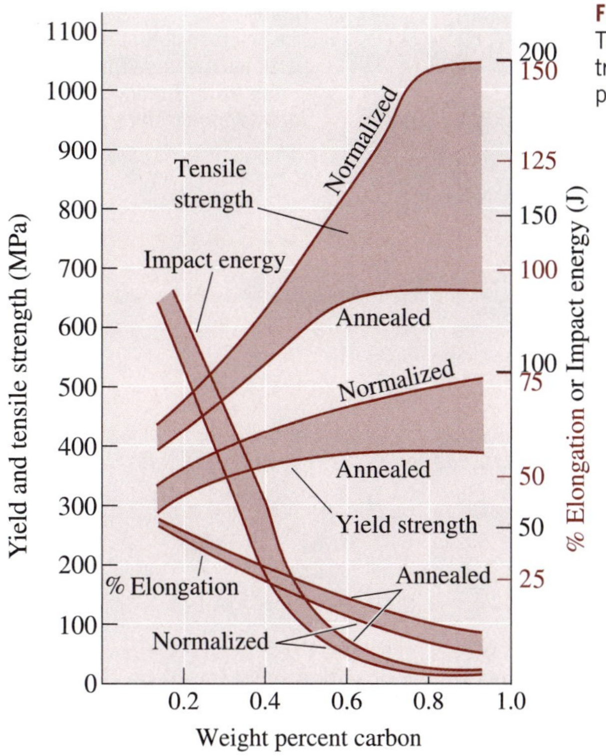

Figure 13-4 The effect of carbon and heat treatment on the properties of plain-carbon steels.

For annealing, austenitizing of hypoeutectoid steels is conducted about 30°C above the A_3, producing 100% γ. However, austenitizing of a hypereutectoid steel is done at about 30°C above the A_1, producing austenite and Fe_3C. This process prevents the formation of a brittle, continuous film of Fe_3C at the grain boundaries that occurs on slow cooling from the 100% γ region. In both cases, the slow furnace cool and coarse pearlite provide relatively low strength and good ductility.

For normalizing, austenitizing is done at about 55°C above the A_3 or A_{cm}; the steel is then removed from the furnace and cooled in air. The faster cooling gives fine pearlite and provides higher strength.

Spheroidizing—Improving Machinability Steels, which contain a large concentration of Fe_3C, have poor machining characteristics. It is possible to transform the morphology of Fe_3C using *spheroidizing*. During the spheroidizing treatment, which requires several hours at about 30°C below the A_1, the Fe_3C phase morphology changes into large, spherical particles in order to reduce boundary area. The microstructure, known as **spheroidite**, has a continuous matrix of soft, machinable ferrite (Figure 13-5). After machining, the steel is given a more sophisticated heat treatment to produce the required properties. A similar microstructure occurs when martensite is tempered just below the A_1 for long periods of time. The difference between tempered martensite and spheroidite is that in formation of spheroidite, martensite is never formed. As noted before, alloying elements such as Pb and S are also added to improve machinability of steels and, more recently, lead-free "green steels" that have very good machinability have been developed.

The following example shows how different heat treatment conditions can be developed for a given composition of steel.

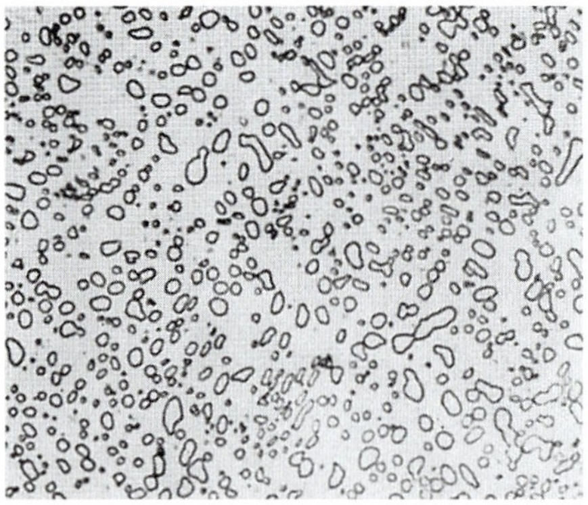

Figure 13-5
The microstructure of spheroidite, with Fe_3C particles dispersed in a ferrite matrix (×850). (*From* ASM Handbook, *Vol. 7, (1972), ASM International, Materials Park, OH 44073.*)

EXAMPLE 13-2 *Determination of Heat Treating Temperatures*

Recommend temperatures for the process annealing, annealing, normalizing, and spheroidizing of 1020, 1077, and 10120 steels.

SOLUTION

From Figure 13-1, we find the critical A_1, A_3, or A_{cm} temperatures for each steel. We can then specify the heat treatment based on these temperatures.

Steel Type	1020	1077	10120
Critical temperatures	$A_1 = 727°C$ $A_3 = 830°C$	$A_1 = 727°C$	$A_1 = 727°C$ $A_{cm} = 895°C$
Process annealing	727 – (80 to 170) = 557°C to 647°C	Not done	Not done
Annealing	830 + 30 = 860°C	727 + 30 = 757°C	727 + 30 = 757°C
Normalizing	830 + 55 = 885°C	727 + 55 = 782°C	895 + 55 = 950°C
Spheroidizing	Not done	727 – 30 = 697°C	727 – 30 = 697°C

13-3 Isothermal Heat Treatments

The effect of transformation temperature on the properties of a 1080 (eutectoid) steel was discussed in Chapter 12. As the isothermal transformation temperature decreases, pearlite becomes progressively finer before bainite begins to form. At very low temperatures, martensite is obtained.

Austempering and Isothermal Annealing The isothermal transformation heat treatment used to produce bainite, called **austempering**, simply involves austenitizing the steel, quenching to some temperature below the nose of the TTT curve, and holding at that temperature until all of the austenite transforms to bainite (Figure 13-6).

Annealing and normalizing are usually used to control the fineness of pearlite. However, pearlite formed by an **isothermal anneal** (Figure 13-6) may give more uniform properties, since the cooling rates and microstructure obtained during annealing and normalizing vary across the cross section of the steel. *Note that the TTT diagrams only*

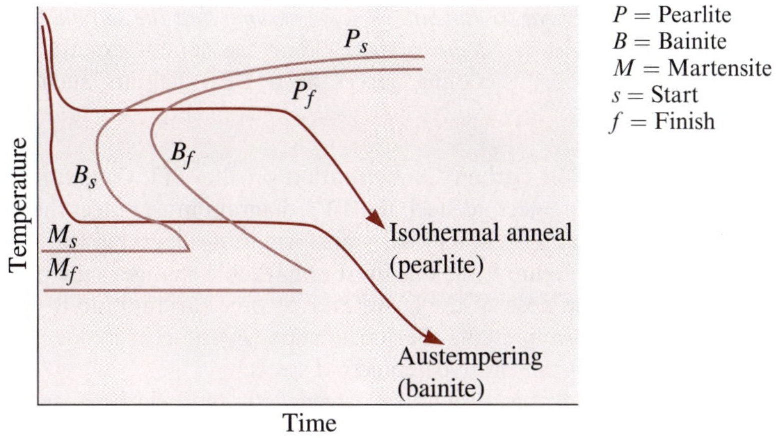

Figure 13-6 The austempering and isothermal anneal heat treatments in a 1080 steel.

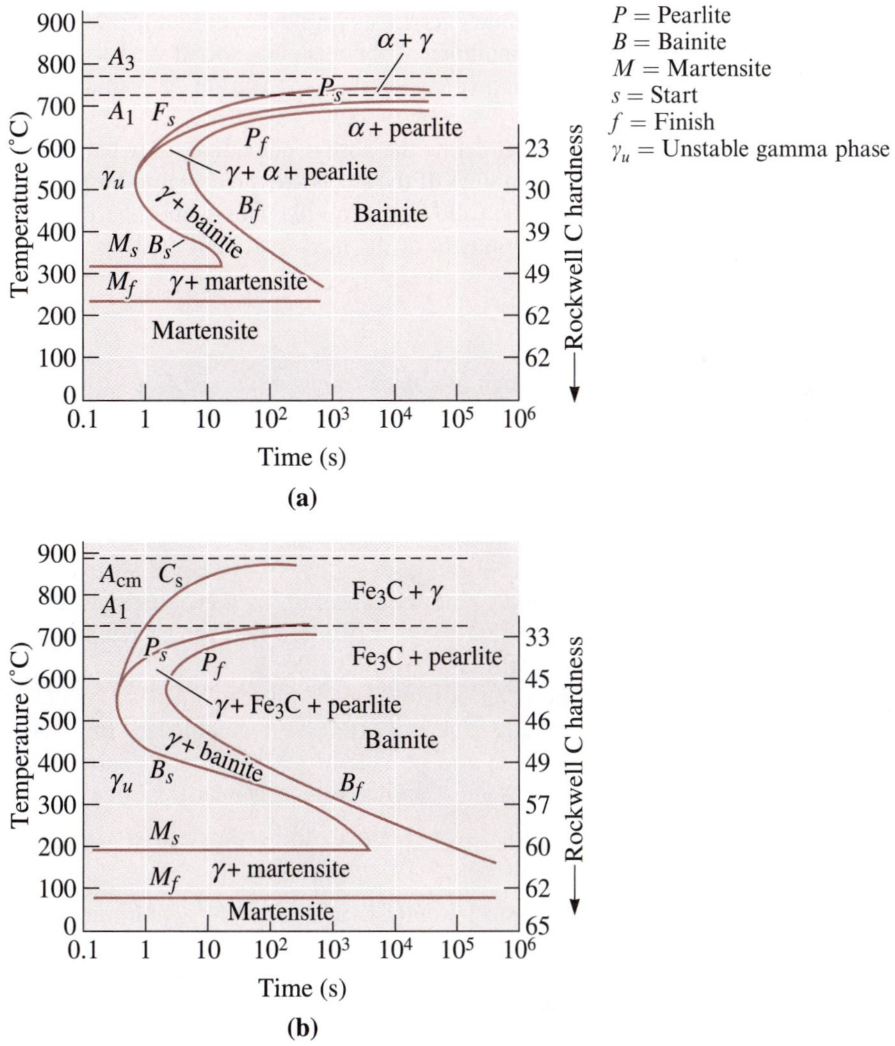

Figure 13-7 The TTT diagrams for (a) a 1050 and (b) a 10110 steel.

describe isothermal heat treatments (i.e., we assume that the sample begins and completes heat treatment at a given temperature). Thus, we cannot exactly describe heat treatments by superimposing cooling curves on a TTT diagram such as those shown in Figure 13-6.

Effect of Changes in Carbon Concentration on the TTT Diagram In either a hypoeutectoid or a hypereutectoid steel, the TTT diagram must reflect the possible formation of a primary phase. The isothermal transformation diagrams for a 1050 and a 10110 steel are shown in Figure 13-7. The most remarkable change is the presence of a "wing" which begins at the nose of the curve and becomes asymptotic to the A_3 or A_{cm} temperature. The wing represents the ferrite start (F_s) time in hypoeutectoid steels or the cementite start (C_s) time in hypereutectoid steels.

When a 1050 steel is austenitized, quenched, and held between the A_1 and the A_3, primary ferrite nucleates and grows. Eventually, an equilibrium amount of ferrite and austenite result. Similarly, primary cementite nucleates and grows to its equilibrium amount in a 10110 steel held between the A_{cm} and A_1 temperatures.

If an austenitized 1050 steel is quenched to a temperature between the nose and the A_1 temperatures, primary ferrite again nucleates and grows until reaching the equilibrium amount. The remainder of the austenite then transforms to pearlite. A similar situation, producing primary cementite and pearlite, is found for the hypereutectoid steel.

If we quench the steel below the nose of the curve, only bainite forms, regardless of the carbon content of the steel. If the steels are quenched to temperatures below the M_s, martensite will form. The following example shows how the phase diagram and TTT diagram can guide development of the heat treatment of steels.

EXAMPLE 13-3 *Design of a Heat Treatment for an Axle*

A heat treatment is needed to produce a uniform microstructure and hardness of HRC 23 in a 1050 steel axle.

SOLUTION

We might attempt this task in several ways. We could austenitize the steel, then cool at an appropriate rate by annealing or normalizing to obtain the correct hardness. By doing this, however, we find that the structure and hardness vary from the surface to the center of the axle.

A better approach is to use an isothermal heat treatment. From Figure 13-7, we find that a hardness of HRC 23 is obtained by transforming austenite to a mixture of ferrite and pearlite at 600°C. From Figure 13-1, we find that the A_3 temperature is 770°C. Therefore, our heat treatment is:

1. Austenitize the steel at 770 + (30 to 55) = 805°C to 825°C, holding for 1 h and obtaining 100% γ.

2. Quench the steel to 600°C and hold for a minimum of 10 s. Primary ferrite begins to precipitate from the unstable austenite (γ_u) after about 1.0 s. After 1.5 s, pearlite begins to grow, and the austenite is completely transformed to ferrite and pearlite after about 10 s. After this treatment, the microconstituents present are:

$$\text{Primary } \alpha = \left[\frac{(0.77 - 0.5)}{(0.77 - 0.0218)}\right] \times 100 = 36\%$$

$$\text{Pearlite} = \left[\frac{(0.5 - 0.0218)}{(0.77 - 0.0218)}\right] \times 100 = 64\%$$

3. Cool in air to room temperature, preserving the equilibrium amounts of primary ferrite and pearlite. The microstructure and hardness are uniform because of the isothermal anneal.

Interrupting the Isothermal Transformation Complicated microstructures are produced by interrupting the isothermal heat treatment. For example, we could austenitize the 1050 steel (Figure 13-8) at 800°C, quench to 650°C and hold for 10 s (permitting some ferrite and pearlite to form), then quench to 350°C and hold for 1 h (3600 s). Whatever unstable austenite remained before quenching to 350°C transforms to bainite. The final structure is ferrite, pearlite, and bainite. We could complicate the treatment further by interrupting the treatment at 350°C after 1 min (60 s) and quenching. Any austenite remaining after 1 min at 350°C forms martensite. The final structure now contains ferrite, pearlite, bainite, and martensite. Note that each time we change the temperature, we start at zero time! In practice, temperatures can not be changed instantaneously (i.e., we cannot go instantly from 800 to 650 or 650 to 350°C). This is why it is better to use the continuous cooling transformation (CCT) diagrams.

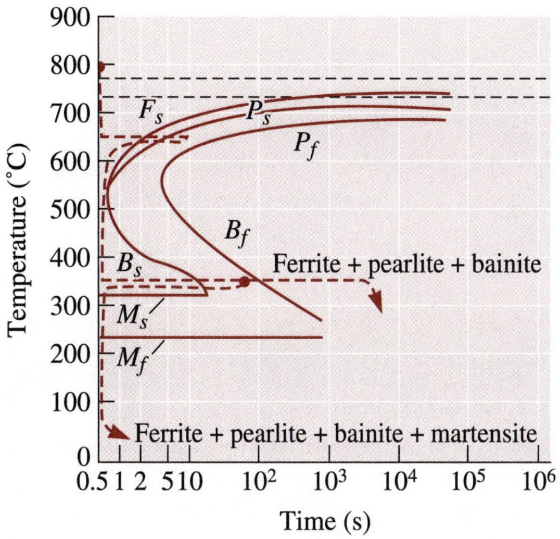

Figure 13-8
Producing complicated structures by interrupting the isothermal heat treatment of a 1050 steel.

13-4 Quench and Temper Heat Treatments

Quenching hardens most steels and tempering increases the toughness. This has been known for perhaps thousands of years. For example, a series of such heat treatments has been used for making Damascus steel and Japanese Samurai swords. We can obtain an exceptionally fine dispersion of Fe_3C and ferrite (known as tempered martensite) if

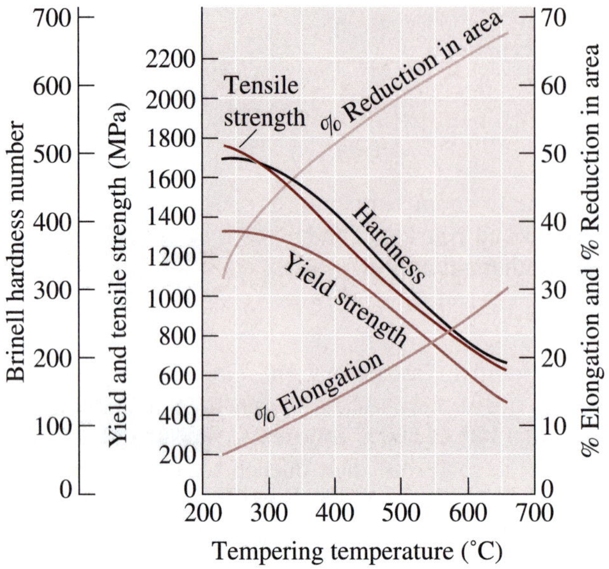

Figure 13-9 The effect of tempering temperature on the mechanical properties of a 1050 steel.

we first quench the austenite to produce martensite, then temper. During tempering, an intimate mixture of ferrite and cementite forms from the martensite, as discussed in Chapter 12. The tempering treatment controls the final properties of the steel (Figure 13-9). Note that this is different from a spheroidizing heat treatment (Figure 13-5). The following example shows how a combination of heat treatments is used to obtain steels with desired properties.

EXAMPLE 13-4 *Design of a Quench and Temper Treatment*

A rotating shaft that delivers power from an electric motor is made from a 1050 steel. Its yield strength should be at least 1000 MPa, yet it should also have at least 15% elongation in order to provide toughness. Design a heat treatment to produce this part.

SOLUTION

We are not able to obtain this combination of properties by annealing or normalizing (Figure 13-4). However a quench and temper heat treatment produces a microstructure that can provide both strength and toughness. Figure 13-9 shows that the yield strength exceeds 1000 MPa if the steel is tempered below 460°C, whereas the elongation exceeds 15% if tempering is done above 425°C. The A_3 temperature for the steel is 770°C. A possible heat treatment is:

1. Austenitize above the A_3 temperature of 770°C for 1 h. An appropriate temperature may be $770 + 55 = 825$°C.
2. Quench rapidly to room temperature. Since the M_f is about 250°C, martensite will form.
3. Temper by heating the steel to 440°C. Normally, 1 h will be sufficient if the steel component is not too thick.
4. Cool to room temperature.

Figure 13-10
Retained austenite (white) trapped between martensite needles (black) (×1000). (*From* ASM Handbook, *Vol. 8, (1973), ASM International, Materials Park, OH 44073.*)

Retained Austenite There is a large volume expansion when martensite forms from austenite. As the martensite plates form during quenching, they surround and isolate small pools of austenite (Figure 13-10), which deform to accommodate the lower-density martensite. However, for the remaining pools of austenite to transform, the surrounding martensite must deform. Because the strong martensite resists the transformation, either the existing martensite cracks or the austenite remains trapped in the structure as **retained austenite**. Retained austenite can be a serious problem. Martensite softens and becomes more ductile during tempering. After tempering, the retained austenite cools below the M_s and M_f temperatures and transforms to martensite, since the surrounding **tempered martensite** can deform. But now the steel contains more of the hard, brittle martensite! A second tempering step may be needed to eliminate the martensite formed from the retained austenite. Retained austenite is also more of a problem for high-carbon steels. The martensite start (M_s) and finish (M_f) temperatures are reduced when the carbon content increases (Figure 13-11). High-carbon steels must be refrigerated to produce all martensite.

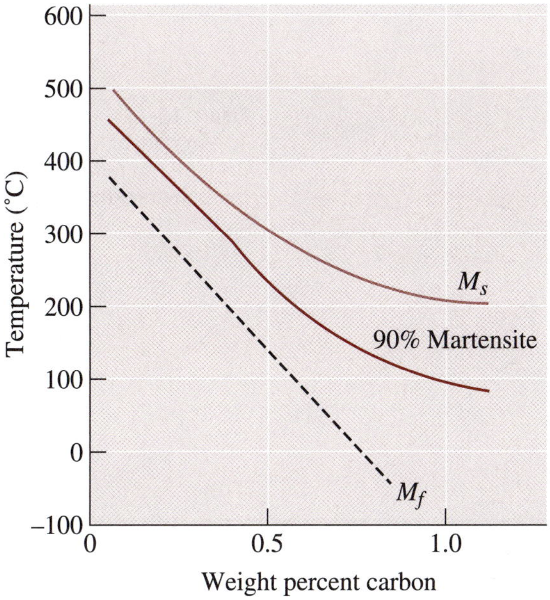

Figure 13-11
Increasing carbon reduces the M_s and M_f temperatures in plain-carbon steels.

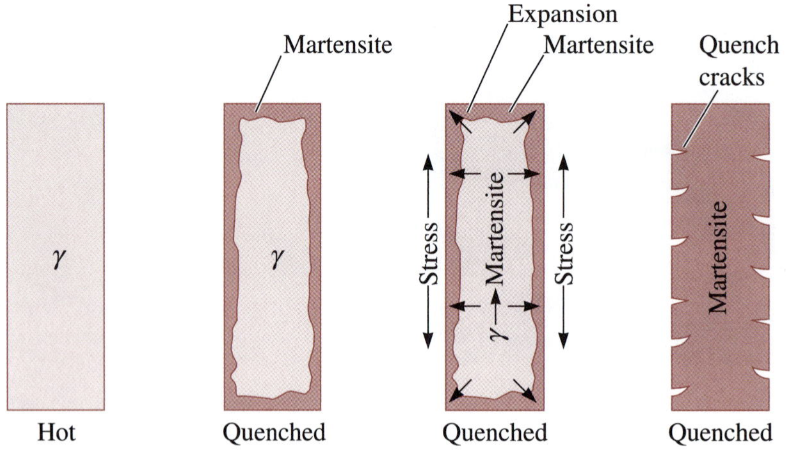

Figure 13-12 Formation of quench cracks caused by residual stresses produced during quenching. The figure illustrates the development of stresses as the austenite transforms to martensite during cooling.

Residual Stresses and Cracking Residual stresses are also produced because of the volume change or because of cold working. A stress-relief anneal can be used to remove or minimize residual stresses due to cold working. Stresses are also induced because of thermal expansion and contraction. In steels, there is one more mechanism that causes stress. When steels are quenched, the surface of the quenched steel cools rapidly and transforms to martensite. When the austenite in the center later transforms, the hard surface is placed in tension, while the center is compressed. If the residual stresses exceed the yield strength, **quench cracks** form at the surface (Figure 13-12). However, if we first cool to just above the M_s and hold until the temperature equalizes in the steel, subsequent quenching permits all of the steel to transform to martensite at about the same time. This heat treatment is called **marquenching** or **martempering** (Figure 13-13). Note that, strictly speaking, the CCT diagrams should be used to examine non-isothermal heat treatments. This will be discussed later in this section.

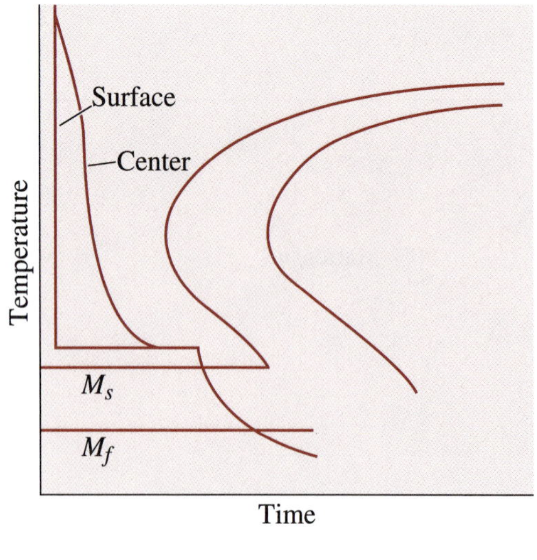

Figure 13-13
The marquenching heat treatment, designed to reduce residual stresses and quench cracking.

TABLE 13-2 ■ ***The H coefficient, or severity of the quench, for several quenching media***

Medium	H Coefficient	Cooling Rate at the Center of a 2.5 cm Bar (°C/s)
Oil (no agitation)	0.25	18
Oil (agitation)	1.0	45
H_2O (no agitation)	1.0	45
H_2O (agitation)	4.0	190
Brine (no agitation)	2.0	90
Brine (agitation)	5.0	230

Quench Rate In using the TTT diagram, we assumed that we could cool from the austenitizing temperature to the transformation temperature instantly. Because this does not occur in practice, undesired microconstituents may form during the quenching process. For example, pearlite may form as the steel cools past the nose of the curve, particularly because the time of the nose is less than one second in plain-carbon steels.

The rate at which the steel cools during quenching depends on several factors. First, the surface always cools faster than the center of the part. In addition, as the size of the part increases, the cooling rate at any location is slower. Finally, the cooling rate depends on the temperature and heat transfer characteristics of the quenching medium (Table 13-2). Quenching in oil, for example, produces a lower H coefficient, or slower cooling rate, than quenching in water or brine. The H coefficient is equivalent to the heat transfer coefficient. Agitation helps break the vapor blanket (e.g., when water is the quenching medium) and improves overall heat transfer rate by bringing cooler liquid into contact with the parts being quenched.

Continuous Cooling Transformation Diagrams We can develop a continuous cooling transformation (CCT) diagram by determining the microstructures produced in the steel at various rates of cooling. The CCT curve for a 1080 steel is shown in Figure 13-14.

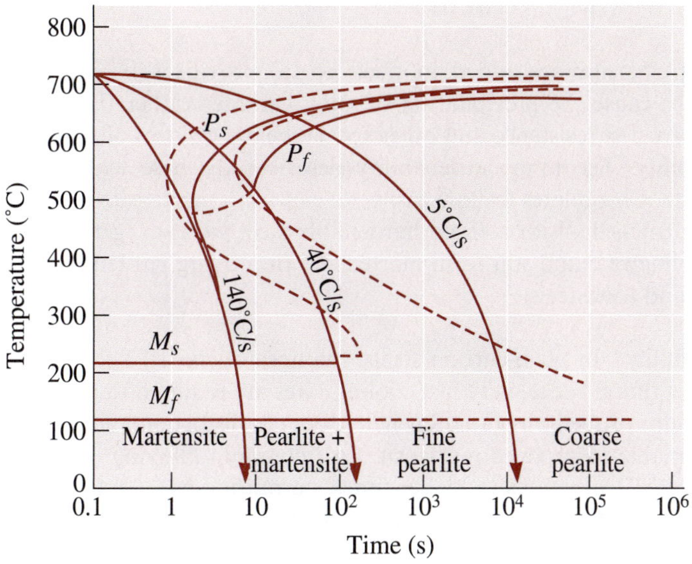

Figure 13-14 The CCT diagram (solid lines) for a 1080 steel compared with the TTT diagram (dashed lines).

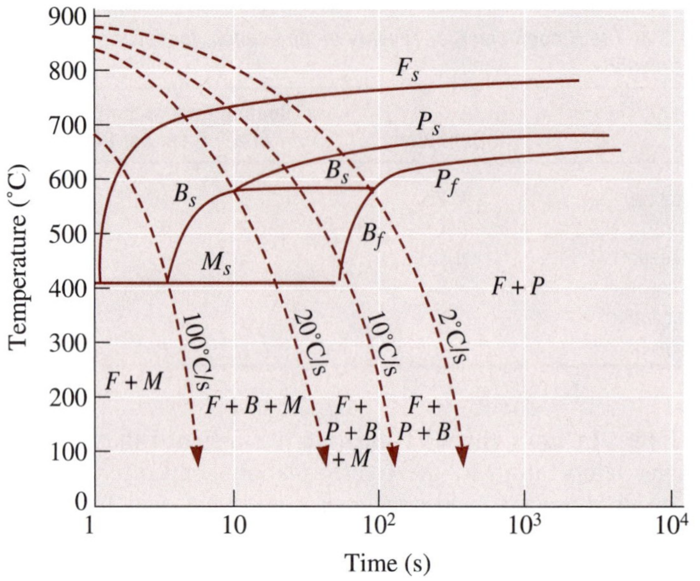

Figure 13-15 The CCT diagram for a low-alloy, 0.2% C steel.

The CCT diagram differs from the TTT diagram in that longer times are required for transformations to begin and no bainite region is observed.

If we cool a 1080 steel at 5°C/s, the CCT diagram tells us that we obtain coarse pearlite; we have annealed the steel. Cooling at 35°C/s gives fine pearlite and is a normalizing heat treatment. Cooling at 100°C/s permits pearlite to start forming, but the reaction is incomplete and the remaining austenite changes to martensite. We obtain 100% martensite and thus are able to perform a quench and temper heat treatment, only if we cool faster than 140°C/s. Other steels, such as the low-carbon steel in Figure 13-15, have more complicated CCT diagrams. In various handbooks, you can find a compilation of TTT and CCT diagrams for different grades of steels.

13-5 Effect of Alloying Elements

Alloying elements are added to steels to (a) provide solid-solution strengthening of ferrite, (b) cause the precipitation of alloy carbides rather than that of Fe_3C, (c) improve corrosion resistance and other special characteristics of the steel, and (d) improve **hardenability**. The term hardenability describes the ease with which steels can form martensite. This relates to how easily we can form martensite in a thick section of steel that is quenched. With a more hardenable steel we can "get away" with a relatively slow cooling rate and still form martensite. Improving hardenability is most important in alloy and tool steels.

Hardenability In plain-carbon steels, the nose of the TTT and CCT curves occurs at very short times; hence, very fast cooling rates are required to produce all martensite. In thin sections of steel, the rapid quench produces distortion and cracking. In thick steels, we are unable to produce martensite. All common alloying elements in steel shift the TTT and CCT diagrams to longer times, permitting us to obtain all martensite even in thick sections at slow cooling rates. Figure 13-16 shows the TTT and CCT curves for a 4340 steel.

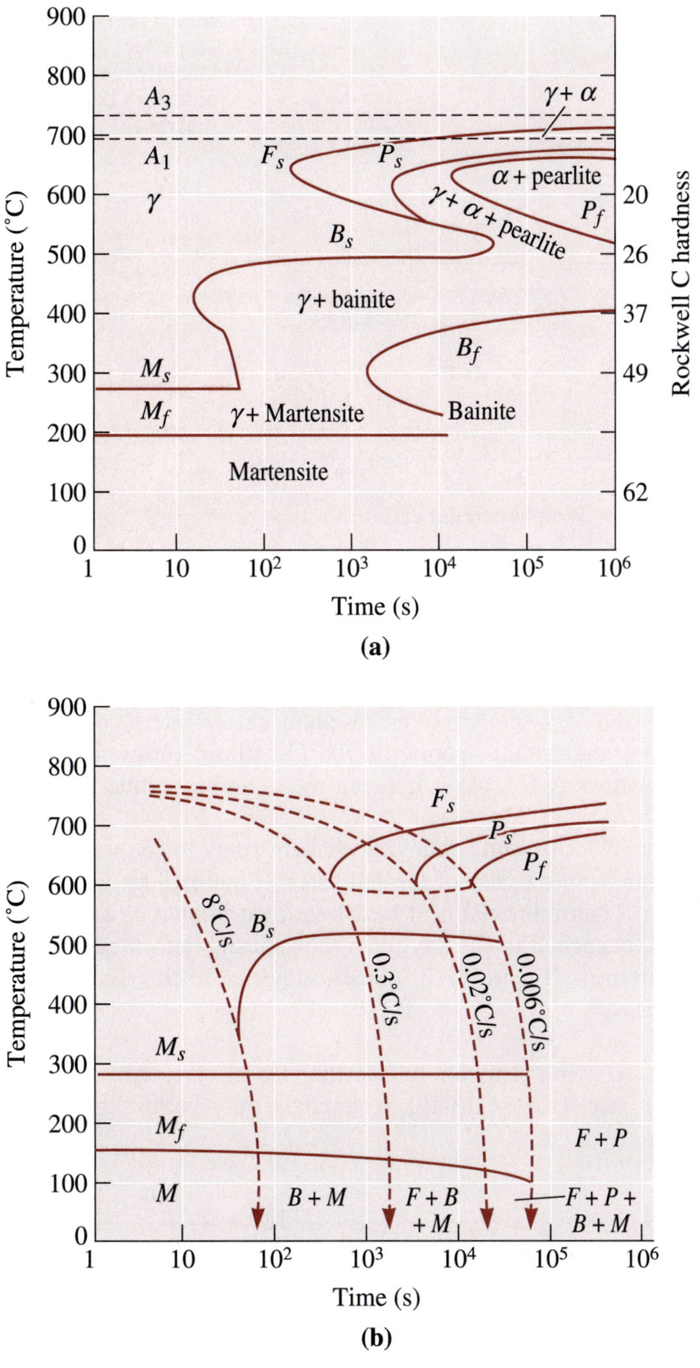

Figure 13-16 (a) TTT and (b) CCT curves for a 4340 steel.

Plain-carbon steels have low hardenability—only very high cooling rates produce all martensite. Alloy steels have high hardenability—even cooling in air may produce martensite. Hardenability does not refer to the hardness of the steel. A low-carbon, high-alloy steel may easily form martensite but, because of the low-carbon content, the martensite formed is not hard.

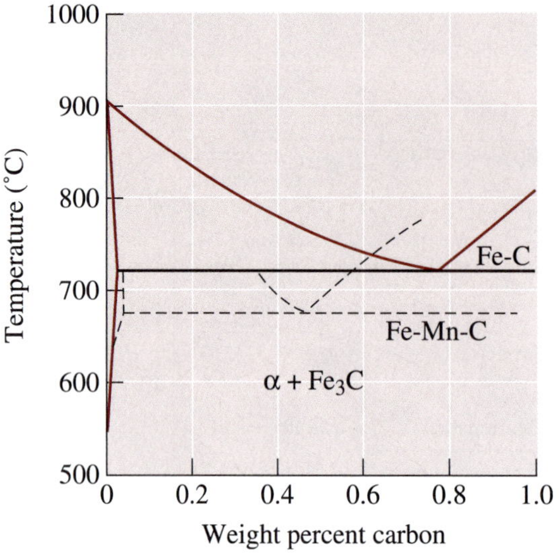

Figure 13-17
The effect of 6% manganese on the stability ranges of the phases in the eutectoid portion of the Fe-Fe_3C phase diagram.

Effect on the Phase Stability When alloying elements are added to steel, the binary Fe-Fe_3C stability is affected and the phase diagram is altered (Figure 13-17). Alloying elements reduce the carbon content at which the eutectoid reaction occurs and change the A_1, A_3, and A_{cm} temperatures. A plain carbon steel containing only 0.6% C is hypoeutectoid and would operate at 700°C without forming austenite; the otherwise same steel containing 6% Mn is hypereutectoid and austenite forms at 700°C.

Shape of the TTT Diagram Alloying elements may introduce a "bay" region into the TTT diagram, as in the case of the 4340 steel (Figure 13-16). The bay region is used as the basis for a thermomechanical heat treatment known as **ausforming**. A steel can be austenitized, quenched to the bay region, plastically deformed, and finally quenched to produce martensite (Figure 13-18). Steels subjected to this treatment are known as *ausformed steels.*

Tempering Alloying elements reduce the rate of tempering compared with that of a plain-carbon steel (Figure 13-19). This effect may permit the alloy steels to operate

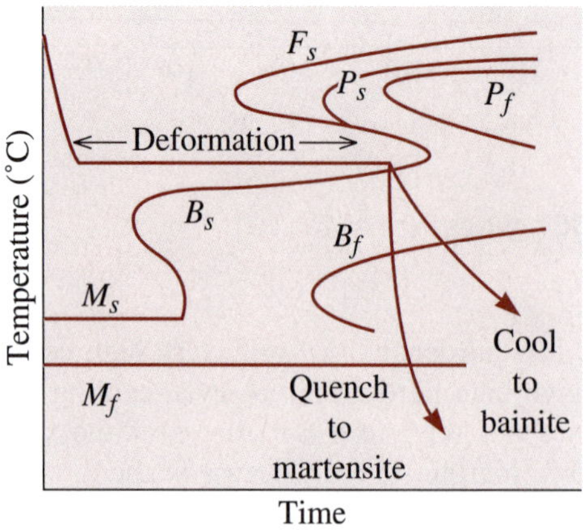

Figure 13-18
When alloying elements introduce a bay region into the TTT diagram, the steel can be ausformed.

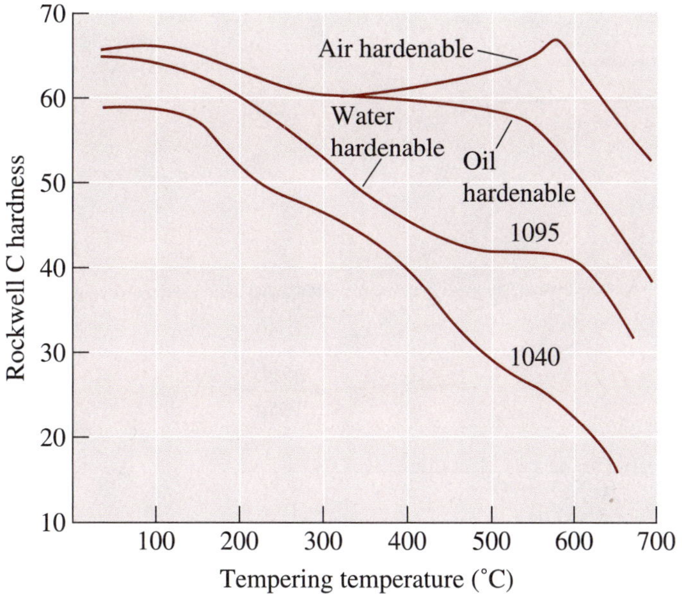

Figure 13-19
The effect of alloying elements on the phases formed during the tempering of steels. The air-hardenable steel shows a secondary hardening peak.

more successfully at higher temperatures than plain-carbon steels since overaging will not occur during service.

13-6 Application of Hardenability

A **Jominy test** (Figure 13-20) is used to compare hardenabilities of steels. A steel bar 10-cm long and 2.5 cm in diameter is austenitized, placed into a fixture, and sprayed at one end with water. This procedure produces a range of cooling rates—very fast at the quenched end, almost air cooling at the opposite end. After the test, hardness measurements are made along the test specimen and plotted to produce a **hardenability curve** (Figure 13-21). The distance from the quenched end is the **Jominy distance** and is related to the cooling rate (Table 13-3).

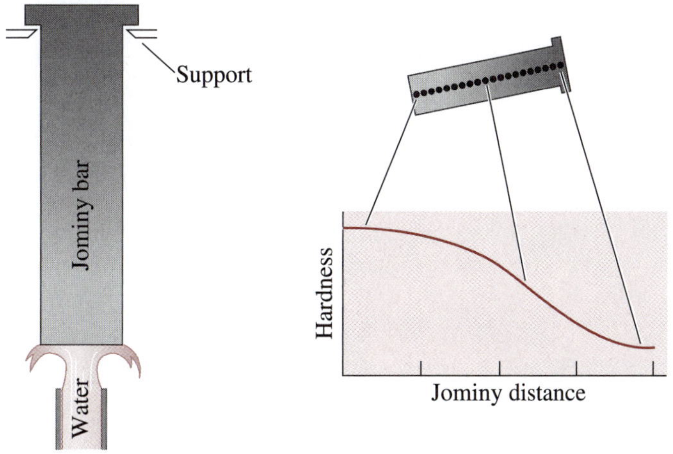

Figure 13-20
The set-up for the Jominy test used for determining the hardenability of a steel.

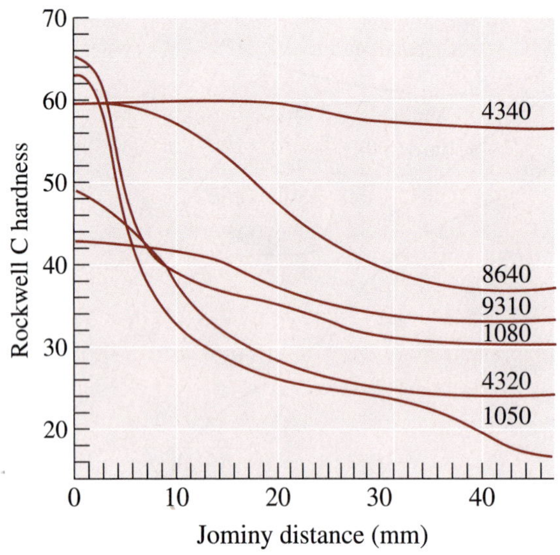

Figure 13-21
The hardenability curves for several steels.

Virtually any steel transforms to martensite at the quenched end. Thus, the hardness at zero Jominy distance is determined solely by the carbon content of the steel. At larger Jominy distances, there is a greater likelihood that bainite or pearlite will form instead of martensite. An alloy steel with a high hardenability (such as 4340) maintains a rather flat hardenability curve; a plain-carbon steel (such as 1050) has a curve that drops off quickly. The hardenability is determined primarily by the alloy content of the steel.

We can use hardenability curves in selecting or replacing steels in practical applications. The fact that two different steels cool at the same rate if quenched under identical conditions helps in this selection process. The Jominy test data are used as shown in the following example.

TABLE 13-3 ■ ***The relationship between cooling rate and Jominy distance***

Jominy Distance (mm)	Cooling Rate (°C/s)
1.6	315
3.1	110
4.7	50
6.3	36
7.8	28
9.4	22
10.9	17
12.5	15
15.6	10
18.8	8
25.0	5
31.3	3
37.5	2.8
43.8	2.5
56.3	2.2

EXAMPLE 13-5 **Design of a Wear-Resistant Gear**

A gear made from 9310 steel, which has an as-quenched hardness at a critical location of HRC 40, wears at an excessive rate. Tests have shown that an as-quenched hardness of at least HRC 50 is required at that critical location. Design a steel that would be appropriate.

SOLUTION

We know that if different steels of the same size are quenched under identical conditions, their cooling rates or Jominy distances are the same. From Figure 13-21, a hardness of HRC 40 in a 9310 steel corresponds to a Jominy distance of 15.6 mm (10°C/s). If we assume the same Jominy distance, the other steels shown in Figure 13-21 have the following hardnesses at the critical location:

1050	HRC 28
1080	HRC 36
4320	HRC 31
8640	HRC 52
4340	HRC 60

Both the 8640 and 4340 steels are appropriate. The 4320 steel has too low a carbon content ever to reach HRC 50; the 1050 and 1080 have enough carbon, but the hardenability is too low. In Table 13-1, we find that the 86xx steels contain less alloying elements than the 43xx steels; thus the 8640 steel is probably less expensive than the 4340 steel and might be our best choice. We must also consider other factors such as durability.

In another simple technique, we utilize the severity of the quench and the Grossman chart (Figure 13-22) to determine the hardness at the *center* of a round bar. The bar diameter and H coefficient, or severity of the quench in Table 13-2, give the Jominy distance at the center of the bar. We can then determine the hardness from the hardenability curve of the steel. (See Example 13-6.)

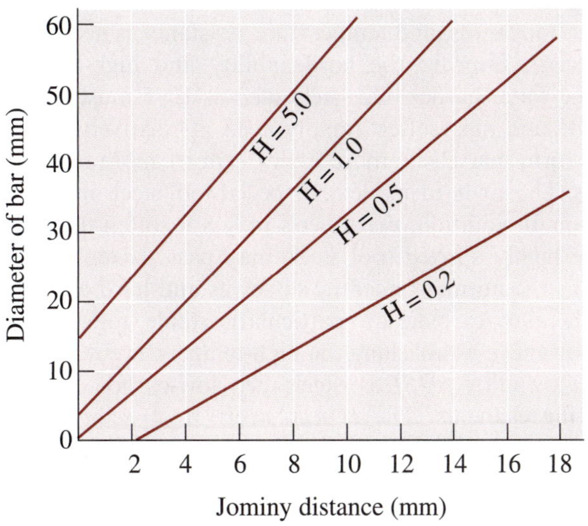

Figure 13-22
The Grossman chart used to determine the hardenability at the center of a steel bar for different quenchants.

EXAMPLE 13-6 Design of a Quenching Process

Design a quenching process to produce a minimum hardness of HRC 40 at the center of a 3.75 cm diameter 4320 steel bar.

SOLUTION

Several quenching media are listed in Table 13-2. We can find an approximate H coefficient for each of the quenching media, then use Figure 13-22 to estimate the Jominy distance in a 3.75 cm diameter bar for each media. Finally, we can use the hardenability curve (Figure 13-21) to find the hardness in the 4320 steel. The results are listed below.

	H Coefficient	Jominy Distance (mm)	HRC
Oil (no agitation)	0.25	17.2	30
Oil (agitation)	1.00	9.4	39
H_2O (no agitation)	1.00	9.4	39
H_2O (agitation)	4.00	6.3	44
Brine (no agitation)	2.00	7.8	42
Brine (agitation)	5.00	4.7	46

The last three methods, based on brine or agitated water, are satisfactory. Using an unagitated brine quenchant might be least expensive, since no extra equipment is needed to agitate the quenching bath. However, H_2O is less corrosive than the brine quenchant.

13-7 Specialty Steels

There are many special categories of steels, including tool steels, interstitial-free steels, high-strength-low-alloy (*HSLA*) steels, dual-phase steels, and maraging steels.

Tool steels are usually high-carbon steels that obtain high hardnesses by a quench and temper heat treatment. Their applications include cutting tools in machining operations, dies for die casting, forming dies, and other uses in which a combination of high strength, hardness, toughness, and temperature resistance is needed.

Alloying elements improve the hardenability and high-temperature stability of the tool steels. The water-hardenable steels such as 1095 must be quenched rapidly to produce martensite and also soften rapidly even at relatively low temperatures. Oil-hardenable steels form martensite more easily, temper more slowly, but still soften at high temperatures. The air-hardenable and special tool steels may harden to martensite while cooling in air. In addition, these steels may not soften until near the A_1 temperature. In fact, the highly alloyed tool steels may pass through a **secondary hardening peak** near 500°C as the normal cementite dissolves and hard alloy carbides precipitate (Figure 13-19). The alloy carbides are particularly stable, resist growth or spheroidization, and are important in establishing the high-temperature resistance of these steels.

High-strength-low-alloy (*HSLA*) steels are low-carbon steels containing small amounts of alloying elements. The *HSLA* steels are specified on the basis of yield strength, with grades up to 552 MPa the steels contain the least amount of alloying element that still provides the proper yield strength without heat treatment. In these steels,

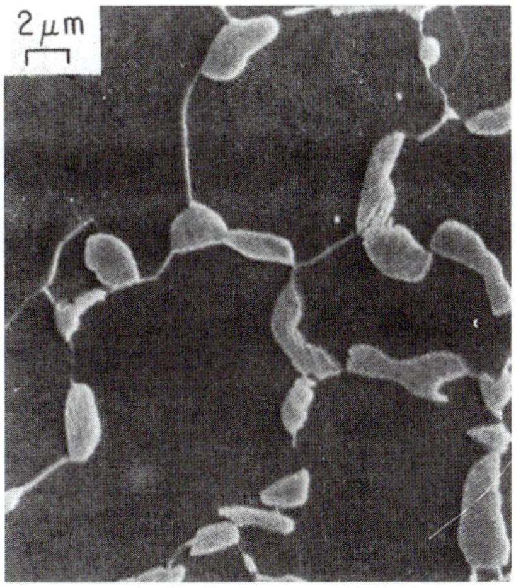

Figure 13-23
Microstructure of a dual-phase steel, showing islands of light martensite in a ferrite matrix (×2500). (*From G. Speich, "Physical Metallurgy of Dual-Phase Steels,"* Fundamentals of Dual-Phase Steels, *The Metallurgical Society of AIME, 1981.*)

careful processing permits precipitation of carbides and nitrides of Nb, V, Ti, or Zr, which provide dispersion strengthening and a fine grain size.

Dual-phase steels contain a uniform distribution of ferrite and martensite, with the dispersed martensite providing yield strengths of 414 to 1000 MPa These low-carbon steels do not contain enough alloying elements to have good hardenability using the normal quenching processes. But when the steel is heated into the ferrite-plus-austenite portion of the phase diagram, the austenite phase becomes enriched in carbon, which provides the needed hardenability. During quenching, only the austenite portion transforms to martensite (Figure 13-23).

Maraging steels are low-carbon, highly alloyed steels. The steels are austenitized and quenched to produce a soft martensite that contains less than 0.3% C. When the martensite is aged at about 500°C, intermetallic compounds such as Ni_3Ti, Fe_2Mo, and Ni_3Mo precipitate.

Interstitial-free steels are steels containing Nb and Ti. They react with C and S to form precipitates of carbides and sulfides. Thus, virtually no carbon remains in the ferrite and hence, the name interstitial-free steels. These steels are very formable and attractive for the automobile industry.

Grain-oriented steels containing silicon are used as soft magnetic materials and are used in transformer cores. Nearly pure iron powder (known as carbonyl iron), obtained by the decomposition of iron pentacarbonyl ($Fe(CO)_5$) and sometimes a reducing heat treatment, is used to make magnetic materials. Pure iron powder is also used as an additive for food supplements in breakfast cereals and other iron-fortified food products under the name reduced iron.

As mentioned before, many steels are also coated, usually to provide good corrosion protection. *Galvanized* steel is coated with a thin layer of zinc, *terne* steel is coated with lead, and other steels are coated with aluminum or tin.

EXAMPLE 13-7 *High Strength Microalloyed Steels for Petroleum Pipelines*

At present, due to worldwide increased demand for oil, there is a need to develop exceptionally strong (yield strength of ~800 MPa) steels for pipelines. Such strong steels allow for operating the pipelines under higher pressures and over a long distance. Such **microalloyed steels** often can contain such elements as Mn, Nb, V, Ti, Mo, Ni, Cr, and Cu. [Ref. Microstructure and High

Strength–Toughness Combination of a New 700 MPa Nb-Microalloyed Pipeline Steel *Materials Science and Engineering:* A.S. Shanmugam, N.K. Ramisetti, R.D.K. Misra, J. Hartmann and S.G. Jansto (in press, 2007).] (a) What other factors must be considered for high-strength steels for oil pipelines? (b) Based on what you have learned in terms of mechanisms of strengthening, how can you choose the best possible alloying elements for this application? Assume that you can conduct whatever experiments are necessary.

SOLUTION

(a) One of the first considerations is that when we increase strength significantly, is there a loss of ductility? The trick to developing such materials is that they provide not just strength but also an adequate level of ductility such that forming (e.g., rolling, extrusion, etc., Chapter 8) is still possible. Another consideration is the possibility of these steels becoming brittle due to incorporation of hydrogen or sulfur. We can also choose such alloying elements as chromium, nickel, and copper, because they have been known to provide adequate strength even in a corrosive environment (Section 13-10). An additional consideration would be the weldability of the steels and the microstructure both of the steel and of the area surrounding the weld region (Section 13-9).

(b) We can get some strengthening due to solid-solution formation (Chapter 10) and the formation of second phases (Chapter 12). We can also process the steel so as to get a fine grain size to get the Hall-Petch strengthening effect (Chapter 4). The primary grain size of the ferrite that will form from austenite is controlled by the grain size of the recrystallized austenite. We can minimize the primary austenite grain size by using niobium. It is known that niobium is a good carbide former. Fine precipitates of niobium carbides will pin the boundaries of austenite and prevent it from grain growth (Chapter 5).

To get exceptionally high strengths, we can resort to the formation of nano-sized precipitates of a second phase in a soft matrix such as ferrite. The choice of alloying elements that could be added can be guided by knowing what forms of precipitates can form so as to get a significant level of precipitation strengthening. This strengthening mechanism can be achieved by additions of not only niobium but also of titanium and vanadium. Care must be taken to ensure that such precipitates are fine (<10 nm), coherent, and high in number concentrations (Chapter 5). We can also study how such an alloy is affected as a result of the exposure to heat from the welding processes that will be used. Will there be a heat-affected zone (Chapter 9) in which any precipitates formed may grow in size and thereby become ineffective in providing strengthening? Another important consideration is the cost and amounts of alloying elements needed; this must be optimized.

This example is meant to illustrate how the principles of what you have learned so far are applied in an integrated fashion for developing a material suitable for a given real-life application. Obviously, even with a team of experienced researchers and engineers from many disciplines, experimentation and field testing are almost always required to develop a suitable and cost-effective material. It is not the intent of this example to expect someone to just develop a material simply by reading about the concepts.

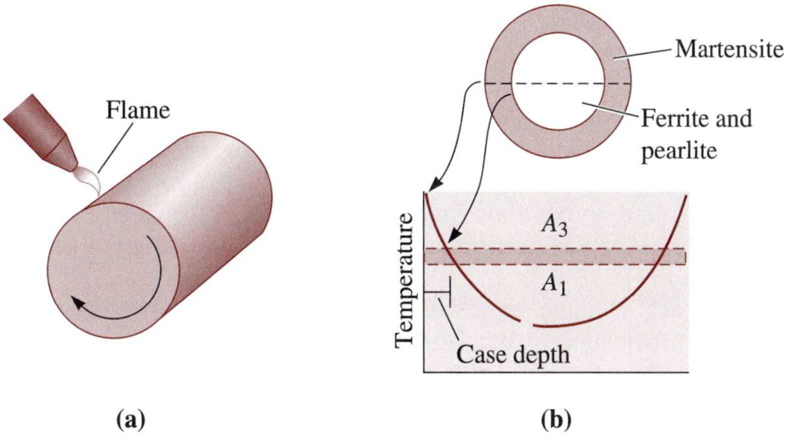

Figure 13-24 (a) Surface hardening by localized heating. (b) Only the surface heats above the A_1 temperature and is quenched to martensite.

13-8 Surface Treatments

We can, by proper heat treatment, produce a structure that is hard and strong at the surface, so that excellent wear and fatigue resistance are obtained, but at the same time gives a soft, ductile, tough core that provides good resistance to impact failure. We have seen principles of carburizing in Chapter 5, when we discussed diffusion. In this section, we see this and other similar processes.

Selectively Heating the Surface We could begin by rapidly heating the surface of a medium-carbon steel above the A_3 temperature (the center remains below the A_1). After the steel is quenched, the center is still a soft mixture of ferrite and pearlite, while the surface is martensite (Figure 13-24). The depth of the martensite layer is the **case depth**. Tempering produces the desired hardness at the surface. We can provide local heating of the surface by using a gas flame, an induction coil, a laser beam, or an electron beam. We can, if we wish, harden only selected areas of the surface that are most subject to failure by fatigue or wear.

Carburizing and Nitriding These techniques involve controlled diffusion of carbon and nitrogen, respectively (Chapter 5). For best toughness, we start with a low-carbon steel. In **carburizing**, carbon is diffused into the surface of the steel at a temperature above the A_3 (Figure 13-25). A high-carbon content is produced at the surface due to rapid diffusion and the high solubility of carbon in austenite. When the steel is quenched and

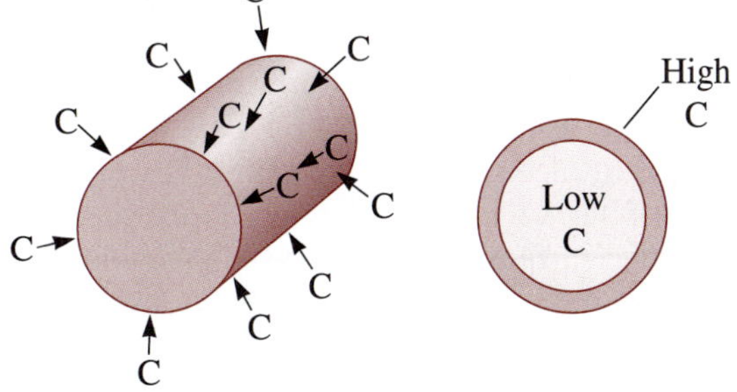

Figure 13-25 Carburizing of a low-carbon steel to produce a high-carbon, wear-resistant surface.

tempered, the surface becomes a high-carbon tempered martensite, while the ferritic center remains soft and ductile. The thickness of the hardened surface, called the case depth, is much smaller in carburized steels than in flame- or induction-hardened steels.

Nitrogen provides a hardening effect similar to that of carbon. In **cyaniding**, the steel is immersed in a liquid cyanide bath which permits both carbon and nitrogen to diffuse into the steel. In **carbonitriding**, a gas containing carbon monoxide and ammonia is generated and both carbon and nitrogen diffuse into the steel. Finally, only nitrogen diffuses into the surface from a gas in **nitriding**. Nitriding is carried out below the A_1 temperature.

In each of these processes, compressive residual stresses are introduced at the surface, providing excellent fatigue resistance (Chapter 8) in addition to the good combination of hardness, strength, and toughness. The following example explains considerations that go into considering heat treatments such as quenching and tempering and surface hardening.

EXAMPLE 13-8 *Design of Surface-Hardening Treatments for a Drive Train*

Design the materials and heat treatments for an automobile axle and drive gear (Figure 13-26).

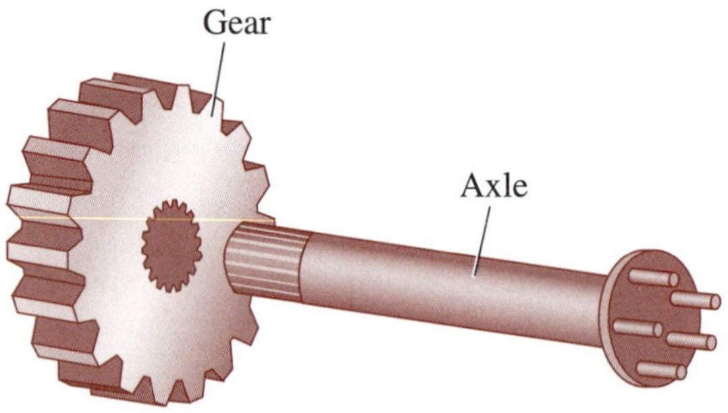

Figure 13-26 Sketch of axle and gear assembly (for Example 13-8).

SOLUTION

Both parts require good fatigue resistance. The gear also should have a good hardness to avoid wear, and the axle should have good overall strength to withstand bending and torsional loads. Both parts should have good toughness. Finally, since millions of these parts will be made, they should be inexpensive.

Quenched and tempered alloy steels might provide the required combination of strength and toughness; however, the alloy steels are expensive. An alternative approach for each part is described next.

The axle might be made from a forged 1050 steel containing a matrix of ferrite and pearlite. The axle could be surface-hardened, perhaps by moving the axle through an induction coil to selectively heat the surface of the steel above the A_3 temperature (about 770°C). After the coil passes any particular location of the axle, the cold interior quenches the surface to martensite. Tempering then softens the martensite to improve ductility. This combination of carbon content and heat treatment meets our requirements. The plain-carbon steel is inexpensive; the core of ferrite and pearlite produces good toughness and strength; and the hardened surface provides good fatigue and wear resistance.

The gear is subject to more severe loading conditions, for which the 1050 steel does not provide sufficient toughness, hardness, and wear resistance.

Instead, we might carburize a 1010 steel for the gear. The original steel contains mostly ferrite, providing good ductility and toughness. By performing a gas carburizing process above the A_3 temperature (about 860°C), we introduce about 1.0% C in a very thin case at the surface of the gear teeth. This high-carbon case, which transforms to martensite during quenching, is tempered to control the hardness. Now we obtain toughness due to the low-carbon ferrite core, wear resistance due to the high-carbon surface, and fatigue resistance due to the high-strength surface containing compressive residual stresses introduced during carburizing. In addition, the plain-carbon 1010 steel is an inexpensive starting material that is easily forged into a near-net shape prior to heat treatment.

13-9 Weldability of Steel

In Chapter 9, we discussed welding and other joining processes. We noted that steels are the most widely used structural materials. In bridges, buildings, and many other applications, steels must be welded. The structural integrity of steel structures not only depends upon the strength of the steel but also the strength of the welded joints. This is why the weldability of steel is always an important consideration.

Many low-carbon steels weld easily. Welding of medium- and high-carbon steels is comparatively more difficult since martensite can form in the heat-affected zone rather easily, thereby causing a weldment with poor toughness (Figure 13-27). Several strategies such as preheating the material or minimizing incorporation of hydrogen have been developed to counter these problems. The incorporation of hydrogen causes the steel to become brittle. In low-carbon steels, the strength of the welded regions in these

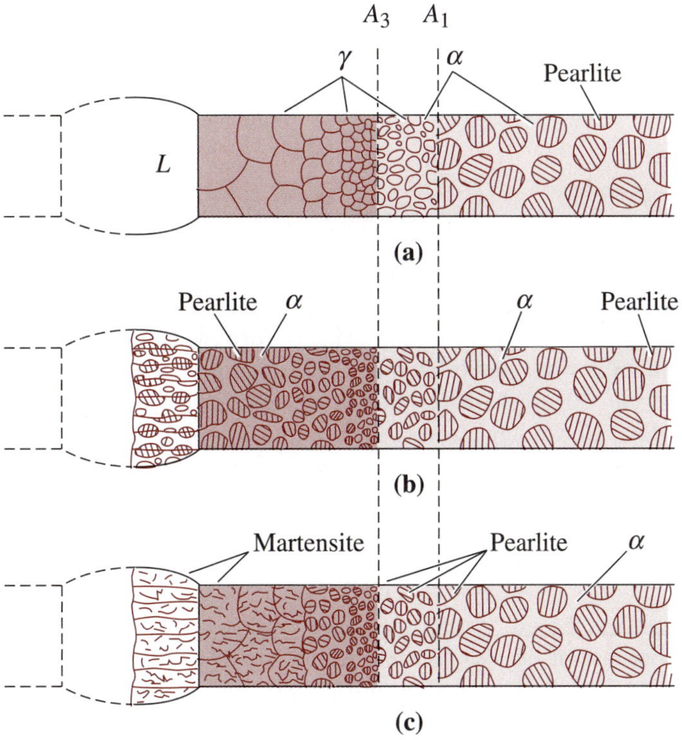

Figure 13-27 The development of the heat-affected zone in a weld: (a) the structure at the maximum temperature, (b) the structure after cooling in a steel of low hardenability, and (c) the structure after cooling in a steel of high hardenability.

materials is higher than the base material. This is due to the finer pearlite microstructure that forms during cooling of the heat-affected zone. Retained austenite along ferrite grain boundaries also limits recrystallization and thus helps retain a fine grain size, which contributes to the strength of the welded region. During welding, the metal nearest the weld heats above the A_1 temperature and austenite forms (Figure 13-27). During cooling, the austenite in this heat-affected zone transforms to a new structure, depending on the cooling rate and the CCT diagram for the steel. Plain low-carbon steels have such a low hardenability that normal cooling rates seldom produce martensite. However, an alloy steel may have to be preheated to slow down the cooling rate or post-heated to temper any martensite that forms.

A steel that is originally quenched and tempered has two problems during welding. First, the portion of the heat-affected zone that heats above the A_1 may form martensite after cooling. Second, a portion of the heat-affected zone below the A_1 may overtemper. Normally, we should not weld a steel in the quenched and tempered condition. The following example shows how the heat-affected zone microstructure can be accounted for using CCT diagrams.

EXAMPLE 13-9 *Structures of Heat-Affected Zones*

Compare the structures in the heat-affected zones of welds in 1080 and 4340 steels if the cooling rate in the heat-affected zone is 5°C/s.

SOLUTION

From the CCT diagrams, Figures 13-14 and 13-16, the cooling rate in the weld produces the following structures:

1080: 100% pearlite

4340: Bainite and martensite

The high hardenability of the alloy steel reduces the weldability, permitting martensite to form and embrittle the weld.

13-10 Stainless Steels

Stainless steels are selected for their excellent resistance to corrosion.

All true stainless steels contain a minimum of about 11% Cr, which permits a thin, protective surface layer of chromium oxide to form when the steel is exposed to oxygen. The chromium is what makes stainless steels stainless. Chromium is also a *ferrite stabilizing element*. Figure 13-28(a) illustrates the effect of chromium on the iron-carbon phase diagram. Chromium causes the austenite region to shrink, while the ferrite region increases in size. For high-chromium, low-carbon compositions, ferrite is present as a single phase up to the solidus temperature.

There are several categories of stainless steels based on crystal structure and strengthening mechanism. Typical properties are included in Table 13-4.

Ferritic Stainless Steels Ferritic stainless steels contain up to 30% Cr and less than 0.12% C. Because of the BCC structure, the ferritic stainless steels have good strengths and moderate ductilities derived from solid-solution strengthening and strain hardening.

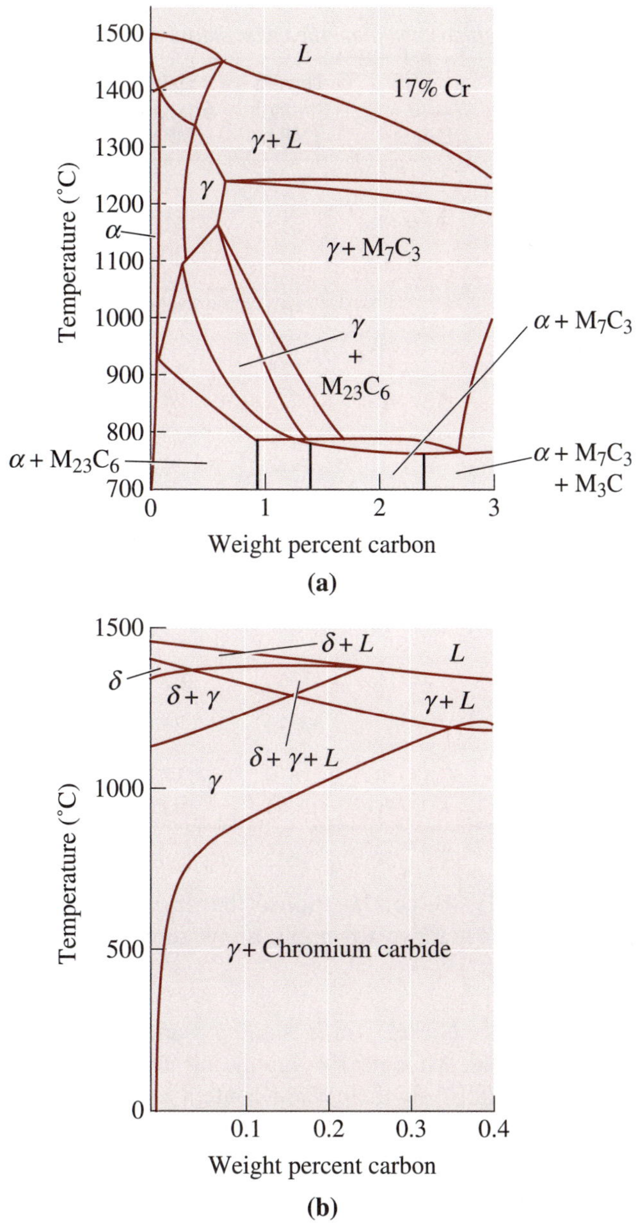

Figure 13-28
(a) The effect of 17% chromium on the iron-carbon phase diagram. At low-carbon contents, ferrite is stable at all temperatures. (b) A section of the iron-chromium-nickel-carbon phase diagram at a constant 18% Cr-8% Ni. At low-carbon contents, austenite is stable at room temperature.

Ferritic stainless steels are magnetic. They are not heat treatable. They have excellent corrosion resistance, moderate formability and are relatively inexpensive.

Martensitic Stainless Steels From Figure 13-28(a), we find that a 17% Cr-0.5% C alloy heated to 1200°C forms 100% austenite, which transforms to martensite on quenching in oil. The martensite is then tempered to produce high strengths and hardnesses [Figure 13-29(a)]. The chromium content is usually less than 17% Cr; otherwise, the austenite field becomes so small that very stringent control over both austenitizing temperature and carbon content is required. Lower chromium contents also permit the carbon content to vary from about 0.1% to 1.0%, allowing martensites of different

TABLE 13-4 ■ *Typical compositions and properties of stainless steels*

Steel	% C	% Cr	% Ni	Others	Tensile Strength (MPa)	Yield Strength (MPa)	% Elongation	Condition
Austenitic:								
201	0.15	17	5	6.5% Mn	655	310	40	Annealed
304	0.08	19	10		517	207	30	Annealed
					1276	965	9	Cold-worked
304L	0.03	19	10		517	207	30	Annealed
316	0.08	17	12	2.5% Mo	517	207	30	Annealed
321	0.08	18	10	0.4% Ti	586	241	55	Annealed
347	0.08	18	11	0.8% Nb	621	241	50	Annealed
Ferritic:								
430	0.12	17			448	207	22	Annealed
442	0.12	20			517	276	20	Annealed
Martensitic:								
416	0.15	13		0.6% Mo	1241	965	18	Quenched and tempered
431	0.20	16	2		1379	1034	16	Quenched and tempered
440C	1.10	17		0.7% Mo	1965	1896	2	Quenched and tempered
Precipitation hardening:								
17-4	0.07	17	4	0.4% Nb	1310	1172	10	Age-hardened
17-7	0.09	17	7	1.0% Al	1655	1586	6	Age-hardened

hardnesses to be produced. The combination of hardness, strength, and corrosion resistance makes the alloys attractive for applications such as high-quality knives, ball bearings, and valves.

Austenitic Stainless Steels Nickel, which is an *austenite stabilizing element*, increases the size of the austenite field, while nearly eliminating ferrite from the iron-chromium-carbon alloys [Figure 13-28(b)]. If the carbon content is below about 0.03%, the car-

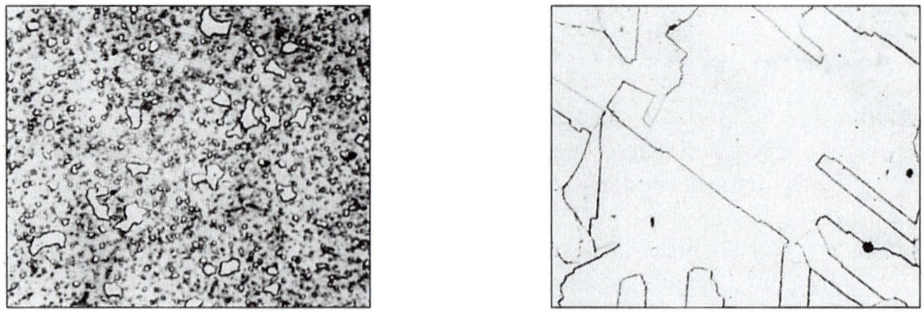

(a) (b)

Figure 13-29 (a) Martensitic stainless steel containing large primary carbides and small carbides formed during tempering (×350). (b) Austenitic stainless steel (×500). (*From* ASM Handbook, *Vols. 7 and 8, (1972, 1973), ASM International, Materials Park, OH 44073.*)

bides do not form and the steel is virtually all austenite at room temperature [Figure 13-29(b)].

The FCC austenitic stainless steels have excellent ductility, formability, and corrosion resistance. Strength is obtained by extensive solid-solution strengthening, and the austenitic stainless steels may be cold worked to higher strengths than the ferritic stainless steels. These are nonmagnetic, which is an advantage for many applications. For example, as seen in Chapter 12, cardiovascular stents are often made from 316 stainless steels. The steels have excellent low-temperature impact properties, since they have no transition temperature. Furthermore, the austenitic stainless steels are not ferromagnetic. Unfortunately, the high-nickel and chromium contents make the alloys expensive. The 304 alloy containing 18% Cr and 8% nickel (also known as 18-8 stainless) is the most widely used grade of stainless steel. Although stainless, this alloy can undergo **sensitization**. When heated to a temperature of ~480 to 860°C, chromium carbides precipitate along grain boundaries rather than within grains. This causes chromium depletion in the interior of the grains and this will cause the stainless steel to corrode very easily. This is known as sensitization.

Precipitation-Hardening (PH) Stainless Steels The precipitation-hardening (or PH) stainless steels contain Al, Nb, or Ta and derive their properties from solid-solution strengthening, strain hardening, age hardening, and the martensitic reaction. The steel is first heated and quenched to permit the austenite to transform to martensite. Reheating permits precipitates such as Ni_3Al to form from the martensite. High-mechanical properties are obtained even with low-carbon contents.

Duplex Stainless Steels In some cases, mixtures of phases are deliberately introduced into the stainless steel structure. By appropriate control of the composition and heat treatment, a **duplex stainless steel** containing approximately 50% ferrite and 50% austenite can be produced. This combination provides a set of mechanical properties, corrosion resistance, formability, and weldability not obtained in any one of the usual stainless steels.

Stainless steels are recyclable, and the following example shows how differences in properties can be used to separate different types of stainless steels.

EXAMPLE 13-10 *Design of a Test to Separate Stainless Steels*

In order to efficiently recycle stainless steel scrap, we wish to separate the high-nickel stainless steel from the low-nickel stainless steel. Design a method for doing this.

SOLUTION

Performing a chemical analysis on each piece of scrap is tedious and expensive. Sorting based on hardness might be less expensive; however, because of the different types of treatments—such as annealing, cold working, or quench and tempering—the hardness may not be related to the steel composition.

The high-nickel stainless steels are ordinarily austenitic, whereas the low-nickel alloys are ferritic or martensitic. An ordinary magnet will be attracted to the low-nickel ferritic and martensitic steels, but will not be attracted to the high-nickel austenitic steel. We might specify this simple and inexpensive magnetic test for our separation process.

13-11 Cast Irons

Cast irons are iron-carbon-silicon alloys, typically containing 2–4% C and 0.5–3% Si, that pass through the eutectic reaction during solidification. The microstructures of the five important types of cast irons are shown schematically in Figure 13-30.

Eutectic Reaction in Cast Irons Based on the Fe-Fe_3C phase diagram (dashed lines in Figure 13-31), the eutectic reaction that occurs in Fe-C alloys at 1140°C is:

$$L \rightarrow \gamma + Fe_3C \tag{13-1}$$

This reaction produces *white cast iron*, with a microstructure composed of Fe_3C and pearlite (formed by the decomposition of austenite). The Fe-Fe_3C system, however, is really a metastable phase diagram. Under truly equilibrium conditions, the eutectic reaction is:

$$L \rightarrow \gamma + \text{graphite} \tag{13-2}$$

The Fe-C phase diagram is shown as solid lines in Figure 13-31. When the stable $L \rightarrow \gamma +$ graphite eutectic reaction occurs at 1146°C, gray, ductile, or compacted-graphite cast iron forms.

In Fe-C alloys, the liquid easily undercools 6°C (the temperature difference between the stable and metastable eutectic temperatures) and white iron forms. Adding about 2% silicon to the iron increases the temperature difference between the eutectics, permitting larger undercoolings to be tolerated and more time for the stable graphite eutectic to nucleate and grow. Silicon, therefore, is a *graphite stabilizing* element. Elements such as chromium and bismuth have the opposite effect and encourage white cast

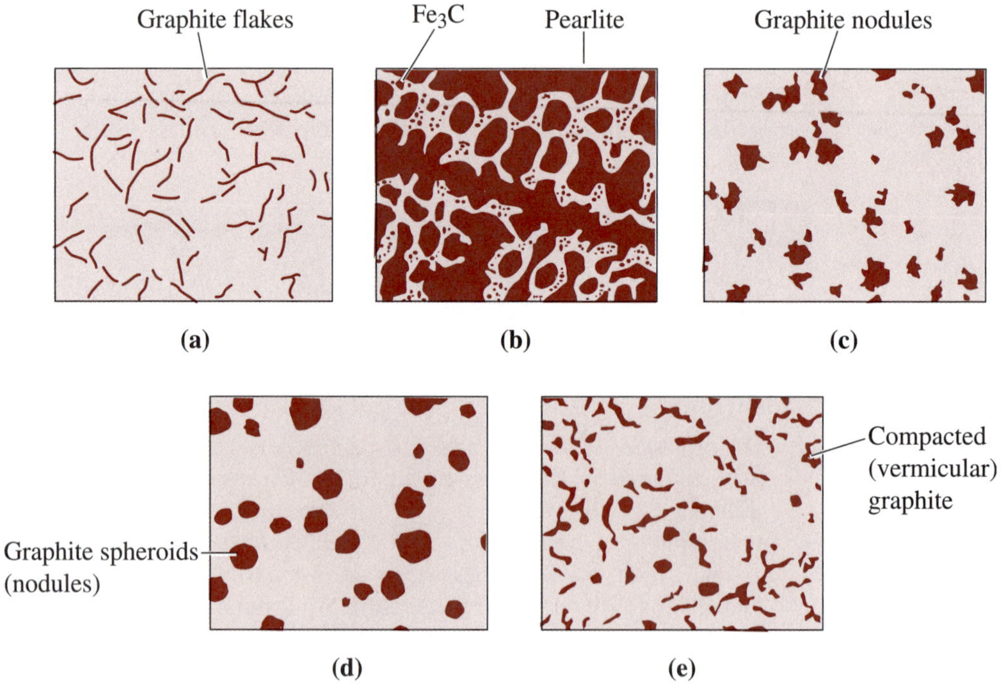

Figure 13-30 Schematic drawings of the five types of cast iron: (a) gray iron, (b) white iron, (c) malleable iron, (d) ductile iron, and (e) compacted graphite iron.

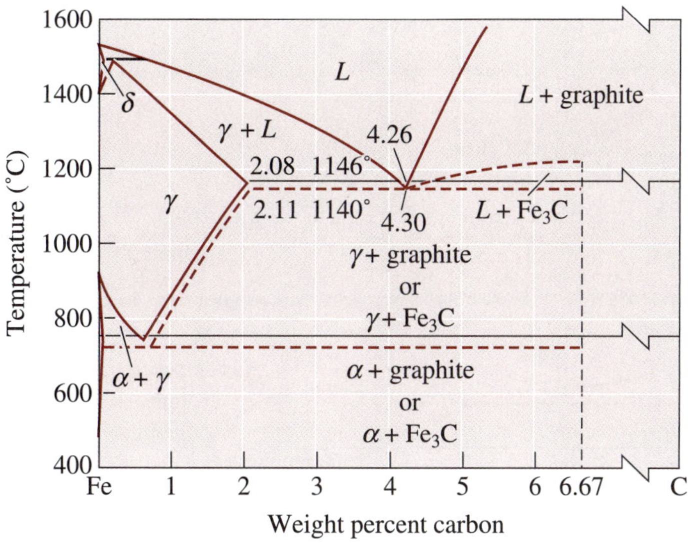

Figure 13-31 The iron-carbon phase diagram showing the relationship between the stable iron-graphite equilibria (solid lines) and the metastable iron-cementite reactions (dashed lines).

iron. We can also introduce *inoculants*, such as silicon (as Fe-Si ferrosilicon), to encourage the nucleation of graphite, or we can reduce the cooling rate of the casting to provide more time for the growth of graphite.

Silicon also reduces the amount of carbon contained in the eutectic. We can take this effect into account by defining the **carbon equivalent (CE)**:

$$CE = \%\ C + \tfrac{1}{3}\%\ Si \tag{13-3}$$

The eutectic composition is always near 4.3% CE. A high-carbon equivalent encourages the growth of the graphite eutectic.

Eutectoid Reaction in Cast Irons The matrix structure and properties of each type of cast iron are determined by how the austenite transforms during the eutectoid reaction. In the Fe-Fe_3C phase diagram used for steels, the austenite transformed to ferrite and cementite, often in the form of pearlite. However, silicon also encourages the *stable* eutectoid reaction:

$$\gamma \rightarrow \alpha + \text{graphite} \tag{13-4}$$

Under equilibrium conditions, carbon atoms diffuse from the austenite to existing graphite particles, leaving behind the low-carbon ferrite. The transformation diagram (Figure 13-32) describes how the austenite might transform during heat treatment. **Annealing** (or furnace cooling) of cast iron gives a soft ferritic matrix (not coarse pearlite as in steels!). Normalizing, or air cooling, gives a pearlitic matrix. The cast irons can also be austempered to produce bainite, or can be quenched to martensite and tempered. Austempered ductile iron, with strengths of up to 1379 MPa, is used for high-performance gears.

Gray cast iron contains small, interconnected graphite flakes that cause low strength and ductility. This is the most widely used cast iron and is named for the dull gray color of the fractured surface. Gray cast iron contains many clusters, or **eutectic cells**, of interconnected graphite flakes (Figure 13-33). The point at which the flakes are connected is the original graphite nucleus. Inoculation helps produce smaller eutectic cells, thus improving strength. The gray irons are specified by a class number of 20 to 80. A class 20 gray iron has a nominal tensile strength of 138 MPa. In thick castings, however,

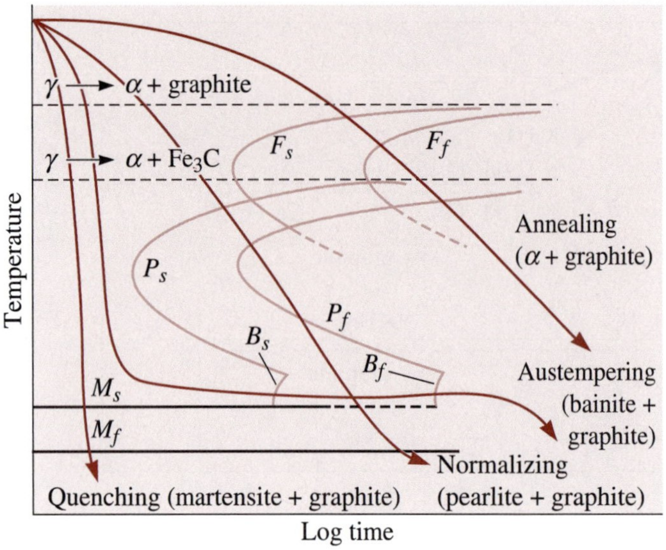

Figure 13-32 The transformation diagram for austenite in a cast iron.

coarse graphite flakes and a ferrite matrix produce tensile strengths as low as 83 MPa (Figure 13-34), whereas in thin castings, fine graphite and pearlite form and give tensile strengths near 276 MPa. Higher strengths are obtained by reducing the carbon equivalent, by alloying, or by heat treatment. The graphite flakes concentrate stresses and cause low strength and ductility, but gray iron has a number of attractive properties, including high compressive strength, good machinability, good resistance to sliding wear, good resistance to thermal fatigue, good thermal conductivity, and good vibration damping.

White cast iron is a hard, brittle alloy containing massive amounts of Fe_3C. A fractured surface of this material appears white, hence the name. A group of highly alloyed white irons are used for their hardness and resistance to abrasive wear. Elements such as chromium, nickel, and molybdenum are added so that, in addition to the alloy carbides formed during solidification, martensite is formed during subsequent heat treatment.

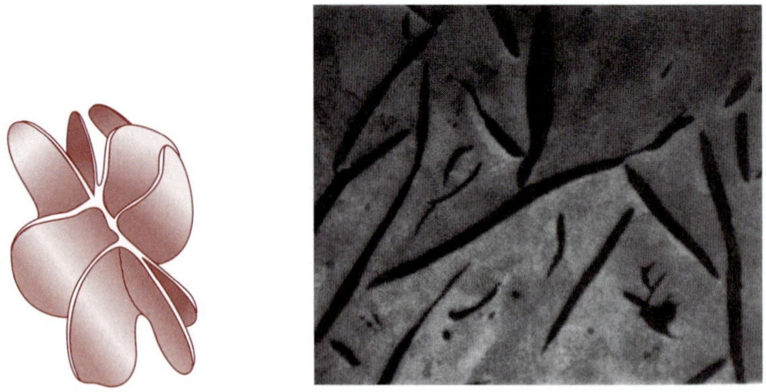

(a) **(b)**

Figure 13-33 (a) Sketch and (b) photomicrograph of the flake graphite in gray cast iron ($\times$100).

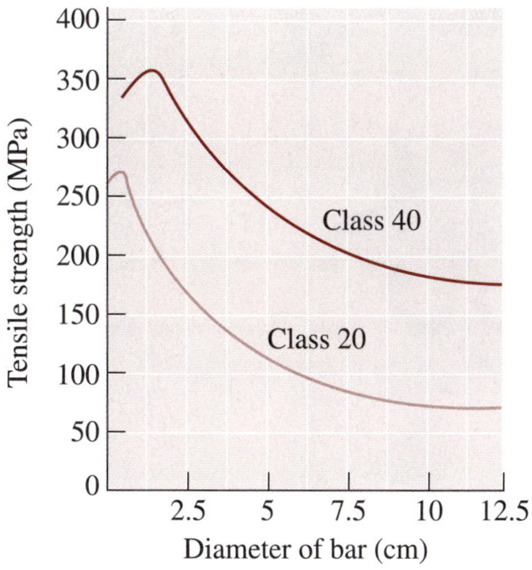

Figure 13-34 The effect of the cooling rate or casting size on the tensile properties of two gray cast irons.

Malleable cast iron, formed by the heat treatment of white cast iron, produces rounded clumps of graphite. It exhibits better ductility than gray or white cast irons. It is also very machinable. Malleable iron is produced by heat treating unalloyed 3% carbon equivalent (2.5% C, 1.5% Si) white iron. During the malleabilizing heat treatment, the cementite formed during solidification is decomposed and graphite clumps, or nodules, are produced. The nodules, or temper carbon, often resemble popcorn. The rounded graphite shape permits a good combination of strength and ductility. The production of malleable iron requires several steps (Figure 13-35). Graphite nodules nucleate as the white iron is slowly heated. During **first stage graphitization** (FSG), cementite decomposes to the stable austenite and graphite phases as the carbon in Fe_3C diffuses to the graphite nuclei. Following FSG, the austenite transforms during cooling. Figure 13-36 shows the microstructures of the original white iron (a) and the two types of malleable iron that can be produced (b and c). To make *ferritic malleable iron*, the casting is cooled

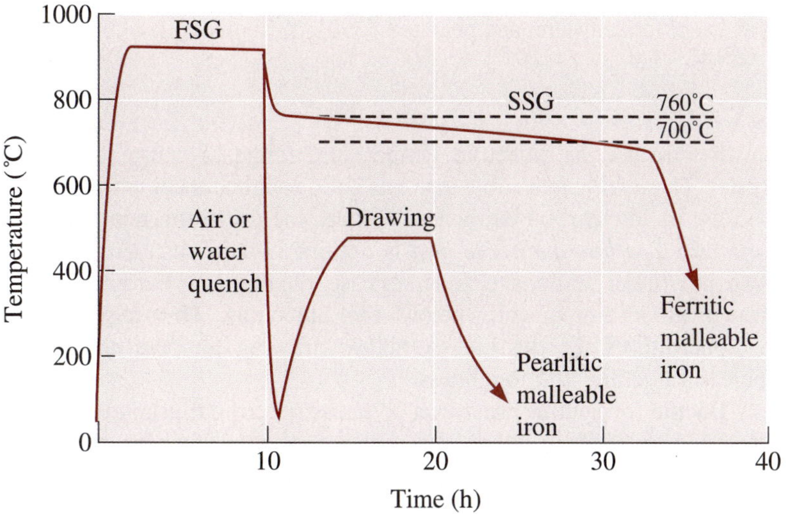

Figure 13-35 The heat treatments for ferritic and pearlitic malleable irons.

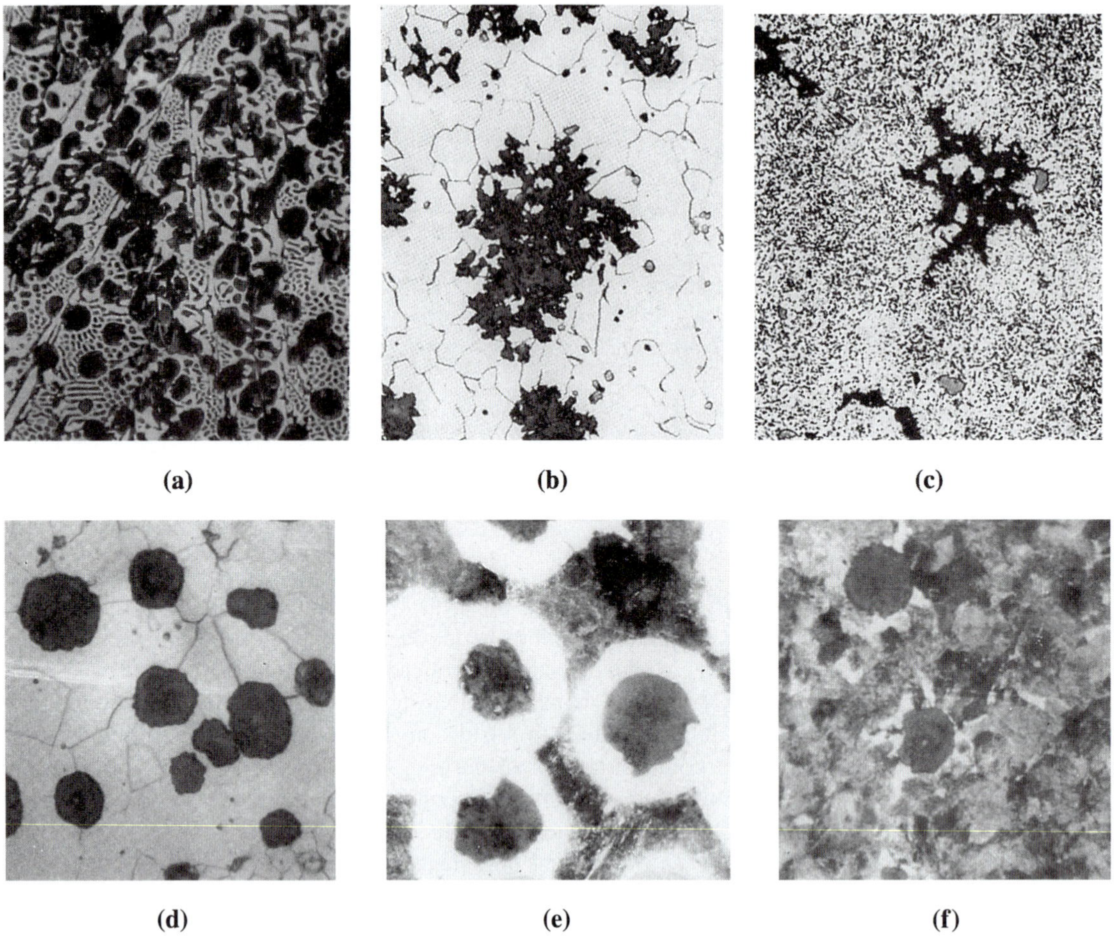

Figure 13-36 (a) White cast iron prior to heat treatment (×100). (b) Ferritic malleable iron with graphite nodules and small MnS inclusions in a ferrite matrix (×200). (c) Pearlitic malleable iron drawn to produce a tempered martensite matrix (×500). (*Images (b) and (c) are from* Metals Handbook, *Vols. 7 and 8, (1972, 1973), ASM International, Materials Park, OH 44073.*) (d) Annealed ductile iron with a ferrite matrix (×250). (e) As-cast ductile iron with a matrix of ferrite (white) and pearlite (×250). (f) Normalized ductile iron with a pearlite matrix (×250).

slowly through the eutectoid temperature range to cause **second stage graphitization** (SSG). The ferritic malleable iron has good toughness compared with that of other irons because its low-carbon equivalent reduces the transition temperature below room temperature. *Pearlitic malleable iron* is obtained when austenite is cooled in air or oil to form pearlite or martensite. In either case, the matrix is hard and brittle. The iron is then drawn at a temperature below the eutectoid. **Drawing** tempers the martensite or spheroidizes the pearlite. A higher drawing temperature decreases strength and increases ductility and toughness.

Ductile (or **nodular**) **cast iron** contains spheroidal graphite particles. Ductile iron is produced by treating liquid iron with a carbon equivalent of near 4.3% with magnesium, which causes spheroidal graphite (called nodules) to grow during solidification, rather than during a lengthy heat treatment. Several steps are required to produce this iron. These include desulfurization, **nodulizing**, and inoculation. In desulfurization, any sulfur and oxygen in the liquid metal is removed by adding desulfurizing agents such as calcium oxide (CaO). In nodulizing, Mg is added, usually in a dilute form such as an MgFeSi alloy. If pure Mg is added, the nodulizing reaction is very violent, since the boiling point of Mg is much lower than the temperature of the liquid iron, and most of the Mg will be lost. A residual of about 0.03% Mg must be in the liquid iron after treatment in order for spheroidal graphite to grow. Finally, **inoculation** with FeSi compounds to cause heterogeneous nucleation of the graphite is essential; if inoculation is not effective, white iron will form instead of ductile iron. The nodulized and inoculated iron must then be poured into molds within a few minutes to avoid fading. **Fading** occurs by the gradual, nonviolent loss of Mg due to vaporization and/or reaction with oxygen, resulting in flake or compacted graphite instead of spheroidal graphite. In addition, the inoculant effect will also fade, resulting in white iron.

Compared with gray iron, ductile cast iron has excellent strength and ductility. Due to the higher silicon content (typically around 2.4%) in ductile irons compared with 1.5% Si in malleable irons, the ductile irons are stronger but not as tough as malleable irons.

Compacted-graphite cast iron contains rounded but interconnected graphite also produced during solidification. The graphite shape in compacted-graphite cast iron is intermediate between flakes and spheres, with numerous rounded rods of graphite that are interconnected to the nucleus of the eutectic cell. This compacted graphite, sometimes called **vermicular graphite**, also forms when ductile iron fades. The compacted graphite permits strengths and ductilities that exceed those of gray cast iron, but allows the iron to retain good thermal conductivity and vibration damping properties. The treatment for the compacted-graphite iron is similar to that for ductile iron; however, only about 0.015% Mg is introduced during nodulizing. A small amount of titanium (Ti) is added to assure the formation of the compacted graphite.

Typical properties of cast irons are given in Table 13-5.

TABLE 13-5 ■ *Typical properties of cast irons*

	Tensile Strength (MPa)	Yield Strength (MPa)	% E	Notes
Gray irons:				
Class 20	83–276	—	—	
Class 40	193–372	—	—	
Class 60	303–455	—	—	
Malleable irons:				
32510	345	224	10	Ferritic
35018	365	241	18	Ferritic
50005	483	345	5	Pearlitic
70003	586	483	3	Pearlitic
90001	724	621	1	Pearlitic
Ductile irons:				
60–40–18	414	276	18	Annealed
65–45–12	448	310	12	As-cast ferritic
80–55–06	552	379	6	As-cast pearlitic
100–70–03	689	483	3	Normalized
120–90–02	827	621	2	Quenched and tempered
Compacted-graphite irons:				
Low strength	276	193	5	90% Ferritic
High strength	448	379	1	80% Pearlitic

SUMMARY

- The properties of steels, determined by dispersion strengthening, depend on the amount, size, shape, and distribution of cementite. These factors are controlled by alloying and heat treatment.
- A process anneal recrystallizes cold-worked steels and reduces stresses.
- Spheroidizing produces large, spheroidal Fe_3C and good machinability in high-carbon steels.
- Annealing, involving a slow furnace cool after austenitizing, gives a coarse pearlitic structure containing lamellar Fe_3C.
- Normalizing, involving an air cool after austenitizing, gives a fine pearlitic structure and higher strength compared with annealing.
- In isothermal annealing, pearlite with a uniform interlamellar spacing is obtained by transforming the austenite at a constant temperature.
- Austempering is used to produce bainite, containing rounded Fe_3C, by an isothermal transformation.
- Quench and temper heat treatments require the formation and decomposition of martensite, providing exceptionally fine dispersions of round Fe_3C.
- We can better understand the mechanics of the heat treatments by use of TTT diagrams, CCT diagrams, and hardenability curves.
- The hardenability curves compare the ease with which different steels transform to martensite.
- Alloying elements increase the times required for transformations in the TTT diagrams, reduce the cooling rates necessary to produce martensite in the CCT diagrams, and improve the hardenability of the steel.
- Specialty steels and heat treatments provide unique properties or combinations of properties. Of particular importance are surface-hardening treatments, such as carburizing, that produce an excellent combination of fatigue and impact resistance. Stainless steels, which contain a minimum of 12% Cr, have excellent corrosion resistance.
- Cast irons, by definition, undergo the eutectic reaction during solidification. Depending on the composition and treatment, either γ and Fe_3C or γ and graphite form during solidification.

GLOSSARY

Annealing (cast iron) A heat treatment used to produce a ferrite matrix in a cast iron by austenitizing, then furnace cooling.

Annealing (steel) A heat treatment used to produce a soft, coarse pearlite in steel by austenitizing, then furnace cooling.

Ausforming A thermomechanical heat treatment in which austenite is plastically deformed below the A_1 temperature, then permitted to transform to bainite or martensite.

Austempering The isothermal heat treatment by which austenite transforms to bainite.

Austenitizing Heating a steel or cast iron to a temperature where homogeneous austenite can form. Austenitizing is the first step in most of the heat treatments for steel and cast irons.

Carbon equivalent (CE) Carbon plus one-third of the silicon in a cast iron.

Carbonitriding Hardening the surface of steel with carbon and nitrogen obtained from a special gas atmosphere.

Carburizing A group of surface-hardening techniques by which carbon diffuses into steel.

Case depth The depth below the surface of a steel at which hardening occurs by surface hardening and carburizing processes.

Cast iron Ferrous alloys containing sufficient carbon so that the eutectic reaction occurs during solidification.

Compacted-graphite cast iron A cast iron treated with small amounts of magnesium and titanium to cause graphite to grow during solidification as an interconnected, coral-shaped precipitate, giving properties midway between gray and ductile iron.

Cyaniding Hardening the surface of steel with carbon and nitrogen obtained from a bath of liquid cyanide solution.

Drawing Reheating a malleable iron in order to reduce the amount of carbon combined as cementite by spheroidizing pearlite, tempering martensite, or graphitizing both.

Dual-phase steels Special steels treated to produce martensite dispersed in a ferrite matrix.

Ductile cast iron Cast iron treated with magnesium to cause graphite to precipitate during solidification as spheres, permitting excellent strength and ductility. Also known as nodular cast iron.

Duplex stainless steel A special class of stainless steels containing a microstructure of ferrite and austenite.

Electric arc furnace A furnace used to melt steel scrap using electricity. Often, specialty steels are made using electric arc furnaces.

Eutectic cell A cluster of graphite flakes produced during solidification of gray iron that are all interconnected to a common nucleus.

Fading The loss of the nodulizing or inoculating effect in cast irons as a function of time, permitting undesirable changes in microstructure and properties.

First stage graphitization The first step in the heat treatment of a malleable iron, during which the massive carbides formed during solidification are decomposed to graphite and austenite.

Gray cast iron Cast iron which, during solidification, contains graphite flakes, causing low strength and poor ductility. This is the most widely used type of cast iron.

Hardenability The ease with which a steel can be quenched to form martensite. Steels with high hardenability form martensite even on slow cooling.

Hardenability curves Graphs showing the effect of the cooling rate on the hardness of as-quenched steel.

Hot metal The molten iron produced in a blast furnace, also known as pig iron. It contains about 95% iron, 4% carbon, 0.3–0.9% silicon, 0.5% Mn, and 0.025–0.05% each of sulfur, phosphorus, and titanium.

Inoculation The addition of an agent to the molten cast iron that provides nucleation sites at which graphite precipitates during solidification.

Interstitial free steels These are steels containing Nb and Ti. They react with C and S to form precipitates of carbides and sulfides, leaving the ferrite nearly free of interstial elements.

Isothermal annealing Heat treatment of a steel by austenitizing, cooling rapidly to a temperature between the A_1 and the nose of the TTT curve, and holding until the austenite transforms to pearlite.

Jominy distance The distance from the quenched end of a Jominy bar. The Jominy distance is related to the cooling rate.

Jominy test The test used to evaluate hardenability. An austenitized steel bar is quenched at one end only, thus producing a range of cooling rates along the bar.

Malleable cast iron Cast iron obtained by a lengthy heat treatment, during which cementite decomposes to produce rounded clumps of graphite. Good strength, ductility, and toughness are obtained as a result of this structure.

Maraging steels A special class of alloy steels that obtain high strengths by a combination of the martensitic and age-hardening reactions.

Marquenching Quenching austenite to a temperature just above the M_S and holding until the temperature is equalized throughout the steel before further cooling to produce martensite. This process reduces residual stresses and quench cracking. Also known as *martempering*.

Microalloyed steels High-strength steels containing relatively low carbon but small amounts of alloying elements such as Mn, Nb, Ti, and V added to enhance strength and hardenability.

Nitriding Hardening the surface of steel with nitrogen obtained from a special gas atmosphere.

Nodulizing The addition of magnesium to molten cast iron to cause the graphite to precipitate as spheres rather than as flakes during solidification.

Normalizing A simple heat treatment obtained by austenitizing and air cooling to produce a fine pearlitic structure. This can be done for steels and cast irons.

Pig iron The molten iron produced in a blast furnace also known as hot metal. It contains about 95% iron, 4% carbon, 0.3–0.9% silicon, 0.5% Mn, and 0.025–0.05% each of sulfur, phosphorus, and titanium.

Process anneal A low-temperature heat treatment used to eliminate all or part of the effect of cold working in steels.

Quench cracks Cracks that form at the surface of a steel during quenching due to tensile residual stresses that are produced because of the volume change that accompanies the austenite-to-martensite transformation.

Retained austenite Austenite that is unable to transform into martensite during quenching because of the volume expansion associated with the reaction.

Second stage graphitization The second step in the heat treatment of malleable irons that are to have a ferritic matrix. The iron is cooled slowly from the first stage graphitization temperature so that austenite transforms to ferrite and graphite rather than to pearlite.

Secondary hardening peak Unusually high hardness in a steel tempered at a high temperature caused by the precipitation of alloy carbides.

Sensitization When heated to a temperature of ~480–860°C, chromium carbides precipitate along grain boundaries rather than within grains, causing chromium depletion in the interior. This causes the stainless steel to corrode very easily.

Spheroidite A microconstituent containing coarse spheroidal cementite particles in a matrix of ferrite, permitting excellent machining characteristics in high-carbon steels.

Stainless steels A group of ferrous alloys that contain at least 11% Cr, providing extraordinary corrosion resistance.

Tempered martensite The microconstituent of ferrite and cementite formed when martensite is tempered.

Tool steels A group of high-carbon steels that provide combinations of high hardness, toughness, or resistance to elevated temperatures.

Vermicular graphite The rounded, interconnected graphite that forms during the solidification of cast iron. This is the intended shape in compacted-graphite iron, but it is a defective shape in ductile iron.

White cast iron Cast iron that produces cementite rather than graphite during solidification. The white irons are hard and brittle.

PROBLEMS

Section 13-1 Designations and Classification of Steels

13-1 What is the difference between cast iron and steels?

13-2 What do A_1, A_3, and A_{cm} temperatures refer to? Are these temperatures constant?

13-3 Calculate the amounts of ferrite, cementite, primary microconstituent, and pearlite in the following steels:

(a) 1015 **(b)** 1035 **(c)** 1095 **(d)** 10130

13-4 Estimate the AISI-SAE number for steels having the following microstructures:

(a) 38% pearlite-62% primary ferrite
(b) 93% pearlite-7% primary cementite
(c) 97% ferrite-3% cementite
(d) 86% ferrite-14% cementite

13-5 Complete the following table:

	1035 Steel	10115 Steel
A_1 temperature		
A_3 or A_{cm} temperature		
Full annealing temperature		
Normalizing temperature		
Process annealing temperature		
Spheroidizing temperature		

13-6 What do the terms low-, medium-, and high-carbon steels mean?

Section 13-2 Simple Heat Treatments

Section 13-3 Isothermal Heat Treatments

13-7 Explain the following heat treatments: **(a)** process anneal, **(b)** austenitizing, **(c)** annealing, **(d)** normalizing, and **(e)** quenching.

13-8 Explain why, strictly speaking, TTT diagrams can be used for isothermal treatments only.

13-9 In a pearlitic 1080 steel, the cementite platelets are 4×10^{-5} cm thick, and the ferrite platelets are 14×10^{-5} cm thick. In a spheroidized 1080 steel, the cementite spheres are 4×10^{-3} cm in diameter. Estimate the total interface area between the ferrite and cementite in a cubic centimeter of each steel. Determine the percentage reduction in surface area when the pearlitic steel is spheroidized. The density of ferrite is 7.87 g/cm^3 and that of cementite is 7.66 g/cm^3.

13-10 Describe the microstructure present in a 1050 steel after each step in the following heat treatments:

(a) heat at 820°C, quench to 650°C and hold for 90 s, and quench to 25°C
(b) heat at 820°C, quench to 450°C and hold for 90 s, and quench to 25°C
(c) heat at 820°C, and quench to 25°C
(d) heat at 820°C, quench to 720°C and hold for 100 s, and quench to 25°C
(e) heat at 820°C, quench to 720°C and hold for 100 s, quench to 400°C and hold for 500 s, and quench to 25°C

(f) heat at 820°C, quench to 720°C and hold for 100 s, quench to 400°C and hold for 10 s, and quench to 25°C

(g) heat at 820°C, quench to 25°C, heat to 500°C and hold for 10^3 s, and air cool to 25°C

13-11 Describe the microstructure present in a 10110 steel after each step in the following heat treatments

(a) heat to 900°C, quench to 400°C and hold for 10^3 s, and quench to 25°C

(b) heat to 900°C, quench to 600°C and hold for 50 s, and quench to 25°C

(c) heat to 900°C, and quench to 25°C

(d) heat to 900°C, quench to 300°C and hold for 200 s, and quench to 25°C

(e) heat to 900°C, quench to 675°C and hold for 1 s, and quench to 25°C

(f) heat to 900°C, quench to 675°C and hold for 1 s, quench to 400°C and hold for 900 s, and slowly cool to 25°C

(g) heat to 900°C, quench to 675°C and hold for 1 s, quench to 300°C and hold for 10^3 s, and air cool to 25°C

(h) heat to 900°C, quench to 300°C and hold for 100 s, quench to 25°C, heat to 450°C for 3600 s, and cool to 25°C

13-12 Recommend appropriate isothermal heat treatments to obtain the following, including appropriate temperatures and times:

(a) an isothermally annealed 1050 steel with HRC 23,

(b) an isothermally annealed 10110 steel with HRC 40,

(c) an isothermally annealed 1080 steel with HRC 38,

(d) an austempered 1050 steel with HRC 40,

(e) an austempered 10110 steel with HRC 55, and

(f) an austempered 1080 steel with HRC 50.

13-13 Compare the minimum times required to isothermally anneal the following steels at 600°C. Discuss the effect of the carbon content of the steel on the kinetics of nucleation and growth during the heat treatment.

(a) 1050 **(b)** 1080 **(c)** 10110

Section 13-4 Quench and Temper Heat Treatments

13-14 Explain the following terms: **(a)** quenching, **(b)** tempering, **(c)** retained austenite, and **(d)** marquenching/martempering.

13-15 We wish to produce a 1050 steel that has a Brinell hardness of at least 330 and an elongation of at least 15%.

(a) Recommend a heat treatment, including appropriate temperatures, that permits this to be achieved. Determine the yield strength and tensile strength that are obtained by this heat treatment.

(b) What yield and tensile strength would be obtained in a 1080 steel by the same heat treatment?

(c) What yield strength, tensile strength and % elongation would be obtained in the 1050 steel if it were normalized?

13-16 We wish to produce a 1050 steel that has a tensile strength of at least 1207 MPa and a reduction in area of at least 50%.

(a) Recommend a heat treatment, including appropriate temperatures, that permits this to be achieved. Determine the Brinell hardness number, % elongation, and yield strength that are obtained by this heat treatment.

(b) What yield strength and tensile strength would be obtained in a 1080 steel by the same heat treatment?

(c) What yield strength, tensile strength, and % elongation would be obtained in the 1050 steel if it were annealed?

13-17 A 1030 steel is given an improper quench and temper heat treatment, producing a final structure composed of 60% martensite and 40% ferrite. Estimate the carbon content of the martensite and the austenitizing temperature that was used. What austenitizing temperature would you recommend?

13-18 A 1050 steel should be austenitized at 820°C, quenched in oil to 25°C, and tempered at 400°C for an appropriate time.

(a) What yield strength, hardness, and % elongation would you expect to obtain from this heat treatment?

(b) Suppose the actual yield strength of the steel is found to be 862 MPa. What might have gone wrong in the heat treatment to cause this low strength?

(c) Suppose the hardness is found to be HB 525. What might have gone wrong in the heat treatment to cause this high hardness?

13-19 A part produced from a low-alloy, 0.2% C steel (Figure 13-15) has a microstructure containing ferrite, pearlite, bainite, and martensite after quenching. What microstructure would be obtained if we used a 1080 steel? What microstructure would be obtained if we used a 4340 steel?

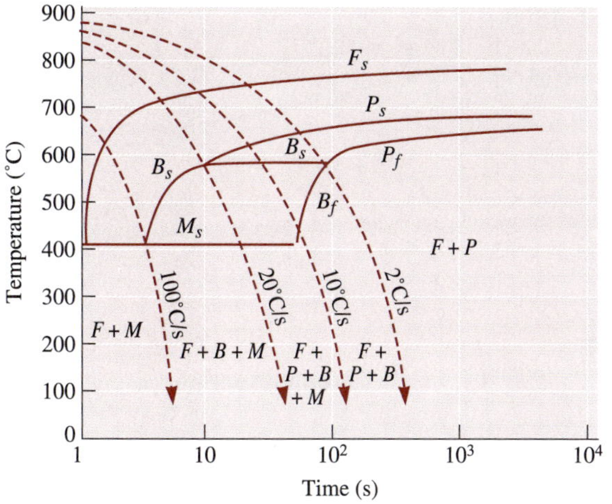

Figure 13-15 (Repeated for Problem 13-19.) The CCT diagram for a low-alloy, 0.2% C steel.

13-20 Fine pearlite and a small amount of martensite are found in a quenched 1080 steel. What microstructure would be expected if we used a low-alloy, 0.2% C steel? What microstructure would be expected if we used a 4340 steel?

Section 13-5 Effect of Alloying Elements

Section 13-6 Application of Hardenability

13-21 Explain the difference between hardenability and hardness. Explain using a sketch how hardenability of steels is measured.

13-22 We have found that a 1070 steel, when austenitized at 750°C, forms a structure containing pearlite and a small amount of grain-boundary ferrite that gives acceptable strength and ductility. What changes in the microstructure, if any, would be expected if the 1070 steel contained an alloying element such as Mo or Cr? Explain.

13-23 Using the TTT diagrams, compare the hardenabilities of 4340 and 1050 steels by determining the times required for the isothermal transformation of ferrite and pearlite (F_s, P_s, and P_f) to occur at 650°C.

13-24 We would like to obtain a hardness of HRC 38 to 40 in a quenched steel. What range of cooling rates would we have to obtain for the following steels? Are some steels inappropriate for achieving these levels of hardness?

(a) 4340 **(b)** 8640 **(c)** 9310
(d) 4320 **(e)** 1050 **(f)** 1080

13-25 A steel part must have an as-quenched hardness of HRC 35 in order to avoid excessive-wear rates during use. When the part is made from 4320 steel, the hardness is only HRC 32. Determine the hardness if the part were made under identical conditions, but with the following steels. Which, if any, of these steels would be better choices than 4320?

(a) 4340 **(b)** 8640 **(c)** 9310
(d) 1050 **(e)** 1080

13-26 A part produced from a 4320 steel has a hardness of HRC 35 at a critical location after quenching. Determine

(a) the cooling rate at that location, and
(b) the microstructure and hardness that would be obtained if the part were made of a 1080 steel.

13-27 A 1080 steel is cooled at the fastest possible rate that still permits all pearlite to form. What is this cooling rate? What Jominy distance, and hardness are expected for this cooling rate?

13-28 Determine the hardness and microstructure at the center of a 3.75-cm-diameter 1080 steel bar produced by quenching in

(a) unagitated oil,
(b) unagitated water, and
(c) agitated brine.

13-29 A 5-cm-diameter bar of 4320 steel is to have a hardness of at least HRC 35. What is the minimum severity of the quench (H coefficient)? What type of quenching medium would you recommend to produce the desired hardness with the least chance of quench cracking?

13-30 A steel bar is to be quenched in agitated water. Determine the maximum diameter of the bar that will produce a minimum hardness of HRC 40 if the bar is:

(a) 1050 **(b)** 1080 **(c)** 4320
(d) 8640 **(e)** 4340

13-31 The center of a 2.5-cm-diameter bar of 4320 steel has a hardness of HRC 40. Determine the hardness and microstructure at the center of a 5-cm bar of 1050 steel quenched in the same medium.

Section 13-7 Specialty Steels

Section 13-8 Surface Treatments

Section 13-9 Weldability of Steel

13-32 What is the principle of the surface hardening of steels using carburizing and nitriding?

13-33 A 1010 steel is to be carburized using a gas atmosphere that produces 1.0% C at the surface of the steel. The case depth is defined as the distance below the surface that contains at least 0.5% C. If carburizing is done at 1000°C, determine the time required to produce a case depth of 0.025 cm (See Chapter 5 for a review.)

13-34 A 1015 steel is to be carburized at 1050°C for 2 h using a gas atmosphere that produces 1.2% C at the surface of the steel. Plot the percent carbon versus the distance from the surface of the steel. If the steel is slowly cooled after carburizing, determine the amount of each phase and microconstituent at 0.005-cm intervals from the surface. (See Chapter 5.)

13-35 Why is it that the strength of the heat-affected zone is higher for low-carbon steels? What is the role of retained austenite, in this case?

13-36 Why is it easier to weld low-carbon steels, however, it is difficult to weld high-carbon steels?

13-37 A 1050 steel is welded. After cooling, hardnesses in the heat-affected zone are obtained at various locations from the edge of the fusion zone. Determine the hardnesses expected at each point if a 1080 steel were welded under the same conditions. Predict the microstructure at each location in the as-welded 1080 steel.

Distance from Edge of Fusion Zone	Hardness in 1050 Weld
0.05 mm	HRC 50
0.10 mm	HRC 40
0.15 mm	HRC 32
0.20 mm	HRC 28

Section 13-10 Stainless Steels

13-38 What is a stainless steel? Why are stainless steels stainless?

13-39 We wish to produce a martensitic stainless steel containing 17% Cr. Recommend a carbon content and austenitizing temperature that would permit us to obtain 100% martensite during the quench. What microstructure would be produced if the martensite were then tempered until the equilibrium phases formed?

13-40 Occasionally, when an austenitic stainless steel is welded, the weld deposit may be slightly magnetic. Based on the Fe-Cr-Ni-C phase diagram [Figure 13-28(b)], what phase would you expect is causing the magnetic behavior? Why might this phase have formed? What could you do to restore the nonmagnetic behavior?

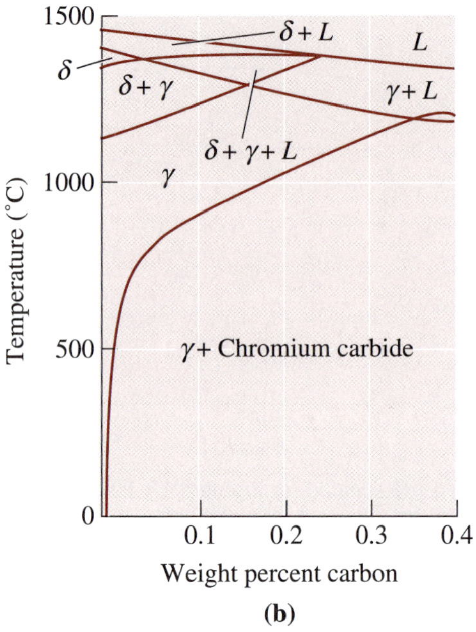

Figure 13-28 (Repeated for Problem 13-40) (b) A section of the iron-chromium-nickel-carbon phase diagram at a constant 18% Cr-8% Ni. At low-carbon contents, austenite is stable at room temperature.

Section 13-11 Cast Irons

13-41 Define cast iron using the Fe-Fe_3C phase diagram.

13-42 What are the different types of cast irons? Explain using a sketch.

13-43 A tensile bar of a class 40 gray iron casting is found to have a tensile strength of 345 MPa. Why is the tensile strength greater than that given by the class number? What do you think is the diameter of the test bar?

13-44 You would like to produce a gray iron casting that freezes with no primary austenite or graphite. If the carbon content in the iron is 3.5%, what percentage of silicon must you add?

Design Problems

13-45 We would like to produce a 5-cm-thick steel wear plate for a rock-crushing unit. To avoid frequent replacement of the wear plate, the hardness should exceed HRC 38 within 0.625 cm of the steel surface. However, the center of the plate should have a hardness of no more than HRC 32 to assure some toughness. We have only a water quench available to us. Design the plate, assuming that we only have the steels given in Figure 13-21 available to us.

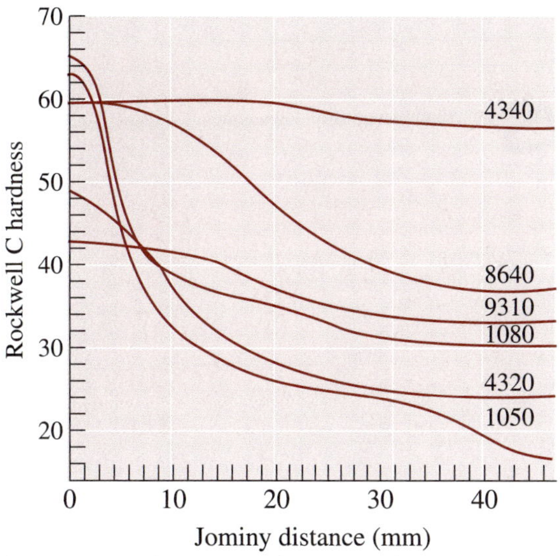

Figure 13-21 (Repeated for Problem 13-45.) The hardenability curves for several steels.

13-46 A quenched and tempered 10110 steel is found to have surface cracks that cause the heat-treated part to be rejected by the customer. Why did the cracks form? Design a heat treatment, including appropriate temperatures and times that will minimize these problems.

13-47 Design a corrosion-resistant steel to use for a pump that transports liquid helium (He) at 4 K in a superconducting magnet.

13-48 Design a heat treatment for a hook made of 2.5-cm-diameter steel rod having a microstructure containing a mixture of ferrite, bainite, and martensite after quenching. Estimate the mechanical properties of your hook.

13-49 Design an annealing treatment for a 1050 steel. Be sure to include details of temperatures, cooling rates, microstructures, and properties.

13-50 Design a process to produce a 0.5-cm-diameter steel shaft having excellent toughness, yet excellent wear and fatigue resistance. The surface hardness should be at least HRC 60, and the hardness 0.01 cm beneath the surface should be approximately HRC 50. Describe the process, including details of the heat-treating atmosphere, the composition of the steel, temperatures, and times.

14

Nonferrous Alloys

Have You Ever Wondered?

- *In the history of mankind, which came first: copper or steel?*
- *What materials are used to manufacture biomedical implants for hip prostheses?*
- *Would it be possible to make aluminum cars?*
- *What materials are used as "catalysts" in the automobile catalytic converter? What do they "convert"?*

Nonferrous alloys (i.e., alloys of elements other than iron) include, but are not limited to, alloys based on aluminum, copper, nickel, cobalt, zinc, precious metals (such as Pt, Au, Ag, Pd), and other metals (e.g., Nb, Ta, W). In this chapter, we will briefly explore the properties and applications of Cu, Al, and Ti alloys in load-bearing applications. We will not discuss the electronic, magnetic, and other applications of nonferrous alloys.

In many applications, weight is a critical factor. To relate the strength of the material to its weight, a *specific strength*, or strength-to-weight ratio, is defined:

$$\text{Specific strength} = \frac{\text{strength}}{\text{density}} \qquad (14\text{-}1)$$

Table 14-1 compares the specific strength of steel, some high-strength nonferrous alloys, and polymer-matrix composites. Note that these are just examples of properties and these should not be used for design purposes. The actual values for any given family of alloy will vary.

Another factor to consider in designing with nonferrous metals is their cost, which also varies considerably. Table 14-1 gives the approximate price of different materials. One should note, however, that the price of the material is only a small portion of the cost of a part. Fabrication and finishing, not to mention marketing and distribution, often contribute much more to the overall cost of a part.

Composites based on carbon and other fibers also have significant advantages with respect to their specific strength. However, their properties could be anisotropic and the temperature at which they can be used is limited. In practice, to overcome the anisotropy, composites are often made in many layers. The directions of fibers are changed in different layers so as to minimize the anisotropy in properties.

TABLE 14-1 ■ ***Specific strength and cost of nonferrous alloys, steels, and polymer composites***

Metal	Density g/cm^3	Tensile Strength (MPa)	Specific Strength (cm)	Cost per kg (\$)[c]
Aluminum	2.70	572	21.5×10^5	1.32
Beryllium composites	1.85	379	20.5×10^5	771.60
Copper	8.93	207–483	11.8×10^5	1.57
Lead	11.36	69	0.5×10^5	0.99
Magnesium	1.74	379	21.8×10^5	3.31
Nickel	8.90	1241	14.0×10^5	9.04
Titanium	4.51	1103	24.5×10^5	8.82
Tungsten	19.25	1034	5.5×10^5	8.82
Zinc	7.13	517	7.3×10^5	0.88
Steels	~7.87	1379	17.5×10^5	0.22
Aramid/epoxy (Kevlar, vol. fraction of fibers 0.6, longitudinal tension)	1.4	1379	10.0×10^6	—
Aramid/epoxy (Kevlar, vol. fraction of fibers 0.6, transverse tension)[a]	1.4	30	2.15×10^4	—
Glass/epoxy (Vol. fraction of E-glass fibers 0.6, longitudinal tension)[b]	2.1	1034	5.0×10^6	—
Glass/epoxy (Vol. fraction of E-glass fibers 0.6, transverse tension)	2.1	48	23.3×10^4	—

[a] *Data for composites from Harper, C.A.,* Handbook of Materials Product Design, *3rd ed. 2001: McGraw-Hill. Commodity composites are relatively inexpensive; high-performance composites are expensive.*
[b] *Properties of composites are highly anisotropic. This is taken care of during fabrication though.*
[c] *Costs based on average prices for the years 1998 to 2002.*

14-1 Aluminum Alloys

Aluminum is the third most plentiful element on earth (next to oxygen and silicon), but, until the late 1800s, was expensive and difficult to produce. The six-pound cap installed on the top of the Washington Monument in 1884 was one of the largest aluminum parts made at that time.

General Properties and Uses of Aluminum Aluminum has a density of 2.70 g/cm^3, or one-third the density of steel, and a modulus of elasticity of 69×10^3 MPa. Although aluminum alloys have lower tensile properties compared with those of steel, their specific strength (or strength-to-weight ratio) is excellent. The Wright brothers used an Al-Cu alloy for their engine for this very reason. Aluminum can be formed easily, it has high thermal and electrical conductivity, and does not show a ductile-to-brittle transition at low temperatures. It is nontoxic and can be recycled with only about 5% of the energy that was needed to make it from alumina (Al_2O_3). This is why the recycling of aluminum is so successful. Aluminum's beneficial physical properties include nonmagnetic behavior and its resistance to oxidation and corrosion. However, aluminum does not display a true endurance limit, so failure by fatigue eventually may occur, even at low stresses. Because of its low-melting temperature, aluminum does not perform well at elevated temperatures. Finally, aluminum alloys have low hardness, leading to poor wear resistance. Aluminum responds readily to strengthening mechanisms. Table 14-2 compares the strength of pure annealed aluminum with that of alloys strengthened by various techniques. The alloys may be 30 times stronger than pure aluminum.

TABLE 14-2 ■ ***The effect of strengthening mechanisms in aluminum and aluminum alloys***

Material	Tensile Strength (MPa)	Yield Strength (MPa)	% Elongation	Ratio of Alloy-to-Metal Yield Strengths
Pure Al	45	17	60	1.0
Commercially pure Al (at least 99% pure)	90	35	45	2.0
Solid-solution-strengthened Al alloy	110	41	35	2.4
Cold-worked Al	166	152	15	8.8
Dispersion-strengthened Al alloy	290	152	35	8.8
Age-hardened Al alloy	572	503	11	29.2

About 25% of the aluminum produced today is used in the transportation industry, another 25% is used for the manufacture of beverage cans and other packaging, about 15% is used in construction, 15% in electrical applications, and 20% in other applications. About 200 pounds of aluminum was used in an average car made in the United States in 1996. Aluminum reacts with oxygen, even at room temperature, to produce an extremely thin aluminum-oxide layer that protects the underlying metal from many corrosive environments. We should be careful, though, not to generalize this behavior. For example, aluminum powder (because it has a high surface area), when present in the form of an oxidizer, such as ammonium perchlorate and iron oxide as catalysts, serves as the fuel for *solid rocket boosters* (SRBs). These boosters use ~91,000 kg of atomized aluminum powder every time the space shuttle takes off and can generate enough force for the shuttle to reach a speed of ~4800 km per hour. New developments related to aluminum include the development of aluminum alloys containing higher Mg concentrations for use in making automobiles.

Designation Aluminum alloys can be divided into two major groups: wrought and casting alloys, depending on their method of fabrication. **Wrought alloys**, which are shaped by plastic deformation, have compositions and microstructures significantly different from casting alloys, reflecting the different requirements of the manufacturing process. Within each major group we can divide the alloys into two subgroups: heat-treatable and non-heat treatable alloys.

Aluminum alloys are designated by the numbering system shown in Table 14-3. The first number specifies the principle alloying elements, and the remaining numbers refer to the specific composition of the alloy.

TABLE 14-3 ■ *Designation system for aluminum alloys*

Wrought alloys:		
1xxx[a]	Commercially pure Al (>99% Al)	Not age-hardenable
2xxx	Al-Cu and Al-Cu-Li	Age-hardenable
3xxx	Al-Mn	Not age-hardenable
4xxx	Al-Si and Al-Mg-Si	Age-hardenable if magnesium is present
5xxx	Al-Mg	Not age-hardenable
6xxx	Al-Mg-Si	Age-hardenable
7xxx	Al-Mg-Zn	Age-hardenable
8xxx	Al-Li, Sn, Zr, or B	Age-hardenable
9xxx	Not currently used	
Casting alloys:		
1xx.x.[b]	Commercially pure Al	Not age-hardenable
2xx.x.	Al-Cu	Age-hardenable
3xx.x.	Al-Si-Cu or Al-Mg-Si	Some are age-hardenable
4xx.x.	Al-Si	Not age-hardenable
5xx.x.	Al-Mg	Not age-hardenable
7xx.x.	Al-Mg-Zn	Age-hardenable
8xx.x.	Al-Sn	Age-hardenable
9xx.x.	Not currently used	

[a] *The first digit shows the main alloying element, the second digit shows modification, and the last two digits show the decimal % of the Al concentration (e.g., 1060: will be 99.6% Al alloy).*
[b] *Last digit indicates product form, 1 or 2 is ingot (depends upon purity) and 0 is for casting.*

The degree of strengthening is given by the **temper designation** T or H, depending on whether the alloy is heat-treated or strain-hardened (Table 14-4). Other designations indicate whether the alloy is annealed (O), solution-treated (W), or used in the as-fabricated condition (F). The numbers following the T or H indicate the amount of strain hardening, the exact type of heat treatment, or other special aspects of the processing of the alloy. Typical alloys and their properties are included in Table 14-5.

Wrought Alloys The 1xxx, 3xxx, 5xxx, and most of the 4xxx wrought alloys are not age-hardenable. The 1xxx and 3xxx alloys are single-phase alloys except for the presence of small amounts of inclusions or intermetallic compounds (Figure 14-1). Their properties are controlled by strain hardening, solid-solution strengthening, and grain-size control. However, because the solubilities of the alloying elements in aluminum are

TABLE 14-4 ■ *Temper designations for aluminum alloys*

F	As-fabricated (hot-worked, forged, cast, etc.)
O	Annealed (in the softest possible condition)
H	Cold-worked
	H1x—cold-worked only. (x refers to the amount of cold work and strengthening.)
	H12—cold work that gives a tensile strength midway between the O and H14 tempers.
	H14—cold work that gives a tensile strength midway between the O and H18 tempers.
	H16—cold work that gives a tensile strength midway between the H14 and H18 tempers.
	H18—cold work that gives about 75% reduction.
	H19—cold work that gives a tensile strength greater than 14 MPa of that obtained by the H18 temper.
	H2x—cold-worked and partly annealed.
	H3x—cold-worked and stabilized at a low temperature to prevent age hardening of the structure.
W	Solution-treated
T	Age-hardened
	T1—cooled from the fabrication temperature and naturally aged.
	T2—cooled from the fabrication temperature, cold-worked, and naturally aged.
	T3—solution-treated, cold-worked, and naturally aged.
	T4—solution-treated and naturally aged.
	T5—cooled from the fabrication temperature and artificially aged.
	T6—solution-treated and artificially aged.
	T7—solution-treated and stabilized by overaging.
	T8—solution-treated, cold-worked, and artificially aged.
	T9—solution-treated, artificially aged, and cold-worked.
	T10—cooled from the fabrication temperature, cold-worked, and artificially aged.

small at room temperature, the degree of solid-solution strengthening is limited. The 5xxx alloys contain two phases at room temperature—α, a solid solution of magnesium in aluminum, and Mg_2Al_3, a hard, brittle intermetallic compound (Figure 14-2). The aluminum-magnesium alloys are strengthened by a fine dispersion of Mg_2Al_3, as well as by strain hardening, solid-solution strengthening, and grain-size control. However,

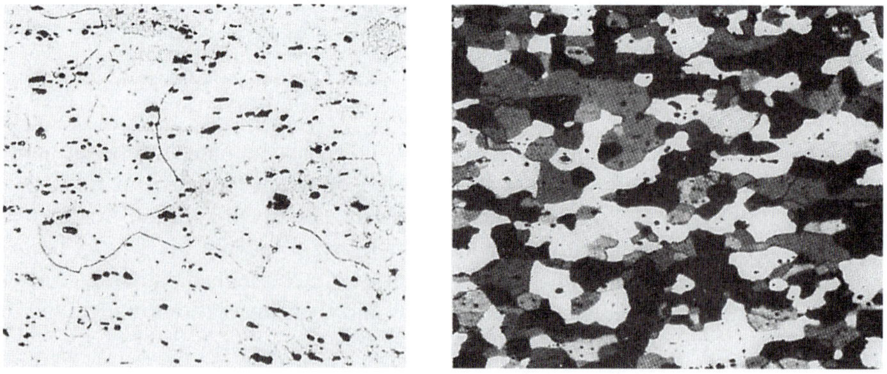

(a) **(b)**

Figure 14-1 (a) $FeAl_3$ inclusions in annealed 1100 aluminum (×350). (b) Mg_2Si precipitates in annealed 5457 aluminum alloy (×75). (*From* ASM Handbook, *Vol. 7, (1972), ASM International, Materials Park, OH 44073.*)

TABLE 14-5 ■ ***Properties of typical aluminum alloys***

Alloy		Tensile Strength (MPa)	Yield Strength (MPa)	% Elongation	Applications
Non heat-treatable wrought alloys:					
1100-O	>99% Al	90	35	40	Electrical components, foil,
1100-H18		166	152	10	food processing,
3004-O	1.2% Mn-1.0% Mg	179	69	25	beverage can bodies,
3004-H18		283	248	9	architectural uses,
4043-O	5.2% Si	145	69	22	filler metal for welding,
4043-H18		283	269	1	beverage can tops, and
5182-O	4.5% Mg	290	131	25	marine components
5182-H19		421	393	4	
Heat-treatable wrought alloys:					
2024-T4	4.4% Cu	469	324	20	Truck wheels, aircraft skins,
2090-T6	2.4% Li-2.7% Cu	552	517	6	pistons, canoes, railroad
4032-T6	12% Si-1% Mg	379	317	9	cars, and aircraft frames
6061-T6	1% Mg-0.6% Si	310	276	15	
7075-T6	5.6% Zn-2.5% Mg	572	503	11	
Casting alloys:					
201-T6	4.5% Cu	483	434	7	Transmission housings,
319-F	6% Si-3.5% Cu	186	124	2	general purpose castings,
356-T6	7% Si-0.3% Mg	228	166	3	aircraft fittings, motor
380-F	8.5% Si-3.5% Cu	317	159	3	housings, automotive
390-F	17% Si-4.5% Cu	283	241	1	engines, food-handling
443-F	5.2% Si (sand cast)	131	55	8	equipment, and marine
	(permanent mold)	159	62	10	fittings
	(die cast)	228	110	9	

because Mg_2Al_3 is not coherent, age-hardening treatments are not possible. The 4xxx series alloys also contain two phases, α and nearly pure silicon, β (Chapter 11).

Alloys that contain both silicon and magnesium can be age hardened by permitting Mg_2Si to precipitate. The 2xxx, 6xxx, and 7xxx alloys are age-hardenable alloys. Although excellent specific strengths are obtained for these alloys, the amount of precipitate that can form is limited. In addition, they cannot be used at temperatures

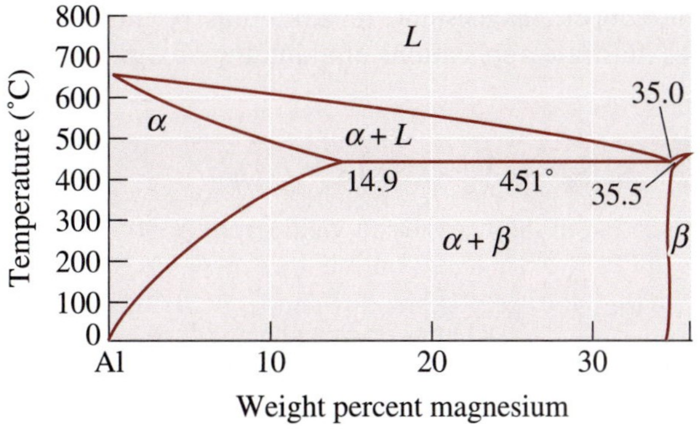

Figure 14-2
Portion of the aluminum-magnesium phase diagram.

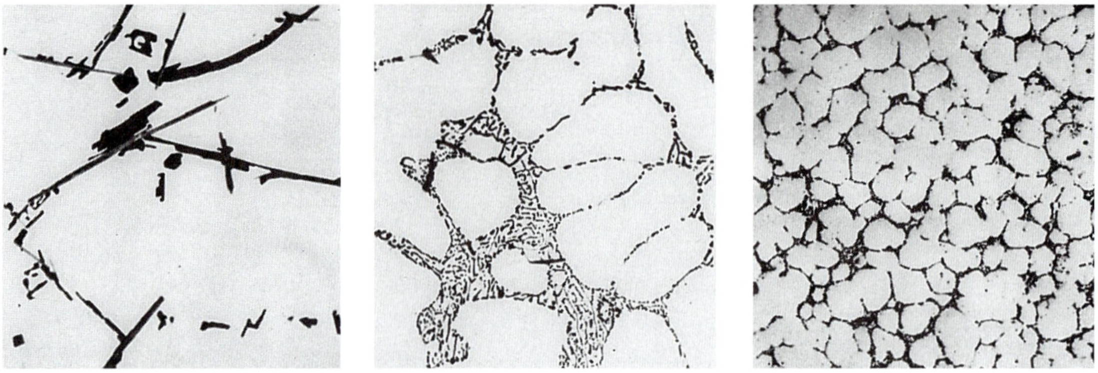

(a) **(b)** **(c)**

Figure 14-3 (a) Sand-cast 443 aluminum alloy containing coarse silicon and inclusions. (b) Permanent-mold 443 alloy containing fine dendrite cells and fine silicon due to faster cooling. (c) Die-cast 443 alloy with a still finer microstructure (×350). (*From* ASM Handbook, *Vol. 7, (1972), ASM International, Materials Park, OH 44073.*)

above approximately 175°C in the aged condition. The most widely used aircraft aluminum alloy is 2024. There is also an interest in the development of precipitation-hardened Al-Li alloys due to their high Young's modulus and low density. However, high-processing costs, anisotropic properties, and lower fracture toughness have proved to be limiting factors. Al-Li alloys are used to make space shuttle fuel tanks.

Casting Alloys Many of the common aluminum casting alloys shown in Table 14-5 contain enough silicon to cause the eutectic reaction, giving the alloys low melting points, good fluidity, and good castability. **Fluidity** is the ability of the liquid metal to flow through a mold without prematurely solidifying, and **castability** refers to the ease with which a good casting can be made from the alloy.

The properties of the aluminum-silicon alloys are controlled by solid-solution strengthening of the α aluminum matrix, dispersion strengthening by the β phase, and solidification, which controls the primary grain size and shape as well as the nature of the eutectic microconstituent. Fast cooling obtained in die casting or permanent-mold casting (Chapter 9) increases strength by refining grain size and the eutectic microconstituent (Figure 14-3). Grain refinement using boron and titanium additions, modification using sodium or strontium to change the eutectic structure, and hardening with phosphorus to refine the primary silicon (Chapter 9) are all done in certain alloys to improve the microstructure and, thus, the degree of dispersion strengthening. Many alloys also contain copper, magnesium, or zinc, thus permitting age hardening. The following examples illustrate applications of aluminum alloys.

EXAMPLE 14-1 *Strength-to-Weight Ratio in Design*

A steel cable 1.25 cm in diameter has a yield strength of 483 MPa. The density of steel is about 7.87 g/cm^3. Based on the data in Table 14-5, determine (a) the maximum load that the steel cable can support, (b) the diameter of a cold-worked aluminum-manganese alloy (3004-H 18) required to support the same load as the steel, and (c) the weight per meter of the steel cable versus the aluminum alloy cable.

SOLUTION

a. $\text{Load} = F = (\sigma_y \times A) = 483 \text{ MPa}\left(\frac{\pi}{4}\right)(1.25 \text{ cm})^2 = 59{,}297 \text{ N}$

b. The yield strength of the aluminum alloy is 248 MPa. Thus:

$$A = \frac{\pi}{4}d^2 = \frac{F}{\sigma_y} = \frac{59{,}297 \text{ N}}{248 \text{ MPa}} = 2.39 \text{ cm}^2$$

$$\therefore d = 1.744 \text{ cm}$$

c. Density of steel $= \rho = 7.87 \text{ g/cm}^3$

Density of aluminum $= \rho = 2.70 \text{ g/cm}^3$

$$\text{Weight of steel} = Al\rho = \frac{\pi}{4}(1.25 \text{ cm})^2(100)(7.87) = 966.183 \text{ g/m}$$

$$\text{Weight of aluminum} = Al\rho = \frac{\pi}{4}(1.744)^2(100)(2.7) = 645.24 \text{ g/m}$$

Although the yield strength of the aluminum alloy is lower than that of the steel and the cable must be larger in diameter, the aluminum alloy cable weighs only about half as much as the steel cable. When comparing materials, a proper factor-of-safety should also be included during design.

EXAMPLE 14-2 *Design of an Aluminum Recycling Process*

Design a method for recycling aluminum alloys used for beverage cans.

SOLUTION

Recycling aluminum is advantageous because only a fraction (about 5%) of the energy required to produce aluminum from Al_2O_3 is required. However, recycling beverage cans does present several difficulties.

The beverage cans are made of two aluminum alloys (3004 for the main body, and 5182 for the lids) having different compositions (Table 14-5). The 3004 alloy has the exceptional formability needed to perform the deep-drawing process; the 5182 alloy is harder and permits the pull-tops to function properly. When the cans are remelted, the resulting alloy contains both Mg and Mn and is not suitable for either application.

One approach to recycling the cans is to separate the two alloys from the cans. The cans are shredded, then heated to remove the lacquer that helps protect the cans during use. We could then further shred the material at a temperature where the 5182 alloy begins to melt. The 5182 alloy has a wider freezing range than the 3004 alloy and breaks into very small pieces; the more ductile 3004 alloy remains in larger pieces. The small pieces of 5182 can therefore be separated by passing the material through a screen. The two separated alloys can then be melted, cast, and rolled into new can stock.

An alternative method would be to simply remelt the cans. Once the cans have been remelted, we could bubble chlorine gas through the liquid alloy. The chlorine reacts selectively with the magnesium, removing it as a chloride. The remaining liquid can then be adjusted to the proper composition and be recycled as 3004 alloy.

EXAMPLE 14-3 ***Design/Materials Selection for a Cryogenic Tank***

Design the material to be used to contain liquid hydrogen fuel for the space shuttle.

SOLUTION

Liquid hydrogen is stored below −253°C; therefore, this tank must have good cryogenic properties. The tank is subjected to high stresses, particularly when the space shuttle enters into orbit, and it should have good fracture toughness to minimize the chances of catastrophic failure. Finally, it should be light in weight to permit higher payloads or less fuel consumption.

A lightweight aluminum alloy would appear to be a good choice. Aluminum does not show a ductile to brittle transition. Because of its good ductility, we expect aluminum to also have good fracture toughness, particularly when the alloy is in the annealed condition.

One of the most common cryogenic aluminum alloys is 5083-O. Aluminum-lithium alloys are also being considered for low-temperature applications to take advantage of their even lower density.

14-2 Magnesium and Beryllium Alloys

Magnesium, which is often extracted electrolytically from concentrated magnesium chloride in seawater, is lighter than aluminum with a density of 1.74 g/cm^3, and it melts at a slightly lower temperature than aluminum. In many environments, the corrosion resistance of magnesium approaches that of aluminum; however, exposure to salts, such as that near a marine environment, can cause rapid deterioration. Alloys of magnesium that are resistant to corrosion have been developed. Although magnesium alloys are not as strong as aluminum alloys, their specific strengths are comparable. Consequently, magnesium alloys are used in aerospace applications, high-speed machinery, and transportation and materials handling equipment.

Magnesium, however, has a low modulus of elasticity (44.8×10^3 MPa) and poor resistance to fatigue, creep, and wear. Magnesium also poses a hazard during casting and machining, since it combines easily with oxygen and burns. Finally, the response of magnesium to strengthening mechanisms is relatively poor.

Structure and Properties Magnesium, which has the HCP structure, is less ductile than aluminum. However, the alloys do have some ductility because alloying increases the number of active slip planes. Some deformation and strain hardening can be accomplished at room temperature, and the alloys can be readily deformed at elevated temperatures. Strain hardening produces a relatively small effect in pure magnesium because of the low strain-hardening coefficient (Chapters 6 and 8).

As in aluminum alloys, the solubility of alloying elements in magnesium at room temperature is limited, causing only a small degree of solid-solution strengthening. However the solubility of many alloying elements increases with temperature, as shown in the Mg-Al phase diagram (Figure 14-4). Therefore, alloys may be strengthened

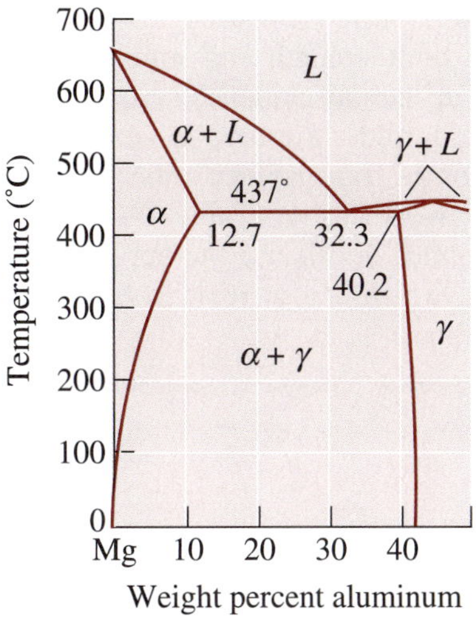

Figure 14-4
The magnesium-aluminum phase diagram.

by either dispersion strengthening or age hardening. Some age-hardened magnesium alloys, such as those containing Zr, Th, Ag, or Ce, have good resistance to overaging at temperatures as high as 300°C. Alloys containing up to 9% Li have exceptionally light weight. Properties of typical magnesium alloys are listed in Table 14-6.

Advanced magnesium alloys include those with very low levels of impurities and those containing large amounts (>5%) of cerium and other rare earths. These alloys form a protective MgO film that improves corrosion resistance. Rapid solidification processing permits larger amounts of alloying elements to be dissolved in the magnesium, further improving corrosion resistance. Improvements in strength, particularly at high temperatures, can be obtained by introducing ceramic particles or fibers such as silicon carbide into the metal.

Beryllium is lighter than aluminum, with a density of 1.848 g/cm^3, yet it is stiffer than steel, with a modulus of elasticity of 290×10^3 MPa. Beryllium alloys are rather

TABLE 14-6 ■ ***Properties of typical magnesium alloys***

Alloy	Composition	Tensile Strength (MPa)	Yield Strength (MPa)	% Elongation
Pure Mg:				
Annealed		159	90	3–15
Cold-worked		179	117	2–10
Casting alloys:				
AM 100-T6	10% Al-0.1% Mn	276	152	1
AZ81A-T4	7.6% Al-0.7% Zn	276	83	15
ZK61A-T6	6% Zn-0.7% Zr	310	193	10
Wrought alloys:				
AZ80A-T5	8.5% Al-0.5% Zn	379	276	7
ZK40A-T5	4% Zn-0.45% Zr	276	255	4
HK31A-H24	3% Th-0.6% Zr	262	207	8

limited. Composites based on Be and Al have yield strengths of 207 to 345 MPa, have high specific strengths and maintain both strength and stiffness to high temperatures. Instrument grade beryllium is used in inertial guidance systems where the elastic deformation must be minimal; structural grades are used in aerospace applications; and nuclear applications take advantage of the transparency of beryllium to electromagnetic radiation. Unfortunately, beryllium is expensive (Table 14-1), brittle, reactive, and toxic. Its production is quite complicated and hence the applications of Be alloys are very limited. Beryllium oxide (BeO), which is also toxic *in a powder form,* is used to make high-thermal conductivity ceramics.

EXAMPLE 14-4 *Magnesium Alloys for Orthopedic Implant Applications*

Currently, titanium alloys, stainless steels, and cobalt chromium alloys are used as biomedical implants. It has been suggested that magnesium alloys have mechanical properties similar to those of bones and will make suitable implant materials. One of the most commonly used magnesium alloy is AZ91, which has 9 wt.% Al, and 1 wt.% Zn. This material can be prepared by the so-called squeeze-casting process, which allows for faster cooling and, hence, a finer grain size (Chapter 9). The effect of grain size on the strength of AZ91 on the yield strength of AZ91 alloy is shown in Figure 14-5.

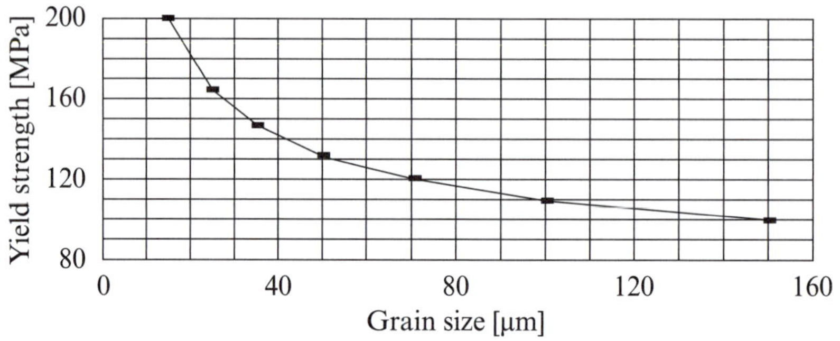

Figure 14-5 The dependence of yield strength of Mg alloy AZ91 with grain size. (*Source: Data from Kainer, K.U. and Benzler, T.U.,* Magnesium Alloys and Technology, *Ed. K.U. Kainer, 2003, p. 60.*)

Assume that these data will fit the Hall-Petch equation (Chapter 4) and calculate the value of constants σ_0 and K for this alloy. What will be the yield strength of the 10 μm alloy?

SOLUTION

The Hall-Petch equation (Chapter 4) is

$$\sigma = \sigma_0 + Kd^{-1/2}$$

We use a spreadsheet program (in this case Excel™) to fit these data to a straight line. To do this, we plot the yield stress as a function of inverse of the square root of the grain size. These data are shown in the following table.

Grain Size (d) Micrometers	Inverse of Grain Size Square Root (Micrometers)$^{1/2}$	Yield Stress (MPa)
15	0.25819889	200
25	0.2	165
35	0.169030851	150
50	0.141421356	130
70	0.119522861	120
100	0.1	110
150	0.081649658	100

When the yield stress (σ) and $d^{-1/2}$ values are fitted to a straight line equation, we get the slope (K) as 567.53 MPa(μm)$^{1/2}$ and the intercept (σ_0) = 52.54 MPa. Thus, in this case, the Hall-Petch equation is

$$\sigma = 52.54 \text{ MPa} + 567.53(\text{MPa} - \mu\text{m}^{1/2}) \times d^{-1/2}$$

From this, for a grain size of 10 μm, the predicted yield strength for the Mg alloy will be

$$\sigma = 52.54 \text{ MPa} + 567.53(\text{MPa} - \mu\text{m}^{1/2}) \times (10)^{-1/2}$$

Therefore, the predicted yield strength of an Mg-alloy casting with a 10 μm grain size will be 232 MPa.

14-3 Copper Alloys

Copper occurs in nature as sulfides and also as elemental copper. Copper was extracted successfully from rock long before iron, since the relatively lower temperatures required for copper extraction could be achieved more easily. Copper is typically produced by a pyrometallurgical (high-temperature) process. The copper ore containing high-sulfur contents is concentrated, then converted into a molten immiscible liquid containing copper sulfide-iron sulfide and is known as a copper matte. This is done in a flash smelter. In a separate reactor, known as a copper converter, oxygen introduced to the matte converts the iron sulfide to iron oxide and the copper sulfide to an impure copper called **blister copper**, which is then purified electrolytically. Other methods for copper extraction include leaching copper from low-sulfur ores with an acid, then electrolytically extracting the copper from the solution.

Copper-based alloys have higher densities than that for steels. Although the yield strength of some alloys is high, their specific strength is typically less than that of aluminum or magnesium alloys. These alloys have better resistance to fatigue, creep, and wear than the lightweight aluminum and magnesium alloys. Many of these alloys have excellent ductility, corrosion resistance, electrical and thermal conductivity, and most can easily be joined or fabricated into useful shapes. Applications for copper-based alloys include electrical components (such as wire), pumps, valves, and plumbing parts, where these properties are used to advantage.

Copper alloys are also unusual in that they may be selected to produce an appropriate decorative color. Pure copper is red; zinc additions produce a yellow color, and nickel produces a silver color. Copper can corrode easily; forming a basic copper sulfate

TABLE 14-7 ■ *Properties of typical copper alloys obtained by different strengthening mechanisms*

Material	Tensile Strength (MPa)	Yield Strength (MPa)	% Elongation	Strengthening Mechanism
Pure Cu, annealed	209	33	60	None
Commercially pure Cu, annealed to coarse grain size	221	69	55	Solid solution
Commercially pure Cu, annealed to fine grain size	234	76	55	Grain size
Commercially pure Cu, cold-worked 70%	393	365	4	Strain hardening
Annealed Cu-35% Zn	324	103	62	Solid solution
Annealed Cu-10% Sn	455	193	68	Solid solution
Cold-worked Cu-35% Zn	676	434	3	Solid solution + strain hardening
Age-hardened Cu-2% Be	1310	1207	4	Age hardening
Quenched and tempered Cu-Al	759	414	5	Martensitic reaction
Cast manganese bronze	490	193	30	Eutectoid reaction

($CuSO_4 \cdot 3Cu(OH)_2$). This is a green compound that is insoluble in water (but soluble in acids). This green patina provides an attractive finish for many applications. The Statue of Liberty is green because of the green patina of the oxidized copper skin that covers the steel structure.

The wide variety of copper-based alloys take advantage of all of the strengthening mechanisms that we have discussed. The effects of these strengthening mechanisms on the mechanical properties are summarized in Table 14-7.

Copper containing less than 0.1% impurities is used for electrical and microelectronics applications. Small amounts of cadmium, silver, and Al_2O_3 improve their hardness without significantly impairing conductivity. The single-phase copper alloys are strengthened by cold working. Examples of this effect are shown in Table 14-7. The FCC structure of copper provides for excellent ductility and a high strain-hardening coefficient.

Solid-Solution-Strengthened Alloys A number of copper-based alloys contain large quantities of alloying elements, yet remain single phase. Important binary phase diagrams are shown in Figure 14-6. The copper-zinc, or **brass**, alloys with less than 40% Zn form single-phase solid solutions of zinc in copper. The mechanical properties—even elongation—increase as the zinc content increases (Figure 10-8). These alloys can be cold formed into rather complicated yet corrosion-resistant components. **Bronzes** are alloys of copper containing tin and can certainly contain other elements. Manganese bronze is a particularly high-strength alloy containing manganese as well as zinc for solid-solution strengthening.

Tin bronzes, often called phosphor bronzes, may contain up to 10% Sn and remain single phase. The phase diagram predicts that the alloy will contain the Cu_3Sn (ε) compound. However, the kinetics of the reaction are so slow that the precipitate particles often do not form.

Alloys containing less than about 9% Al or less than 3% Si are also single phase. These aluminum bronzes and silicon bronzes have good forming characteristics and are often selected for their good strength and excellent toughness.

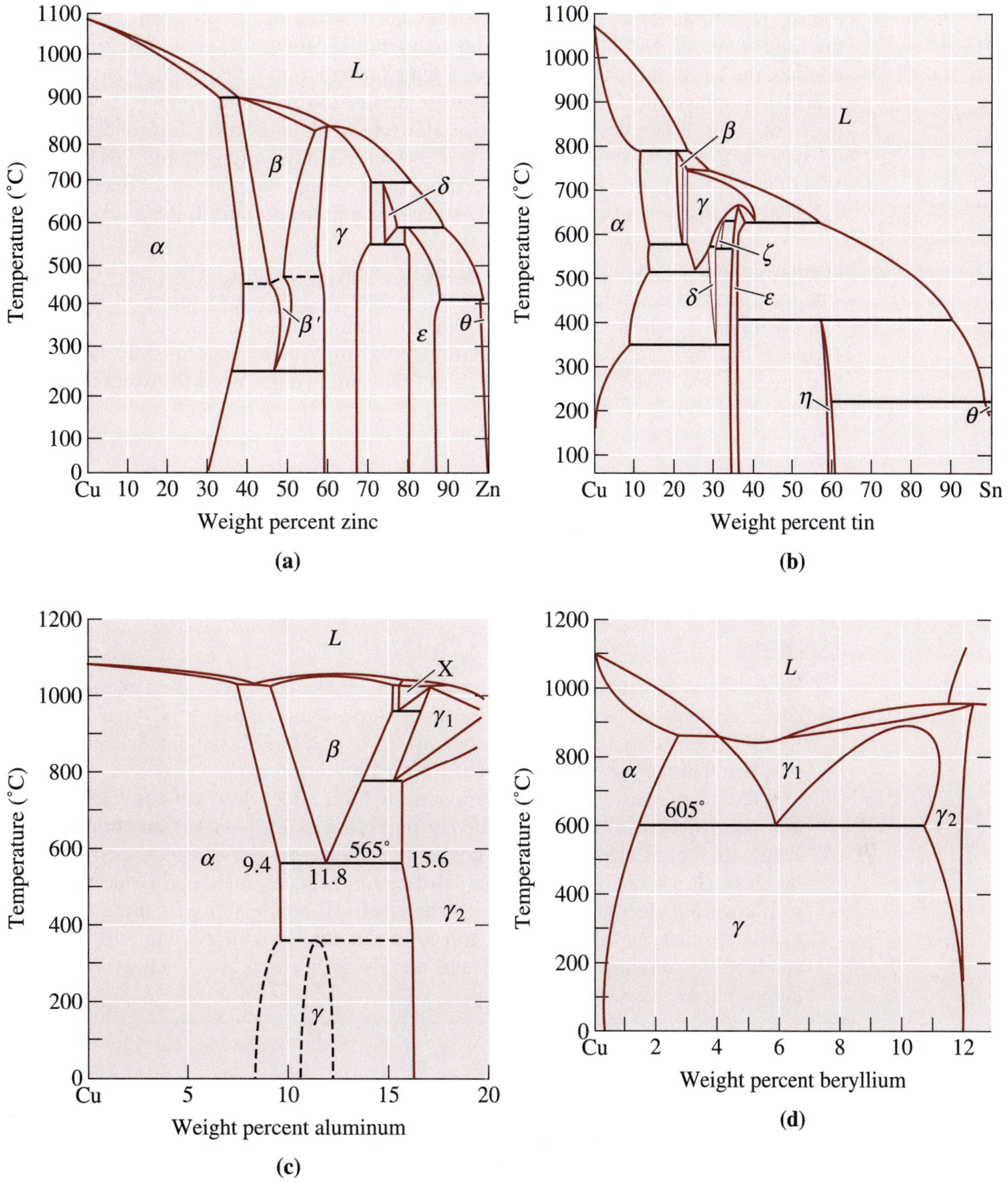

Figure 14-6 Binary phase diagrams for the (a) copper-zinc, (b) copper-tin, (c) copper-aluminum, and (d) copper-beryllium systems.

Age-Hardenable Alloys A number of copper-base alloys display an age-hardening response, including zirconium-copper, chromium-copper, and beryllium-copper. The copper-beryllium alloys are used for their high strength, their high stiffness (making them useful as springs and fine wires), and their nonsparking qualities (making them useful for tools to be used near flammable gases and liquids).

Phase Transformations Aluminum bronzes that contain over 9% Al can form β phase on heating above 565°C, the eutectoid temperature [Figure 14-6(c)]. On subsequent cooling, the eutectoid reaction produces a lamellar structure (like pearlite) that contains a brittle γ_2 compound. The low-temperature peritectoid reaction, $\alpha + \gamma_2 \rightarrow \gamma$, normally does not occur. The eutectoid product is relatively weak and brittle, but we can rapidly quench the β to produce martensite, or β', which has high strength and low ductility. When β' is subsequently tempered, a combination of high strength, good ductility, and excellent toughness is obtained as fine platelets of α precipitate from the β'.

Leaded-Copper Alloys Virtually any of the wrought alloys may contain up to 4.5% Pb. The lead forms a monotectic reaction with copper and produces tiny lead spheres as the last liquid to solidify. The lead improves machining characteristics. Use of leaded-copper alloys, however, has a major environmental impact and, consequently, new alloys that are lead free have been developed. The following two examples illustrate the use of copper-based alloys.

EXAMPLE 14-5 *Design/Materials Selection for an Electrical Switch*

Design the contacts for a switch or relay that opens and closes a high-current electrical circuit.

SOLUTION

When the switch or relay opens and closes, contact between the conductive surfaces can cause wear and result in poor contact and arcing. A high hardness would minimize wear, but the contact materials must allow the high current to pass through the connection without overheating or arcing.

Therefore, our design must provide for both good electrical conductivity and good wear resistance. A relatively pure copper alloy dispersion strengthened with a hard phase that does not disturb the copper lattice would, perhaps, be ideal. In a Cu-Al_2O_3 alloy, the hard ceramic-oxide particles provide wear resistance but do not interfere with the electrical conductivity of the copper matrix. In fact, as the oxygen dissolved in the copper matrix is removed, the conductivity actually increases. There are also other materials such as W-Ag alloys that are also used.

EXAMPLE 14-6 *Design of a Heat Treatment for a Cu-Al Alloy Gear*

Design the heat treatment required to produce a high-strength aluminum-bronze gear containing 10% Al.

SOLUTION

The aluminum bronze can be strengthened by a quench and temper heat treatment. We must heat above 900°C to obtain 100% β for a Cu-10% Al alloy [Figure 14-6(c)]. The eutectoid temperature for the alloy is 565°C. Therefore, our recommended heat treatment is:

1. Heat the alloy to 950°C and hold to produce 100% β.
2. Quench the alloy to room temperature to cause β to transform to martensite, β', which is supersaturated in copper.
3. Temper below 565°C; a temperature of 400°C might be suitable. During tempering, the martensite transforms to α and γ_2. The amount of the γ_2 that forms at 400°C is:

$$\%\,\gamma_2 = \frac{10 - 9.4}{15.6 - 9.4} \times 100 = 9.7\%$$

4. Cool rapidly to room temperature so that the equilibrium γ does not form.

Note that if tempering were carried out below about 370°C, γ would form rather than γ_2.

14-4 Nickel and Cobalt Alloys

Nickel and cobalt alloys are used for corrosion protection and for high-temperature resistance, taking advantage of their high melting points and high strengths. Nickel is FCC and has good formability; cobalt is an allotropic metal, with an FCC structure above 417°C and an HCP structure at lower temperatures. Special cobalt alloys are used for exceptional wear resistance and, because of biocompatibility for prosthetic devices. Typical alloys and their applications are listed in Table 14-8.

TABLE 14-8 ■ *Compositions, properties, and applications for selected nickel and cobalt alloys*

Material	Tensile Strength (MPa)	Yield Strength (MPa)	% Elongation	Strengthening Mechanism	Applications
Pure Ni (99.9% Ni)	345	110	45	Annealed	Corrosion resistance
	655	621	4	Cold-worked	Corrosion resistance
Ni-Cu alloys:					
Monel 400 (Ni-31.5% Cu)	538	269	37	Annealed	Valves, pumps, heat exchangers
Monel K-500 (Ni-29.5% Cu-2.7% Al-0.6% Ti)	1034	759	30	Aged	Shafts, springs, impellers
Ni superalloys:					
Inconel 600 (Ni-15.5% Cr-8% Fe)	621	200	49	Carbides	Heat-treatment equipment
Hastelloy B-2 (Ni-28% Mo)	896	414	61	Carbides	Corrosion resistance
DS-Ni (Ni-2% ThO_2)	490	331	14	Dispersion	Gas turbines
Fe-Ni superalloys:					
Incoloy 800 (Ni-46% Fe-21% Cr)	614	283	37	Carbides	Heat exchangers
Co superalloys:					
Stellite 6B (60% Co-30% Cr-4.5% W)	1220	710	4	Carbides	Abrasive wear resistance

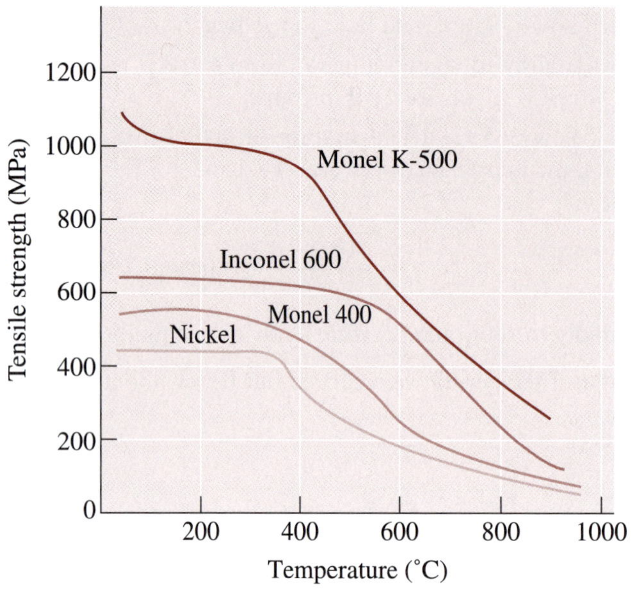

Figure 14-7
The effect of temperature on the tensile strength of several nickel-based alloys.

In Chapter 9, we saw how rapidly solidified powders of nickel- and cobalt-based superalloys can be formed using spray atomization followed by hot isostatic pressing. These materials are used to make the rings that retain turbine blades, as well as for turbine blades for aircraft engines. In Chapter 12, we mentioned shape-memory alloys based on Ni-Ti. Iron, nickel and cobalt are magnetic. Certain Fe-Ni- and Fe-Co-based alloys form very good magnetic materials. A Ni-36% Fe alloy (Invar) displays practically no expansion during heating; this effect is exploited in producing bimetallic composite materials. Cobalt is used as a tough phase that absorbs shocks and vibrations in WC-Co cutting tools.

Nickel and Monel Nickel and its alloys have excellent corrosion resistance and forming characteristics. When copper is added to nickel, the maximum strength is obtained near 60% Ni. A number of alloys, called **monels**, with approximately this composition are used for their strength and corrosion resistance in salt water and at elevated temperatures. Some of the monels contain small amounts of aluminum and titanium. These alloys show an age-hardening response by the precipitation of γ', a coherent Ni_3Al or Ni_3Ti precipitate which nearly doubles the tensile properties. The precipitates resist overaging at temperatures up to 425°C (Figure 14-7).

Superalloys Nickel, iron-nickel, and cobalt alloys that contain large amounts of alloying elements intended to produce a combination of high strength at elevated temperatures, resistance to creep at temperatures up to 1000°C, and resistance to corrosion. These **superalloys** excellent high-temperature properties are obtained even though the melting temperatures of the alloys are about the same as that for steels. Typical applications include vanes and blades for turbine and jet engines, heat exchangers, chemical reaction vessel components, and heat-treating equipment.

To obtain high strength and creep resistance, the alloying elements must produce a stable microstructure at high temperatures. Solid-solution strengthening, dispersion strengthening, and precipitation hardening generally are employed.

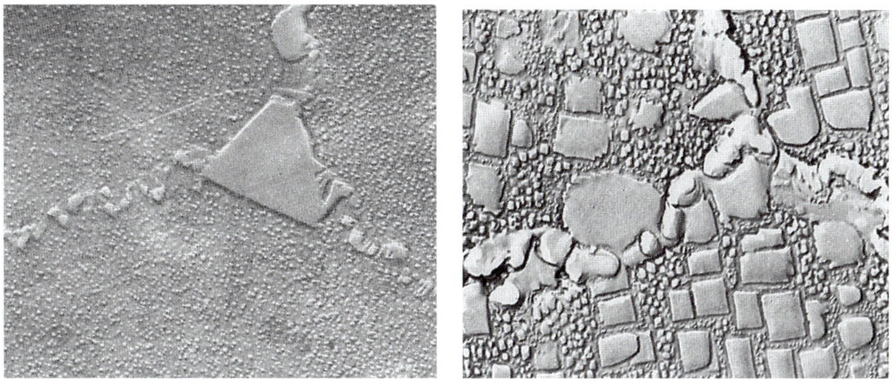

Figure 14-8 (a) Microstructure of a superalloy, with carbides at the grain boundaries and γ' precipitates in the matrix ($\times$15,000). (b) Microstructure of a superalloy aged at two temperatures, producing both large and small cubical γ' precipitates ($\times$10,000). (ASM Handbook, *Vol. 9, Metallography and Microstructure (1985), ASM International, Materials Park, OH 44073.*)

Solid-Solution Strengthening Large additions of chromium, molybdenum, and tungsten and smaller additions of tantalum, zirconium, niobium, and boron provide solid-solution strengthening. The effects of solid-solution strengthening are stable and, consequently, make the alloy resistant to creep, particularly when large atoms such as molybdenum and tungsten (which diffuse slowly) are used.

Carbide-Dispersion Strengthening All alloys contain a small amount of carbon which, by combining with other alloying elements, produces a network of fine, stable carbide particles. The carbide network interferes with the dislocation movement and prevents grain boundary sliding. The carbides include TiC, BC, ZrC, TaC, Cr_7C_3, $Cr_{23}C_6$, Mo_6C, and W_6C, although often they are more complex and contain several alloying elements. Stellite 6B, a cobalt-based superalloy, has unusually good wear resistance at high temperatures due to these carbides.

Precipitation Hardening Many of the nickel and nickel-iron superalloys that contain aluminum and titanium form the coherent precipitate γ'(Ni_3Al or Ni_3Ti) during aging. The γ' particles (Figure 14-8) have a crystal structure and lattice parameter similar to that of the nickel matrix; this similarity leads to a low-surface energy and minimizes overaging of the alloys, providing good strength and creep resistance even at high temperatures.

By varying the aging temperature, precipitates of various sizes can be produced. Small precipitates, formed at low-aging temperatures, can grow between the larger precipitates produced at higher temperatures, therefore increasing the volume percentage of the γ' and further increasing the strength [Figure 14-8(b)].

The high-temperature use of the superalloys can be improved when a ceramic coating is used. The next example shows the application of a nickel-based superalloy.

EXAMPLE 14-7 ***Design/Materials Selection for a High-Performance Jet Engine Turbine Blade***

Design a nickel-based superalloy for producing turbine blades for a gas turbine aircraft engine that will have a particularly long creep-rupture time at temperatures approaching 1100°C.

SOLUTION

First, we need a temperature stable microstructure. Addition of aluminum or titanium permits the precipitation of up to 60 vol% of the γ' phase during heat treatment and may permit the alloy to operate at temperatures approaching 0.85 times the absolute melting temperature. Addition of carbon and alloying elements such as Ta and Hf permits the precipitation of alloy carbides that prevent grain boundaries from sliding at high temperatures. Other alloying elements, including molybdenum and tungsten, provide solid-solution strengthening.

Second, we might produce a directionally solidified or even single-crystal turbine blade (Chapter 9). In directional solidification, only columnar grains form during freezing, eliminating transverse grain boundaries that might nucleate cracks. In a single crystal, no grain boundaries are present. We might use the investment casting process, being sure to pass the liquid superalloy through a filter to trap any tiny inclusions before the metal enters the ceramic investment mold.

We would then heat treat the casting to assure that the carbides and γ' precipitate with the correct size and distribution. Multiple aging temperatures might be used to assure that the largest possible volume percent γ' is formed.

Finally, the blade might contain small cooling channels along its length. Air for combustion in the engine can pass through these channels, providing active cooling to the blade, before reacting with fuel in the combustion chamber.

14-5 Titanium Alloys

Titanium is traditionally produced from TiO_2 by the Kroll process. The TiO_2 is converted to $TiCl_4$ (titanium tetra chloride, also informally known as *tickle*!), which is subsequently reduced to titanium metal by sodium or magnesium. The resultant titanium sponge is then consolidated, alloyed as necessary, and processed using vacuum arc melting. Recently, a new electrochemical process for the production of titanium sponge directly from TiO_2 has been reported. Titanium provides excellent corrosion resistance, high specific strength, and good high-temperature properties. Strengths up to 1380 MPa, coupled with a density of 4.505 g/cm^3, provide excellent mechanical properties. An adherent, protective TiO_2 film provides excellent resistance to corrosion and contamination below 535°C. Above 535°C, the oxide film breaks down and small atoms such as carbon, oxygen, nitrogen, and hydrogen embrittle the titanium.

Titanium's excellent corrosion resistance provides applications in chemical processing equipment, marine components, and biomedical implants such as hip prostheses. Titanium is an important aerospace material, finding applications as airframe and jet engine components. When it is combined with niobium, a superconductive intermetallic compound is formed; when it is combined with nickel, the resulting alloy displays the

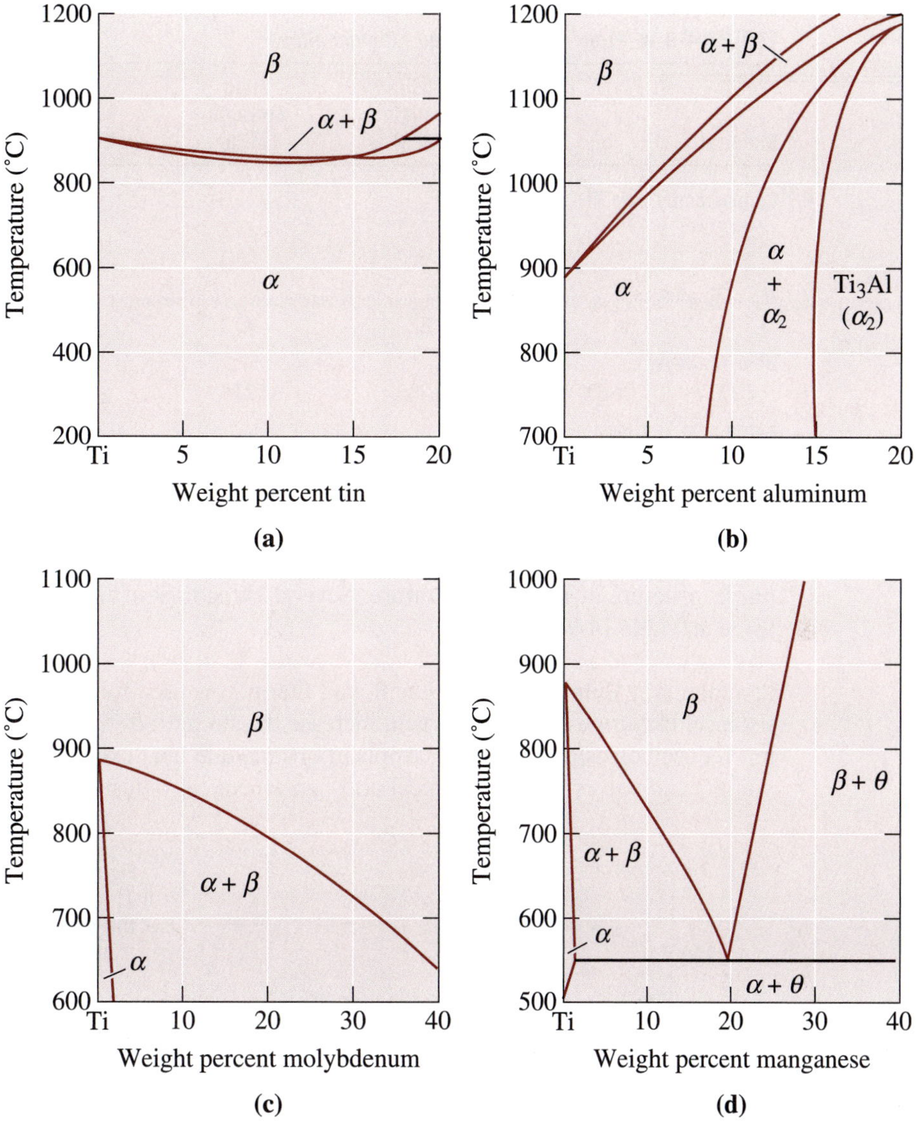

Figure 14-9 Portions of the phase diagrams for (a) titanium-tin, (b) titanium-aluminum, (c) titanium-molybdenum, and (d) titanium-manganese.

shape-memory effect; when it is combined with aluminum, a new class of intermetallic alloys is produced, as discussed in Chapter 11. Titanium alloys are used for sports equipment such as the heads of golf clubs.

Titanium is allotropic, with the HCP crystal structure (α) at low temperatures and a BCC structure (β) above 882°C. Alloying elements provide solid-solution strengthening and change the allotropic transformation temperature. The alloying elements can be divided into four groups (Figure 14-9). Additions such as tin and zirconium provide solid-solution strengthening without affecting the transformation temperature. Aluminum, oxygen, hydrogen, and other alpha-stabilizing elements increase the temperature at which α transforms to β. Beta stabilizers such as vanadium, tantalum, molybdenum, and niobium lower the transformation temperature, even causing β to be stable at room temperature. Finally, manganese, chromium, and iron produce a eutectoid reaction,

TABLE 14-9 ■ *Properties of selected titanium alloys*

Material	Tensile Strength (MPa)	Yield Strength (MPa)	% Elongation
Commercially pure Ti:			
99.5% Ti	241	172	24
99.0% Ti	552	483	15
Alpha Ti alloys:			
5% Al-2.5% Sn	862	779	15
Beta Ti alloys:			
13% V-11% Cr-3% Al	1289	1214	5
Alpha-beta Ti alloys:			
6% Al-4% V	1034	965	8

reducing the temperature at which the α–β transformation occurs and producing a two-phase structure at room temperature. Several categories of titanium and its alloys are listed in Table 14-9.

Commercially Pure Titanium Unalloyed titanium is used for its superior corrosion resistance. Impurities, such as oxygen, increase the strength of the titanium (Figure 14-10) but reduce corrosion resistance. Applications include heat exchangers, piping, reactors, pumps, and valves for the chemical and petrochemical industries.

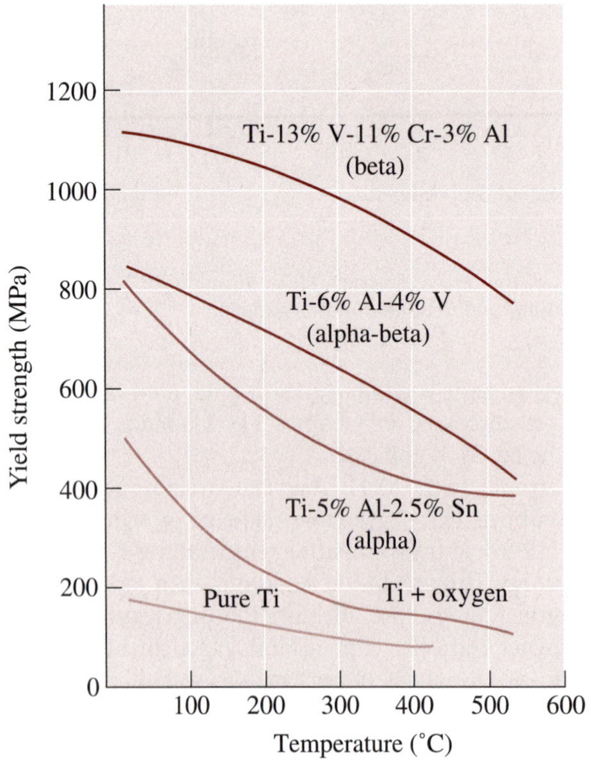

Figure 14-10
The effect of temperature on the yield strength of selected titanium alloys.

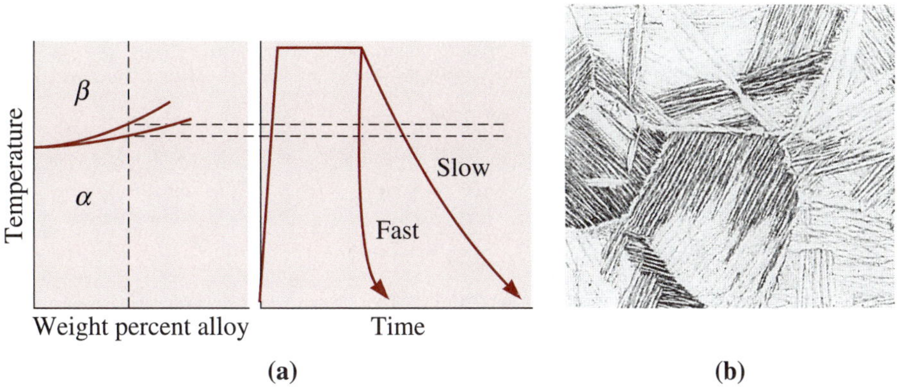

Figure 14-11 (a) Annealing and (b) microstructure of rapidly cooled alpha titanium (×100). Both the grain boundary precipitate and the Widmanstðtten plates are alpha. (*From* ASM Handbook, *Vol. 7, (1972), ASM International, Materials Park, OH 44073.*)

Alpha Titanium Alloys The most common of the all-alpha alloys contains 5% Al and 2.5% Sn, which provide solid-solution strengthening of the HCP alpha phase. The alpha alloys are annealed at high temperatures in the β region. Rapid cooling gives an acicular, or Widmanstðtten, α-grain structure (Figure 14-11) that provides good resistance to fatigue. Furnace cooling gives a more plate-like α structure that provides better creep resistance.

Beta Titanium Alloys Although large additions of vanadium or molybdenum produce an entirely β structure at room temperature, none of the beta alloys are actually alloyed to that extent. Instead, they are rich in β stabilizers, so that rapid cooling produces a metastable structure composed of all β. Strengthening is obtained both from the large amount of solid-solution-strengthening alloying elements and by aging the metastable β structure to permit α to precipitate. Applications include high-strength fasteners, beams, and other fittings for aerospace applications.

Alpha-Beta Titanium Alloys With proper balancing of the α and β stabilizers, a mixture of α and β is produced at room temperature. Ti-6% Al-4%V, an example of this approach, is by far the most common of all the titanium alloys. Because the alloys contain two phases, heat treatments can be used to control the microstructure and properties.

Annealing provides a combination of high ductility, uniform properties, and good strength. The alloy is heated just below the β-transition temperature, permitting a small amount of α to remain and prevent grain growth (Figure 14-12). Slow cooling causes equiaxed α grains to form; the equiaxed structure provides good ductility and formability while making it difficult for fatigue cracks to nucleate. Faster cooling, particularly from above the α-β transus temperature, produces an acicular—or "basketweave"—alpha phase (Figure 14-12). Although fatigue cracks may nucleate more easily in this structure, cracks must follow a tortuous path along the boundaries between α and β. This condition results in a low-fatigue crack growth rate, good fracture toughness, and good resistance to creep.

Two possible microstructures can be produced when the β phase is quenched from a high temperature. The phase diagram in Figure 14-13 includes a dashed martensite start line, which provides the basis for a quench and temper treatment. The β transforms

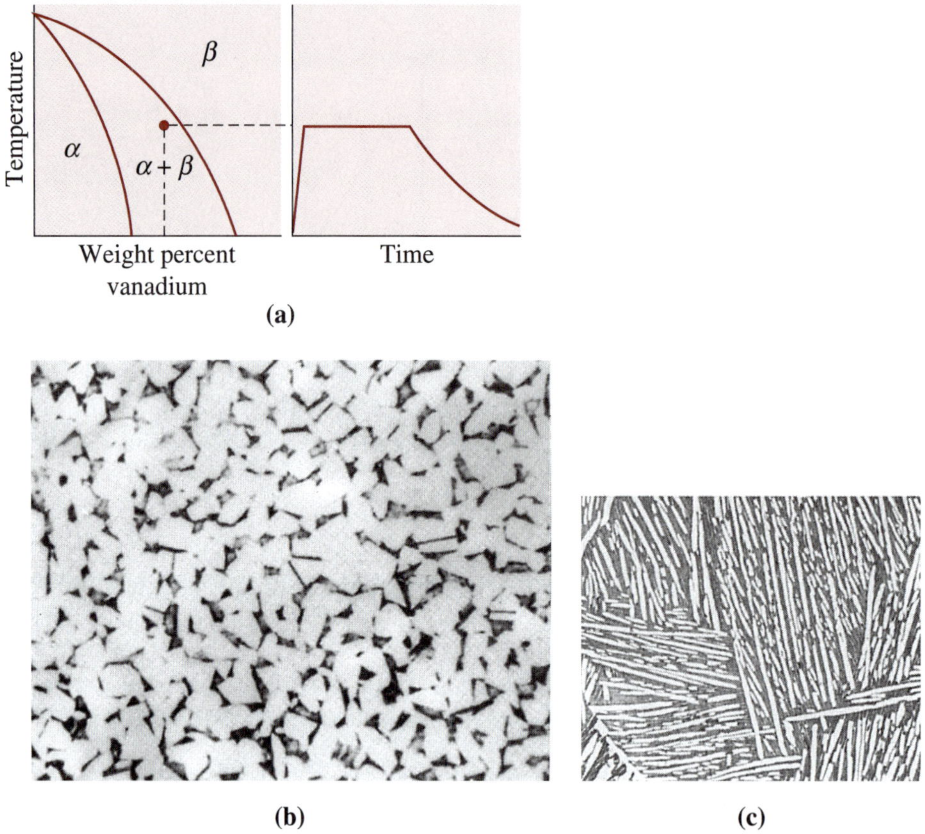

Figure 14-12 Annealing of an alpha-beta titanium alloy. (a) Annealing is done just below the α–β transformation temperature, (b) slow cooling gives equiaxed α grains ($\times$250), and (c) rapid cooling yields acicular α grains ($\times$2500). (*From* Metals Handbook, *Vol. 7, (1972), ASM International, Materials Park, OH 44073.*)

to titanium martensite (α') in an alloy that crosses the M_s line on cooling. The titanium martensite is a relatively soft supersaturated phase. When α' is reheated, tempering occurs by the precipitation of β from the supersaturated α':

$$\alpha' \rightarrow \alpha + \beta \text{ precipitates}$$

Fine β precipitates initially increase the strength compared with the α', opposite to what is found when a steel martensite is tempered. However, softening occurs when tempering is done at too high a temperature.

More highly alloyed α-β compositions are age-hardened. When the β phase in these alloys is quenched, β_{ss}, which is supersaturated in titanium, remains. When β_{ss} is aged, α precipitates in a Widmanstðtten structure, (Figure 14-13):

$$\beta_{ss} \rightarrow \beta + \alpha \text{ precipitates}$$

The formation of this structure leads to improved strength and fracture toughness. Components for airframes, rockets, jet engines, and landing gear are typical applications for the heat-treated alpha-beta alloys. Some alloys, including the Ti-6% Al-4%V

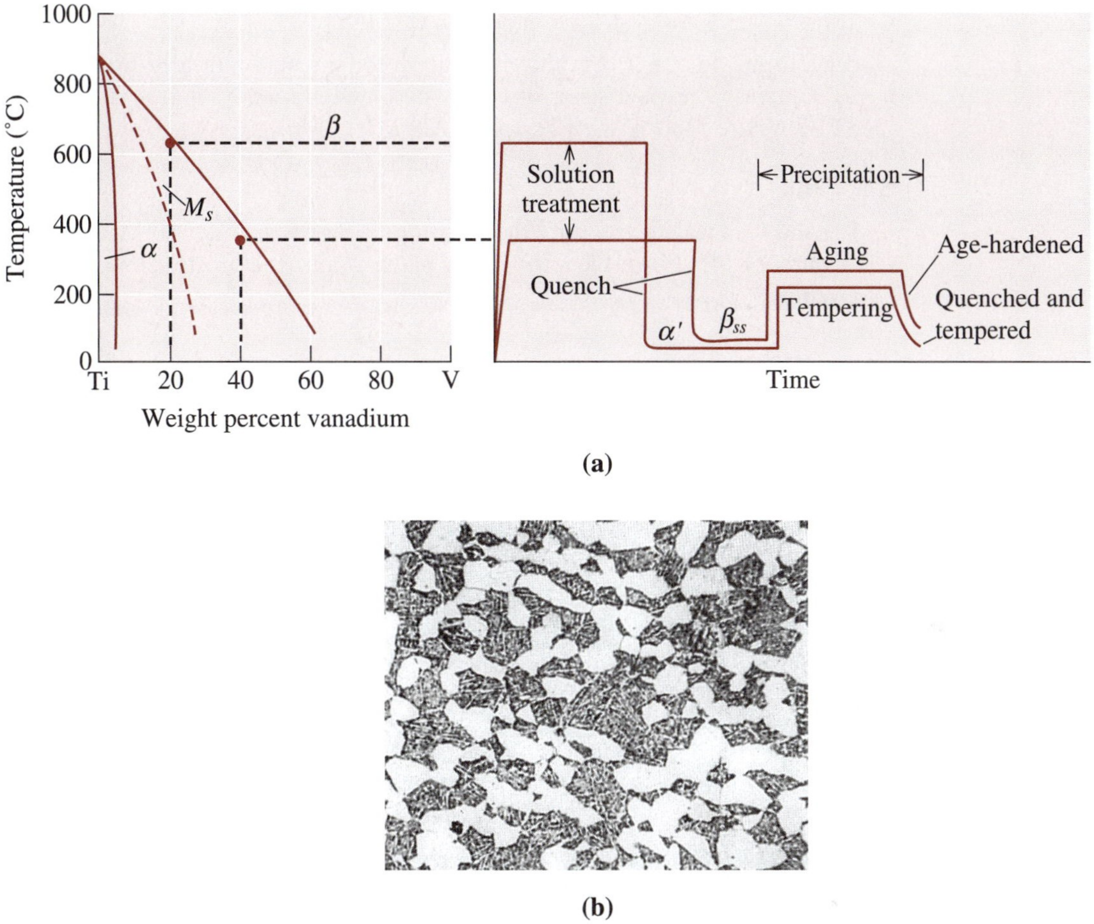

Figure 14-13 (a) Heat treatment and (b) microstructure of the alpha-beta titanium alloys. The structure contains primary α (large white grains) and a dark β matrix with needles of α formed during aging (×250). (*From* ASM Handbook, *Vol. 7, (1972), ASM International, Materials Park, OH 44073.*)

alloy, are superplastic and can be deformed as much as 1000%. This alloy is also used for making implants for human bodies. Titanium alloys are considered **biocompatible** (i.e., they are not rejected by the body). By developing porous coatings of bone-like ceramic compositions known as hydroxyapatite, it may be possible to make titanium implants **bioactive** (i.e., the natural bone can grow into the hydroxyapatite coating). The following three examples illustrate applications of titanium alloys.

EXAMPLE 14-8 ***Fracture Toughness of Titanium 6246 Alloy***

The fracture toughness of different titanium alloys is shown in Figure 14-14. The short transverse (ST) direction is through thickness direction, while the length (L) is the longitudinal direction or the rolling direction. A titanium-alloy 6246 plate is exposed to a stress of 200 MPa. Calculate the flaw size for a crack at the edge of the plate such that after this crack size the crack will grow at a catastrophic rate. Perform this calculation for the beta annealed titanium and ST samples. The yield stress of β-annealed, bimodal alloy, β-processed L, and ST alloys are 1180, 1200, 1275, and 1280 MPa, respectively.

SOLUTION

For the β-annealed structure, the fracture toughness is about 60 MPa $- \text{m}^{-1/2}$ (Figure 14-14). From Chapter 7, the plane-strain fracture toughness (K_{Ic}) is given by

$$K_{Ic} = f \times \sigma \times \sqrt{\pi \times a}$$

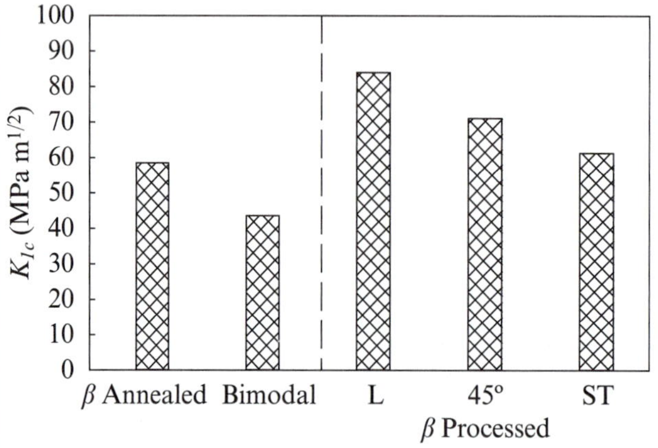

Figure 14-14 Fracture Toughness of Titanium 6246 Alloy. (*Source:* Titanium, *2007, page 309, K.U. Kainer, Gerd Lutjering and James C. Williams, Figure 7.27. With kind permission of Springer Science and Business Media.*)

The value of f is 1.12, since the crack or notch is on the edge of the sample. Therefore,

$$60 \text{ MPa} - \text{m}^{-1/2} = 1.12 \times (200 \text{ MPa}) \times \sqrt{\pi \times a}$$

This gives us a value of a, where the crack size will be 2.28 cm (i.e., any crack that is larger than about 2.28 cm will grow catastrophically).

For the L sample, the fracture toughness is about 80 MPa $- \text{m}^{-1/2}$. Therefore, the critical flaw length for a tensile stress of 200 MPa will be given by

$$80 \text{ MPa} - \text{m}^{-1/2} = 1.12 \times (200 \text{ MPa}) \times \sqrt{\pi \times a}$$

This works out to about 4 cm. Thus, the L sample titanium alloy with its higher fracture toughness will be able to withstand a much larger sized crack before the crack will grow catastrophically.

Later in Chapter 15 we will see that ceramic materials have much lower fracture toughness, so that the critical flaw size is very small compared to that for titanium and other alloys with very high fracture toughness.

EXAMPLE 14-9 *Design of a Connecting Rod*

Design a high-performance connecting rod for the engine of a racing automobile (Figure 14-15).

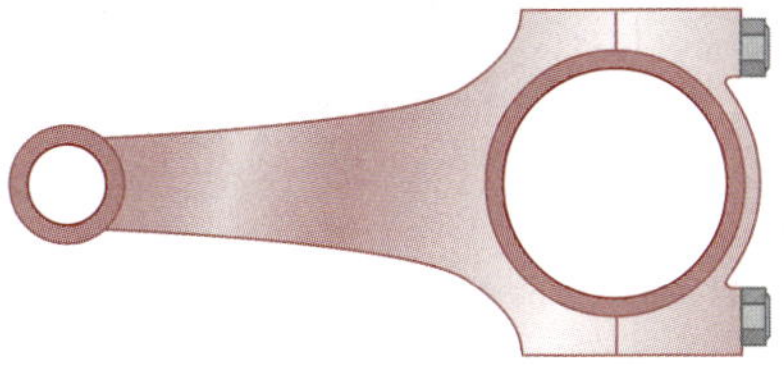

Figure 14-15
Sketch of connecting rod (for Example 14-9).

SOLUTION

A high-performance racing engine requires materials that can operate at high temperatures and stresses while minimizing the weight of the engine. In normal automobiles, the connecting rods are often a forged steel or a malleable cast iron. We might be able to save considerable weight by replacing these parts with titanium.

To achieve high strengths, we might consider an alpha-beta titanium alloy. Because of its availability, the Ti-6% Al-4% V alloy is a good choice. The alloy is heated to about 1065°C, which is in the all-β portion of the phase diagram. On quenching, a titanium martensite forms; subsequent tempering produces a microstructure containing β precipitates in an α matrix.

When the heat treatment is performed in the all-β region, the tempered martensite has an acicular structure, which reduces the rate of growth of any fatigue cracks that might develop.

EXAMPLE 14-10 *Materials for Hip Prosthesis*

What type of a material would you choose for an implant to be used for a total hip replacement implant?

SOLUTION

A hip prosthesis is intended to replace part of the worn out or damaged femur bone. The implant has a metal head and fits down the cavity of the femur. We need to consider the following factors: biocompatibility, corrosion resistance, high-fracture toughness, excellent fatigue life (so that implants last for many years since it is difficult to do the surgery as patients get older), and wear resistance. We also need to consider the stiffness. If the alloy chosen is too stiff compared to the bone, most of the stress will be carried by the implant. This leads to weakening of the remaining bone and, in turn, can make the implant loose. Thus, we need a material that has a high tensile strength, corrosion resistance, biocompatibility, and fracture toughness. These requirements suggest 316 stainless steel or Ti-6%Al-4% V. Neither of these materials are magnetic and both are opaque to x-rays. This means no interference for magnetic resonance and x-ray imaging. Titanium alloys are not very hard and can wear out. Stainless steels are harder, but they are much stiffer than bone. Titanium is biocompatible and would be a better choice. Perhaps a composite material in which the stem is made from a Ti-6%Al-4% V alloy and a head that is made from a wear-resistant, corrosion resistant, and fractured tough ceramic, such as alumina, may be an answer. The inside of the socket could be made from an ultra-high-density (ultra-high molecular weight) polyethylene that has a very low-friction coefficient. The surface of the implant could be made porous so as to encourage the bone to grow. Another option is to coat the implant with a material like porous hydroxyapatite to encourage bone growth.

14-6 Refractory and Precious Metals

The **refractory metals**, which include tungsten, molybdenum, tantalum, and niobium (or columbium), have exceptionally high-melting temperatures (above 1925°C) and, consequently, have the potential for high-temperature service. Applications include filaments for light bulbs, rocket nozzles, nuclear power generators, tantalum- and niobium-based electronic capacitors, and chemical processing equipment. These metals, however, have a high density, limiting their specific strengths (Table 14-10).

TABLE 14-10 ■ *Properties of some refractory metals*

			T = 1000°C		
Metal	Melting Temperature (°C)	Density (g/cm^3)	Tensile Strength (MPa)	Yield Strength (MPa)	Transition Temperature (°C)
Nb	2468	8.57	117	55	−140
Mo	2610	10.22	345	207	30
Ta	2996	16.6	186	166	−270
W	3410	19.25	455	103	300

Oxidation The refractory metals begin to oxidize between 200 and 425°C and are rapidly contaminated or embrittled. Consequently, special precautions are required during casting, hot working, welding, or powder metallurgy. The metals must also be protected during service at elevated temperatures. For example, the tungsten filament in a light bulb is protected by a vacuum.

For some applications, the metals may be coated with a silicide or aluminide coating. The coating must (a) have a high melting temperature, (b) be compatible with the refractory metal, (c) provide a diffusion barrier to prevent contaminants from reaching the underlying metal, and (d) have a coefficient of thermal expansion similar to that of the refractory metal. Coatings are available that protect the metal to about 1650°C. In some applications, such as capacitors for cellular phones, the formation of oxides is useful since we want to make use of the oxide as a nonconducting material.

Forming Characteristics The refractory metals, which have the BCC crystal structure, display a ductile-to-brittle transition temperature. Because the transition temperatures for niobium and tantalum are well below room temperature (Table 14-10), these two metals can readily be formed. However, annealed molybdenum and tungsten normally have a transition temperature above room temperature, causing them to be brittle at room temperature. Fortunately, if these metals are hot worked to produce a fibrous microstructure, the transition temperature is lowered and the forming characteristics are improved.

Alloys Large increases in both room-temperature and high-temperature mechanical properties are obtained by alloying. Tungsten alloyed with hafnium, rhenium, and carbon can operate up to 2100°C. These alloys typically are solid-solution strengthened; in fact, tungsten and molybdenum form a complete series of solid solutions, much like copper and nickel. Some alloys, such as W-2% ThO_2, are dispersion strengthened by oxide particles during their manufacture by powder metallurgy processes. Composite materials, such as niobium reinforced with tungsten fibers, may also improve high-temperature properties.

Precious Metals These include gold (Au), silver (Ag), palladium (Pd), platinum (Pt), and rhodium (Rh). As their name suggests, these are precious and expensive. From an engineering viewpoint, these materials resist corrosion and make conductors of electricity. As a result, alloys of these materials are often used as electrodes for devices. These electrodes are formed using a thin-film deposition (e.g., sputtering or electroplating) or screen printing of metal powder dispersions/pastes. Nano-sized particles of Pt/Rh/Pd (loaded onto a ceramic support) are also used as catalysts in automobiles. These metals facilitate the oxidation of CO to CO_2 and NO_x to N_2 and O_2. They are also used as catalysts in petroleum refining.

SUMMARY

- The "light metals" include low-density alloys based on aluminum, magnesium, and beryllium. Aluminum alloys have a high specific strength due to their low density and, as a result, find many aerospace applications. Excellent corrosion resistance and electrical conductivity of aluminum also provide for a vast number of applications. Aluminum and magnesium are limited to use at low temperatures because of the loss of their mechanical properties as a result of overaging or recrystallization. Copper alloys (brasses and bronzes) are also used in many structural and other applications. Titanium alloys have intermediate densities and temperature resistance, along with excellent corrosion resistance, leading to applications in aerospace, chemical processing, and biomedical devices.
- Nickel and cobalt alloys, including superalloys, provide good properties at even higher temperatures. Combined with their good corrosion resistance, these alloys find many applications in aircraft engines and chemical processing equipment.

GLOSSARY

Bioactive A material that is not rejected by the human body and eventually becomes part of the body (e.g., hydroxyapatite).

Biocompatible A material that is not rejected by the human body.

Blister copper An impure form of copper obtained during the copper refining process.

Brass A group of copper-based alloys, normally containing zinc as the major alloying element.

Bronze Generally, copper alloys containing tin but can contain other elements.

Castability The ease with which a metal can be poured into a mold to make a casting without producing defects or requiring unusual or expensive techniques to prevent casting problems.

Fluidity The ability of liquid metal to fill a mold cavity without prematurely freezing.

Monel The copper-nickel alloy, containing approximately 60% Ni, that gives the maximum strength in the binary alloy system.

Nonferrous alloy An alloy based on some metal other than iron.

Refractory metals Metals having a melting temperature above 1925°C.

Specific strength The ratio of strength to density. Also called the strength-to-weight ratio.

Superalloys A group of nickel, iron-nickel, and cobalt-based alloys that have exceptional heat resistance, creep resistance, and corrosion resistance.

Temper designation A shorthand notation using letters and numbers to describe the processing of an alloy. H tempers refer to cold-worked alloys; T tempers refer to age-hardening treatments.

Wrought alloys Alloys that are shaped by a deformation process.

PROBLEMS

14-1 In some cases, we may be more interested in cost per unit volume than in cost per unit weight. Rework Table 14-1 to show the cost in terms of $/cm^3. Does this change/alter the relationship between the different materials?

Section 14-1 Aluminum Alloys

14-2 Assuming that the density remains unchanged, compare the specific strength of the 2090-T6 aluminum alloy to that of a die-cast 443-F aluminum alloy. If you considered the actual density, do you think the difference between the specific strengths would increase or become smaller? Explain.

14-3 Explain why aluminum alloys containing more than about 15% Mg are not used.

14-4 Would you expect a 2024-T9 aluminum alloy to be stronger or weaker than a 2024-T6 alloy? Explain.

14-5 Estimate the tensile strength expected for the following aluminum alloys:

(a) 1100-H14 **(b)** 5182-H12 **(c)** 3004-H16

14-6 Suppose, by rapid solidification from the liquid state, that a supersaturated Al-7% Li alloy can be produced and subsequently aged. Compare the amount of β that will form in this alloy with that formed in a 2090 alloy.

14-7 Determine the amount of Mg_2Al_3 (β) expected to form in a 5182-O aluminum alloy (See Figure 14-2).

14-8 Based on the phase diagrams, which of the following alloys would be most suited for thixocasting? Explain your answer. (See Figure 14-2 and phase diagrams from Chapters 11 and 12.)

(a) Al-12% Si **(b)** Al-1% Cu **(c)** Al-10% Mg

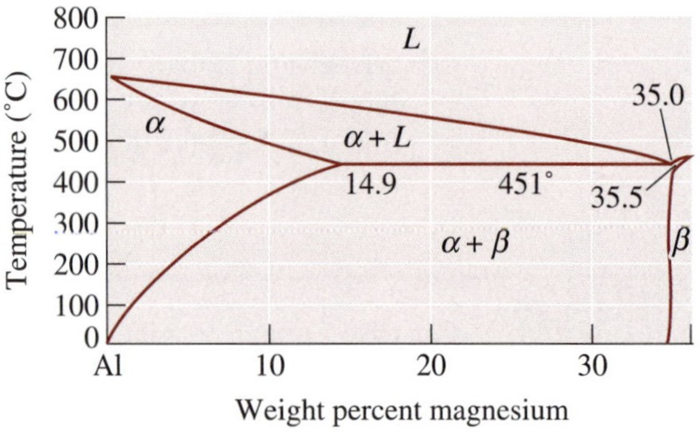

Figure 14-2 (Repeated for Problems 14-7 and 14-8) Portion of the aluminum-magnesium phase diagram.

Section 14-2 Magnesium and Beryllium Alloys

14-9 From the data in Table 14-6, estimate the ratio by which the yield strength of magnesium can be increased by alloying and heat treatment and compare with that of aluminum alloys.

14-10 Suppose a 60-cm-long round bar is to support a load of 181 kg without any permanent deformation. Calculate the minimum diameter of the bar if it is made of

(a) AZ80A-T5 magnesium alloy, and
(b) 6061-T6 aluminum alloy.

Calculate the weight of the bar and the approximate cost (based on pure Al and Mg) in each case.

14-11 A 10-m rod 0.5 cm in diameter must elongate no more than 2 mm under load. Determine the maximum force that can be applied if the rod is made of:

(a) aluminum **(b)** magnesium **(c)** beryllium

14-12 For the Mg alloy AZ91, calculate the grain size from the Hall-Petch equation for a casting that had strength of 250 MPa (See Example 14-4).

Section 14-3 Copper Alloys

14-13 **(a)** Explain how pure copper is made. **(b)** What are some of the important properties of copper? **(c)** What is brass? **(d)** What is bronze? **(e)** Why does the Statue of Liberty appear green?

14-14 We say that copper can contain up to 40% Zn or 9% Al and still be single phase. How do we explain this statement in view of the phase diagrams for the Cu-Zn system? [See Figure 14-6(a).]

14-15 Compare the percentage increase in the yield strength of commercially pure annealed aluminum, magnesium, and copper by strain hardening. Explain the differences observed.

14-16 We would like to produce a quenched and tempered aluminum bronze containing 13% Al. Recommend a heat treatment, including appropriate temperatures. Calculate the amount of each phase after each step of the treatment.

14-17 A number of casting alloys have very high lead contents; however, the Pb content in wrought alloys is comparatively low. Why isn't more lead added to the wrought alloys? What precautions must be taken when a leaded wrought alloy is hot worked or heat treated?

14-18 Would you expect the fracture toughness of quenched and tempered aluminum bronze to be high or low? Would there be a difference in the resistance of the alloy to crack nucleation compared with crack growth? Explain.

Section 14-4 Nickel and Cobalt Alloys

14-19 Based on the photomicrograph in Figure 14-8(a), would you expect the γ' precipitate or the carbides to provide a greater strengthening effect in superalloys at low temperatures? Explain.

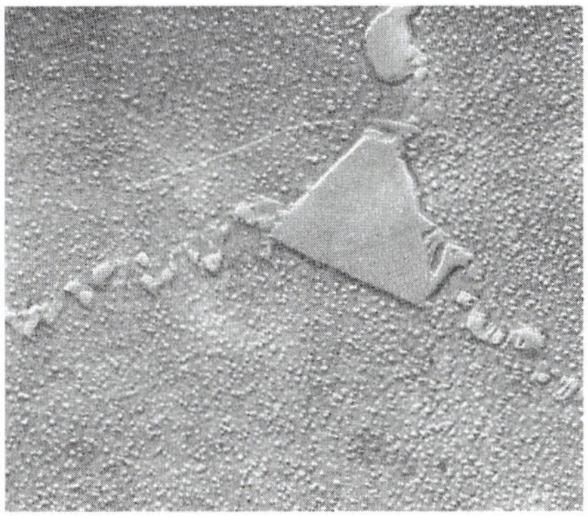

(a)

Figure 14-8 (Repeated for Problem 14-19)
(a) Microstructure of a superalloy, with carbides at the grain boundaries and γ' precipitates in the matrix ($\times$15,000).

14-20 The density of Ni_3Al is 7.5 g/cm^3. Suppose a Ni-5 wt% Al alloy is heat treated so that all of the aluminum reacts with nickel to produce Ni_3Al. Determine the volume percentage of the Ni_3Al precipitate in the nickel matrix.

Section 14-5 Titanium Alloys

14-21 When steel is joined using arc welding, only the liquid fusion zone must be protected by a gas or flux. However, when titanium is welded, both the front and back sides of the welded metal must be protected. Why must these extra precautions be taken when joining titanium?

14-22 Both a Ti-15% V alloy and a Ti-35% V alloy are heated to a temperature at which all β just forms. They are then quenched and reheated to 300°C. Describe the changes in microstructure during the heat treatment for each alloy, including the amount of each phase. What is the matrix and what is the precipitate in each case? Which is an age-hardening process? Which is a quench and temper process? [See Figure 14-13(a)].

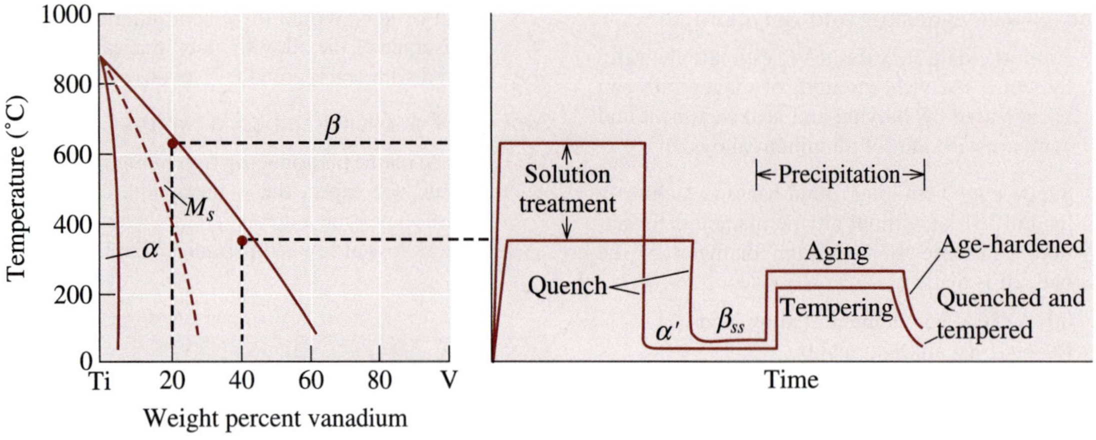

(a)

Figure 14-13 (Repeated for Problem 14-22) (a) Heat treatment of the alpha-beta titanium alloys.

14-23 Determine the specific strength of the strongest Al, Mg, Cu, Ti, and Ni alloys. Use the densities of the pure metals in your calculations. Try to explain their order.

14-24 Based on the phase diagrams, estimate the solubilities of Ni, Zn, Al, Sn, and Be in copper at room temperature. Are these solubilities expected in view of Hume-Rothery's conditions for solid solubility (Chapter 10)? Explain.

14-25 A titanium 6246 alloy plate of ST type was found to have an edge crack of size 3 mm. What will be the highest level of tensile stress (σ) that can be supported on this plate without causing catastrophic failure? [See Figure 14-14.]

14-26 A titanium 6246 alloy plate of a bimodal microstructure was found to have an edge crack of size 4 mm. What will be the highest level of tensile stress (σ) that can be supported on this plate without causing catastrophic failure? [See Figure 14-14.]

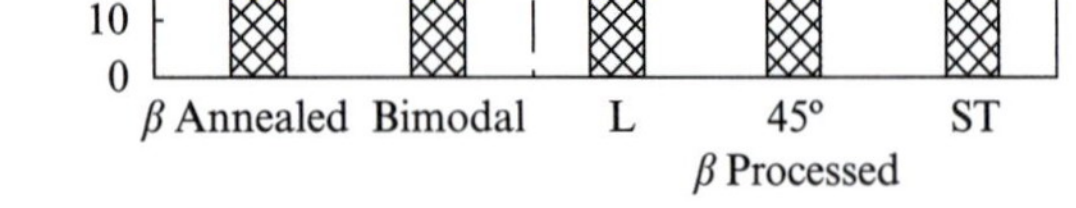

Figure 14-14 (Repeated for Problems 14-25 and 14-26) Fracture Toughness of Titanium 6246 Alloy. *(Source: Titanium, Lutjering, G. and Williams, J.C., Figure 7.27, p. 309.)*

Section 14-6 Refractory and Precious Metals

14-27 What is a refractory metal or an alloy? What is a precious metal?

14-28 The temperature of a coated tungsten part is increased. What happens when the protective coating on a tungsten part expands more than the tungsten? What happens when the protective coating on a tungsten part expands less than the tungsten?

14-29 For what applications are Pt, Rh, Pd, Ag used?

Design Problems

14-30 A part for an engine mount for a private aircraft must occupy a volume of 60 cm^3 with a minimum thickness of 0.5 cm and a minimum width of 4 cm. The load on the part during service may be as much as 75,000 N. The part is expected to remain below 100°C during service. Design a material and its treatment that will perform satisfactorily in this application.

14-31 You wish to design the rung on a ladder. The ladder should be light in weight so that it can be easily transported and used. The rungs on the ladder should be 0.625 cm $\times$ 2.5 cm and are 30 cm long. Design a material and its processing for the rungs.

14-32 We have determined that we need an alloy having a density of 2.3 $\pm$ 0.05 g/cm^3 that must be strong, yet still have some ductility. Design a material and its processing that might meet these requirements.

14-33 We wish to design a mounting device that will position and aim a laser for precision cutting of a composite material. What design requirements might be important? Design a material and its processing that might meet these requirements.

14-34 Design a nickel-titanium alloy that will produce 60 volume percent Ni_3Ti precipitate in a pure-nickel matrix.

14-35 An actuating lever in an electrical device must open and close almost instantly and carry a high current when closed. What design requirements would be important for this application? Design a material and its processing to meet these requirements.

14-36 A fan blade in a chemical plant must operate at temperatures as high as 400°C under rather corrosive conditions. Occasionally, solid material is ingested and impacts the fan. What design requirements would be important? Design a material and its processing for this application.

15

Ceramic Materials

Have You Ever Wondered?

- *What is the magnetic strip on a credit card made from?*
- *What material is used to protect the space shuttle from high temperatures during re-entry?*
- *What ceramic material is commonly added to paints?*
- *What ceramic material is found in bone and teeth?*
- *What are spark plugs made from?*

The goal of this chapter is to examine more closely the synthesis, processing, and applications of ceramic materials. Ceramics have been used for many thousands of years. Most ceramics exhibit good strength under compression; however, typically they exhibit virtually no ductility under tension. The family of ceramic materials includes polycrystalline and single-crystal inorganic materials, amorphous inorganic glasses, and glass-ceramics.

In Chapters 2 and 3, we learned about the bonding in ceramic materials, the crystal structures of technologically useful ceramics, and the arrangements of ions in glasses.

This chapter focuses on the synthesis, processing, and applications of ceramics. We will also recapitulate the processing and applications of inorganic glasses and glass-ceramics. We begin with a discussion that summarizes the classification and applications of ceramics.

15-1 Applications of Ceramics

One way to classify ceramics is based on their class of chemical compounds (e.g., oxides, carbides, nitrides, sulfides, fluorides, etc.). Another way, which we will use here, is to classify ceramics by their major function.

Ceramics are used in a wide range of technologies such as refractories, spark plugs, dielectrics in capacitors, sensors, abrasives, magnetic recording media, etc. The space shuttle makes use of ~25,000 reusable, lightweight, highly porous ceramic tiles that protect the aluminum frame from the heat generated during re-entry into the Earth's atmosphere. These tiles are made from high-purity silica fibers and colloidal silica coated with a borosilicate glass. Ceramics also appear in nature as oxides and in natural materials; the human body has the amazing ability of making hydroxyapatite, a ceramic found in bones and teeth. Ceramics are also used as coatings. **Glazes** are ceramic coatings applied to glass objects; **enamels** are ceramic coatings applied to metallic objects. Let's follow the classification shown in Table 15-1 and take note of different applications. Alumina and silica are the most widely used ceramic materials and, as you will notice, there are numerous applications listed in Table 15-1 that depend upon the use of these two ceramics.

The following is a brief summary of applications of some of the more widely used ceramic materials:

- *Alumina* (Al_2O_3) is used to contain molten metal or in applications where a material must operate at high temperatures, but where high strength is also required. Alumina is also used as a low dielectric constant substrate for electronic packaging that houses silicon chips. One classical application is for insulators in spark plugs. Some unique applications are also being found in dental and medical use. Chromium-doped alumina is used for making lasers. Fine particles of alumina are used as catalyst supports.

- *Diamond* (C) is the hardest naturally occurring material. Industrial diamonds are used as abrasives for grinding and polishing. Diamond and diamond-like coatings prepared using chemical vapor deposition processes are used to make abrasion-resistant coatings for many different applications (e.g., cutting tools). It is, of course, also used in jewelry.

- *Silica* (SiO_2) is probably the most widely used ceramic material. Silica is an essential ingredient in glasses and many glass ceramics. Silica-based materials are used in thermal insulation, refractories, abrasives, fiber-reinforced composites, laboratory glassware, etc. In the form of long continuous fibers, silica is used to make optical fibers for communications. Powders made using fine particles of silica are used in tires, paints, and many other applications.

- *Silicon carbide* (SiC) provides outstanding oxidation resistance at temperatures even above the melting point of steel. SiC often is used as a coating for metals,

TABLE 15-1 ■ *Functional classification of ceramics**

Function	Application	Examples of Ceramics
Electrical	Capacitor dielectrics	$BaTiO_3$, $SrTiO_3$, Ta_2O_5
	Microwave dielectrics	$Ba(Mg_{1/3}Ta_{2/3})O_3$, $Ba(Zn_{1/3}Ta_{2/3})O_3$
		$BaTi_4O_9$, $Ba_2Ti_9O_{20}$, $Zr_xSn_{1-x}TiO_4$, Al_2O_3
	Conductive oxides	In-doped SnO_2 (*ITO*)
	Superconductors	$YBa_2Cu_3O_{7-x}$ (*YBCO*)
	Electronic packaging	Al_2O_3
	Insulators	Porcelain
	Solid-oxide fuel cells	ZrO_2, $LaCrO_3$, $LaMnO_3$
	Piezoelectric	$Pb(Zr_xTi_{1-x})O_3$(*PZT*), $Pb(Mg_{1/3}Nb_{2/3})O_3$ (PMN)
	Electro-optical	*PLZT*, $LiNbO_3$
Magnetic	Recording media	γ-Fe_2O_3, CrO_2 ("chrome" cassettes)
	Ferrofluids, credit cards	Fe_3O_4
	Circulators, isolators,	Nickel zinc ferrite
	Inductors, magnets	Manganese zinc ferrite
Optical	Fiber optics	Doped SiO_2
	Glasses	SiO_2 based
	Lasers	Al_2O_3, yttrium aluminum garnate (*YAG*)
	Lighting	Al_2O_3, glasses.
Automotive	Oxygen sensors, fuel cells	ZrO_2
	Catalyst support	Cordierite
	Spark plugs	Al_2O_3
	Tires	SiO_2
	Windshields/windows	SiO_2 based glasses
Mechanical/Structural	Cutting tools	WC-Co cermets
		Sialon
		Al_2O_3
	Composites	SiC, Al_2O_3, silica glass fibers
	Abrasives	SiC, Al_2O_3, diamond, BN, $ZrSiO_4$
Biomedical	Implants	Hydroxyapatite
	Dentistry	Porcelain, Al_2O_3
	Ultrasound imaging	*PZT*
Construction	Buildings	Concrete
		Glass
		Sanitaryware
Others	Defense applications	*PZT*, B_4C
	Armor materials	
	Sensors	SnO_2
	Nuclear	UO_2
		Glasses for waste disposal
	Metals processing	Alumina and silica-based refractories, oxygen sensors, casting molds, etc.
Chemical	Catalysis	Various oxides (Al_2O_3, ZrO_2, ZnO, TiO_2)
	Air, liquid filtration	
	Sensors	
	Paints, rubber	
Domestic	Tiles, sanitaryware,	Clay, alumina, and silica-based ceramics, glass-ceramics, diamond, ruby, cubic zirconia and other crystals
	Whiteware, kitchenware,	
	Pottery, art, jewelry	

**Acronyms are indicated in italics.*

carbon-carbon composites, and other ceramics to provide protection at these extreme temperatures. SiC is also used as an abrasive in grinding wheels and as particulate and fibrous reinforcement in both metal matrix and ceramic matrix composites. It is also used to make heating elements for furnaces. SiC is a semiconductor and is a very good candidate for high-temperature electronics.

- *Silicon nitride* (Si_3N_4) has properties similar to those of SiC, although its oxidation resistance and high-temperature strength are somewhat lower. Both silicon nitride and silicon carbide are likely candidates for components for automotive and gas turbine engines, permitting higher operating temperatures and better fuel efficiencies with less weight than traditional metals and alloys.
- *Titanium Dioxide* (TiO_2) is used to make electronic ceramics such as $BaTiO_3$. The largest use, though, is as a white pigment to make paints. Titania is used in certain glass ceramics as a nucleating agent. Fine particles of TiO_2 are used to make suntan lotions that provide protection against ultraviolet rays.
- *Zirconia* (ZrO_2) is used to make many other ceramics such as zircon. Zirconia is also used to make oxygen gas sensors that are used in automotives and to measure dissolved oxygen in molten steels. Zirconia is used as an additive in many electronic ceramics as well as a refractory material. The cubic form of zirconia single crystals is used to make jewelry items. Fuel cells based on zirconia will likely appear in cars by the year 2015.

15-2 Properties of Ceramics

The properties of some ceramics are summarized in Table 15-2. Mechanical properties of some structural ceramics are summarized in Table 15-3.

Take note of the high-melting temperatures and high-compressive strengths of ceramics. We should also remember that the tensile and flexural strength values show considerable variation since the strength of ceramics is dependent on the distribution of flaw sizes and is not affected by dislocation motion. We discussed the Weibull distribution and the strength of ceramics and glasses in Chapter 7. Also note that, contrary to common belief, ceramics are not always brittle. Under smaller strain rates and at high temperatures, many ceramics with a very fine grain size indeed show superplastic behavior.

TABLE 15-2 ■ *Properties of commonly encountered polycrystalline ceramics*

Material	Melting Point (°C)	Thermal Expansion Coefficient ($\times 10^{-6}$ cm/cm)/°C	Knoop Hardness (HK) (100 g)
Al_2O_3	2000	~6.8	2100
BN	2732	0.57[a], −0.46[b]	5000
SiC	2700	~3.7	2500
Diamond		1.02	7000
Mullite	1810	4.5	—
TiO_2	1840	8.8	—
Cubic ZrO_2	2700	10.5	—

[a] *Perpendicular to pressing direction.*
[b] *Parallel to pressing direction.*

TABLE 15-3 ■ ***Mechanical properties of selected advanced ceramics***

Material	Density (g/cm^3)	Tensile Strength (MPa)	Flexural Strength (MPa)	Compressive Strength (MPa)	Young's Modulus (MPa)	Fracture Toughness ($MPa\sqrt{m}$)
Al_2O_3	3.98	207	552	3025	386×10^3	5.5
SiC (sintered)	3.1	172	552	3860	414×10^3	4.4
Si_3N_4 (reaction bonded)	2.5	138	241	1030	207×10^3	3.3
Si_3N_4 (hot pressed)	3.2	552	896	3450	310×10^3	5.5
Sialon	3.24	414	965	3450	310×10^3	9.9
ZrO_2 (partially stabilized)	5.8	448	690	1860	207×10^3	11.0
ZrO_2 (transformation toughened)	5.8	345	793	1725	200×10^3	12.1

15-3 Synthesis and Processing of Ceramic Powders

Ceramic materials melt at high temperatures and they usually exhibit a brittle behavior under tension. As a result, the casting and thermomechanical processing, used widely for metals, alloys, and thermoplastics, cannot be applied when processing ceramics. Inorganic glasses, though, make use of lower melting temperatures due to the formation of eutectics and are made using the float-glass process. Since melting, casting, and thermomechanical processing is not a viable option for polycrystalline ceramics, we typically process ceramics into useful shapes starting with ceramic powders. A *powder* is a collection of fine particles. The step of making a ceramic powder is defined here as the **synthesis** of ceramics. We begin with a ceramic powder and get it ready for shaping by crushing, grinding, separating impurities, blending different powders, and **spray drying** to form soft agglomerates. Different techniques such as compaction, **tape casting**, extrusion, and **slip casting** are then used to convert properly processed powders into a desired shape to form what is known as a **green ceramic**. A green ceramic is a ceramic that has not yet been sintered. The steps of converting a ceramic powder (or mixture of powders) into a useful shape are known as **powder processing**. The green ceramic is then consolidated further using a high-temperature treatment known as sintering or firing. In this process, the green ceramic is heated to a high temperature, using a controlled heat treatment and atmosphere, so that a dense material is obtained. The ceramic may be then subjected to additional operations such as grinding, polishing, or machining as needed for the final application. In some cases, leads will be attached, electrodes will be deposited, or coatings may have to be deposited. These general steps encountered in the synthesis and processing of ceramics are summarized in Figure 15-1.

Ceramic powders prepared using conventional or chemical techniques are shaped using the techniques shown in Figure 15-2. We emphasize that very similar processes are used for processing metal and alloy powders, a route known as **powder metallurgy**. Powders consist of particles that are loosely bonded, and powder processing involves the consolidation of these powders into a desired shape. Often, the ceramic powders prepared need to be converted into soft agglomerates by spraying a slurry of the powder through a nozzle into a chamber (spray dryer) in the presence of hot air. This process leads to the formation of soft agglomerates that flow into the dies used for powder compaction; this is known as *spray drying*.

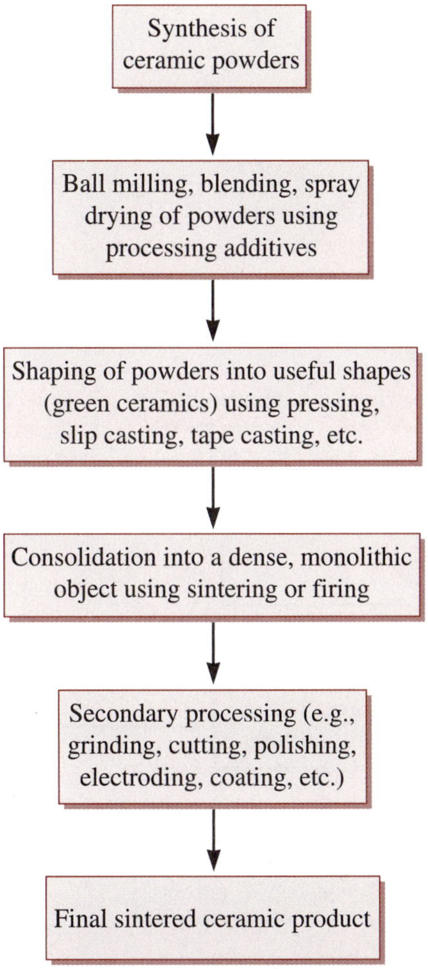

Figure 15-1
Typical steps encountered in the processing of ceramics.

Compaction and Sintering One of the most cost-effective ways to produce thousands of relatively small pieces ($\sim<15$ cm) of simple shapes is compaction and sintering. Many electronic and magnetic ceramics, WC-Co (**cermet**) cutting tool bits, and other materials are processed using this technique. The driving force for sintering is the reduction in the surface area of a powder (Chapter 5). Fine powders can be spray dried, forming soft agglomerates that flow and compact well. The different steps of uniaxial compaction, in which the compacting force is applied in one direction, are shown in Figure 15-3(a) on page 475. As an example, the microstructure of a barium magnesium tantalate ceramic prepared using compaction and sintering is shown in Figure 15-3(b). Sintering involves different mass transport mechanisms [Figure 15-3(c)]. With sintering, the grain boundary and bulk (volume) diffusion contribute to densification (increase in density). Surface diffusion and evaporation condensation can cause grain growth, but they do not cause densification.

The compaction process can be completed within one minute for smaller parts; thus, uniaxial compaction is well suited for making a large number of smaller and simple shapes. Compaction is used to create what we call "green ceramics"; these have respectable strengths and can be handled and machined. In some cases, very large pieces (up to 100 cm in diameter and 200 cm to 240 cm long) can be produced using a process called **cold isostatic pressing (CIP)** where pressure is applied using oil. Such large pieces

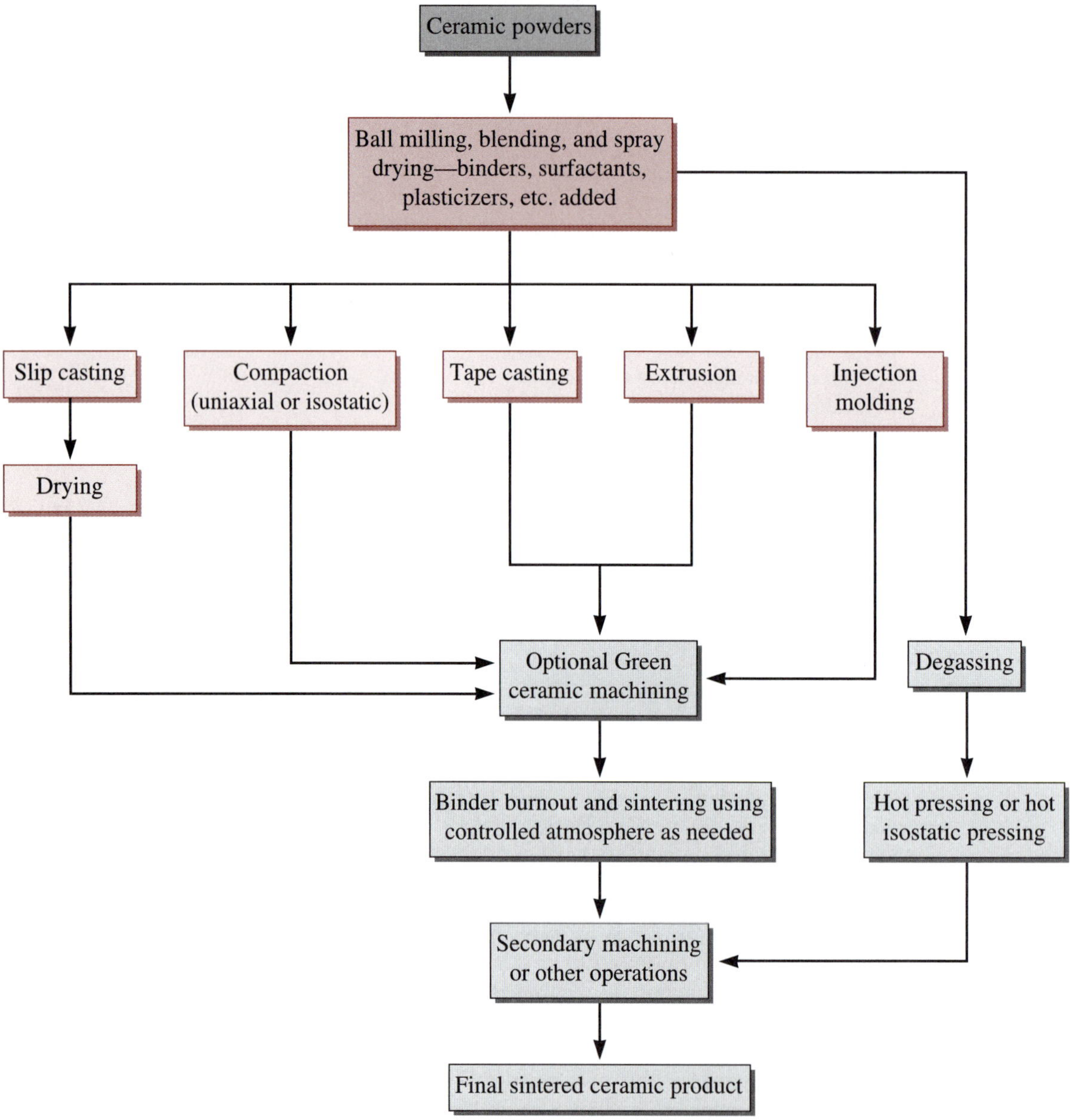

Figure 15-2 Different techniques for processing of advanced ceramics.

are then sintered with or without pressure. Cold isostatic pressing is used for achieving a higher green ceramic density or where the compaction of more complex shapes is required.

In some cases, parts may be produced under conditions in which sintering is conducted using applied pressure. This technique, known as **hot pressing**, is used for refractory and covalently bonded ceramics that do not show good pressureless sintering behavior. Similarly, large pieces of metals and alloys compacted using CIP can be sintered under pressure in a process known as **hot isostatic pressing (HIP)**. In hot pressing or HIP, the applied pressure acts against the internal pore pressure and enhances densification without causing grain growth. Hot pressing or hot isostatic pressing also are

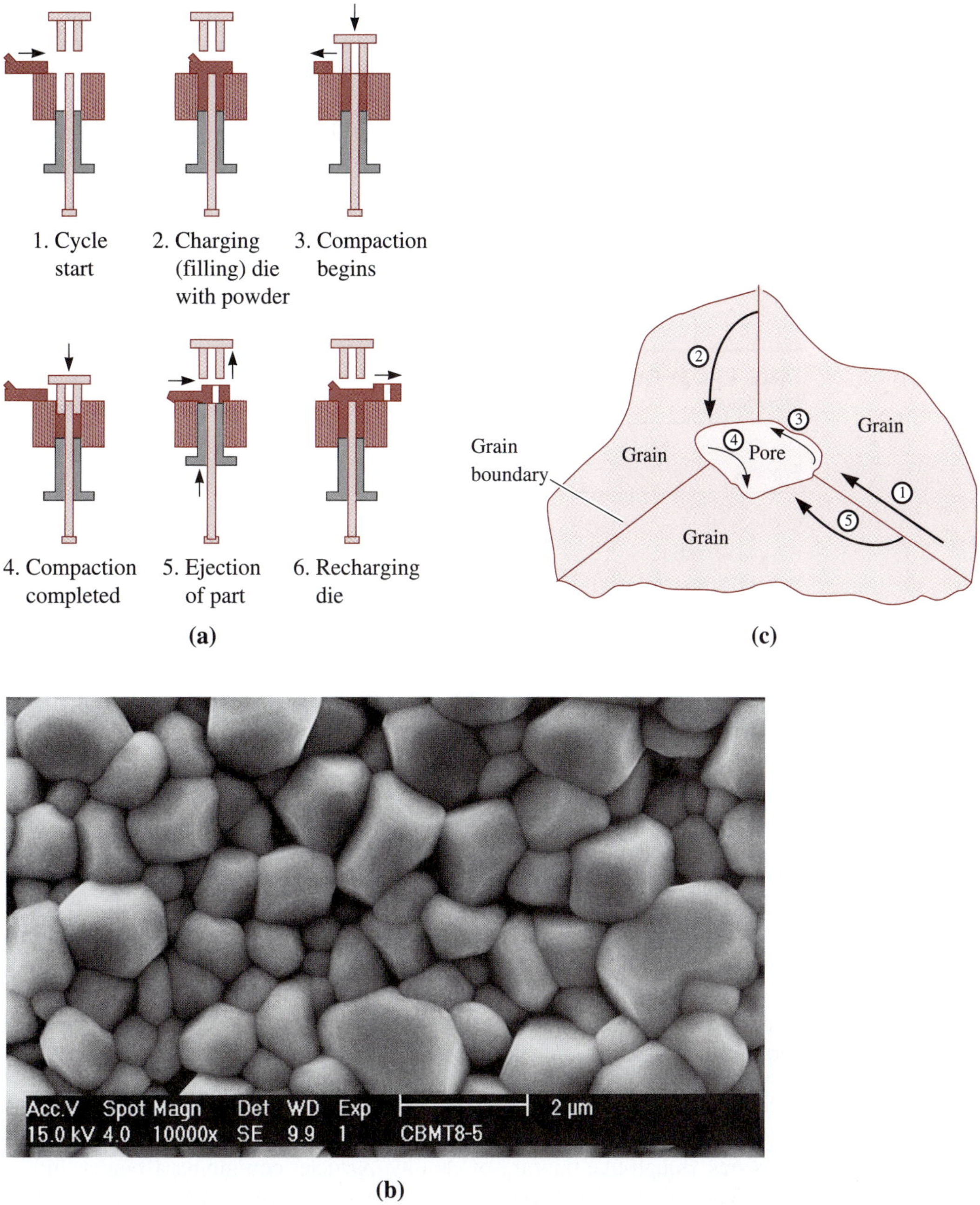

Figure 15-3 (a) Uniaxial powder compaction showing the die-punch assembly during different stages. Typically, for small parts these stages are completed in less than a minute. (*Source: From* Materials and Processes in Manufacturing, *Eighth Edition, by E.P. DeGarmo, J.T. Black, and R.A. Koshe, Fig. 16-4. Copyright © 1997 Prentice Hall. Reprinted with permission from John Wiley & Sons, Inc.*) (b) Microstructure of a barium magnesium tantalate (BMT) ceramic prepared using compaction and sintering. (*Photo Courtesy Schott North America.*) (c) Different diffusion mechanisms involved in sintering. The grain boundary and bulk diffusion (1, 2 and 5) to the neck contribute to densification. Evaporation-condensation (4) and surface diffusion (3) do not contribute to densification. (*Source: From* Physical Ceramics: Principles for Ceramic Science and Engineering, *by Y.M. Chiang, D. Birnie, and W.D. Kingery, Fig. 5-40. Copyright 1997 John Wiley & Sons, Inc. This material is used by John Wiley & Sons, Inc.*)

used for making ceramics or metallic alloys where very little or almost no porosity is required. Some recent innovative processes that make use of microwaves (similar to the way food gets heated in microwave oven) have also been developed for the drying and sintering of ceramic materials.

Some ceramics, such as silicon nitride (Si_3N_4), are produced by **reaction bonding**. Silicon is formed into a desired shape and then reacted with nitrogen to form the nitride. Reaction bonding, which can be done at lower temperatures, provides better dimensional control compared with hot pressing; however, lower densities and mechanical properties are obtained. As a comparison, the effect of processing on silicon nitride ceramics is shown in Table 15-4.

TABLE 15-4 ■ ***Properties of Si_3N_4 processed using different techniques***

Process	Compressive Strength (MPa)	Flexural Strength (MPa)
Slip casting	138	69
Reaction bonding	772	207
Hot pressing	345	862

Tape Casting A technique known as *tape casting* is used for the production of thin ceramic tapes (~3 to 100 μm). The tape is subjected to sintering. Many commercially important electronic packages based on alumina substrates and millions of barium titanate capacitors are made using this type of tape casting process.

Slip Casting This technique typically uses an aqueous slurry of ceramic powder. The slurry, known as the **slip**, is poured into a plaster of Paris ($CaSO_4:2H_2O$) mold (Figure 15-4). As the water from the slurry begins to move out by capillary action, a thick mass builds along the mold wall. When sufficient product thickness is built, the rest of the slurry is poured out (this is called *drain casting*). It is also possible to continue to pour more slurry in to form a solid piece (this is called *solid casting*) (Figure 15-4). Pressure may also be used to inject the slurry into polymer molds. The green ceramic is then dried and "fired" or sintered at a high temperature. Slip casting is widely used to make ceramic art (figurines and statues), sinks and other ceramic sanitaryware.

Extrusion and Injection Molding These are popular techniques used for making furnace tubes, bricks, tiles, and insulators. The idea behind the extrusion process is to use a viscous, dough-like mixture of ceramic particles containing a binder and other additives. This mixture has a clay-like consistency, which is then fed to an extruder where it is mixed well in a pug mill, sheared, deaerated, and then injected into a die where a continuous shape of green ceramic is produced by the extruder. This material is cut at appropriate lengths and then dried and sintered. Cordierite ceramics used for making catalyst honeycomb structures are also made using the extrusion process.

Injection molding of ceramics is similar to injection molding of polymers (Chapter 16). Ceramic powder is mixed with a thermoplastic plasticizer and other additives. The mixture is then taken through an extruder and injected into a die. Ceramic injection molding is better suited for complex shapes. The polymer contained in the injection-molded ceramic is burnt off and the rest of the ceramic body is sintered at a high temperature. This technique is especially useful for manufacturing complex shapes.

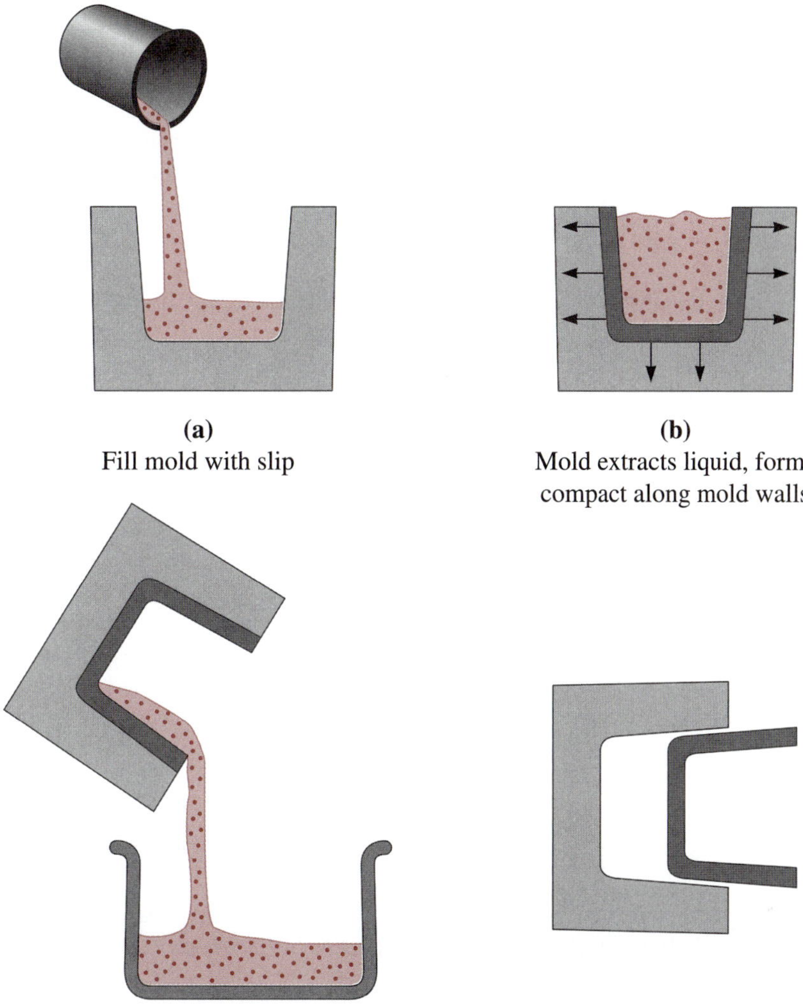

(a) Fill mold with slip

(b) Mold extracts liquid, forms compact along mold walls

(c) Excess slip drained

(d) Casting removed after partial drying

Figure 15-4 Steps in slip casting of ceramics. (*Source: From* Modern Ceramic Engineering, *by D.W. Richerson, p. 462, Fig. 10-34. Copyright © 1992 Marcel Dekker. Reprinted by permission.*)

15-4 Characteristics of Sintered Ceramics

For sintered ceramics, the average grain size, grain size distribution, and the level and type of porosity are important. Similarly, depending upon the application, second phases in the microstructure could occur as separate grains of components dissolved in solid solutions of the matrix. Therefore, second phases at grain boundaries also become important. In the case of extruded ceramics, orientation effects also can be important.

Grains and Grain Boundaries The average grain size is often closely related to the primary particle size. An exception to this is if there is grain growth due to long sintering times or exaggerated or abnormal grain growth (Chapter 5). Typically, ceramics with a small grain size are stronger than coarse-grained ceramics. Finer grain sizes help reduce stresses that develop at grain boundaries due to anisotropic expansion and contraction. Normally, starting with fine ceramic raw materials produces a fine grain size. Magnetic, dielectric, and optical properties of ceramic materials depend upon the average grain size and, in these applications, grain size must be controlled properly. Although we have not discussed this here in detail, in certain applications, it is important to use single crystals of ceramic materials so as to avoid the deleterious grain boundaries that are always present in polycrystalline ceramics.

Porosity Pores represent the most important defect in polycrystalline ceramics. The presence of pores is usually detrimental to the mechanical properties of bulk ceramics, since pores provide a pre-existing location from which a crack can grow. The presence of pores is one of the reasons why ceramics show such brittle behavior under tensile loading. Since there is a distribution of pore sizes and the overall level of porosity changes, the mechanical properties of ceramics vary. This variability is measured using the Weibull statistics (Chapter 7). The presence of pores, on the other hand, may be useful for increasing the resistance to thermal shock. In certain applications, such as filters for hot metals and alloys or for liquids or gases, the presence of interconnected pores is desirable.

Pores in a ceramic may be either interconnected or closed. The **apparent porosity** measures the interconnected pores and determines the permeability, or the ease with which gases and fluids seep through the ceramic component. The apparent porosity is determined by weighing the dry ceramic (W_d), then reweighing the ceramic both when it is suspended in water (W_s) and after it is removed from the water (W_w). Using units of grams and cm^3:

$$\text{Apparent porosity} = \frac{W_w - W_d}{W_w - W_s} \times 100 \tag{15-1}$$

The **true porosity** includes both interconnected and closed pores. The true porosity, which better correlates with the properties of the ceramic, is:

$$\text{True porosity} = \frac{\rho - B}{\rho} \times 100 \tag{15-2}$$

where

$$B = \frac{W_d}{W_w - W_s} \tag{15-3}$$

B is the **bulk density** and ρ is the true density or specific gravity of the ceramic. The bulk density is the weight of the ceramic divided by its volume. The following example illustrates how porosity levels in ceramics are determined.

EXAMPLE 15-1 Silicon Carbide Ceramics

Silicon carbide particles are compacted and fired at a high temperature to produce a strong ceramic shape. The specific gravity of SiC is 3.2 g/cm^3. The ceramic shape subsequently is weighed when dry (360 g), after soaking in water (385 g), and while suspended in water (224 g). Calculate the apparent porosity, the true porosity, and the fraction of the pore volume that is closed.

SOLUTION

$$\text{Apparent porosity} = \frac{W_w - W_d}{W_w - W_s} \times 100 = \frac{385 - 360}{385 - 224} \times 100 = 15.5\%$$

$$\text{Bulk density} = B = \frac{W_d}{W_w - W_s} = \frac{360}{385 - 224} = 2.24$$

$$\text{True porosity} = \frac{\rho - B}{\rho} \times 100 = \frac{3.2 - 2.24}{3.2} \times 100 = 30\%$$

The closed-pore percentage is the true porosity minus the apparent porosity, or $30 - 15.5 = 14.5\%$. Thus,

$$\text{Fraction closed pores} = \frac{14.5}{30} = 0.483$$

15-5 Inorganic Glasses

In Chapter 3, we discussed amorphous materials such as glasses. We also discussed the concepts of short- versus long-range order in terms of atomic or ionic arrangements in noncrystalline materials. The most important of the noncrystalline materials are glasses, especially those based on silica. Of course, there are glasses based on other compounds (e.g., sulfides, fluorides, and different alloys). A **glass** is a metastable material that has hardened and become rigid without crystallizing. A glass, in some ways, resembles an undercooled liquid. Below the **glass temperature** (T_g) (Figure 15-5), the rate of volume contraction on cooling is reduced and the material can be considered a "glass" rather than an "undercooled liquid." Joining silica tetrahedra or other ionic groups produces a solid, but noncrystalline framework structure produces the glassy structures (Chapter 3).

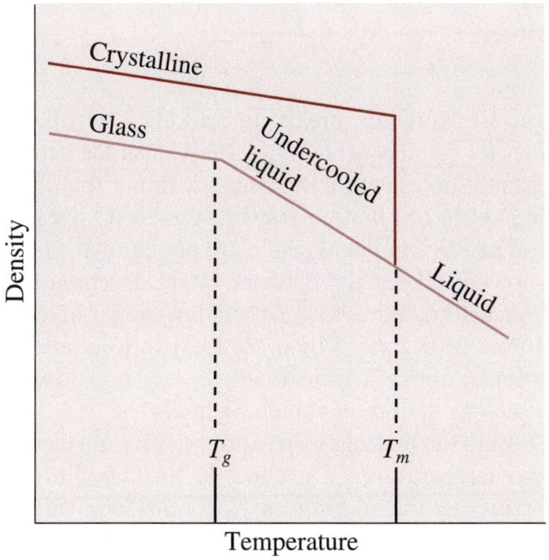

Figure 15-5
When silica crystallizes on cooling, an abrupt change in the density is observed. For silica glass, however, the change in slope at the glass temperature indicates the formation of a glass from the undercooled liquid. Glass does not have a fixed T_m or T_g. Crystalline materials have a fixed T_m and they do not have a T_g.

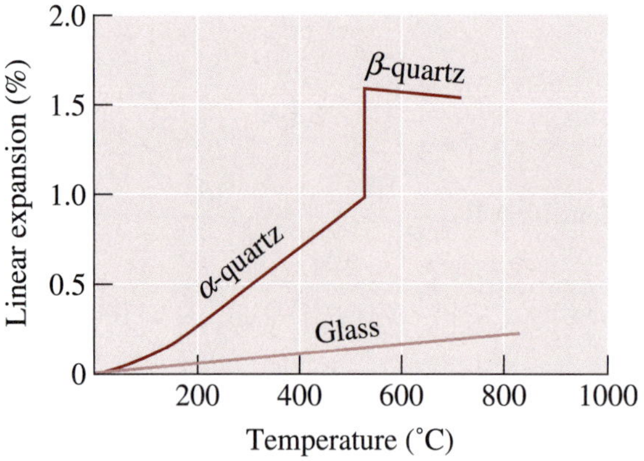

Figure 15-6
The expansion of quartz. In addition to the regular—almost linear—expansion, a large, abrupt expansion accompanies the α- to β-quartz transformation. However, glasses expand uniformly.

Silicate Glasses The silicate glasses are the most widely used. *Fused silica*, formed from pure SiO_2, has a high-melting point, and the dimensional changes during heating and cooling are small (Figure 15-6). Generally, however, the silicate glasses contain additional oxides (Table 15-5). While oxides such as silica behave as **glass formers**, an **intermediate** oxide (such as lead oxide or aluminum oxide) does not form a glass by itself but is incorporated into the network structure of the glass formers. A third group of oxides, the *modifiers*, break up the network structure and eventually cause the glass to devitrify, or crystallize.

TABLE 15-5 ■ ***Division of the oxides into glass formers, intermediates, and modifiers***

Glass Formers	Intermediates	Modifiers
B_2O_3	TiO_2	Y_2O_3
SiO_2	ZnO	MgO
GeO_2	PbO_2	CaO
P_2O_5	Al_2O_3	PbO
V_2O_3	BeO	Na_2O

Modified Silicate Glasses Modifiers break up the silica network if the oxygen-to-silicon ratio (O:Si) increases significantly. When Na_2O is added to silica glass, for example, the sodium ions enter holes within the network rather than becoming part of the network. However, the oxygen ion that enters with the Na_2O does become part of the network (Figure 15-7). When this happens, there are not enough silicon ions to combine with the extra oxygen ions and keep the network intact. Eventually, a high O:Si ratio causes the remaining silica tetrahedra to form chains, rings, or compounds, and the silica no longer transforms to a glass. When the O:Si ratio is above about 2.5, silica glasses are difficult to form; above a ratio of three, a glass forms only when special precautions are taken, such as the use of rapid cooling rates.

Modification also lowers the melting point and viscosity of silica, making it possible to produce glass at lower temperatures. The effect of Na_2O additions to silica is shown in Figure 15-8. As you can see, the addition of Na_2O produces eutectics with very low melting temperatures. Adding CaO, which reduces the solubility of the glass in water, further modifies these glasses. The example that follows shows how to design a glass.

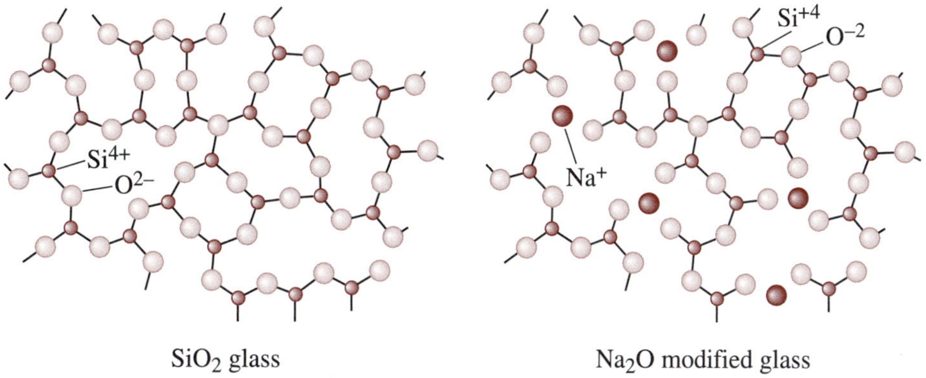

Figure 15-7 The effect of Na_2O on the silica glass network. Sodium oxide is a modifier, disrupting the glassy network and reducing the ability to form a glass.

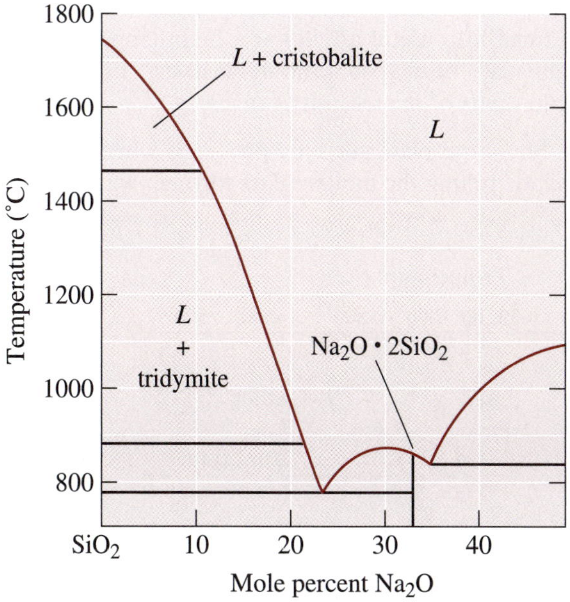

Figure 15-8 The SiO_2-Na_2O phase diagram. Additions of soda (Na_2O) to silica dramatically reduce the melting temperature of silica by forming eutectics.

EXAMPLE 15-2 *Design of a Glass*

We produce good chemical resistance in a glass when we introduce B_2O_3 into silica. To assure that we have good glass-forming tendencies, we wish the O:Si ratio to be no more than 2.5, but we also want the glassware to have a low-melting temperature to make the glass-forming process easier and more economical. Design such a glass.

SOLUTION

Because B_2O_3 reduces the melting temperature of silica, we would like to add as much as possible. We also, however, want to assure that the O:Si ratio is no more than 2.5, so the amount of B_2O_3 is limited. As an example, let us determine the amount of B_2O_3 we must add to obtain exactly an O:Si ratio of 2.5. Let f_B be the mole fraction of B_2O_3 added to the glass, and $(1 - f_B)$ be the mole fraction of SiO_2:

$$\frac{O}{Si} = \frac{\left(3\frac{\text{O ions}}{B_2O_3}\right)(f_B) + \left(2\frac{\text{O ions}}{SiO_2}\right)(1 - f_B)}{\left(1\frac{\text{Si ion}}{SiO_2}\right)(1 - f_B)} = 2.5$$

$$3f_B + 2 - 2f_B = 2.5 - 2.5f_B \quad \text{or} \quad f_B = 0.143$$

Therefore, we must produce a glass containing no more than 14.3 mol% B_2O_3. In weight percent:

$$\text{wt\% } B_2O_3 = \frac{(f_B)(69.62 \text{ g/mol})}{(f_B)(69.62 \text{ g/mol}) + (1 - f_B)(60.08 \text{ g/mol})} \times 100$$

$$\text{wt\% } B_2O_3 = \frac{(0.143)(69.62)}{(0.143)(69.62) + (0.857)(60.08)} \times 100 = 16.2$$

Glasses are manufactured into useful articles at a high temperature with viscosity controlled so that the glass can be shaped without breaking. Figure 15-9 helps us understand the processing in terms of the viscosity ranges.

1. *Liquid range*. Sheet and plate glass are produced when the glass is in the molten state. Techniques include rolling the molten glass through water-cooled rolls or floating

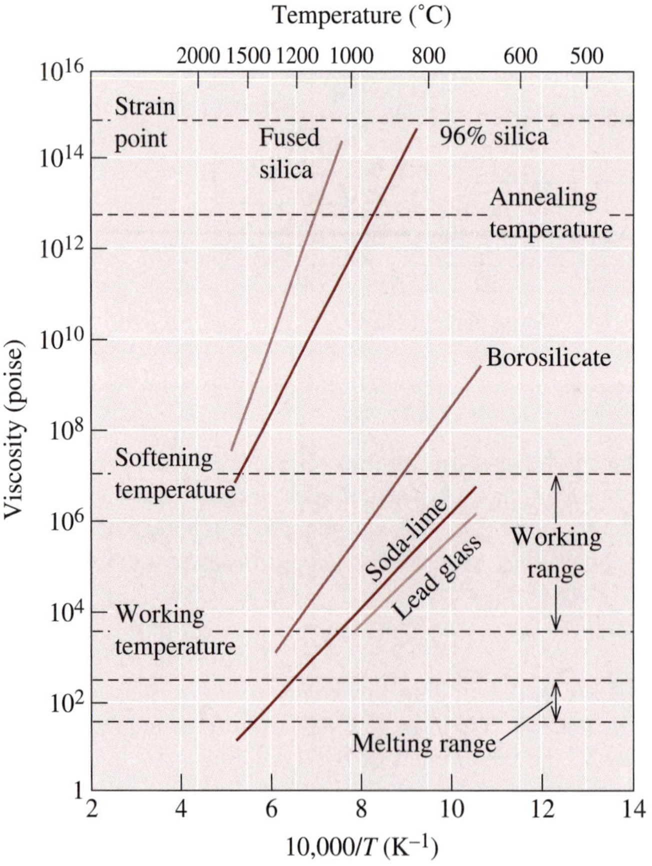

Figure 15-9 The effect of temperature and composition on the viscosity of glass.

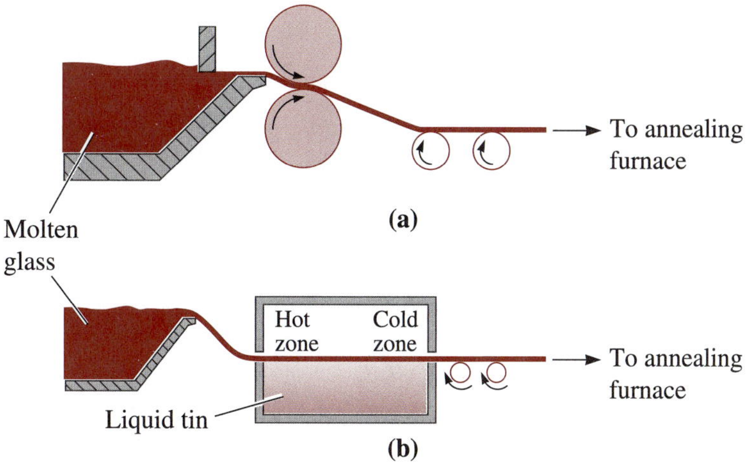

Figure 15-10 Techniques for manufacturing sheet and plate glass: (a) rolling and (b) floating the glass on molten tin.

the molten glass over a pool of liquid tin (Figure 15-10). This float-glass process produces an exceptionally smooth surface on the glass. The development of the float-glass process was a genuine breakthrough in the area of glass processing. Basic float-glass composition has been essentially unchanged for many years (Table 15-6). Glasses for automotive applications contain small amounts of iron oxide to control absorption of light and infrared radiation.

Some glass shapes, including large optical mirrors, are produced by casting the molten glass into a mold, then assuring that cooling is as slow as possible to minimize residual stresses and avoid cracking of the glass part. Glass fibers may be produced by drawing the liquid glass through small openings in a platinum die [Figure 15-11(c)]. Typically, many fibers are produced simultaneously for a single die.

2. *Working range*. Shapes such as those of containers or light bulbs can be formed by pressing, drawing, or blowing glass into molds (Figure 15-11). A hot *gob* of liquid

TABLE 15-6 ■ ***Compositions of typical glasses (in weight percent)***

Glass	SiO_2	Al_2O_3	CaO	Na_2O	B_2O_3	MgO	PbO	Others
Fused silica	99							
Vycor™	96				4			
Pyrex™	81	2		4	12			
Glass jars	74	1	5	15		4		
Window glass	72	1	10	14		2		
Plate glass/Float glass	73	1	13	13				
Light bulbs	74	1	5	16		4		
Fibers	54	14	16		10	4		
Thermometer	73	6		10	10			
Lead glass or crystal	67			6			17	10% K_2O
Optical flint	50			1			19	13% BaO, 8% K_2O, ZnO
Optical crown	70			8		10		2% BaO, 8% K_2O
E-glass fibers	55	15	20		10			
S-glass fibers	65	25				10		

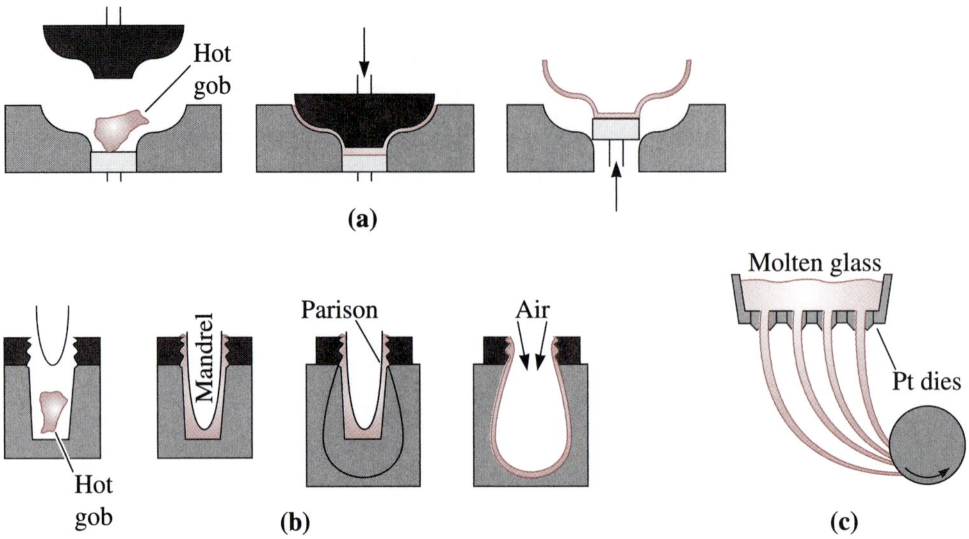

Figure 15-11 Techniques for forming glass products: (a) pressing, (b) press and blow process, and (c) drawing of fibers.

glass may be preformed into a crude shape (**parison**), then pressed or blown into a heated die to produce the final shape. The glass is heated to the working range so that the glass is formable, but not "runny."

3. *Annealing range*. Some glass parts may be annealed to reduce residual stresses introduced during forming. Large glass castings, for example, are often annealed and slowly cooled to prevent cracking. Some glasses may be heat treated to cause **devitrification**, or the precipitation of a crystalline phase from the glass.

Tempered glass is produced by quenching the surface of plate glass with air, causing the surface layers to cool and contract. When the center cools, its contraction is restrained by the already rigid surface, which is placed in compression (Figure 15-12). Tempered glass is capable of withstanding much higher tensile stresses and impact blows than untempered glass. Tempered glass is used in car and home windows, shelving for refrigerators, ovens, furniture, and many other applications where safety is important. **Laminated glass**, consisting of two annealed glass pieces with a polymer (such as polyvinyl butyral or PVB) in between, is shatterproof and used to make car windshields.

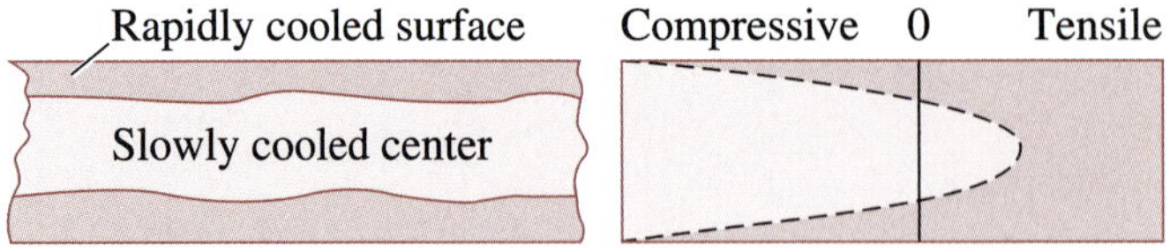

Figure 15-12 Tempered glass is cooled rapidly to produce compressive residual stresses at the surface.

Glass Compositions Pure SiO_2 must be heated to very high temperatures to obtain viscosities that permit economical forming. Most commercial glasses are based on silica (Table 15-6); modifiers such as soda (Na_2O) are added to break down the network structure and form eutectics with low melting temperatures, whereas lime (CaO) is added to reduce the solubility of the glass in water. The most common commercial glass contains approximately 75% SiO_2, 15% Na_2O, and 10% CaO and is known as soda-lime glass.

Borosilicate glasses, which contain about 15% B_2O_3, have excellent chemical and dimensional stability. Their uses include laboratory glassware (Pyrex^{TM}) and containers for the disposal of high-level radioactive nuclear waste. Calcium aluminoborosilicate glass—or E-glass—is used as a general-purpose fiber for composite materials, such as fiberglass. Aluminosilicate glass, with 20% Al_2O_3 and 12% MgO, and high-silica glasses, with 3% B_2O_3, are excellent for high-temperature resistance and for protection against heat or thermal shock. The S-glass, a magnesium aluminosilicate, is used to produce high-strength fibers for composite materials. Fused silica, or virtually pure SiO_2, has the best resistance to high temperature, thermal shock, and chemical attack, although it is also expensive.

Special optical qualities can also be obtained, including sensitivity to light. Photochromic glass, which is darkened by the ultraviolet portion of sunlight, is used for sunglasses. Photosensitive glass darkens permanently when exposed to ultraviolet light; if only selected portions of the glass are exposed and then immersed in hydrofluoric acid, etchings can be produced. Polychromatic glasses are sensitive to all light, not just ultraviolet radiation. Similarly, nanosized crystals of semiconductors such as cadmium sulfide (CdS) are nucleated in silicate glasses in a process known as *striking*. These glasses exhibit lively colors and also have useful optical properties.

15-6 Glass-Ceramics

Glass-ceramics are crystalline materials that are derived from amorphous glasses. Usually, glass-ceramics have a substantial level of crystallinity (~>70–99%). The formation of glass-ceramics was discovered serendipitously by Don Stookey. With glass-ceramics, we can take advantage of the formability of glass. Also a product that contains very low porosity can be obtained by producing a shape with conventional glass-forming techniques, such as pressing or blowing. Glass, however, has poor creep resistance. We then crystallize the glass using heterogeneous nucleation by such oxides as TiO_2 and/or ZrO_2. These oxides react with the glass and with each other and provide the nuclei that ultimately lead to glass crystallization. Phase separation of glasses plays an important role in formation of the nuclei. In some commercial glass-ceramics (e.g., Visionware^{TM}) nano-sized crystallites are formed and the resultant material remains optically transparent.

The first step in producing a glass-ceramic is to assure that crystallization does not occur during cooling from the forming temperature. A continuous and isothermal cooling transformation diagram, much like the CCT and TTT diagrams for steels, can be seen for silicate-based glasses. Figure 15-13(a) shows a TTT diagram for a glass. If glass cools too slowly, a transformation line is crossed; nucleation and growth of the crystals begin, but in an uncontrolled manner. Addition of modifying oxides to glass, much like addition of alloying elements to steel, shifts the transformation curve to longer times and prevents devitrification even at slow cooling rates. As noted in previous chapters, strictly speaking we should make use of CCT (and not TTT) diagrams for this discussion.

Nucleation of the crystalline phase is controlled in two ways. First, the glass contains agents, such as TiO_2, that react with other oxides and form phases that provide the nucleation sites. Second, a heat treatment is designed to provide the appropriate number of nuclei; the temperature should be relatively low in order to maximize the rate

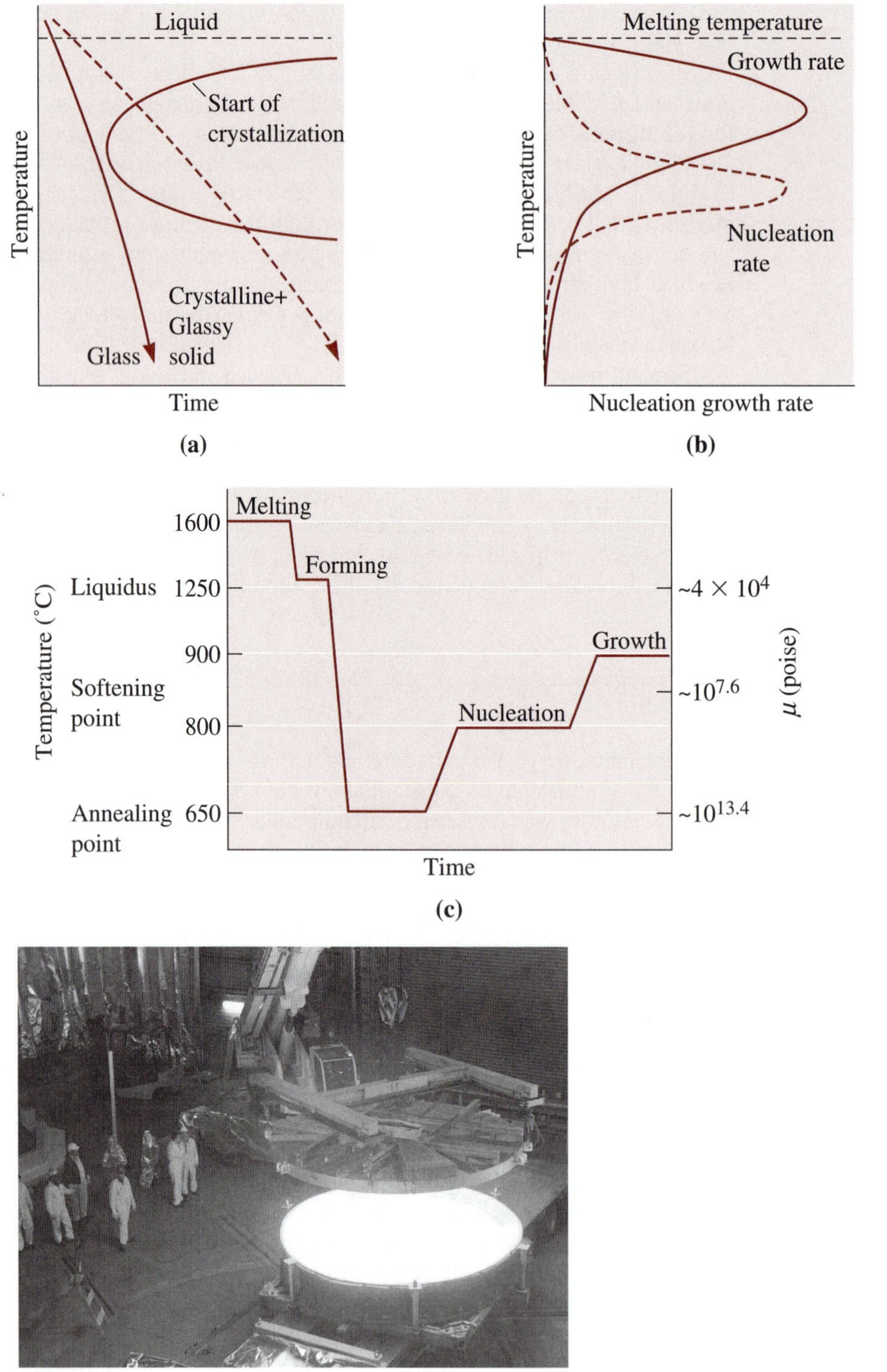

Figure 15-13 Producing a glass-ceramic: (a) Cooling must be rapid to avoid the start of crystallization. Isothermal and continuous cooling curves for lunar glass. (b) The rate of nucleation of precipitates is high at low temperatures, whereas the rate of growth of the precipitates is high at higher temperatures. (c) A typical heat-treatment profile for glass-ceramics fabrication, illustrated here for Li_2O-Al_2O_3-SiO_2 glasses. (d) A large mirror blank of the Zerodur™ made by Schott Glass. *(Source: Photo Courtesy of Schott Glass.)*

of nucleation [Figure 15-13(b)]. However, the overall rate of crystallization depends on the growth rate of the crystals once nucleation occurs; higher temperatures are required to maximize the growth rate. Consequently, a heat-treatment schedule similar to that shown in Figure 15-13(c) for the Li_2O-Al_2O_3-SiO_2 glass-ceramics (Pyroceram^{TM}) can be used. The low-temperature step provides nucleation sites, and the high-temperature step speeds the rate of growth of the crystals; as much as 99% of the part may crystallize.

This special structure of glass-ceramics can provide good mechanical strength and toughness, often with a low coefficient of thermal expansion and high-temperature corrosion resistance. Perhaps the most important glass-ceramic is based on the Li_2O-Al_2O_3-SiO_2 system. These materials are used for cooking utensils (Corning Ware^{TM}) and ceramic tops for stoves. Other glass-ceramics are used in communication, computer, and optical applications. Figure 15-13(d) shows the processing of a large telescope mirror being prepared using the Zerodur^{TM} glass-ceramic material. This glass-ceramic is especially suited for telescope mirror blanks because it has a very small coefficient of thermal expansion (Chapter 2).

15-7 Processing and Applications of Clay Products

Crystalline ceramics are often manufactured into useful articles by preparing a shape, or compact, composed of the raw materials in a fine powder form. The powders are then bonded by chemical reaction, partial or complete **vitrification** (melting), or sintering.

Clay products form a group of traditional ceramics used for producing pipe, brick, cooking ware, and other common products. Clay, such as kaolinite, and water serve as the initial binder for the ceramic powders, which are typically silica. Other materials, such as feldspar [$(K, Na)_2O \cdot Al_2O_3 \cdot 6SiO_2$] serve as fluxing (glass-forming) agents during later heat treatment.

Forming Techniques for Clay Products The powders, clay, **flux**, and water are mixed and formed into a shape (Figures 15-3 and 15-4). Dry or semi-dry mixtures are mechanically pressed into "green" (unbaked) shapes of sufficient strength to be handled. For more uniform compaction of complex shapes, isostatic pressing may be done; the powders are placed into a rubber mold and subjected to high pressures through a gas or liquid medium. Higher moisture contents permit the powders to be more plastic or formable. **Hydroplastic forming** processes, including extrusion, jiggering, and hand working, can be applied to these plastic mixes. Ceramic slurries containing large amounts of organic plasticizers, rather than water, can be injected into molds.

Still higher moisture contents permit the formation of a slip, or pourable slurry, containing fine ceramic powder. The slip is poured into a porous mold. The water in the slip nearest to the mold wall is drawn into the mold, leaving behind a soft solid which has a low-moisture content. When enough water has been drawn from the slip to produce a desired thickness of solid, the remaining liquid slip is poured from the mold, leaving behind a hollow shell (Figure 15-4). Slip casting is used in manufacturing washbasins and other commercial products. After forming, the ceramic bodies—or greenware—are still weak, contain water or other lubricants, and are porous, and subsequent drying and firing are required.

Drying and Firing of Clay Products During drying, excess moisture is removed and large dimensional changes occur. Initially, the water between the clay platelets—the interparticle water—evaporates and provides most of the shrinkage. Relatively little

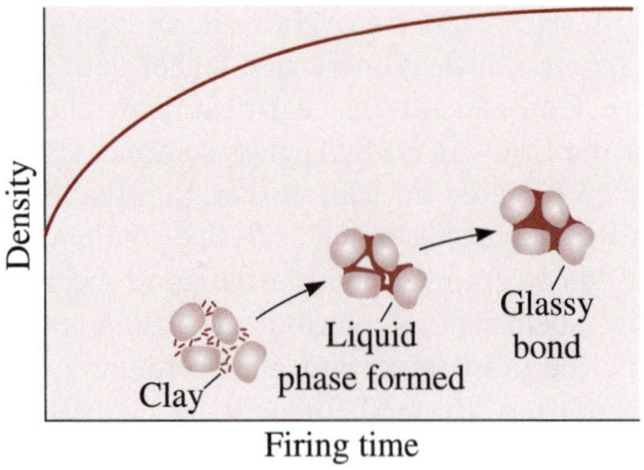

Figure 15-14
During firing, clay and other fluxing materials react with coarser particles to produce a glassy bond and reduce porosity.

dimensional change occurs as the remaining water between the pores evaporates. The temperature and humidity are carefully controlled to provide uniform drying throughout the part, thus minimizing stresses, distortion, and cracking.

The rigidity and strength of a ceramic part are obtained during **firing**. During heating, the clay dehydrates, eliminating the hydrated water that is part of the kaolinite crystal structure, and vitrification begins (Figure 15-14). Impurities and the fluxing agent react with the ceramic particles (SiO_2) and clay, producing a low-melting-point liquid phase at the grain surfaces. The liquid helps eliminate porosity and, after cooling, changes to a rigid glass that binds the ceramic particles. This glassy phase provides a **ceramic bond**, but it also causes additional shrinkage of the entire ceramic body.

The grain size of the final part is determined primarily by the size of the original powder particles. Furthermore, as the amount of flux increases, the melting temperature decreases; more glass forms, and the pores become rounder and smaller. A smaller initial grain size accelerates this process by providing more surface area at which vitrification can occur.

Applications of Clay Products Many structural clay products and whitewares are produced using these processes. Brick and tile used for construction are pressed or extruded into shape, dried, and fired to produce the ceramic bond. Higher firing temperatures or finer original particle sizes produce more vitrification, less porosity, and higher density. The higher density improves mechanical properties but reduces the insulating qualities of the brick or tile.

Earthenware are porous clay bodies fired at relatively low temperatures. Little vitrification occurs; the porosity is very high and interconnected, and earthenware ceramics may leak. Consequently, these products must be covered with an impermeable glaze.

Higher firing temperatures, which provide more vitrification and less porosity, produce stoneware. The stoneware, which is used for drainage and sewer pipe, contains only 2% to 4% porosity. Ceramics known as china and porcelain require even higher firing temperatures to cause complete vitrification and virtually no porosity.

15-8 Refractories

Refractory materials are important components of the equipment used in the production, refining, and handling of metals and glasses, for constructing heat-treating furnaces, and for other high-temperature processing equipment. The **refractories** must survive at high temperatures without being corroded or weakened by the surrounding environ-

TABLE 15-7 ■ *Compositions of typical refractories (weight percents)*

Refractory	SiO_2	Al_2O_3	MgO	Fe_2O_3	Cr_2O_3
Acidic					
Silica	95–97				
Superduty firebrick	51–53	43–44			
High-alumina firebrick	10–45	50–80			
Basic					
Magnesite			83–93	2–7	
Olivine	43		57		
Neutral					
Chromite	3–13	12–30	10–20	12–25	30–50
Chromite-magnesite	2–8	20–24	30–39	9–12	30–50

(From Ceramic Data Book, *Cahners Publishing Co., 1982.)*

ment. Typical refractories are composed of coarse oxide particles bonded by a finer refractory material. The finer material melts during firing, providing bonding. In some cases, refractory bricks contain about 20 to 25% apparent porosity to provide improved thermal insulation.

Refractories are often divided into three groups—acid, basic, and neutral—based on their chemical behavior (Table 15-7).

Acid Refractories Common acidic refractories include silica, alumina, and fireclay (an impure kaolinite). Pure silica is sometimes used to contain molten metal. In some applications, the silica may be bonded with small amounts of boron oxide, which melts and produces the ceramic bond. When a small amount of alumina is added to silica, the refractory contains a very low-melting-point eutectic microconstituent (Figure 15-15) and is not suited for refractory applications at temperatures above about 1600°C, a temperature often required for steel making. However, when larger amounts of alumina are added, the microstructure contains increasing amounts of mullite, $3Al_2O_3 \cdot 2SiO_2$, which has a high melting temperature. Thcsc fireclay refractories are generally relatively weak, but they are inexpensive. Alumina concentrations above about 50% constitute the high-alumina refractories.

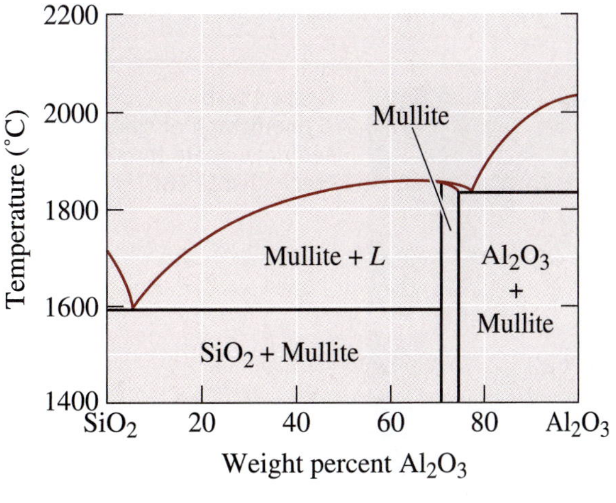

Figure 15-15
A simplified SiO_2-Al_2O_3 phase diagram, the basis for alumina silicate refractories.

Basic Refractories A number of refractories are based on MgO (magnesia, or periclase). Pure MgO has a high melting point, good refractoriness, and good resistance to attack by the basic environments often found in steel-making processes. Olivine refractories contain forsterite, or Mg_2SiO_4, and also have high melting points. Other magnesia refractories may include CaO or carbon. Typically, the basic refractories are more expensive than the acid refractories.

Neutral Refractories These refractories, which include chromite and chromite-magnesite, might be used to separate acid and basic refractories, preventing them from attacking one another.

Special Refractories Carbon, or graphite, is used in many refractory applications, particularly when oxygen is not present. Other refractory materials include zirconia (ZrO_2), zircon ($ZrO_2 \cdot SiO_2$), and a variety of nitrides, carbides, and borides. Most of the carbides, such as TiC and ZrC, do not resist oxidation well, and their high-temperature applications are best suited to reducing conditions. However, silicon carbide (SiC) is an exception; when SiC is oxidized at high temperatures, a thin layer of SiO_2 forms at the surface, protecting the SiC from further oxidation up to about 1500°C. Nitrides and borides also have high-melting temperatures and are less susceptible to oxidation. Some of the oxides and nitrides are candidates for use in jet engines.

15-9 Other Ceramic Materials

In addition to their use in producing construction materials, appliances, structural materials, and refractories, ceramics find a host of other applications, including the following.

Cements Ceramic raw materials are joined using a binder that does not require firing or sintering in a process called **cementation**. A chemical reaction converts a liquid resin to a solid that joins the particles. In the case of sodium silicate, the introduction of CO_2 gas acts as a catalyst to dehydrate the sodium silicate solution into a glassy material:

$$x\text{Na}_2\text{O} \cdot y\text{SiO}_2 \cdot \text{H}_2\text{O} + \text{CO}_2 \rightarrow \text{glass (not balanced)}$$

Figure 15-16 shows silica sand grains used to produce molds for metal casting. The liquid sodium silicate coats the sand grains and provides bridges between the sand grains. Introduction of the CO_2 converts the bridges to a solid, joining the sand grains.

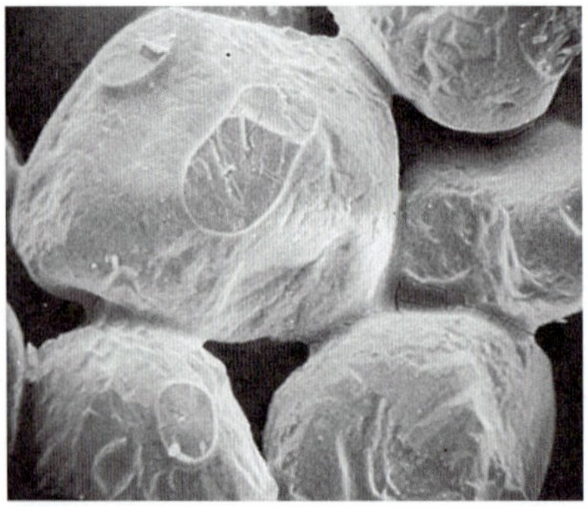

Figure 15-16
A photograph of silica sand grains bonded with sodium silicate through the cementation mechanism (×60).

Fine alumina-powder solutions catalyzed with phosphoric acid produce an aluminum phosphate cement:

$$Al_2O_3 + 2H_3PO_4 \rightarrow 2AlPO_4 + 3H_2O$$

When alumina particles are bonded with the aluminum phosphate cement, refractories capable of operating at temperatures as high as 1650°C are produced.

Plaster of paris, or gypsum, is another material that is hardened by a cementation reaction:

$$CaSO_4 \cdot \tfrac{1}{2}H_2O + \tfrac{3}{2}H_2O \rightarrow CaSO_4 \cdot 2H_2O$$

When the liquid slurry reacts, interlocking solid crystals of gypsum ($CaSO_4 \cdot 2H_2O$) grow, with very small pores between the crystals. Larger amounts of water in the original slurry provide more porosity, but they also decrease the strength of the final plaster. One of the important uses of this material is for construction of walls in buildings.

The most common and important of the cementation reactions occurs in Portland cement, which is used to produce concrete.

Coatings Ceramics are often used to provide protective coatings to other materials. Common commercial coatings include glazes and enamels. Glazes are applied to the surface of a ceramic material to seal a permeable clay body, to provide protection and decoration, or for special purposes. Enamels are applied to metal surfaces. The enamels and glazes are clay products that vitrify easily during firing. A common composition is $CaO \cdot Al_2O_3 \cdot 2SiO_2$.

Special colors can be produced in glazes and enamels by the addition of other minerals. Zirconium silicate gives a white glaze, cobalt oxide makes the glaze blue, chromium oxide produces green, lead oxide gives a yellow color, and a red glaze may be produced by adding a mixture of selenium and cadmium sulfides.

One of the problems encountered with a glaze or enamel is surface cracking, or crazing, which occurs when the glaze has a coefficient of thermal expansion different than that of the underlying material. This is frequently the most important factor in determining the composition of the coating.

Special coatings are used for advanced ceramics and high-service temperature metals. SiC coatings are applied to carbon-carbon composite materials to improve their oxidation resistance. Zirconia coatings are applied to nickel-based superalloys to provide thermal barriers that protect the metal from melting or adverse reactions.

Thin Films and Single Crystals Thin films of many complex and multi-component ceramics are produced using different techniques such as sputtering, sol-gel, and chemical-vapor deposition (CVD). Usually, the thickness of such films is less than 0.05 to 10 μm, and more likely greater than 2 μm. Many functional electronic ceramic thin films are prepared and integrated onto silicon wafers, glasses, and other substrates. For example, the magnetic strips on credit cards use iron oxide (γ-Fe_2O_3 or Fe_3O_4) thin films for storing data. Indium tin oxide (ITO), a conductive and transparent material, is coated on glass and used in applications such as touch-screen displays (Table 15-1). Many other coatings are used on glass to make the glass energy efficient. Recently, a self-cleaning glass using a TiO_2 coating has been developed. Similarly, thin films of ceramics, such as PZT, PLZT, and $BaTiO_3$-$SrTiO_3$ solid solutions, can be prepared and used. Often, the films develop an orientation, or texture, which may be advantageous for a given application. Single crystals of ceramics (e.g., SiO_2 or quartz, lithium niobate ($LiNbO_3$), sapphire, or yttrium aluminum garnate) are used in many electrical and electro-optical applications. These crystals are grown from melts using techniques similar to those described in Chapter 9.

Fibers Fibers are produced from ceramic materials for several uses: as a reinforcement in composite materials, for weaving into fabrics, or for use in fiber-optic systems. Borosilicate glass fibers, the most commonly produced fibers, provide strength and stiffness in fiberglass. Fibers can be produced from a variety of other ceramics, including alumina, silicon carbide, silica, and boron carbide. The sol-gel process is also used to produce commercial fibers for many applications.

A special type of fibrous material is the silica tile used to provide the thermal protection system for NASA's space shuttle. Silica fibers are bonded with colloidal silica to produce an exceptionally lightweight tile with densities as low as 0.144 g/cm^3; the tile is coated with special high-emissivity glazes to permit protection up to 1300°C.

Joining and Assembly of Ceramic Components Ceramics are often made as monolithic components rather than assemblies of numerous components. When two ceramic parts are placed in contact under a load, stress concentrations at the brittle surface lead to an increased probability of failure. In addition, methods for joining ceramic parts into a larger assembly are limited. The brittle ceramics cannot be joined by fusion welding or deformation bonding processes. At low temperatures, adhesive bonding using polymer materials may be accomplished; ceramic cements may be used at higher temperatures. Diffusion bonding and brazing can be used to join ceramics and to join ceramics to metals.

SUMMARY

- Ceramics are inorganic materials that have high hardness and high melting points. These include single crystal and polycrystalline ceramics, glasses, and glass-ceramics. Typical ceramics are electrical and thermal insulators with good chemical stability and good strength in compression.

- Polycrystalline ceramics typically exhibit a brittle behavior, partly because of porosity. Because most polycrystalline ceramics cannot plastically deform (unless special conditions with respect to temperature and strain rates are created), the porosity limits the ability of a ceramic material to withstand a tensile load.

- Ceramics play a critical role in a wide array of technologies related to electronic, magnetic, optical, and energy. Many advanced ceramics play a very important role in providing thermal insulation and high-temperature properties. Applications of advanced ceramics range from credit cards, tiles for the space shuttle, medical imaging, optical fibers that enable communication, and safe, energy-efficient glasses. Traditional ceramics play a very important role as refractories for metals processing and consumer applications.

- Ceramic-powder synthesis involves the treatment of ores using processes such as crushing and grinding, followed by ball milling, and calcination. These processes are inexpensive and widely used for making many ceramic powders.

- Ceramic processing is commonly conducted using compaction and sintering. For specialized applications, isostatic compaction, hot pressing, and hot isostatic pressing (HIP) are used, especially to achieve higher-densification levels.

- Tape casting, slip casting, extrusion, and injection molding are some of the other techniques used to form green ceramics into different shapes. These processes are then followed by a burnout step in which binders and plasticizers are burnt off, and the resultant ceramic is sintered.

- Many silicates and other ceramics form glasses rather easily, since the kinetics of crystallization are sluggish. Glasses can be formed in the form of sheets using float-glass or fibers and other shapes. Silicate glasses are used in a significant number of applications that include window glass, windshields, fiber optics, and fiberglass.
- Glass-ceramics are formed using controlled crystallization of inorganic glasses. These materials are used widely for kitchenware and many other applications.
- Ceramics, in the form of fibers, thin films, coatings, and single crystals, have many different applications.

GLOSSARY

Apparent porosity The percentage of a ceramic body that is composed of interconnected porosity.

Bulk density The mass of a ceramic body per unit volume, including closed and interconnected porosity.

Cementation Bonding ceramic raw materials into a useful product, using binders that form a glass or gel without firing at high temperatures.

Ceramic An inorganic material with high melting temperature. Usually hard and brittle.

Ceramic bond Bonding ceramic materials by permitting a glassy product to form at high firing temperatures.

Cermet A ceramic-metal composite (e.g., WC-Co) providing a good combination of hardness with other properties such as toughness.

Cold isostatic pressing (CIP) A powder-shaping technique in which hydrostatic pressure is applied during compaction. This is used for achieving a higher green ceramic density or compaction of more complex shapes.

Devitrification The crystallization of glass.

Enamel A ceramic coating on metal.

Firing Heating a ceramic body at a high temperature to cause a ceramic bond to form.

Flux Additions to ceramic raw materials that reduce the melting temperature.

Glass An amorphous material derived by cooling of a melt.

Glass-ceramics Polycrystalline ceramics formed initially in the glassy state and later crystallized during heat treatment to achieve improved strength and toughness.

Glass formers Oxides with a high bond strength that easily produce a glass during processing.

Glass temperature (T_g) The temperature below which an undercooled liquid becomes a glass. This is not a fixed temperature.

Glaze A ceramic coating applied to glass. The glaze contains glassy and crystalline ceramic phases.

Green ceramic A ceramic that has been shaped into a desired form but has not yet been sintered.

Hot isostatic pressing (HIP) A powder-processing technique in which large pieces of metals, alloys and ceramics can be produced using sintering under a hydrostatic pressure generated by a gas.

Hot pressing A processing technique in which sintering is conducted under uniaxial pressure.

Hydroplastic forming A number of processes by which a moist ceramic clay body is formed into a useful shape.

Injection molding A processing technique in which a thermoplastic mass (loaded with ceramic powder) is mixed in an extruder-like setup and then injected into a die to form complex parts. In the case of ceramics, the thermoplastic is burnt off.

Intermediates Oxides that, when added to a glass, help to extend the glassy network, although the oxides normally do not form a glass themselves.

Laminated glass Annealed glass with a polymer (e.g., polyvinyl butyral, *PVB*) sandwiched in between, used for car windshields.

Parison A crude glassy shape that serves as an intermediate step in the production of glassware. The parison is later formed into a finished product.

Powder metallurgy Powder processing routes used for converting metal and alloy powders into useful shapes.

Powder processing Unit operations conducted to convert powders into useful shapes (e.g., pressing, tape casting, etc.).

Reaction bonding A ceramic processing technique by which a shape is made using one material that is later converted into a ceramic material by reaction with a gas.

Refractories A group of ceramic materials capable of withstanding high temperatures for prolonged periods of time.

Slip A liquid slurry that is poured into a mold. When the slurry begins to harden at the mold surface, the remaining liquid slurry is decanted, leaving behind a hollow ceramic casting.

Slip casting Forming a hollow ceramic part by introducing a pourable slurry into a mold. The water in the slurry is extracted into the porous mold, leaving behind a drier surface. Excess slurry can then be decanted.

Spray drying A slurry of a ceramic powder is sprayed into a large chamber in the presence of hot air. This leads to the formation of soft agglomerates that can flow well into the dies used during powder compaction.

Synthesis Steps conducted to make a ceramic powder.

Tape casting A process for making thin sheets of ceramics using a ceramic slurry consisting of binders, plasticizers, etc. The slurry is cast with the help of a blade onto a plastic substrate. The resultant green tape is then dried, cut, and machined and used to make electronic ceramic and other devices.

Tempered glass A high-strength glass that has a surface layer where the stress is compressive, induced thermally during cooling or by the chemical diffusion of ions.

True porosity The percentage of a ceramic body that is composed of both closed and interconnected porosity.

Vitrification Melting, or formation of a glass.

PROBLEMS

Section 15-1 Applications of Ceramics

15-1 What are the primary types of atomic bonds in ceramics?

15-2 Explain the meaning of following terms: ceramics, inorganic glasses, and glass-ceramics.

15-3 Explain why ceramics typically are processed as powders. How is this similar to or different from the processing of metals?

15-4 What do the terms "glaze" and "enamel" mean?

15-5 What material is used to make the tiles that provide thermal protection in NASA's space shuttle?

15-6 Which ceramic materials are most widely used?

15-7 Explain how ceramic materials can be classified in different ways.

15-8 State any one application of the following ceramics: **(a)** alumina, **(b)** silica, **(c)** barium titanate, **(d)** zirconia, **(e)** boron carbide, and **(f)** diamond.

Section 15-2 Properties of Ceramics

15-9 What are some of the typical characteristics of ceramic materials?

15-10 Why is the tensile strength of ceramics much lower than their compressive strength?

15-11 Plastic deformation due to dislocation motion is important in metals; however, this is not a very important consideration for the properties of ceramics and glasses. Explain.

15-12 Can ceramic materials show superplastic behavior or are they always brittle? Explain.

15-13 Explain why the strength of ceramics tends to show a wide scatter in their mechanical properties.

Section 15-3 Synthesis and Processing of Ceramic Powders

15-14 What is the driving force to sintering?

15-15 What is the driving force to grain growth?

15-16 What mechanisms of diffusion play the most important role in the solid-state sintering of ceramics?

15-17 Explain the use of the following processes (use a sketch as needed): **(a)** uniaxial compaction and sintering, **(b)** hot pressing, **(c)** HIP, and **(d)** tape casting.

Section 15-4 Characteristics of Sintered Ceramics

15-18 What are some of the important characteristics of sintered ceramics?

15-19 What typical density levels are obtained in sintered ceramics?

15-20 What do the terms "apparent porosity" and "true porosity" of ceramics mean?

15-21 The specific gravity of Al_2O_3 is 3.96 g/cm^3. A ceramic part is produced by sintering alumina powder. It weighs 80 g when dry, 92 g after it has soaked in water, and 58 g when suspended in water. Calculate the apparent porosity, the true porosity, and the closed pores.

15-22 Silicon carbide (SiC) has a specific gravity of 3.1 g/cm^3. A sintered SiC part is produced, occupying a volume of 500 cm^3 and weighing 1200 g. After soaking in water, the part weighs 1250 g. Calculate the bulk density, the true porosity, and the volume fraction of the total porosity that consists of closed pores.

Section 15-5 Inorganic Glasses

15-23 What is the main reason why glass formation is easy in silicate systems?

15-24 Can glasses be formed using metallic materials?

15-25 Define the terms "glass formers", "intermediates", and "modifiers".

15-26 What does the term "glass temperature" mean? Is this a fixed temperature for a given composition of glass?

15-27 How many grams of BaO can be added to 1 kg of SiO_2 before the O:Si ratio exceeds 2.5 and glass-forming tendencies are poor? Compare this with the case when Li_2O is added to SiO_2.

15-28 Calculate the O:Si ratio when 30 wt% Y_2O_3 is added to SiO_2. Will this material provide good glass-forming tendencies?

Section 15-6 Glass-Ceramics

15-29 How is a glass-ceramic different from a glass and a ceramic?

15-30 What are the advantages of using glass-ceramics as compared to either glasses or ceramics?

15-31 Draw a typical heat-treatment profile encountered in processing of glass-ceramics.

15-32 What are some of the important applications of glass-ceramics?

16

Polymers

Have You Ever Wondered?

- *What are compact disks (CDs) made from?*
- *What is Silly Putty® made from?*
- *What polymer is used in chewing gum?*
- *Which was the first synthetic fiber ever made?*
- *Why are some plastics "dishwasher safe" and some not?*
- *What are bulletproof vests made from?*
- *What polymer is used for non-stick cookware?*

The word *mer* means a "unit." In this context, the term mer refers to a unit group of atoms or molecules that defines a characteristic arrangement for a polymer. A polymer can be thought of as a material made by combining several mers or units. *Polymers* are materials consisting of giant or macromolecules, chain-like molecules having average molecular weights from 10,000 to more than 1,000,000 g/mol. They are built by joining many mers or units through chemical bonding. Molecular weight is defined as the sum of atomic masses in each molecule. Most polymers, solids or liquids, are carbon-based; however, they can be inorganic (e.g., silicones based on a Si-O network). **Plastics** are materials that are composed principally of polymers containing additives such as glass fibers, fillers, pigments, and the like that further enhance their properties. Plastics include thermoplastics (commodity and engineering), thermoset materials, and elastomers (natural or synthetic). In this book, we use the terms plastics and polymers interchangably. *Polymerization* is the process by which small molecules consisting of one unit (known as a **monomer**) or a few units (known as **oligomers**) are chemically joined to create these giant molecules. Polymerization normally begins with the production of long chains in which the atoms are strongly joined by covalent bonding. Plastics are used in an amazing number of applications including clothing, toys, home appliances, structural and decorative items, coatings, paints, adhesives, automobile tires, biomedical devices, car bumpers and interiors, foams, and packaging. Polymers are often used in composites, both as fibers and as a matrix. Liquid crystal displays (LCDs) are based on polymers. We also use polymers to make eyeglasses with photochromic lenses. Plastics are often used to make electronic components because of their insulating ability and low dielectric constant. More recently, significant developments have occurred in the area of flexible electronic devices based on the useful piezoelectricity, semiconductivity, optical and electro-optical properties seen in some polymers. Polymers such as polyvinyl acetate (PVA) are water-soluble. Many such polymers can be dissolved in water or organic solvents to be used as binders, surfactants, or plasticizers in processing ceramics and semiconductors, and as additives to many consumer products. Polyvinyl butyral (PVB), a polymer, makes up part of the laminated glass used for car windshields (Chapter 15). Polymers are probably used in more technologies than any other class of materials.

Commercial—or standard commodity—polymers are lightweight, corrosion-resistant materials with low strength and stiffness, and they are not suitable for use at high temperatures. These polymers are, however, relatively inexpensive and are readily formed into a variety of shapes, ranging from plastic bags to mechanical gears to bathtubs. *Engineering polymers* are designed to give improved strength or better performance at elevated temperatures. These materials are produced in relatively small quantities and often are expensive. Some of the engineering polymers can perform at temperatures as high as 350°C; others—usually in a fiber form—have strengths that are greater than that of steel.

Polymers also have many useful physical properties. Some polymers such as acrylics like Plexiglas™ and Lucite™ are transparent and can substitute for glasses. Although most polymers are electrical insulators, special polymers (such as the acetals) and polymer-based composites possess useful electrical conductivity. Teflon™ has a low coefficient of friction and is the coating for nonstick cookware. Polymers also resist corrosion and chemical attack.

16-1 Classification of Polymers

Polymers are classified in several ways: by how the molecules are synthesized, by their molecular structure, or by their chemical family. One way to classify polymers is to state if the polymer is a **linear polymer** or a **branched polymer** (Figure 16-1). A linear polymer consists of spaghetti-like molecular chains. In a branched polymer, there are primary polymer chains and secondary offshoots of smaller chains that stem from these main chains. Note that even though we say "linear", the chains are actually not in the form of straight lines. A better method to describe polymers is in terms of their mechanical and thermal behavior. Table 16-1 compares the three major polymer categories.

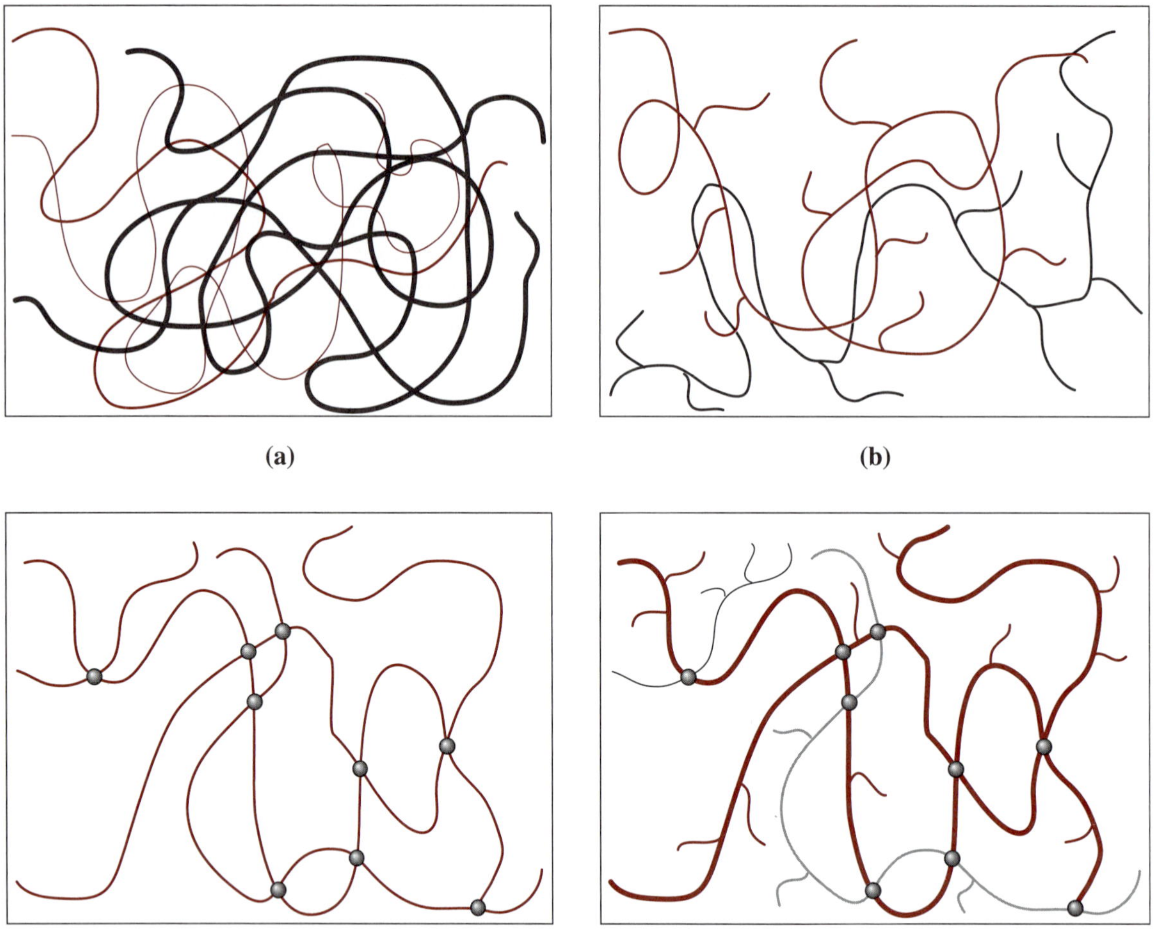

(a) (b) (c) (d)

Figure 16-1 Schematic showing linear and branched polymers. Note that branching can occur in any type of polymer (e.g., thermoplastics, thermosets, and elastomers). (a) Linear unbranched polymer: notice chains are not straight lines and not connected. Different polymer chains are shown using different shades and design to show clearly that each chain is not connected to another. (b) Linear branched polymer: chains are not connected, however they have branches. (c) Thermoset polymer without branching: chains are connected to one another by covalent bonds but they do not have branches. Joining points are highlighted with solid circles. (d) Thermoset polymer that has branches and chains that are interconnected via covalent bonds. Different chains and branches are shown in different shades for better contrast. Places where chains are actually chemically bonded are shown with filled circles.

TABLE 16-1 ■ ***Comparison of the three polymer categories***

Behavior	General Structure	Example
Thermoplastic	Flexible linear chains (straight or branched)	Polyethylene
Thermosetting	Rigid three-dimensional network (chains may be linear or branched)	Polyurethanes
Elastomers	Thermoplastics or lightly cross-linked thermosets, consist of spring-like molecules	Natural rubber

Thermoplastics are composed of long chains produced by joining together monomers; they typically behave in a ductile manner. The chains may or may not have branches. Individual chains are intertwined. There are relatively weak van der Waals bonds between atoms of different chains. This is somewhat similar to a few trees that are tangled up together. The trees may or may not have branches, each tree is on its own and not connected to another. The chains in thermoplastics can be untangled by application of a tensile stress. Thermoplastics can be amorphous or crystalline. Upon heating, thermoplastics soften and melt. They are processed into shapes by heating to elevated temperatures. Thermoplastics are easily recycled.

Thermosetting polymers are composed of long chains (linear or branched) of molecules that are strongly cross-linked to one another to form three-dimensional network structures. Network or thermosetting polymers are like a bunch of strings that are knotted to one another in several places and not just tangled up. Each string may have other side strings attached to it. Thermosets are generally stronger, but more brittle, than thermoplastics. Thermosets do not melt upon heating but begin to decompose. They cannot easily be reprocessed after the cross-linking reaction has occurred and hence recycling is difficult.

Elastomers These are known as rubbers. They have an elastic deformation >200%. These may be thermoplastics or lightly cross-linked thermosets. The polymer chains consist of coil-like molecules that can reversibly stretch by applying a force.

Thermoplastic elastomers are a special group of polymers. They have the processing ease of thermoplastics and the elastic behavior of elastomers.

Representative Structures Figure 16-2 shows three ways we could represent a segment of polyethylene, the simplest of the thermoplastics. The polymer chain consists of a

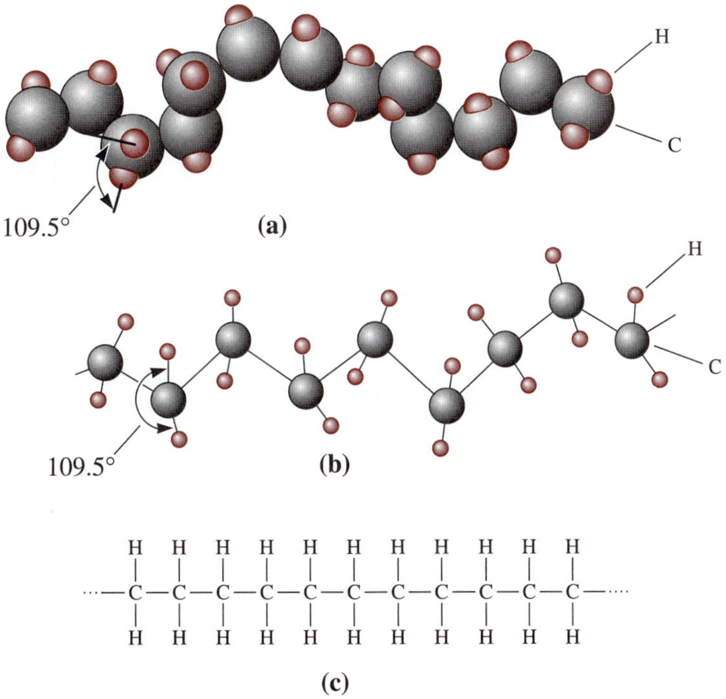

Figure 16-2 Three ways to represent the structure of polyethylene: (a) a solid three-dimensional model, (b) a three-dimensional "space" model, and (c) a simple two-dimensional model.

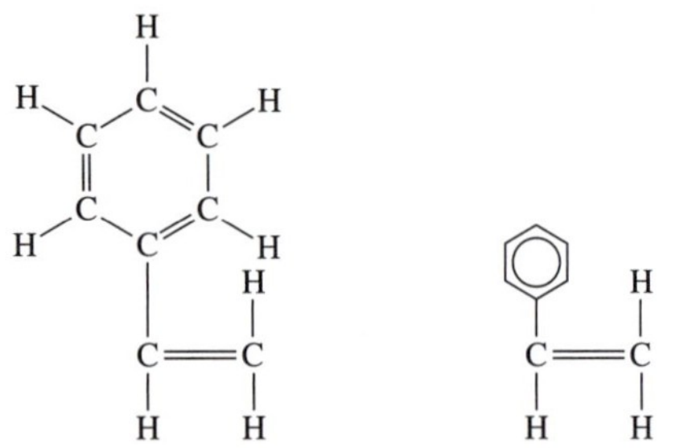

Figure 16-3
Two ways to represent the benzene ring. In this case, the benzene ring is shown attached to a pair of carbon atoms, producing styrene.

backbone of carbon atoms; two hydrogen atoms are bonded to each carbon atom in the chain. The chain twists and turns throughout space. In Figure 16-2, polyethylene has no branches and hence is a linear thermoplastic. The simple two-dimensional model in Figure 16-2(c) includes the essential elements of the polymer structure and will be used to describe the various polymers. The single lines (—) between the carbon atoms and between the carbon and hydrogen atoms represent a single covalent bond. Two parallel lines (=) represent a double covalent bond between atoms. A number of polymers include ring structures, such as the benzene ring found in polystyrene and other polymers (Figure 16-3).

In the structure shown in Figure 16-2(c), if we replace one of the hydrogen atoms in CH_2 with a methyl group (CH_3), a benzene ring, or chlorine, we get the structures of polypropylene, polystyrene, and polyvinyl chloride (PVC), respectively. If we replaced all H in CH_2 groups with fluorine (F), we get the structure of polytetrafluoroethylene or Teflon™. Like many other discoveries Teflon™ was also discovered by accident. Many polymer structures can thus be derived from the structure of polyethylene. The following example shows how different types of polymers are used.

EXAMPLE 16-1 *Design/Materials Selection for Polymer Components*

Design the type of polymer material you might select for the following applications: a surgeon's glove, a beverage container and a pulley.

SOLUTION

The glove must be capable of stretching a great deal in order to slip onto the surgeon's hand, yet it must conform tightly to the hand to permit the maximum sensation of touch during surgery. A material that undergoes a large amount of elastic strain—particularly with relatively little applied stress—might be appropriate; this requirement describes an elastomer.

The beverage container should be easily and economically produced. It should have some ductility and toughness so that it does not accidentally shatter and leak the contents. If the beverage is carbonated, diffusion of CO_2 is a major concern (Chapter 5). A thermoplastic such as polyethylene terephthalate (PET) will have the necessary formability and ductility needed for this application.

The pulley will be subjected to some stress and wear as a belt passes over it. A relatively strong, rigid, hard material is required to prevent wear, so a thermosetting polymer might be most appropriate. It may also be possible to use a material like nylon.

16-2 Addition and Condensation Polymerization

Addition polymerization and **condensation polymerization** are the two main ways to conduct "polymerization" (creating a polymer). The polymers derived from these processes are known as addition and condensation polymers, respectively. The formation of the most common polymer, polyethylene (PE), from ethylene molecules is an example of addition or chain-growth polymerization. Ethylene, a gas, is the monomer (single unit) and has the formula C_2H_4. The two carbon atoms are joined by a double covalent bond. Each carbon atom shares two of its electrons with the second carbon atom, and two hydrogen atoms are bonded to each of the carbon atoms (Figure 16-4).

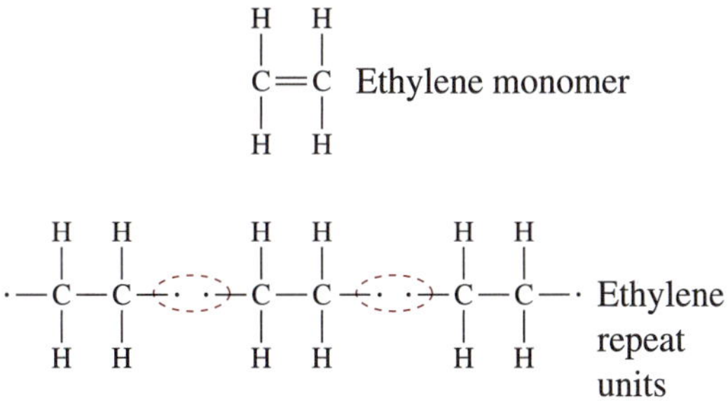

Figure 16-4
The addition reaction for producing polyethylene from ethylene molecules. The unsaturated double bond in the monomer is broken to produce active sites, which then attract additional repeat units to either end to produce a chain.

In presence of an appropriate combination of heat, pressure, and catalysts, the double bond between the carbon atoms is broken and replaced with a single covalent bond. The double bond is an **unsaturated bond**. After changing to a single bond, the carbon atoms are still joined, but they become active; other **repeat units** or mers can be added to produce the polymer chain.

We need polymers that have a designated average molecular weight and a molecular weight distribution. Thus, the polymerization reactions must have an "off" switch as well! The polymerization of chains may be terminated by two mechanisms. First, the ends of two growing chains may be joined. This process, called *combination*, creates a single large chain from two smaller chains. Second, the active end of one chain may remove a hydrogen atom from a second chain by a process known as *disproportionation*. This reaction terminates two chains, rather than combining two chains into one larger chain. Sometimes, compounds known as terminators are added to end the polymerization reactions. In general, for thermoplastics, the higher the average molecular weight the higher will be the melting temperature and the higher will be the Young's modulus of the polymer (Section 16-3). The following example illustrates an addition polymerization reaction.

EXAMPLE 16-2 Calculation of Initiator Required

Calculate the amount of benzoyl peroxide (BPO) initiator required to produce 1 kg of polyethylene with an average molecular weight of 200,000 g/mol. What is the degree of polymerization? Assume that 20% of the initiator is actually effective and that all termination occurs by the combination mechanism.

SOLUTION

For 100% efficiency, we need one molecule of benzoyl peroxide for each polyethylene chain. One of the free radicals would initiate one chain, the other free radical a second chain; then, the two chains combine into one larger one. The molecular weight of ethylene is $= (2\ \text{C})(12) + (4\ \text{H})(1) = 28$ g/mol. Therefore, the degree of polymerization is:

$$\frac{200{,}000\ \text{g/mol}}{28\ \text{g/mol}} = 7143 \text{ ethylene molecules per average chain}$$

$$\frac{(1000\ \text{g polyethylene})(6.02 \times 10^{23}\ \text{monomers/mol})}{28\ \text{g/mol}} = 215 \times 10^{23} \text{ monomers}$$

The combination mechanism requires the number of benzoyl peroxide molecules to be:

$$\frac{215 \times 10^{23} \text{ ethylene molecules}}{7143 \text{ ethylenes/chain}} = 0.03 \times 10^{23} \text{ chains}$$

The molecular weight of benzoyl peroxide is $(14\ \text{C})(12) + (10\ \text{H})(1) + (4\ \text{O})(16) = 242$ g/mol. Therefore, the amount of initiator needed to form the ends of the chains is:

$$\frac{(0.03 \times 10^{23})(242\ \text{g/mol})}{6.02 \times 10^{23}} = 1.206\ \text{g}$$

However, only 20% of the initiator actually is effective; the rest recombines or combines with other molecules and does not cause initiation of a chain. Thus, we need five times this amount, or 6.03 g of benzoyl peroxide per kilogram of polyethylene.

Condensation Polymerization Polymer chains can also form by condensation reactions, or *step-growth* polymerization, producing structures and properties that resemble those of addition polymers. In condensation polymerization, a relatively small molecule (such as water, ethanol, methanol etc.) is formed as a result of the polymerization reaction. This mechanism may often involve different monomers as starting or precursor molecules. The polymerization of dimethyl terephthalate and ethylene glycol (also used as radiator coolant) to produce *polyester* is an important example (Figure 16-5).

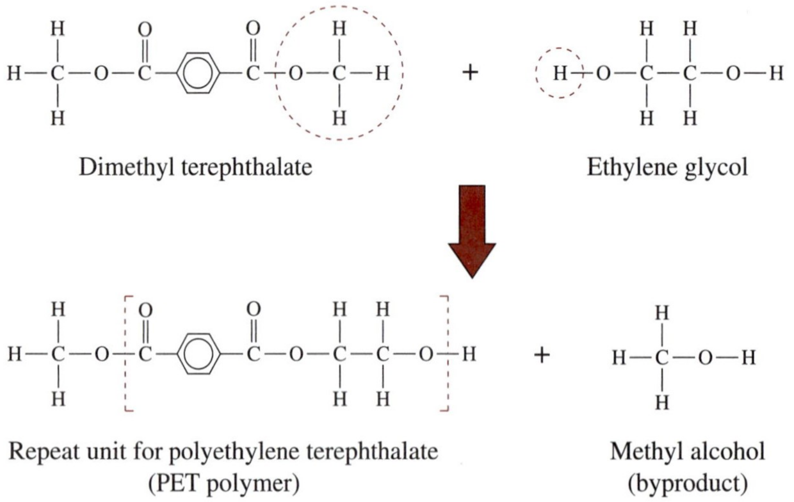

Figure 16-5 The condensation reaction for polyethylene terephthalate (PET), a common polyester. The OCH_3 group and a hydrogen atom are removed from the monomers, permitting the two monomers to join and producing methyl alcohol as a byproduct.

During polymerization, a hydrogen atom on the end of the ethylene glycol monomer combines with an OCH_3 (methoxy) group from the dimethyl terephthalate. A byproduct, methyl alcohol (CH_3OH), is "condensed" off and the two monomers combine to produce a larger molecule. Each of the monomers in this example are bifunctional, and the condensation polymerization can continue by the same reaction. Eventually, a long polymer chain—a polyester—is produced. The repeat unit for this polyester consists of two original monomers: ethylene glycol and dimethyl terephthalate. The length of the polymer chain depends on the ease with which the monomers can diffuse to the ends and undergo the condensation reaction. Chain growth ceases when no more monomers reach the end of the chain to continue the reaction. Condensation polymerization reactions also occur in sol-gel processing of ceramic materials.

The following example describes the discovery of nylon and calculations related to condensation polymerization.

EXAMPLE 16-3 *Condensation Polymerization of 6,6-Nylon*

Nylon was first reported by Wallace Hume Carothers, of du Pont in about 1934. In 1939, du Pont's Charles Stine reported the discovery of this first synthetic fiber to a group of 3000 women gathered for the New York World's Fair. The first application was nylon stockings that were strong. Today nylon is used in hundreds of applications. Prior to nylon, Carothers had discovered neoprene (an elastomer).

The linear polymer 6,6-nylon is to be produced by combining 1000 g of hexamethylene diamine (HMDA) with adipic acid. A condensation reaction then produces the polymer. Show how this reaction occurs and determine the byproduct that forms. How many grams of adipic acid are needed, and how much 6,6-nylon is produced, assuming 100% efficiency?

SOLUTION

The molecular structures of the monomers are shown below. The linear nylon chain is produced when a hydrogen atom from the hexamethylene diamine combines with an OH group from adipic acid to form a water molecule.

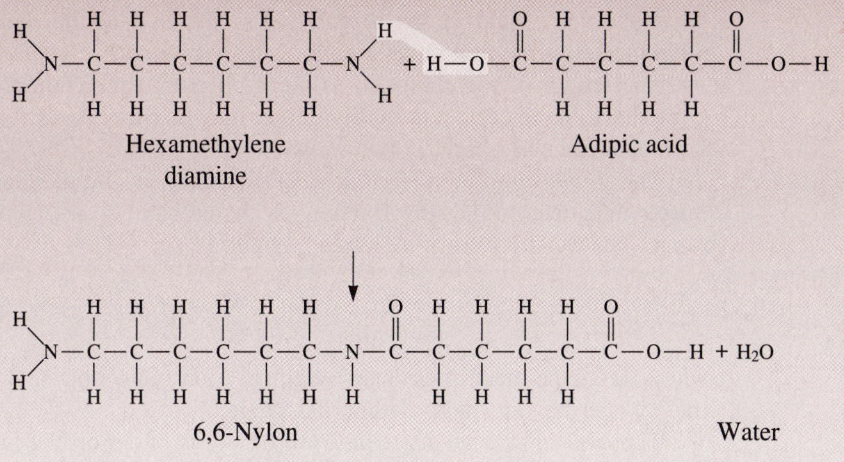

Note that the reaction can continue at both ends of the new molecule; consequently, long chains may form. This polymer is called 6,6-nylon because both monomers contain six carbon atoms.

We can determine that the molecular weight of hexamethylene diamine (HMDA) is 116 g/mol, of adipic acid is 146 g/mol, and of water is 18 g/mol. The number of moles of hexamethylene diamine added (calculated below) is equal to the number of moles of adipic acid:

$$\frac{1000\text{ g}}{116\text{ g/mol}} = 8.621\text{ moles} = \frac{x\text{ g}}{146\text{ g/mol}}$$

$$x = 1259\text{ g of adipic acid required}$$

The number of moles of water lost is also 8.621:

$$y = (8.621\text{ moles})(18\text{ g/mol}) = 155.2\text{ g }H_2O$$

But each time one more monomer is attached, another H_2O is released. Therefore, the total amount of nylon produced is

$$1000\text{ g} + 1259\text{ g} - 2(155.2\text{ g}) = 1948.6\text{ g}.$$

16-3 Degree of Polymerization

Polymers, unlike organic or inorganic compounds, do not have a fixed molecular weight. For example, polyethylene may have a molecular weight that ranges from ~25,000 to 6 million! The average length of a linear polymer is represented by the **degree of polymerization**, or the number of repeat units in the chain. The degree of polymerization can also be defined as:

$$\text{Degree of polymerization} = \frac{\text{average molecular weight of polymer}}{\text{molecular weight of repeat unit}} \quad (16\text{-}1)$$

If the polymer contains only one type of monomer, the molecular weight of the repeat unit is that of the monomer. If the polymer contains more than one type of monomer, the molecular weight of the repeat unit is the sum of the molecular weights of the monomers, less the molecular weight of the byproduct.

The length of the chains in a linear polymer varies considerably. Some may be quite short due to early termination; other may be exceptionally long. We can define an average molecular weight in two ways.

The *weight average molecular weight* is obtained by dividing the chains into size ranges and determining the fraction of chains having molecular weights within that range. The weight average molecular weight $\overline{M}_w$ is

$$\overline{M}_w = \sum f_i M_i \quad (16\text{-}2)$$

where M_i is the mean molecular weight of each range and f_i is the weight fraction of the polymer having chains within that range.

The *number average molecular weight* $\overline{M}_n$ is based on the number fraction, rather than the weight fraction, of the chains within each size range. It is always smaller than the weight average molecular weight and is given by

$$\overline{M}_n = \sum x_i M_i \quad (16\text{-}3)$$

where M_i is again the mean molecular weight of each size range, but x_i is the fraction of the total number of chains within each range. Either $\overline{M}_x$ or $\overline{M}_n$ can be used to calculate the degree of polymerization.

The two following examples illustrate these concepts.

EXAMPLE 16-4 Degree of Polymerization for 6,6-Nylon

Calculate the degree of polymerization if 6,6-nylon has a molecular weight of 120,000 g/mol.

SOLUTION

The reaction by which 6,6-nylon is produced was described in Example 16-3. Hexamethylene diamine and adipic acid combine and release a molecule of water. When a long chain forms, there is, on an average, one water molecule released for each reacting molecule. The molecular weights are 116 g/mol for hexamethylene diamine, 146 g/mol for adipic acid, and 18 g/mol for water. The repeat unit for 6,6-nylon is:

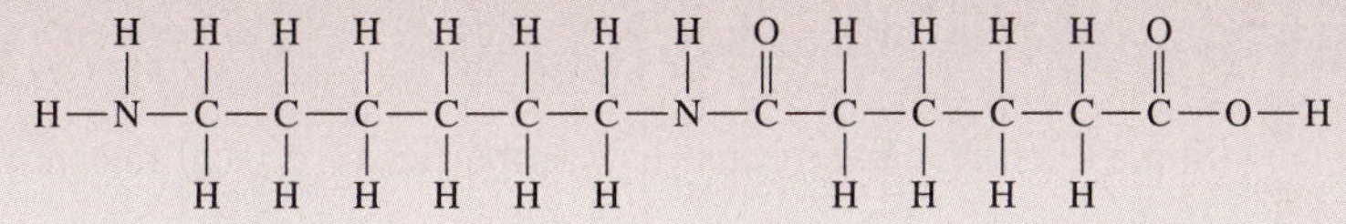

The molecular weight of the repeat unit is the sum of the molecular weights of the two monomers, minus that of the two water molecules that are evolved:

$$M_{\text{repeat unit}} = 116 + 146 - 2(18) = 226 \text{ g/mol}$$

$$\text{Degree of polymerization} = \frac{120{,}000}{226} = 531$$

The degree of polymerization refers to the total number of repeat units in the chain. The chain contains 531 hexamethylene diamine and 531 adipic acid molecules.

EXAMPLE 16-5 Number and Weight Average Molecular Weights

We have a polyethylene sample containing 4000 chains with molecular weights between 0 and 5000 g/mol, 8000 chains with molecular weights between 5000 and 10,000 g/mol, 7000 chains with molecular weights between 10,000 and 15,000 g/mol, and 2000 chains with molecular weights between 15,000 and 20,000 g/mol. Determine both the number and weight average molecular weights.

SOLUTION

First we need to determine the number fraction x_i and weight fraction f_i for each of the four ranges. For calculating x_i, we simply divide the number in each range by 21,000, the total number of chains. To find f_i, we first multiply the number of chains by the mean molecular weight of the chains in each range, giving the "weight" of each group, then find f_i by dividing by the total weight of 192.5×10^6. We can then use Equations 16-2 and 16-3 to find the molecular weights.

Number of Chains	Mean M per Chain	x_i	x_iM_i	Weight	f_i	f_iM_i
4000	2500	0.191	477.5	10×10^6	0.0519	129.75
8000	7500	0.381	2857.5	60×10^6	0.3118	2338.50
7000	12,500	0.333	4162.5	87.5×10^6	0.4545	5681.25
2000	17,500	0.095	1662.5	35×10^6	0.1818	3181.50
$\sum = 21{,}000$		$\sum = 1.00$	$\sum = 9160$	$\sum = 192.5 \times 10^6$	$\sum = 1$	$\sum = 11{,}331$

$$\overline{M}_n = \sum x_iM_i = 9160 \text{ g/mol}$$

$$\overline{M}_w = \sum f_iM_i = 11{,}331 \text{ g/mol}$$

The weight average molecular weight is larger than the number average molecular weight.

16-4 Typical Thermoplastics

Some of the typical mechanical properties of typical thermoplastics are shown in Table 16-2.

TABLE 16-2 ■ ***Properties of selected thermoplastics***

	Tensile Strength (MPa)	% Elongation	Elastic Modulus (MPa)	Density (g/cm^3)	Izod Impact (J/cm)
Polyethylene (PE):					
Low-density	21	800	276	0.92	4.9
High-density	38	130	1241	0.96	2.2
Ultrahigh molecular weight	48	350	690	0.934	16.2
Polyvinyl chloride (PVC)	62	100	4140	1.40	
Polypropylene (PP)	41	700	1517	0.90	0.5
Polystyrene (PS)	55	60	3103	1.06	0.2
Polyacrylonitrile (PAN)	62	4	4000	1.15	2.6
Polymethyl methacrylate (PMMA) (acrylic, Plexiglas)	83	5	3100	1.22	0.3
Polychlorotrifluoroethylene	41	250	2070	2.15	1.4
Polytetrafluoroethylene (PTFE, Teflon)	48	400	550	2.17	1.6
Polyoxymethylene (POM) (acetal)	83	75	3590	1.42	1.2
Polyamide (PA) (nylon)	83	300	3450	1.14	1.1
Polyester (PET)	72	300	4140	1.36	0.3
Polycarbonate (PC)	76	130	2760	1.20	8.6
Polyimide (PI)	117	10	2070	1.39	0.8
Polyetheretherketone (PEEK)	70	150	3790	1.31	0.9
Polyphenylene sulfide (PPS)	66	2	3310	1.30	0.3
Polyether sulfone (PES)	84	80	2410	1.37	0.9
Polyamide-imide (PAI)	186	15	5030	1.39	2.2

Because bonding within the chains is stronger, rotation and sliding of the chains is more difficult, leading to higher strengths, higher stiffnesses, and higher melting points than the simpler addition polymers (Table 16-3). In some cases, good impact properties can be gained from these complex chains, with polycarbonates being particularly remarkable. Polycarbonates (Lexan™, Merlon™, and Sparlux™) are used to make bulletproof windows, compact disks for data storage, and in many other applications.

TABLE 16-3 ■ ***Repeat units and applications for selected addition thermoplastics***

Polymer	Repeat Unit	Application	Polymer	Repeat Unit	Application
Polyethylene (PE)	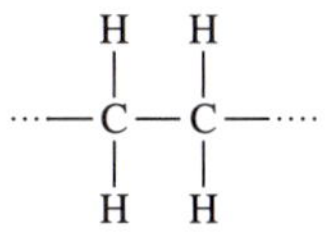	Packing films, wire insulation, squeeze bottles, tubing, household items	Polyacrylonitrile (PAN)	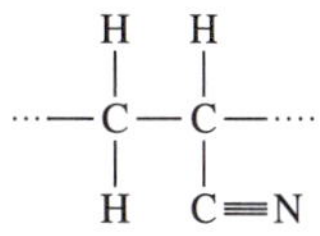	Textile fibers, precursor for carbon fibers, food container
Polyvinyl chloride (PVC)	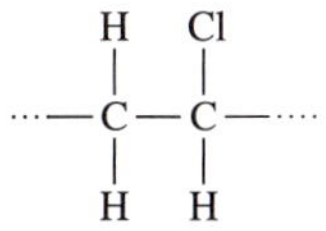	Pipe, valves, fittings, floor tile, wire insulation, vinyl automobile roofs	Polymethyl methacrylate (PMMA) (acrylic-Plexiglas)	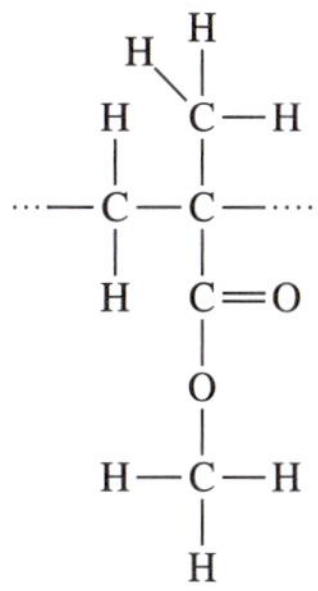	Windows, windshields, coatings, hard contact lenses, lighted signs
Polypropylene (PP)	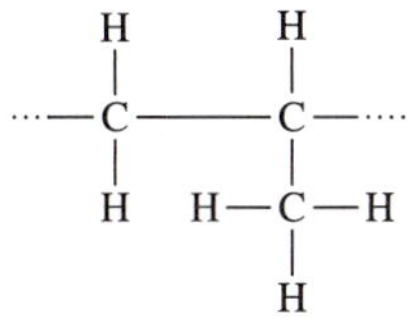	Tanks, carpet fibers, rope, packaging			
Polystyrene (PS)	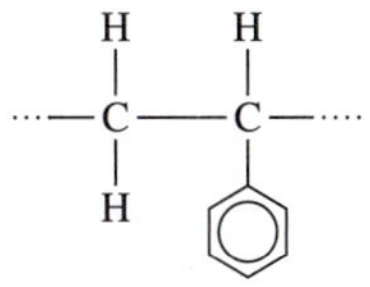	Packaging and insulation foams, lighting panels, appliance components, egg cartons	Polychlorotri-fluoroethylene	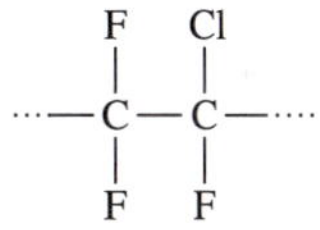	Valve components, gaskets, tubing, electrical insulation
			Polytetrafluoro-ethylene (Teflon) (PTFE)	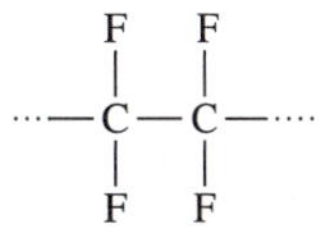	Seals, valves, nonstick coatings

Thermoplastics with Complex Structures A large number of polymers, that typically are used for special applications and in relatively small quantities, are formed from complex monomers, often by the condensation mechanism. Oxygen, nitrogen, sulfur, and benzene rings (or aromatic groups) may be incorporated into the chain. Table 16-4 shows the repeat units and typical applications for a number of these complex polymers. Polyoxymethylene, or acetal, is a simple example in which the backbone of the polymer chain contains alternating carbon and oxygen atoms. A number of these polymers, including polyimides and polyetheretherketone (PEEK), are important aerospace materials.

TABLE 16-4 ■ *Repeat units and applications for complex thermoplastics*

Polymer	Repeat Unit	Applications
Polyoxymethylene (acetal)(POM)	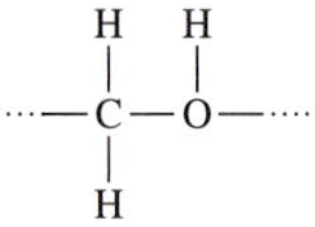	Plumbing fixtures, pens, bearings, gears, fan blades
Polyamide (nylon) (PA)	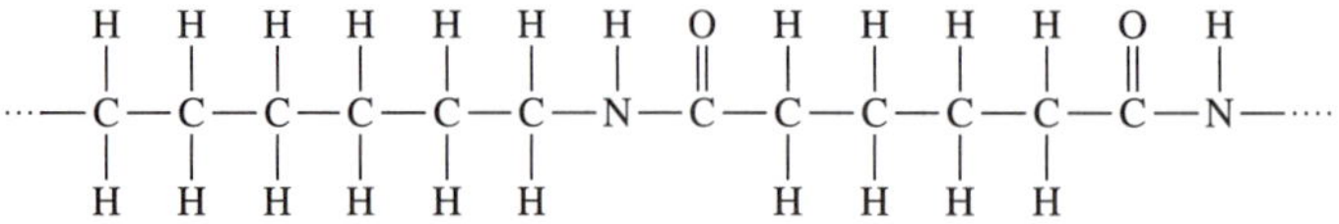	Bearings, gears, fibers, rope, automotive components, electrical components
Polyester (PET)	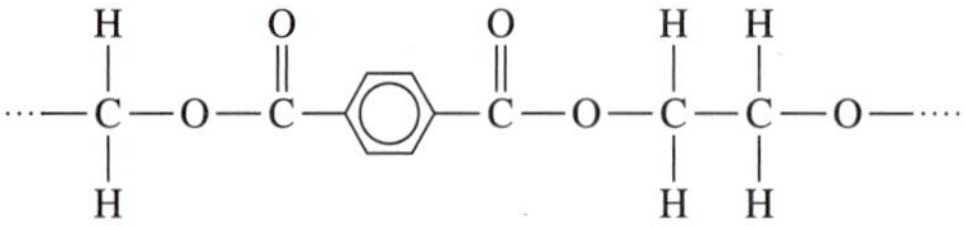	Fibers, photographic film, recording tape, boil-in-bag containers, beverage containers
Polycarbonate (PC)	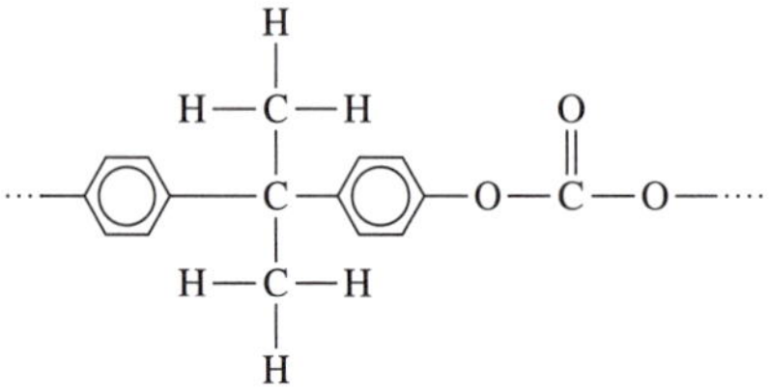	Electrical and appliance housings, automotive components, football helmets, returnable bottles
Polyimide (PI)	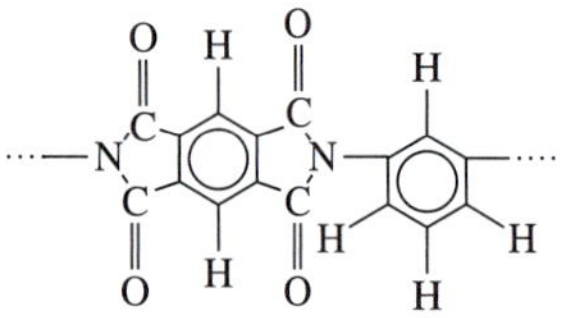	Adhesives, circuit boards, fibers for space shuttle
Polyetheretherketone (PEEK)	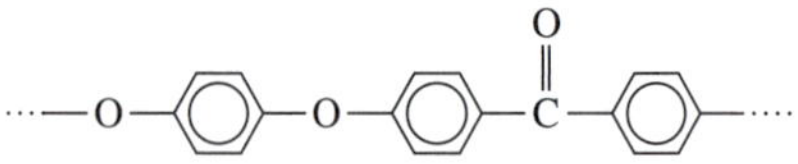	High-temperature electrical insulation and coatings
Polyphenylene sulfide (PPS)	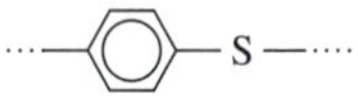	Coatings, fluid-handling components, electronic components, hair dryer components
Polyether sulfone (PES)	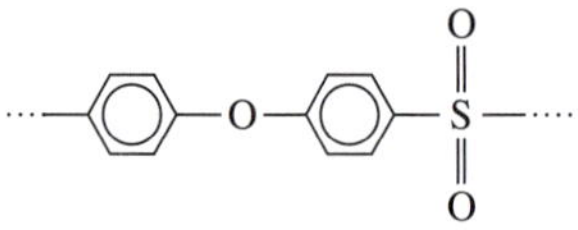	Electrical components, coffeemakers, hair dryers, microwave oven components
Polyamide-imide (PAI)	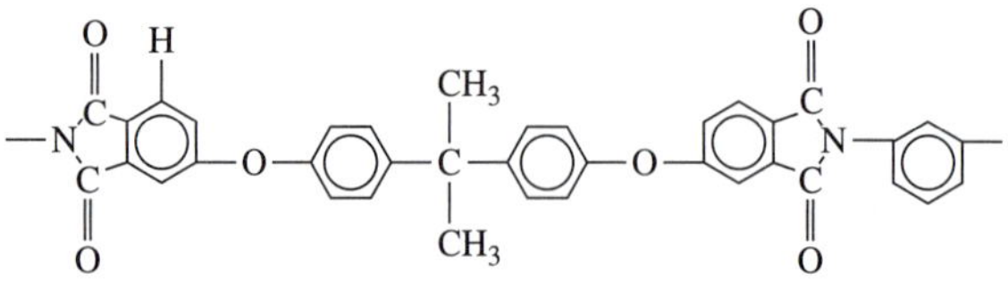	Electronic components, aerospace and automotive applications

16-5 Structure–Property Relationships in Thermoplastics

Degree of Polymerization In general, for a given type of thermoplastic (e.g., polyethylene) the tensile strength, creep resistance, impact toughness, wear resistance, and melting temperature all increase with increasing average molecular weight or degree of polymerization. The increases in these properties are not linear. As the average molecular weight increases, the melting temperature increases and this makes the processing more difficult. In fact, we can make use of a bimodal molecular weight distribution in polymer processing. One component has lower molecular weight and helps melting. In certain applications (e.g., polystyrene for coffee cups), it is important to ensure that the residual monomer concentration is extremely small (<1 ppm) because of the toxicity concerns.

Effect of Side Groups In polyethylene, the linear chains easily rotate and slide when stress is applied, and no strong polar bonds are formed between the chains; thus, polyethylene has a low strength.

Vinyl compounds have one of the hydrogen atoms replaced with a different atom or atom group. When *R* in the side group is chlorine, we produce polyvinyl chloride (PVC); when the side group is CH_3, we produce polypropylene (PP); addition of a benzene ring as a side group gives polystyrene (PS); and a CN group produces polyacrylonitrile (PAN). Generally, a head-to-tail arrangement of the repeat units in the polymers is obtained (Figure 16-6). When two of the hydrogen atoms are replaced, the monomer is a *vinylidene compound,* important examples of which include polyvinylidene chloride (the basis for Saran Wrap™) and polymethyl methacrylate (acrylics such as Lucite™ and Plexiglas™).

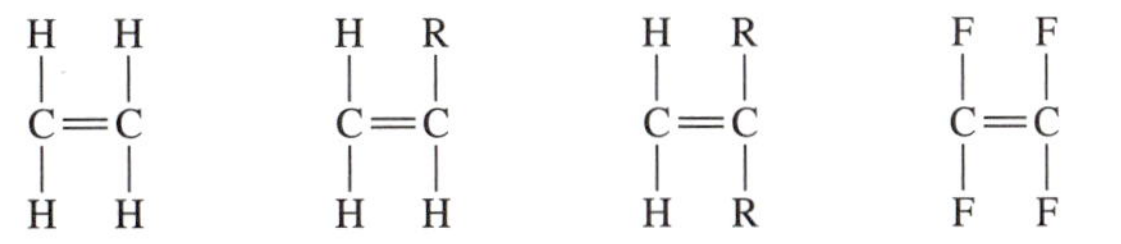

Ethylene | Vinyl compound | Vinylidene compound | Tetrafluoro-ethylene

The effects of adding other atoms or atom groups to the carbon backbone in place of hydrogen atoms are illustrated by the typical properties given in Table 16-3. Larger atoms such as chlorine or groups of atoms such as methyl (CH_3) and benzene make it more difficult for the chains to rotate, uncoil, disentangle, and deform by viscous flow when a stress is applied or when the temperature is increased. This condition leads to higher strength, stiffness, and melting temperature than those for polyethylene. The chlorine atom in PVC and the carbon-nitrogen group in PAN are strongly attracted by hydrogen bonding to hydrogen atoms on adjacent chains. This, for example, is the reason why PVC is more rigid than many other thermoplastics. The way to get around the rigidity of PVC is to add low molecular weight compounds such as phthalate esters,

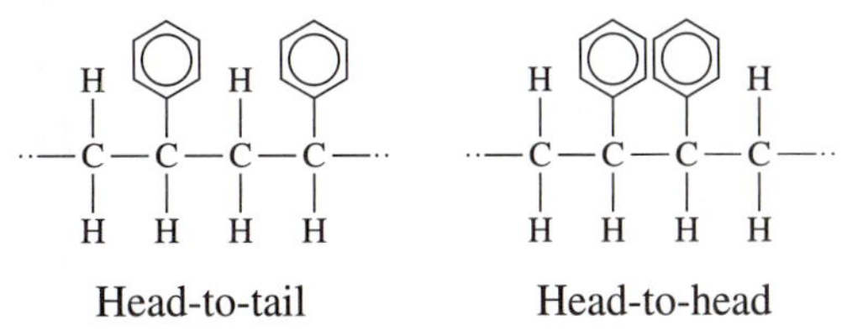

Figure 16-6
Head-to-tail versus head-to-head arrangement of repeat units. The head-to-tail arrangement is most typical.

known as plasticizers. When PVC contains these compounds, the glass temperature is lowered. This makes PVC more ductile and workable and is known as vinyl (not to be confused with the vinyl group mentioned here and in other places). PVC is used to make three-ring binders, pipes, tiles, and clear Tygon™ tubing.

In polytetrafluoroethylene (PTFE or Teflon™), all four hydrogen atoms in the polyethylene structure are replaced by fluorine. The monomer again is symmetrical, and the strength of the polymer is not much greater than that of polyethylene. However, the C-F bond permits PTFE to have a high melting point with the added benefit of low friction, nonstick characteristics that make the polymer useful for bearings and cookware. Teflon™ was invented by accident by Roy Plunkett, who was working with tetrafluoroethylene gas. He found a tetrafluoroethylene gas cylinder that had no pressure (and, thus, seemed empty) but was still heavy. The gas inside had polymerized into solid Teflon™!

Branching prevents dense packing of the chains, thereby reducing the density, stiffness, and strength of the polymer. Low-density (LD) polyethylene, which has many branches, is weaker than high-density (HD) polyethylene, which has virtually no branching (Table 16-2).

Crystallization and Deformation *Crystallinity* is important in polymers since it affects mechanical and optical properties. Crystallinity evolves in the processing of polymers as a result of temperature changes and applied stress (e.g., formation of PET bottles). If crystalline regions become too large, they begin to scatter light and make the plastic translucent. Of course, in certain special polymers, localized regions crystallize in response to an applied electric field and this is the principle by which the liquid crystal displays work. As we have discussed previously, encouraging crystallization of the polymer also helps to increase density, resistance to chemical attack, and mechanical properties—even at higher temperatures—because of the stronger bonding between the chains. In addition, deformation straightens and aligns the chains, producing a preferred orientation. Deformation of a polymer is often used in producing fibers having mechanical properties in the direction of the fiber that exceed those of many metals and ceramics. This texture strengthening (Chapter 8), in fact, played a key role in the discovery of nylon fibers. During their processing, PET bottles develop a biaxial texture and strength along the radial and length direction.

Tacticity When a polymer is formed from nonsymmetrical repeat units, the structure and properties are determined by the location of the nonsymmetrical atoms or atom groups. This condition is called **tacticity**, or stereoisomerism. In the syndiotactic arrangement, the atoms or atom groups alternatively occupy positions on opposite sides of the linear chain. The atoms are all on the same side of the chain in *isotactic* polymers, whereas the arrangement of the atoms is random in *atactic* polymers (Figure 16-7).

The atactic structure, which is the least regular and least predictable, tends to give poor packing, low density, low strength and stiffness, and poor resistance to heat or chemical attack. Atactic polymers are more likely to have an amorphous structure with a relatively high glass temperature. An important example of the importance of tacticity occurs in polypropylene. Atactic polypropylene is an amorphous wax-like polymer with poor mechanical properties, whereas isotactic polypropylene may crystallize and is one of the most widely used commercial polymers.

Copolymers Similar to the concept of solid solutions or the idea of composites, linear addition chains composed of two or more types of molecules can be arranged to form **copolymers**. This is a very powerful way to blend properties of different polymers. The arrangement of the monomers in a copolymer may take several forms (Figure 16-8).

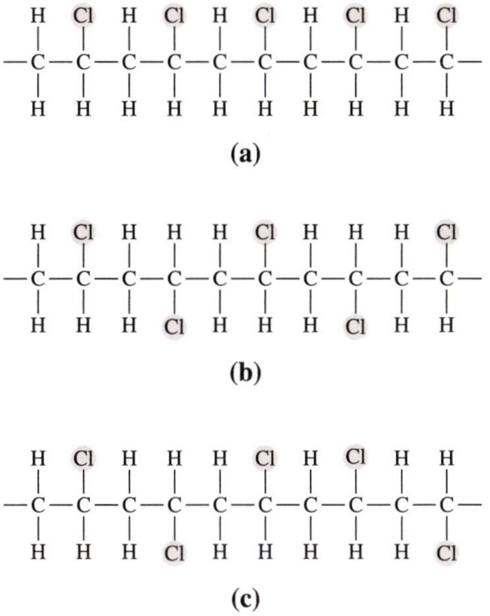

Figure 16-7
Three possible arrangements of nonsymmetrical monomers: (a) isotactic, (b) syndiotactic, and (c) atactic.

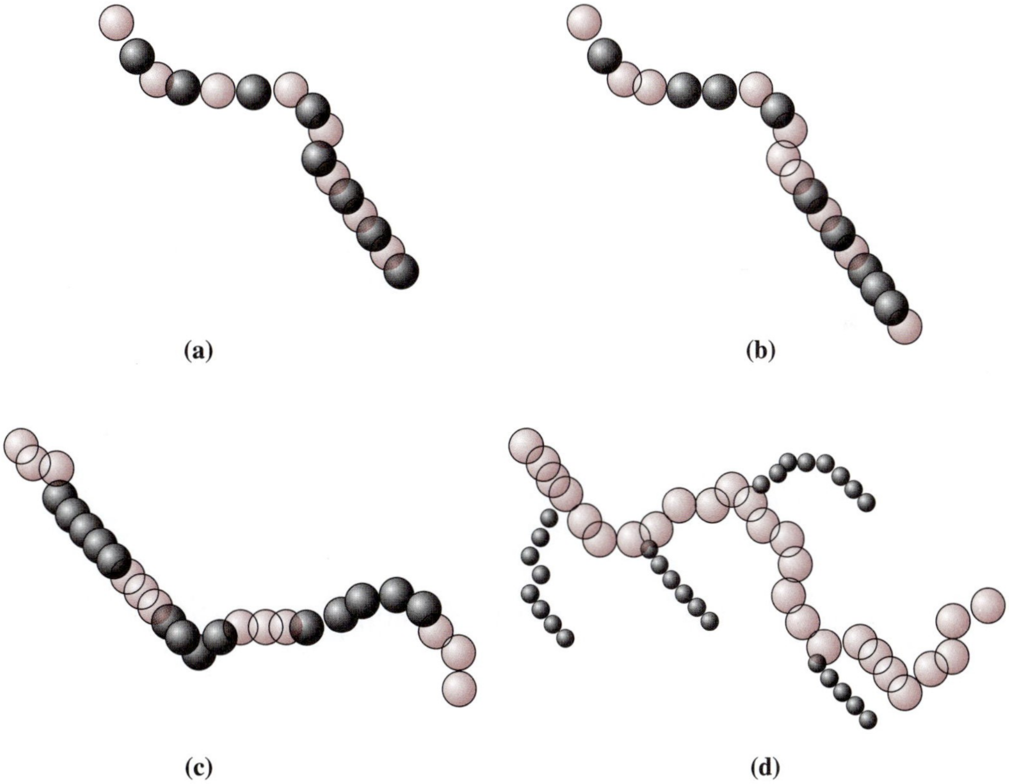

Figure 16-8 Four types of copolymers: (a) alternating monomers, (b) random monomers, (c) block copolymers, and (d) grafted copolymers. Circles of different colors or sizes represent different monomers.

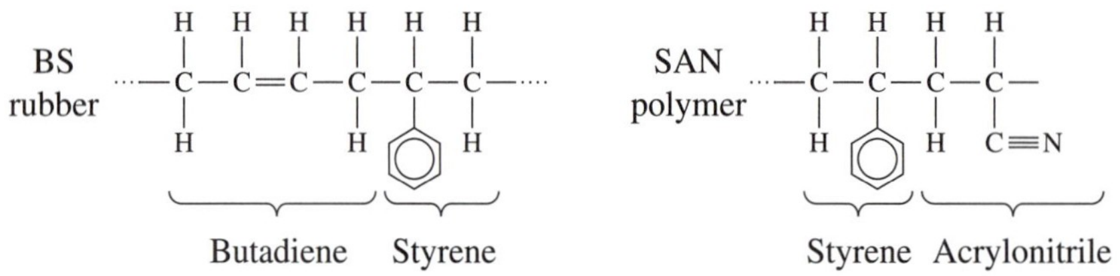

Figure 16-9 Copolymerization produces the polymer ABS, which is really made up of two copolymers, SAN and BS, grafted together.

These include alternating, random, block, and grafted copolymers. ABS, composed of acrylonitrile, butadiene (a synthetic elastomer), and styrene, is one of the most common polymer materials (Figure 16-9). Styrene and acrylonitrile form a linear copolymer (SAN) that serves as a matrix. Styrene and butadiene also form a linear copolymer, BS rubber, which acts as the filler material. The combination of the two copolymers gives ABS an excellent combination of strength, rigidity, and toughness. Another common copolymer contains repeat units of ethylene and propylene. Whereas polyethylene and polypropylene are both easily crystallized, the copolymer remains amorphous. When this copolymer is cross-linked, it behaves as an elastomer. Dylark™ is a copolymer of maleic anhydride and styrene. Styrene provides toughness, while maleic anhydride provides high-temperature properties. Carbon black (for protection from ultraviolet rays and enhancing stiffness), rubber (for toughness), and glass fibers (for stiffness) are added to the Dylark™ copolymer. It is used to make instrument panels for car dashboards. The Dylark™ plastic is then coated with vinyl that provides a smooth and soft finish.

Blending and Alloying We can improve the mechanical properties of many of the thermoplastics by blending or alloying. By mixing an immiscible elastomer with the thermoplastic, we produce a two-phase polymer, as we found in ABS. The elastomer does not enter the structure as a copolymer but, instead, helps to absorb energy and improve toughness. Polycarbonates used to produce transparent aircraft canopies are also toughened by elastomers in this manner.

Liquid Crystalline Polymers Some of the complex thermoplastic chains become so stiff that they act as rigid rods, even when heated above the melting point. These materials are **liquid crystalline polymers** (LCPs). Some aromatic polyesters and aromatic polyamides (or **aramids**) are examples of liquid crystalline polymers and are used as high-strength fibers (as to be discussed in Chapter 17). Kevlar™, an aromatic polyamide, is the most familiar of the LCPs and is used as a reinforcing fiber for aerospace applications and for bulletproof vests. Liquid crystal polymers are, of course, used to make electronic displays.

16-6 Effect of Temperature on Thermoplastics

Properties of thermoplastics change depending upon temperature. We need to know how these changes occur because this can help us (a) better design components, and (b) guide the type of processing techniques that need to be used. Several critical temperatures and structures, summarized in Figures 16-10 and 16-11, may be observed.

Thermoplastics can be amorphous or crystalline once they cool below the melting temperature, (Figure 16-10). Most often, engineered thermoplastics consist of regions

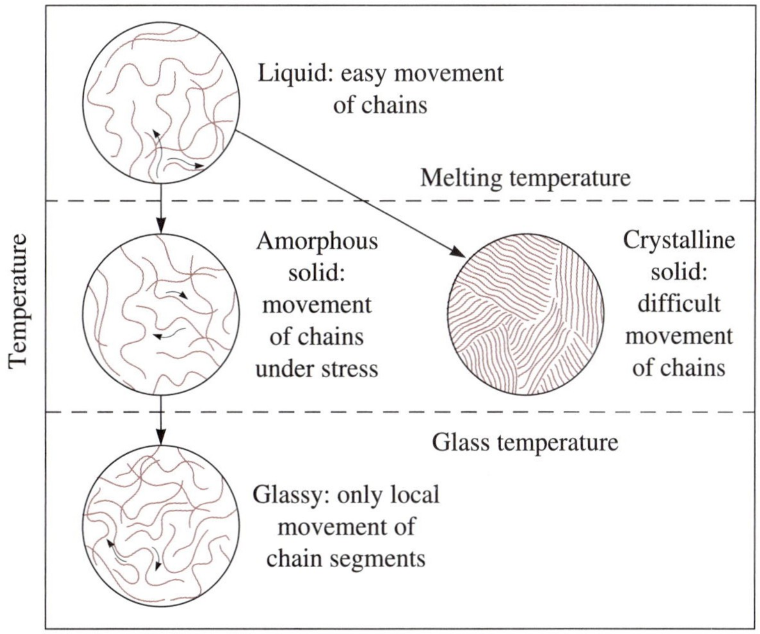

Figure 16-10 The effect of temperature on the structure and behavior of thermoplastics.

that are amorphous and crystalline. The crystallinity in thermoplastics can be introduced by temperature (slow cooling) or by **stress-induced crystallization** (the application of stress that can untangle chains, Chapter 7). Similar to dispersion strengthening of metallic materials, the formation of crystalline regions in an otherwise amorphous matrix helps increase the strength of thermoplastics. In typical thermoplastics, bonding within the chains is covalent, but the long coiled chains are held to one another by weak van der Waals bonds and by entanglement. When a tensile stress is applied to the thermoplastic, the weak bonding between the chains can be overcome and the chains can rotate and slide relative to one another. The ease with which the chains slide depends on both temperature and the polymer structure.

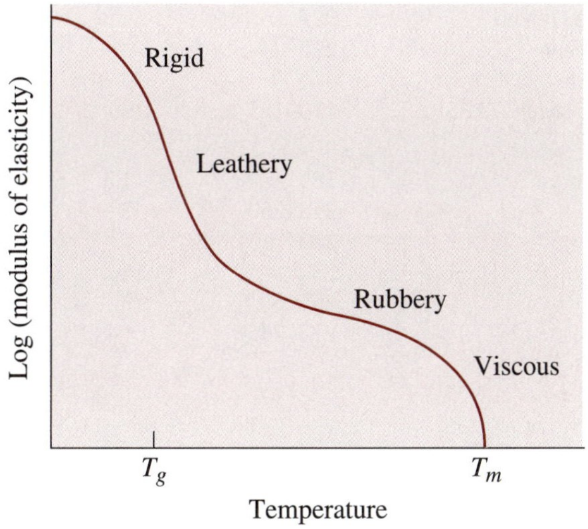

Figure 16-11
The effect of temperature on the modulus of elasticity for an amorphous thermoplastic. Note that T_g and T_m are not fixed.

Degradation Temperature At very high temperatures, the covalent bonds between the atoms in the linear chain may be destroyed, and the polymer may burn or char. In thermoplastics decomposition occurs in the liquid state, in thermosets the decomposition occurs in the solid state. This temperature T_d (not shown in Figure 16-11), is the **degradation** (or decomposition) **temperature**. When plastics burn, they create smoke, and that is dangerous. Some materials (such as limestone, talc, alumina etc.) added to thermoplastics are thermal or heat stabilizers. They absorb heat and protect the polymer matrix. Fire retardant additives, such as hydrated alumina, antimony compounds, or halogen compounds (e.g., MgBr, PCl_5) are added to retard the flammability of polymers. Some additives retard fire by excluding oxygen but generate dangerous gases and are not appropriate for certain applications.

Exposure to other forms of chemicals or energy (e.g., oxygen, ultraviolet radiation, and attack by bacteria) also cause a polymer to degrade or **age** slowly, even at low temperatures. Carbon black (up to ~3%) is one of the commonly used additives that helps improve the resistance of plastics to ultraviolet degradation.

Liquid Polymers Thermoplastics usually do not melt at a precise temperature. Instead there is usually a range of temperatures over which melting occurs. The approximate melting ranges of typical polymers are included in Table 16-5. At or above the melting temperature T_m, bonding between the twisted and intertwined chains is weak. If a force is applied, the chains slide past one another and the polymer flows with virtually no

TABLE 16-5 ■ ***Melting, glass, and processing temperature ranges (°C) for selected thermoplastics and elastomers***

Polymer	Melting Temperature Range	Glass Temperature Range (T_g)	Processing Temperature Range
Addition polymers			
Low-density (LD) polyethylene	98–115	−90 to −25	149–232
High-density (HD) polyethylene	130–137	−110	177–260
Polyvinyl chloride	175–212	87	
Polypropylene	160–180	−25 to −20	190–288
Polystyrene	240	85–125	
Polyacrylonitrile	320	107	
Polytetrafluoroethylene (Teflon)	327		
Polychlorotrifluoroethylene	220		
Polymethyl methacrylate (acrylic)		90–105	
Acrylonitrile butadiene styrene (ABS)	110–125	100	177–260
Condensation polymers			
Acetal	181	−85	
6,6-nylon	243–260	49	260–327
Cellulose acetate	230		
Polycarbonate	230	149	271–300
Polyester	255	75	
Polyethylene terephthalate (PET)	212–265	66–80	227–349
Elastomers			
Silicone		−123	
Polybutadiene	120	−90	
Polychloroprene	80	−50	
Polyisoprene	30	−73	

elastic strain. The strength and modulus of elasticity are nearly zero and the polymer is suitable for casting and many forming processes. Most thermoplastic melts are shear thinning (i.e., their apparent viscosity decreases within an increase in the steady-state shear rate).

Rubbery and Leathery States Below the melting temperature, the polymer chains are still twisted and intertwined. These polymers have an amorphous structure. Just below the melting temperature, the polymer behaves in a *rubbery* manner. When stress is applied, both elastic and plastic deformation of the polymer occurs. When the stress is removed, the elastic deformation is quickly recovered, but the polymer is permanently deformed due to the movement of the chains. Some of this deformation is recovered over a period of time. Thus, many polymers exhibit a viscoelastic behavior (Chapter 6). Large permanent elongations can be achieved, permitting the polymer to be formed into useful shapes by molding and extrusion.

At lower temperatures, bonding between the chains is stronger, the polymer becomes stiffer and stronger, and a *leathery* behavior is observed. Many of the commercial polymers, including polyethylene, have a useable strength in this condition.

Glassy State Below the **glass temperature** T_g, the linear amorphous polymer becomes hard, brittle, and glass-like. This is again not a fixed temperature but a range of temperatures. When the polymer cools below the glass temperature, certain properties—such as density or modulus of elasticity—change at a different rate (Figure 16-12).

Although glassy polymers have poor ductility and formability, they do have good strength, stiffness, and creep resistance. A number of important polymers, including polystyrene and polyvinyl chloride, have glass temperatures above room temperature (Table 16-5).

The glass temperature is typically about 0.5 to 0.75 times the absolute melting temperature T_m. Polymers such as polyethylene, which have no complicated side groups attached to the carbon backbone, have low glass temperatures (even below room temperature) compared with polymers such as polystyrene, which have more complicated side groups.

As pointed out in Chapter 6, many thermoplastics become brittle at lower temperatures. The brittleness of the polymer used for some of the O-rings ultimately caused the 1986 *Challenger* disaster. The lower temperatures that existed during the launch time caused the embrittlement of the rubber O-rings used for the booster rockets.

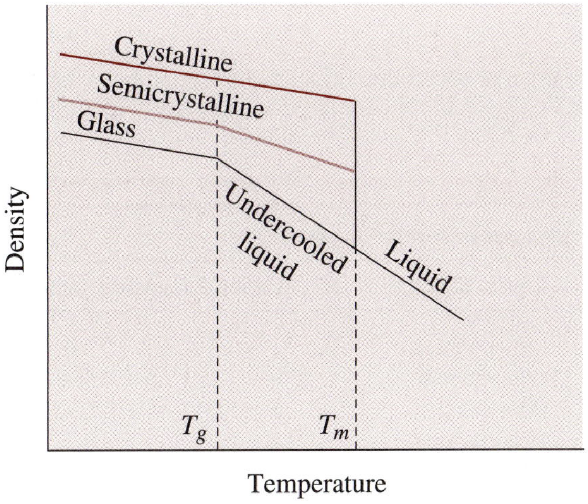

Figure 16-12
The relationship between the density and the temperature of the polymer shows the melting and glass temperatures. Note that T_g and T_m are not fixed; rather, they are ranges of temperatures.

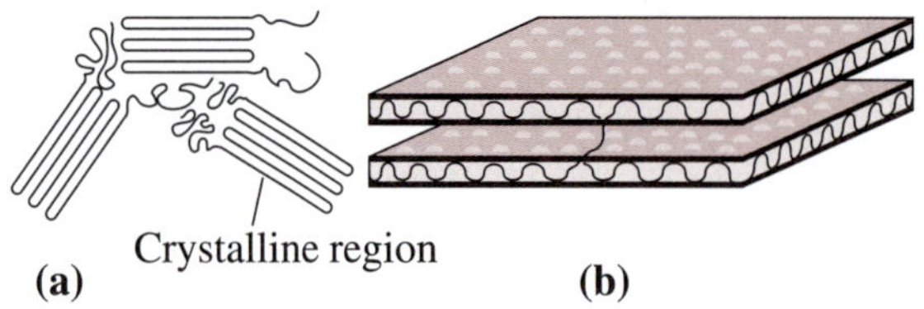

Figure 16-13 The folded chain model for crystallinity in polymers, shown in (a) two dimensions and (b) three dimensions.

Observing and Measuring Crystallinity in Polymers Many thermoplastics partially crystallize when cooled below the melting temperature, with the chains becoming closely aligned over appreciable distances. A sharp increase in the density occurs as the coiled and intertwined chains in the liquid are rearranged into a more orderly, close-packed structure (Figure 16-12).

One model describing the arrangement of the chains in a crystalline polymer is shown in Figure 16-13. In this *folded chain* model, the chains loop back on themselves, with each loop being approximately 100 carbon atoms long. The folded chain extends in three dimensions, producing thin plates or lamellae. The crystals can take various forms, with the spherulitic shape shown in Figure 16-14(a) being particularly common. The crystals have a unit cell that describes the regular packing of the chains. The crystal structure for polyethylene, shown in Figure 16-14(b), describes one such unit cell. Crystal structures for several polymers are described in Table 16-6. Some polymers are polymorphic, having more than one crystal structure.

Even in crystalline polymers, there are always thin regions between the lamellae, as well as between spherulites, that are amorphous transition zones. The weight percentage

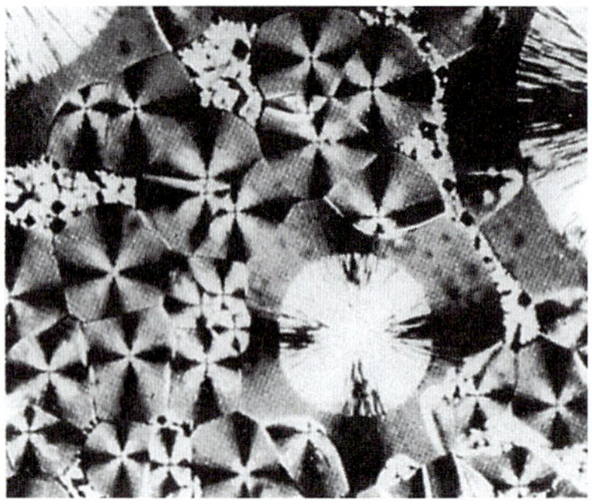

(a)

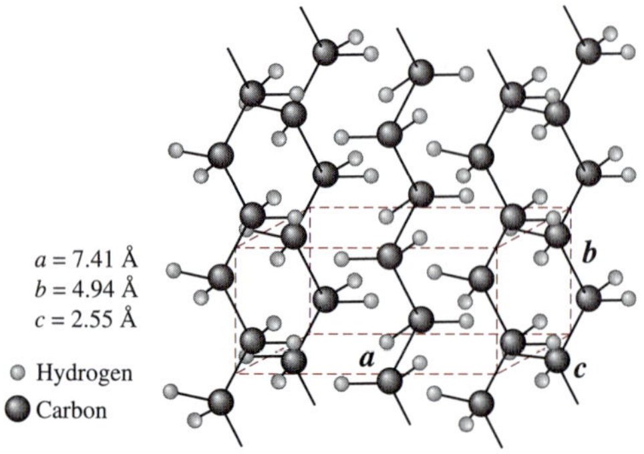

(b)

Figure 16-14 (a) Photograph of spherulitic crystals in an amorphous matrix of nylon (×200). (*Source: From R. Brick, A. Pense and R. Gordon,* Structure and Properties of Engineering Materials, *4th Ed., McGraw-Hill, 1997.*) (b) The unit cell of crystalline polyethylene.

TABLE 16-6 ■ *Crystal structures of several polymers*

Polymer	Crystal Structure	Lattice Parameters (nm)		
Polyethylene	Orthorhombic	$a = 0.742$	$b = 0.495$	$c = 0.255$
Polypropylene	Orthorhombic	$a = 1.450$	$b = 0.569$	$c = 0.740$
Polyvinyl chloride	Orthorhombic	$a = 1.040$	$b = 0.530$	$c = 0.510$
Polyisoprene (cis)	Orthorhombic	$a = 1.246$	$b = 0.886$	$c = 0.810$

of the structure that is crystalline can be calculated from the density of the polymer:

$$\% \text{ Crystalline} = \frac{\rho_c}{\rho}\frac{(\rho - \rho_a)}{(\rho_c - \rho_a)} \times 100 \tag{16-4}$$

where ρ is the measured density of the polymer, ρ_a is the density of amorphous polymer, and ρ_c is the density of completely crystalline polymer. Similarly, x-ray diffraction (XRD) can be used to measure the level of crystallinity and determination of lattice constants for single crystal polymers.

As the side groups get more complex, it becomes harder to crystallize thermoplastics. For example, polyethylene (H as side group) can be crystallized more easily than polystyrene (benzene ring as side group). High-density polyethylene (HDPE) has a higher level of crystallinty and, therefore, a higher density (0.97 g/cc). Low-density polyethylene (LDPE) has a density of 0.92 g/cc. The crystallinity and, hence, the density in LDPE is lower, since the polymer is branched. Thus, branched polymers show lower levels of crystallinity. A completely crystalline polymer would not display a glass temperature; however, the amorphous regions in semicrystalline polymers do change to a glassy material below the glass temperature (Figure 16-12). Such polymers as acetal, nylon, HDPE, and polypropylene are referred to as crystalline even though the level of crystallinity may be moderate. Sometimes, the trade names can be confusing. For example, what is described as a "crystal polystyrene" is actually an amorphous material. It looks transparent and shiny though, and hence, the trade name crystal polystyrene. The following examples show how properties of plastics can be accounted for in different applications.

EXAMPLE 16-6 *Design of a Polymer Insulation Material*

A storage tank for liquid hydrogen will be made of metal, but we wish to coat the metal with a 3-mm thickness of a polymer as an intermediate layer between the metal and additional insulation layers. The temperature of the intermediate layer may drop to −80°C. Design a material for this layer.

SOLUTION

We want the material to have reasonable ductility. As the temperature of the tank changes, stresses develop in the coating due to differences in thermal expansion, and we do not want the polymer to fail due to these stresses. A material that has good ductility and/or can undergo large elastic strains is needed. We therefore would prefer either a thermoplastic that has a glass temperature below −80°C or an elastomer, also with a glass temperature below −80°C. Of the polymers listed in Table 16-2, thermoplastics such as polyethylene and acetal are satisfactory. Suitable elastomers include silicone and polybutadiene.

We might prefer one of the elastomers, for they can accommodate thermal stress by elastic, rather than plastic, deformation.

EXAMPLE 16-7 *Impact-Resistant Polyethylene*

A new grade of flexible, impact-resistant polyethylene for use as a thin film requires a density of 0.88 to 0.915 g/cm^3. Design the polyethylene required to produce these properties. The density of amorphous polyethylene is about 0.87 g/cm^3.

SOLUTION

To produce the required properties and density, we must control the percent crystallinity of the polyethylene. We can use Equation 16-4 to determine the crystallinity that corresponds to the required density range. To do so, however, we must know the density of completely crystalline polyethylene. We can use the data in Table 16-3 to calculate this density if we recognize that there are two polyethylene repeat units in each unit cell:

$$\rho_c = \frac{(4\text{ C})(12) + (8\text{ H})(1)}{(7.42)(4.95)(2.55)(10^{-24})(6.02 \times 10^{23})} = 0.9932\text{ g/cm}^3$$

We know that $\rho_a = 0.87$ g/cm^3 and that ρ varies from 0.88 to 0.915 g/cm^3. The required crystallinity then varies from:

$$\%\text{ crystalline} = \frac{(0.9932)(0.88 - 0.87)}{(0.88)(0.9932 - 0.87)} \times 100 = 9.2$$

$$\%\text{ crystalline} = \frac{(0.9932)(0.915 - 0.87)}{(0.915)(0.9932 - 0.87)} \times 100 = 39.6$$

Therefore, we must be able to process the polyethylene to produce a range of crystallinity between 9.2 and 39.6%.

16-7 Mechanical Properties of Thermoplastics

Most thermoplastics (molten and solid) exhibit a non-Newtonian and **viscoelastic** behavior. The behavior is non-Newtonian (i.e., the stress and strain are not linearly related for most parts of the stress-strain curve). The viscoelastic behavior means when an external force is applied to a thermoplastic polymer, both elastic and plastic (or viscous) deformation occurs. The mechanical behavior is closely tied to the manner in which the polymer chains move relative to one another under load. Deformation is more complicated in thermoplastics. The deformation process depends on both time and the rate at which the load is applied. Figure 16-15 shows a stress-strain curve for 6,6-nylon.

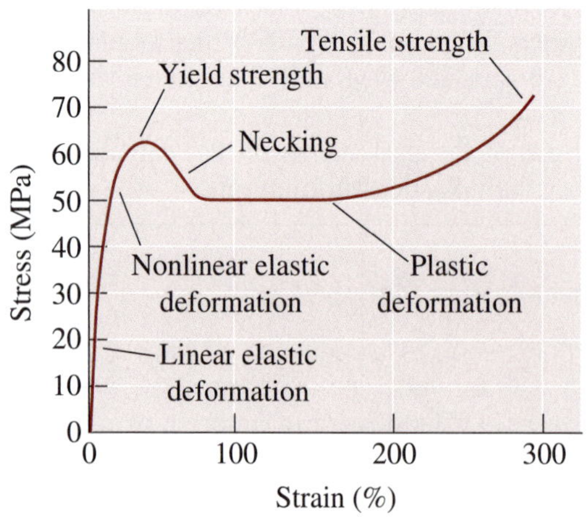

Figure 16-15 The stress-strain curve for 6,6-nylon, a typical thermoplastic polymer.

Elastic Behavior Elastic deformation in thermoplastics is the result of two mechanisms. An applied stress causes the covalent bonds within the chain to stretch and distort, allowing the chains to elongate elastically. When the stress is removed, recovery from this distortion is almost instantaneous. This behavior is similar to that in metals and ceramics, which also deform elastically by the stretching of metallic, ionic, or covalent bonds. But in addition, entire segments of the polymer chains may be distorted; when the stress is removed, the segments move back to their original positions only over a period of time—often hours or even months. This time-dependent, or viscoelastic, behavior may contribute to some nonlinear elastic behavior.

Plastic Behavior of Amorphous Thermoplastics These polymers deform plastically when the stress exceeds the yield strength. Unlike deformation in the case of metals, however, plastic deformation is not a consequence of dislocation movement. Instead, chains stretch, rotate, slide, and disentangle under load to cause permanent deformation. The drop in the stress beyond the yield point can be explained by this phenomenon. Initially, the chains may be highly tangled and intertwined. When the stress is sufficiently high, the chains begin to untangle and straighten. Necking also occurs, permitting continued sliding of the chains at a lesser stress. Eventually, however, the chains become almost parallel and close together; stronger van der Waals bonding between the more closely aligned chains requires higher stresses to complete the deformation and fracture process (Figure 16-16). This type of crystallization due to orientation played an important role in the discovery of nylon as a material to make strong fibers.

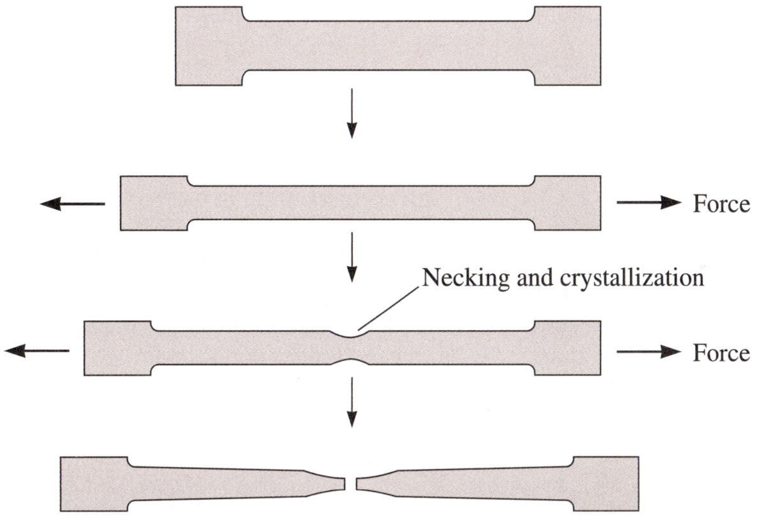

Figure 16-16 Necks are not stable in amorphous polymers, because local alignment strengthens the necked region and reduces its rate of deformation.

EXAMPLE 16-8 *Comparing Mechanical Properties of Thermoplastics*

Compare the mechanical properties of LD polyethylene, HD polyethylene, polyvinyl chloride, polypropylene, and polystyrene, and explain their differences in terms of their structures.

SOLUTION

Let us look at the maximum tensile strength and modulus of elasticity for each polymer.

Polymer	Tensile Strength (MPa)	Modulus of Elasticity (MPa)	Structure
LD polyethylene	21	276	Highly branched, amorphous structure with symmetrical monomers
HD polyethylene	38	1241	Amorphous structure with symmetrical monomers but little branching
Polypropylene	41	1517	Amorphous structure with small methyl side groups
Polystyrene	55	3103	Amorphous structure with benzene side groups
Polyvinyl chloride	62	4137	Amorphous structure with large chlorine atoms as side groups

We can conclude that:

1. Branching, which reduces the density and close packing of chains, reduces the mechanical properties of polyethylene.

2. Adding atoms or atom groups other than hydrogen to the chain increases strength and stiffness. The methyl group in polypropylene provides some improvement, the benzene ring of styrene provides higher properties, and the chlorine atom in polyvinyl chloride provides a large increase in properties.

Creep and Stress Relaxation Thermoplastics also exhibit creep, a time-dependent permanent deformation with constant stress or load (Figures 16-17 and 16-18).

They also show **stress relaxation** (i.e., under a constant strain the stress level decreases with time) (Chapter 6). Stress relaxation, like creep, is a consequence of the viscoelastic behavior of the polymer. Perhaps the most familiar example of this behavior is a rubber band (an elastomer) stretched around a pile of books. Initially, the tension in the rubber band is high, when the rubber band is taut. After several weeks, the strain in the rubber band is unchanged (it still completely encircles the books), but the stress will have decreased—that is, the band is no longer taut. Similarly, the nylon strings in tennis rackets are pulled at a higher tension initially since this tension (i.e., stress) decreases with time.

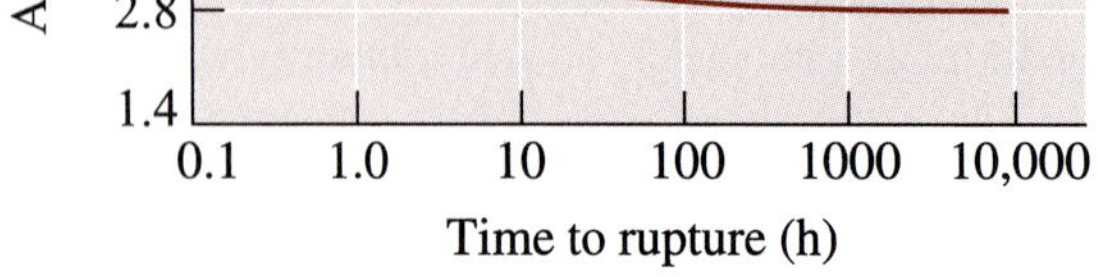

Figure 16-17
The effect of temperature on the stress-rupture behavior of high-density polyethylene.

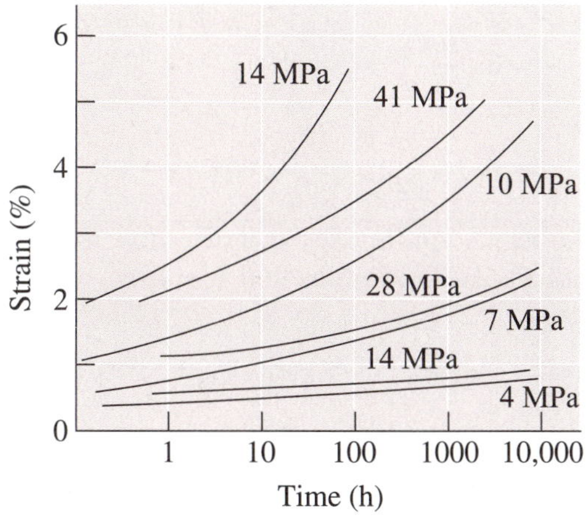

Figure 16-18
Creep curves for acrylic (PMMA) (colored lines) and polypropylene (black lines) at 20°C and several applied stresses.

In a simple model, the rate at which stress relaxation occurs is related to the **relaxation time** λ, which is considered a property of the polymer (more complex models consider a distribution of relaxation times). The stress after time t is given by

$$\sigma = \sigma_0 \exp(-t/\lambda) \tag{16-5}$$

where σ_0 is the original stress. The *relaxation time*, in turn, depends on the viscosity and, thus, the temperature:

$$\lambda = \lambda_0 \exp(Q/RT) \tag{16-6}$$

where σ_0 is a constant and Q is the activation energy related to the ease with which polymer chains slide past each other. Stress relaxation occurs more *rapidly* at *higher temperatures* and for polymers with a low viscosity.

The following example shows how stress relaxation can be accounted for while designing with polymers.

EXAMPLE 16-9 *Design of Initial Stress in a Polymer*

A band of polyisoprene is to hold together a bundle of steel rods for up to one year. If the stress on the band is less than 10 MPa, the band will not hold the rods tightly. Design the initial stress that must be applied to a polyisoprene band when it is slipped over the steel. A series of tests showed that an initial stress of 7 MPa decreased to 6.8 MPa after six weeks.

SOLUTION

Although the strain of the elastomer band may be constant, the stress will decrease over time due to stress relaxation. We can use Equation 16-5 and our initial tests to determine the relaxation time for the polymer:

$$\sigma = \sigma_0 \exp\left(-\frac{t}{\lambda}\right)$$

$$6.8 = 7 \exp\left(-\frac{6}{\lambda}\right)$$

$$-\frac{6}{\lambda} = \ln\left(\frac{6.8}{7}\right) = \ln(0.97) = -0.0305$$

$$\lambda = \frac{6}{0.0305} = 197 \text{ weeks}$$

Now that we know the relaxation time, we can determine the stress that must be initially placed onto the band in order that it still be stressed to 10 MPa after 1 year (52 weeks).

$$10 = \sigma_0 \exp(-52/197) = \sigma_0 \exp(-0.264) = 0.768\sigma_0$$

$$\sigma_0 = \frac{10}{0.768} = 13 \text{ MPa}$$

The polyisoprene band must be made significantly undersized so it can slip over the materials it is holding together with a tension of 13 MPa. After one year, the stress will still be 10 MPa.

One more practical measure for high temperature and creep properties of a polymer is the **heat deflection temperature** or **heat distortion temperature** under load, which is the temperature at which a given deformation of a beam occurs for a standard load. A high-heat deflection temperature indicates good resistance to creep and permits us to compare various polymers. The deflection temperatures for several polymers are shown in Table 16-7, which gives the temperature required to cause a 0.025 cm deflection for a 1.8 MPa load at the center of a bar resting on supports 10 cm apart. A polymer is "dishwasher safe" if it has a heat distortion temperature greater than ~50°C.

TABLE 16-7 ■ *Deflection temperatures for selected polymers for a 1.8 MPa load*

Polymer	Deflection Temperature (°C)
Polyester	40
Polyethylene (ultra-high density)	40
Polypropylene	60
Phenolic	80
Polyamide (6,6-nylon)	90
Polystyrene	100
Polyoxymethylene (acetal)	130
Polyamide-imide	280
Epoxy	290

Impact Behavior Viscoelastic behavior also helps us understand the impact properties of polymers. At very high rates of strain, as in an impact test, there is insufficient time for the chains to slide and cause plastic deformation. For these conditions, the thermoplastics behave in a brittle manner and have poor impact values. Polymers may have a transition temperature. At low temperatures, brittle behavior is observed in an impact test, whereas more ductile behavior is observed at high temperatures, where the chains move more easily. These effects of temperature and strain rate are similar to those seen in metals that exhibit a ductile-to-brittle transition temperature; however, the mechanisms by which ductile to brittle transition occurs are different.

Deformation of Crystalline Polymers A number of polymers are used in the crystalline state. As we discussed earlier, however, the polymers are never completely crystalline. Instead, small regions—between crystalline lamellae and between crystalline spherulites—are amorphous transition regions. Polymer chains in the crystalline region extend into these amorphous regions as tie chains. When a tensile load is applied to the polymer, the crystalline lamellae within the spherulites slide past one another and begin to separate as the tie chains are stretched. The folds in the lamellae tilt and become aligned with the direction of the tensile load. The crystalline lamellae break into smaller units and slide past one another, until eventually the polymer is composed of small aligned crystals joined by tie chains and oriented parallel to the tensile load. The spherulites also change shape and become elongated in the direction of the applied stress. With continued stress, the tie chains disentangle or break, causing the polymer to fail.

Crazing Crazing occurs in thermoplastics when localized regions of plastic deformation occur in a direction perpendicular to that of the applied stress. In transparent thermoplastics, such as some of the glassy polymers, the craze produces a translucent or opaque region that looks like a crack. The craze can grow until it extends across the entire cross-section of the polymer part. However, the craze is not a crack, and, in fact, it can continue to support an applied stress. The process is similar to that for the plastic deformation of the polymer, but the process can proceed even at a low stress over an extended length of time. Crazing can lead to brittle fracture of the polymer and is often assisted by the presence of a solvent (known as solvent crazing).

Blushing Blushing or whitening refers to failure of a plastic because of localized crystallization (due to repeated bending, for example) that ultimately causes voids to form.

16-8 Elastomers [Rubbers]

A number of natural and synthetic polymers called elastomers display a large amount (>200%) of elastic deformation when a force is applied. Rubber bands, automobile tires, O-rings, hoses, and insulation for electrical wires are common uses for these materials. Crude natural rubber could erase pencil marks; hence, elastomers got the name rubber.

Geometric Isomers Some monomers that have different structures, even though they have the same composition, are called **geometric isomers**. Isoprene, or natural rubber, is an important example (Figure 16-19). The monomer includes two double bonds between carbon atoms; this type of monomer is called a **diene**. Polymerization occurs by breaking both double bonds, creating a new double bond at the center of the molecule and active sites at both ends.

In the *trans* form of isoprene, the hydrogen atom and the methyl group at the center of the repeat unit are located on opposite sides of the newly formed double bond. This arrangement leads to relatively straight chains; the polymer crystallizes and forms a hard rigid polymer called *gutta percha*. This is used to make golf balls and shoe soles.

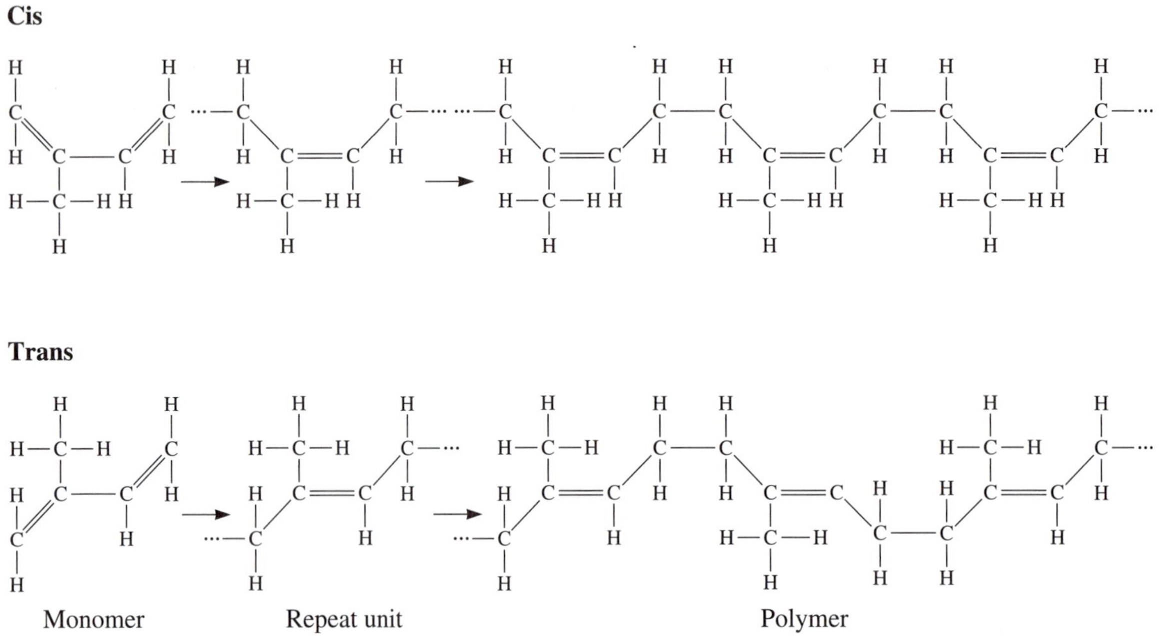

Figure 16-19 The cis and trans structures of isoprene. The cis form is useful for producing the isoprene elastomer.

In the *cis* form, however, the hydrogen atom and the methyl group are located on the same side of the double bond. This different geometry causes the polymer chains to develop a highly coiled structure, preventing close packing and leading to an amorphous, rubbery polymer. If a stress is applied to the cis-isoprene, the polymer behaves in a viscoelastic manner. The chains uncoil and bonds stretch, producing elastic deformation, but the chains also slide past one another, producing nonrecoverable plastic deformation. The polymer behaves as a thermoplastic rather than an elastomer (Figure 16-20).

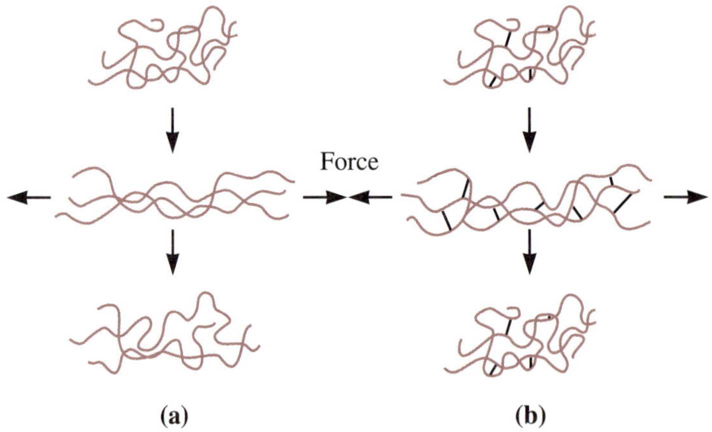

Figure 16-20 (a) When the elastomer contains no cross-links, the application of a force causes both elastic and plastic deformation; after the load is removed, the elastomer is permanently deformed. (b) When cross-linking occurs, the elastomer still may undergo large elastic deformation; however, when the load is removed, the elastomer returns to its original shape.

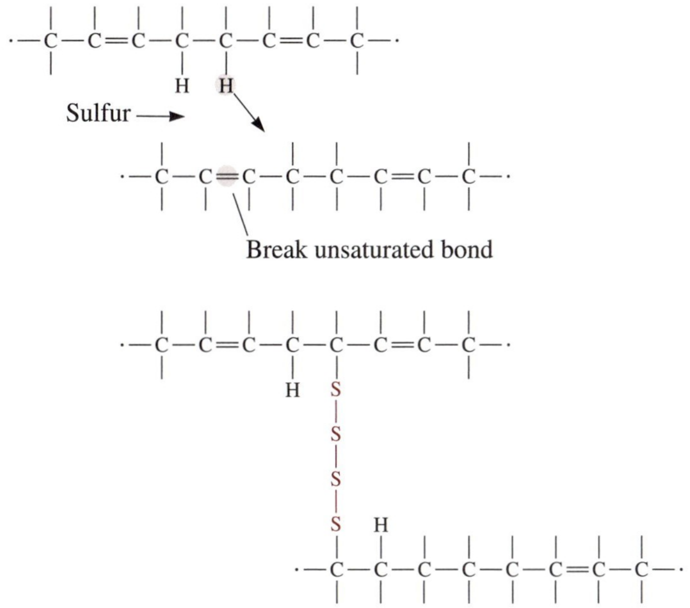

Figure 16-21
Cross-linking of polyisoprene chains may occur by introducing strands of sulfur atoms. Sites for attachment of the sulfur strands occur by rearrangement or loss of a hydrogen atom and the breaking of an unsaturated bond.

Cross-Linking We prevent viscous plastic deformation while retaining large elastic deformation by **cross-linking** the chains. **Vulcanization**, which uses sulfur atoms, is a common method for cross-linking. Vulcanization describes how strands of sulfur atoms can link the polymer chains as the polymer is processed and shaped at temperatures of about 120 to 180°C (Figure 16-21). The cross-linking steps may include rearranging a hydrogen atom and replacing one or more of the double bonds with single bonds. The cross-linking process is not reversible; consequently, the elastomer cannot be recycled easily.

The stress-strain curve for an elastomer is shown in Figure 16-22. Virtually all of the curve represents elastic deformation; thus, elastomers display a nonlinear elastic behavior. Initially, the modulus of elasticity decreases because of the uncoiling of the chains. However, after the chains have been extended, further elastic deformation occurs by the stretching of the bonds, leading to a higher modulus of elasticity.

The number of cross-links determines the elasticity of the rubber or the amount of sulfur added to the material. Low sulfur additions leave the rubber soft and flexible, as in elastic bands or rubber gloves. Increasing the sulfur content restricts the uncoiling of

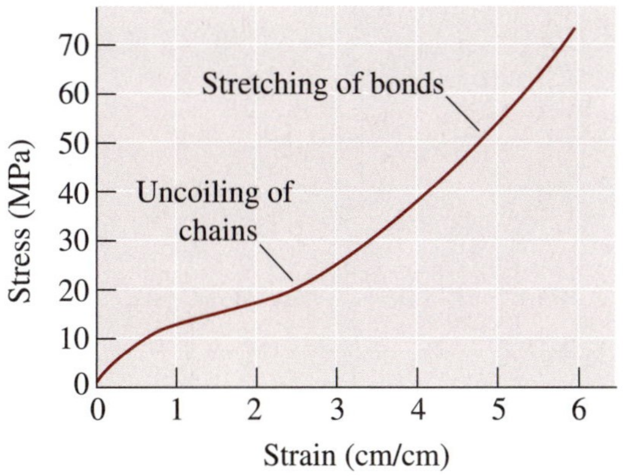

Figure 16-22
The stress-strain curve for an elastomer. Virtually all of the deformation is elastic; therefore, the modulus of elasticity varies as the strain changes.

the chains and the rubber becomes harder, more rigid, and brittle (as in rubber used for motor mounts). Typically, 0.5 to 5% sulfur is added to provide cross-linking in elastomers. Many more *efficient vulcanizing* (EV) systems, which are sulfur free, also have been developed and used in recent years.

Typical Elastomers Elastomers, which are amorphous polymers, do not easily crystallize during processing. They have a low glass temperature, and chains can easily be deformed elastically when a force is applied. The typical elastomers (Tables 16-8 and 16-9) meet these requirements. Polyisoprene is a natural rubber. Polychloroprene, or

TABLE 16-8 ■ *Repeat units and applications for selected elastomers*

Polymer	Repeat Unit	Applications
Polyisoprene	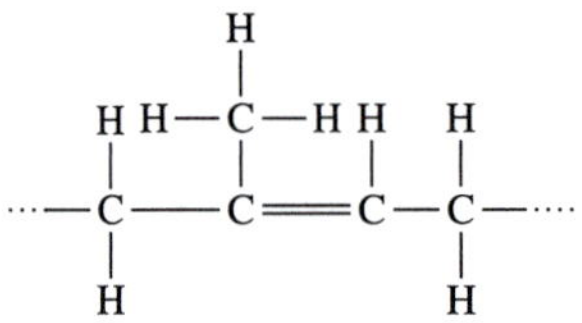	Tires, golf balls, shoe soles
Polybutadiene (or butadiene rubber or Buna-S)	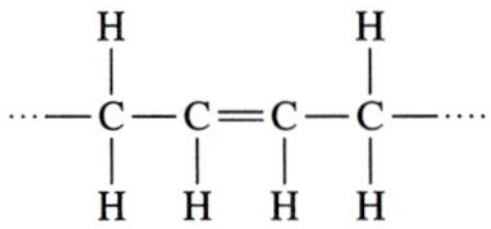	Industrial tires, toughening other elastomers, inner tubes of tires, weatherstripping, steam hoses
Polyisobutylene (or butyl rubber)	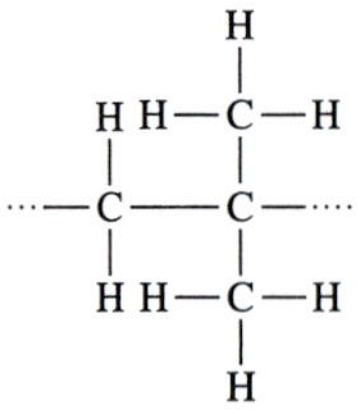	
Polychloroprene (Neoprene)	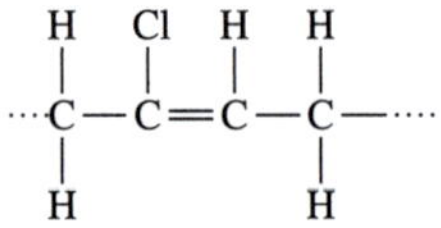	Hoses, cable sheathing
Butadiene-styrene (BS or SBR rubber)	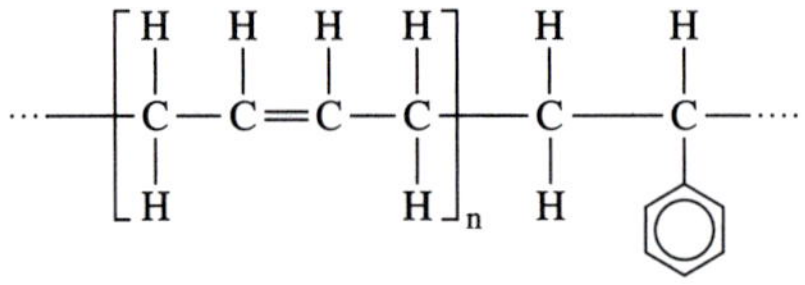	Tires
Butadiene-acrylonitrile (Buna-N)	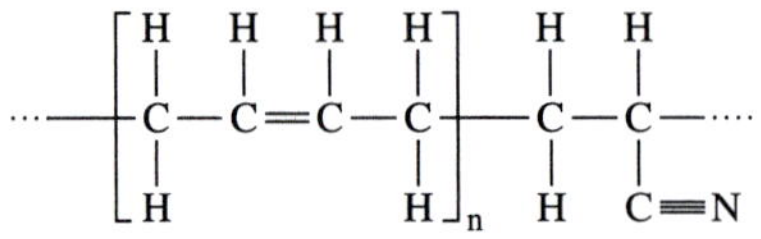	Gaskets, fuel hoses
Silicones	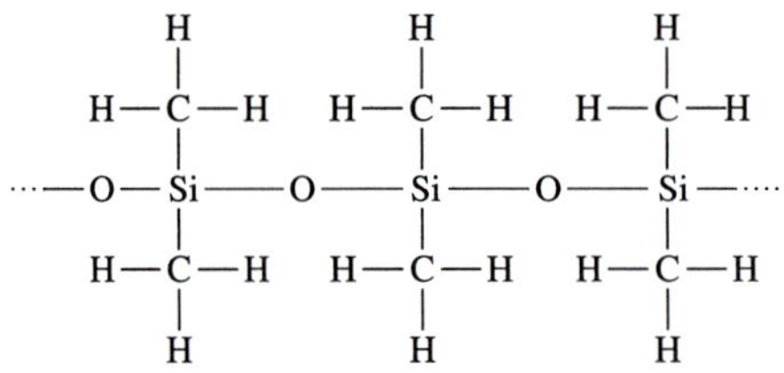	Gaskets, seals

TABLE 16-9 ■ ***Properties of selected elastomers***

	Tensile Strength (MPa)	% Elongation	Density (g/cm^3)
Polyisoprene	21	800	0.93
Polybutadiene	24		0.94
Polyisobutylene	28	350	0.92
Polychloroprene (Neoprene)	24	800	1.24
Butadiene-styrene (BS or SBR rubber)	21	2000	1.0
Butadiene-acrylonitrile	5	400	1.0
Silicones	7	700	1.5
Thermoplastic elastomers	35	1300	1.06

Neoprene, is a common material for hoses and electrical insulation. Many of the important synthetic elastomers are copolymers. Polybutadiene (butadiene rubber or Buna-S) is similar to polyisoprene, but the repeat unit has four carbon atoms consisting of one double bond. This is a relatively low-cost rubber, though the resistance to solvents is poor. As a result, it is used as a toughening material to make other elastomers. Butadiene-styrene rubber (BSR or BS), which is also one of the components of ABS (Figure 16-9), is used for automobile tires. Butyl rubber is different from polybutadiene (butadiene rubber). Butyl rubber, or polyisobutadiene, is used to make inner tubes for tires, vibration mounts, and as weather-stripping material. Silicones are another important elastomer based on chains composed of silicon and oxygen atoms. Silly Putty® was invented by James Wright of General Electric. It is made using hydroxyl terminated polydimethyl siloxane, boric oxide, and some other additives. At slow strain rates you can stretch it significantly, while if you pull it fast it snaps. The silicone rubbers (also known as polysiloxanes) provide high-temperature resistance, permitting use of the elastomer at temperatures as high as 315°C. Low molecular weight silicones form liquids and are known as silicon oils. Silicones can also be purchased as a two-part system that can be molded and cured. Chewing gum contains a base that is made from natural rubber, styrene butadiene, or polyvinyl acetate (PVA).

Thermoplastic Elastomers (TPEs) This is a special group of polymers that do not rely on cross-linking to produce a large amount of elastic deformation. Figure 16-23 shows the structure of a styrene-butadiene block copolymer engineered so that the styrene repeat units are located only at the ends of the chains. Approximately 25% of the chain is composed of styrene. The styrene ends of several chains form spherical-shaped domains. The styrene has a high glass temperature; consequently, the domains are strong and rigid and tightly hold the chains together. Rubbery areas containing butadiene repeat units are located between the styrene domains; these portions of the polymer have a glass temperature below room temperature and therefore behave in a soft, rubbery manner. Elastic deformation occurs by recoverable movement of the chains; however, sliding of the chains at normal temperatures is prevented by the styrene domains.

The styrene-butadiene block copolymers differ from the BS rubber discussed earlier in that cross-linking of the butadiene monomers is not necessary and, in fact, is undesirable. When the thermoplastic elastomer is heated, the styrene heats above the glass temperature, the domains are destroyed, and the polymer deforms in a viscous manner—that is, it behaves as any other thermoplastic, making fabrication very easy. When the polymer cools, the domains reform and the polymer reverts to its elastomer

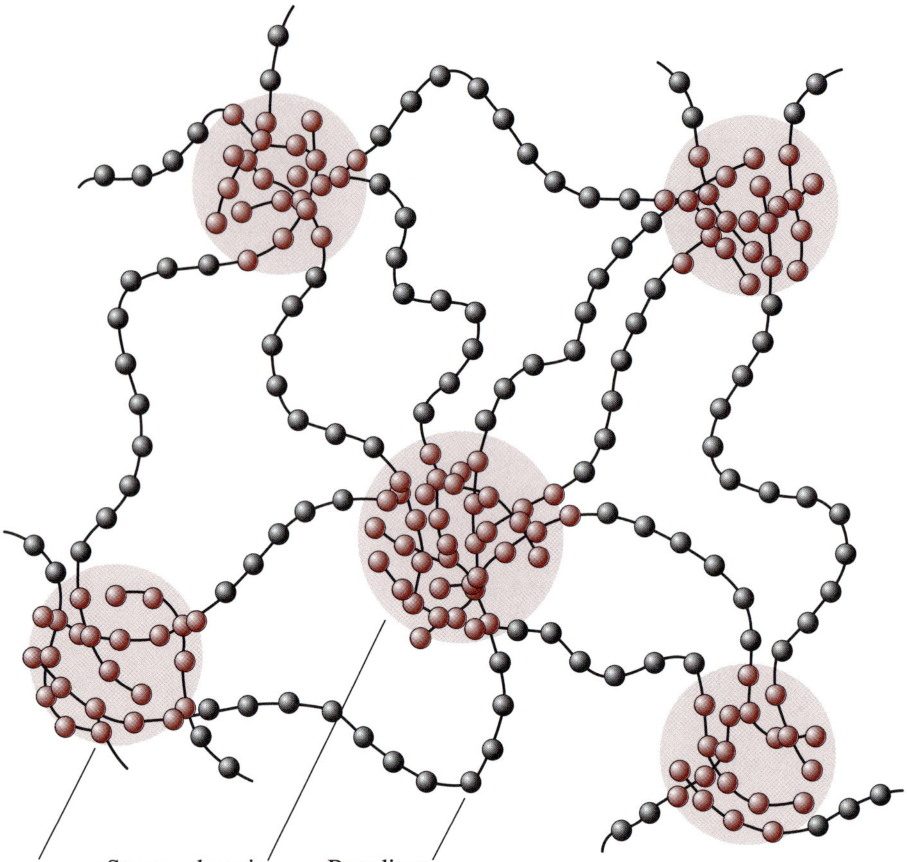

Figure 16-23 The structure of the styrene-butadiene (SB) copolymer in a thermoplastic elastomer. The glassy nature of the styrene domains provides elastic behavior without cross-linking of the butadiene.

characteristics. The thermoplastic elastomers consequently behave as ordinary thermoplastics at elevated temperatures and as elastomers at low temperatures. This behavior also permits the thermoplastic elastomers to be more easily recycled than conventional elastomers. A useful fluoroelastomer for high temperature and corrosive environments is Viton^{TM}. It is used for seals, O-rings, etc.

16-9 Thermosetting Polymers

Thermosets are highly cross-linked polymer chains that form a three-dimensional network structure. Because the chains cannot rotate or slide, these polymers possess good strength, stiffness, and hardness. However, thermosets also have poor ductility and impact properties and a high glass temperature. In a tensile test, thermosetting polymers display the same behavior as a brittle metal or a ceramic.

Thermosetting polymers often begin as linear chains. Depending on the type of repeat units and the degree of polymerization, the initial polymer may be either a solid or a liquid resin; in some cases, a two- or three-part liquid resin is used (as in the case of

TABLE 16-10 ■ *Functional units and applications for selected thermosets*

Polymer	Functional Units	Typical Applications
Phenolics	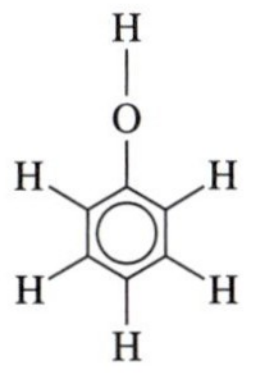	Adhesives, coatings, laminates
Amines	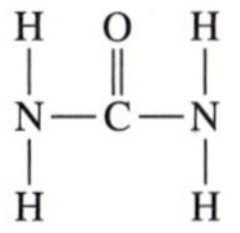	Adhesives, cookware, electrical moldings
Polyesters	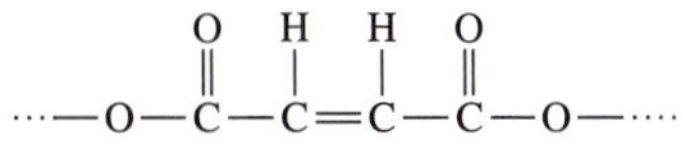	Electrical moldings, decorative laminates, polymer matrix in fiberglass
Epoxies	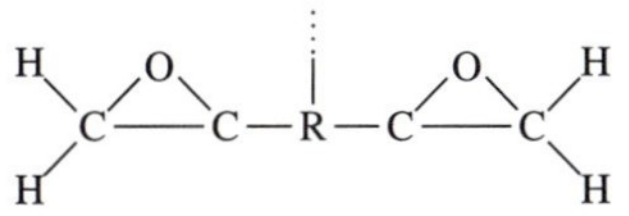	Adhesives, electrical moldings, matrix for composites
Urethanes	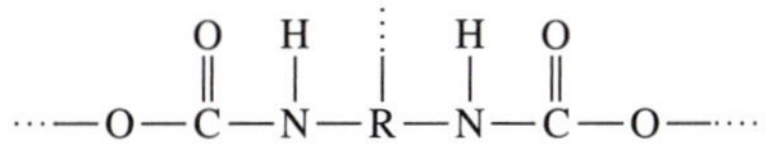	Fibers, coatings, foams, insulation
Silicone	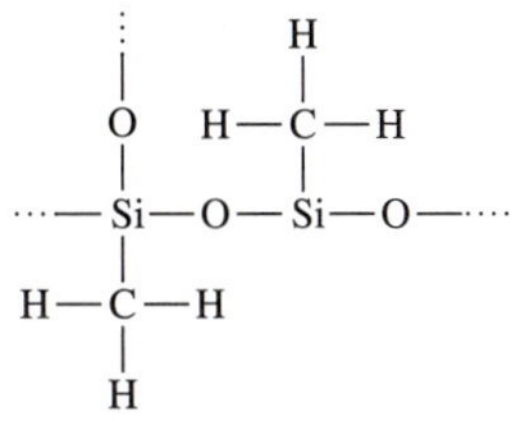	Adhesives, gaskets, sealants

the two tubes of epoxy glue that we often use). Heat, pressure, mixing of the various resins, or other methods initiate the cross-linking process. Cross-linking is not reversible; once formed, the thermosets cannot be reused or recycled conveniently.

The functional groups for a number of common thermosctting polymers are summarized in Table 16-10, and representative properties are given in Table 16-11.

TABLE 16-11 ■ *Properties of typical thermosetting polymers*

	Tensile Strength (MPa)	% Elongation	Elastic Modulus (MPa)	Density (g/cm^3)
Phenolics	62	2	9	1.27
Amines	69	1	11	1.50
Polyesters	90	3	5	1.28
Epoxies	103	6	4	1.25
Urethanes	69	6		1.30
Silicone	28	0	8	1.55

Phenolics Phenolics, the most commonly used thermosets, are often used as adhesives, coatings, laminates, and molded components for electrical or motor applications.

Bakelite™ is one of the common phenolic thermosets. A condensation reaction joining phenol and formaldehyde molecules produces the initial linear phenolic resin.

This process continues until a linear phenol-formaldehyde chain is formed. However, phenol is trifunctional. After the chain has formed, there is a third location on each phenol ring that provides a site for cross-linking with the adjacent chains.

Amines Amino resins, produced by combining urea or melamine monomers with formaldehyde, are similar to the phenolics. The monomers are joined by a formaldehyde link to produce linear chains. Excess formaldehyde provides the cross-linking needed to give strong, rigid polymers suitable for adhesives, laminates, molding materials for cookware, and electrical hardware such as circuit breakers, switches, outlets, and wall plates.

Urethanes Depending on the degree of cross-linking, the urethanes behave as thermosetting polymers, thermoplastics, or elastomers. These polymers find application as fibers, coatings, and foams for furniture, mattresses, and insulation.

Polyesters Polyesters form chains from acid and alcohol molecules by a condensation reaction, giving water as a byproduct. When these chains contain unsaturated bonds, a styrene molecule may provide cross-linking. Polyesters are used as molding or casting materials for a variety of electrical applications, decorative laminates, boats and other marine equipment, and as a matrix for composites such as fiberglass.

Epoxies Epoxies are thermosetting polymers formed from molecules containing a tight C—O—C ring. During polymerization, the C—O—C rings are opened and the bonds are rearranged to join the molecules. The most common of the commercial epoxies is based on bisphenol A, to which have been added two epoxide units. These molecules are polymerized to produce chains and then co-reacted with curing agents that provide cross-linking. Epoxies are used as adhesives, rigid molded parts for electrical applications, automotive components, circuit boards, sporting goods and a matrix for high-performance fiber-reinforced composite materials for aerospace.

Polyimides Polyimides display a ring structure that contains a nitrogen atom. One special group, the bismaleimides (BMI), is important in the aircraft and aerospace industry. They can operate continuously at temperatures of 175°C and do not decompose until reaching 460°C.

Interpenetrating Polymer Networks Some special polymer materials can be produced when linear thermoplastic chains are intertwined through a thermosetting framework, forming **interpenetrating polymer networks**. For example, nylon, acetal, and polypropylene chains can penetrate into a cross-linked silicone thermoset. In more advanced systems, two interpenetrating thermosetting framework structures can be produced.

16-10 Adhesives

Adhesives are polymers used to join other polymers, metals, ceramics, composites, or combinations of these materials. The adhesives are used for a variety of applications. The most critical of these are the "structural adhesives," which find use in the automotive, aerospace, appliance, electronics, construction, and sporting equipment areas.

Chemically Reactive Adhesives These adhesives include polyurethane, epoxy, silicone, phenolics, anaerobics, and polyimides. One-component systems consist of a single polymer resin cured by exposure to moisture, heat, or—in the case of anaerobics—the absence of oxygen. Two-component systems (such as epoxies) cure when two resins are combined.

Evaporation or Diffusion Adhesives The adhesive is dissolved in either an organic solvent or water and is applied to the surfaces to be joined. When the carrier evaporates, the remaining polymer provides the bond. Water-base adhesives are preferred from the standpoint of environmental and safety considerations. The polymer may be completely dissolved in water or may consist of latex, or a stable dispersion of polymer in water. A number of elastomers, vinyls, and acrylics are used.

Hot-Melt Adhesives These thermoplastics and thermoplastic elastomers melt when heated. On cooling, the polymer solidifies and joins the materials. Typical melting temperatures of commercial hot-melts are about 80°C to 110°C, which limits the elevated-temperature use of these adhesives. High-performance hot-melts, such as polyamides and polyesters, can be used up to 200°C.

Pressure-Sensitive Adhesives These adhesives are primarily elastomers or elastomer copolymers produced as films or coatings. Pressure is required to cause the polymer to stick to the substrate. They are used to produce electrical and packaging tapes, labels, floor tiles, wall coverings, and wood-grained textured films.

Conductive Adhesives A polymer adhesive may contain a filler material such as silver, copper, or aluminum flakes or powders to provide electrical and thermal conductivity. In some cases, thermal conductivity is desired but electrical conductivity is not wanted; alumina, boron nitride, and silica may be used as fillers to provide this combination of properties.

16-11 Polymer Processing and Recycling

There are a number of methods for producing polymer shapes, including molding, extrusion, and manufacture of films and fibers. The techniques used to form the polymers depend to a large extent on the nature of the polymer—in particular, whether it is thermoplastic or thermosetting. The greatest variety of techniques are used to form the thermoplastics. The polymer is heated to near or above the melting temperature so that it becomes rubbery or liquid. The polymer is then formed in a mold or die to produce the required shape. Thermoplastic elastomers can be formed in the same manner. In these processes, scrap can be easily recycled and waste is minimized. Fewer forming techniques are used for the thermosetting polymers because, once cross-linking has occurred, the thermosetting polymers are no longer capable of being formed. Elastomers are processed in high-shear equipment such as a Banbury mixer. Carbon black and other additives are added. The heating from viscoelastic deformation can begin to cross-link the material prematurely. After the mixing step, a curing agent (e.g., zinc oxide) is added. The material discharged from the mixer is pliable and is processed using a short extruder, molded using a two-roll mill, or applied on parts by dip coating. This processing of elastomers is known as **compounding** of rubber.

The following are some of the techniques mainly used for processing of polymers; most of these, you will note, apply only to thermoplastics.

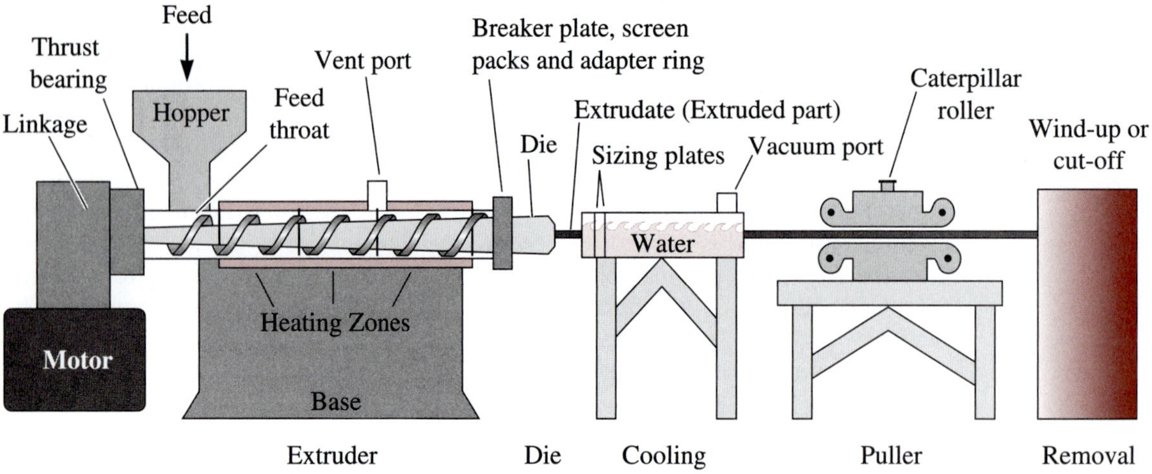

Figure 16-24 Schematic of an extruder used for polymer processing. (*Source: Strong, A. Brent,* Plastics: Materials and Processing, *2nd, © 2000. Electronically reproduced by permission of Pearson Education, Inc., Upper Saddle River, New Jersey.*)

Extrusion This is the most widely used technique for processing thermoplastics. Extrusion can serve two purposes. First, it provides a way to form certain simple shapes continuously (Figure 16-24). Second, extrusion provides an excellent mixer for additives (e.g., carbon black, fillers, etc.) when processing polymers (Figure 16-25) that ultimately may be processed using some other process. A screw mechanism consisting of one or

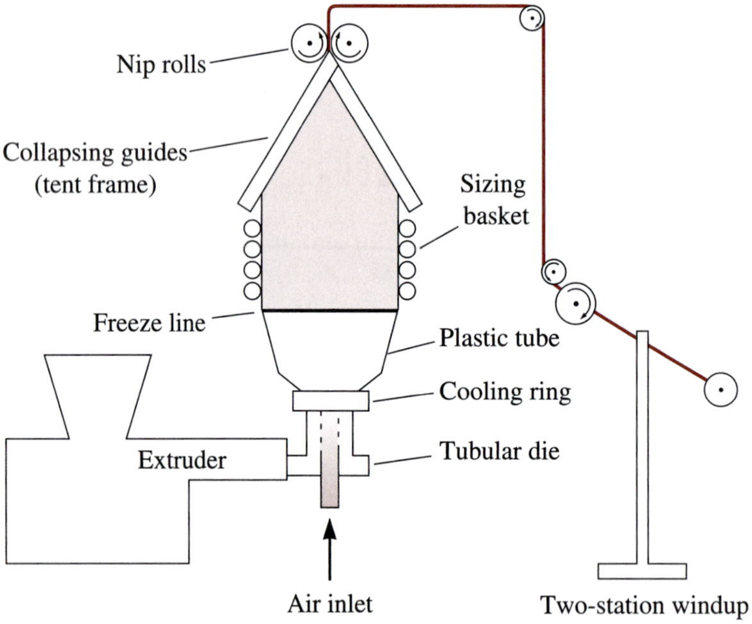

Figure 16-25 One technique by which polymer films (used in the manufacture of garbage bags, for example) can be produced. The film is extruded in the form of a bag, which is separated by air pressure until the polymer cools. (*Source: Strong, A. Brent,* Plastics: Materials and Processing, *2nd, © 2000. Electronically reproduced by permission of Pearson Education, Inc., Upper Saddle River, New Jersey.*)

a pair of screws (twin screw) forces heated thermoplastic (either solid or liquid) and additives through a die opening to produce solid shapes, films, sheets, tubes, pipes, and even plastic bags (Figure 16-24). An industrial extruder can be up to 18 m to 21 m long, 60 cm in diameter, and consist of different heating or cooling zones. Since thermoplastics show shear thinning behavior and are viscoelastic, the control of both temperature and viscosity is critical in polymer extrusion. One special extrusion process for producing films is illustrated in Figure 16-25. Extrusion also can be used to coat wires and cables with either thermoplastics or elastomers.

Blow Molding A hollow preform of a thermoplastic called a **parison** is introduced into a die by gas pressure and expanded against the walls of the die (Figure 16-26). This process is used to produce plastic bottles, containers, automotive fuel tanks, and other hollow shapes.

Injection Molding Thermoplastics heated above the melting temperature using an extruder are forced into a closed die to produce a molding. This process is similar to die casting of molten metals. A plunger or a special screw mechanism applies pressure

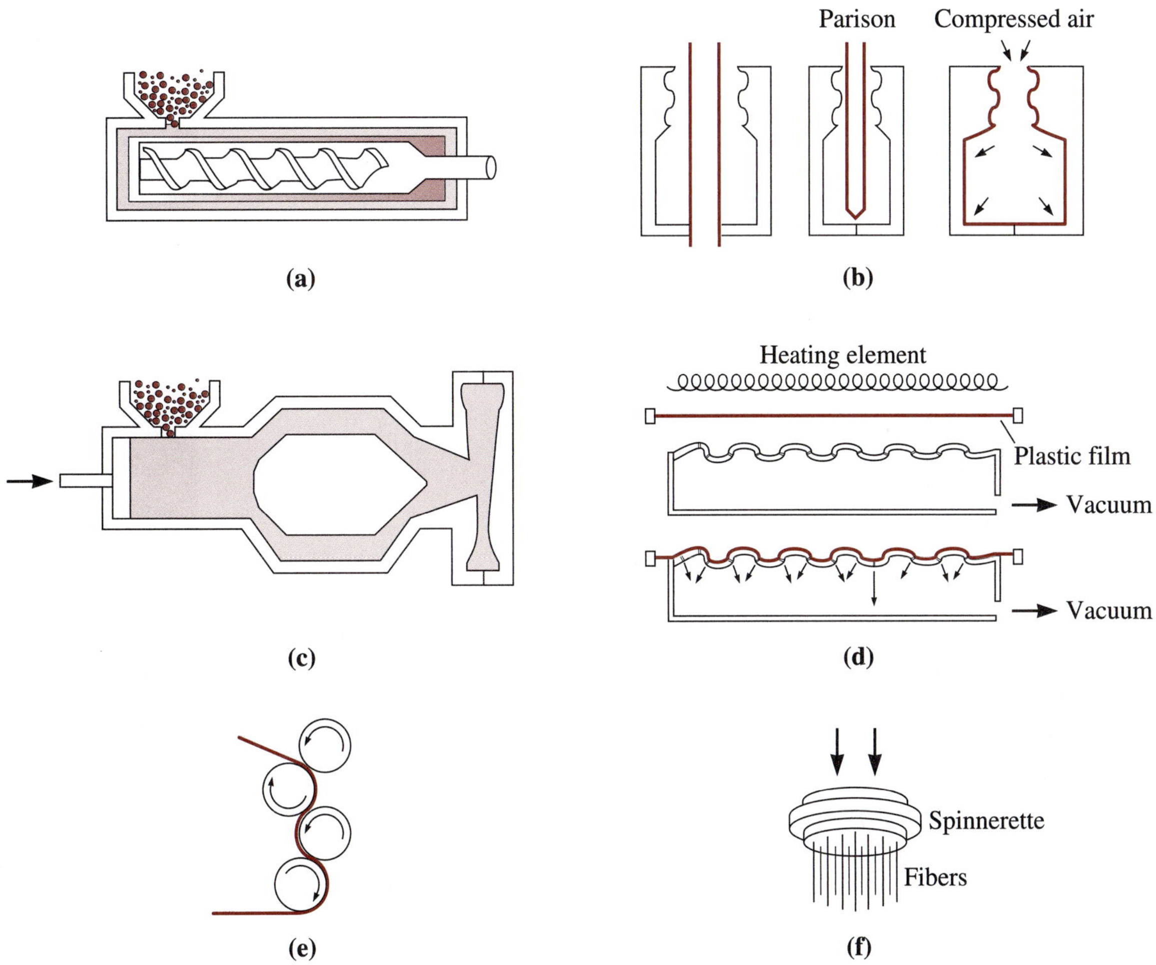

Figure 16-26 Typical forming processes for thermoplastic: (a) extrusion, (b) blow molding, (c) injection molding, (d) thermoforming, (e) calendaring, and (f) spinning.

to force the hot polymer into the die. A wide variety of products, ranging from cups, combs, and gears to garbage cans, can be produced in this manner.

Thermoforming Thermoplastic polymer sheets heated to the plastic region can be formed over a die to produce such diverse products as egg cartons and decorative panels. The forming can be done using matching dies, a vacuum, or air pressure.

Calendaring In a calendar, molten plastic is poured into a set of rolls with a small opening. The rolls, which may be embossed with a pattern, squeeze out a thin sheet of the polymer—often, polyvinyl chloride. Typical products include vinyl floor tile and shower curtains.

Spinning Filaments, fibers, and yarns may be produced by spinning. The molten thermoplastic polymer is forced through a die containing many tiny holes. The die, called a **spinnerette**, can rotate and produce a yarn. For some materials, including nylon, the fiber may subsequently be stretched to align the chains parallel to the axis of the fiber; this process increases the strength of the fibers.

Casting Many polymers can be cast into molds and permitted to solidify. The molds may be plate glass for producing individual thick plastic sheets or moving stainless steel belts for continuous casting of thinner sheets. *Rotational molding* is a special casting process in which molten polymer is poured into a mold rotating about two axes. Centrifugal action forces the polymer against the walls of the mold, producing a thin shape such as a camper top.

Compression Molding Thermoset moldings are most often formed by placing the solid material before cross-linking into a heated die (Figure 16-27a). Application of high

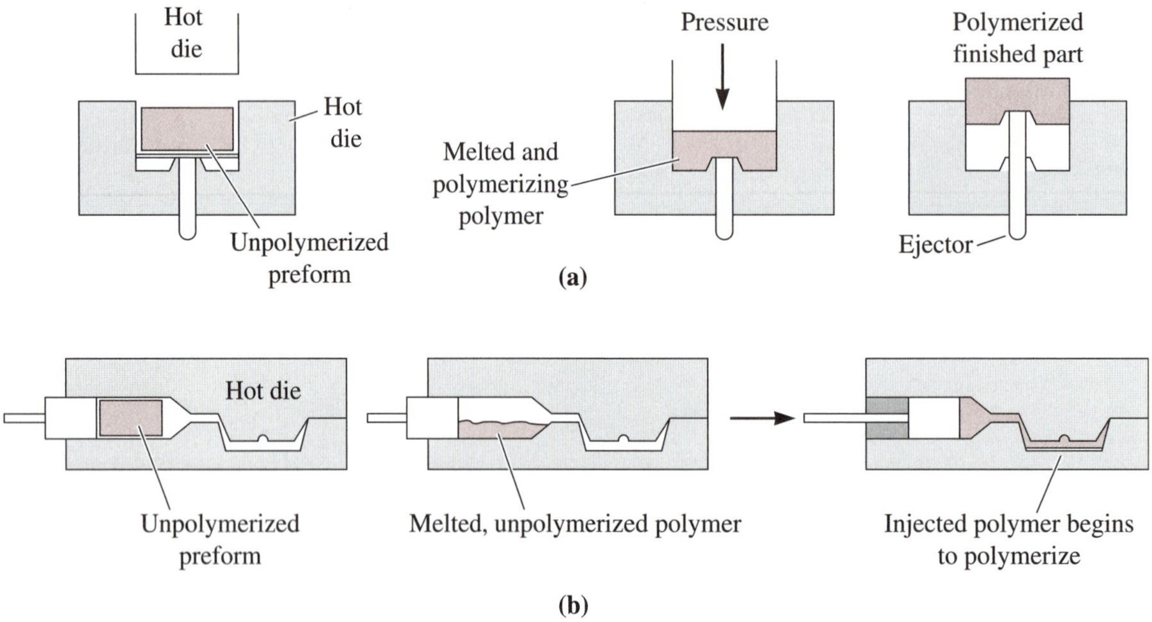

Figure 16-27 Typical forming processes for thermosetting polymers: (a) compression molding and (b) transfer molding.

pressure and temperature causes the polymer to melt, fill the die, and immediately begin to harden. Small electrical housings as well as fenders, hoods, and side panels for automobiles can be produced by this process.

Transfer Molding A double chamber is used in the transfer molding of thermosetting polymers. The polymer is heated under pressure in one chamber. After melting, the polymer is injected into the adjoining die cavity. This process permits some of the advantages of injection molding to be used for thermosetting polymers (Figure 16-27b).

Reaction Injection Molding (RIM) Thermosetting polymers in the form of liquid resins are first injected into a mixer and then directly into a heated mold to produce a shape. Forming and curing occur simultaneously in the mold. In reinforced-reaction injection molding (RRIM), a reinforcing material consisting of particles or short fibers is introduced into the mold cavity and is impregnated by the liquid resins to produce a composite material. Automotive bumpers, fenders, and furniture parts are made using this process.

Foams Foamed products can be produced in polystyrene, urethanes, polymethyl methacrylate, and a number of other polymers. The polymer is produced in the form of tiny beads, often containing a blowing agent such as pentane. During the pre-expansion process, the bead increases in diameter by as many as 50 times. The pre-expanded beads are then injected into a die, with the individual beads fusing together, using steam, to form exceptionally lightweight products with densities of perhaps only 0.02 g/cm^3. Expandable polystyrene (EPS) cups, packaging, and insulation are some of the applications for foams. Engine blocks for many automobiles are made using a pattern made from expanded polystyrene beads.

EXAMPLE 16-10 *Insulation Boards for Houses*

You want to design a material that can be used for making insulation boards that are approximately 120 cm wide and 240 cm tall. The material must provide good thermal insulation. What material would you choose?

SOLUTION

Glasses tend to be good insulators of heat. However, they will be heavy, more expensive, and prone to fracture. Polymers are lightweight, can be produced inexpensively, and they can be good thermal insulators. We can use foamed polystyrene since the air contained in the beads adds significantly to their effectiveness as thermal insulators. For better mechanical properties, we may want to produce foams that have relatively high density (compared to foams that are used to make coffee cups). Finally, from a safety viewpoint, we want to be sure that some fire and flame retardants are added to the foams. Such panels are made using expanded polystyrene beads containing pentane. A molding process is used to make the foams. The sheets can be cut into required sizes using a heated metal wire.

EXAMPLE 16-11 *Thermoset Polymers by Compression Molding*

Thermosets are commonly processed using a process known as compression molding. This process is similar to sintering of ceramics in that resin (like ceramic powder) is fed into mold cavities, that depend upon the shape of the object being made, and then compressed between platens. The resin is then heated while being pressed. Under the application of heat and pressure the resin crosslinks (Figure 16-27). Additional pressure is needed if the component being made is deeper (thicker). The force required to form a part of certain size by compression molding is given by:

$$F = (A)[(P_A) + (\rho \times d_e)]$$

In this, F: force, A: projected area of the part which is exposed to pressure, P_A is the cavity pressure (usually experimentally developed) necessary for a given thermoset, ρ: depth factor, i.e., pressure per unit length needed beyond a certain depth d_e (Ref. A.B. Strong, *Plastics: Materials and Processing, Second Edn.*, Publisher Prentice Hall). Calculate the force needed to form a thermoset shown below. Assume that for a thin part (i.e., less than 3 cm thick part) the value of P_A is 25 MPa, ρ is 2 MPa/cm beyond a depth of 3 cm. Assume the projected width and length are 10 cm and 20 cm, respectively. Assume that the depth of the part is 7 cm.

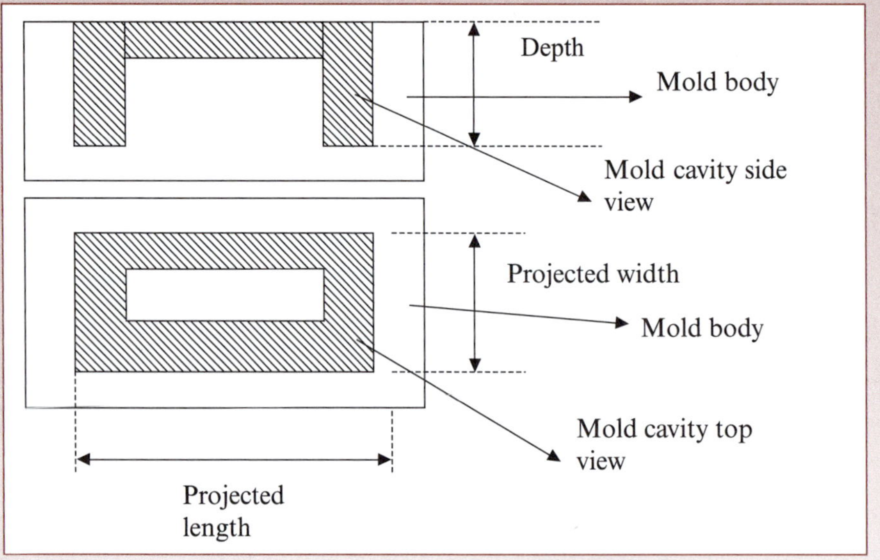

SOLUTION

The total projected area for the cavity is 20 × 10 cm = 200 cm^2. In this case, there is only one cavity that is in the mold that will be filled with the resin. The depth of the part is 7 cm; this is 4 cm beyond what is considered thin here. Thus, the total force necessary will be:

$$F = (A)[(P_A) + (\rho \times d_e)]$$

$$F = (200\ \text{cm}^2)[25\ \text{MPa} + (2\ \text{MPa/cm} \times (7 - 3)\text{cm})]$$

$$F = 200\ \text{cm}^2[33\ \text{MPa}]$$

$$F = (200 \times 10^{-4}\ \text{m}^2) \times 33\ \text{MN/m}^2 = 0.66\ \text{MN} = 660\ \text{kN}$$

In practice, it can be assumed that a factor of safety of 1.3 is necessary. This means, we should have equipment that can deliver a necessary force of 660 + 198 kN. Thus, a force of 858 kN or approximately 900 kN will be required.

Recycling of Plastics Recycling is a very important issue and a full discussion of the entire process is outside the scope of this book. But recycling plays an important role in our everyday lives. Material is recycled in many ways. For example, part of the polymer that is scrap from a manufacturing process (known as regrind) is used by recycling plants. The recycling of thermoplastics is relatively easy and practiced widely. Note that many of the everyday plastic products you encounter (bags, soda bottles, yogurt containers, etc.) have numbers stamped on them. For PET products (recycling symbol "PETE", because of trademark issues), the number is 1. For HDPE vinyl (recycling symbol V), LDPE, PP, and PS the numbers are 2, 3, 4, 5 and 6, respectively. Other plastics are marked number 7.

Thermosets and elastomers are more difficult to recycle, although they can still be used. For example, tires can be shredded and used to make safer playground surfaces or roads.

Despite enormous recycling efforts, a large portion of the materials in a landfill today are plastics (the largest is that of paper). Given the limited amount of petroleum, the threat of global warming, and a need for a cleaner and safer environment, careful use and recycling makes sense for all materials.

SUMMARY

- Polymers are made from large macromolecules produced by the joining of smaller molecules, called monomers, using addition or condensation polymerization reactions. *Plastics* are materials that are based on polymeric compounds, and they contain many other additives that improve their properties. Compared with most metals and ceramics, plastics have low strength, stiffness, and melting temperatures; however, they also have a low density and good chemical resistance. Plastics are used in a most diverse number of technologies.

- Thermoplastics have chains that are not chemically bonded to each other, permitting the material to be easily formed into useful shapes, to have good ductility, and to be economically recycled. Thermoplastics can have an amorphous structure, which provides low strength and good ductility when the ambient temperature is above the glass temperature. The polymers are more rigid and brittle when the temperature falls below the glass temperature. Many thermoplastics can partially crystallize during cooling or by application of a stress. This increases their strength.

- The thermoplastic chains can be made more rigid and stronger by using nonsymmetrical monomers that increase the bonding strength between the chains and make it more difficult for the chains to disentangle when stress is applied. In addition, many monomers produce more rigid chains containing atoms or groups of atoms other than carbon; this structure also produces high-strength thermoplastics.

- Elastomers are thermoplastics or lightly cross-linked thermosets that exhibit greater than 200% elastic deformation. Chains are eventually cross-linked using vulcanization. The cross-linking makes it possible to obtain very large elastic deformations without permanent plastic deformation. Increasing the number of cross-links increases the stiffness and reduces the amount of elastic deformation of the elastomers.

- Thermoplastic elastomers combine features of both thermoplastics and elastomers. At high temperatures, these polymers behave as thermoplastics and are plastically formed into shapes; at low temperatures, they behave as elastomers.
- Thermosetting polymers are highly cross-linked into a three-dimensional network structure. Typically, high glass temperatures, good strength, and brittle behavior are found. Once cross-linking occurs, these polymers cannot be easily recycled.
- Manufacturing processes used for polymers depend on their behavior. Processes such as extrusion, injection molding, thermoforming, casting, drawing, and spinning are made possible by the viscoelastic behavior of the thermoplastics. The nonreversible behavior of bonding in thermosetting polymers limits their processing to fewer techniques, such as compression molding, transfer molding and reaction-injection molding.

GLOSSARY

Addition polymerization Process by which polymer chains are built up by adding monomers together without creating a byproduct.

Aging Slow degradation of polymers as a result of exposure to low levels of heat, oxygen, bacteria, or ultraviolet rays.

Aramids Polyamide polymers containing aromatic groups of atoms in the linear chain.

Blushing A thermoplastic bent repeatedly leads to small volumes of the material crystallizing; this leads to voids that ultimately cause the material to fail.

Branched polymer Any polymer consisting of chains that consist of a main chain and secondary chains that branch off from the main chain.

Compounding Processing of elastomers in device known as a Banbury mixer followed by forming using extrusion, molding, or dip coating.

Condensation polymerization A polymerization mechanism in which a small molecule (e.g., water, methanol, etc.) is condensed out as a byproduct.

Copolymer An addition polymer produced by joining more than one type of monomer.

Crazing Localized plastic deformation in a polymer. A craze may lead to the formation of cracks in the material.

Cross-linking Attaching chains of polymers together to produce a three-dimensional network polymer.

Degree of polymerization The average molecular weight of the polymer divided by the molecular weight of the monomer.

Diene A group of monomers that contain two double-covalent bonds. These monomers are often used in producing elastomers.

Elastomers These are polymers (thermoplastics or lightly cross-linked thermosets) that have an elastic deformation $> 200\%$.

Geometric isomer A molecule that has the same composition as, but a structure different from, a second molecule.

Glass temperature (T_g) The temperature range below which the amorphous polymer assumes a rigid glassy structure.

Heat-deflection temperature The temperature at which a polymer will deform a given amount under a standard load (also called distortion temperature).

Heat-degradation temperature The temperature above which a polymer burns, chars, or decomposes.

Interpenetrating polymer networks Polymer structures produced by intertwining two separate polymer structures or networks.

Linear polymer Any polymer in which molecules are in the form of spaghetti-like chains.

Liquid-crystalline polymers Exceptionally stiff polymer chains that act as rigid rods, even above their melting point.

Mer A unit group of atoms and molecules that defines a characteristic arrangement for a polymer. A polymer can be thought off as a material made by combining several mers or units.

Monomer The molecule from which a polymer is produced.

Oligomer Low molecular weight molecules, these may contain two (dimers) or three (trimers) mers.

Parison A hot glob of soft or molten polymer that is blown or formed into a useful shape.

Plastic A predominantly polymeric material made with some additives.

Polymer Polymers are materials made from giant (or macromolecular), chain-like molecules having average molecular weights from 10,000 to more than 1,000,000 g/mol built by the joining of many mers or units by chemical bonds. Polymers are usually, but not always, carbon based.

Relaxation time A property of a polymer that is related to the rate at which stress relaxation occurs.

Repeat unit The repeating structural unit from which a polymer is built. Also called a *mer*.

Spinnerette An extrusion die containing many small openings through which hot or molten polymer is forced to produce filaments. Rotation of the spinnerette twists the filaments into a yarn.

Stress-induced crystallization The process of forming crystals by the application of an external stress. Typically, a significant fraction of many amorphous plastics can be crystallized in this fashion, making them stronger.

Stress relaxation A reduction of the stress acting on a material over a period of time at a constant strain due to viscoelastic deformation.

Tacticity Describes the location in the polymer chain of atoms or atom groups in nonsymmetrical monomers.

Thermoplastic elastomers Polymers that behave as thermoplastics at high temperatures, but as elastomers at lower temperatures.

Thermoplastics Linear or branched polymers in which chains of molecules are not interconnected to one another.

Thermosetting polymers Polymers that are heavily cross-linked to produce a strong three dimensional network structure.

Unsaturated bond The double- or even triple-covalent bond joining two atoms together in an organic molecule. When a single covalent bond replaces the unsaturated bond, polymerization can occur.

Viscoelasticity The deformation of a material by elastic deformation and viscous flow of the material when stress is applied.

Vulcanization Cross-linking elastomer chains by introducing sulfur or other chemicals.

PROBLEMS

Section 16-1 Classification of Polymers

16-1 What are linear and branched polymers? Can thermoplastics be branched?

16-2 Define **(a)** a thermoplastic, **(b)** thermosetting plastics, **(c)** elastomers, and **(d)** thermoplastic elastomers.

16-3 What electrical and optical applications are polymers used for? Explain using examples.

16-4 What are the major advantages associated with plastics compared to ceramics, glasses, and metallic materials?

Sections 16-2 Addition and Condensation Polymerization

16-5 What do the terms condensation polymerization, addition polymerization, initiator, and terminator mean?

Section 16-3 Degree of Polymerization

Section 16-4 Typical Thermoplastics

16-6 Explain why low-density polyethylene is good to make grocery bags, however, super high molecular weight polyethylene must be used where strength and very high wear resistance is needed.

16-7 The molecular weight of polymethyl methacrylate (see Table 16-3) is 250,000 g/mol. If all of the polymer chains are the same length,

(a) calculate the degree of polymerization, and
(b) the number of chains in 1 g of the polymer.

16-8 Calculate **(a)** the degree of polymerization if polytetrafluoroethylene (see Table 16-3) is 7500. **(b)** If all of the polymer chains are the same length, calculate

(i) the molecular weight of the chains, and
(ii) the total number of chains in 1000 g of the polymer.

16-9 A polyethylene rope weighs 15.12 g/cm. If each chain contains 7000 repeat units,

(a) calculate the number of polyethylene chains in a 3-m length of rope, and
(b) the total length of chains in the rope, assuming that carbon atoms in each chain are approximately 0.15 nm apart.

16-10 Analysis of a sample of polyacrylonitrile (see Table 16-3) shows that there are six lengths of chains, with the following number of chains of each length. Determine

(a) the weight average molecular weight and degree of polymerization, and
(b) the number average molecular weight and degree of polymerization.

Number of Chains	Mean Molecular Weight of Chains (g/mol)
10,000	3,000
18,000	6,000
17,000	9,000
15,000	12,000
9,000	15,000
4,000	18,000

Section 16-5 Structure–Property Relationships in Thermoplastics

Section 16-6 Effect of Temperature on Thermoplastics

Section 16-7 Mechanical Properties of Thermoplastics

16-11 Explain what the following terms mean: decomposition temperature, heat distortion temperature, glass temperature, and melting temperature. Why is it that thermoplastics do not have a fixed melting or glass temperature?

16-12 Using Table 16-5, plot the relationship between the glass temperatures and the melting temperatures of the addition thermoplastics. What is the approximate relationship between these two critical temperatures? Do the condensation thermoplastics and the elastomers also follow the same relationship?

16-13 List the addition polymers in Table 16-5 that might be good candidates for making the bracket that holds the rearview mirror onto the outside of an automobile, assuming that temperatures frequently fall below zero degrees Celsius. Explain your choices.

16-14 Based on Table 16-5, which of the elastomers might be suited for use as a gasket in a pump for liquid CO_2 at $-78°C$? Explain.

16-15 How do the glass temperatures of polyethylene, polypropylene, and polymethyl methacrylate compare? Explain their differences, based on the structure of the monomer.

16-16 Which of the addition polymers in Table 16-5 are used in their leathery condition at room temperature? How is this condition expected to affect their mechanical properties compared with those of glassy polymers?

16-17 What factors influence the crystallinity of polymers? Explain the development and role of crystallinity in PET and nylon.

16-18 Describe the relative tendencies of the following polymers to crystallize. Explain your answer.

- **(a)** branched polyethylene versus linear polyethylene,
- **(b)** polyethylene versus polyethylene-polypropylene copolymer,
- **(c)** isotactic polypropylene versus atactic polypropylene, and
- **(d)** polymethyl methacrylate versus acetal (polyoxymethylene).

16-19 Explain the meaning of these terms: creep, stress relaxation, crazing, blushing, environmental stress cracking, and aging of polymers.

16-20 A stress of 17 MPa is applied to a polymer serving as a fastener in a complex assembly. At a constant strain, the stress drops to 16.6 MPa after 100 h. If the stress on the part must remain above 14.5 MPa in order for the part to function properly, determine the life of the assembly.

16-21 A stress of 7 MPa is applied to a polymer that operates at a constant strain; after six months, the stress drops to 5.9 MPa. For a particular application, a part made of the same polymer must maintain a stress of 6.2 MPa after 12 months. What should be the original stress applied to the polymer for this application?

16-22 Data for the rupture time of polyethylene are shown in Figure 16-17. At an applied stress of 4.8 MPa, the figure indicates that the polymer ruptures in 0.2 h at 90°C but survives 10,000 h at 65°C. Assuming that the rupture time is related to the viscosity, calculate the activation energy for the viscosity of the polyethylene and estimate the rupture time at 23°C.

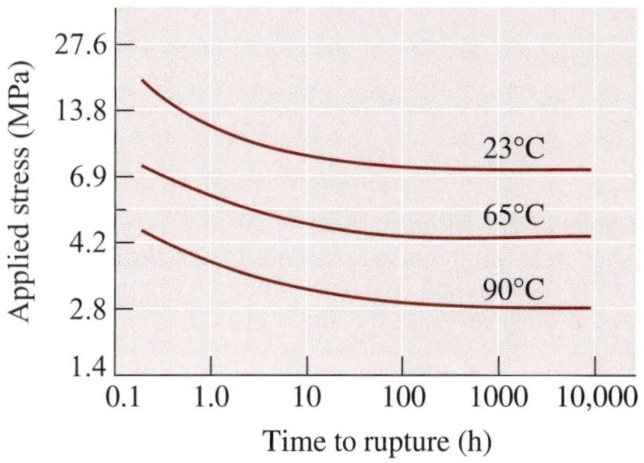

Figure 16-17 (Repeated for Problem 16-22) The effect of temperature on the stress-rupture behavior of high-density polyethylene.

16-23 For each of the following pairs, recommend the one that will most likely have the better impact properties at 25°C. Explain each of your choices.

- **(a)** polyethylene versus polystyrene,
- **(b)** low-density polyethylene versus high-density polyethylene, and
- **(c)** polymethyl methacrylate versus polytetrafluoroethylene.

Section 16-8 Elastomers (Rubbers)

16-24 The polymer ABS can be produced with varying amounts of styrene, butadiene, and acrylonitrile monomers, which are present in the form of two copolymers: BS rubber and SAN.

- **(a)** How would you adjust the composition of ABS if you wanted to obtain good impact properties?
- **(b)** How would you adjust the composition if you wanted to obtain good ductility at room temperature?
- **(c)** How would you adjust the composition if you wanted to obtain good strength at room temperature?

16-25 Figure 16-22 on the next page shows the stress-strain curve for an elastomer. From the curve, calculate and plot the modulus of elasticity versus strain and explain the results.

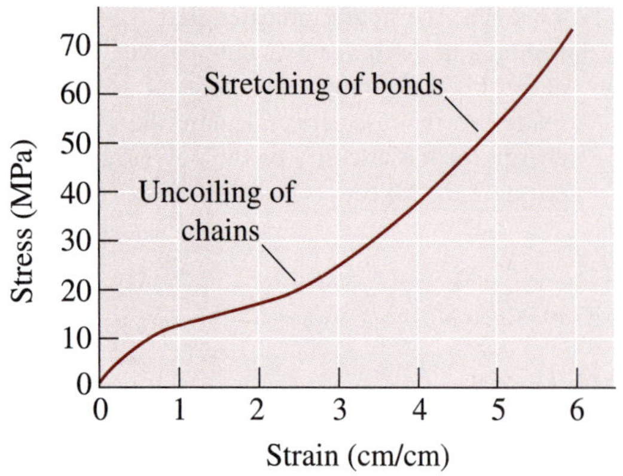

Figure 16-22 (Repeated for Problem 16-25)
The stress-strain curve for an elastomer. Virtually all of the deformation is elastic; therefore, the modulus of elasticity varies as the strain changes.

Section 16-9 Thermosetting Polymers

16-26 Explain the term "thermosetting polymer". Why can't a thermosetting polymer be produced using just adipic acid and ethylene glycol?

16-27 Explain why the degree of polymerization is not usually used to characterize thermosetting polymers.

16-28 Defend or contradict the choice to use the following materials as hot-melt adhesives for an application in which the assembled part is subjected to impact-type loading:

(a) polyethylene,
(b) polystyrene,
(c) styrene-butadiene thermoplastic elastomer,
(d) polyacrylonitrile, and
(e) polybutadiene.

16-29 Compare and contrast properties of thermoplastics, thermosetting materials, and elastomers.

17

Composites: Teamwork and Synergy in Materials

Have You Ever Wondered?

- *What are some of the naturally occurring composite materials?*
- *Why is abalone shell, made primarily of calcium carbonate, so much stronger than chalk, which is also made of calcium carbonate?*
- *What sporting gear applications make use of composites?*
- *Why does the new 787* Dreamliner *plane have 50% composite materials?*

Composites are produced when two or more materials or phases are used together to give a combination of properties that cannot be attained otherwise. Composite materials may be selected to give unusual combinations of stiffness, strength, weight, high-temperature performance, corrosion resistance, hardness, or conductivity. Composites highlight how different materials can work in synergy. Abalone shell, wood, bone, and teeth are examples of naturally occurring composites. Microstructures of selected composites are shown in Figure 17-1. An example of a material that is a composite at a macro-scale would be steel-reinforced concrete. Composites that are at a

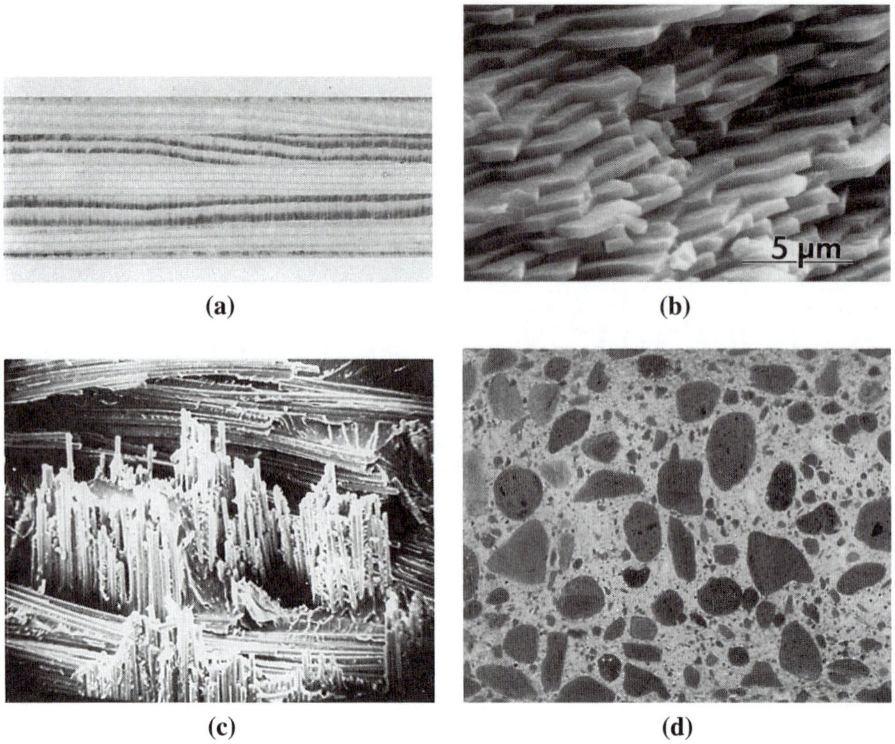

Figure 17-1 Some examples of composite materials: (a) plywood is a laminar composite of layers of wood veneer, (b) abalone shell, (c) fiberglass is a fiber-reinforced composite containing stiff, strong glass fibers in a softer polymer matrix ($\times 175$), and (d) concrete is a particulate composite containing coarse sand or gravel in a cement matrix (reduced 50%).

micro-scale include such materials as carbon or glass-fiber reinforced plastics (CFRP or GFRP). These composites offer significant gains in specific strengths and are finding increasing uses in airplanes, electronic components, automotives and sporting equipment. For example, the latest Boeing 787 *Dreamliner* plane has 50% composite materials. The resultant lightweight structure will cause 20% increase in its fuel efficiency.

As mentioned in Chapters 12 and 13, dispersion-strengthened (like steels) and precipitation-hardened alloys are examples of traditional materials that are **nanocomposites**. In a nanocomposite, the **dispersed phase** consists of nanoscale particles and is distributed in a *matrix phase*. Essentially, the same concept has been applied in developing what are described as **hybrid organic-inorganic nanocomposites**. These are materials in which the molecular, or microstructure, of the composites consists of an inorganic part or block and an organic block. The idea is not too different from the formation of block copolymers (Chapter 16). These are often made using the sol-gel process. These and other functional composites can provide unusual combinations of electronic, magnetic, or optical properties. For example, a porous dielectric material prepared using phase-separated inorganic glasses exhibits a dielectric constant that is lower than that for the same material with no porosity. A transparent polymer that is responsive to magnetic field can also be prepared using composites. Space shuttle tiles made from silica fibers are lightweight because they consists of air and silica fibers and exhibit a very small thermal conductivity. The two phases in these examples would be ceramic and air. Many glass-ceramics are nano-scale composites of different ceramic phases. Many plastics can be considered composites as well. For example, Dylark™ is a composite of maleic anhydride-styrene copolymer. It contains carbon black for stiffness and protection against ultraviolet rays. It also contains glass fibers for increased Young's modulus and rubber for toughness. Silver-filled epoxies provide thermal conductivities higher than those for expoxies. Some dielectric materials are made using multiple phases such that the overall dielectric properties of interest (e.g., the dielectric constant) do not change appreciably with temperature (within a certain range). Similarly, many composites are prepared using magnetic and optical materials. Some composite structures may consist of different materials arranged in different layers. This leads to what are known as functionally graded materials and structures. For example, yttria stabilized zirconia (YSZ) coating on a turbine blade will have other layers in between that provide bonding with the turbine blade material. The YSZ coating itself is made using a plasma spray or other technique and contains certain levels of porosity which are essential for providing protection against high temperatures. Similarly, different coatings on glass are examples of composite structures. Thus, the *concept* of using composites is a generic one and can be applied at the macro, micro, and nano length-scales.

In composites, the properties and volume fractions of individual phases are important. The connectivity of phases is also very important. Usually the **matrix phase** is the continuous phase and the other phase is said to be the dispersed phase. Thus, terms such as "metal-matrix" indicate a metallic material used to form the continuous phase. Connectivity describes how the two or more phases are connected in the composite. Newnham has described a connectivity model for describing connectivities for functional composites. Composites are often classified based on the shape or nature of the dispersed phase (e.g., particle-reinforced, whisker-reinforced, or fiber-reinforced composites). **Whiskers** are like fibers; but their length is much smaller. The bonding between the particles, whiskers, or fibers and the matrix is also very important. In structural composites, polymeric molecules known as "coupling agents" are used. These molecules form bonds with the dispersed phase and become integrated into the continuous matrix phase as well.

In this chapter, we will primarily focus on composites used in structural or mechanical applications. Composites can be placed into three categories—particulate, fiber, and laminar—based on the shapes of the materials (Figure 17-1). Concrete, a mixture of cement and gravel, is a particulate composite; fiberglass, containing glass fibers embedded in a polymer, is a fiber-reinforced composite; and plywood, having alternating layers of wood veneer, is a laminar composite. If the reinforcing particles are uniformly distributed, particulate composites have isotropic properties; fiber composites may be either isotropic or anisotropic; laminar composites always display anisotropic behavior.

17-1 Dispersion-Strengthened Composites

A special group of dispersion-strengthened nanocomposite materials containing particles 10 to 250 nm in diameter is classified as particulate composites. These **dispersoids**, usually a metallic oxide, are introduced into the matrix by means other than traditional phase transformations (Chapters 12 and 13). Even though the small particles are not coherent with the matrix, they block the movement of dislocations and produce a pronounced strengthening effect.

At room temperature, the dispersion-strengthened composites may be weaker than traditional age-hardened alloys, which contain a coherent precipitate. However, because the composites do not catastrophically soften by overaging, overtempering, grain

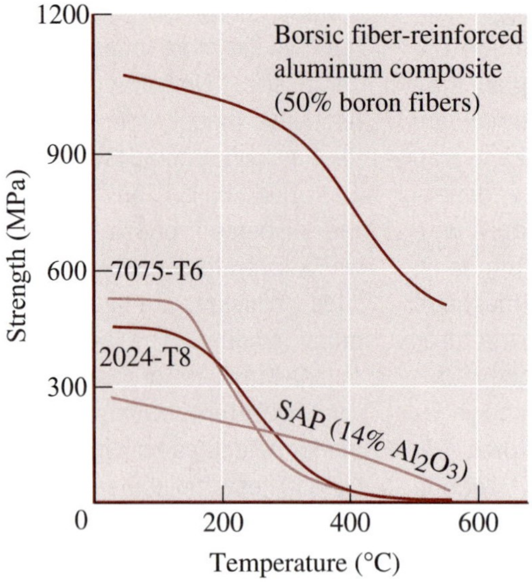

Figure 17-2
Comparison of the yield strength of dispersion-strengthened sintered aluminum powder (SAP) composite with that of two conventional two-phase high-strength aluminum alloys. The composite has benefits above about 300°C. A fiber-reinforced aluminum composite is shown for comparison.

growth, or coarsening of the dispersed phase, the strength of the composite decreases only gradually with increasing temperature (Figure 17-2). Furthermore, their creep resistance is superior to that of metals and alloys.

The dispersoid must have a low solubility in the matrix and must not chemically react with the matrix, but a small amount of solubility may help improve the bonding between the dispersant and the matrix. Copper oxide (Cu_2O) dissolves in copper at high temperatures; thus, the Cu_2O-Cu system would not be effective. However, Al_2O_3 does not dissolve in aluminum; the Al_2O_3-Al system does give an effective dispersion-strengthened material.

Examples of Dispersion-Strengthened Composites Table 17-1 lists some materials of interest. Perhaps the classic example is the sintered aluminum powder (SAP) composite. SAP has an aluminum matrix strengthened by up to 14% Al_2O_3. The composite is formed by powder metallurgy. In one method, aluminum and alumina powders are blended, compacted at high pressures, and sintered. In a second technique, the aluminum powder is treated to add a continuous-oxide film on each particle. When the powder is compacted, the oxide film fractures into tiny flakes that are surrounded by the aluminum metal during sintering.

TABLE 17-1 ■ *Examples and applications of selected dispersion-strengthened composites*

System	Applications
Ag-CdO	Electrical contact materials
Al-Al_2O_3	Possible use in nuclear reactors
Be-BeO	Aerospace and nuclear reactors
Co-ThO_2, Y_2O_3	Possible creep-resistant magnetic materials
Ni-20% Cr-ThO_2	Turbine engine components
Pb-PbO	Battery grids
Pt-ThO_2	Filaments, electrical components
W-ThO_2, ZrO_2	Filaments, heaters

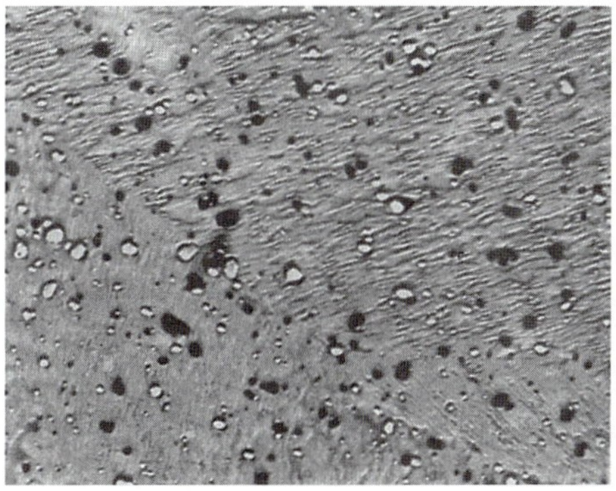

Figure 17-3
Electron micrograph of TD-nickel. The dispersed ThO_2 particles have a diameter of 300 nm or less (×2000). (*From* Oxide Dispersion Strengthening, *p. 714, Gordon and Breach, 1968. © AIME.*)

Another important group of dispersion-strengthened composites includes thoria-dispersed (ThO_2) metals such as TD-nickel (Figure 17-3). TD-nickel can be produced by internal oxidation. Thorium is present in nickel as an alloying element. After a powder compact is made, oxygen is allowed to diffuse into the metal, react with the thorium, and produce thoria (ThO_2). The following example illustrates calculations related to a dispersion-strengthened composite.

EXAMPLE 17-1 ***TD-Nickel Composite***

Suppose 2 wt% ThO_2 is added to nickel. Each ThO_2 particle has a diameter of 1000 A. How many particles are present in each cubic centimeter?

SOLUTION

The densities of ThO_2 and nickel are 9.69 and 8.9 g/cm^3, respectively. The volume fraction is:

$$f_{ThO_2} = \frac{\frac{2}{9.69}}{\frac{2}{9.69} + \frac{98}{8.9}} = 0.0184$$

Therefore, there is 0.0184 cm^3 of ThO_2 per cm^3 of composite. The volume of each ThO_2 sphere is:

$$V_{ThO_2} = \tfrac{4}{3}\pi r^3 = \tfrac{4}{3}\pi(0.5 \times 10^{-5}\text{ cm})^3 = 0.52 \times 10^{-15}\text{ cm}^3$$

$$\text{Concentration of ThO}_2\text{ particles} = \frac{0.0184}{0.52 \times 10^{-15}} = 35.4 \times 10^{12}\text{ particles/cm}^3.$$

17-2 Particulate Composites

The particulate composites contain large amounts of coarse particles that do not effectively block slip. The particulate composites are designed to produce unusual combinations of properties rather than to improve strength.

Rule of Mixtures Certain properties of a particulate composite depend only on the relative amounts and properties of the individual constituents. The **rule of mixtures** can accurately predict these properties. The density of a particulate composite, for example, is

$$\rho_c = \sum(f_i \cdot \rho_i) = f_1\rho_1 + f_2\rho_2 + \cdots + f_n\rho_n \tag{17-1}$$

where ρ_c is the density of the composite, $\rho_1, \rho_2, \ldots, \rho_n$ are the densities of each constituent in the composite, and $f_1, f_2, \ldots, f_n$ are the volume fractions of each constituent. Note that the connectivity of different phases (i.e., how the dispersed phase is arranged with respect to the continuous phase) is also very important for many properties.

Cemented Carbides **Cemented carbides**, or **cermets**, contain hard ceramic particles dispersed in a metallic matrix (Chapter 1). Tungsten carbide (WC) inserts used for cutting tools in machining operations are typical of this group. Tungsten carbide is a hard, stiff, high-melting-temperature ceramic. To improve toughness, tungsten carbide particles are combined with cobalt powder and pressed into powder compacts. The compacts are heated above the melting temperature of the cobalt. The liquid cobalt surrounds each of the solid tungsten carbide particles (Figure 17-4). After solidification, the cobalt serves as the binder for tungsten carbide and provides good impact resistance. Other carbides, such as TaC and TiC, may also be included in the cermet. The following example illustrates the calculation of density for cemented carbide.

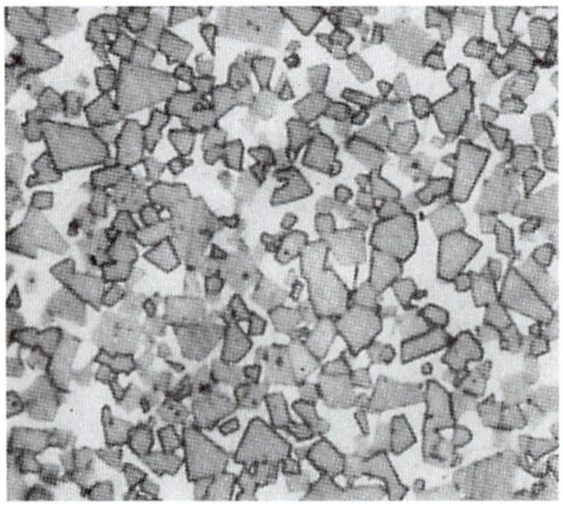

Figure 17-4
Microstructure of tungsten carbide—20% cobalt-cemented carbide (×1300). (*From* Metals Handbook, *Vol. 7, 8th Ed., American Society for Metals, 1972.*)

EXAMPLE 17-2 *Cemented Carbides*

A cemented carbide cutting tool used for machining contains 75 wt% WC, 15 wt% TiC, 5 wt% TaC, and 5 wt% Co. Estimate the density of the composite.

SOLUTION

First, we must convert the weight percentages to volume fractions. The densities of the components of the composite are:

$$\rho_{WC} = 15.77 \text{ g/cm}^3 \quad \rho_{TiC} = 4.94 \text{ g/cm}^3$$
$$\rho_{TaC} = 14.5 \text{ g/cm}^3 \quad \rho_{Co} = 8.90 \text{ g/cm}^3$$

$$f_{WC} = \frac{75/15.77}{75/15.77 + 15/4.94 + 5/14.5 + 5/8.9} = \frac{4.76}{8.70} = 0.547$$

$$f_{TiC} = \frac{15/4.94}{8.70} = 0.349$$

$$f_{TaC} = \frac{5/14.5}{8.70} = 0.040$$

$$f_{Co} = \frac{5/8.90}{8.70} = 0.064$$

From the rule of mixtures, the density of the composite is

$$\rho_c = \sum (f_i \rho_i) = (0.547)(15.77) + (0.349)(4.94) + (0.040)(14.5) + (0.064)(8.9)$$
$$= 11.50 \text{ g/cm}^3$$

Abrasives Grinding and cutting wheels are formed from alumina (Al_2O_3), silicon carbide (SiC), and cubic boron nitride (CBN). To provide toughness, the abrasive particles are bonded by a glass or polymer matrix. Diamond abrasives are typically bonded with a metal matrix. As the hard particles wear, they fracture or pull out of the matrix, exposing new cutting surfaces.

Electrical Contacts Materials used for electrical contacts in switches and relays must have a good combination of wear resistance and electrical conductivity. Otherwise, the contacts erode, causing poor contact and arcing. Tungsten-reinforced silver provides this combination of characteristics. A tungsten powder compact is made using conventional powder metallurgical processes (Figure 17-5) to produce high interconnected

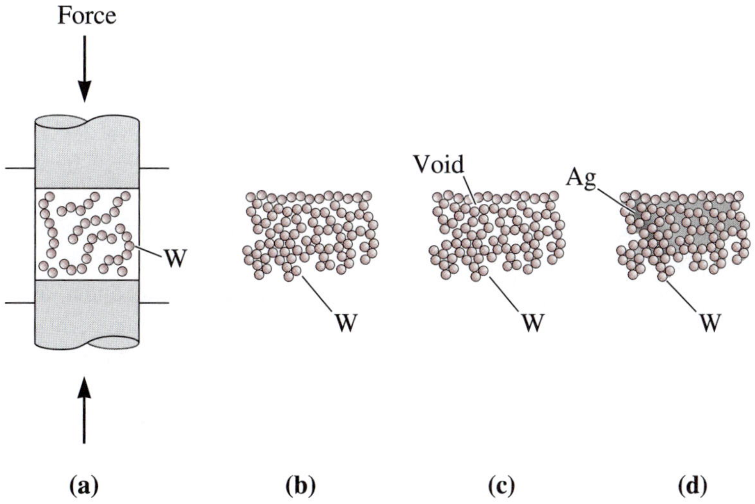

Figure 17-5 The steps in producing a silver-tungsten electrical composite: (a) Tungsten powders are pressed, (b) a low-density compact is produced, (c) sintering joins the tungsten powders, and (d) liquid silver is infiltrated into the pores between the particles.

porosity. Liquid silver is then vacuum infiltrated to fill the interconnected voids. Both the silver and the tungsten are continuous. Thus, the pure silver efficiently conducts current while the hard tungsten provides wear resistance.

EXAMPLE 17-3 *Silver-Tungsten Composite*

A silver-tungsten composite for an electrical contact is produced by first making a porous tungsten powder metallurgy compact, then infiltrating pure silver into the pores. The density of the tungsten compact before infiltration is 14.5 g/cm^3. Calculate the volume fraction of porosity and the final weight percent of silver in the compact after infiltration.

SOLUTION

The densities of pure tungsten and pure silver are 19.3 g/cm^3 and 10.49 g/cm^3. We can assume that the density of a pore is zero, so from the rule of mixtures:

$$\rho_c = f_W \rho_W + f_{pore}\rho_{pore}$$

$$14.5 = f_W(19.3) + f_{pore}(0)$$

$$f_W = 0.75$$

$$f_{pore} = 1 - 0.75 = 0.25$$

After infiltration, the volume fraction of silver equals the volume fraction of pores:

$$f_{Ag} = f_{pore} = 0.25$$

$$\text{wt\% Ag} = \frac{(0.25)(10.49)}{(0.25)(10.49) + (0.75)(19.3)} \times 100 = 15.3\%$$

This solution assumes that all of the pores are open, or interconnected.

Polymers Many engineering polymers that contain fillers and extenders are particulate composites. A classic example is carbon black in vulcanized rubber. Carbon black consists of tiny carbon spheroids only 5 to 500 nm in diameter. The carbon black improves the strength, stiffness, hardness, wear resistance, resistance to degradation due to ultraviolet rays, and heat resistance of the rubber. As mentioned in Chapter 16, nanoparticles of silica are added to rubber tires to enhance their stiffness.

Extenders, such as calcium carbonate ($CaCO_3$), solid glass spheres, and various clays, are added so that a smaller amount of the more expensive polymer is required. The extenders may stiffen the polymer, increase the hardness and wear resistance, increase thermal conductivity, or improve resistance to creep; however, strength and ductility normally decrease (Figure 17-6). Introducing hollow glass spheres may impart the same changes in properties while significantly reducing the weight of the composite. Other special properties can be obtained. Elastomer particles are introduced into polymers to improve toughness (e.g., Dylark^{TM}). Polyethylene may contain metallic powders, such as lead, to improve the absorption of fission products in nuclear applications. The design of a polymer composite is illustrated in the example that follows.

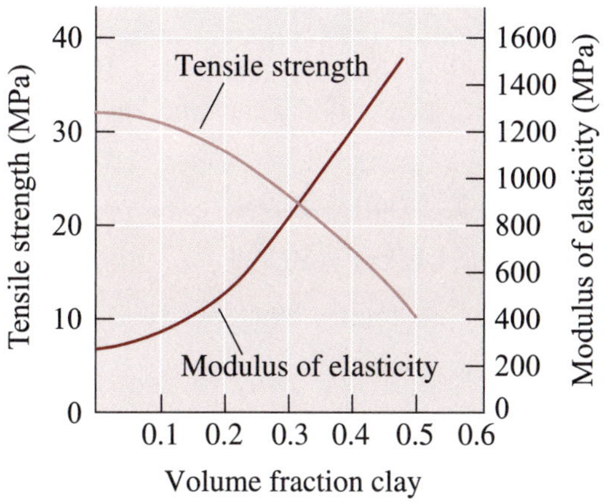

Figure 17-6
The effect of clay on the properties of polyethylene.

EXAMPLE 17-4 *Design of a Particulate Polymer Composite*

Design a clay-filled polyethylene composite suitable for injection molding of inexpensive components. The final part must have a tensile strength of at least 21 MPa and a modulus of elasticity of at least 552 MPa. Polyethylene costs approximately \$1.1 per kg and clay costs approximately \$0.1 per kg. The density of polyethylene is 0.95 g/cm^3 and that of clay is 2.4 g/cm^3.

SOLUTION

From Figure 17-6, a volume fraction of clay below 0.35 is required to maintain a tensile strength greater than 21 MPa, whereas a volume fraction of at least 0.2 is needed for the minimum modulus of elasticity. For lowest cost, we use the maximum allowable clay, or a volume fraction of 0.35.

In 1000 cm^3 of composite parts, there are 350 cm^3 of clay and 650 cm^3 of polyethylene in the composite, or:

$$(350 \text{ cm}^3)(2.4 \text{ g/cm}^3) = 840 \text{ g clay}$$

$$(650 \text{ cm}^3)(0.95 \text{ g/cm}^3) = 617.50 \text{ g polyethylene}$$

The cost of materials is:

$$(840 \text{ g clay})(\$0.1/\text{kg}) = \$0.084$$

$$(617.50 \text{ g PE})/(\$1.1/\text{kg}) = \$0.68$$

$$\text{total} = \$0.764 \text{ per } 1000 \text{ cm}^3$$

Suppose that weight is critical. The composite's density is:

$$\rho_c = (0.35)(2.4) + (0.65)(0.95) = 1.46 \text{ g/cm}^3$$

We may wish to sacrifice some of the economic savings in order to obtain lighter weight. If we use only 0.2 volume fraction clay, then (using the same method as above) we find that we need 480 g clay and 760 g polyethylene. The cost of materials is now:

$$(480\text{ g clay})(\$0.1/\text{kg}) = \$0.048$$
$$(760\text{ g PE})(\$1.1/\text{kg}) = \$0.836$$
$$\text{total} = \$0.88 \text{ per } 1000\text{ cm}^3$$

The density of the composite is:

$$\rho_c = (0.2)(2.4) + (0.8)(0.95) = 1.24\text{ g/cm}^3$$

The material costs about 10% more, but there is a weight savings of more than 10%.

Cast Metal Particulate Composites Aluminum castings containing dispersed SiC particles for automotive applications, including pistons and connecting rods, represent an important commercial application for particulate composites (Figure 17-7). With special processing, the SiC particles can be wet by the liquid, helping to keep the ceramic particles from sinking during freezing.

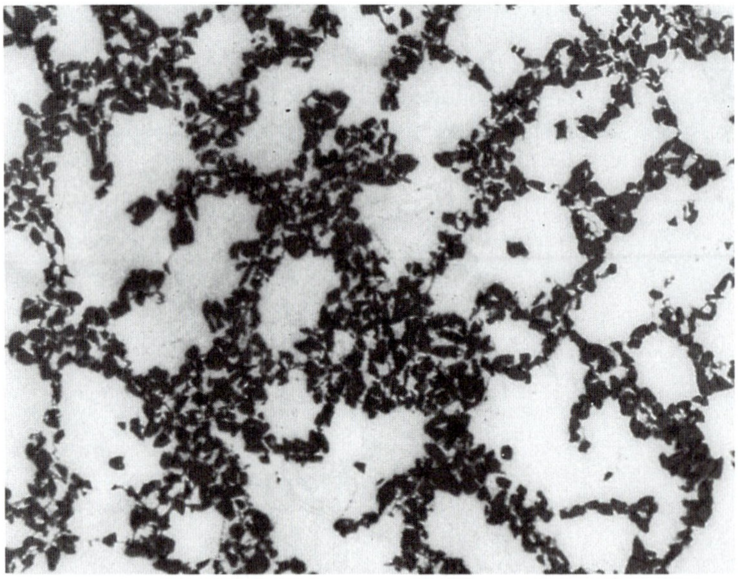

Figure 17-7 Microstructure of an aluminum casting alloy reinforced with silicon carbide particles. In this case, the reinforcing particles have segregated to interdendritic regions of the casting (×125). (*Courtesy of David Kennedy, Lester B. Knight Cost Metals Inc.*)

17-3 Fiber-Reinforced Composites

Most fiber-reinforced composites provide improved strength, fatigue resistance, Young's modulus, and strength-to-weight ratio by incorporating strong, stiff, but brittle fibers into a softer, more ductile matrix. The matrix material transmits the force to the fibers, which carry most of the applied force. The matrix also provides protection for the fiber surface and minimizes diffusion of species such as oxygen or moisture that can degrade the mechanical properties of fibers. The strength of the composite may be high at both room temperature and elevated temperatures (Figure 17-2).

Many types of reinforcing materials are employed. Straw has been used to strengthen mud bricks for centuries. Steel-reinforcing bars are introduced into concrete structures. Glass fibers in a polymer matrix produce fiberglass for transportation and aerospace applications. Fibers made of boron, carbon, polymers (e.g., aramids, Chapter 16), and ceramics provide exceptional reinforcement in advanced composites based on matrices of polymers, metals, ceramics, and even intermetallic compounds.

The Rule of Mixtures in Fiber-Reinforced Composites As for particulate composites, the rule of mixtures always predicts the density of fiber-reinforced composites:

$$\rho_c = f_m \rho_m + f_f \rho_f \tag{17-2}$$

where the subscripts m and f refer to the matrix and the fiber. Note that $f_m = 1 - f_f$. The subscript c refers to the composite.

In addition, the rule of mixtures accurately predicts the electrical and thermal conductivity of fiber-reinforced composites along the fiber direction if the fibers are *continuous* and *unidirectional*:

$$K_c = f_m K_m + f_f K_f \tag{17-3}$$

$$\sigma_c = f_m \cdot \sigma_m + f_f \cdot \sigma_f \tag{17-4}$$

where K is the thermal conductivity and σ is the electrical conductivity. Thermal or electrical energy can be transferred through the composite at a rate that is proportional to the volume fraction of the conductive material. In a composite with a metal matrix and ceramic fibers, the bulk of the energy would be transferred through the matrix; in a composite consisting of a polymer matrix containing metallic fibers, energy would be transferred through the fibers.

When the fibers are not continuous or unidirectional, the simple rule of mixtures may not apply. For example, in a metal fiber-polymer matrix composite, electrical conductivity would be low and would depend on the length of the fibers, the volume fraction of the fibers, and how often the fibers touch one another. This is expressed using the concept of connectivity of phases.

Modulus of Elasticity The rule of mixtures is used to predict the modulus of elasticity when the fibers are continuous and unidirectional. Parallel to the fibers, the modulus of elasticity may be as high as:

$$E_{c,\|} = f_m \cdot E_m + f_f \cdot E_f \tag{17-5}$$

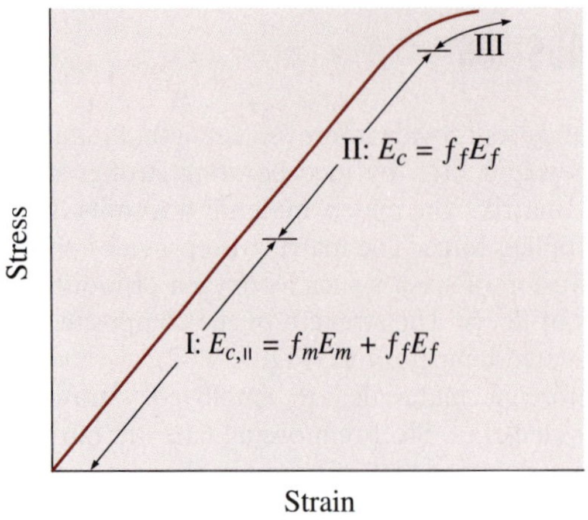

Figure 17-8 The stress-strain curve for a fiber-reinforced composite. At low stresses (region I), the modulus of elasticity is given by the rule of mixtures. At higher stresses (region II), the matrix deforms and the rule of mixtures is no longer obeyed.

However, when the applied stress is very large, the matrix begins to deform and the stress-strain curve is no longer linear (Figure 17-8). Since the matrix now contributes little to the stiffness of the composite, the modulus can be approximated by:

$$E_{c,\parallel} = f_f \cdot E_f \tag{17-6}$$

When the load is applied perpendicular to the fibers, each component of the composite acts independently of the other. The modulus of the composite is now:

$$\frac{1}{E_{c,\perp}} = \frac{f_m}{E_m} + \frac{f_f}{E_f} \tag{17-7}$$

Again, if the fibers are not continuous and unidirectional, the rule of mixtures does not apply.

The following examples further illustrate these concepts.

EXAMPLE 17-5 *Rule of Mixtures for Composites: Stress Parallel to Fibers*

Derive the rule of mixtures (Equation 17-5) for the modulus of elasticity of a fiber-reinforced composite when a stress (σ) is applied along the axis of the fibers.

SOLUTION

The total force acting on the composite is the sum of the forces carried by each constituent:

$$F_c = F_m + F_f$$

Since $F = \sigma A$:

$$\sigma_c A_c = \sigma_m A_m + \sigma_m A_f$$

$$\sigma_c = \sigma_m \left(\frac{A_m}{A_c}\right) + \sigma_f \left(\frac{A_f}{A_c}\right)$$

If the fibers have a uniform cross-section, the area fraction equals the volume fraction f:

$$\sigma_c = \sigma_m \cdot f_m + \sigma_f \cdot f_f$$

From Hooke's law, $\sigma = \varepsilon E$. Therefore:

$$E_{c,\parallel}\varepsilon_c = E_m\varepsilon_m f_m + E_f\varepsilon_f f_f$$

If the fibers are rigidly bonded to the matrix, both the fibers and the matrix must stretch equal amounts (iso-strain conditions):

$$\varepsilon_c = \varepsilon_m = \varepsilon_f$$

$$E_{c,\parallel} = f_m E_m + f_f E_f$$

EXAMPLE 17-6 Modulus of Elasticity for Composites: Stress Perpendicular to Fibers

Derive the equation for the modulus of elasticity of a fiber-reinforced composite when a stress is applied perpendicular to the axis of the fiber (Equation 17-7).

SOLUTION

In this example, the strains are no longer equal; instead, the weighted sum of the strains in each component equals the total strain in the composite, whereas the stresses in each component are equal (iso-stress conditions):

$$\varepsilon_c = f_m\varepsilon_m + f_f\varepsilon_f$$

$$\frac{\sigma_c}{E_c} = f_m\left(\frac{\sigma_m}{E_m}\right) + f_f\left(\frac{\sigma_f}{E_f}\right)$$

Since $\sigma_c = \sigma_m = \sigma_f$:

$$\frac{1}{E_{c,\perp}} = \frac{f_m}{E_m} + \frac{f_f}{E_f}$$

Strength of Composites The tensile strength of a fiber-reinforced composite (TS_c) depends on the bonding between the fibers and the matrix. However, the rule of mixtures is sometimes used to approximate the tensile strength of a composite containing continuous, parallel fibers:

$$TS_c = f_f TS_f + f_m\sigma_m \tag{17-8}$$

where TS_f is the tensile strength of the fiber and σ_m is the stress acting on the matrix when the composite is strained to the point where the fiber fractures. Thus, σ_m is *not* the actual tensile strength of the matrix. Other properties, such as ductility, impact properties, fatigue properties, and creep properties, are difficult to predict even for unidirectionally aligned fibers.

EXAMPLE 17-7 *Boron Aluminum Composites*

Boron coated with SiC (or Borsic) reinforced aluminum containing 40 vol% fibers is an important high-temperature, lightweight composite material. Estimate the density, modulus of elasticity, and tensile strength parallel to the fiber axis. Also estimate the modulus of elasticity perpendicular to the fibers.

SOLUTION

The properties of the individual components are shown below.

Material	Density (g/cm^3)	Modulus of Elasticity (MPa)	Tensile Strength (MPa)
Fibers	2.36	379×10^3	2760
Aluminum	2.70	69×10^3	35

From the rule of mixtures:

$$\rho_c = (0.6)(2.7) + (0.4)(2.36) = 2.56 \text{ g/cm}^3$$

$$E_{c,\|} = (0.6)(69 \times 10^3) + (0.4)(379 \times 10^3) = 193 \times 10^3 \text{ MPa}$$

$$TS_c = (0.6)(35) + (0.4)(2760) = 1125 \text{ MPa}$$

Perpendicular to the fibers:

$$\frac{1}{E_{c,\perp}} = \frac{0.6}{69 \times 10^3} + \frac{0.4}{379 \times 10^3} = 0.00976 \times 10^{-3}$$

$$E_{c,\perp} = 102.5 \times 10^3 \text{ MPa}$$

The actual modulus and strength parallel to the fibers are shown in Figure 17-9. The calculated modulus of elasticity (193×10^3 MPa) is exactly the same as the measured modulus. However, the estimated strength (1125 MPa) is substantially higher than the actual strength (about 896 MPa). We also note that the modulus of elasticity is very anisotropic, with the modulus perpendicular to the fiber being only half the modulus parallel to the fibers.

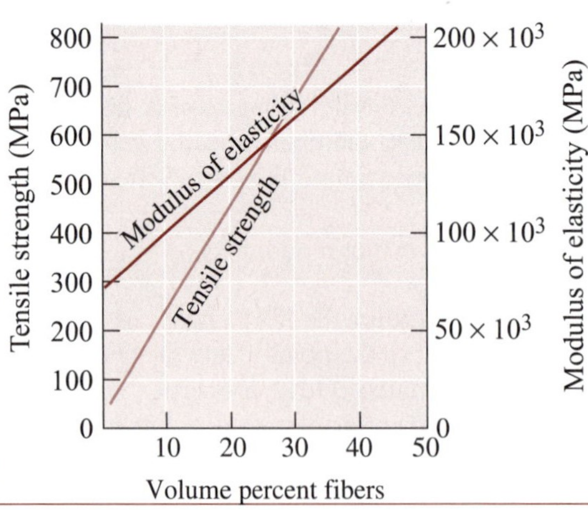

Figure 17-9 The influence of volume percent boron-coated SiC (Borsic) fibers on the properties of Borsic-reinforced aluminum parallel to the fibers (for Example 17-7).

EXAMPLE 17-8 ***Nylon-Glass Fiber Composites***

Glass fibers in nylon provide reinforcement. If the nylon contains 30 vol% E-glass, what fraction of the applied force is carried by the glass fibers?

SOLUTION

The modulus of elasticity for each component of the composite is:

$$E_{\text{glass}} = 72.4 \times 10^3 \text{ MPa} \quad E_{\text{nylon}} = 2.8 \times 10^3 \text{ MPa}$$

Both the nylon and the glass fibers have equal strain if bonding is good, so:

$$\varepsilon_c = \varepsilon_m = \varepsilon_f$$

$$\varepsilon_m = \frac{\sigma_m}{E_m} = \varepsilon_f = \frac{\sigma_f}{E_f}$$

$$\frac{\sigma_f}{\sigma_m} = \frac{E_f}{E_m} = \frac{72.4 \times 10^3}{2.8 \times 10^3} = 25.86$$

$$\text{Fraction} = \frac{F_f}{F_f + F_m} = \frac{\sigma_f A_f}{\sigma_f A_f + \sigma_m A_m} = \frac{\sigma_f(0.3)}{\sigma_m(0.3) + \sigma_m(0.7)}$$

$$= \frac{0.3}{0.3 + 0.7(\sigma_m/\sigma_f)} = \frac{0.3}{0.3 + 0.7(1/25.86)} = 0.92$$

Almost all of the load is carried by the glass fibers.

17-4 Characteristics of Fiber-Reinforced Composites

Many factors must be considered when designing a fiber-reinforced composite, including the length, diameter, orientation, amount, and properties of the fibers; the properties of the matrix; and the bonding between the fibers and the matrix.

Fiber Length and Diameter Fibers can be short, long, or even continuous. Their dimensions are often characterized by the **aspect ratio** defined as l/d, where l is the fiber length and d is the diameter. Typical fibers have diameters varying from 10 μm (10×10^{-4} cm) to 150 μm (150×10^{-4} cm).

The strength of a composite improves when the aspect ratio is large. Fibers often fracture because of surface imperfections. Making the diameter as small as possible gives the fiber less surface area and, consequently, fewer flaws that might propagate during processing or under a load. We also prefer long fibers. The ends of a fiber carry less of the load than the remainder of the fiber; consequently, the fewer the ends, the higher the load-carrying ability of the fibers (Figure 17-10).

In many fiber-reinforced systems, discontinuous fibers with an aspect ratio greater than some critical value are used to provide an acceptable compromise between processing ease and properties. A critical fiber length l_c, for any given fiber diameter d can be determined:

$$l_c = \frac{TS_f d}{2\tau_i} \tag{17-9}$$

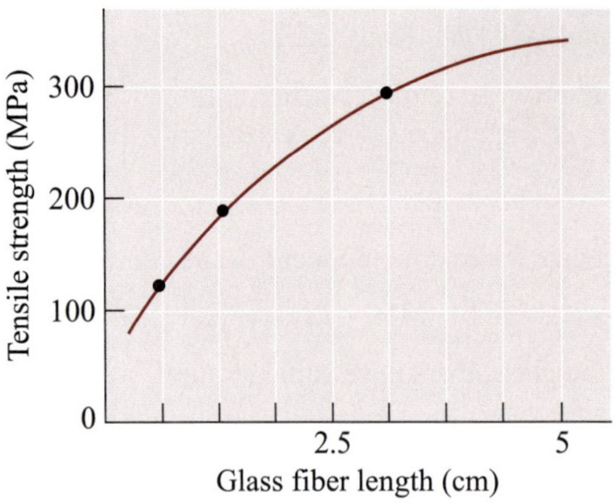

Figure 17-10
Increasing the length of chopped E-glass fibers in an epoxy matrix increases the strength of the composite. In this example, the volume fraction of glass fibers is about 0.5.

where TS_f is the strength of the fiber and τ_i is related to the strength of the bond between the fiber and the matrix, or the stress at which the matrix begins to deform. If the fiber length l is smaller than l_c, little reinforcing effect is observed; if l is greater than about $15l_c$, the fiber behaves almost as if it were continuous. The strength of the composite can be estimated from

$$\sigma_c = f_f TS_f\left(1 - \frac{l_c}{2l}\right) + f_m\sigma_m \tag{17-10}$$

where σ_m is the stress on the matrix when the fibers break.

Volume Fraction of Fiber A greater volume fraction of fibers increases the strength and stiffness of the composite, as we would expect from the rule of mixtures. However, the maximum volume fraction is about 80%, beyond which fibers can no longer be completely surrounded by the matrix.

Orientation of Fibers The reinforcing fibers may be introduced into the matrix in a number of orientations. Short, randomly oriented fibers having a small aspect ratio—typical of fiberglass—are easily introduced into the matrix and give relatively isotropic behavior in the composite.

Long, or even continuous, unidirectional arrangements of fibers produce anisotropic properties, with particularly good strength and stiffness parallel to the fibers. These fibers are often designated as 0° plies, indicating that all of the fibers are aligned with the direction of the applied stress. However, unidirectional orientations provide poor properties if the load is perpendicular to the fibers (Figure 17-11).

One of the unique characteristics of fiber-reinforced composites is that their properties can be tailored to meet different types of loading conditions. Long, continuous fibers can be introduced in several directions within the matrix (Figure 17-12); in orthogonal arrangements (0°/90° plies), good strength is obtained in two perpendicular directions. More complicated arrangements (such as 0°/±45°/90° plies) provide reinforcement in multiple directions.

Fibers can also be arranged in three-dimensional patterns. In even the simplest of fabric weaves, the fibers in each individual layer of fabric have some small degree of orientation in a third direction. Better three-dimensional reinforcement occurs when the fabric layers are knitted or stitched together. More complicated three-dimensional weaves (Figure 17-13) can also be used.

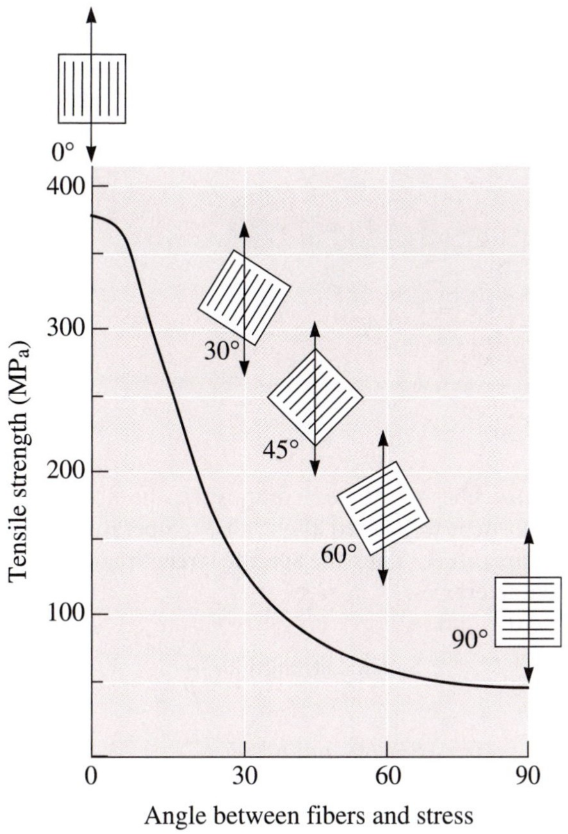

Figure 17-11
Effect of fiber orientation on the tensile strength of E-glass fiber-reinforced epoxy composites.

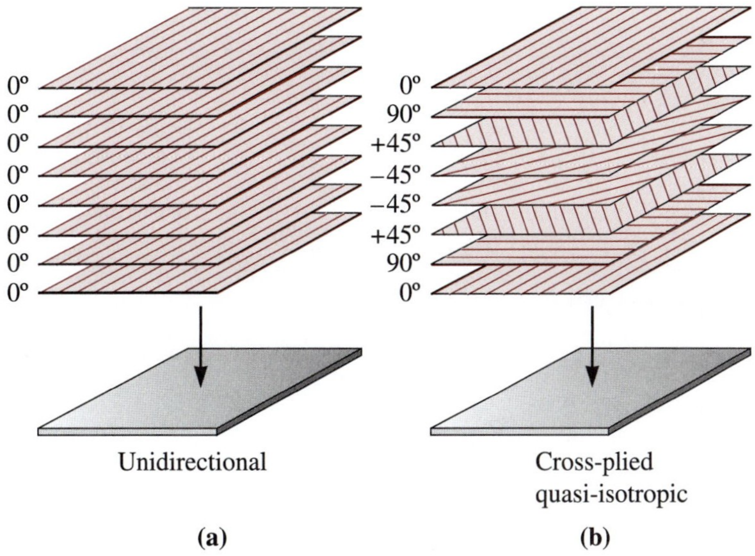

Figure 17-12 (a) Tapes containing aligned fibers can be joined to produce a multi-layered unidirectional composite structure. (b) Tapes containing aligned fibers can be joined with different orientations to produce a quasi-isotropic composite. In this case, a 0°/±45°/90° composite is formed.

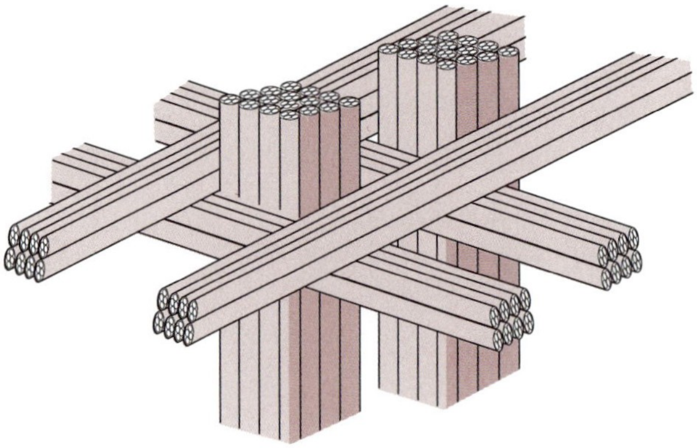

Figure 17-13
A three-dimensional weave for fiber-reinforced composites.

Fiber Properties In most fiber-reinforced composites, the fibers are strong, stiff, and lightweight. If the composite is to be used at elevated temperatures, the fiber should also have a high melting temperature. Thus the **specific strength** and **specific modulus** of the fiber are important characteristics:

$$\text{Specific strength} = \frac{TS}{\rho} \tag{17-11}$$

$$\text{Specific modulus} = \frac{E}{\rho} \tag{17-12}$$

where TS is the tensile strength, ρ is the density, and E is the modulus of elasticity.

Properties of typical fibers are shown in Table 17-2 and Figure 17-14. Note in Table 17-2, the density is in g/cm^3. The highest specific modulus is usually found in materials having a low atomic number and covalent bonding, such as carbon and boron. These two elements also have a high strength and melting temperature.

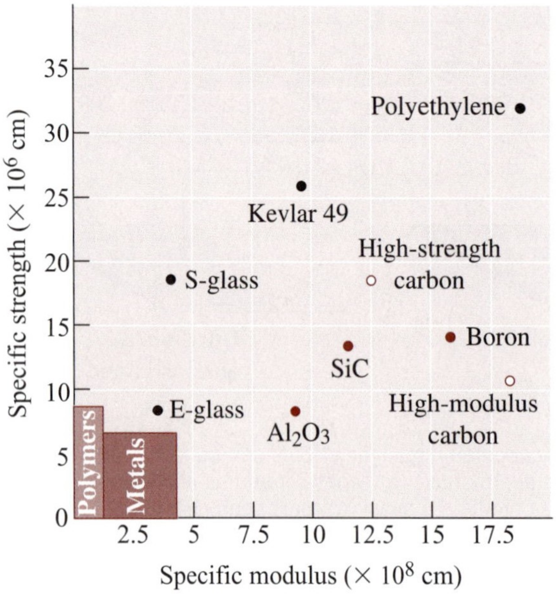

Figure 17-14
Comparison of the specific strength and specific modulus of fibers versus metals and polymers.

TABLE 17-2 ■ ***Properties of selected reinforcing materials****

Material	Density (g/cm^3)	Tensile Strength (MPa)	Modulus of Elasticity ($\times 10^3$ MPa)	Melting Temperature (°C)	Specific Modulus ($\times 10^7$ cm)	Specific Strength ($\times 10^6$ cm)
Polymers:						
Kevlar™	1.44	4480	124	500	86.8	31.3
Nylon	1.14	83	3.5	249	2.5	7.3
Polyethylene	0.97	21–48	0.3–0.7	147	17.8	34.3
Metals:						
Be composites	1.83	276–345	303	1277	193.8	7.0
Boron	2.36	3450	379	2030	161.8	11.8
W	19.40	4000	407	3410	21.3	2.0
Glass:						
E-glass	2.55	3450	72.4	<1725	28.5	14.0
S-glass	2.50	4480	86.9	<1725	35.0	18.0
Carbon:						
HS (high strength)	1.75	5650	276	3700	158.8	32.5
HM (high modulus)	1.90	1860	531	3700	280.0	9.8
Ceramics:						
Al_2O_3	3.95	2070	379	2015	97.0	5.3
B_4C	2.36	2280	483	2450	206.0	9.8
SiC	3.00	3930	483	2700	118.3	13.3
ZrO_2	4.84	2070	345	2677	71.5	4.3
Whiskers:						
Al_2O_3	3.96	20700	428	1982	108.5	52.5
Cr	7.20	8900	241	1890	33.5	12.3
Graphite	1.66	20700	703	3700	425.0	125.5
SiC	3.18	20700	483	2700	152.0	65.5
Si_3N_4	3.18	13800	379		119.5	43.8

Aramid fibers, of which Kevlar™ is the best known example, are aromatic polyamide polymers strengthened by a backbone containing benzene rings (Figure 17-15) and are examples of liquid-crystalline polymers in that the polymer chains are rod-like and very stiff. Specially prepared polyethylene fibers are also available. Both the aramid and polyethylene fibers have excellent strength and stiffness but are limited to low-temperature use. Because of their lower density, polyethylene fibers have superior specific strength and specific modulus.

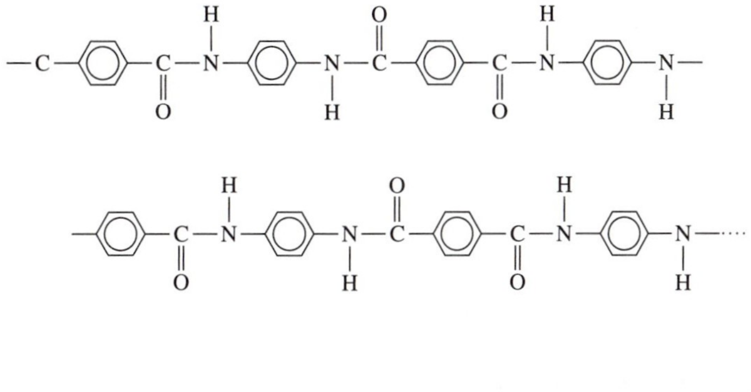

Figure 17-15 The structure of Kevlar™. The fibers are joined by secondary bonds between oxygen and hydrogen atoms on adjoining chains.

Ceramic fibers and whiskers, including alumina, glass, and silicon carbide, are strong and stiff. Glass fibers, which are the most commonly used, include pure silica, S-glass (25% Al_2O_3, 100% MgO, balance SiO_2), and E-glass (18% CaO, 15% Al_2O_3, balance SiO_2). Although they are considerably denser than the polymer fibers, the ceramics can be used at much higher temperatures. Beryllium and tungsten, although metallically bonded, have a high modulus that makes them attractive fiber materials for certain applications. The following example discusses issues related to designing with composites.

EXAMPLE 17-9 *Design of an Aerospace Composite*

We are now using a 7075-T6 aluminum alloy (modulus of elasticity of 69×10^3 MPa to make a 227 kg panel on a commercial aircraft. Experience has shown that each kg reduction in weight on the aircraft reduces the fuel consumption by 4200 litres each year. Design a material for the panel that will reduce weight, yet maintain the same specific modulus, and will be economical over a 10-year lifetime of the aircraft.

SOLUTION

There are many possible materials that might be used to provide a weight savings. As an example, let's consider using a boron fiber-reinforced Al-Li alloy in the T6 condition. Both the boron fiber and the lithium alloying addition increase the modulus of elasticity; the boron and the Al-Li alloy also have densities less than that of typical aluminum alloys.

The specific modulus of the current 7075-T6 alloy is:

$$\text{Specific modulus} = \frac{(69 \times 10^3 \text{ MPa})}{\left(2.7 \frac{\text{g}}{\text{cm}^3}\right)}$$

$$= 2.61 \times 10^8 \text{ cm}$$

The density of the boron fibers is approximately 2.36 g/cm^3 and that of a typical Al-Li alloy is approximately 2.5 g/cm^3. If we use 0.6 volume fraction boron fibers in the composite, then the density, modulus of elasticity, and specific modulus of the composite are:

$$\rho_c = (0.6)(2.36) + (0.4)(2.5) = 2.416 \text{ g/cm}^3$$

$$E_c = (0.6)(379 \times 10^3 \text{ MPa}) + (0.4)(76 \times 10^3 \text{ MPa}) = 258 \times 10^3 \text{ MPa}$$

$$\text{Specific modulus} = \frac{258 \times 10^3 \text{ MPa}}{2.416 \text{ g/cm}^3} = 10.9 \times 10^8 \text{ cm}$$

If the specific modulus is the only factor influencing the design of the component, the thickness of the part might be reduced by 75%, giving a component weight of 56.8 kg rather than 227 kg. The weight savings would then be 170.2 kg, or (4200 litres/kg)(170.2 kg) = 714,840 litres per year. At about \$.50 per litre, about \$357,000 in fuel savings could be realized each year, or \$3.57 million over the 10-year aircraft lifetime.

This is certainly an optimistic comparison, since strength or fabrication factors may not permit the part to be made as thin as suggested. In addition, the high cost of boron fibers (over $660/kg) and higher manufacturing costs of the composite compared with those of 7075 aluminum would reduce cost savings. As mentioned before, Boeing has used 50% carbon-fiber-reinforced plastic in its latest 787 *Dreamliner* airplane to achieve 20% increase in fuel efficiency.

Matrix Properties The matrix supports the fibers and keeps them in the proper position, transfers the load to the strong fibers, protects the fibers from damage during manufacture and use of the composite, and prevents cracks in the fiber from propagating throughout the entire composite. The matrix usually provides the major control over electrical properties, chemical behavior, and elevated-temperature use of the composite.

Polymer matrices are particularly common. Most polymer materials—both thermoplastics and thermosets—are available in short glass fiber-reinforced grades. These composites are formed into useful shapes by the processes described in Chapter 16. Sheet-molding compounds (SMCs) and bulk-molding compounds (BMCs) are typical of this type of composite. Thermosetting aromatic polyimides are used for somewhat higher temperature applications.

Metal-matrix composites include aluminum, magnesium, copper, nickel, and intermetallic compound alloys reinforced with ceramic and metal fibers. A variety of aerospace and automotive applications are satisfied by the MMCs. The metal matrix permits the composite to operate at high temperatures, but producing the composite is often more difficult and expensive than producing the polymer-matrix materials.

The ceramic-matrix composites (CMCs) have good properties at elevated temperatures and are lighter in weight than the high-temperature metal-matrix composites. In a later section, we discuss how to develop toughness in CMCs.

Bonding and Failure Particularly in polymer and metal-matrix composites, good bonding must be obtained between the various constituents. The fibers must be firmly bonded to the matrix material if the load is to be properly transmitted from the matrix to the fibers. In addition, the fibers may pull out of the matrix during loading, reducing the strength and fracture resistance of the composite if bonding is poor. Figure 17-16 on the next page illustrates poor bonding of carbon fibers in a copper matrix. In some cases, special coatings may be used to improve bonding. Glass fibers are coated with a silane coupling or "keying" agent (called **sizing**) to improve bonding and moisture resistance in fiberglass composites. Carbon fibers are similarly coated with an organic material to improve bonding. Boron fibers have been coated with silicon carbide or boron nitride to improve bonding with an aluminum matrix; in fact, these fibers have been called Borsic fibers to reflect the presence of the silicon carbide (SiC) coating.

Another property that must be considered when combining fibers into a matrix is the similarity between the coefficients of thermal expansion for the two materials. If the fiber expands and contracts at a rate much different from that of the matrix, fibers may break or bonding can be disrupted, causing premature failure.

In many composites, individual plies or layers of fabric are joined. Bonding between these layers must also be good or another problem—**delamination**—may occur. Delamination has been suspected to be a cause in some accidents involving airplanes using composite-based structures. The layers may tear apart under load and cause failure. Using composites with a three-dimensional weave will help prevent delamination.

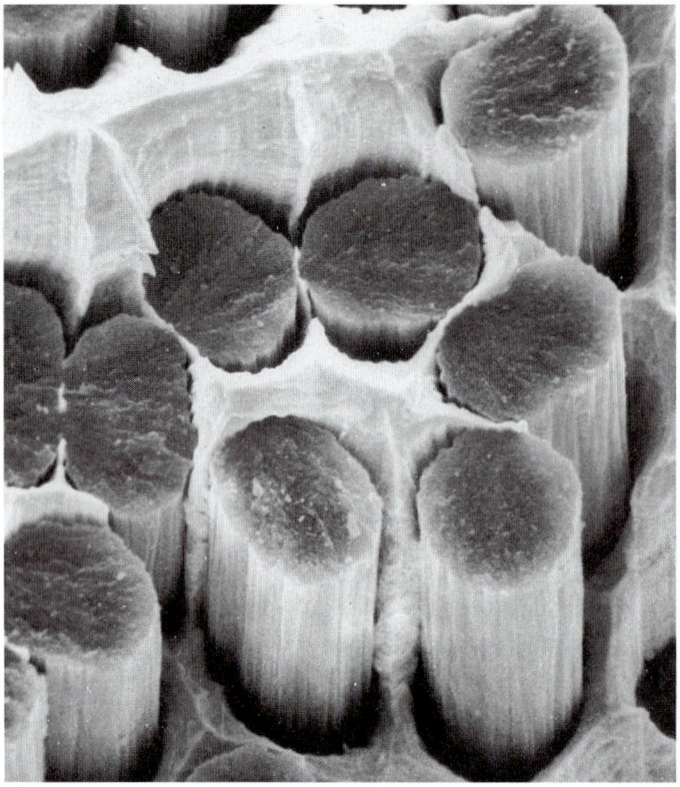

Figure 17-16 Scanning electron micrograph of the fracture surface of a silver-copper alloy reinforced with carbon fibers. Poor bonding causes much of the fracture surface to follow the interface between the metal matrix and the carbon tows (×3000). (*From* Metals Handbook, *American Society for Metals, Vol. 9, 9th Ed., 1985.*)

17-5 Manufacturing Fibers and Composites

Producing a fiber-reinforced composite involves several steps, including producing the fibers, arranging the fibers into bundles or fabrics, and introducing the fibers into the matrix.

Making the Fiber Metallic fibers, glass fibers, and many polymer fibers (including nylon, aramid, and polyacrylonitrile) can be formed by drawing processes, as described in Chapter 8 (wire drawing of metal) and Chapter 16 (using the spinnerette for polymer fibers).

Boron, carbon, and ceramics are too brittle and reactive to be worked by conventional drawing processes. Boron fiber is produced by **chemical vapor deposition** (CVD) [Figure 17-17(a)]. A very fine, heated tungsten filament is used as a substrate, passing through a seal into a heated chamber. Vaporized boron compounds such as BCl_3 are introduced into the chamber, decompose, and permit boron to precipitate onto the tungsten wire (Figure 17-18). SiC fibers are made in a similar manner, with carbon fibers as the substrate for the vapor deposition of silicon carbide.

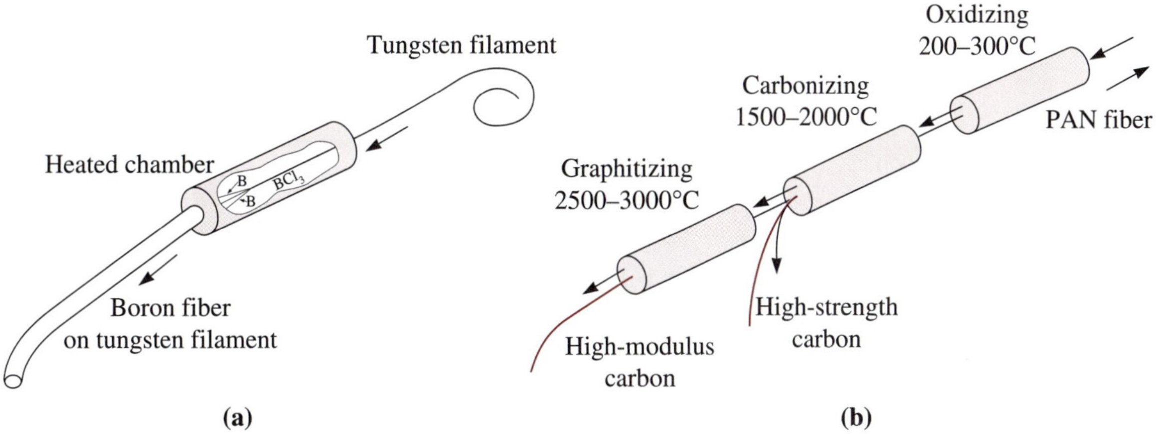

Figure 17-17 Methods for producing (a) boron and (b) carbon fibers.

Carbon fibers are made by **carbonizing**, or pyrolizing, an organic filament, which is more easily drawn or spun into thin, continuous lengths [Figure 17-17(b)]. The organic filament, known as a **precursor**, is often rayon (a cellulosic polymer), polyacrylonitrile (PAN), or pitch (various aromatic organic compounds). High temperatures decompose the organic polymer, driving off all of the elements but carbon. As the carbonizing temperature increases from 1000°C to 3000°C, the tensile strength decreases while the modulus of elasticity increases (Figure 17-19 on the next page). Drawing the carbon filaments at critical times during carbonizing may produce desirable preferred orientations in the final carbon filament.

Whiskers are single crystals with aspect ratios of 20 to 1000. Because the whiskers contain no mobile dislocations, slip cannot occur and they have exceptionally high strengths. Because of the complex processing required to produce whiskers, their cost may be quite high.

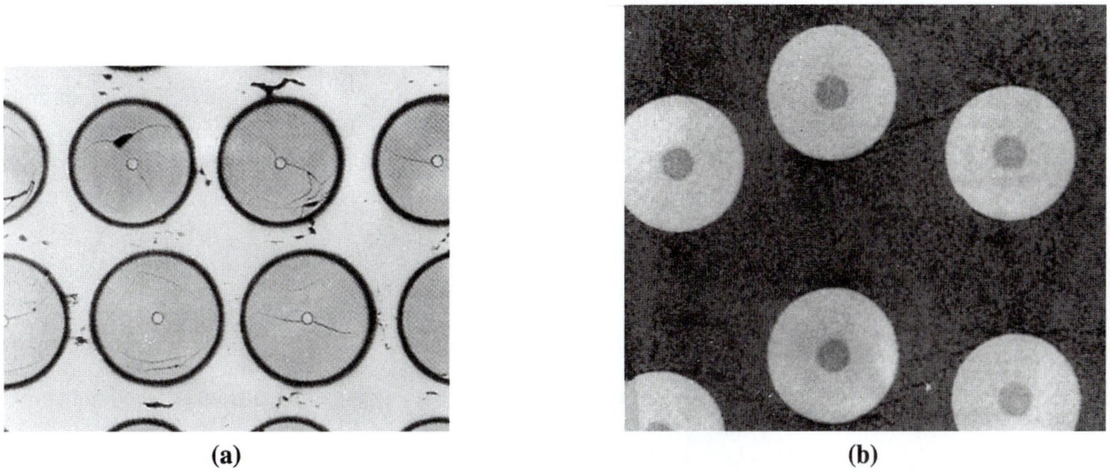

Figure 17-18 Photomicrographs of two fiber-reinforced composites: (a) In Borsic fiber-reinforced aluminum, the fibers are composed of a thick layer of boron deposited on a small-diameter tungsten filament (×1000). (*From* Metals Handbook, *American Society for Metals, Vol. 9, 9th Ed., 1985.*) (b) In this microstructure of a ceramic-fiber–ceramic-matrix composite, silicon carbide fibers are used to reinforce a silicon nitride matrix. The SiC fiber is vapor-deposited on a small carbon precursor filament (×125). (*Courtesy of Dr. R.T. Bhatt, NASA Lewis Research Center.*)

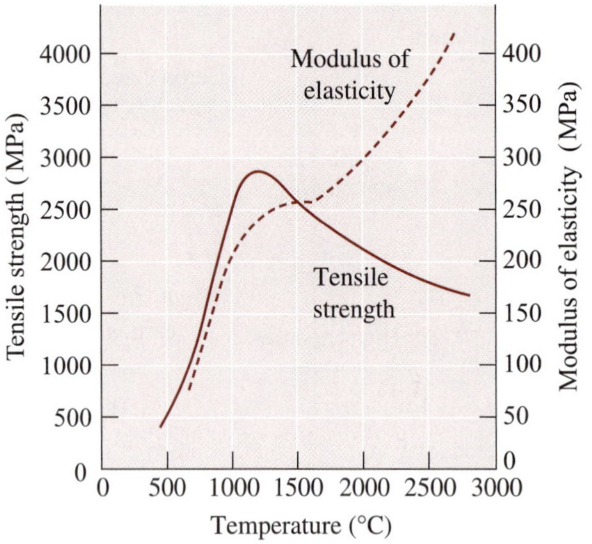

Figure 17-19
The effect of heat-treatment temperature on the strength and modulus of elasticity of carbon fibers.

Arranging the Fibers Exceptionally fine filaments are bundled together as rovings, yarns, or tows. In **yarns**, as many as 10,000 filaments are twisted together to produce the fiber. A **tow** contains a few hundred to more than 100,000 untwisted filaments (Figure 17-20). **Rovings** are untwisted bundles of filaments, yarns, or tows.

Often, fibers are chopped into short lengths of 1 cm or less. These fibers, also called **staples**, are easily incorporated into the matrix and are typical of the sheet-molding and bulk-molding compounds for polymer-matrix composites. The fibers often are present in the composite in a random orientation.

Long or continuous fibers for polymer-matrix composites can be processed into mats or fabrics. *Mats* contain non-woven, randomly oriented fibers loosely held together by a polymer resin. The fibers can also be woven, braided, or knitted into two-dimensional or three-dimensional fabrics. The fabrics are then impregnated with a polymer resin. The resins at this point in the processing have not yet been completely polymerized; these mats or fabrics are called **prepregs**.

When unidirectionally aligned fibers are to be introduced into a polymer matrix, **tapes** may be produced. Individual fibers can be unwound from spools onto a mandrel, which determines the spacing of the individual fibers, and prepregged with a polymer resin. These tapes, only one fiber diameter thick, may be up to 120 cm wide. Figure 17-21 illustrates that tapes can also be produced by covering the fibers with upper and lower layers of metal foil that are then joined by diffusion bonding.

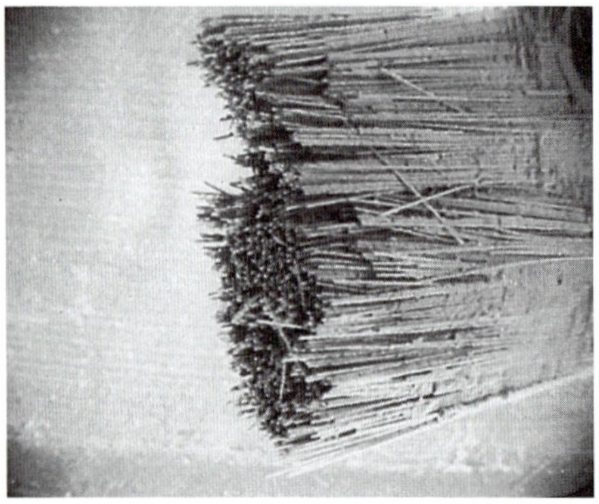

Figure 17-20
A scanning electron micrograph of a carbon tow containing many individual carbon filaments (×200).

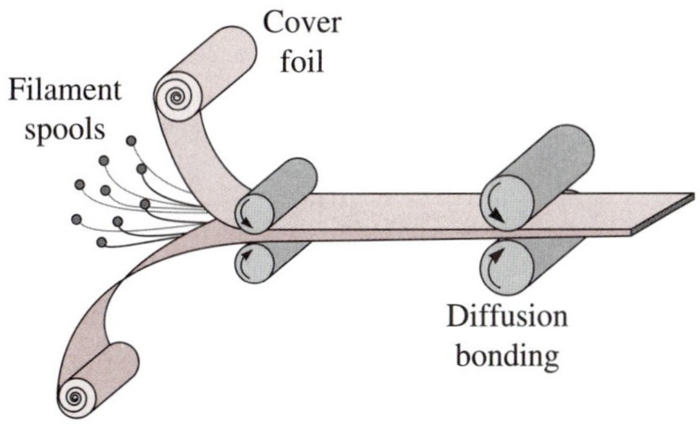

Figure 17-21
Production of fiber tapes by encasing fibers between metal cover sheets by diffusion bonding.

Producing the Composite A variety of methods for producing composite parts are used, depending on the application and materials. Short fiber-reinforced composites are normally formed by mixing the fibers with a liquid or plastic matrix, then using relatively conventional techniques such as injection molding for polymer-base composites or casting for metal-matrix composites. Polymer matrix composites can also be produced by a spray-up method, in which short fibers mixed with a resin are sprayed against a form and cured.

Special techniques, however, have been devised for producing composites using continuous fibers, either in unidirectionally aligned, mat, or fabric form (Figure 17-22). In hand lay-up techniques, the tapes, mats, or fabrics are placed against a form, saturated with a polymer resin, rolled to assure good contact and freedom from porosity, and finally cured. Fiberglass car and truck bodies might be made in this manner, which is generally slow and labor intensive.

Tapes and fabrics can also be placed in a die and formed by bag molding. High-pressure gases or a vacuum are introduced to force the individual plies together so that good bonding is achieved during curing. Large polymer matrix components for the skins of military aircraft have been produced by these techniques. In matched die

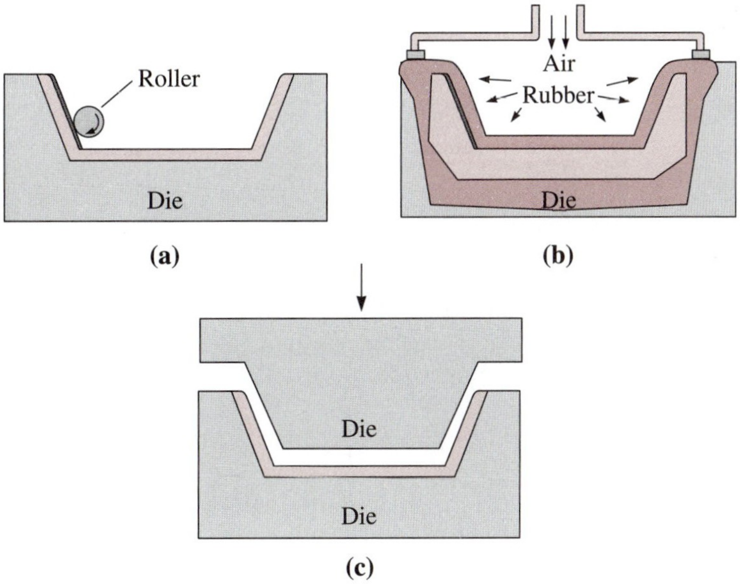

Figure 17-22 Producing composite shapes in dies by (a) hand lay-up, (b) pressure bag molding, and (c) matched die molding.

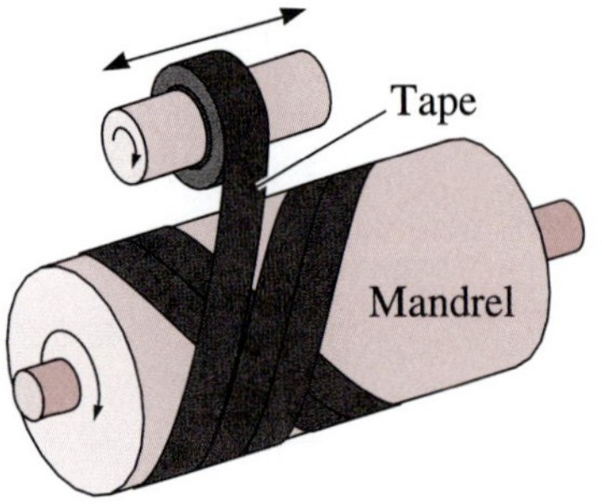

Figure 17-23
Producing composite shapes by filament winding.

molding, short fibers or mats are placed into a two-part die; when the die is closed, the composite shape is formed.

Filament winding is used to produce products such as pressure tanks and rocket motor castings (Figure 17-23). Fibers are wrapped around a form or mandrel to gradually build up a hollow shape that may be even several feet in thickness. The filament can be dipped in the polymer-matrix resin prior to winding, or the resin can be impregnated around the fiber during or after winding. Curing completes the production of the composite part.

Pultrusion is used to form a simple-shaped product with a constant cross section, such as round, rectangular, pipe, plate, or sheet shapes (Figure 17-24). Fibers or mats are drawn from spools, passed through a polymer resin bath for impregnation, and gathered together to produce a particular shape before entering a heated die for curing. Curing of the resin is accomplished almost immediately, so a continuous product is produced. The pultruded stock can subsequently be formed into somewhat more complicated shapes, such as fishing poles, golf club shafts, and ski poles.

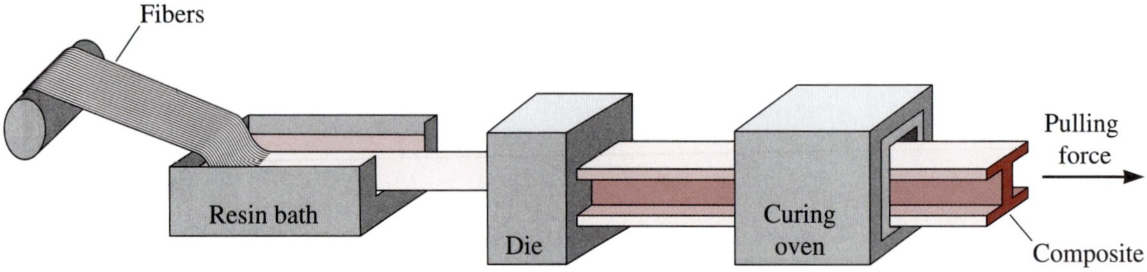

Figure 17-24 Producing composite shapes by pultrusion.

Metal-matrix composites with continuous fibers are more difficult to produce than are the polymer-matrix composites. Casting processes that force liquid around the fibers using capillary rise, pressure casting, vacuum infiltration, or continuous casting are used. Various solid-state compaction processes can also be used.

17-6 Fiber-Reinforced Systems and Applications

Before completing our discussion of fiber-reinforced composites, let's look at the behavior and applications of some of the most common of these materials. Figure 17-25 compares the specific modulus and specific strength of several composites with those of

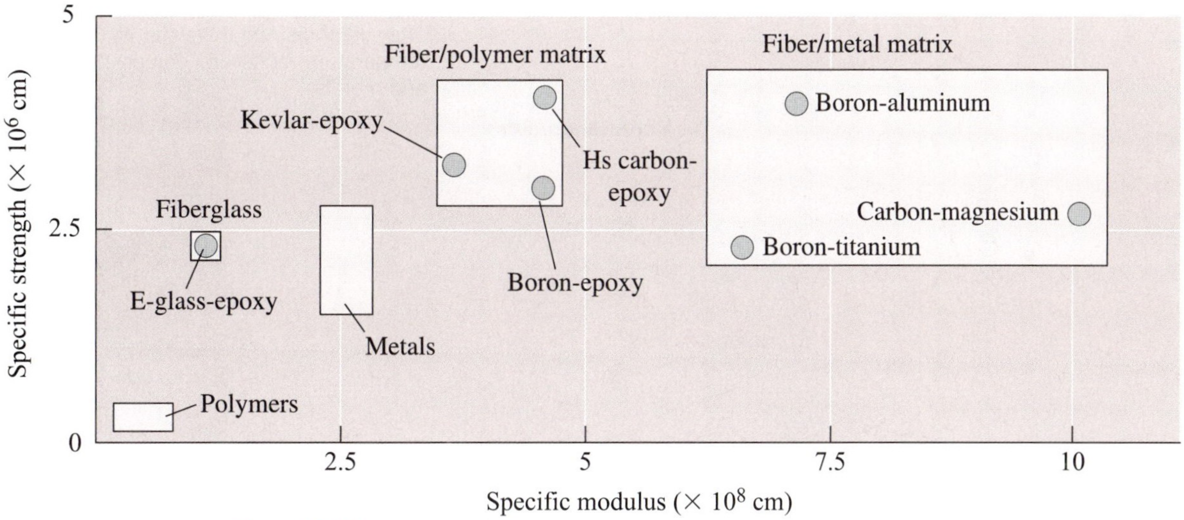

Figure 17-25 A comparison of the specific modulus and specific strength of several composite materials with those of metals and polymers.

metals and polymers. Note that the values in this figure are lower than those in Figure 17-14, since we are now looking at the composite, not just the fiber.

Advanced Composites The term advanced composites is often used when the composite is intended to provide service in very critical applications, as in the aerospace industry (Table 17-3). The advanced composites normally are polymer–matrix composites reinforced with high-strength polymer, metal, or ceramic fibers. Carbon fibers are used extensively where particularly good stiffness is required; aramid—and, to an even greater extent, polyethylene—fibers are better suited to high-strength applications in which toughness and damage resistance are more important. Unfortunately, the polymer fibers lose their strength at relatively low temperatures, as do all of the polymer matrices (Figure 17-26 on the next page).

The advanced composites are also frequently used for sporting goods. Tennis rackets, golf clubs, skis, ski poles, and fishing poles often contain carbon or aramid fibers because the higher stiffness provides better performance. In the case of golf clubs, carbon fibers allow less weight in the shaft and therefore more weight in the head. Fabric reinforced with polyethylene fibers is used for lightweight sails for racing yachts.

A unique application for aramid fiber composites is armor. Tough Kevlar™ composites provide better ballistic protection than other materials, making them suitable for lightweight, flexible bulletproof clothing.

TABLE 17-3 ■ *Examples of fiber-reinforced materials and applications*

Material	Applications
Borsic aluminum	Fan blades in engines, other aircraft and aerospace applications
Kevlar™-epoxy and Kevlar™-polyester	Aircraft, aerospace applications (including space shuttle), boat hulls, sporting goods (including tennis rackets, golf club shafts, fishing rods), flak jackets
Graphite-polymer	Aerospace and automotive applications, sporting goods
Glass-polymer	Lightweight automotive applications, water and marine applications, corrosion-resistant applications, sporting goods equipment, aircraft and aerospace components

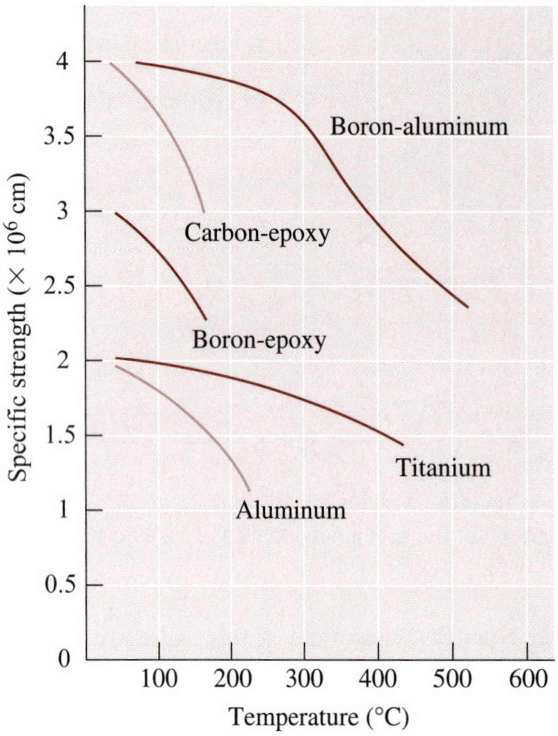

Figure 17-26
The specific strength versus temperature for several composites and metals.

Hybrid composites are composed of two or more types of fibers. For instance, Kevlar™ fibers may be mixed with carbon fibers to improve the toughness of a stiff composite, or Kevlar™ may be mixed with glass fibers to improve stiffness. Particularly good tailoring of the composite to meet specific applications can be achieved by controlling the amounts and orientations of each fiber.

Tough composites can also be produced if careful attention is paid to the choice of materials and processing techniques. Better fracture toughness in the rather brittle composites can be obtained by using long fibers, amorphous (such as PEEK and PPS) rather than crystalline or cross-linked matrices, thermoplastic-elastomer matrices, or interpenetrating network polymers.

Metal-Matrix Composites These materials, strengthened by metal or ceramic fibers, provide high-temperature resistance. Aluminum reinforced with borsic fibers has been used extensively in aerospace applications, including struts for the space shuttle. Copper-based alloys have been reinforced with SiC fibers, producing high-strength propellers for ships.

Aluminum is commonly used in metal-matrix composites. Al_2O_3 fibers reinforce the pistons for some diesel engines; SiC fibers and whiskers are used in aerospace applications, including stiffeners and missile fins; and carbon fibers provide reinforcement for the aluminum antenna mast of the Hubble telescope. Polymer fibers, because of their low melting or degradation temperatures, are not normally used in a metallic matrix. *Polymets*, however, are produced by hot extruding aluminum powder and high-melting-temperature liquid-crystalline polymers. A reduction of 1000 to 1 during the extrusion process elongates the polymer into aligned filaments and bonds the aluminum powder particles into a solid matrix.

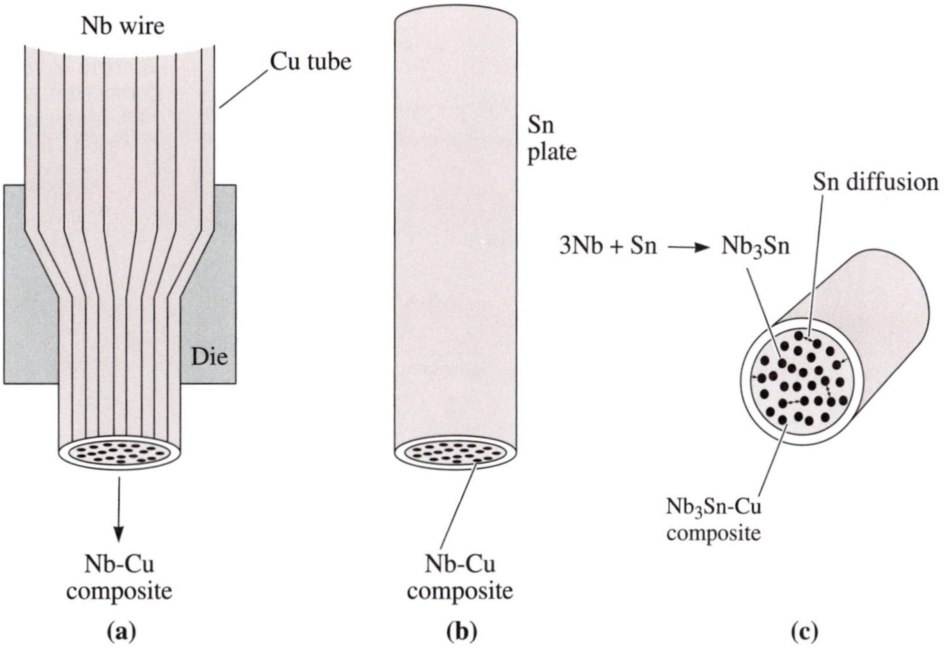

Figure 17-27 The manufacture of composite superconductor wires: (a) Niobium wire is surrounded with copper during forming. (b) Tin is plated onto Nb-Cu composite wire. (c) Tin diffuses to niobium to produce the Nb_3Sn-Cu composite.

Metal-matrix composites may find important applications in components for rocket or aircraft engines. Superalloys reinforced with metal fibers (such as tungsten) or ceramic fibers (such as SiC or B_4N) maintain their strength at higher temperatures, permitting jet engines to operate more efficiently. Similarly, titanium and titanium aluminides reinforced with SiC fibers are considered for turbine blades and disks.

A unique application for metal-matrix composites is in the superconducting wire required for fusion reactors. The intermetallic compound Nb_3Sn has good superconducting properties but is very brittle. To produce Nb_3Sn wire, pure niobium wire is surrounded by copper as the two metals are formed into a wire composite (Figure 17-27). The niobium-copper composite wire is then coated with tin. The tin diffuses through the copper and reacts with the niobium to produce the intermetallic compound. Niobium-titanium systems are also used.

Ceramic-Matrix Composites Composites containing ceramic fibers in a ceramic matrix are also finding applications. Two important uses will be discussed to illustrate the unique properties that can be obtained with these materials.

Carbon-carbon (C—C) composites are used for extraordinary temperature resistance in aerospace applications. Carbon-carbon composites can operate at temperatures of up to 3000°C and, in fact, are stronger at high temperatures than at low temperatures (Figure 17-28). Carbon-carbon composites are made by forming a polyacrylonitrile or carbon fiber fabric into a mold, then impregnating the fabric with an organic resin, such as a phenolic. The part is pyrolyzed to convert the phenolic resin to carbon. The composite, which is still soft and porous, is impregnated and pyrolyzed several more times, continually increasing the density, strength, and stiffness. Finally, the part is coated with silicon carbide to protect the carbon-carbon composite from oxidation. Strengths of 2070 MPa and a Young's modulus (measure of stiffness) of 345×10^3 MPa can be

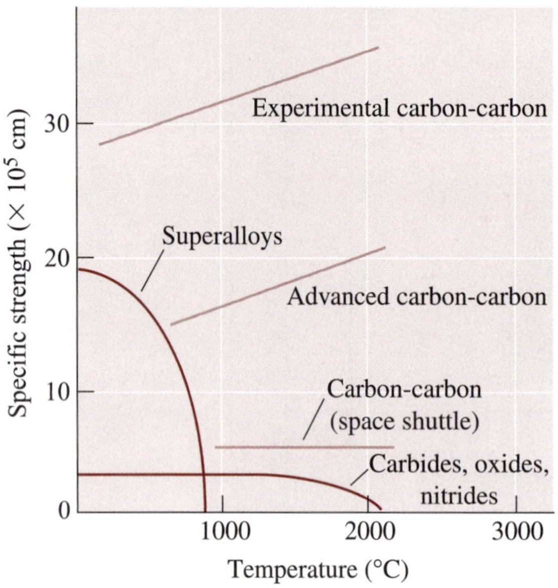

Figure 17-28
A comparison of the specific strength of various carbon-carbon composites with that of other high-temperature materials relative to temperature.

obtained. Carbon-carbon composites have been used as nose cones and leading edges of high-performance aerospace vehicles such as the space shuttle, and as brake discs on racing cars and commercial jet aircraft.

Ceramic-fiber–ceramic-matrix composites provide improved strength and fracture toughness compared with conventional ceramics (Table 17-4). Fiber reinforcements improve the toughness of the ceramic matrix in several ways. First, a crack moving through the matrix encounters a fiber; if the bonding between the matrix and the fiber is poor, the crack is forced to propagate around the fiber in order to continue the fracture process. In addition, poor bonding allows the fiber to begin to pull out of the matrix [Figure 17-29(a)]. Both processes consume energy, thereby increasing fracture toughness. Finally, as a crack in the matrix begins, unbroken fibers may bridge the crack, providing a compressive stress that helps keep the crack from opening [Figure 17-29(b)].

TABLE 17-4 ■ ***Effect of SiC-reinforcement fibers on the properties of selected ceramic materials***

Material	Flexural Strength (MPa)	Fracture Toughness ($MPa\sqrt{m}$)
Al_2O_3	550	5.5
Al_2O_3/SiC	790	8.8
SiC	500	4.4
SiC/SiC	760	25.3
ZrO_2	210	5.5
ZrO_2/SiC	450	22.2
Si_3N_4	470	4.4
Si_3N_4/SiC	790	56.0
Glass	60	1.1
Glass/SiC	830	18.7
Glass ceramic	210	2.2
Glass ceramic/SiC	830	17.6

(a)

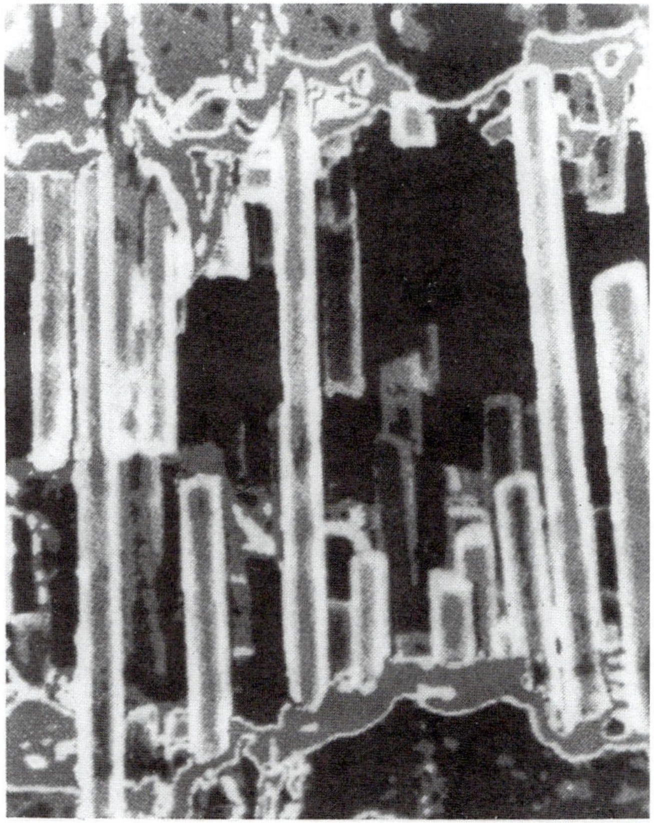

(b)

Figure 17-29 Two failure modes in ceramic-ceramic composites: (a) Extensive pull-out of SiC fibers in a glass matrix provides good composite toughness (×20). (*From* Metals Handbook, *American Society for Metals, Vol. 9, 9th Ed., 1985.*) (b) Bridging of some fibers across a crack enhances the toughness of a ceramic-matrix composite (unknown magnification). (*From* Journal of Metals, *May 1991.*)

Unlike polymer and metal matrix composites, poor bonding—rather than good bonding—is required! Consequently, control of the interface structure is crucial. In a glass-ceramic (based on $Al_2O_3 \cdot SiO_2 \cdot Li_2O$) reinforced with SiC fibers, an interface layer containing carbon and NbC is produced that makes debonding of the fiber from the matrix easy. If, however, the composite is heated to a high temperature, the interface is oxidized; the oxide occupies a large volume, exerts a clamping force on the fiber, and prevents easy pull-out. Fracture toughness is then decreased. The following example illustrates some of the cost and property issues that come up while working with composites.

EXAMPLE 17-10 *Design of a Composite Strut*

Design a unidirectional fiber-reinforced epoxy-matrix strut having a round cross-section. The strut is 3 m long and, when a force of 2224 N is applied, it should stretch no more than 0.25 cm. We want to assure that the stress acting on the strut is less than the yield strength of the epoxy matrix, 83 MPa. If the fibers should happen to break, the strut will stretch an extra amount but may not catastrophically fracture. Epoxy costs about \$1.8/kg and has a modulus of elasticity of 3450 MPa.

SOLUTION

Suppose that the strut were made entirely of epoxy (that is, no fibers):

$$\varepsilon_{\text{max}} = \frac{0.25\text{ cm}}{300} = 0.00083\text{ cm/cm}$$

$$\sigma_{\text{max}} = E\varepsilon = (3450)(0.00083) = 2.9\text{ MPa}$$

$$A_{\text{strut}} = \frac{F}{\sigma} = \frac{2224\text{ N}}{2.9\text{ MPa}} = 7.67\text{ cm}^2 \quad \text{or} \quad d = 3.12\text{ cm}$$

Since $\rho_{\text{epoxy}} = 1.25\text{ g/cm}^3$,

$$\text{Weight}_{\text{strut}} = (1.25)(\pi)(3.12/2)^2(300) = 2.87\text{ kg}$$

$$\text{Cost}_{\text{strut}} = (2.87\text{ kg})(\$1.8/\text{kg}) = \$5.17$$

With no reinforcement, the strut is large and heavy; the materials cost is high due to the large amount of epoxy needed.

In a composite, the maximum strain is still 0.00083 cm/cm If we make the strut as small as possible—that is, it operates at 83 MPa—then the minimum modulus of elasticity E_c of the composite is:

$$E_c > \frac{\sigma}{\varepsilon_{\text{max}}} = \frac{83}{0.00083} = 100 \times 10^3\text{ MPa}$$

Let's look at several possible composite systems. The modulus of glass fibers is less than 100×10^3 MPa; therefore, glass reinforcement is not a possible choice.

For high modulus carbon fibers, $E = 531 \times 10^3$ MPa; the density is 1.9 g/cm^3, and the cost is about \$66/kg. The minimum volume fraction of carbon fibers needed to give a composite modulus of 100×10^3 MPa is:

$$E_c = f_C(531 \times 10^3) + (1 - f_C)(3.45 \times 10^3) > 100 \times 10^3$$

$$f_C = 0.183$$

The volume fraction of epoxy remaining is 0.817. An area of 0.817 times the total cross-sectional area of the strut must support a 2224 N load with no more than 83 MPa if all of the fibers should fail:

$$A_{epoxy} = 0.817 A_{total} = \frac{F}{\sigma} = \frac{2224 \text{ N}}{83 \text{ MPa}} = 0.268 \text{ cm}^2$$

$$A_{total} = \frac{0.268}{0.817} = 0.328 \text{ cm}^2 \quad \text{or} \quad d = 0.646 \text{ cm}$$

$$\text{Volume}_{strut} = A_{total} = (0.328 \text{ cm}^2)(300 \text{ cm}) = 98.4 \text{ cm}^3$$

$$\text{Weight}_{strut} = \rho V = [(1.9)(0.183) + (1.25)(0.817)](98.4)$$
$$= 0.135 \text{ kg}$$

$$\text{Weight fraction carbon} = \frac{(0.183)(1.9 \text{ g/cm}^3)}{(0.183)(1.9) + (0.817)(1.25)} = 0.254$$

$$\text{Weight carbon} = (0.254)(0.135 \text{ kg}) = 0.034 \text{ kg}$$

$$\text{Weight epoxy} = (0.746)(0.135 \text{ kg}) = 0.101 \text{ kg}$$

$$\text{Cost of each strut} = (0.034)(\$66) + (0.101)(\$1.80) = \$2.42$$

The carbon-fiber reinforced strut is less than one-quarter the diameter of an all-epoxy structure, with only 5% of the weight and half of the cost.

We might also repeat these calculations using Kevlar™ fibers, with a modulus of 124×10^3 MPa, a density of 1.44 g/cm^3, and a cost of about \$44/kg. By doing so, we would find that a volume fraction of 0.8 fibers is required. Note that 0.8 volume fraction is at the maximum of fiber volume that can be incorporated into a matrix. We would also find that the required diameter of the strut is 1.288 cm and that the strut weighs 0.573 kg and costs \$20.94. The modulus of the Kevlar™ is not high enough to offset its high cost.

Although the carbon fibers are the most expensive, they permit the lightest weight and the lowest material cost strut. (This calculation does not, however, take into consideration the costs of manufacturing the strut.) Our design, therefore, is to use a 0.646-cm-diameter strut containing 0.183 volume fraction high-modulus carbon fiber.

17-7 Laminar Composite Materials

Laminar composites include very thin coatings, thicker protective surfaces, claddings, bimetallics, laminates, and a host of other applications. In addition, the fiber-reinforced composites produced from tapes or fabrics can be considered partly laminar. Many laminar composites are designed to improve corrosion resistance while retaining low cost, high strength, or light weight. Other important characteristics include superior wear or abrasion resistance, improved appearance, and unusual thermal expansion characteristics.

Rule of Mixtures Some properties of the laminar composite materials parallel to the lamellae are estimated from the rule of mixtures. The density, electrical and thermal conductivity, and modulus of elasticity parallel to the lamellae can be calculated with little error using following formulas:

$$\text{Density} = \rho_{c,\|} = \sum (f_i \rho_i) \tag{17-13}$$

$$\left.\begin{aligned} \text{Electrical conductivity} &= \sigma_{c,\|} = \sum (f_i \sigma_i) \\ \text{Thermal conductivity} &= K_{c,\|} = \sum (f_i K_i) \\ \text{Modulus of elasticity} &= E_{c,\|} = \sum (f_i E_i) \end{aligned}\right\} \tag{17-14}$$

The laminar composites are very anisotropic. The properties perpendicular to the lamellae are:

$$\left.\begin{aligned} \text{Electrical conductivity} &= \frac{1}{\sigma_{c,\perp}} = \sum \left(\frac{f_i}{\sigma_i}\right) \\ \text{Thermal conductivity} &= \frac{1}{K_{c,\perp}} = \sum \left(\frac{f_i}{K_i}\right) \\ \text{Modulus of elasticity} &= \frac{1}{E_{c,\perp}} = \sum \left(\frac{f_i}{E_i}\right) \end{aligned}\right\} \tag{17-15}$$

However, many of the really important properties, such as corrosion and wear resistance, depend primarily on only one of the components of the composite, so the rule of mixtures is not applicable.

Producing Laminar Composites Several methods are used to produce laminar composites, including a variety of deformation and joining techniques used primarily for metals. (Figure 17-30).

Individual plies are often joined by *adhesive bonding*, as is the case in producing plywood. Polymer-matrix composites built up from several layers of fabric or tape prepregs are also joined by adhesive bonding; a film of unpolymerized material is placed between each layer of prepreg. When the layers are pressed at an elevated temperature, polymerization is completed and the prepregged fibers are joined to produce composites that may be dozens of layers thick.

Most of the metallic laminar composites, such as claddings and bimetallics, are produced by *deformation bonding*, such as hot- or cold-roll bonding. The pressure exerted by the rolls breaks up the oxide film at the surface, brings the surfaces into

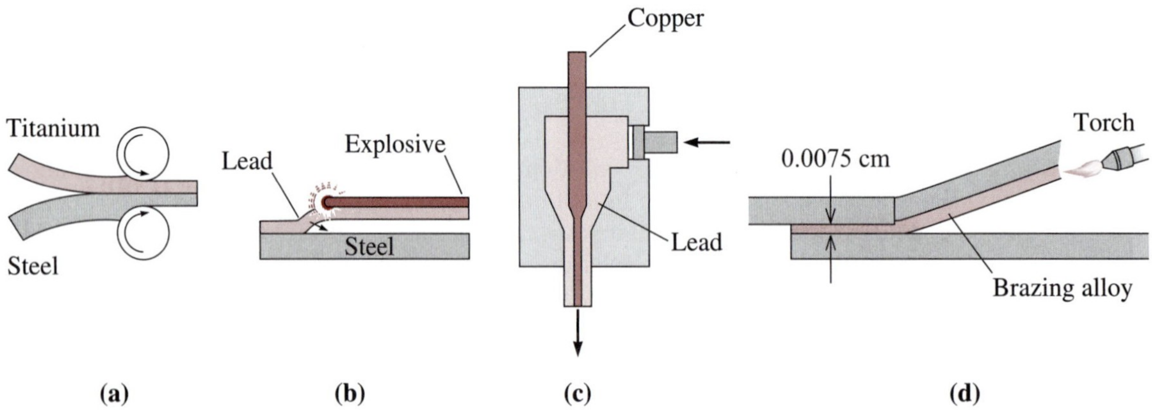

Figure 17-30 Techniques for producing laminar composites: (a) roll bonding, (b) explosive bonding, (c) coextrusion, and (d) brazing.

atom-to-atom contact, and permits the two surfaces to be joined. Explosive bonding can also be used. An explosive charge provides the pressure required to join metals. This process is particularly well suited for joining very large plates that will not fit into a rolling mill. Very simple laminar composites, such as coaxial cable, are produced by coextruding two materials through a die in such a way that the soft material surrounds the harder material. Metal conductor wire can be coated with an insulating thermoplastic polymer in this manner.

Brazing can join composite plates (Chapter 9). The metallic sheets are separated by a very small clearance—preferably, about 0.0075 cm—and heated above the melting temperature of the brazing alloy. The molten brazing alloy is drawn into the thin joint by capillary action.

17-8 Examples and Applications of Laminar Composites

The number of laminar composites is so varied and their applications and intentions are so numerous that we cannot make generalizations concerning their behavior. Instead we will examine the characteristics of a few commonly used examples.

Laminates Laminates are layers of materials joined by an organic adhesive. In laminated safety glass, a plastic adhesive, such as polyvinyl butyral (PVB), joins two pieces of glass; the adhesive prevents fragments of glass from flying about when the glass is broken (Chapter 15). Laminates are used for insulation in motors, for gears, for printed circuit boards, and for decorative items such as Formica® countertops and furniture.

Microlaminates include composites composed of alternating layers of aluminum sheet and fiber-reinforced polymer. *Arall* (aramid-aluminum laminate) and *Glare* (glass-aluminum laminate) have been developed as possible skin materials for aircraft. In Arall, an aramid fiber such as Kevlar™ is prepared as a fabric or unidirectional tape, impregnated with an adhesive, and laminated between layers of aluminum alloy (Figure 17-31). The composite laminate has an unusual combination of strength, stiffness, corrosion resistance, and light weight. Fatigue resistance is improved, since the interface between the layers may block cracks. Compared with polymer-matrix composites, the microlaminates have good resistance to lightning-strike damage (which is important in aerospace applications), are formable and machinable, and are easily repaired.

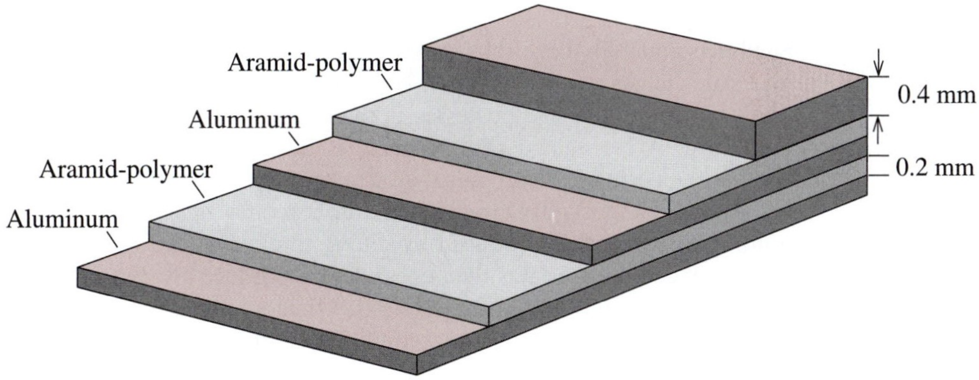

Figure 17-31 Schematic diagram of an aramid-aluminum laminate, Arall, which has potential for aerospace applications.

Clad Metals Clad materials are metal-metal composites. A common example of **cladding** is United States silver coinage. A Cu-80% Ni alloy is bonded to both sides of a Cu-20% Ni alloy. The ratio of thicknesses is about 1/6:2/3:1/6. The high-nickel alloy is a silver color, while the predominantly copper core provides low cost.

Clad materials provide a combination of good corrosion resistance with high strength. *Alclad* is a clad composite in which commercially pure aluminum is bonded to higher-strength aluminum alloys. The pure aluminum protects the higher-strength alloy from corrosion. The thickness of the pure aluminum layer is about 1% to 15% of the total thickness. Alclad is used in aircraft construction, heat exchangers, building construction, and storage tanks, where combinations of corrosion resistance, strength, and light weight are desired.

Bimetallics Temperature indicators and controllers take advantage of the different coefficients of thermal expansion of the two metals in laminar composite. If two pieces of metal are heated, the metal with the higher coefficient of thermal expansion becomes longer. If the two pieces of metal are rigidly bonded together, the difference in their coefficients causes the strip to bend and produce a curved surface. The amount of movement depends on the temperature. By measuring the curvature or deflection of the strip, we can determine the temperature. Likewise, if the free end of the strip activates a relay, the strip can turn on or off a furnace or air conditioner to regulate temperature. Metals selected for **bimetallics** must have (a) very different coefficients of thermal expansion, (b) expansion characteristics that are reversible and repeatable, and (c) a high modulus of elasticity, so that the bimetallic device can do work. Often the low-expansion strip is made from Invar, an iron-nickel alloy, whereas the high-expansion strip may be brass, Monel, or pure nickel.

Bimetallics can act as circuit breakers as well as thermostats; if a current passing through the strip becomes too high, heating causes the bimetallic to deflect and break the circuit.

Multilayer capacitors Similar geometry is used to make billions of multi-layer capacitors. Their structure is comprisesd of thin sheets of $BaTiO_3$-based ceramics separated by Ag/Pd or Ni electrodes.

17-9 Sandwich Structures

Sandwich materials have thin layers of a facing material joined to a lightweight filler material, such as a polymer foam. Neither the filler nor the facing material is strong or rigid, but the composite possesses both properties. A familiar example is corrugated cardboard. A corrugated core of paper is bonded on either side to flat, thick paper. Neither the corrugated core nor the facing paper is rigid, but the combination is.

Another important example is the honeycomb structure used in aircraft applications. A **honeycomb** is produced by gluing thin aluminum strips at selected locations. The honeycomb material is then expanded to produce a very low-density cellular panel that, by itself, is unstable (Figure 17-32). When an aluminum facing sheet is adhesively bonded to either side of the honeycomb, however, a very stiff, rigid, strong, and exceptionally lightweight sandwich with a density as low as 0.04 g/cm^3 is obtained.

The honeycomb cells can have a variety of shapes, including hexagonal, square, rectangular, and sinusoidal, and they can be made from aluminum, fiberglass, paper,

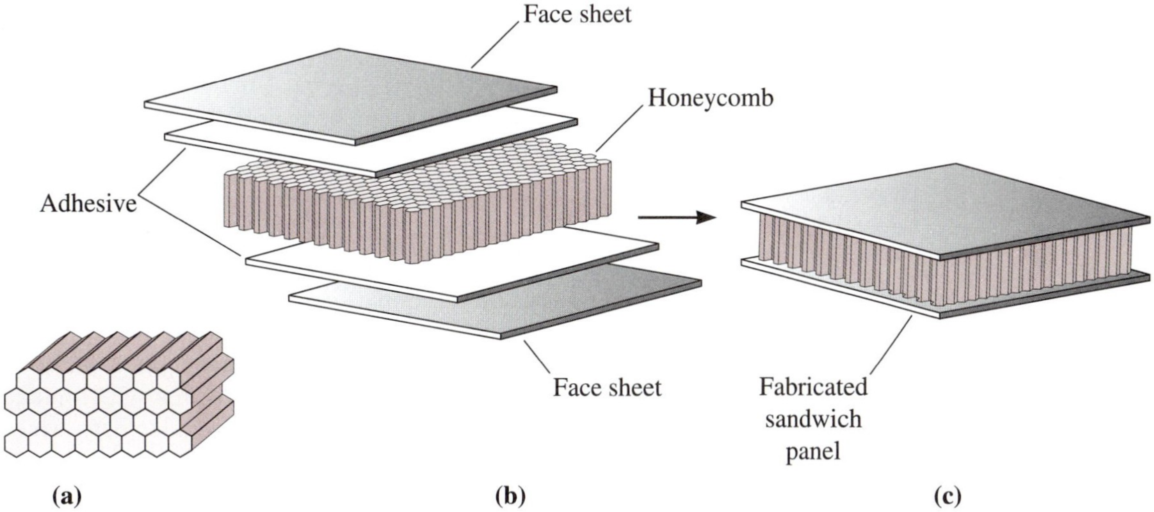

Figure 17-32 (a) A hexagonal cell honeycomb core, (b) can be joined to two face sheets by means of adhesive sheets, (c) producing an exceptionally lightweight yet stiff, strong honeycomb sandwich structure.

aramid polymers, and other materials. The honeycomb cells can be filled with foam or fiberglass to provide excellent sound and vibration absorption. Figure 17-33 describes one method by which the honeycomb can be fabricated.

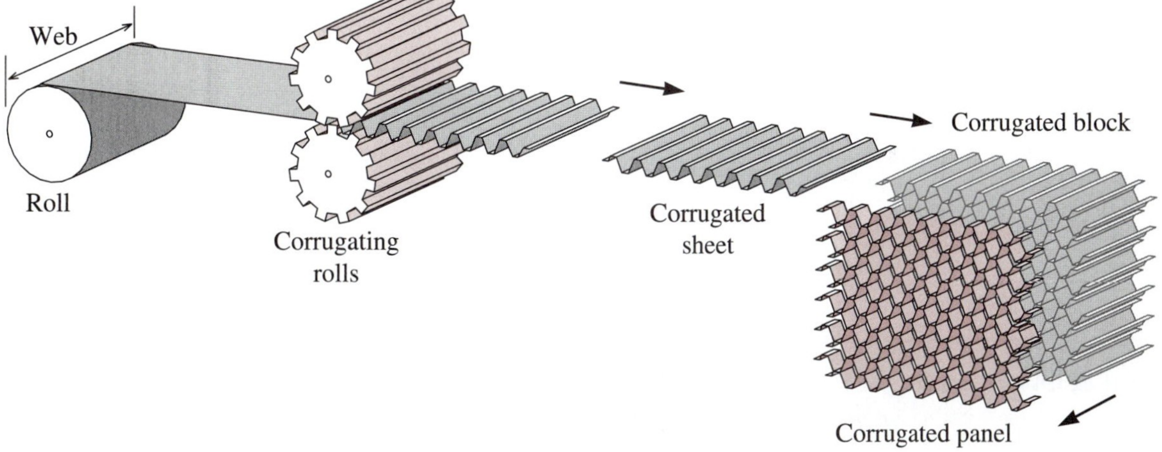

Figure 17-33 In the corrugation method for producing a honeycomb core, the material (such as aluminum) is corrugated between two rolls. The corrugated sheets are joined together with adhesive and then cut to the desired thickness.

SUMMARY

- Composites are composed of two or more materials or phases joined or connected in such a way so as to give a combination of properties that cannot be attained otherwise. The volume fraction and the connectivity of the phases or materials in a composite and the nature of the interface between the dispersed phase and matrix are very important in determining the properties.

- Virtually any combination of metals, polymers, and ceramics is possible. In many cases, the rule of mixtures can be used to estimate some of the properties of a composite.
- Composites have many applications in construction, aerospace, automotive, sports, microelectronics and other industries.
- Dispersion-strengthened materials contain exceptionally small oxide particles in a metal matrix. The small stable dispersoids interfere with slip, providing good mechanical properties at elevated temperatures.
- Particulate composites contain particles that impart combinations of properties to the composite. Metal-matrix composites contain ceramic or metallic particles that provide improved strength and wear resistance and assure good electrical conductivity, toughness, or corrosion resistance. Polymer-matrix composites contain particles that enhance stiffness, heat resistance, or electrical conductivity while maintaining light weight, ease of fabrication, or low cost.
- Fiber-reinforced composites provide improvements in strength, stiffness, or high-temperature performance in metals and polymers and impart toughness to ceramics. Fibers typically have low densities, giving high specific strength and specific modulus, but they often are very brittle. Fibers can be continuous or discontinuous. Discontinuous fibers with a high aspect ratio l/d produce better reinforcement.
- Fibers are introduced into a matrix in a variety of orientations. Random orientations and isotropic behavior are obtained using discontinuous fibers; unidirectionally aligned fibers produce composites with anisotropic behavior, with large improvements in strength and stiffness parallel to the fiber direction. Properties can be tailored to meet the imposed loads by orienting the fibers in multiple directions.
- Laminar composites are built of layers of different materials. These layers may be sheets of different metals, with one metal providing strength, and the other providing hardness or corrosion resistance. Layers may also include sheets of fiber-reinforced material bonded to metal or polymer sheets or even to fiber-reinforced sheets having different fiber orientations. The laminar composites are always anisotropic.
- Sandwich materials, including honeycombs, are exceptionally lightweight laminar composites, with solid facings joined to an almost hollow core.

GLOSSARY

Aramid fibers Polymer fibers, such as Kevlar™, formed from polyamides, which contain the benzene ring in the backbone of the polymer.

Aspect ratio The length of a fiber divided by its diameter.

Bimetallic A laminar composite material produced by joining two strips of metal with different thermal expansion coefficients, making the material sensitive to temperature changes.

Brazing A process in which a liquid filler metal is introduced by capillary action between two solid-based materials that are to be joined. Solidification of the brazing alloy provides the bond.

Carbonizing Driving off the non-carbon atoms from a polymer fiber, leaving behind a carbon fiber of high strength. Also known as pyrolizing.

Cemented carbides or cermets Particulate composites containing hard ceramic particles bonded with a soft metallic matrix. The composite combines high hardness and cutting ability, yet still has good shock resistance.

Chemical vapor deposition (CVD) Method for manufacturing materials by condensing the material from a vapor onto a solid substrate.

Cladding A laminar composite produced when a corrosion-resistant or high-hardness layer of a laminar composite is formed onto a less expensive or higher-strength backing.

Delamination Separation of individual plies of a fiber-reinforced composite.

Dispersed phase The phase or phases that are dispersed in a continuous matrix of the composite.

Dispersoids Tiny oxide particles formed in a metal matrix that interfere with dislocation movement and provide strengthening, even at elevated temperatures.

Filament winding Process for producing fiber-reinforced composites in which continuous fibers are wrapped around a form or mandrel. The fibers may be prepregged or the filament-wound structure may be impregnated to complete the production of the composite.

Honeycomb A lightweight but stiff assembly of aluminum strip joined and expanded to form the core of a sandwich structure.

Hybrid organic-inorganic composites Nanocomposites consisting of structures that are partly organic and partly inorganic, often made using sol-gel processing.

Matrix phase The continuous phase in a composite. The composites are named after the continuous phase (e.g., polymer-matrix composites).

Nanocomposite A material in which the dispersed phase is nano-sized and is distributed in the continuous matrix.

Precursor A starting chemical (e.g., a polymer fiber) that is carbonized to produce carbon fibers.

Prepregs Layers of fibers in unpolymerized resins. After the prepregs are stacked to form a desired structure, polymerization joins the layers together.

Pultrusion A method for producing composites containing mats or continuous fibers.

Rovings Bundles of less than 10,000 filaments.

Rule of mixtures The statement that the properties of a composite material are a function of the volume fraction of each material in the composite.

Sandwich A composite material constructed of a lightweight, low-density material surrounded by dense, solid layers. The sandwich combines overall light weight with excellent stiffness.

Sizing Coating glass fibers with an organic material to improve bonding and moisture resistance in fiberglass.

Specific modulus The modulus of elasticity divided by the density.

Specific strength The tensile or yield strength of a material divided by the density.

Staples Fibers chopped into short lengths.

Tapes Single-filament-thick strips of prepregs, with the filaments either unidirectional or woven fibers. Several layers of tapes can be joined to produce composite structures.

Tow A bundle of more than 10,000 filaments.

Whiskers Very fine fibers grown in a manner that produces single crystals with no mobile dislocations, thus giving nearly theoretical strengths.

Yarns Continuous fibers produced from a group of twisted filaments.

PROBLEMS

Section 17-1 Dispersion-Strengthened Composites

17-1 What is a composite?

17-2 What do the properties of composite materials depend upon?

17-3 Give examples of applications where composites are used for load bearing applications.

17-4 Give two examples of applications where composites are used for non-structural applications.

17-5 What is a dispersion-strengthened composite? How is it different from a particle-reinforced composite?

17-6 What is a nanocomposite? How can steels containing ferrite and martensite be described as composites? Explain.

17-7 Nickel containing 2 wt% thorium is produced in powder form, consolidated into a part, and sintered in the presence of oxygen, causing all of the thorium to produce ThO_2 spheres 80 nm in diameter. Calculate the number of spheres per cm^3. (*Note:* The density of ThO_2 is 9.86 g/cm^3.)

17-8 Spherical aluminum powder (SAP) 0.002 mm in diameter is treated to create a thin oxide layer and is then used to produce a SAP dispersion-strengthened material containing 10 vol% Al_2O_3. Calculate the average thickness of the oxide film prior to compaction and sintering of the powders into the part.

17-9 Yttria (Y_2O_3) particles 750 A in diameter are introduced into tungsten by internal oxidation. Measurements using an electron microscope show that there are 5×10^{14} oxide particles per cm^3. Calculate the wt% Y originally in the alloy. (*Note:* The density of Y_2O_3 is 5.01 g/cm^3.)

17-10 With no special treatment, aluminum is typically found to have an Al_2O_3 layer that is 3 nm thick. If spherical aluminum powder prepared with a total diameter of 0.01 mm is used to produce the SAP dispersion-strengthened material, calculate the volume percent Al_2O_3 in the material and the number of oxide particles per cm^3. Assume that the oxide breaks into disk-shaped flakes 3 nm thick and 3×10^{-4} mm in diameter. Compare the number of oxide particles per cm^3 with the number of solid solution atoms per cm^3 when 3 at% of an alloying element is added to aluminum.

Section 17-2 Particulate Composites

17-11 What is a particulate composite?

17-12 What is a cermet? What is the role of WC and Co in a cermet?

17-13 Calculate the density of a cemented carbide, or cermet, based on a titanium matrix if the composite contains 50 wt% WC, 22 wt% TaC, and 14 wt% TiC. (See Example 17-2 for densities of the carbides.)

17-14 A typical grinding wheel is 22.5 cm in diameter, 2.5 cm thick, and weighs 2.7 kg. The wheel contains SiC (density of 3.2 g/cm^3) bonded by silica glass (density of 2.5 g/cm^3); 5 vol% of the wheel is porous. (*Note:* The SiC is in the form of 0.04 cm cubes.) Calculate

(a) the volume fraction of SiC particles in the wheel and

(b) the number of SiC particles lost from the wheel after it is worn to a diameter of 20 cm.

17-15 An electrical contact material is produced by infiltrating copper into a porous tungsten carbide (WC) compact. The density of the final composite is 12.3 g/cm^3. Assuming that all of the pores are filled with copper, calculate

(a) the volume fraction of copper in the composite,

(b) the volume fraction of pores in the WC compact prior to infiltration, and

(c) the original density of the WC compact before infiltration.

17-16 An electrical contact material is produced by first making a porous tungsten compact that weighs 125 g. Liquid silver is introduced into the compact; careful measurement indicates that 105 g of silver is infiltrated. The final density of the composite is 13.8 g/cm^3. Calculate the volume fraction of the original compact that is interconnected porosity and the volume fraction that is closed porosity (no silver infiltration).

17-17 How much clay must be added to 10 kg of polyethylene to produce a low-cost composite having a modulus of elasticity greater than 827 MPa and a tensile strength greater than 14 MPa? (*Note:* The density of the clay is 2.4 g/cm^3 and that of polyethylene is 0.92 g/cm^3.)

17-18 We would like to produce a lightweight epoxy part to provide thermal insulation. We have available hollow glass beads for which the outside diameter is 0.16 cm and the wall thickness is

0.0025 cm. Determine the weight and number of beads that must be added to the epoxy to produce a one-kg composite with a density of 0.65 g/cm^3. (*Note:* The density of the glass is 2.5 g/cm^3 and that of the epoxy is 1.25 g/cm^3.)

Section 17-3 Fiber-Reinforced Composites

Section 17-4 Characteristics of Fiber-Reinforced Composites

17-19 What is a fiber-reinforced composite?

17-20 What fiber-reinforcing materials are commonly used?

17-21 In a fiber-reinforced composite, what is the role of the matrix?

17-22 What do the terms CFRP and GFRP mean?

17-23 Explain briefly how the volume of fiber, fiber orientation, and fiber strength and modulus affect the properties of fiber-reinforced composites.

17-24 Five kg of continuous boron fibers are introduced in a unidirectional orientation into 8 kg of an aluminum matrix. Calculate

(a) the density of the composite,
(b) the modulus of elasticity parallel to the fibers, and
(c) the modulus of elasticity perpendicular to the fibers.

17-25 We want to produce 4.5 kg of a continuous unidirectional fiber-reinforced composite of HS carbon in a polyimide matrix that has a modulus of elasticity of at least 172×10^3 MPa parallel to the fibers. How many kg of fibers are required? (See Chapter 16 for properties of polyimide.)

17-26 We produce a continuous unidirectionally reinforced composite containing 60 vol% HM carbon fibers in an epoxy matrix. The epoxy has a tensile strength of 103 MPa. What fraction of the applied force is carried by the fibers?

17-27 A polyester matrix with a tensile strength of 90 MPa is reinforced with Al_2O_3 fibers. What vol% fibers must be added to insure that the fibers carry 75% of the applied load?

17-28 An epoxy matrix is reinforced with 40 vol% E-glass fibers to produce a 2-cm-diameter composite that is to withstand a load of 25,000 N. Calculate the stress acting on each fiber.

17-29 A titanium alloy with a modulus of elasticity of 110×10^3 MPa is used to make a 454-kg part for a manned space vehicle. Determine the weight of a part having the same modulus of elasticity parallel to the fibers, if the part is made of

(a) aluminum reinforced with boron fibers and
(b) polyester reinforced with high-modulus carbon fibers (with a modulus of 4482 MPa).
(c) Compare the specific modulus for all three materials.

17-30 Short, but aligned, Al_2O_3 fibers with a diameter of 20 μm are introduced into a 6,6-nylon matrix. The strength of the bond between the fibers and the matrix is estimated to be 7 MPa. Calculate the critical fiber length and compare with the case when 1-μm alumina whiskers are used instead of the coarser fibers. What is the minimum aspect ratio in each case?

17-31 We prepare several epoxy-matrix composites using different lengths of 3-μm-diameter ZrO_2 fibers and find that the strength of the composite increases with increasing fiber length up to 5 mm. For longer fibers, the strength is virtually unchanged. Estimate the strength of the bond between the fibers and the matrix.

Section 17-5 Manufacturing Fibers and Composites

17-32 Explain briefly how boron and carbon fibers are made.

17-33 Explain briefly how continuous-glass fibers are made.

17-34 What is a coupling agent? What is "sizing" as it relates to the production of glass fibers?

17-35 What is the difference between a fiber and a whisker?

17-36 In one polymer-matrix composite, as produced, discontinuous glass fibers are introduced directly into the matrix; in a second case, the fibers are first "sized." Discuss the effect this difference might have on the critical fiber length and the strength of the composite.

17-37 Explain why bonding between carbon fibers and an epoxy matrix should be excellent, whereas bonding between silicon nitride fibers and a silicon carbide matrix should be poor.

Section 17-6 Fiber-Reinforced Systems and Applications

17-38 Explain briefly in what sporting equipment composite materials are used. What is the main reason why composites are used in these applications?

17-39 A polyimide matrix is to be reinforced with 70 vol% carbon fibers to give a minimum modulus of elasticity of 276×10^3 MPa. Recommend a process for producing the carbon fibers required. Estimate the tensile strength of the fibers that are produced.

17-40 What are the advantages of using ceramic-matrix composites?

Section 17-7 Laminar Composite Materials

Section 17-8 Examples and Applications of Laminar Composites

Section 17-9 Sandwich Structures

17-41 What is a laminar composite?

17-42 A microlaminate, Arall, is produced using 5 sheets of 0.4-mm-thick aluminum and 4 sheets of 0.2-mm-thick epoxy reinforced with unidirectionally aligned Kevlar™ fibers. The volume fraction of Kevlar™ fibers in these intermediate sheets is 55%. Calculate the modulus of elasticity of the microlaminate parallel and perpendicular to the unidirectionally aligned Kevlar™ fibers. What are the principle advantages of the Arall material compared with those of unreinforced aluminum?

17-43 A laminate composed of 0.1-mm-thick aluminum sandwiched around a 2-cm-thick layer of polystyrene foam is produced as an insulation material. Calculate the thermal conductivity of the laminate parallel and perpendicular to the layers. The thermal conductivity of aluminum is

$$0.57 \frac{\text{cal}}{\text{cm} \cdot \text{s} \cdot \text{K}}$$

and that of the foam is

$$0.000077 \frac{\text{cal}}{\text{cm} \cdot \text{s} \cdot \text{K}}.$$

17-44 A 0.01-cm-thick sheet of a polymer with a modulus of elasticity of 4.8×10^3 MPa is sandwiched between two 4-mm-thick sheets of glass with a modulus of elasticity of 83×10^3 MPa. Calculate the modulus of elasticity of the composite parallel and perpendicular to the sheets.

17-45 A U.S. quarter is 2.3 cm in diameter and is about 0.16 cm thick. Assuming copper costs about \$2.43 per kg and nickel costs about \$9 per kg. Compare the material cost in a composite quarter versus a quarter made entirely of nickel.

17-46 Calculate the density of a honeycomb structure composed of the following elements: The two 2-mm-thick cover sheets are produced using an epoxy matrix prepreg containing 55 vol% E-glass fibers. The aluminum honeycomb is 2 cm thick; the cells are in the shape of 0.5 cm squares and the walls of the cells are 0.1 mm thick. Estimate the density of the structure. Compare the weight of a 1 m × 2 m panel of the honeycomb compared with a solid aluminum panel of the same dimensions.

Design Problems

17-47 Design the materials and processing required to produce a discontinuous, but aligned, fiber-reinforced fiberglass composite that will form the hood of a sports car. The composite should provide a density of less than 1.6 g/cm^3 and a strength of 138 MPa. Be sure to list all of the assumptions you make in creating your design.

17-48 Design an electrical-contact material and a method for producing the material that will result in a density of no more than 6 g/cm^3, yet at least 50 vol% of the material will be conductive.

17-49 What factors will have to be considered in designing a bicycle frame using an aluminum frame and a frame made using C—C composite?

Appendix A: Selected Physical Properties of Some Elements

Metal		Atomic Number	Crystal Structure	Lattice Parameters (A)	Atomic Mass g/mol	Density (g/cm^3)	Melting Temperature (°C)
Aluminum	Al	13	FCC	4.04958	26.981	2.699	660.4
Antimony	Sb	51	hex	$a = 4.307$ $c = 11.273$	121.75	6.697	630.7
Arsenic	As	33	hex	$a = 3.760$ $c = 10.548$	74.9216	5.778	816
Barium	Ba	56	BCC	5.025	137.3	3.5	729
Beryllium	Be	4	hex	$a = 2.2858$ $c = 3.5842$	9.01	1.848	1290
Bismuth	Bi	83	hex	$a = 4.546$ $c = 11.86$	208.98	9.808	271.4
Boron	B	5	rhomb	$a = 10.12$ $\alpha = 65.5°$	10.81	2.3	2300
Cadmium	Cd	48	HCP	$a = 2.9793$ $c = 5.6181$	112.4	8.642	321.1
Calcium	Ca	20	FCC	5.588	40.08	1.55	839
Cerium	Ce	58	HCP	$a = 3.681$ $c = 11.857$	140.12	6.6893	798
Cesium	Cs	55	BCC	6.13	132.91	1.892	28.6
Chromium	Cr	24	BCC	2.8844	51.996	7.19	1875
Cobalt	Co	27	HCP	$a = 2.5071$ $c = 4.0686$	58.93	8.832	1495
Copper	Cu	29	FCC	3.6151	63.54	8.93	1084.9
Gadolinium	Gd	64	HCP	$a = 3.6336$ $c = 5.7810$	157.25	7.901	1313
Gallium	Ga	31	ortho	$a = 4.5258$ $b = 4.5186$ $c = 7.6570$	69.72	5.904	29.8
Germanium	Ge	32	diamond cubic	5.6575	72.59	5.324	937.4
Gold	Au	79	FCC	4.0786	196.97	19.302	1064.4
Hafnium	Hf	72	HCP	$a = 3.1883$ $c = 5.0422$	178.49	13.31	2227
Indium	In	49	tetra	$a = 3.2517$ $c = 4.9459$	114.82	7.286	156.6
Iridium	Ir	77	FCC	3.84	192.9	22.65	2447
Iron	Fe	26	BCC FCC (>912°C) BCC (>1394°C)	2.866 3.589	55.847	7.87	1538
Lanthanum	La	57	HCP	$a = 3.774$ $c = 12.17$	138.91	6.146	918
Lead	Pb	82	FCC	4.9489	207.19	11.36	327.4
Lithium	Li	3	BCC	3.5089	6.94	0.534	180.7

(continued)

Metal		Atomic Number	Crystal Structure	Lattice Parameters (A)	Atomic Mass g/mol	Density (g/cm^3)	Melting Temperature (°C)
Magnesium	Mg	12	HCP	$a = 3.2087$	24.312	1.738	650
				$c = 5.209$			
Manganese	Mn	25	cubic	8.931	54.938	7.47	1244
Mercury	Hg	80	rhomb		200.59	13.546	−38.9
Molybdenum	Mo	42	BCC	3.1468	95.94	10.22	2610
Nickel	Ni	28	FCC	3.5167	58.71	8.902	1453
Niobium	Nb	41	BCC	3.294	92.91	8.57	2468
Osmium	Os	76	HCP	$a = 2.7341$	190.2	22.57	2700
				$c = 4.3197$			
Palladium	Pd	46	FCC	3.8902	106.4	12.02	1552
Platinum	Pt	78	FCC	3.9231	195.09	21.45	1769
Potassium	K	19	BCC	5.344	39.09	0.855	63.2
Rhenium	Re	75	HCP	$a = 2.760$	186.21	21.04	3180
				$c = 4.458$			
Rhodium	Rh	45	FCC	3.796	102.99	12.41	1963
Rubidium	Rb	37	BCC	5.7	85.467	1.532	38.9
Ruthenium	Ru	44	HCP	$a = 2.6987$	101.07	12.37	2310
				$c = 4.2728$			
Selenium	Se	34	hex	$a = 4.3640$	78.96	4.809	217
				$c = 4.9594$			
Silicon	Si	14	diamond cubic	5.4307	28.08	2.33	1410
Silver	Ag	47	FCC	4.0862	107.868	10.49	961.9
Sodium	Na	11	BCC	4.2906	22.99	0.967	97.8
Strontium	Sr	38	FCC	6.0849	87.62	2.6	768
			BCC (>557°C)	4.84			
Tantalum	Ta	73	BCC	3.3026	180.95	16.6	2996
Technetium	Tc	43	HCP	$a = 2.735$	98.9062	11.5	2200
				$c = 4.388$			
Tellurium	Te	52	hex	$a = 4.4565$	127.6	6.24	449.5
				$c = 5.9268$			
Thorium	Th	90	FCC	5.086	232	11.72	1775
Tin	Sn	50	FCC	6.4912	118.69	5.765	231.9
Titanium	Ti	22	HCP	$a = 2.9503$	47.9	4.507	1668
				$c = 4.6831$			
			BCC (>882°C)	3.32			
Tungsten	W	74	BCC	3.1652	183.85	19.254	3410
Uranium	U	92	ortho	$a = 2.854$	238.03	19.05	1133
				$b = 5.869$			
				$c = 4.955$			
Vanadium	V	23	BCC	3.0278	50.941	6.1	1900
Yttrium	Y	39	HCP	$a = 3.648$	88.91	4.469	1522
				$c = 5.732$			
Zinc	Zn	30	HCP	$a = 2.6648$	65.38	7.133	420
				$c = 4.9470$			
Zirconium	Zr	40	HCP	$a = 3.2312$	91.22	6.505	1852
				$c = 5.1477$			
			BCC (>862°C)	3.6090			

Appendix B: The Atomic and Ionic Radii of Selected Elements

Element	Atomic Radius (A)	Valence	Ionic Radius (A)
Aluminum	1.432	+3	0.51
Antimony	1.45	+5	0.62
Arsenic	1.15	+5	2.22
Barium	2.176	+2	1.34
Beryllium	1.143	+2	0.35
Bismuth	1.60	+5	0.74
Boron	0.46	+3	0.23
Bromine	1.19	−1	1.96
Cadmium	1.49	+2	0.97
Calcium	1.976	+2	0.99
Carbon	0.77	+4	0.16
Cerium	1.84	+3	1.034
Cesium	2.65	+1	1.67
Chlorine	0.905	−1	1.81
Chromium	1.249	+3	0.63
Cobalt	1.253	+2	0.72
Copper	1.278	+1	0.96
Fluorine	0.6	−1	1.33
Gallium	1.218	+3	0.62
Germanium	1.225	+4	0.53
Gold	1.442	+1	1.37
Hafnium	1.55	+4	0.78
Hydrogen	0.46	+1	1.54
Indium	1.570	+3	0.81
Iodine	1.35	−1	2.20
Iron	1.241 (BCC)	+2	0.74
	1.269 (FCC)	+3	0.64
Lanthanum	1.887	+3	1.016
Lead	1.75	+4	0.84
Lithium	1.519	+1	0.68
Magnesium	1.604	+2	0.66
Manganese	1.12	+2	0.80
		+3	0.66
Mercury	1.55	+2	1.10
Molybdenum	1.363	+4	0.70
Nickel	1.243	+2	0.69

(*continued*)

Element	Atomic Radius (A)	Valence	Ionic Radius (A)
Niobium	1.426	+4	0.74
Nitrogen	0.71	+5	0.15
Oxygen	0.60	−2	1.32
Palladium	1.375	+4	0.65
Phosphorus	1.10	+5	0.35
Platinum	1.387	+2	0.80
Potassium	2.314	+1	1.33
Rubidium	2.468	+1	0.70
Selenium	1.15	−2	1.91
Silicon	1.176	+4	0.42
Silver	1.445	+1	1.26
Sodium	1.858	+1	0.97
Strontium	2.151	+2	1.12
Sulfur	1.06	−2	1.84
Tantalum	1.43	+5	0.68
Tellurium	1.40	−2	2.11
Thorium	1.798	+4	1.02
Tin	1.405	+4	0.71
Titanium	1.475	+4	0.68
Tungsten	1.371	+4	0.70
Uranium	1.38	+4	0.97
Vanadium	1.311	+3	0.74
Yttrium	1.824	+3	0.89
Zinc	1.332	+2	0.74
Zirconium	1.616	+4	0.79

Note that 1 A = 10^{-8} cm = 0.1 nanometer (nm)

Answers to Selected Problems

CHAPTER 2

2-6 **(a)** 1.037×10^{21} atoms.

2-7 **(a)** 1.078×10^{28} atoms/ton.

2-8 **(a)** 5.70×10^{23} atoms. **(b)** 0.948 mol.

2-10 3.

2-16 0.086.

2-27 Al_2O_3, with strong ionic bonding, has lower coefficient.

2-29 Weak secondary bonds between chains.

CHAPTER 3

3-3 **(a)** 1.426×10^{-8} cm. **(b)** 1.4447×10^{-8} cm.

3-5 **(a)** 5.3355 A. **(b)** 2.3103 A.

3-7 FCC.

3-9 BCT.

3-10 **(a)** 8 atoms/cell. **(b)** 0.387.

3-15 0.6% contraction.

3-19 A: $[00\bar{1}]$. B: $[1\bar{2}0]$. C: $[\bar{1}11]$. D: $[2\bar{1}\bar{1}]$.

3-21 A: $(1\bar{1}1)$. B: (030). C: $(10\bar{2})$.

3-25 $[\bar{1}10]$, $[1\bar{1}0]$, $[101]$, $[\bar{1}0\bar{1}]$, $[011]$, $[0\bar{1}\bar{1}]$.

3-27 Tetragonal—4; orthorhombic—2; cubic—12.

3-29 **(a)** (111). **(c)** $(0\bar{1}2)$.

3-31 [100]: 0.35089 nm, 2.85 nm^{-1}, 0.866. [110]: 0.496 nm, 2.015 nm^{-1}, 0.612. [111]: 0.3039 nm, 3.291 nm^{-1}, 1.

3-32 (100): $1.1617 \times 10^{15}/cm^2$, packing fraction 0.7854. (110): $1.144 \times 10^{15}/cm^2$, packing fraction 0.555. (111): $1.867 \times 10^{15}/cm^2$, 0.907.

3-35 **(a)** 0.2797 A. **(b)** 0.629 A.

3-44 Fluorite. **(a)** 5.2885 A. **(b)** 12.13 g/cm^3. **(c)** 0.624.

3-46 Cesium chloride. **(a)** 4.1916 A. **(b)** 4.8 g/cm^3. **(c)** 0.693.

CHAPTER 4

4-1 4.97×10^{19} vacancies/cm^3.

4-3 **(a)** 0.002375. **(b)** 1.61×10^{20} vacancies/cm^3.

4-5 **(a)** 1.157×10^{20} vacancies/cm^3. **(b)** 0.532 g/cm^3.

4-7 0.345.

4-9 8.265 g/cm^3.

4-11 **(a)** 0.0081. **(b)** one H atom per 123.5 unit cells.

4-13 **(a)** 0.0534 defects/unit cell. **(b)** 2.52×10^{20} defects/cm^3.

4-14 **(a)** $[0\bar{1}1]$, $[01\bar{1}]$, $[\bar{1}10]$, **(b)** $[1\bar{1}0]$, $[\bar{1}01]$, $[10\bar{1}]$.

4-16 $(1\bar{1}0)$, $(\bar{1}10)$, $(0\bar{1}1)$, $(01\bar{1})$, $(10\bar{1})$, $(\bar{1}01)$.

4-18 Expected: $\mathbf{b} = 2.863$ A, $d = 2.338$ A. (110)[111]: $\mathbf{b} = 7.014$ A, $d = 2.863$ A. Ratio = 0.44.

4-26 **(a)** $K = 0.13$ MPa/$\sqrt{m}$, $\sigma_0 = 415.7$ MPa. **(b)** 705.7 MPa.

4-28 **(a)** 128 grains/in^2. **(b)** 1,280,000 grains/in^2.

4-30 3.6.

CHAPTER 5

5-4 1.08×10^9 jumps/s.

5-6 **(a)** 59,230 cal/mol. **(b)** 0.055 cm^2/s.

5-12 **(a)** −0.02495 at% Sb/cm. **(b)** -1.246×10^{19} Sb/$cm^3 \cdot$ cm.

5-14 **(a)** -1.969×10^{11} H atoms/$cm^3 \cdot$ cm. **(b)** 3.3×10^7 H atoms/$cm^2 \cdot$ s.

5-16 1.245×10^{-3} g/h.

5-18 −198°C.

5-28 0.01 cm: 0.87% C. 0.05 cm: 0.43% C. 0.10 cm: 0.18% C.

5-30 907°C.

5-32 0.53% C.

5-34 2.9 min.

5-36 12.8 min.

5-39 667°C.

5-41 51,286 cal/mol; yes.

CHAPTER 6

6-3 **(a)** Deforms. **(b)** Does not neck.

6-5 **(b)** 4740 N.

6-7 82,750 N.

6-9 **(a)** 5 cm.

6-10 15.2 m.

6-14 **(a)** 80 MPa. **(b)** 90.1 MPa. **(c)** 4267.7 MPa. **(d)** 4.6%. **(e)** 3.4%. **(f)** 80 MPa. **(g)** 83.2 MPa. **(h)** 0.53 MPa.

6-16 **(a)** 274 MPa. **(b)** 417 MPa. **(c)** 172 GPa. **(d)** 18.55%. **(e)** 15.8%. **(f)** 397.9 MPa. **(g)** 473 MPa. **(h)** 0.17 MPa.

6-18 **(a)** $e_f = 30.0077$ cm. **(b)** $d_f = 0.9999$ cm.

6-20 **(a)** 546.6 MPa. **(b)** 159, 430 MPa.

6-22 **(a)** 41 mm; will not fracture.

6-25 29.8.

6-30 No transition temperature.

6-33 Not notch-sensitive; poor toughness.

CHAPTER 7

7-4 0.99 MPa $\sqrt{m}$.

7-5 No; test will not be sensitive enough.

7-11 68 N.

7-13 $d = 4$ cm.

7-17 22 MPa; max = +22 MPa, min = −22 MPa, mean = 0 MPa; reduce fatigue strength due to heating.

CHAPTER 8

8-1 $n = 0.12$; BCC.

8-4 $n = 0.15$.

8-7 0.39 cm.

8-9 166 MPa tensile, 152 MPa yield, 5% elongation.

8-11 First step: 36% CW giving 179 MPa tensile, 159 MPa yield, 6% elongation. Second step: 64% CW giving 207 MPa tensile, 186 MPa yield, 3% elongation. Third step: 84% CW giving 221 MPa tensile, 200 MPa yield, 2% elongation.

8-12 1.95 to 2.4 cm.

8-13 48% CW: 193 MPa tensile, 172 MPa yield, 4% elongation.

8-16 **(a)** 6100 N. **(b)** Will not break.

8-18 **(a)** 550°C, 750°C, 950°C. **(b)** 700°C. **(c)** 900°C. **(d)** 2285°C.

8-21 Slope = 0.4.

8-23 CW 75% from 5 to 25 cm, anneal. CW 75% from 2.5 to 1.25 cm, anneal. CW 72.3% from 1.25 to 0.66 cm, anneal. CW 42% from 0.66 to 0.5 cm or hot work 98.3% from 5 to 0.66 cm, then CW 42% from 0.66 to 0.5 cm.

CHAPTER 9

9-2 **(a)** 6.65 A. **(b)** 109 atoms.

9-4 1.136×10^6 atoms.

9-6 **(a)** 0.0333. **(b)** 0.333. **(c)** All.

9-8 1265°C.

9-11 31.15 s.

9-13 $B = 305$ s/cm^2, $n = 1.58$.

9-17 $c = 0.0032$ s, $m = 0.34$.

9-19 **(a)** 900°C. **(b)** 420°C. **(c)** 480°C. **(d)** 312°C/min. **(e)** 9.7 min. **(f)** 8.1 min. **(g)** 60°C. **(h)** Zinc. **(i)** 14 min/cm^2.

9-22 $D \geq 16.69$ cm, $H \geq 25$ cm, $V \geq 5459$ cm^3.

9-26 V/A (riser) = 0.68, V/A (middle) = 1.13, V/A (end) = 0.89; not effective.

9-28 $D_{Cu} = 3.7$ cm $D_{Fe} = 3.24$ cm.

9-30 **(a)** 46 cm^3. **(b)** 4.1%.

9-32 23.04 cm.

CHAPTER 10

10-9 **(a)** Yes. **(c)** No. **(e)** No. **(g)** No.

10-11 *Cd* should give smallest decrease in conductivity; none should give unlimited solid solubility.

10-17 **(b)** 44.1 at% Cu – 55.9 at% Al.

10-20 750 g Ni, Ni/Cu = 1.62.

10-22 332 g MgO.

10-24 64.1 wt% FeO.

10-26 **(a)** 49 wt% W in L, 70 wt% W in α. **(b)** Not possible.

10-28 313.2 kg; 761 kg.

10-33 **(a)** 2900°C, 2690°C, 210°C. **(b)** 60% L containing 49% W, 40% α containing 70% W.

10-35 **(a)** 55% W. **(b)** 18% W.

10-39 **(a)** 2900°C. **(b)** 2710°C. **(c)** 190°C. **(d)** 2990°C. **(e)** 90°C. **(f)** 300 s. **(g)** 340 s. **(h)** 60% W.

CHAPTER 11

11-7 **(a)** θ. **(b)** α, β, γ, η. **(c)** 1100°C: peritectic. 900°C: monotectic. 680°C: eutectic. 600°C: peritectoid. 300°C: eutectoid.

11-10 $SnCu_3$.

11-11 $SiCu_4$.

11-13 **(a)** 2.5% Mg. **(b)** 600°C, 470°C, 400°C, 130°C. **(c)** 74% α containing 7% Mg, 26% L containing 26% Mg. **(d)** 100% α containing 12% Mg. **(e)** 67% α containing 1% Mg, 33% β containing 34% Mg.

11-15 **(a)** Hypereutectic. **(b)** 98% Sn. **(c)** 22.8% β containing 97.5% Sn, 77.2% L containing 61.9% Sn. **(d)** 35% α containing 19% Sn, 65% β containing 97.5% Sn. **(e)** 22.8% primary β containing 97.5% Sn, 77.2% eutectic containing 61.9% Sn. **(f)** 30% α containing 2% Sn, 70% β containing 100% Sn.

11-17 **(a)** Hypoeutectic. **(b)** 1% Si. **(c)** 78.5% α containing 1.65% Si, 21.5% L containing 12.6% Si. **(d)** 97.6% α containing 1.65% Si, 2.4% β containing 99.83% Si. **(e)** 78.5 primary α containing 1.65% Si, 21.5% eutectic containing 12.6% Si. **(f)** 96% α containing 0% Si, 4% β containing 100% Si.

11-19 Hypoeutectic.

11-21 52% Sn.

11-23 Hypereutectic. **(b)** 64% α, 36% β.

11-25 0.54.

11-27 **(a)** 1150°C. **(b)** 150°C. **(c)** 1000°C. **(d)** 577°C. **(e)** 423°C. **(f)** 10.5 min. **(g)** 11.5 min. **(h)** 45% Si.

CHAPTER 12

12-2 $c = 6.47 \times 10^{-6}$, $n = 2.89$.

12-8 For Al – 4% Mg: solution treat between 210 and 451°C, quench, age below 210°C. For Al – 12% Mg: solution treat between 390 and 451°C, quench, age below 390°C.

12-18 **(a)** Solution treat between 290 and 400°C, quench, age below 290°C. **(c)** Not good candidate. **(e)** Not good candidate.

12-26 **(a)** 795°C. **(b)** Primary ferrite. **(c)** 56.1% ferrite containing 0.0218% C and 43.9% austenite containing 0.77% C. **(d)** 95.1% ferrite containing 0.0218% C and 4.9% cementite containing 6.67% C. **(e)** 56.1% primary ferrite containing 0.0218% C and 43.9% pearlite containing 0.77% C.

12-28 0.53% C, hypoeutectoid.

12-30 0.156% C, hypoeutectoid.

12-32 0.281% C.

12-34 760°C, 0.212% C.

12-45 **(a)** 615°C. **(b)** 1.67×10^{-5} cm.

12-47 Bainite with HRC 47.

12-49 Martensite with HRC 66.

12-51 **(a)** 37.2% martensite with 0.77% C and HRC 65. **(b)** 84.8% martensite with 0.35% C and HRC 58.

12-53 **(a)** 750°C. **(b)** 0.455% C.

12-59 3.06% expansion.

12-61 Austenitize at 750°C, quench, temper above 330°C.

CHAPTER 13

13-3 **(a)** 97.8% ferrite, 2.2% cementite, 82.9% primary ferrite, 17.1% pearlite. **(c)** 85.8% ferrite, 14.2% cementite, 3.1% primary cementite, 96.9% pearlite.

13-5 For 1035: $A_1 = 727°C$; $A_3 = 790°C$; anneal = 820°C; normalize = 845°C; process anneal = 557–647°C; not usually spheroidized.

13-10 **(a)** Ferrite and pearlite. **(c)** Martensite. **(e)** Ferrite and bainite. **(g)** Tempered martensite.

13-12 **(a)** Austenitize at 820°C, hold at 600°C for 10 s, cool. **(c)** Austenitize at 780°C, hold at 600°C for 10 s, cool. **(e)** Austenitize at 900°C, hold at 320°C for 5000 s, cool.

13-15 **(a)** Austenitize at 820°C, quench, temper between 420 and 480°C; 1000 to 1300 MPa tensile, 900 to 1100 MPa yield. **(b)** 1200 to 1300 MPa tensile, 890 to 940 MPa yield. **(c)** 690 MPa tensile, 449 MPa yield, 20% elongation.

13-17 0.48% C in martensite; austenitized at 770°C; should austenitize at 860°C.

13-19 1080: fine pearlite. 4340: martensite.

13-22 May become hypereutectoid, with grain boundary cementite.

13-24 Not applicable. **(c)** 8 to 10°C/s. **(e)** 32 to 36°C/s.

13-26 **(a)** 16°C/s. **(b)** Pearlite with HRC 38.

13-28 **(a)** Pearlite with HRC 36. **(c)** Pearlite and martensite with HRC 46.

13-30 **(a)** 3.3 cm. **(c)** 4.8 cm. **(e)** greater than 6.3 cm.

13-33 0.25 h.

13-37 0.05 mm: pearlite and martensite with HRC 53. 0.15 mm: medium pearlite with HRC 38.

13-40 δ-ferrite; nonequilibrium freezing; quench anneal.

13-44 2.4% Si.

CHAPTER 14

14-3 Eutectic microconstituent contains 97.6% β.

14-6 27% β versus 2.2% β.

14-8 Al – 10% Mg.

14-10 **(a)** 0.28 cm, 6.58 g, $0.02. **(b)** 0.28 cm, 10.21 g, $0.013.

14-15 Al: 440%. Mg: 130%. Cu: 1100%.

14-17 Lead may melt during hot working.

14-19 γ' more at low temperature.

14-22 Ti-15% V: 100% β transforms to 100% α', which then transforms to 24% β precipitate in an α matrix. Ti-35% V: 100% β transforms to 100% β', which then transforms to 27% α precipitate in a β matrix.

14-23 Al: 18.5×10^5 cm Cu: 13.4×10^5 cm Ni: 8.5×10^5 cm.

14-28 Spalls off; cracks.

CHAPTER 15

15-22 $B = 2.4$; true = 22.58%; fraction = 0.44.

15-27 1.257 kg BaO; 0.245 kg Li_2O.

CHAPTER 16

16-7 **(a)** 2500. **(b)** 2.4×10^{18}.

16-10 **(a)** 211. **(b)** 175.

16-14 Polybutadiene and silicone.

16-16 Polyethylene and polypropylene.

16-21 8.59 MPa.

16-23 **(a)** PE. **(b)** LDPE. **(c)** PTFE.

16-25 At $\varepsilon = 1$, $E = 5.8$ MPa; at $\varepsilon = 4$, $E = 14$ MPa.

CHAPTER 17

17-7 7.65×10^{13} per cm^3.

17-9 2.47%.

17-13 9.408 g/cm^3.

17-15 **(a)** 0.507. **(b)** 0.507. **(c)** 7.775 g/cm^3.

17-17 11.18 to 22.2 kg.

17-24 **(a)** 2.53 g/cm^3. **(b)** 201,100 MPa. **(c)** 107,500 MPa.

17-26 0.964.

17-28 188 MPa.

17-30 For $d = 20$ μm, $l_c = 0.30$ cm, $l_c/d = 150$.

17-36 Sizing improves strength.

17-39 Pyrolize at 2500°C; 1720 MPa.

17-42 $E_{parallel} = 69{,}920$ MPa; $E_{perpendicular} = 20{,}830$ MPa.

17-44 $E_{parallel} = 82{,}040$ MPa; $E_{perpendicular} = 69{,}600$ MPa.

17-46 0.417 g/cm^3; 20.0 kg versus 129.6 kg.

Index

A

D

E

F

G

H

I

N

O

P

T

U

V

W

X

Y

Z